Methods in Molecular Biology

Series Editor
John M. Walker
School of Life and Medical Sciences
University of Hertfordshire
Hatfield, Hertfordshire, UK

For further volumes:
http://www.springer.com/series/7651

For over 35 years, biological scientists have come to rely on the research protocols and methodologies in the critically acclaimed *Methods in Molecular Biology* series. The series was the first to introduce the step-by-step protocols approach that has become the standard in all biomedical protocol publishing. Each protocol is provided in readily-reproducible step-by-step fashion, opening with an introductory overview, a list of the materials and reagents needed to complete the experiment, and followed by a detailed procedure that is supported with a helpful notes section offering tips and tricks of the trade as well as troubleshooting advice. These hallmark features were introduced by series editor Dr. John Walker and constitute the key ingredient in each and every volume of the *Methods in Molecular Biology* series. Tested and trusted, comprehensive and reliable, all protocols from the series are indexed in PubMed.

Pichia pastoris

Methods and Protocols

Edited by

Xavier Garcia-Ortega

Department of Chemical, Biological and Environmental Engineering, Engineering School, Universitat Autònoma de Barcelona, Bellaterra, Barcelona, Spain

Anton Glieder

Christian Doppler Laboratory for Innovative Pichia pastoris Host and Vector Systems, Institute of Molecular Biotechnology, Graz University of Technology, Graz, Austria

Karin Kovar

daspool, Wädenswil, Switzerland

Lukas Rieder

Christian Doppler Laboratory for Innovative Pichia pastoris Host and Vector Systems, Institute of Molecular Biotechnology, Graz University of Technology, Graz, Austria

Humana Press

Editors
Xavier Garcia-Ortega
Department of Chemical, Biological and Environmental Engineering
Engineering School,
Universitat Autònoma de Barcelona
Bellaterra, Barcelona, Spain

Anton Glieder
Christian Doppler Laboratory for Innovative Pichia pastoris Host and Vector Systems,
Institute of Molecular Biotechnology
Graz University of Technology
Graz, Austria

Karin Kovar
daspool
Wädenswil, Switzerland

Lukas Rieder
Christian Doppler Laboratory for Innovative Pichia pastoris Host and Vector Systems,
Institute of Molecular Biotechnology
Graz University of Technology
Graz, Austria

ISSN 1064-3745 ISSN 1940-6029 (electronic)
Methods in Molecular Biology
ISBN 978-1-0716-4778-3 ISBN 978-1-0716-4779-0 (eBook)
https://doi.org/10.1007/978-1-0716-4779-0

This Humana imprint is published by the registered company Springer Science+Business Media, LLC, part of Springer Nature.
The registered company address is: 1 New York Plaza, New York, NY 10004, U.S.A.

Dedication

This volume is dedicated to the memory of James M. (Jim) Cregg, PhD, whose pioneering work established *Pichia pastoris* as a powerful system for recombinant protein expression. Jim's vision, creativity, and generosity transformed the field, providing a platform that has enabled discoveries across academic and industrial biotechnology worldwide. His contributions as a scientist, mentor, and colleague continue to inspire, and his legacy lives on in every experiment that relies on *Pichia*.

Anton Glieder, Editor in Chief, with Tom Chappell, Ilya Tostorukov and Knut Madden, co-founders of BioGrammatics with Jim (Sept. 11, 2025).

Preface

Pichia pastoris, an Evolving Microbial Platform for Recombinant Gene Expression

Pichia pastoris, a methylotrophic yeast, has undoubtedly become one of the most versatile and widely adopted microbial cell factories for recombinant production of proteins, as well as, chemicals and for research in biotechnology and molecular biology. Due to its wide use in research and industry for more than 40 years, the name *Pichia pastoris* is still commonly used despite the recognition of *Pichia pastoris* as a new genus named *Komagataella* in 1995.

This new genus includes the most frequently used species *Komagataella phaffii* and *Komagataella pastoris*. Many protocols of this book will be equally applicable to *K. phaffii*, as well as *K. pastoris*, commonly summarized as *P. pastoris*; however, most recent scientific publications reporting specific differences between the closely related strain lines within the species *K. phaffii* and compared to *K. pastoris*, also adopted the current correct scientific names.

Remarkably, over the past decades, this expression system has been adopted for very diverse applications from bulk products such as SCP (single-cell protein) to recombinant proteins with complex, humanized glycosylation patterns, and natural product biosynthesis by metabolic engineering and synthetic biology. The implementation of these *Pichia* bioprocesses at an industrial scale may provide sustainable solutions to many of today's unresolved and even unaddressed problems in the near future.

Following the initial wave of industrial applications for enzyme production and the early FDA approvals of pharmaceutical proteins, we are now witnessing a resurgence of interest in the use of *Komagataella phaffii* (reclassified from *Pichia pastoris*) in the human food sector. Pioneering work by Impossible Foods, which introduced leghemoglobin into the market over the past decade, has catalyzed this trend. More recently, the European Food Safety Authority (EFSA) issued safety opinions regarding both the leghemoglobin itself and the genetic modifications used for its production. Additionally, a series of new Generally Recognized as Safe (GRAS) notices for other proteins intended for human food and nutrition have been published. This growing momentum is further supported by an increasing number of new products, start-ups, and research laboratories, all building upon the robust foundation of *P. pastoris* expression systems and related technologies.

Would you like to implement your next *Pichia* project efficiently and effectively?

Practitioners may strive to obtain all established knowledge on the *Pichia* expression and production system from one source. An international cohort of authors herewith provides a useful and comprehensive resource of laboratory protocols for working with *Pichia pastoris*. These protocols cover basic and newer laboratory methods, starting literally from the gene and genetic construct to mature yet reasonable amounts of a recombinant product being produced, purified, and characterized. However, although this volume provides the most comprehensive collection of *P. pastoris* protocols, the need for an additional complementary volume is obvious, given the growing large international community contributing new methods continuously

This volume in the Methods in Molecular Biology series is organized into seven parts:

- Part I: Introduction
- Part II: Bioinformatics
- Part III: Cell and Systems Biology
- Part IV: Gene Expression Cassettes and Vectors
- Part V: *Pichia pastoris* Expression Strains
- Part VI: Upstream Bioprocessing
- Part VII: Downstream Processing and Product Analysis

We sincerely hope that this book will become a practical guide and a joy for you, and we look forward to receiving your feedback, proposals for complementary protocols, and your support for the *Pichia* community.

Bellaterra, Barcelona, Spain ***Xavier Garcia-Ortega***
Graz, Austria ***Anton Glieder***
Wädenswil, Switzerland ***Karin Kovar***
Graz, Austria ***Lukas Rieder***

Contents

Contributors

NÚRIA ADELANTADO • *Lonza AG, Visp, Switzerland*

FARZANA ALAM • *Department of Chemical Engineering, Imperial College London, London, UK; Imperial College Centre for Synthetic Biology, Imperial College London, London, UK; Department of Genetic Engineering and Biotechnology, Jagannath University, Dhaka, Bangladesh*

JOAN ALBIOL • *Department of Chemical Engineering, Autonomous University of Barcelona, Barcelona, Catalonia, Spain; Department of Chemical, Biological and Environmental Engineering, Universitat Autònoma de Barcelona, Bellaterra, Spain*

MIGUEL ALCALDE • *Department of Biocatalysis, Institute of Catalysis, CSIC, Madrid, Spain*

S. ALJOSCHA WAHL • *Department of Biotechnology, Delft University of Technology, Delft, The Netherlands; Lehrstuhl für Bioverfahrenstechnik, Friedrich Alexander Universität Erlangen, Nürnberg, Germany*

MARTIN ALTVATER • *Department of Biotechnology, Institute of Microbiology and Microbial Biotechnology, BOKU University, Vienna, Austria*

ROCHELLE AW • *Department of Chemical Engineering, Imperial College London, London, UK; Imperial College Centre for Synthetic Biology, Imperial College London, London, UK; Department of Bioengineering, Stanford University, Stanford, CA, USA*

THOMAS BECKER • *Technical University of Munich, TUM School of Life Science, Chair of Brewing and Beverage Technology, Freising, Germany*

CECILIA BERGES • *TBI, Université de Toulouse, CNRS, INRAE, INSA, Toulouse, France*

NÚRIA BERNAT-CAMPS • *Department of Chemical, Biological and Environmental Engineering, School of Engineering, Universitat Autònoma de Barcelona, Bellaterra, Barcelona, Spain; Austrian Centre of Industrial Biotechnology (ACIB), Graz, Austria*

LUKAS BERNAUER • *Institute of Molecular Biotechnology, Graz University of Technology, NAWI Graz, Graz, Austria; BioTechMed-Graz, Graz, Austria*

JULIO BERRIOS • *School of Biochemical Engineering, Pontificia Universidad Católica de Valparaíso, Valparaíso, Chile*

MORGANE BOONE • *Department of Biochemistry and Microbiology, Ghent University, Ghent, Belgium; VIB Center for Medical Biotechnology, Ghent, Belgium; Altos Labs, Inc., Bay Area Institute of Science, Redwood City, CA, USA*

JASON BROWN • *Thermo Fisher Scientific, Logan, UT, USA*

KAI BÜCHNER • *Technical University of Munich, TUM School of Life Science, Chair of Brewing and Beverage Technology, Freising, Germany*

NICO CALLEWAERT • *Department of Biochemistry and Microbiology, Ghent University, Ghent, Belgium; VIB Center for Medical Biotechnology, Ghent, Belgium*

MARC CARNICER • *Department of Chemical Engineering, Autonomous University of Barcelona, Barcelona, Catalonia, Spain; Bioengineering Department, IQS, Universitat Ramon Llull, Barcelona, Catalonia, Spain*

ANTONI CASABLANCAS • *Department of Chemical, Biological and Environmental Engineering, Engineering School, Universitat Autònoma de Barcelona, Bellaterra, Barcelona, Spain*

ROCÍO CUEVAS CASADO • *Leanbio S.L., Barcelona, Spain*

THOMAS G. CHAPPELL • *BioGrammatics Inc., Carlsbad, CA, USA*

GESINE CORNELISSEN • *Hamburg University of Applied Sciences, Faculty Life Sciences, Department Biotechnology, Hamburg, Germany*
JOAN CORTADA-GARCIA • *Lonza AG, Visp, Switzerland*
JAMES M. CREGG • *Biogrammatics Inc., Carlsbad, CA, USA; Keck Graduate Institute of Applied Life Sciences, Claremont, CA, USA*
ANA G. DEL HIERRO • *Bioprocess R&D Pilot Plant, Bionet, Fuente Álamo, Spain; BiOrbic Bioeconomy SFI Research Centre, University College Dublin, Dublin, Ireland*
MIKEL DOLZ • *EvoEnzyme S.L., Parque Científico de Madrid, Madrid, Spain*
KATHARINA EBNER • *bisy GmbH, Wuenschendorf, Hofstätten an der Raab, Austria*
HANNAH EECKHAUT • *Department of Biochemistry and Microbiology, Ghent University, Ghent, Belgium; VIB Center for Medical Biotechnology, Ghent, Belgium*
ANITA EMMERSTORFER-AUGUSTIN • *Institute of Molecular Biotechnology, Graz University of Technology, NAWI Graz, Graz, Austria; BioTechMed-Graz, Graz, Austria; acib—Austrian Centre of Industrial Biotechnology, Graz, Austria*
JULIA FEICHTINGER • *Division of Cell Biology, Histology and Embryology, Gottfried Schatz Research Center for Cell Signaling, Metabolism and Aging, Medical University of Graz, Graz, Austria; Christian Doppler Laboratory for Innovative Pichia pastoris Host and Vector Systems, Graz, Austria*
PAU FERRER • *Department of Chemical Engineering, Autonomous University of Barcelona, Barcelona, Catalonia, Spain; Department of Chemical, Biological and Environmental Engineering, Universitat Autònoma de Barcelona, Bellaterra, Spain*
PATRICK FICKERS • *Microbial Processes and Interactions, TERRA Teaching and Research Centre, Gembloux Agro-Bio Tech, University of Liège, Gembloux, Belgium*
ALBERT FINA • *Department of Chemical, Biological and Environmental Engineering, Universitat Autònoma de Barcelona, Bellaterra, Spain*
KARLHEINZ FLICKER • *Lonza AG, Visp, Switzerland*
SARAH GANGL • *acib GmbH, Krenngasse, Graz, Austria; Austrian Centre of Industrial Biotechnology (ACIB), Graz, Austria*
XAVIER GARCIA-ORTEGA • *Department of Chemical, Biological and Environmental Engineering, Engineering School, Universitat Autònoma de Barcelona, Bellaterra, Barcelona, Spain*
BRIGITTE GASSER • *Department of Biotechnology and Food Science, BOKU University, Institute of Microbiology and Microbial Biotechnology, Vienna, Austria; Austrian Centre of Industrial Biotechnology (ACIB GmbH), Vienna, Austria*
ARNAU GASSET • *Department of Chemical, Biological and Environmental Engineering, School of Engineering, Universitat Autònoma de Barcelona, Bellaterra (Barcelona), Spain*
THOMAS GASSLER • *Institute of Microbiology, Department of Biology, ETH Zurich, Zurich, Switzerland*
KURT R. GEHLSEN • *Gehlsen Consulting, Palm Beach County, FL, USA*
MARTINA GEIER • *bisy GmbH, Wuenschendorf, Hofstätten an der Raab, Austria*
ANTON GLIEDER • *Christian Doppler Laboratory for Innovative Pichia pastoris Host and Vector Systems, Institute of Molecular Biotechnology, Graz University of Technology, Graz, Austria; bisy GmbH, Hofstätten an der Raab, Austria*
CARLA MAGDALENA AGUILAR GOMEZ • *Institute of Molecular Biotechnology, Graz University of Technology, Petersgasse, Graz, Austria; Christian Doppler Laboratory for innovative Pichia pastoris host and vector systems, Graz University of Technology, Petersgasse, Graz, Austria*

David Gonzalez-Perez • *Department of Biocatalysis, Institute of Catalysis, CSIC, Madrid, Spain*
Alexandra Charlotte Ilse Grebe • *Institute of Molecular Biotechnology, Graz University of Technology, Petersgasse, Graz, Austria*
Valentina Haag • *daspool, Wädenswil, Switzerland*
Amina Hasanagic • *Christian Doppler Laboratory for Innovative Pichia pastoris Host and Vector Systems, Graz University of Technology, Petersgasse, Graz, Austria*
Sven Heimhilcher • *bisy GmbH, Wünschendorf, Hofstätten a.d. Raab, Austria*
Stéphanie Heux • *TBI, Université de Toulouse, CNRS, INRAE, INSA, Toulouse, France*
Maik Hilgarth • *Roche Diagnostics GmbH, Penzberg, Germany*
Hauke Holm • *Lonza AG, Visp, Switzerland*
Andrea K. Hönikl • *Institute of Molecular Biotechnology, Graz University of Technology, Petersgasse, Graz, Austria*
Albert Font Inglés • *Leanbio S.L., Barcelona, Spain*
Christian Kaiser • *Hamburg University of Applied Sciences, Faculty Life Sciences, Department Biotechnology, Hamburg, Germany*
Roland Kerpes • *Technical University of Munich, TUM School of Life Science, Chair of Brewing and Beverage Technology, Freising, Germany*
Merve Keser • *Department of Biocatalysis, Institute of Catalysis, CSIC, Madrid, Spain*
David Keßler • *Christian Doppler Laboratory for Innovative Pichia pastoris Host and Vector Systems, Graz University of Technology, Petersgasse, Graz, Austria*
Shaun A. Kirkpatrick • *Research Corporations Technologies, Tucson, AZ, USA*
Julian Kopp • *Institute of Chemical, Environmental and Bioscience Engineering, Research Group Integrated Bioprocess Development, TU Wien, Vienna, Austria*
Karin Kovar • *daspool, Wädenswil, Switzerland*
Julian Christopher Krappinger • *Division of Cell Biology, Histology and Embryology, Gottfried Schatz Research Center for Cell Signaling, Metabolism and Aging, Medical University of Graz, Graz, Austria; Christian Doppler Laboratory for Innovative Pichia pastoris Host and Vector Systems, Graz, Austria*
Leonie Lehmayer • *Institute of Molecular Biotechnology, Graz University of Technology, NAWI Graz, Graz, Austria; BioTechMed-Graz, Graz, Austria*
Geoff P. Lin-Cereghino • *Department of Biological Sciences, University of the Pacific, Stockton, CA, USA*
Joan Lin-Cereghino • *Department of Biological Sciences, University of the Pacific, Stockton, CA, USA*
Knut Madden • *Biogrammatics Inc., Carlsbad, CA, USA*
Carlos Martínez • *School of Biochemical Engineering, Pontificia Universidad Católica de Valparaíso, Valparaíso, Chile*
Diethard Mattanovich • *Department of Biotechnology and Food Science, BOKU University, Institute of Microbiology and Microbial Biotechnology, Vienna, Austria; Austrian Centre of Industrial Biotechnology (ACIB GmbH), Vienna, Austria*
Magdalena Merkaš • *Institute of Molecular Biotechnology, Graz University of Technology, NAWI Graz, Graz, Austria*
Lisa Michel • *Hamburg University of Applied Sciences, Faculty Life Sciences, Department Biotechnology, Hamburg, Germany*
Pierre Millard • *TBI, Université de Toulouse, CNRS, INRAE, INSA, Toulouse, France*

José Luis Montesinos-Seguí • *Department of Chemical, Biological and Environmental Engineering, School of Engineering, Universitat Autònoma de Barcelona, Bellaterra (Barcelona), Spain*

Juan A. Moreno-Cid • *Bioprocess R&D Pilot Plant, Bionet, Fuente Álamo, Spain*

Charles Moritz • *Department of Biotechnology, Institute of Microbiology and Microbial Biotechnology, BOKU University, Vienna, Austria; Austrian Center of Industrial Biotechnology, Vienna, Austria*

Holger Morschett • *Lonza AG, Visp, Switzerland*

Miguel Angel Nieto-Taype • *Department of Chemical, Biological and Environmental Engineering, School of Engineering, Universitat Autònoma de Barcelona, Barcelona, Spain*

Ruth Ordóñez • *Bioprocess R&D Pilot Plant, Bionet, Fuente Álamo, Spain*

Melanie Ottinger • *daspool, Wädenswil, Switzerland*

Odyssefs-Ioannis Pantelakis • *Institute of Molecular Biotechnology, Graz University of Technology, NAWI Graz, Graz, Austria*

Alexander Pekarsky • *Institute of Chemical, Environmental and Bioscience Engineering, Research Group Integrated Bioprocess Development, TU Wien, Vienna, Austria*

Carsten Pichler • *Christian Doppler Laboratory for Innovative Pichia pastoris Host and Vector Systems, Institute of Molecular Biotechnology, Graz University of Technology, Graz, Austria; Institute of Molecular Biotechnology, Graz University of Technology, Graz, Austria*

Karen M. Polizzi • *Department of Chemical Engineering, Imperial College London, London, UK; Imperial College Centre for Synthetic Biology, Imperial College London, London, UK*

Verena Puxbaum • *Department of Biotechnology and Food Science, BOKU University, Institute of Microbiology and Microbial Biotechnology, Vienna, Austria; Austrian Centre of Industrial Biotechnology (ACIB GmbH), Vienna, Austria*

Astrid Radkohl • *Institute of Molecular Biotechnology, Graz University of Technology, NAWI Graz, Graz, Austria; BioTechMed-Graz, Graz, Austria*

Hana Raschmanová • *University of Chemistry and Technology Prague, Department of Biotechnology, Prague, Czech Republic; Department of Biotechnology, University of Chemistry and Technology Prague, Dejvice, Czech Republic*

Pierre D. M. Reinprecht • *bisy GmbH, Wuenschendorf, Hofstätten a. d. Raab, Austria*

Lukas Rieder • *Christian Doppler Laboratory for Innovative Pichia pastoris Host and Vector Systems, Institute of Molecular Biotechnology, Graz University of Technology, Graz, Austria*

Albert Sales-Vallverdú • *Department of Chemical, Biological and Environmental Engineering, School of Engineering, Universitat Autònoma de Barcelona, Barcelona, Spain*

Lena Schaffer • *bisy GmbH, Wünschendorf, Hofstätten a. d. Raab, Austria; Institute of Biomedical Informatics, Graz University of Technology, Stremayrgasse, Graz, Austria*

Lia Jade Schallengruber • *Institute of Molecular Biotechnology, Graz University of Technology, Petersgasse, Graz, Austria*

Tanja Michelle Schleichert • *bisy GmbH, Hofstätten an der Raab, Austria*

Marina Schmid • *Roche Diagnostics GmbH, Penzberg, Germany*

Veronika Schusterbauer • *bisy GmbH, Wünschendorf, Hofstätten a. d. Raab, Austria; Institute of Biomedical Informatics, Graz University of Technology, Stremayrgasse, Graz, Austria*

REZA MALEKI SEIFAR • *Bioengineering Department, IQS, Universitat Ramon Llull, Barcelona, Catalonia, Spain; Nofalab B.V., Schiedam, The Netherlands*
KIRILL SMIRNOV • *Christian Doppler Laboratory for Innovative Pichia pastoris Host and Vector Systems, Graz University of Technology, Petersgasse, Graz, Austria*
OLIVER SPADIUT • *Institute of Chemical, Environmental and Bioscience Engineering, Research Group Integrated Bioprocess Development, TU Wien, Vienna, Austria*
ALEX J. SPICE • *Department of Chemical Engineering, Imperial College London, London, UK; Imperial College Centre for Synthetic Biology, Imperial College London, London, UK*
MARCEL STRAUMANN • *Molecular Partners AG, Schlieren, Switzerland*
GERHARD G. THALLINGER • *Institute of Biomedical Informatics, Graz University of Technology, Stremayrgasse, Graz, Austria*
ILYA TOLSTORUKOV • *Keck Graduate Institute, Claremont, CA, USA; BioGrammatics Inc., Carlsbad, CA, USA*
FRANCISCO VALERO • *Department of Chemical, Biological and Environmental Engineering, School of Engineering, Universitat Autònoma de Barcelona, Bellaterra (Barcelona), Spain*
WALTER VAN GULIK • *Department of Biotechnology, Delft University of Technology, Delft, The Netherlands*
FUENSANTA VERDÚ-NAVARRO • *Bioprocess R&D Pilot Plant, Bionet, Fuente Álamo, Spain; Instituto de Biotecnología Vegetal, Universidad Politécnica de Cartagena, Cartagena, Spain*
JAVIER VIÑA-GONZALEZ • *EvoEnzyme S.L., Parque Científico de Madrid, Madrid, Spain*
LISA WASSERER • *bisy GmbH, Hofstätten an der Raab, Austria*
FLORIAN WEISS • *Christian Doppler Laboratory for Innovative Pichia pastoris Host and Vector Systems, Institute of Molecular Biotechnology, Graz University of Technology, Graz, Austria*
ASTRID WENINGER • *Graz University of Technology, Institute of Molecular Biotechnology, Graz, Austria*
CHRISTINE WINKLER • *Institute of Molecular Biotechnology, Graz University of Technology, Graz, Austria*
LEON WOLBANK • *Christian Doppler Laboratory for Innovative Pichia pastoris Host and Vector Systems, Institute of Molecular Biotechnology, Graz University of Technology, Graz, Austria*
IWO ZAMORA • *Infors HT, Bottmingen, Switzerland*

Part I

Introduction

Chapter 1

A Brief History of the *Pichia pastoris* Expression System

James M. Cregg, Thomas G. Chappel, Kurt R. Gehlsen, Shaun A. Kirkpatrick, and Knut Madden

Abstract

The origin and development of the *Pichia pastoris* expression system is described from its roots as an organism for the production of single-celled protein (SCP) to its evolution into a major host for the synthesis of recombinant proteins. Throughout its history, the ability of *P. pastoris* to grow on methanol as a sole source of carbon and energy to extremely high densities in fermenter cultures has been a central theme. The expression system would not be what it is today without the efforts of numerous talented researchers in fields from molecular biology to bioprocess engineering. As products made using *P. pastoris* take their places in world markets, it is clear that the future of the system is bright.

Key words *Pichia pastoris*, *Komagataella phaffii*, Recombinant protein production, Foreign gene expression, Alcohol oxidase promoter, DNA-mediated transformation, High-density fermentation, β-galactosidase, Invertase, Glycosylation

1 Yeast Systematics, the Discovery and Re-Classification of *Pichia pastoris*

The genus *Pichia* was originally established by Hansen in 1904 and named in honor of the nineteenth-century scientist P. Pichi. In 1919, Marie Antoine Alexandre Guilliermond isolated a wild yeast from a French chestnut tree and classified it as *Zygosaccharomyces pastori*, later renamed *Saccharomyces pastori* [1]. Several years later, multiple wild yeasts isolated from Yosemite Park oak trees by Herman Phaff were classified as a single, new species of *Pichia*. Phaff could not distinguish these new isolates from Guilliermond's 1919 isolate, so he proposed moving *Saccharomyces pastori*, along with his isolates, into a new species named *Pichia pastoris*. Phaff continued to isolate additional samples of *P. pastoris* from cacti in the Sonoran Desert.

More modern molecular biological approaches in yeast taxonomy enabled researchers to divide some of these isolates into new species. According to partial sequence analysis of 18S and 26S

Xavier Garcia-Ortega et al. (eds.), *Pichia pastoris: Methods and Protocols*, Methods in Molecular Biology, vol. 2697,
https://doi.org/10.1007/978-1-0716-4779-0_1,

ribosomal RNAs, several strains that were originally classified as *P. pastoris* were found to be more heterogeneous. Based on these findings, Yamada et al. proposed a new genus for these strains, *Komagataella,* with a new name, *Komagataella pastoris,* chosen for the type strain, ATCC 28485 [2]. Subsequently, the biotechnological strains of *P. pastoris* were also genotyped. Kurtzman found them to be different enough from *K. pastoris* to warrant becoming their own species of *Komagataella* and named them *Komagataella phaffii*; the type strain for *K. phaffii* is NRRL Y7556 [3]. Other strains also reclassified as *K. phaffii* include (1) NRRL Y-11430, optimized for a decade to grow efficiently on methanol to a high cell density for SCP in animal feed at Phillips Petroleum (Bartlesville, OK), and (2) the histidine auxotroph, GS115, isolated to support selection for a heterologous protein expression system developed during the '80s at the Salk Institutes Biotech Industrial Associate (SIBIA, La Jolla, CA). GS115 and related strains were engineered into "kits" that were distributed worldwide starting in 1994 by Invitrogen (Carlsbad, CA) under license from Research Corporation Technologies (Tucson, AZ).

Thus, from a strict yeast-taxonomy point of view, *Pichia pastoris* is a yeast species that no longer exists. However, since the name is so well known among researchers, it continues to be used and is still accepted. Even though *K. pastoris* and *K. phaffii* have become the norm in peer-reviewed manuscripts, *P. pastoris* (or just "Pichia") is still in the common lexicon of labs generating *Komagataella* data. More importantly, "Pichia" is predominantly used by the larger population of scientists only using the "Pichia" system for protein production, likely due to the marketing efforts for these kits (Fig. 1).

The two *Komagataella* species, *K. pastoris* and *K. phaffii,* are the most cited for academic and industrial biotechnology applications and will be the focus of this publication; most of the protocols can be used with both strains equally well. Note from the editor, *P. pastoris* will be used in cases where the experimental protocols are expected to work for both species.

2 The Single-Cell Protein Years

The existence of yeast that can grow on methanol as a sole source of carbon and energy was unknown until 1969 [4]. The discovery of methylotrophic yeast was quickly recognized by the petroleum industry as a potential opportunity for their industry. In addition to petroleum, oil wells were known to generate large quantities of natural gas, far more than could profitably be sold at that time. As a result, much of this gas was flared off at the wellhead. The companies recognized that this burning process was a waste of a potentially valuable resource. With the discovery of methylotrophic

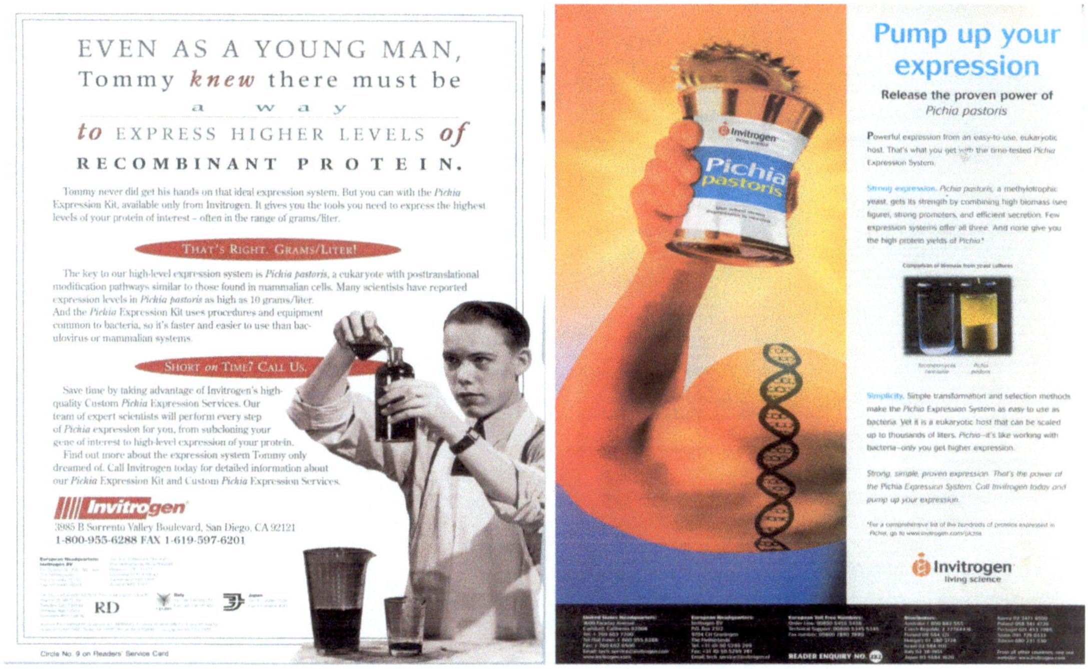

Fig. 1 Early advertisements for Pichia "kits" from Invitrogen (1996/7, Carlsbad, CA, USA)

microorganisms, they conceived of the idea of chemically converting the excess natural gas (mainly methane) to methanol and then feeding the methanol to yeast. The resulting yeast biomass or single-cell protein (SCP) would then be sold as a high-protein animal feed supplement. In this way, a petroleum industry waste product would be converted into a marketable product. During much of the 1970s, research on the SCP-generation idea was very popular and was heavily supported by petroleum companies and government funding agencies around the world.

Of interest to the *Pichia pastoris* expression system were the SCP research and development efforts of the Phillips Petroleum Company of Bartlesville, Oklahoma. During much of the 1970s, Phillips pursued two lines of SCP development research. One was the development of a bioprocessing method for growing methylotrophic yeast continuously (for periods of a month or greater) and at extremely high cell densities (<130 grams/liter dry cell weight or 400–500 OD_{600} units). In their SCP-generation process, a steady-state culture was achieved by continuously feeding methanol and nutrients needed for growth into the bioreactor and simultaneously harvesting yeast culture. The density of these cultures was limited by the ability to provide sufficient oxygen to support a significant rate of growth. In the SCP process, yeast cultures emerged from the fermenters as a viscous paste that was then spray-dried to remove remaining liquid. The resulting dried yeast was the SCP product (Fig. 2) which could be used as a high-protein

Fig. 2 Images of inside and outside the Phillips Petroleum pilot SCP production facility

animal feed supplement. The objective of Phillip's second line of research was to identify the methylotrophic yeast strain that was most compatible with their continuous high-density fermentation process. Because yeasts had never been examined for methylotrophy, Phillips' researchers had to obtain large numbers of yeast species samples from culture collections all over the world and test them for growth on methanol. The few that did grow on methanol were then tested for compatibility with the SCP fermentation process. The yeast species found to be most compatible with the SCP bioprocess was *Pichia pastoris* NRRL Y-1603. This strain was then subjected to long periods of continuous growth on methanol to select strains with optimal fermentation growth characteristics for the SCP process, and the best of these was deposited in the Northern Regional Research Laboratories as NRRL Y-11430.

Despite the intense SCP development efforts and the spending of millions of dollars by Phillips Petroleum, no methanol-based SCP products ever made it to market. During the 1970s, major shortages in world markets for petroleum products (the "oil crisis") occurred, and the price of these products, including the ultimate feedstock for the SCP process, natural gas, increased dramatically. As a result, Phillips and the other SCP producers could not make

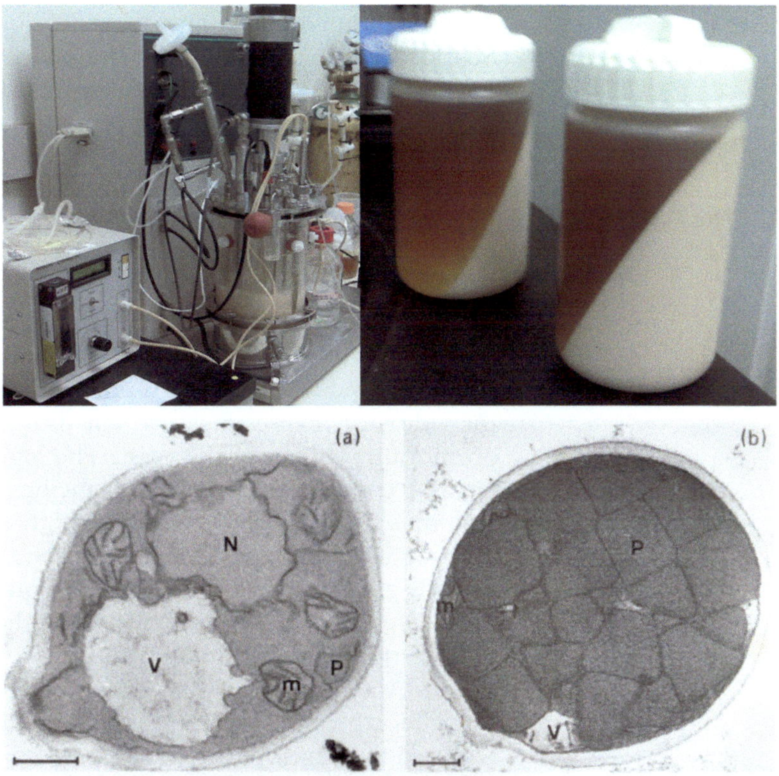

Fig. 3 High cell density culture of *P. pastoris*. Top image, standard fermenter set-up and bottles, post centrifugation, from a *P. pastoris* culture grown in a fermenter to a density of $>$ 200 g/liter wet cell weight. Bottom Image, yeast cells grown on (**a**) glucose, (**b**) methanol (N = nucleus, V = vacuole, m = mitochondria, and P = peroxisome). (Republished with permission from Martin veenhuis, 1989)

SCP at a price competitive with other sources of high-protein animal feed, which in the United States was mainly from soybeans.

Although Phillip's SCP development efforts did not result in a marketable product, they did provide key materials and knowledge needed for the development of a recombinant protein production system, based on *P. pastoris*. First, the base strain from which most other *P. pastoris* expression strains are derived, NRRL Y-11430, was identified. A critical feature of this strain that was important for both the SCP process and recombinant protein production processes was its ability to readily grow to high densities without the generation of ethanol (Fig. 3). Many yeasts like to ferment sugars and other carbon sources, the product of which is ethanol, which, at high cell densities, quickly accumulates to levels that are toxic to the cells and inhibit further growth and recombinant protein synthesis. Second, methods for inducing and growing *P. pastoris* at high cell densities on methanol were defined. Third, high cell density culture media recipes were developed and optimized for the SCP process (e.g., FM22 and the trace element solution PTM1). Each of these SCP-related developments continues to be used for the expression of recombinant proteins in *P. pastoris* today.

3 The Beginnings of *P. pastoris* as an Expression System

By the early 1980s, it was clear to Phillips that the sale of their SCP product was not likely to be profitable; at the same time, the biotech "revolution" had begun. Therefore, to improve the potential market value of *P. pastoris*-produced SCP, Phillips conceived of the idea of expressing growth hormone or other growth-promoting proteins in the yeast. The logic was that young animals fed SCP containing a growth-enhancing protein might grow faster, and thus, Phillips could justify a higher price for their growth-promoting SCP. To create *P. pastoris* strains that synthesized growth-promoting proteins, Phillips contracted with a small start-up biotechnology company owned by the Salk Institute called the Salk Institute Biotechnology/Industrial Associates (SIBIA) in La Jolla, CA, to perform the required research and development.

The expression of a growth-promoting protein in *P. pastoris* required the development of the same components and procedures needed to express any recombinant protein in the yeast, namely the isolation of a transcriptional promoter to control expression of the foreign gene and strains, vector components, and methods to facilitate the transfer of foreign DNA into *P. pastoris* strains. To control expression of foreign proteins, SIBIA researchers isolated the promoter from the *P. pastoris AOX1* gene, which encodes alcohol oxidase (AOX), the first enzyme in the methanol utilization pathway, and is strongly regulated by methanol [5]. In *P. pastoris* cells growing on glucose or most other carbon sources, almost no AOX is present. However, after shifting to methanol, up to 30% of total soluble protein in the yeast can be AOX [6] (Fig. 3). The *AOX1* gene and its promoter were isolated by screening libraries of *P. pastoris* DNA for clones that strongly hybridized to labeled RNA extracted from methanol-grown cells but not from ethanol-grown cells (ethanol, like glucose, is known to be a strong repressor of *AOX1* transcription). Inserts in plasmids that appeared by this assay to contain a methanol-regulated gene fragment were then sequenced, and sequences with the potential to encode the N-terminus of AOX, which was known, were identified. Once the *AOX1* gene and its promoter had been isolated, Northern blot analysis was employed to confirm that, as hoped, the *AOX1* gene was tightly regulated at the transcriptional level by the *AOX1* promoter (Fig. 4). More recently, less than one *AOX1* mRNA/cell is seen in glucose grown Pichia by transcriptome sequencing (BioGrammatics, Inc. unpublished data).

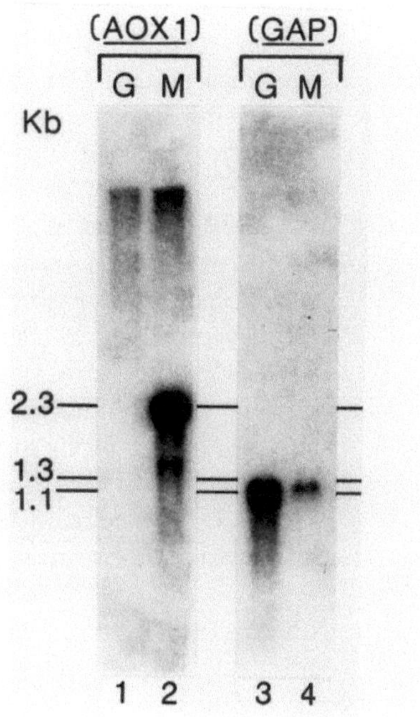

Fig. 4 Northern blot of *AOX* RNA. Total RNAs extracted from glucose (G) and methanol (M) grown cells of *P. pastoris* were separated by agarose gel electrophoresis and blotted onto a nitrocellulose filter. The RNAs in the left-hand blot were then hybridized with P^{32}-labeled DNA from a plasmid containing the *AOX1* gene. As a control, a second blot with the same RNAs was hybridized with labeled DNA from a plasmid containing the constitutively expressed glyceraldehyde dehydrogenase gene (*GAP*)

To transfer and maintain foreign DNA in the yeast, an auxotrophic host/biosynthetic gene transformation system was developed [7]. A suitable host was isolated by mutagenizing cultures of *P. pastoris* strain NRRL Y-11430 and screening for strains that required histidine supplementation for growth (His^- phenotype). Histidine auxotrophs were selected simply because they were the most frequent type of auxotroph identified in populations of mutagenized cells. To determine the specific step in the histidine biosynthetic pathway that was affected, a cell-free extract of each His^- strain was prepared and assayed for activity of each enzyme in the yeast histidine biosynthetic pathway. The first strain identified as specifically defective in activity for a histidine biosynthetic pathway enzyme was defective in histidinol dehydrogenase, the last enzyme in the pathway. In *Saccharomyces cerevisiae*, this enzyme is encoded by the *HIS4* gene, and therefore, the *P. pastoris* gene was also named *HIS4*, and this His^- strain was designated GS115 (*his4*). Once this strain was in hand, a *HIS4* gene was needed to transform the strain. To isolate the *P. pastoris HIS4* gene, SIBIA constructed a *P. pastoris* genomic DNA library in an *S. cerevisiae* shuttle vector, transformed that library into a known *his4* mutant strain of *S. cerevisiae*, and selected for His⁺ transformants on minimal medium plates. The *P. pastoris HIS4* gene was then recovered by extracting total DNA from several of the resulting *S. cerevisiae* His⁺ transformants and transforming that DNA, still in its shuttle vector, into *E. coli*. The gene was then inserted into an *E. coli* plasmid pBR322 to create a *P. pastoris/E. coli* shuttle vector. This vector, along with one containing the *S. cerevisiae HIS4* gene, was employed to transform GS115 (*his4*) to histidine prototrophy. Both vectors worked well. SIBIA went on to characterize the fate of vector DNA transformed into *P. pastoris* and observed that it was similar to *S. cerevisiae*. A vector could be maintained either as an independently replicating element in transformed cells through the presence of a *P. pastoris*-specific autonomous replication sequence (ARS) on the vector, or it could be integrated into the genome of the yeast. Importantly, SIBIA showed that by linearizing a vector within sequences shared by the vector and host genomes (e.g., the *P. pastoris HIS4* gene), the vector could be precisely targeted to integrate into the *P. pastoris* genome at that locus by site-specific recombination. Finally, SIBIA demonstrated that, as in *S. cerevisiae*, DNA sequences could be precisely added or deleted from any specific locus of the *P. pastoris* genome through transformation of *P. pastoris* with a properly constructed segment of DNA (a.k.a., knock-in/knock-out construction) [8].

4 Expression of the First Foreign Gene

With the *AOX1* promoter and a transcriptional terminator (also obtained from the *AOX1* gene) and methods and materials needed to transform and maintain DNA in *P. pastoris*, SIBIA was ready to express a foreign gene in the yeast. As a good reporter gene for initial test experiments, SIBIA employed the *E. coli* β-galactosidase gene (*lacZ*) [9]. The *lacZ* gene was well studied and had the further advantage that the presence of its product (β-gal) in transformed colonies could be easily observed by their blue color when plated on X-gal agar. Finally, *P. pastoris* lacked its own β-gal, thus eliminating the possibility of an endogenous β-gal complicating efforts to measure levels of the foreign β-gal product. Results with *lacZ* in *P. pastoris* were very satisfying. Expression of *lacZ* was regulated almost exactly like that of *AOX1*, and the β-gal product appeared to be a fully active enzyme. Furthermore, in methanol-grown cells of a *lacZ*-expression strain, approximately 20% of total soluble protein was β-gal (Fig. 5).

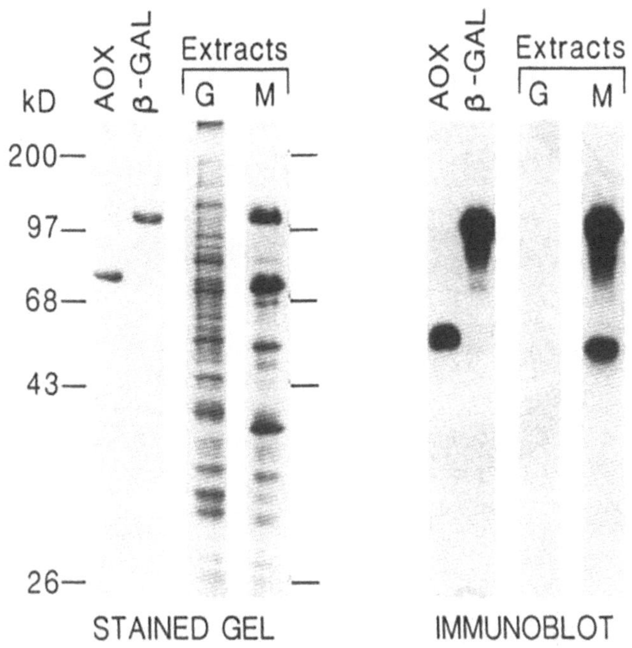

Fig. 5 Expression of the *E. coli lacZ* gene in *P. pastoris*. Cell-free extracts were prepared from glucose (G) and methanol (M) grown cells of a *P. pastoris* strain transformed with a vector containing the *lacZ* gene under the control of the *AOX1* promoter. On the left is an SDS-PA gel through which extract samples were electrophoresed and then stained with Coomassie Blue. As a control, samples of purified β-gal and AOX were electrophoresed in parallel. On the right is a Western blot in which the same samples were electrophoresed through an SDS-PA gel, blotted onto a nitrocellulose filter, and then reacted with radioactively labeled antibodies specific for β-gal and AOX

5 *S. cerevisiae* Invertase, the First Secreted Recombinant Protein

An important advantage of expressing foreign proteins in yeast is the potential for secretion of the protein product. The secretory pathway compartments of eukaryotes are a different environment relative to the cytoplasm and include special proteins called chaperones that aid in the folding of nascent proteins. In addition, it is only in the secretory pathway that proteins are subjected to post-translational modification events, such as proteolytic processing, disulfide bridge formation, and glycosylation. An added advantage of secretion is that yeasts typically do not secrete large amounts of their own proteins. Thus, the successful secretion of a foreign protein into the yeast culture medium amounts to a major purification step, separating the protein product away from the bulk of intracellular yeast proteins. (Note: although secretion, relative to intracellular expression, may seem to be the most attractive method of expressing a foreign protein, the researcher typically does have a

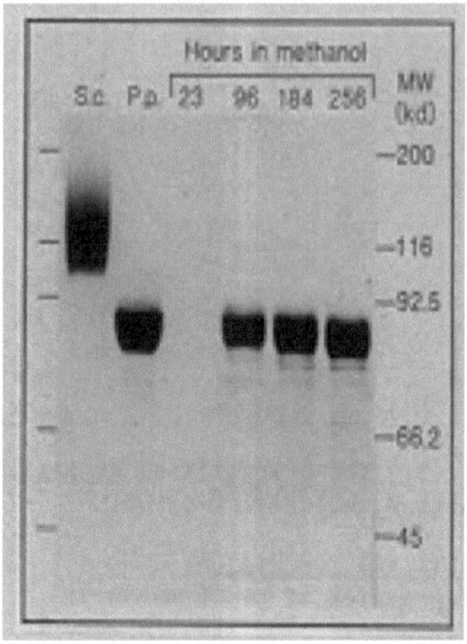

Fig. 6 Secretion of invertase from *P. pastoris*. A *P. pastoris* strain harboring an AOX1-promoter-SUC2 expression cassette was grown in fermenter culture on glycerol and then shifted to methanol. Samples of the culture supernatant were collected periodically and then electrophoresed through SDS-PA gel and stained with Coomassie Blue. The lane on the far left contains approximately 2 ug of invertase purified from S. cerevisiae. The lanes to the right each contain 25 ul of culture supernatant mixed with loading dye. The numbers across the top indicate the time in hours after the culture was shifted from glycerol to methanol

choice in the matter since proteins almost always must be synthesized in the same environment as in their native hosts.) As a first test of the secretion of a foreign protein in *P. pastoris*, SIBIA chose the *S. cerevisiae* invertase encoded by the *SUC2* gene of that yeast [10]. Like β-gal, invertase is a well-characterized enzyme that, in *S. cerevisiae*, is known to undergo signal sequence processing, multiple disulfide bridge formations, and glycosylation at most of its numerous (fourteen) potential N-linked sites. And like β-gal, *P. pastoris* does not have its own invertase, the presence of which might complicate results with the recombinant enzyme. The *SUC2* gene was inserted into a *P. pastoris* shuttle vector behind the *AOX1* promoter, and the construct was transformed into *P. pastoris*. Shake-flask cultures of the strain were induced on methanol, and samples were taken and examined for invertase. As hoped, invertase was present in the culture medium. The invertase-secreting strain was then grown and methanol induced in a fermenter culture (the first recombinant *P. pastoris* strain to be grown this way), and supernatant samples were harvested and examined for invertase activity and protein (Fig. 6). Although the final cell density of the culture was only a modest 40 G/L dry cell weight, results from both activity assays and SDS-PAGE indicated the presence of invertase in the culture medium at approximately 2.5 G/L.

The presence of invertase in the culture medium came as a surprise to SIBIA researchers since in *S. cerevisiae*, invertase, although a product of the yeast secretory pathway, remains largely trapped in the cell wall matrix of the yeast. An explanation for this came with further characterization of the *P. pastoris*-made enzyme. SDS-PAGE visualization revealed a single relatively tight band of approximately 80 kDa, whereas *S. cerevisiae*-made invertase is a very heterogeneous population of molecules ranging in size from approximately 100–140 kDa. Comparison of enzymatically de-glycosylated samples of invertase from both yeasts revealed that the protein portion of the molecules was identical in size. Furthermore, amino-terminal amino acid sequencing of the *P. pastoris*-made invertase was identical to that from *S. cerevisiae*, indicating that the signal sequence had been properly processed. Thus, the difference in size between the two enzymes was most likely due to a difference in glycosylation [11]. In *S. cerevisiae*, N-linked glycosylation typically results in carbohydrates containing a hundred or more mannose residues per chain. However, analysis of the carbohydrates on *P. pastoris*-made invertase by one- and two-dimensional NMR analysis revealed much smaller structures with an average of 9–14 mannose residues [12]. The significantly smaller size of the carbohydrate chains added to *P. pastoris*-made invertase provided an obvious explanation for the ability of *P. pastoris* to secrete the enzyme freely into the culture medium.

6 Alternative Host Strains Developed at SIBIA

Two types of *P. pastoris* host strains that significantly improve the levels of expression of certain recombinant proteins were also developed at SIBIA during the 1980s and continue to be in wide use today. In one of these host strains, the *AOX1* gene has been deleted (e.g., KM71 *aox1Δ AOX2 his4*) [8]. As a result, this strain grows only very slowly on methanol due to the presence of a second functional but transcriptionally weak *AOX2* gene. With this host, SIBIA researchers observed a 3- to 5-fold increase in expression of several recombinant proteins, including β-gal, invertase, and the hepatitis B surface antigen [8, 10, 13]. Since that time, numerous other researchers have reported similar results with *aox1Δ* (a.k.a. Mut slow) hosts. However, examples in which no improvement or even reduced expression have also been reported. The reasons for improved expression with *aox1Δ* hosts have never been determined. One possibility is that *aox1Δ* strains are strongly carbon limited on methanol due to low levels of AOX and respond to this starvation by hyper-induction of the *AOX1* promoter. Since the recombinant gene is under the control of this promoter, transcription of the recombinant gene is instead hyperinduced. Another possibility is that the much-reduced metabolic rate of these hosts allows more time for the cells' translational machinery and/or post-translational protein-modifying machinery to correctly perform their essential functions on recombinant proteins. Thus, a greater portion of the synthesized recombinant protein achieves its final stable and active form.

A second set of *P. pastoris* host strains still in common use are deleted for either or both of the two proteinase genes, *PEP4* and *PRB1* (SMD1168 *pep4 his4*; SMD1165 *prb1 his4*; SMD1163 *pep4 prb1 his4*) [14, 15]. For a certain small set of recombinant proteins, deletion of these genes, particularly *PEP4*, results in a major improvement in the levels of fully intact protein. PEP4 is a proteinase responsible for activating several other secretory pathway hydrolases [16]. Thus, deletion of *PEP4* helps in the expression of proteins that are sensitive to degradation by one or more of these enzymes. For sensitive proteins, the results can be dramatic with an expressed protein going from highly degraded and nearly undetectable to intact and in good yield. As a word of caution, it is not a good idea to use one of these proteinase-deficient strains prophylactically as the only host, since the strains are generally less healthy relative to wild-type strains with the result that many recombinant proteins may be poorly expressed unless the system is further optimized.

7 Phillips Licenses the System to Welcome Biotechnology

In 1988, Phillips Petroleum issued its first commercial license for use of the *P. pastoris* expression system to Wellcome Research Laboratories in the United Kingdom. In addition to being the first licensee and expressing several recombinant proteins in *P. pastoris*, Wellcome researchers contributed two very useful general advances to the system. The first was the construction and use of *P. pastoris* expression vectors encoding the *S. cerevisiae* alpha mating type pre-pro signal sequence (αMF) for directing recombinant proteins through the yeast secretory pathway [17, 18]. The αMF signal has been the most commonly used and generally applicable signal for secretion of recombinant proteins from *P. pastoris*. Although successful at secreting most proteins, the αMF signal can sometimes be problematic with regard to its proper removal from the N-terminus of certain recombinant proteins by the resident KEX2 and STE13 signal peptidases of *P. pastoris*. Thus, N-terminal sequencing of proteins secreted via the αMF signal is usually warranted to confirm that the signal has been properly processed.

Wellcome's second general contribution to the *P. pastoris* system was the development of a method to enrich populations of transformed cells for ones that contained multiple integrated copies of an expression vector [17, 18]. Previously, it was known that strains with additional copies of an expression vector usually (but not always) produced more product. However, a method to easily construct such strains did not exist. The Wellcome method involved the incorporation of a bacterial neomycin (G418) resistance gene into a *P. pastoris* expression vector. The expression vector, after insertion of a recombinant gene, was transformed into a *his4 P. pastoris* host by selection for the *HIS4* gene on the vector. Large numbers of His^+ transformants were then spread on agar plates containing a high level (1–4 mg/ml) of G418. A number (10–100) of highly G418-resistant colonies were recovered and examined for vector copy number by DNA dot blot. Approximately 10% of the recovered G418-resistant colonies were resistant due to an increase in the number of vector copies. Most contained just a few additional vector copies, but a few (~1%) "jack-pot" strains could be found that contained 10–30 or more copies.

Finally, Wellcome engineered a generalized *P. pastoris* expression vector that included segments of DNA encoding both the αMF signal and the G418-resistance gene. In these vectors, the αMF signal was inserted properly downstream of the *AOX1* promoter and immediately followed by a multiple cloning site for insertion of the foreign gene, which was in turn followed by the *AOX1* transcriptional terminator and the *P. pastoris HIS4* gene. The most popular of these multi-copy enrichment vectors was pPIC9K and is still in common use today.

8 The *P. pastoris* Expression System Goes Public

Despite the early success of *P. pastoris* as an expression system, Phillips Petroleum lost interest in the project since growth-promoting proteins expressed in the yeast did not have the "hoped for" effect of stimulating animal growth when incorporated into SCP preparations. In light of this and other strategic factors, in 1993, Phillips sold the *Pichia pastoris* technology and business to Research Corporation Technologies (RCT) of Tucson, AZ. At about the same time, Phillips also contracted with a small new biotechnology company in San Diego, CA, called Invitrogen (more recently Life Technologies and now ThermoFisher), to prepare and sell kits containing *P. pastoris* expression system components, mainly vectors and host strains, along with instructions on how to express genes in the system.

RCT continued and expanded this arrangement with Invitrogen which continues to the present day. Under this arrangement, commercial organizations can purchase a system "kit" and use it under a limited use research license for a trial period of one year. If a kit purchaser is an academic institution performing only non-commercial research, the academic researchers can continue to use the expression system kit for longer than one year. The ready availability of the *P. pastoris* system unleashed a wave of interest and research in the system. Most importantly, large numbers of publications began to appear in the literature with new advances in vector components, host strains, and methods for manipulating the system for maximum yields. Now, 30 plus years later, more than 13,000 publications citing "*Pichia*" are listed in PubMed [19]. The large number of researchers worldwide using the *P. pastoris* system, or wishing to use the system, generated a need for current information about Pichia expression systems, and the function of this volume, and for rapid communication between works in the field, a need met by international *Pichia* meetings held every other year (most recently Pichia 2024 in San Diego).

9 Major Advances Since 1994

At least two major advances in the *P. pastoris* expression field have occurred since 1994. One involves the modification of carbohydrates added to N-linked glycosylation sites on proteins that enter the secretory pathway of the yeast. Like other eukaryotic organisms, *P. pastoris* adds sugars to the amino group of asparagine residues that are part of the N-linked glycosylation recognition sequence Asp-X-Ser/Thr. However, unlike higher eukaryotes, which add a mixture of different sugars to N-linked sites, the sugars added by *P. pastoris* are mostly mannoses. In addition, these mannose structures are typical of yeasts and fungi and differ significantly from higher eukaryotic structures. The differences in glycosylation between yeasts and humans are important since the human immune system recognizes these differences and can mount a strong allergic reaction. Thus, *P. pastoris*-made glycoproteins cannot be directly used as human injectable pharmaceuticals. Starting in 1997, Dr. Roland Contreras of the University of Ghent, Belgium, and colleagues reported on the construction of *P. pastoris* strains with modified N-linked structures, ones that were closer to those of higher eukaryotes [20]. Continuing the work of Drs. Contreras and Callewaert, RCT launched the *Pichia* GlycoSwitch® system through Biogrammatics in 2014. The *Pichia* GlycoSwitch® system consists of the SuperMan5 strains and a series of Pichia GlycoSwitch® vectors that allow researchers to custom-tailor the glycosylation pattern and thus, explore the effects of different glycoforms on the biology of the protein of interest. In a parallel effort, Dr. Tillman Gerngross and colleagues took *Pichia* glycosylation engineering to another level and founded the company Glycofi, later acquired by Merck. In 2004, Glycofi reported the construction of a *P. pastoris* strain that was capable of synthesizing recombinant glycoproteins with fully human carbohydrate structures [21].

The other major advance was the publication of a *P. pastoris* genome nucleotide sequence by the laboratory of Dr. Nico Callewaert, University of Ghent, Belgium [22]. The availability of the *P. pastoris* GS115 genomic sequence allowed researchers to easily isolate and amplify any gene in the yeast's genome by simply identifying the gene in the database, synthesizing appropriate oligonucleotide primers, and amplifying the gene by polymerase chain reaction (PCR). The availability of the *P. pastoris* genomic sequence also made it easier to create specific knock-out mutations. Genome sequences of other *P. pastoris* strains have followed, including the *K. phaffii* and *K. pastoris* type strains.

10 A Few Firsts for the *P. pastoris* System

The first product on the market made by recombinant methods in *P. pastoris* was bovine enterokinase made by Invitrogen starting in 1995 (Table 1). Invitrogen's EnterokinaseMax™ is the catalytic subunit of the enzyme, a serine protease used to digest fusion proteins to release the fusion partner, or tag, from recombinant fusion proteins. The first product launched for human use produced by *P. pastoris* is a hepatitis-B vaccine called Shanvac-B, sold by Shantha Biomechanics of Hyderabad, India, starting in 1997. The first human pharmaceutical produced by P. pastoris and approved for sale in the United States is KALBITOR from Dyax Corp. KALBITOR (ecallantide), approved by the Food and Drug Administration of the United States in 2009 is an inhibitor of plasma kallikrein for the treatment of acute attacks of hereditary angioedema [23]. The second human pharmaceutical produced by *P. pastoris* and approved in the United States is JETRA™, an antithrombolytic drug from ThromboGenics, NV, launched in 2012. In addition to the *P. pastoris*-made products on the market, many others are at various stages of development and testing. More recently, an antibody drug developed by Alder Biopharmaceuticals and subsequently acquired by Lundbeck Pharmaceuticals, was made commercially available under the name Eptinezumab (brand name Vyepti®). Vyepti® is a full-length "humanized" monoclonal antibody for relief from episodic and chronic migraine headaches. Interestingly, although antibodies are normally N-glycosylated at a single site on each of the two heavy chains of the tetrameric molecule, the sequence encoding the glycosylation site is mutated in the Alder construct to remove the site. Thus, glycosylation is not essential to the structure or function of their antibodies.

Table 1
Commercial examples of recombinant products produced by *P. pastoris* (Pichia)

Product	Company	Indication/use	Status
Alkaline Phosphatase	ArticZymes/Biotec Pharmacon ASA	DNA preparation	On the market
Alkaline Phosphatase, DNase, & Proteinase K, Trypsin	Roche Diagnostics GmbH	DNA Preparation	On the market
Allergy Immunogens (house dust mite, cat, mouse and cockroach)	INDOOR Biotechnologies	Research reagent	On the market (US)
Allergy Immunogens (Various)	Phadia/ ThermoFisher	Research and diagnostic reagents	On the market
AquaVac®IPN (vaccine against infectious pancreatic necrosis virus in salmon)	Schering Plough Animal Health	Vaccines for infectious pancreatic necrosis in salmon	On the market (Chile and Norway)
Corbevax® (Protein sub-unit COVID-19 Vaccine)	Biological E. Limited	R_x—COVID-19 vaccine	On the market (India & other countries)
Enterokinase	Invitrogen (Thermo Fisher)	Research reagent	On the market
Gaussia luciferase	Prolume	Bioluminescent applications, e.g. biological markers and BioToys	On the market (US)
Humalog® (human insulin lispro)	Eli Lilly	R_x—Diabetes	On the market
Human Cystatin C	BBI Solutions, Crumlin UK	Research reagent	On the market (EU)
Human type III collagen	Fibrogen	Medical research reagents/dermal filler	On the market (US)
Humulin® (human insulin)	Eli Lilly	R_x—Diabetes	On the market
Indovac (COVID-19 Vaccine)	Biofarma	R_x—COVID-19 vaccine	On the market (Indonesia)
Insugen® (human insulin)	Biocon	R_x—Diabetes	On the market
JETREA™ (Microplasmin, anti-thrombolytic protein)	ThromboGenics Ltd/Alcon	R_x—Vitreomacular adhesion R_x—Diabetic retinopathy	Launched (US & EU) but subsequently withdrawn from the market
Kalbitor® (DX-88 ecallantide, kallikrein inhibitor protein)	Dyax (Shire)	R_x—Hereditary angioedema	On the market (US)

(continued)

Table 1 (continued)

Product	Company	Indication/use	Status
LysN—(endopeptidase)	Immunoprecise Antibodies	Research reagent	On the market
Medway (human serum albumin)	Mitsubishi Tanabe Pharma Corp.	R_x—Blood volume expander	Launched (Japan) but subsequently withdrawn from the market
OptiPhos® Plus (phytase)	Huvepharma	Animal feed additive	On the market
Purifine® (phospholipase C)	DSM/Verenium	Degumming of high phosphorus oils	On the market
Shanferon (interferon-alpha 2b)	Shantha Biotech (Sanofi)	R_x—Hepatitis C/ Cancer	On the market (India)
Shanvac™-B (hepatitis B vaccine)	Shantha Biotech (Sanofi)	R_x—Hepatitis B	On the market (India)
Soy leghemoglobin	Impossible Foods	Food additive (flavor and color)	On the market
Vyepti® (Eptinezumab, anti-CGRP monoclonal antibody)	Alder Pharmaceuticals/ Lundbeck	R_x—Episodic and chronic migraines.	On the market

Additionally, generally regarded as safe (GRAS) notices have been published for recombinant proteins produced by *P. pastoris*, including phospholipase C for use in food processing (BASF, San Diego, CA), and recombinant soy leghemoglobin as a food and color ingredient (e.g., in the Impossible Burger, Impossible Foods, Redwood City, CA). Interestingly, leghemoglobin is expressed intracellularly to obtain full heme "loading," and 6–9% of the host, *Pichia*, proteins are added to the food product with the recombinant leghemoglobin. These "food" protein examples are only the "tip of the iceberg"; currently, egg, dairy, meat, and other food and flavor proteins are generated with Pichia all around the world making new applications possible. Future demand will likely utilize *P. pastoris* in the production of many more proteins for food and feed production and processing, as well as for any protein production where a eukaryotic microbe, with a 2 hour doubling time, excels.

And finally, the speed at which new products can be produced by a Pichia expression system, at a large scale, was demonstrated during the COVID-19 crisis. Corbevax® from Biological E. (India) and Indovac from BioFarma (Indonesia) were both rapidly produced by *Pichia* and introduced into markets around the world, and researchers used *Pichia* to express various proteins associated with COVID-19 in their efforts to end the pandemic [24–36]. It has been 30 years since the first *P. pastoris* products were made at SIBIA, and because of the "kits" launched by Invitrogen/RCT, it is safe to say, *P. pastoris* has become a mainstream tool to produce recombinant proteins with a tremendous amount of ongoing research to further improve the *P. pastoris* expression system. We look forward to the next 30 years of *P. pastoris* innovation.

Competing Interest Statement

The authors have no conflicts of interest to declare that are relevant to the content of this chapter.

References

1. Lodder J, Kreger-van Rij NJW (1952) The yeasts: a taxonomic study. North-Holland Publishing, Amsterdam
2. Yamada Y, Matsuda M, Maeda K, Mikata K (1995) The phylogenetic relationships of methanol-assimilating yeasts based on the partial sequences of 18S and 26S ribosomal RNAs: the proposal of *Komagataella* gen. nov. (Saccharomycetaceae). Biosci Biotechnol Biochem 59:439–444
3. Kurtzman CP (2009) Biotechnological strains of *Komagataella* (*Pichia*) *pastoris* are *Komagataella phaffii* as determined from multigene sequence analysis. J Ind Microbiol Biotechnol 36:1435–1438
4. Ogata K, Nishikawa H, Ohsugi M (1969) A yeast capable of utilizing methanol. Agric Biol Chem 33:1519–1520
5. Ellis SB, Brust PF, Koutz PJ et al (1985) Isolation of alcohol oxidase and two other methanol regulatable genes from the yeast Pichia pastoris. Mol Cell Biol 5:1111–1121
6. Veenhuis M, van Dijken JP, Harder W (1983) The significance of peroxisomes in the metabolism of one-carbon compounds. Adv Microb Physiol 24:1–82
7. Cregg JM, Barringer KJ, Hessler AY, Madden KR (1985) *Pichia pastoris* as a host for transformation. Mol Cell Biol 5:3376–3385
8. Cregg JM, Madden KR, Barringer KJ et al (1989) Functional characterization of the two alcohol oxidase genes from the yeast *Pichia pastoris*. Mol Cell Biol 9:1316–1423
9. Tschopp JF, Brust PF, Cregg JM et al (1987) Expression of the *lacZ* gene from two methanol-regulated promoters in *Pichia pastoris*. Nucleic Acids Res 15:3859–3875
10. Tschopp JF, Sverlow G, Kosson R et al (1987) High level secretion of glycosylated invertase in the methylotrophic yeast, Pichia pastoris. Bio/Technology 5:1305–1308
11. Grinna LS, Tschopp JF (1989) Size distribution and general structural features of N-linked oligosaccharides from the methylotrophic yeast, Pichia pastoris. Yeast 5:107–115
12. Trimble RB, Atkinson PH, Tschopp JG et al (1991) Structure of oligosaccharides on saccharomyces SUC2 invertase secreted by the methylotrophic yeast Pichia pastoris. J Biol Chem 266:22807–22817
13. Cregg JM, Tschopp JF, Stillman C et al (1987) High-level expression and efficient assembly of hepatitis B surface antigen in the methylotrophic yeast *Pichia pastoris*. Bio/Technology 5:479–485
14. White CE, Hunter MJ, Meininger DP et al (1995) Large scale expression, purification and characterization of the smallest active fragment of thrombomodulin: the roles of the sixth domain and of methionine-388. Protein Eng 8:1177–1187
15. Gleeson MAG, White CE, Meininger DP, Komives EA (1998) Generation of protease-deficient strains and their use in heterologous protein expression. Methods Mol Biol 103:81–94
16. Jones EW, Zubenko GS, Parker RR (1982) PEP4 gene function is required for expression of several vacuolar hydrolases in *Saccharomyces cerevisiae*. Genetics 102:665–677

17. Clare JJ, Romanos MA, Rayment FB (1991) Production of mouse epidermal growth factor in yeast: high-level secretion using *Pichia pastoris* strains containing multiple gene copies. Gene 105:205–212
18. Scorer CA, Clare JJ, McCombie WR et al (1994) Rapid selection using G418 of high copy number transformants of *Pichia pastoris* for high-level foreign gene expression. Bio/-Technology 12:181–184
19. www.pubmed.gov; key word: Pichia
20. Jacobs PP, Geysens S, Vervecken W et al (2009) Engineering complex-type N-glycosylation in Pichia pastoris using GlycoSwitch technology. Nat Protoc 4:58–70
21. Hamilton SR, Davidson RC, Sethuraman N et al (2006) Humanization of yeast to produce complex terminally sialylated glycoproteins. Science 313:1441–1443
22. De Schutter K, Lin YC, Tiels P et al (2009) Genome sequence of the recombinant protein production host Pichia pastoris. Nat Biotechnol 27:561–566
23. Zuraw B, Yasothan U, Kirkpatrick P (2010) Ecallantide. Nat Rev Drug Discov 9:189–190
24. Lin-Cereghino J, Cregg JM (2000) Heterologous protein expression in the methylotrophic yeast *Pichia pastoris*. FEMS Microbiol Rev 24:45–66
25. Pollet J, Chen WH, Versteeg L et al (2021) SARS-CoV-2 RBD219-N1C1: a yeast-expressed SARS-CoV-2 recombinant receptor-binding domain candidate vaccine stimulates virus neutralizing antibodies and t-cell immunity in mice. Hum Vaccin Immunother. 17(8):2356–2366
26. Lee J, Liu Z, Chen WH et al (2021) Process development and scale-up optimization of the SARS-CoV-2 receptorbinding domain-based vaccine candidate, RBD219-N1C1. Appl Microbiol Biotechnol. 105(10):4153–4165
27. Argentinian AntiCovid Consortium (2020) Structural and functional comparison of SARS-CoV-2-spike receptor bindingdomain produced in Pichia pastoris and mammalian cells. Sci Rep. 10(1):21779
28. Limonta-Fernández M, Chinea-Santiago G, Martín-Dunn AM et al (2022) An engineered SARS-CoV-2 receptor-binding domain produced in Pichiapastoris as a candidate vaccine antigen. New Biotechnol 72:11–21
29. Kalyoncu S, Yilmaz S, Kuyucu AZ et al (2023) Process development for an effective COVID-19 vaccine candidate harboring recombinant SARS-CoV-2 delta plus receptor binding domain produced by Pichia pastoris. Sci Rep 13(1):5224
30. Nagar G, Jain S, Rajurkar M et al (2023) Large-scale purification and characterization of recombinant receptor-binding domain (RBD) of SARS-CoV-2 spike protein expressed in yeast. Vaccines (Basel). 11(10):1602
31. Liu B, Yin Y, Liu Y et al (2022) A vaccine based on the receptor-binding domain of the spike protein expressed in glycoengineered Pichia pastoris targeting SARS-CoV-2 stimulates neutralizing and protective antibody responses. Engineering (Beijing) 13:107–115
32. Dalvie NC, Rodriguez-Aponte SA, Hartwell BL et al (2021) Engineered SARS-CoV-2 receptor binding domain improves manufacturability in yeast and immunogenicity in mice. Proc Natl Acad Sci USA. 118(38): e2106845118
33. Ahmed S, Khan MS, Gayathri S et al (2021) A stabilized, monomeric, receptor binding domain elicits high-titer neutralizing antibodies against all SARS-CoV-2 variants of concern. Front Immunol. 12:765211
34. Nurdin A, Nency YM, Maddeppungeng M et al (2024) Immunogenicity and safety of SARS-CoV-2 recombinant protein subunit vaccine (IndoVac) adjuvanted with alum and CpG 1018 in Indonesian adults: A phase 3, randomized, active-controlled, multicenter trial. Vaccine. 42(12):3009–3017
35. Schepen B, Van Schie L et al (2021) An affinity-enhanced, broadly neutralizing heavy chain–only antibody protects against SARS-CoV-2 infection in animal models. Sci Transl Med 13(621):eabi7826
36. van Schie L, Van Breedam W, Roels C et al (2023) GlycoVHH: optimal sites for introducing N-glycans on the camelid VHH antibody scaffold and use for macrophage delivery. MAbs 15(1):2210709

Part II

Bioinformatics

Chapter 2

Pichia pastoris Genomic Resources: Accessibility, Recommendations, and Pitfalls

Veronika Schusterbauer, Lena Schaffer, and Gerhard G. Thallinger

Abstract

The availability of reliable genomic data is one of the major drivers in advancing genome engineering. The different databases and annotations contain information from nucleotide sequences up to proteins and their functions. However, not all genomic resources are equally helpful or complete. This chapter aims to provide an overview of available genome references and annotations for the industrially most interesting strains of *Pichia pastoris* (now reclassified to *Komagataella phaffii*, *Komagataella pastoris*, and *Komagataella kurtzmanii*), including a short guideline on which genomic reference to use for which task. Additionally, we will cover the most important databases and data types and show methods on how to access them, based on the strain *K. phaffii* CBS 7435.

Key words *Pichia pastoris*, *Komagataella phaffii*, Genome assembly, Nucleic acid databases, Data retrieval, Data visualization

1 Introduction

A high-quality reference sequence is the key to facilitate genetic engineering, as it enables the reliable design of primers or integration vectors. With the emergence of next-generation sequencing (NGS), the availability of nucleotide sequencing data increased exponentially, and so did the number of genomic assemblies. Unfortunately, *de novo* assemblies based on data from short-read sequencing methods, such as Roche 454 [1, 2] or Illumina (previously Solexa) [3], are often fragmented and need substantial manual postprocessing to be fully resolved. The development of third-generation sequencing methods, including Pacific Biosciences (PacBio) [4, 5] and Oxford Nanopore Technologies (ONT) [6], largely improved the quality and completeness of *de novo* assemblies [7, 8]. With this initial bottleneck solved, the currently most important step is the improvement of available reference genomes

Xavier Garcia-Ortega et al. (eds.), *Pichia pastoris: Methods and Protocols*, Methods in Molecular Biology, vol. 2697, https://doi.org/10.1007/978-1-0716-4779-0_2,

by generating reliable genome annotations, including the position of coding sequences (CDSs) and their putative function.

Although accessing genomic resources and visualizing genomic data are not classical methods performed in the laboratory, the presented chapter sets the foundation for many strain-engineering and protein-expression studies and is thus of major practical relevance. In the presented chapter, we explore why certain genome assemblies are more suitable than others for a specific task and provide detailed information on how to get access to the required genomic data. For readers, not used to working with genomic data, we introduce the most relevant file types, the information they contain, and how to handle them. Importantly, we also provide an overview of tools for nucleotide sequence visualization that allow simplified handling of these large datasets.

1.1 Pichia pastoris *Reference Assemblies*

In 1995, *Pichia pastoris* was reclassified into the species *Komagataella phaffii* and *Komagataella pastoris* based on their 18S and 26S sequences [9], and the sibling *Komagataella kurtzmanii* followed in 2013 [10]. However, due to its popularity, the original species name *Pichia pastoris* is still commonly used. The very first genome assembly of *Pichia pastoris*, published in 2009, is based on the *his4*-deficient strain *P. pastoris* GS115 and was facilitated by Roche 454 sequencing. This was an important step toward simplifying fast and accurate genome engineering in *P. pastoris*, a yeast that had gained high importance in the field of recombinant protein production [11], since its introduction in the early 1990s. Shortly after the assembly for *P. pastoris* GS115, the first genome assemblies for *P. pastoris* DSMZ 70382, which is the *K. pastoris* type strain [12], as well as for the strain *K. phaffii* CBS 7435 [13], were published. All three assemblies are based on short-read sequencing techniques, including Illumina and Roche 454, leaving some gaps in the assembly sequence, especially in repetitive regions. Furthermore, the prediction of protein-coding genes relied solely on the DNA sequence, resulting in 5313, 5450, and 5007 CDSs, respectively. Nonetheless, these genome assemblies allowed for the precise description of the methanol utilization and protein secretion pathways [11–13]. The emergence of third-generation, long-read sequencing led to a new round of high-quality genome assemblies, including assemblies for *K. phaffii* CBS 7435 (NRRL Y-11430), a mutant of the *K. phaffii* type strain, NRRL Y-4290 (NRRL Y-7556, CBS 2612) [14, 15], *K. phaffii* GS115 (NRRL Y-15851) [14], and *K. pastoris* DSMZ 70382 (NRRL Y-1603, CBS 704) [14], which facilitated studies for genome comparisons [14, 16] and analysis of meiotic recombination [17]. Furthermore, the availability of increasing amounts of RNA sequencing (RNA-seq) data led to improved gene annotation as well as the characterization of non-coding RNAs such as rRNAs [18]. These annotations set the basis for further projects, such as the *K. phaffii* proteome atlas

[19]. With all these *P. pastoris* genome assemblies available, the question arises: Which assembly should be used for a given task at hand?

1.2 Databases

The most important databases for nucleotide sequences and related metadata are GenBank [20] at the National Center for Biotechnology Information (NCBI), National Library of Medicine, National Institutes of Health in Bethesda, Maryland, USA; the European Nucleotide Archive (ENA) [21] at the European Molecular Biology Laboratory's European Bioinformatics Institute (EMBL-EBI) in Hinxton, UK; and the DNA Data Bank of Japan (DDBJ) [22] at the National Institute of Genetics in Mishima, Japan. All the mentioned institutes have been collaborating for more than 30 years to foster the International Nucleotide Sequence Database Collaboration (INSDC; http://www.insdc.org/), which enables a daily synchronization of data between the resources to keep them all at the same level of information [23]. The NCBIs Reference Sequence (RefSeq) database [24] aims to represent DNA, RNA, and protein molecules of taxonomically diverse, non-redundant, and richly annotated sequences and therefore only contains a subset of complete genomes of the GenBank database. Usually, RefSeq contains one genome assembly per species, which is referred to as either the reference or the representative genome [25].

Besides the common sources for nucleotide sequences mentioned above, yeast-specific resources for nucleotide sequences and annotations are available. The most important database is the Saccharomyces Genome Database (SGD; www.yeastgenome.org) [26], which is, as the name implies, dedicated to the model organism of yeasts, *Saccharomyces cerevisiae*. The SGD is a collection of *S. cerevisiae* assembly data and gene annotations but also contains extensive additional information, such as a functional description of each gene, phenotype of knockout strains, and interactions between genes. Furthermore, it contains specialized databases and tools for biochemical pathways and gene ontology (GO [27]) annotations, including a GO-term mapper and the yeast GO-slim database, which is reduced to GO-terms applicable to all yeast species [26, 28]. Similarly, a project dedicated to *P. pastoris* is available online (www.pichiagenome.org). It allows to browse assemblies and annotations for *K. phaffii*, *K. pastoris,* and *S. cerevisiae* and also provides an interactive representation of the *K. phaffii* proteome atlas [18, 19].

2 Materials

To be able to work with genomic data, it is important to know the characteristics of the most common file formats and the information they contain. Here we will give an overview of the formats later used to work with genome assemblies and their respective annotations.

2.1 File Formats

2.1.1 FASTA

The FASTA format is used to store either nucleotide (NT) or amino acid (AA) sequences. A single file can contain a single sequence or multiple sequences (extensions: .fa/.fasta/.fna for nucleotide and .fasta/.faa for AA sequences) [29, 30]. In any case, each entry starts with a header line (string starting with ">") which states the sequence ID, often an accession number, and additional information about the sequence. The lines following the header line denote the actual sequence, where each letter stands for either an NT or an AA, conforming to the respective IUPAC standard [31] (example shown in Table 2).

2.1.2 GFF/GTF

Both the Generic Feature Format (GFF, extensions: .gff/.gff2/.gff3) and the Gene Transfer Format (GTF, extensions: .gtf) are formats for feature annotation of nucleotide sequences. They are very similar, tab-delimited file formats, containing nine columns. The first entry states the reference accession of the nucleotide sequence they refer to, followed by seven columns defining the type and position of the feature. The 9th column of the file contains additional annotations for each feature. In general, GFF version 3 should be preferred over the GTF format, since it is more precisely defined [32, 33] (example of a GFF shown in Table 3, *see* **Note 1**).

2.1.3 GenBank Flat File

The GenBank flat file format (GBF, file extensions: .gbff/.gb/.gbk) is commonly used to store information of sequences and annotations for genomes ranging from viruses up to complex eukaryotes. It is favored over other formats because of its extensive abilities to store the NT or AA sequences together with detailed annotations. Additionally, the GBF is designed to be easy to read for humans rather than machines. The format can contain multiple annotated NT or AA sequences, wherein each entry starts with meta information of the sequences, followed by individual feature annotations, and ending with the origin section, which comprises the actual nucleotide or protein sequence (example shown in Table 4) [34, 35].

2.2 *Pichia pastoris* Assemblies

There are a number of different genome assemblies available for *K. phaffii* as well as *K. pastoris*, all of which are available in the NCBI datasets database (https://www.ncbi.nlm.nih.gov/datasets/genome). As of May 2024, 175 genomes of *K. phaffii* and *K. pastoris* strains are accessible via NCBI. Most of them stem from comparative studies [36–39] and either suffer from a low level of completeness or have no genome annotation. Hence, only a low number of assemblies are suitable as a reference genome for experimental design or bioinformatics analysis (Table 1).

Table 1
Overview of selected, historically especially relevant, and strain-specific genome assemblies available on NCBI

Species	Strain	Assembly	Completeness	Additional contigs	Submitter	Year	Ann	Ref.
K. phaffii	CBS 7435 (NRRL Y-11430)	KP_7435-4	Complete Genome	–	ACIB, Sturmberger	2016	2022	*(15)*
		ASM170808v1	Chromosome	1 plasmid 2 unplaced	MIT, Love	2016	2016	*(14)*
		PicPas_Mar2011	Chromosome	mitochondrion	TU Graz, Thallinger	2011	2016	*(13, 18)*
	GS115 (NRRL Y-15851)	ASM174695v1	Chromosome	1 plasmid 5 unplaced	MIT, Love	2016	2016	*(14)*
		ASM2700v1	Chromosome	11 unplaced	UGent, De Schutter	2007	2007	*(11)*
	Pichia Pink strain2 (NRRL Y-48124)	PP2_HC	Scaffold	33 unplaced (no CHRs)	CEBITEC, Wibberg	2018	–	*(40)*
K. pastoris	Isolate: EAMORG09 (see **Note 2**)	ASM2963296v1	Chromosome	11 unplaced	Nanjing Tech University	2023	–	–
	ATCC 28485 (NRRL Y-1603)	ASM170810v1	Chromosome	7 unplaced	MIT, Love	2016	2016	*(14)*
	DSMZ-70382 (NRRL Y-1603)	ASM18278v1	Contig	326 unplaced (no CHRs)	UVienna, Mattanovich	2009	–	*(12)*

The "Additional contigs" field lists sequences included in the assembly, which do not refer to chromosomes. "No CHRs" means that contigs were not mapped to chromosomes. "Ann." denotes the year in which the annotation of the respective assembly was last updated. The representative genomes per species are marked in blue, whereas the recommended assemblies are marked in green if they differ. Assembly ASM2963296v1 is additionally marked with orange stripes as there seem to be inconsistencies regarding the assigned species (*see* **Note 2**). "Completeness" refers to the assembly level of the respective assembly, whereas the most complete is complete genome, which means the assembly contains only gapless chromosomes. Chromosome level means that the sequences may still contain gaps, and Scaffold and Contig assemblies contain sequences that were not placed on any chromosome

2.3 Visualization Tools

Although most data formats in genomics are text-based and can be viewed with any viable text editor, it is more convenient to use specialized tools to view and manipulate genomic data (*see* **Note 4**).

Table 2
Example of an AA sequence in FASTA format, showing the protein Alcohol oxidase (AOX1, accession: CCA40305.1) of strain *K. phaffii* CBS 7435

Format	Example
FASTA	>CCA40305.1 Alcohol oxidase [*Komagataella phaffii* CBS 7435] MAIPEEFDILVLGGGSSGSCIAGRLANLDHSLKVGLIEAGENNLNNPWVYLPGIYPRNMKLDSKTASFYT SNPSPHLNGRRAIVPCANVLGGGSSINFMMYTRGSASDYDDFQAEGWKTKDLLPLMKKTETYQRACNNPD IHGFEGPIKVSFGNYTYPVCQDFLRASESQGIPYVDDLEDLVTAHGAEHWLKWINRDTGRRSDSAHAFVH STMRNHDNLYLICNTKVDKIIVEDGRAAAVRTVPSKPLNPKKPSHKIYRARKQIVLSCGTISSPLVLQRS GFGDPIKLRAAGVKPLVNLPGVGRNFQDHYCFFSPYRIKPQYESFDDFVRGDAEIQKRVFDQWYANGTGP LATNGIEAGVKIRPTPEELSQMDESFQEGYREYFEDKPDKPVMHYSIIAGFFGDHTKIPPGKYMTMFHFL EYPFSRGSIHITSPDPYAAPDFDPGFMNDERDMAPMVWAYKKSRETARRMDHFAGEVTSHHPLFPYSSEA RALEMDLETSNAYGGPLNLSAGLAHGSWTQPLKKPTAKNEGHVTSNQVELHPDIEYDEEDDKAIENYIRE HTETTWHCLGTCSIGPREGSKIVKWGGVLDHRSNVYGVKGLKVGDLSVCPDNVGCNTYTTALLIGEKTAT LVGEDLGYSGEALDMTVPQFKLGTYEKTGLARF

Table 3
Example of feature annotations in GFF format, showing the first four features of chromosome 1 (Accession: CP014715.1) of *K. phaffii* GS115. The 9th column carrying the additional annotation and info for each feature is shown in bold and was split up into multiple lines for ease of reading

<table>
<tr><th>Format</th><th>Example</th></tr>
<tr><td>GFF</td><td>##gff-version 3#!gff-spec-version 1.21
#!processor NCBI annotwriter
#!genome-build ASM174695v1
#!genome-build-accession NCBI_Assembly:GCA_001746955.1
##sequence-region CP014715.1 1 2887587
##species https://www.ncbi.nlm.nih.gov/Taxonomy/Browser/wwwtax.cgi?id=
644223
CP014715.1 Genbank gene 5075 6556 . + .
ID=gene-AT250_GQ6802991;Name=AT250_GQ6802991;end_range=6556,.;
gbkey=Gene;gene_biotype=protein_coding;locus_tag=AT250_GQ6802991;
partial=true; start_range=.,5075
CP014715.1 Genbank mRNA 5075 6556 . + .
ID=rna-gnl|LoveMIT|mrna.GQ68_02991T0;Parent=gene-T250_GQ6802991;
end_range=6556,.;gbkey=mRNA;locus_tag=AT250_GQ6802991;
orig_protein_id=gnl|LoveMIT|GQ68_02991T0;orig_transcript_id=gnl|
LoveMIT|mrna.GQ68_02991T0;partial=true; product=GQ68_02991T0;start_
range=.,5075
CP014715.1 Genbank exon 5075 6556 . + .
ID=exon-gnl|LoveMIT|mrna.GQ68_02991T0-1;Parent=rna-gnl|LoveMIT|mrna.
GQ68_02991T0;
end_range=6556,.;gbkey=mRNA;locus_tag=AT250_GQ6802991; orig_pr-tein_id
=gnl|LoveMIT|GQ68_02991T0;
orig_transcript_id=gnl|LoveMIT|mrna.GQ68_02991T0; partial=true;product
=GQ68_02991T0;start_range=.,5075
CP014715.1 Genbank CDS 5075 6556 . + 0
ID=cds-AOA65887.1;Parent=rna-gnl|LoveMIT|mrna.GQ68_02991T0;
Dbxref=NCBI_GP:AOA65887.1;Name=AOA65887.1;gbkey=CDS;locus_tag=AT250_
GQ6802991;
orig_transcript_id=gnl|LoveMIT|mrna.GQ68_02991T0;product=GQ68_02991T0;
protein_id=AOA65887.1</td></tr>
</table>

Especially for downstream tasks such as primer design for DNA amplification or the design of plasmids and integration cassettes, tools, such as SnapGene (snapgene.com, Insightful Science, San Diego, CA, USA) or the CLC Main Workbench (QIAGEN, Hilden, Germany), are of great use. Both mentioned tools allow the visualization of whole genomes within their free access versions. A completely open-source alternative, which is more directed toward viewing whole genomes as well as NGS data, is the Integrative Genomics Viewer (IGV, software.broadinstitute.org/software/igv) [41] (*see* **Note 5**).

Table 4
Example of an AA sequence including annotations in GBF format. The example shows the Alcohol oxidase coded by the AOX1 gene from *K. phaffii* GS115 (*see* Note 3)

Format	Example
GBF	LOCUS AOA70169 663 aa linear PLN 12-AUG-2016 DEFINITION GQ68_05196T0 [Komagataella phaffii GS115]. ACCESSION AOA70169 VERSION AOA70169.1 DBLINK BioProject: PRJNA304976 BioSample: SAMN04311980 DBSOURCE accession CP014718.1 KEYWORDS . SOURCE Komagataella phaffii GS115 ORGANISM Komagataella phaffii GS115 Eukaryota; Fungi; Dikarya; Ascomycota; Saccharomycotina; Saccharomycetes; Saccharomycetales; Phaffomycetaceae; Komagataella. REFERENCE 1 (residues 1 to 663) AUTHORS Love,K.R., Shah,K.A., Whittaker,C.A., Wu,J., Bartlett,M.C., Ma,D., Leeson,R.L., Priest,M., Young,S.K. and Love,J.C. TITLE Comparative genomic and transcriptomic foundation for *Pichia pastoris* JOURNAL Unpublished REFERENCE 2 (residues 1 to 663) AUTHORS Love,K.R., Shah,K.A., Whittaker,C.A., Wu,J., Bartlett,M.C., Ma,D., Leeson,R.L., Priest,M., Young,S.K. and Love,J.C. TITLE Direct Submission JOURNAL Submitted (10-MAR-2016) The David H. Koch Institute for Integrative Cancer, Massachusetts Institute of Technology, 500 Main Street, Cambridge, MA 02139, USA FEATURES Location/Qualifiers source 1..663 /organism="Komagataella phaffii GS115" /strain="GS115" /culture_collection="ATCC:20864" /db_xref="taxon:644223" /chromosome="4" /country="USA" /collection_date="2014" Protein 1..663 /product="GQ68_05196T0" Region 5..632 /region_name="PRK02106" /note="choline dehydrogenase; Validated" /db_xref="CDD:235000" CDS 1..663 /locus_tag="AT250_GQ6805196" /coded_by="complement(CP014718.1:1590103..1592094)" ORIGIN 1 maipeefdil vlgggssgsc iagrlanldh slkvglieag ennlnnpwvy lpgiyprnmk 61 ldsktasfyt snpsphlngr raivpcanvl gggssinfmm ytrgsasdyd dfqaegwktk 121 dllplmkkte tyqracnnpd ihgfegpikv sfgnytypvc qdflrasesq gipyvddled 181 lvtahgaehw lkwinrdtgr rsdsahafvh stmrnhdnly licntkvdki ivedgraaav 241 rtvpskplnp kkpshkiyra rkqivlscgt issplvlqrs gfgdpiklra agvkplvnlp 301 gvgrnfqdhy cffspyrikp qyesfddfvr gdaeiqkrvf dqwyangtgp latngieagv 361 kirptpeels qmdesfqegy reyfedkpdk pvmhysiiag ffgdhtkipp gkymtmfhfl 421 eypfsrgsih itspdpyaap dfdpgfmnde rdmapmvway kksretarrm dhfagevtsh 481 hplfpyssea ralemdlets nayggplnls aglahgswtq plkkptakne ghvtsnqvel 541 hpdieydeed dkaienyire htettwhclg tcsigpregs kivkwggvld hrsnvygvkg 601 lkvgdlsvcp dnvgcntytt alligektat lvgedlgysg ealdmtvpqf klgtyektgl 661 arf

3 Methods

This chapter provides a step-by-step guide for identifying, downloading, and visualizing genomic assemblies and annotations for *Pichia pastoris* strains. It is intended for researchers who rely on reference assemblies for tasks such as the design of genomic alterations, resequencing analyses, or RNA-seq. Subheadings 3.2 and 3.3 contain methods for the use of either browser or desktop applications. To make the descriptions easier to understand, all fields or buttons that you should interact with are supported by screenshots and marked with red boxes. Within the text, names of buttons or fields are written in *italics*, and words or phrases that are entered into, e.g., a search box, are marked with "quotes."

3.1 Considerations for the Choice of an Assembly

For certain tasks, like the design of primers for, e.g., the amplification of a *K. phaffii* native gene, but also for a resequencing analysis of engineered strains, it is of utmost importance to have a valid genomic sequence. Thus, it is advisable to use genome assemblies that have as little gaps as possible and represent the most closely related strain. As a general guideline, we advise using the assembly ASM174695v1 for working with the *K. phaffii* GS115 strain or any of its descendants like the X-33 or strains from the *Pichia* Pink system (Thermo Fisher Scientific, Waltham, MA, USA), since it is the most complete and up-to-date assembly for *K. phaffii* GS115. For working with the strain CBS 7435 and its descendants, we recommend using either assembly KP_7435-4 or ASM170808v1. These include the TUGraz FTO *Pichia* Pool platform [42, 43] and the *Pichia* platform strains available from BioGrammatics (BioGrammatics Inc., Carlsbad, CA, USA) and bisy (bisy GmbH, Hofstätten a. d. Raab, Austria), which are based on an NRLL Y-11430 strain derivative cured of its killer plasmids. Similarly, as there is no genome assembly available for the OPEN*Pichia* platform (NCYC 2543 hoc1tr-1 [44]) or the parental strain CBS2612 (=NCYC 2543/NRLL Y-7556) we would advise using the genome assemblies for CBS 7435, which show only a few differences in the genome sequence. For tasks such as RNA-seq analysis that often require more information for downstream analysis, including, e.g., GO tags, it can be an advantage to use the representative genome assembly, since other databases usually relate to the corresponding gene or protein tags. Therefore, in these cases, you should consider using assemblies ASM2700v1 or PicPas_-Mar2011, for working with GS115- or CBS 7435-like strains, respectively (*see* **Notes 6** and **7**). Furthermore, the assembly by Küberl *et al.* (PicPas_Mar2011) is the only assembly that includes the nucleotide sequence and annotation for the mitochondrion. Therefore, if you need the mitochondrial sequence, or if you do whole-genome sequencing or RNA-seq, where it is important that

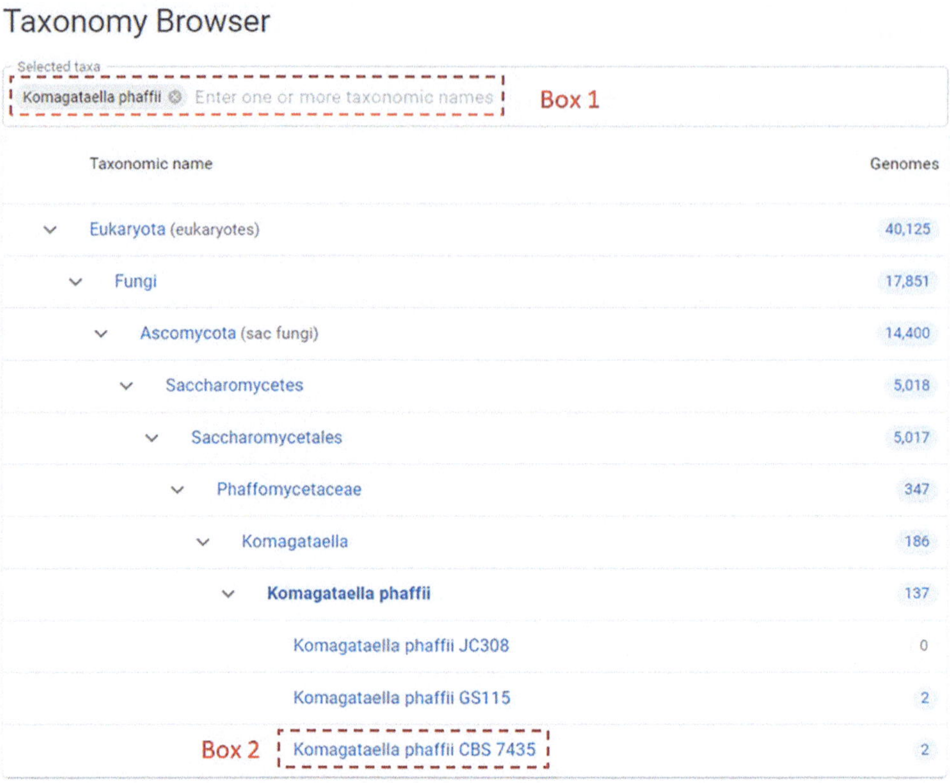

Fig. 1 Overview of available strains of *K. phaffii* on the NCBI taxonomy browser. The number of available genomes per taxonomic subunit is shown on the right in the column *Genome*. Box 1 marks the search field, and Box 2 marks the link to the entry for *K. phaffii* CBS 7435

the reference is as complete as possible, PicPas_Mar2011 is a good choice (*see* **Notes 8–10**). For working with *K. pastoris* strains, you should in any case choose assembly ASM170810v1, as it is the most complete assembly as well as the representative assembly for this species (Table 1).

3.2 NCBI Taxonomy Browser

The NCBI taxonomy browser is especially helpful when trying to get an overview of the different genomic information that is available for a certain species or even a specific strain.

1. Open the following link in a browser of your choice: https://www.ncbi.nlm.nih.gov/datasets/taxonomy/tree/
2. Enter "Komagataella phaffii" in the search field (Fig. 1, Box 1) to obtain an overview of all available *K. phaffii* strains.
3. Click on the link "*Komagataella phaffii* CBS 7435" (Fig. 1, Box 2) to show the lineage and detailed information about available genomic resources on strain CBS 7435 (Fig. 2).

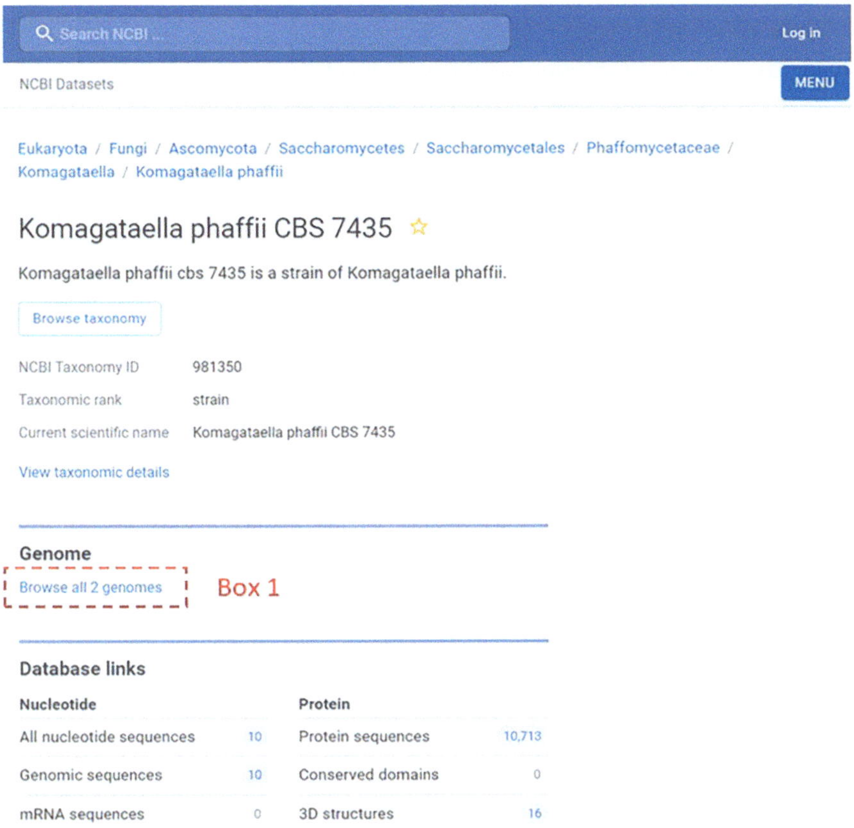

Fig. 2 Detailed view of *K. phaffii* CBS 7435 on the NCBI taxonomy browser, with the lineage details on top and the information box about the different genomic resources available via Entrez at the bottom. Box 1 marks the number of available genomes for *K. phaffii* CBS 7435

4. To show all available genomes for this strain, click the link in the Genomes section, stating "Browse all x genomes" (Fig. 2, Box 1), which will link you to the *datasets* database of NCBI.

3.3 Download of NCBI Sequence and Annotation Information

Each assembly-specific page on NCBI contains detailed information about the assembly and its quality. The most important information provided is the number of contigs, the total length of assembled sequences, and the completion status of the assembly. Besides the assembly characteristics, metadata like the institution that uploaded the assembly and the related publication can be found on the webpage (Fig. 3). The next steps will detail how to download the relevant sequence and annotation files for an assembly, exemplified for *K. phaffii* CBS 7435.

1. To open the assembly page for the latest assembly of *K. phaffii* CBS 7435, click the link to KP_7435-4 in the list of available genomes on NCBI (or directly open https://www.ncbi.nlm.nih.gov/datasets/genome/GCA_900235035.2/). We recommend

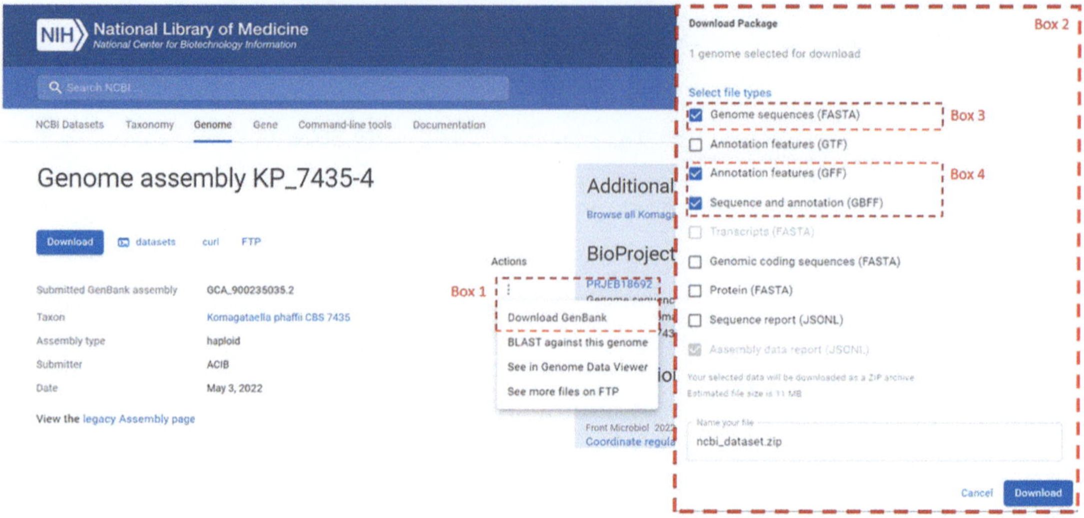

Fig. 3 NCBI genome page for KP_7435-4. On the right-hand side, the opened pop-up menu for the genome download is shown. Box 1 marks the *Actions* dropdown, including the Download GenBank option, to open the pop-up shown in Box 2. Box 3 and Box 4 mark the relevant *File types* for genome sequences and annotations download, respectively

this assembly, as it is the most recent and complete assembly available for *K. phaffii* CBS 7435 (Table 1, *see* **Note 11**).

2. To download the assembly as a GenBank flat file, click on the three dots below *Actions* and select *Download GenBank* in the dropdown menu (Fig. 3, Box 1). Then select *Sequence and annotation (GBFF)* in the pop-up window (Fig. 3, Box 4) and download it.
3. To download the nucleotide sequence without annotations as a FASTA file, select *Genomic sequences (FASTA)* from the pop-up menu and download. (Fig. 3, Box 2, *see* **Notes 12** and **13**).
4. To download the respective features for the genomic sequence separately, choose *Annotation features (GFF)* from the drop-down (Fig. 3, Box 3, *see* **Notes 14** and **15**).
5. All your downloaded files will be compressed into a zip archive. Once extracted, the files can be found in *ncbi_dataset/data/GCA_900235035.2* (.gbff/.fna/.gff) (*see* **Note 16**).

With your assembly data at hand, you can now go on to use it for designing your lab work, but using an accurate assembly as a reference is also an important requirement for NGS data analysis, like whole-genome resequencing or RNA-seq. A guide on RNA-seq in *Pichia pastoris* is provided in the chapter "*Komagataella phaffii* RNA-sequencing and Data Analysis."

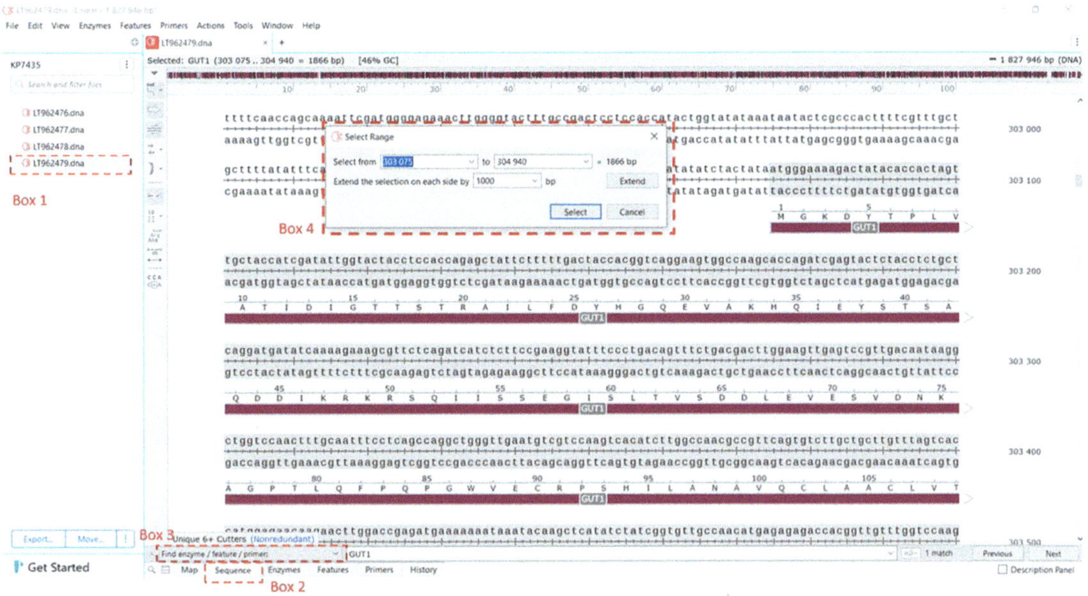

Fig. 4 SnapGene sequence view showing the 5′ end of the *GUT1* gene and mRNA features (gray) and the *GUT1* CDS (dark red). Box 1 marks the DNA file for chromosome 4. The sequence tab is marked as Box 2, and the search field and the respective dropdown menu are marked as Box 3. Box 4 marks the pop-up for extending the selected range

3.4 Visualization of Genomes

In this part, we will describe different ways to interact with the data, we just retrieved in a convenient way. We will view the genome assemblies and extract specific information, which is buried within a massive number of annotations.

3.4.1 SnapGene

We will first load our retrieved genome assembly with SnapGene to view it and then examine the different possibilities to search for specific regions/genes. To illustrate how this is done, we have chosen *GUT1* as an example gene.

1. To load a genome for the first time, simply choose *File > Open Files* and navigate to your GBF file (*see* **Note 17**). This will generate the specified folder, which contains a SnapGene-specific file (extension: .dna) for each contig (*see* **Note 18**).
2. Open file LT962479.dna, which refers to chromosome 4 (Fig. 4, Box 1, *see* **Notes 19** and **20**).
3. Switch to the sequence view by selecting the *Sequence* tab at the bottom of the SnapGene window (Fig. 4, Box 2) and open the search field by pressing <Ctrl>+F (or ⌘+F on macOS).
4. In the sequence view, you can find a gene by its name by choosing *Find enzyme/feature/primer* and typing “GUT1” into the search field (Fig. 4, Box 2). The view will jump to the respective feature (Fig. 4, Box 3, *see* **Note 21**).

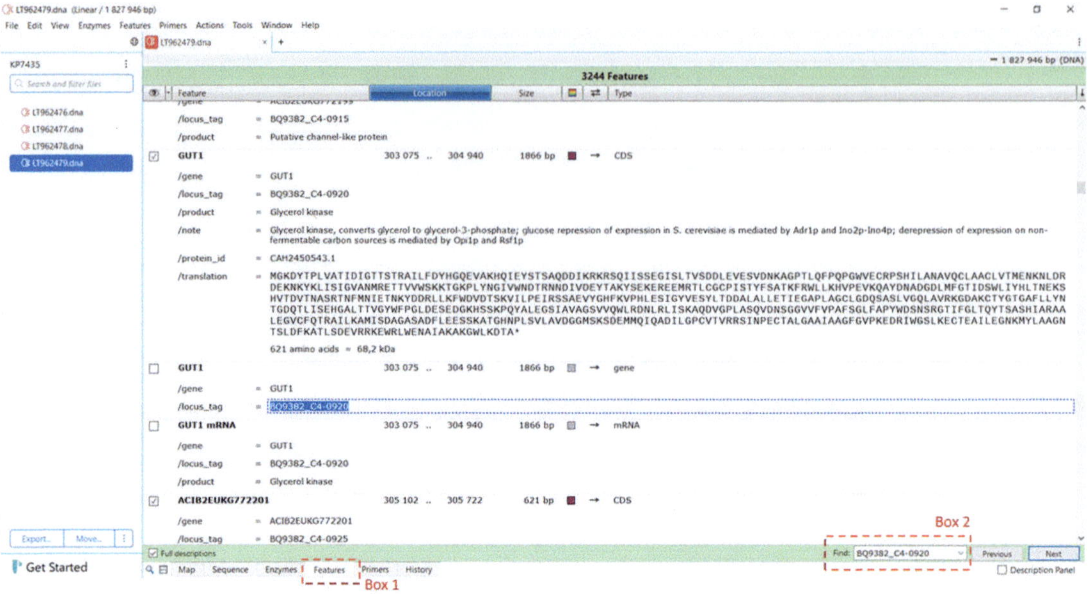

Fig. 5 SnapGene features view, showing the CDS, gene, and mRNA features of *GUT1*. The features tab is marked as Box 1, and the search field is marked as Box 2

5. Once you have identified the gene of interest, you can click on the feature to mark the nucleotide sequence and press <Ctrl>+C (⌘+C on macOS) to copy the sequence (Fig. 4).
6. Double clicking on the feature will show you detailed information, such as the position of the gene on chromosome 4.
7. To copy a feature with overhangs, click on the desired feature to select its sequence, then right click on the selection (sequence with blue/gray background) and choose *Select Range* from the appearing menu. In the *Select Range* menu (Fig. 4, Box 4), choose *1000* for the *Extend selection on each side* option. Subsequently, press the *Extend* button followed by the *Select* button. This will mark the extended region. Now press <Ctrl>+C (⌘+C on macOS) to copy the extended region.
8. To search for a gene by any of its identifiers, such as *gene*, *protein_id*, *locus_tag*, switch to the *Features* view by clicking on the *Features* tab (Fig. 5, Box 1) at the bottom of the SnapGene window and again open the search field by pressing <Ctrl>+F (⌘-F on macOS). Type in the *locus_tag* "BQ9382_C4-0920" to find all feature annotations related to the *GUT1* gene (Fig. 5, Box 2).

3.4.2 IGV Viewer

The following steps explain how to load the genomic resources into the IGV viewer, followed by how to find a certain feature and how to copy its sequence. We will again perform all steps on the example gene *GUT1*.

1. Load the genome reference by going to *Genomes* > *Load genome from file* and navigating to your FASTA file (*see* **Note 22**).
2. Add the annotation data by opening *File* > *Load from file* and selecting your GFF file.
3. To get a better view of the annotations, right click on the annotations track and choose *Expanded* as viewing mode.
4. Go to *File* > *Save* session in order to be able to open both files together the next time.
5. In IGV, you can search for features across all chromosomes and contigs by all their identifiers. To find our example gene *GUT1*, simply type "GUT1" into the search field. Alternatively, if you know the *protein_id*, you can type in "CAH2450543.1".
6. Click on the mRNA feature of the *GUT1* gene to show all additional information, given in the GFF file, such as additional identifiers and description of the gene product (Fig. 6, Box 2 and Box 3).
7. To copy the DNA sequence of the gene, right click on the feature and choose *Copy Sequence.*
8. To copy the genetic sequence with overhangs, you need to define a region of interest (ROI). Click on the button for the *regions navigator* (Fig. 6, Box 4). Once the region's navigator is active, define your ROI by choosing the desired start and end point within the annotation track. This will add a red bar marking the ROI (Fig. 6, Box 5).
9. Right click on the ROI and choose *Copy sequence* to copy the nucleotide sequence of the defined region (Fig. 6).

With the nucleotide sequences of your desired regions, you can now perform further steps such as primer design for DNA amplification or the alignment of Sanger sequencing results in a tool of your choice, e.g., SnapGene or the CLC Genomics Workbench (*see* **Note 23**).

4 Notes

1. When working with very specialized analysis tools, e.g., in RNA-seq analysis, it sometimes happens that tools will only accept either GTF or GFF format, or that they will require specific annotation tags to be present, e.g., the *exon* or *gene* tag.

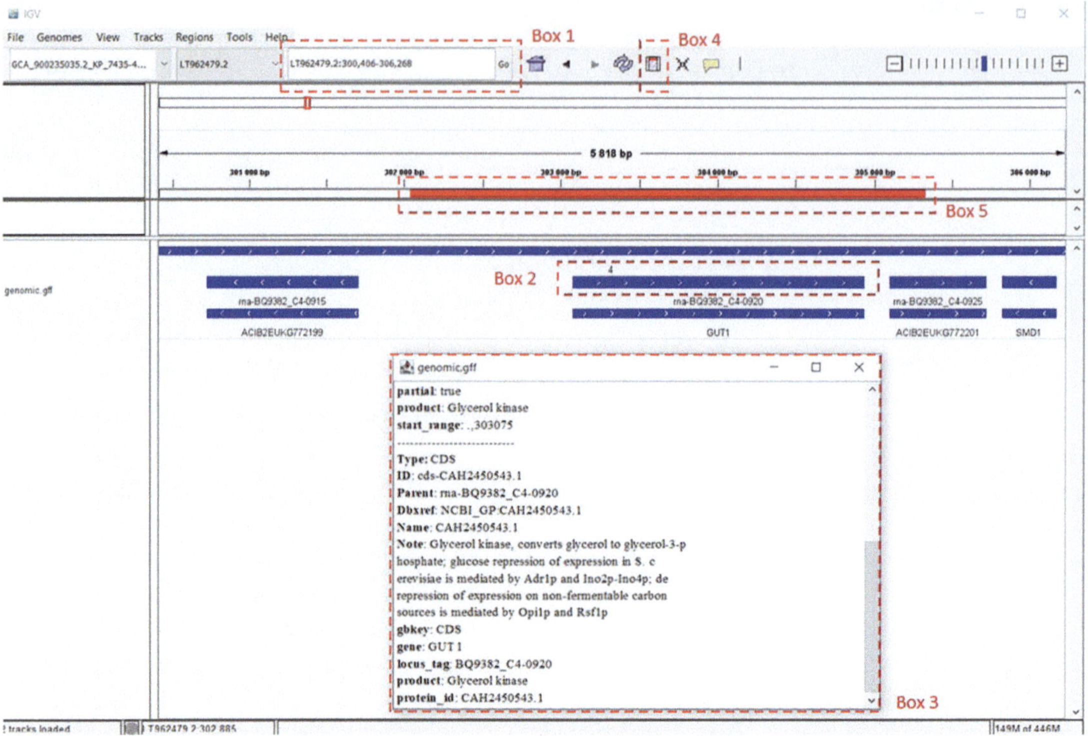

Fig. 6 IGV viewer showing the imported GFF file of *K. phaffii* CBS 7435 as a Features track. Box 1 marks the search field. Box 2 and Box 4 mark the mRNA feature for gene GUT1 and the annotation information for it, respectively. Box 4 marks the button for the *Regions Navigator,* and the red bar above the features track, marked with Box 5, represents the defined ROI

2. The current (08.05.2024) representative of *K. pastoris* in the NCBI RefSeq database is assembly ASM2963296v1, which is stated to have been generated from genomic material of isolate *K. pastoris* EAMORG09 and which was uploaded to NCBI by the Nanjing Tech University, China. However, a comparison of the average nucleotide identity (ANI) of ASM2963296v1 with the published assemblies KP_7435-4 (*K. phaffii* CBS 7435) and ASM170810v1 (*K. pastoris* ATCC 28485) resulted in a two-way ANI of 99.95% and 88.51%, respectively. Hence, it seems very likely that ASM2963296v1 represents a *K. phaffii* strain rather than one from *K. pastoris.* The ANI was calculated using the ANI calculator from the Enveomics Collection [45] (http://enve-omics.ce.gatech.edu/ani/).
3. Note that the protein AOA70169.1 is annotated as "choline dehydrogenase", but the AA sequence is identical to the protein CCA40305.1. A good way to compare annotations between strains and assemblies is to look at the identical protein groups at NCBI.

4. When working with bioinformatics text files, it is advisable to download a programming text editor, such as Sublime [46] (Windows/Mac/Linux) or Notepad++ [47] (Windows). Make sure that your text editor does not automatically add empty lines at the end of the file!
5. Since IGV is mainly designed for viewing mapped reads from NGS reads, you can also import and view mapped sequences in BAM format.
6. The KEGG [48] database of biological pathways of *P. pastoris* also uses the gene tags of ASM2700v1, the *K. phaffii* representative genome.
7. Although PicPas_Mar2011 is not the representative genome of *K. phaffii*, QuickGO references the UniProt IDs and also the locus_tags from this assembly.
8. The assembly and annotation data of *K. phaffii* CBS 7435 by Küberl et al. [13]/Valli et al. [18] (PicPas_Mar2011) is also available via the *Downloads* section of the *Pichia* Genome Browser (www.pichiagenome.org) [18]. Note that the GFF file provided there is slightly more complete than the one offered by NCBI.
9. Alternatively, to get the best out of all genome data, you can download the sequence and annotation for the mitochondrion separately and add it to one of the other, more up-to-date *K. phaffii* assemblies. You can download FASTA, GBF, and GFF files for the mitochondrion of *K. phaffii* CBS 7435 from the NCBI nucleotide section (accession: NC_015384.1, www.ncbi.nlm.nih.gov/nuccore/NC_015384.1). You can simply add the data to your assembly by copying and pasting it to the end of the respective text files.
10. None of the assemblies contains full sequences of the two linear cytoplasmic plasmids. However, the plasmid sequences and annotations of *K. phaffii* CBS7435 can be downloaded from the NCBI nucleotide database via accessions MG491503.1 and MG491504.1 [15]. You can simply add the data to your assembly by copying and pasting it to the end of the respective text files. If you plan to do NGS analysis with strains that still have the killer plasmids, it is advised to add these sequences to the genome assembly.
11. The reference genome is not always the best choice, since it is often the first complete genome sequence published and not the latest or most complete. Therefore, it is advisable to check all available genomes and assemblies and the related literature.
12. Beware that there are additional FASTA entries available named *Genomic CDSs* and *Protein*. Those contain only the

sequences of the (protein) coding features, either as NT or AA sequences, but not the whole chromosomes.

13. If FASTA files are used as a reference for sequence alignment, they should not contain empty lines, since some tools cannot handle them properly.
14. Alternatively, choose *Genomic GTF* to download the features for the genomic sequence in GTF format.
15. GFF3 files downloaded from NCBI often contain HTML tags of special characters within the annotation field (e.g., %2C), probably caused by some format conversion step. This can lead to the loss of information, because some tools will interpret the % character as a comment tag.
16. Appropriate tools for compressing and extracting data are, e.g., 7-Zip (Windows) or iPackr (macOS). Depending on how you download files from NCBI, they sometimes have a .txt file extension by default or no file extension at all. Make sure that you always store files with a commonly used file extension and that the settings of your operating system allow you to do so.
17. If you have the full version of SnapGene and set it as the default for .gbff files, you can simply open the file with a double-click.
18. The SnapGene full version will automatically build a collection with all the contigs. When prompted with the question to import multiple sequences, select "yes." When asked which sequences to display as circular, select plasmid sequences and mitochondrial sequences only. In the case of the CBS 7435, the "NCBI genome" does not contain any complete mitochondrial or plasmid sequences.
19. If your computer has limited RAM, it can be advisable to use IGV for loading whole chromosomes instead of SnapGene.
20. Usually, the first 4 files will refer to the first 4 chromosomes. If you are unsure which accession refers to which chromosome, this information can be found in the headers of each of the downloaded files, or it is also given at the bottom of the genome-specific NCBI page.
21. If you search for sequences in a SnapGene collection, it will only search within the active sequence. Therefore, if you do not know on which chromosome your gene of interest is encoded, you will have to search through each of them separately.
22. You need to change the file ending of the genomic sequence from ".fna" to ".fasta" to open it with IGV.
23. For the alignment of sequences and primer design, you will need the full versions of the mentioned tools.

Acknowledgments

For this work, V.S. and L.S. were supported by the Austrian Research Promotion Agency (FFG).

Competing Interest Statement The authors have no conflicts of interest to declare that are relevant to the content of this chapter.

References

1. Nyrén P (1987) Enzymatic method for continuous monitoring of DNA polymerase activity. Anal Biochem 167:235–238
2. Ronaghi M, Karamohamed S, Pettersson B et al (1996) Real-time DNA sequencing using detection of pyrophosphate release. Anal Biochem 242:84–89
3. Bentley DR, Balasubramanian S, Swerdlow HP et al (2008) Accurate whole human genome sequencing using reversible terminator chemistry. Nature 456:53–59
4. Travers KJ, Chin C-S, Rank DR et al (2010) A flexible and efficient template format for circular consensus sequencing and SNP detection. Nucleic Acids Res 38:e159
5. Eid J, Fehr A, Gray J et al (2009) Real-time DNA Sequencing from Single Polymerase Molecules. Science 323:133–138
6. Clarke J, Wu H-C, Jayasinghe L et al (2009) Continuous base identification for single-molecule nanopore DNA sequencing. Nat Nanotechnol 4:265–270
7. Larsen PA, Heilman AM, Yoder AD (2014) The utility of PacBio circular consensus sequencing for characterizing complex gene families in non-model organisms. BMC Genomics 15:720
8. Wenger AM, Peluso P, Rowell WJ et al (2019) Accurate circular consensus long-read sequencing improves variant detection and assembly of a human genome. Nat Biotechnol 37:1155–1162
9. Yamada Y, Matsuda M, Maeda K et al (1995) The phylogenetic relationships of methanol-assimilating yeasts based on the partial sequences of 18S and 26S ribosomal RNAs: the proposal of *Komagataella* Gen. Nov. (*Saccharomycetaceae*). Biosci Biotechnol Biochem 59:439–444
10. Naumov GI, Naumova ES, Tyurin OV et al (2013) *Komagataella kurtzmanii* sp. nov., a new sibling species of *Komagataella* (*Pichia*) *pastoris* based on multigene sequence analysis. Antonie van Leeuwenhoek Int J Gen Mol Microbiol 104:339–347
11. De SK, Lin YC, Tiels P et al (2009) Genome sequence of the recombinant protein production host *Pichia pastoris*. Nat Biotechnol 27: 561–566
12. Mattanovich D, Graf A, Stadlmann J et al (2009) Genome, secretome and glucose transport highlight unique features of the protein production host *Pichia pastoris*. Microb Cell Fact 8:29
13. Küberl A, Schneider J, Thallinger GG et al (2011) High-quality genome sequence of *Pichia pastoris* CBS7435. J Biotechnol 154: 312–320
14. Love KR, Shah KA, Whittaker CA et al (2016) Comparative genomics and transcriptomics of *Pichia pastoris*. BMC Genomics 17:550
15. Sturmberger L, Chappell T, Geier M et al (2016) Refined *Pichia pastoris* reference genome sequence. J Biotechnol 235:121–131
16. Brady JR, Whittaker CA, Tan MC et al (2020) Comparative genome-scale analysis of *Pichia pastoris* variants informs selection of an optimal base strain. Biotechnol Bioeng 117:543–555
17. Braun-Galleani S, Dias JA, Coughlan AY et al (2019) Genomic diversity and meiotic recombination among isolates of the biotech yeast *Komagataella phaffii* (*Pichia pastoris*). Microb Cell Fact 18:211
18. Valli M, Tatto NE, Peymann A et al (2016) Curation of the genome annotation of *Pichia pastoris* (*Komagataella phaffii*) CBS7435 from gene level to protein function. FEMS Yeast Res 16:fow051
19. Valli M, Grillitsch K, Grünwald-Gruber C et al (2020) A subcellular proteome atlas of the yeast *Komagataella phaffii*. FEMS Yeast Res 20:foaa001
20. Sayers EW, Cavanaugh M, Clark K et al (2020) GenBank. Nucleic Acids Res 48:D84–D86
21. Amid C, Alako BTF, Balavenkataraman Kadhirvelu V et al (2020) The European Nucleotide Archive in 2019. Nucleic Acids Res 48: D70–D76

22. Fukuda A, Kodama Y, Mashima J et al (2021) DDBJ update: streamlining submission and access of human data. Nucleic Acids Res 49: D71–D75
23. Arita M, Karsch-Mizrachi I, Cochrane G (2021) The international nucleotide sequence database collaboration. Nucleic Acids Res 49: D121–D124
24. Pruitt K, Brown G, Tatusova T, et al (2002) The Reference Sequence (RefSeq) Database, In: The NCBI Handbook. National Center for Biotechnology Information (US), Bethesda (MD), pp 1–24
25. National Library of Medicine Prokaryotic RefSeq Genomes, https://www.ncbi.nlm.nih.gov/refseq/about/prokaryotes/. Accessed on 30 April 2024
26. Cherry JM, Hong EL, Amundsen C et al (2012) *Saccharomyces* genome database: the genomics resource of budding yeast. Nucleic Acids Res 40:D700–D705
27. Carbon S, Douglass E, Dunn N et al (2019) The Gene Ontology Resource: 20 years and still GOing strong. Nucleic Acids Res 47: D330–D338
28. Boyle EI, Weng S, Gollub J et al (2004) GO::TermFinder—open source software for accessing Gene Ontology information and finding significantly enriched Gene Ontology terms associated with a list of genes. Bioinformatics 20:3710–3715
29. Stuart A, Nick W, and Tsyganov K (2018) Introduction to RNA-Seq, https://monashdatafluency.github.io/rnaseq-intro/reference-sequence.html. Accessed on 22 June 2022
30. Harrington R (2020) File formats used in bioinformatics – Bioinformatics Notebook, https://rnnh.github.io/bioinfo-notebook/docs/file_formats.html#fasta. Accessed on 22 June 2022
31. NCBI FASTA Format for Nucleotide Sequences, https://www.ncbi.nlm.nih.gov/genbank/fastaformat/. Accessed on 22 June 2022
32. EMBL-EBI GFF/GTF File Format, https://www.ensembl.org/info/website/upload/gff.html. Accessed on 22 June 2022
33. Stein LD Generic Feature Format Version 3 (GFF3), https://github.com/The-Sequence-Ontology/Specifications/blob/master/gff3.md. Accessed on 22 June 2022
34. NCBI Sample GenBank Record, https://www.ncbi.nlm.nih.gov/Sitemap/samplerecord.html. Accessed on 22 June 2022
35. Lee TH, Kim YK, Hie BH (2008) GBParsy: a GenBank flatfile parser library with high speed. BMC Bioinformatics 9:321
36. Offei B, Braun-Galleani S, Venkatesh A et al (2022) Identification of genetic variants of the industrial yeast *Komagataella phaffii* (*Pichia pastoris*) that contribute to increased yields of secreted heterologous proteins. PLoS Biol 20: e3001877
37. Heistinger L, Dohm JC, Paes BG et al (2022) Genotypic and phenotypic diversity among *Komagataella* species reveals a hidden pathway for xylose utilization. Microb Cell Fact 21:70
38. Schwarzhans J-P, Wibberg D, Winkler A et al (2016) Non-canonical integration events in *Pichia pastoris* encountered during standard transformation analysed with genome sequencing. Sci Rep 6:38952
39. Schwarzhans JP, Wibberg D, Winkler A et al (2016) Integration event induced changes in recombinant protein productivity in *Pichia pastoris* discovered by whole genome sequencing and derived vector optimization. Microb Cell Fact 15:84
40. Zirpel B, Degenhardt F, Zammarelli C et al (2018) Optimization of Δ9-tetrahydrocannabinolic acid synthase production in *Komagataella phaffii* via post-translational bottleneck identification. J Biotechnol 272–273:40–47
41. Robinson JT, Thorvaldsdóttir H, Winckler W et al (2011) Integrative genomics viewer. Nat Biotechnol 29:24–26
42. Näätsaari L, Mistlberger B, Ruth C et al (2012) Deletion of the *Pichia pastoris* KU70 homologue facilitates platform strain generation for gene expression and synthetic biology. PLoS One 7:e39720
43. Ahmad M, Hirz M, Pichler H et al (2014) Protein expression in *Pichia pastoris*: recent achievements and perspectives for heterologous protein production. Appl Microbiol Biotechnol 98:5301–5317
44. Claes K, Van HD, Vanluchene R et al (2024) OPEN*Pichia*: licence-free *Komagataella phaffii* chassis strains and toolkit for protein expression. Nat Microbiol 9(3):864–876
45. Rodriguez-R LM, Konstantinidis KT (2016) The enveomics collection: a toolbox for specialized analyses of microbial genomes and metagenomes. PeerJ Preprints 4:e1900v1
46. SUBLIME TEXT Sublime Text – Text Editing, Done Right, https://www.sublimetext.com/. Accessed on 1 Aug 2022
47. Ho D Notepad++, https://notepad-plus-plus.org/. Accessed on 1 Aug 2022
48. Ogata H, Goto S, Sato K et al (1999) KEGG: Kyoto encyclopedia of genes and genomes. Nucleic Acids Res 27:29–34

Chapter 3

Isolation of *Pichia pastoris* Genomic DNA for Long-Read DNA Sequencing and PCR Applications

Sven Heimhilcher, David Keßler, and Lukas Rieder

Abstract

Nanopore sequencing applications allow for the rapid acquisition of genetic information. However, the fast and cost-efficient extraction of high molecular-mass genomic DNA from *Komagataella phaffii* can become a bottleneck in setting up an effective sequencing pipeline. In this chapter, we describe two methods for the extraction of genomic DNA that are suitable for downstream PCR applications and for long-read genomic DNA sequencing via a Nanopore sequencing device, respectively.

Key words *Pichia pastoris*, *Komagataella phaffii*, High molecular mass genomic DNA, Sequencing, PCR, LiAc

1 Introduction

The methylotrophic yeast *Komagataella phaffii* (reclassified from *Pichia pastoris*) has become a relevant production host due to its ability to produce and secrete recombinant proteins with high titers. Ectopic integration of heterologous genes and expression cassettes into the genome of *K. phaffii* is frequently used for the generation of new high-producing strains. Such differences in production are often caused by sites of integration, varying copy numbers, and additional mutations, including InDels or chromosomal rearrangements. Therefore, the extraction of genomic DNA (gDNA) is crucial both to determine the presence of heterologous genes via PCR applications and to characterize the genetic differences of individual production strains by genome sequencing. An important step toward providing easy and direct access to genomic sequence information was the introduction of Nanopore sequencing applications, which enable the sequencing of samples in situ. However, the quality of the sequencing results is largely dependent

Authors Sven Heimhilcher and David Keßler have equally contributed to this chapter.

Xavier Garcia-Ortega et al. (eds.), *Pichia pastoris: Methods and Protocols*, Methods in Molecular Biology, vol. 2697,
https://doi.org/10.1007/978-1-0716-4779-0_3,

on the length of the reads obtained. Hence, the extraction of gDNA as well as the sample preparation has a critical influence, as the occurrence of excessive shear forces, such as by vortexing, excessive pipetting, or the use of beads during cell lysis, must be avoided in order to ensure the integrity of high-molecular-mass DNA. For this reason, methods based on the enzymatic degradation of the cell wall, followed by lysis using a detergent, are the methods of choice for obtaining high molecular weight DNA. Lysis is combined with phenol-chloroform extraction and DNA retrieval via isopropanol-induced precipitation.

In this chapter, we describe two alternative and often complementary methods for the extraction of gDNA from *K. phaffii*, resulting in different qualities of gDNA suitable for various applications. The first method presented is based on the lithium acetate method [1] and is ideal for PCR-based downstream applications, as the method is fast but results in shorter DNA fragments (Fig. 1). In contrast, the second method presented is suitable for obtaining high-molecular-mass genomic DNA, which is crucial for obtaining long reads (e.g., Nanopore sequencing), but it is more labor-intense [2, 3].

2 Materials

2.1 Preparation of K. phaffii Cultures

1. 250 mL baffled shake flasks with appropriate sealing (e.g. cotton cloth) (sterilized).
2. YPD medium: 1% (w/v) Yeast extract, 2% (w/v) peptone, 2% (w/v) glucose. Autoclave and store at room temperature.
3. 10 mg/mL Ampicillin stock solution.
4. YPD-Agar plates with appropriate selection pressure: 1% (w/v) yeast extract, 2% (w/v) peptone, 2% (w/v) glucose, and 1.5% (w/v) agar-agar. Autoclave. When lukewarm, add selection reagents, pour into Petri dishes, and let solidify.

2.2 gDNA Isolation for PCR Applications

1. Table centrifuge.
2. 1.5 mL plastic reaction tubes.
3. YPD-Agar plates with appropriate selection pressure: 1% (w/v) yeast extract, 2% (w/v) peptone, 2% (w/v) glucose, and 1.5% (w/v) agar-agar. Autoclave. When lukewarm, add selection reagents, pour into Petri dishes, and let solidify.
4. Lithium acetate lysis solution: 1% SDS, 0.2 M LiAc. Autoclave and store at room temperature.
5. 99.9% ethanol.
6. Thermoblock.
7. 10 μM primers suitable to amplify the gene of interest.

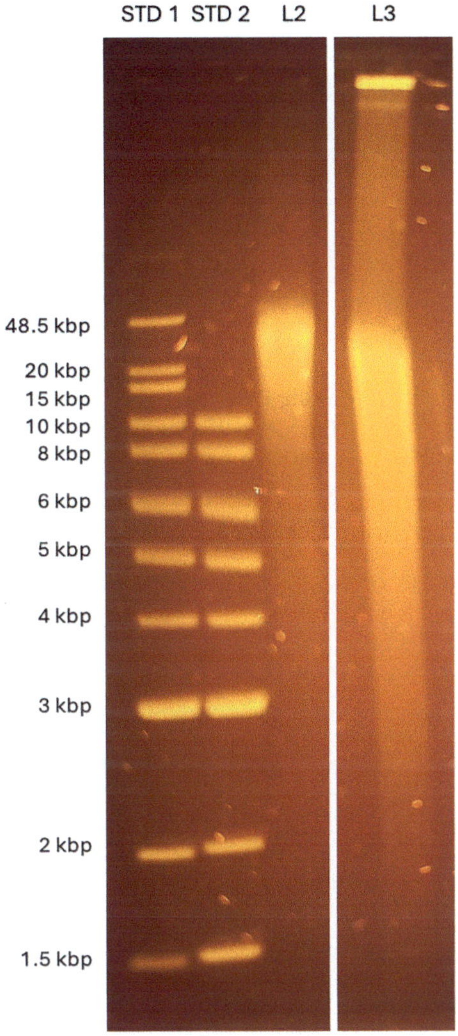

Fig. 1 Pulsed-field gel to compare *K. phaffii* BSYBG11 gDNA prepared using the LiAC (Lane 2) and high-molecular-weight gDNA extraction (Lane 3) methods, respectively. 1 μg of DNA was loaded into each well

8. Phire Plant Direct PCR Master Mix (Thermo Scientific™).
9. Thermocycler.

2.3 gDNA Isolation for Long-Read DNA Sequencing

1. Chemical fume hood.
2. Thermoblock.
3. Cooled centrifuge with rotor for 50 mL tubes.
4. Water bath or heating block with adapter for 50 mL tubes.
5. 50 mL tubes.
6. Nuclease-free water.

Table 1
Components, their concentrations, and volumes for the preparation of resuspension solution

Solution 1			
	Stock solutions	**Final concentrations**	**Mixture 100 mL**
EDTA	0.5 M EDTA pH 8.0	40 mM EDTA	8 mL
2-Mercaptoethanol	≥99% stock	90 mM 2-ME	0.69 mL
Water			Fill to 100 mL

Table 2
Components, their concentrations, and volumes for the preparation of digestion solution

Solution 2			
	Stock solutions	**Final concentrations**	**Mixture 100 mL**
Sorbitol	1 M sorbitol	~1 M sorbitol	100 mL
EDTA	0.5 M EDTA pH 8.0	1 mM EDTA	200 μL
DTT	1 M DTT	3 mM DTT	300 μL
Zymolyase		2 mg/mL	200 mg

Table 3
Components, their concentrations, and volumes for the preparation of spheroplast resuspension solution

Solution 3			
	Stock solutions	**Final concentrations**	**Mixture 100 mL**
Tris-HCl	1 M Tris-HCl pH 8.0	10 mM Tris-HCl pH 8.0	1 mL
NaCl	5 M NaCl	100 mM NaCl	2 mL
$MgCl_2$	1 M $MgCl_2$	10 mM $MgCl_2$	1 mL
2-Mercaptoethanol	≥99% stock	2 mM 2-ME	15 μL
Water			Fill to 100 mL

7. Resuspension Solution: 40 mM EDTA, 90 mM 2-mercaptoethanol; adjust pH to 8.0 with HCl. Store at 4 °C (Table 1).
8. Digestion Solution: 1 M sorbitol, 1 mM EDTA, 3 mM DTT. Store at 4 °C. Add Zymolyase® to a final concentration of 2 mg/mL directly before use (Table 2).

Table 4
Components, their concentrations, and volumes for the preparation of lysis solution

Solution 4			
	Stock solutions	**Final concentrations**	**Mixture 100 mL**
Tris-HCl	1 M tris-HCl pH 8.0	50 mM tris-HCl pH 8.0	5 mL
NaCl	5 M NaCl	1 M NaCl	20 mL
SDS	10% SDS	2% SDS	20 mL
EDTA	0.5 M EDTA pH 8.0	50 mM EDTA	10 mL
Water			Fill to 100 mL

9. Spheroplast Resuspension Solution: 10 mM Tris-base, 100 mM NaCl, 10 mM $MgCl_2$, 2 mM 2-mercaptoethanol; adjust to pH 8.0. Store at 4 °C (Table 3).
10. Lysis Solution: 50 mM Tris-HCl, 1 M NaCl, 2% SDS, 50 mM EDTA. Store at 4 °C (Table 4).
11. 2-Mercaptoethanol stock (>99%).
12. Zymolyase® 100 T (Carl Roth GmbH + Co. KG, 9329.1).
13. TE buffer: 10 mM Tris-HCl, 1 mM EDTA, pH 8.0.
14. 99.9% isopropanol.
15. 99.9% ethanol.
16. 70% ethanol.
17. Phenol–chloroform–isoamyl alcohol mixture for extraction of nucleic acids (Carl Roth GmbH + Co. KG, A156.1).
18. 10 mg/mL RNase A.
19. 20 mg/mL Proteinase K.
20. Qubit™ fluorometer.
21. Qubit™ HS Buffer.
22. Qubit™ HS reagent.
23. Qubit™ dsDNA HS Standards #1 and #2.
24. Nanodrop.
25. Pulse field gel electrophoresis device.
26. Appropriate DNA size standard (e.g., NEB® Quick-Load® 1 kb extend DNA Ladder).

3 Methods

3.1 gDNA Isolation for PCR-Based Applications

1. Resuspend a single *K. phaffii* colony in a 1.5 mL reaction tube containing 100 μL of lithium acetate lysis solution using a small inoculation loop or by vortexing.
2. Incubate at 65 °C for 10 min.
3. Add 300 μL of 99.9% ethanol and mix.
4. Centrifuge at 20,000 rcf and room temperature for 3 min.
5. Discard the supernatant.
6. Wash the pellet with 500 μL of 70% ethanol (*see* **Note 1**).
7. Discard the supernatant.
8. Air-dry the pellet for 10–15 min.
9. Resuspend the pellet in 70 μL of ddH$_2$O.
10. Centrifugation at 20,000 rcf at room temperature for 1 min.
11. Use 1–5 μL of the gDNA preparation for PCR, as exemplified by the amplification of *ku70* (Tables 5 and 6, Fig. 2) (*see* **Note 2**).

3.2 gDNA Isolation for Long-Read DNA Sequencing

1. Inoculate 50 mL of YPD growth medium in a sterile 250 mL baffled shake flask with cell material from a single colony of *K. phaffii* or a few microliters of an overnight culture (ONC).
2. Incubate for 36–48 h at 28 °C or until the early stationary phase is reached.
3. Prepare solutions for gDNA extraction (*see* **Note 3**).
4. Prepare the heating block at 30 °C and the water bath at 45 °C.
5. Incubate Solution 4 at 45 °C for ~20 min in the water bath (*see* **Note 4**).

Table 5
Components and their respective volumes for PCR reaction with gDNA as template

Component	Volume
Phire plant direct PCR master mix	10 μL
Supernatant	X μL
Primer 1: 5'- GATGTAACAACACCTTCCAC-3' (10 μM)	2.5 μL
Primer 2: 5'-GACGTTGCAGATAAAATCCAATC -3'(10 μM)	2.5 μL
ddH$_2$O	5–X μL
Total volume	20 μL

Table 6
PCR cycling profile used for amplification

# of cycles	Time	Temperature
1	1 min	98 °C
35	5 s	98 °C
	5 s	Dependent on primers
	20 s/kb	72 °C
1	1 min	72 °C
1	∞	10 °C

6. Pellet yeast cells in 50 mL tubes by centrifugation at 3000 rcf and room temperature for 5 min.
7. Discard the supernatant.
8. Resuspend the cell pellet in 5 mL of the Resuspension Solution.
9. Incubate the cells for 5 min.
10. Repeat the centrifugation step at 3000 rcf and room temperature for 3 min. Discard the supernatant.
11. Resuspend the cell pellet in 5 mL Digestion Solution and incubate at 100 rpm and 30 °C for 1 h (*see* **Note 5**).
12. Centrifuge samples to pellet spheroplasts at 3000 rcf and room temperature for 5 min.
13. Carefully pour off the supernatant.
14. Gently resuspend spheroplasts in 2 mL of Spheroplast Resuspension Solution by gently moving and flicking the tube. Do not vortex (*see* **Note 6**)!
15. Add 2 mL of Lysis Solution and mix gently by inverting the tube.
16. Incubate the tube in the water bath at 45 °C for 10 min. Gently invert the tube several times during the incubation period.
17. Add 2 mL of the 1:1 phenol/chloroform/isoamyl alcohol mixture and invert gently until the mixture is homogenous (*see* **Note 7**).
18. Centrifuge the tube at 3000 rcf and room temperature for 5 min.
19. Transfer the upper aqueous phase to a fresh 50 mL plastic reaction tube using wide-bore pipette tips. Do not disturb the interface (*see* **Note 8**).
20. Add 0.7 volumes of isopropanol to precipitate the gDNA.
21. Mix gently by inversion.

Fig. 2 Confirmation of the *ku70* gene (4394 bp) via PCR amplification, using the gDNA obtained by employing the LiAc protocol as a template

22. If 50 mL of culture was used, a DNA pellet should be visible and can be scooped out with a glass rod or pipette tip; otherwise, centrifuge the DNA at 15,000 rcf and 4 °C for 15 min. Discard the supernatant (*see* **Note 9**).
23. Wash the pellet with 3 mL 70% ethanol.
24. Centrifuge the tube at 15,000 rcf and 4 °C for 15 min. Discard the supernatant (*see* **Note 10**).
25. Dry the pellet for 5–20 min at room temperature or under laminar flow conditions.
26. Resuspend DNA in TE buffer for several hours or overnight. Do not pipette the DNA (*see* **Note 11**).

27. Add RNase A to a final concentration of 100 μg/mL.
28. Incubate with gentle shaking at 37 °C for 1 h.
29. Add proteinase K to a final concentration of 200 μg/mL.
30. Incubate at 56 °C for 2 h.
31. Repeat steps 17–26.
32. For next-generation sequencing applications, resuspend the DNA in nuclease-free water (*see* **Note 12**).
33. Determine DNA concentration and purity using Nanodrop and Qubit™ measurement (*see* **Note 13**).
34. For sequencing applications, we recommend A260/A280 ratios of 1.8–2.0 and A260/A230 ratios of 2.0–2.2.

4 Notes

1. Gently wash the pellet by pipetting the 70% ethanol mixture up and down, allowing it to flow over the pellet without disturbing it.
2. If higher DNA yields are desired, we observed higher DNA concentrations according to Nanodrop measurements after multiple hours of incubation.
3. For preparation of the solutions, it is recommended to prepare the following stock solutions: 1 M Tris-HCl pH 8.0, 0.5 M EDTA pH 8.0, 1 M sorbitol, 1 M DTT, 5 M NaCl, 1 M $MgCl_2$, and 10% SDS.
4. Solution 4 will show at room temperature a white precipitate. When heated to 45 °C, this precipitate will solubilize. Use solution 4 only when the solution is completely clear.
5. Spheroplast formation can be determined under the microscope by preparing two microscope slides with 2 μL of the sample. To one slide, add 2 μL of 10% SDS solution. In case of successful spheroplast formation, the sample with SDS added should show lysed cells, while the other sample shows intact spheroplasts.
6. Careful handling of the spheroplasts is recommended to ensure the structural integrity of high-molecular-weight gDNA.
7. When using a prepared phenol/chloroform/isoamyl alcohol stock, ensure that the solution is always drawn from the lower phase. Do not shake the bottle.
8. Take only as much of the upper aqueous layer as possible without contaminating the sample with phenol, chloroform, or protein from the interface. The high starting volume of 50 mL cell culture allows for prioritization of DNA purity over yield.

9. Do not pour off the supernatant, as the DNA pellet is often not strongly attached to the inner wall of the plastic reaction tube and therefore might be lost. Instead, use a serological pipette to remove the supernatant.
10. Steps 23 and 24 can be repeated to ensure better removal of co-precipitated salts.
11. If RNase A and Proteinase K digestion is performed afterward, TE buffer is recommended.
12. If gDNA samples are used directly for downstream applications such as PCR or sequencing, ddH_2O is sufficient for elution. For long-term storage, elution in TE buffer is highly recommended.
13. As DNA quantification using the Qubit® Broad Range dsDNA assay is considered more precise than spectrophotometric determination of nucleic acid concentrations via Nanodrop measurement, we recommend using the first method to determine DNA concentration, while the latter can be used to check for the presence of impurities before sequencing.

Acknowledgments

We kindly thank the Christian Doppler Gesellschaft and the industrial partners BioGrammatics, SeSam Biotech, ZETA, and Solar-Biotech for funding of the Christian Doppler Laboratory for Innovative *P. pastoris* Host and Vector Systems, as well as the FFG for support of the bisy GmbH "Basisprogramm" project FO999909039. LR received funding from the EIC Pathfinder program NewCat under grant agreement No. 101046815.

Competing Interest Statement The authors declare that they have no conflicts of interest relevant to the content of this chapter.

References

1. Lõoke M, Kristjuhan K, Kristjuhan A (2011) Extraction of genomic DNA from yeasts for PCR-based applications. BioTechniques 50(5): 325–328. https://doi.org/10.2144/000113672
2. Sherman F, Fink GR, Hicks JB (1986) Yeast DNA isolations. In: Laboratory course manual for methods in yeast genetics. Cold Spring Harbor Laboratory, Cold Spring Harbor, pages 125–126
3. Hoyer LL (2023) Extraction of yeast high-molecular-weight genomic DNA. protocols.io https://doi.org/10.17504/protocols.io.rm7vzb1b4vx1/v1

Chapter 4

Pichia pastoris RNA-Sequencing and Data Analysis

Julian Christopher Krappinger, Astrid Radkohl, Veronika Schusterbauer, Anita Emmerstorfer-Augustin, and Julia Feichtinger

Abstract

RNA sequencing (RNA-seq) allows for transcriptome profiling, usually with the goal of comparing the gene expression between conditions of interest. Transcriptome profiling helped to understand and improve the use of microbes in industrial processes, as exemplified numerous times for the biotechnologically relevant yeast *Komagataella phaffii*. A typical RNA-seq experiment begins with (a) experimental design, (b) RNA extraction, and (c) library preparation and sequencing and continues with its data analysis, including (d) quality control and preprocessing, (e) alignment, (f) quantification, and (g) differential gene expression analysis. In particular, experimental design should be considered carefully, while RNA extraction is relatively straightforward, and library preparation and sequencing are commonly purchased as a service. RNA-seq data analysis can be carried out in a number of ways and comprises a sequence of steps with many aspects to consider. Here, we provide a protocol for RNA extraction, a summary of aspects to consider for experimental design, library preparation, and sequencing, as well as the basics of a data analysis workflow.

Key words *Pichia pastoris*, *Komagataella phaffii*, RNA extraction, RNA-sequencing, Data analysis

1 Introduction

Komagataella phaffii (reclassified from *Pichia pastoris*) is also called the 'biotech' yeast, as it is used in numerous industrial applications. The reasons for that include strong and (methanol) inducible promoters, an efficient protein secretion machinery, and the ease of genetic manipulation [1]. Recently, *K. phaffii* also started to gain attention as an alternative yeast model organism [2], which broadens the fundamental knowledge of yeast cell biology typically exclusively gained from *Saccharomyces cerevisiae*. To better understand yeast physiology, cell fitness, or cell stress at a given moment under given conditions, RNA sequencing (RNA-seq) is a powerful tool. Although RNA-seq is a relatively young method, transcriptome profiling based on high-throughput next-generation sequencing is now widely used in life sciences [3]. RNA-seq allows for a deeper understanding of gene expression and its changes in response to

Xavier Garcia-Ortega et al. (eds.), *Pichia pastoris: Methods and Protocols*, Methods in Molecular Biology, vol. 2697,
https://doi.org/10.1007/978-1-0716-4779-0_4,

different conditions and provides insights into gene structure and regulation [4]. The most common aim of a transcriptomics experiment is the comparison of gene expression between conditions, termed differential gene expression (DGE) analysis [3]. The recent sequencing and annotation of the *K. phaffii* genome but also of other *Pichia pastoris* strains [5–10] were crucial steps in facilitating genome re-sequencing and transcriptomics analyses. In recent years, RNA-seq has served as a tool to study the transcriptome of these strains under different conditions [9–11]. Additionally, RNA-seq was successfully used to optimize growth media [12], screen for strong endogenous promoters [13], conduct metabolic flux analysis (reviewed by [14]), and study the negative effects of growth inhibitors [15].

As a basis for RNA-seq, RNA has to be extracted quickly and efficiently from cells propagated at the conditions to be investigated, e.g., at a certain temperature, in special media, or after being treated with different reagents. In order to generate the most sensitive and biologically relevant results, some important steps must be considered before, during, and after the actual RNA purification. First, cells have to be harvested quickly, since even small changes in cultivation conditions (e.g., a temperature shift) can dramatically influence the whole transcriptome [16, 17]. Second, cells have to be disrupted quickly in the presence of RNase inhibitor reagents (typically strong denaturants like guanidine salts, sodium dodecylsulfate (SDS), or phenol-based compounds that are designed to lower the risk of RNA degradation) [18]. Third, the extracted RNA must be of excellent quality, since RNA integrity is a major concern for gene expression studies. RNA is an unstable molecule that is prone to hydrolysis in the presence of alkaline solutions and rapidly degraded by ribonucleases. Typically, after isolation, RNA is stored at −80 °C in RNase-free water, and freeze-thaw cycles should be avoided in order not to compromise RNA quality [19]. To avoid sequencing of degraded samples, sample quality has to be assessed carefully, which includes comparing the ratio of 25S:18S ribosomal RNA by ethidium bromide-based gel electrophoresis and the determination of the RNA integrity number (RIN) [20].

After RNA extraction, library preparation and sequencing are performed, although these steps are commonly outsourced to a sequencing facility [21]. Nevertheless, the researcher should carefully consider the experimental design, including the number of biological replicates, the sequencing technology, the library preparation method, and the read type, length, and output, depending on the aim and complexity of the planned experiment. The RNA-seq data analysis does not follow a strict/standardized workflow and involves a series of steps [11, 22–24], for which the researcher should also carefully review a number of aspects, such as workflow and tool choice, depending on the goal of the

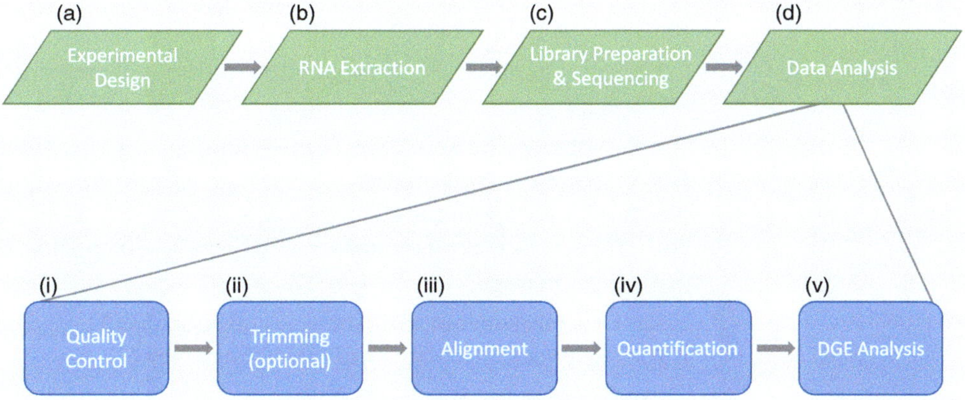

Fig. 1 A basic RNA-sequencing workflow including (**a**) experimental design, (**b**) RNA extraction, (**c**) library preparation and sequencing, and (**d**) data analysis. A closer look at data analysis divides this step into

experiment. For these aspects, we would like to direct the reader to an excellent review by Stark et al., which provides a good overview of possible data analysis routes [3]. In this chapter, we focus on a commonly used workflow, consisting of sequence read alignment to a reference genome followed by a quantification step to generate read counts for each feature, which then can be used as input for the DGE analysis.

With the rising popularity of RNA-seq experiments, expertise in bioinformatics and adequate computing resources are in high demand [25], and protocols help the researcher to understand the principles and guide them through the individual steps. In this protocol, we provide a basic workflow for RNA sequencing and data analysis (Fig. 1), comprising (a) RNA extraction, (b) important aspects considering experimental design, library preparation, and sequencing, (c) quality control and preprocessing, (d) alignment to a reference genome, (e) quantification, and (f) an overview of a DGE analysis. This should serve as a basis for the reader to independently carry out RNA-seq experiments and subsequent data analyses.

2 Materials

2.1 RNA Extraction

1. Host strain *K. phaffii* (here exemplified with the CBS7435 *his4*Δ).
2. Minimal dextrose (MD) medium: 2% dextrose, 1.34% of yeast nitrogen base lacking amino acids, 0.00004% of biotin, with 0.4% of histidine (for histidine auxotrophic strain).
3. 12 mL glass culture tubes for overnight cultures.

4. 250 mL shake flasks with baffles.
5. Shaker (e.g., HT Infors Shaker).
6. Tabletop centrifuge (e.g., Eppendorf Microcentrifuge 5415 R).
7. RNA Isolation System (we recommend using the SV Total RNA Isolation System from Promega Corporation, USA). Certain kit components (RNA wash solution, RNA lysis buffer, and DNase stop solution) have to be prepared according to the provider's manual.
8. RNase surface decontaminant (e.g., RNase AWAY™, Thermo Fisher Scientific Inc., St. Leon-Rot, Germany).
9. Glass beads, 0.25–0.5 mm.
10. Vortex-Genie™ 2.
11. SafeSeal tips professional with incorporated filter, sterile (10, 200, and 1000 μL).
12. RNase-free 1.5 mL reaction tubes with conical bottoms.
13. Ethanol absolute (abs.).
14. Liquid nitrogen.
15. Resuspension solution: 1 M sorbitol, 0.1 M EDTA, 0.1% (v/v) 2-mercaptoethanol (added fresh to the solution prior to use).
16. RNase-free water, Aqua bidest. "Fresenius" (*see* **Note 1**).
17. Rotilab® syringe filters 0.22 μm, sterile.
18. DNase I supplied with the SV Total RNA Isolation System is dissolved in the indicated volume of Fresenius water, aliquoted as 50 μL stocks, and stored at −20 °C (*see* **Note 2**).

2.2 RNA Gel Electrophoresis

1. NanoDrop 2000 UV-Vis Spectrophotometer.
2. Agarose.
3. 1× TAE buffer: 40 mM Tris, 20 mM acetic acid, 1 mM EDTA.
4. 10 × DNA sample loading dye.
5. Agarose gel electrophoresis equipment.
6. Ethidium bromide.
7. 99.5% formamide.
8. GeneRuler DNA Ladder Mix.

2.3 RNA-seq Data Analysis

The following tools are popular choices for the analysis of RNA-seq data (Table 1). Installation guides as well as user manuals can be found at the links stated in Table 1 (*see* **Notes 3** and **4**).

Table 1
A collection of popular RNA-seq data analysis tools

Tools	Analysis step	Link	References
FastQC*	Quality control	https://www.bioinformatics.babraham.ac.uk/projects/fastqc/	[26]
MultiQC*	Quality control	https://multiqc.info/	[27]
Trimmomatic	Preprocessing	http://www.usadellab.org/cms/?page=trimmomatic	[28]
Cutadapt	Preprocessing	https://cutadapt.readthedocs.io/en/stable/	[29]
STAR*	Alignment	https://github.com/alexdobin/STAR	[30]
HISAT2	Alignment	http://daehwankimlab.github.io/hisat2/	[31]
Samtools*	Sorting	http://www.htslib.org/	[32]
IGV*	Visualization	https://software.broadinstitute.org/software/igv/	[33]
Qualimap*	Quality control	http://qualimap.conesalab.org/	[34, 35]
RSeQC	Quality control	http://rseqc.sourceforge.net/	[36]
htseq-count*	Quantification	https://htseq.readthedocs.io/en/master/	[37]
featureCounts	Quantification	http://subread.sourceforge.net/	[38]
R*		https://www.r-project.org/	[39]
DESeq2* (R package)	DGE analysis	https://bioconductor.org/packages/release/bioc/html/DESeq2.html	[40]
limma (R package)	DGE analysis	https://bioconductor.org/packages/release/bioc/html/limma.html	[41]
edgeR (R package)	DGE analysis	https://bioconductor.org/packages/release/bioc/html/edgeR.html	[42]

Tools marked with an asterisk are used in this protocol, while the others represent popular alternatives or tools for optional steps

3 Methods

3.1 RNA Extraction

RNA isolation in this protocol is conducted with the SV Total RNA Isolation System (Promega Corporation, USA) as instructed by the provider's manual. This kit allows for RNA isolation from a broad range of eukaryotic cells. When using a different RNA isolation kit, certain steps might be different. Important adaptations to the provider's protocol are highlighted below, which are required for efficient RNA isolation from the yeast *K. phaffii*.

1. Pre-grow cells overnight at 28 °C, 120 rpm, and 80% humidity to saturation in 2 mL of MD medium in 12 mL glass culture tubes.
2. Dilute cells in 50 mL of MD medium to an OD_{600} of 0.015 in 250 mL shake flasks with baffles and let the culture grow at 28 ° C, 120 rpm, and 80% humidity until it reaches the late exponential growth phase (OD_{600} ~ 4–5) (*see* **Note 5**).
3. Prior to cell harvest, clean the bench, the pipettes, and the centrifuge with RNase surface decontaminate (*see* **Notes 6** and 7), and precool the centrifuge at 4 °C.
4. Harvest 18 OD units (*see* **Note 8**) by transferring the corresponding volume of cells to RNase-free 1.5 mL reaction tubes with conical bottoms and centrifugation at room temperature at 1600 rpm for 30 s (*see* **Note 9**).
5. Discard the supernatant (*see* **Note 10**) and immediately resuspend the cell pellet in 200 μL of precooled (4 °C) resuspension solution.
6. Add glass beads to the cell suspension (glass beads should comprise 50% of the volume of the cell suspension), and let the cell suspension rest on ice before mechanical cell disruption (*see* **Note 11**).
7. For mechanical cell disruption, vortex the 1.5 mL reaction tubes with conical bottoms at max. speed for 30 s. Then, chill the tubes on ice for 30 s. Repeat this step 8 times (*see* **Note 12**).
8. Add 150 μL of precooled (4 °C) RNA lysis solution and mix gently by inverting the tubes 5 times.
9. Centrifuge the mixture at 1600 rpm for 30 s at 4 °C.
10. Transfer 200 μL of the supernatant to a fresh RNase-free 1.5 mL reaction tube filled with 350 μL of RNA dilution buffer and mix gently by inverting the tube 5 times.
11. Centrifuge at max. speed for 10 min at 4 °C.
12. Transfer the cleared lysate into a fresh tube.
13. Add 200 μL of absolute ethanol, and mix by inverting the tube.
14. Transfer mixture to the spin column provided with the SV Total RNA Isolation System and centrifuge at max. speed for 1 min at room temperature.
15. Add 600 μL of RNA wash solution and centrifuge at max. speed for 1 min at room temperature.
16. Pipette 50 μL of DNase incubation onto the filter membrane of the spin column and incubate for 15 min at room temperature.
17. Add 200 μL of DNase Stop Solution and centrifuge at max. speed for 1 min at room temperature.

18. Add 600 μL of RNA wash solution, centrifuge at max. speed for 1 min at room temperature, and discard the flow-through.
19. Add 250 μL of RNA wash solution and centrifuge at max. speed for 2 min at room temperature.
20. Transfer the insert to a fresh RNase-free 1.5 mL reaction tube.
21. Pipette 100 μL of RNase-free Fresenius water onto the membrane of the insert and elute RNA by centrifugation at max. speed for 1 min at room temperature.
22. Take 10 μL aliquots of each RNA sample and store it in a separate tube for the quality assessment.
23. Quickly freeze all RNA isolates with liquid nitrogen and store at—80 °C until continuing the procedure (*see* **Notes 13** and **14**).

3.2 RNA Quality Assessment

3.2.1 Gel Electrophoresis of Isolated RNA

After isolation, the quality of the RNA has to be determined. As a first measure, RNA concentrations can be determined spectrophotometrically. The RNA purity is thereby assessed by comparing A_{260} with A_{280} and A_{230}, respectively. Ideally, $A_{260/280}$ should be between 1.7 and 2.1, and $A_{260/230}$ between 1.7 and 2.0. A NanoDrop also calculates RNA concentrations by following the Lambert-Beer law. Additionally, denaturing gel electrophoresis is recommended. Degradation of RNA cannot be detected spectrophotometrically but instead by visualizing the two distinct RNA bands corresponding to 25S and 18S RNA (Fig. 2). These bands will result in a smear on the denaturing agarose gel if the RNA is degraded. In general, the signal of the 28S RNA band should be 3 times brighter than the signal of the 18S RNA band.

1. Use 1.5 μL of aliquoted RNA for UV-Vis spectrophotometry.
2. When the purity determined with the NanoDrop is of sufficient quality and quantity (*see* **Note 15**), samples are further investigated on a denaturing agarose gel electrophoresis.
3. Prepare a 1.2% (w/v) TAE agarose gel containing ethidium bromide (or any other compound typically used for visualizing DNA/RNA in agarose gels).
4. While waiting until the agarose gel is polymerized, mix ~1000 ng of your RNA samples (*see* **Note 16**) with a final concentration of min. 60% (v/v) formamide and DNA loading dye.
5. Incubate RNA mixtures at 65 °C for 5 min.
6. Afterwards, cool the samples at 4 °C for 5 min.
7. Transfer the solid agarose gel to an electrophoresis chamber and fill the chamber with 1x TAE until the agarose gel is completely covered by the buffer.

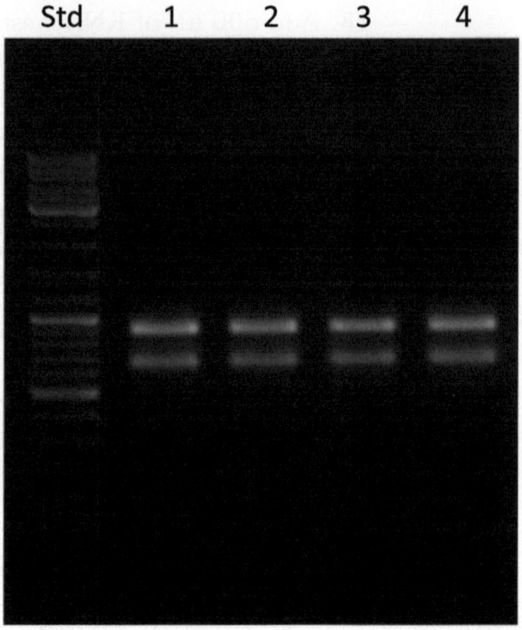

Fig. 2 Denaturing agarose gel of RNA samples. ~ 1000 ng of isolated RNA mixed with 60% (v/v) formamide and DNA loading dye were heat denatured at 65 °C, loaded onto the denaturing ethidium bromide gel, separated electrophoretically, and visualized under UV light. Std: GeneRuler DNA Ladder Mix. 1–4: RNA samples

8. Load a DNA standard and the RNA samples and run the gel at 70 V for 90 min.
9. Visualize RNA bands under UV light.

3.2.2 RNA Integrity Categories

For RNA sequencing, it is recommended to consult with the sequencing company of choice about which level of RNA quality and RNA amounts are needed. This information can usually also be found on the provider's website. All commercial sequencing providers usually perform their own quality controls, which include the determination of the RIN (or the RNA quality number—RQN, which is equivalent to RIN, depending on the Bioanalyzer used) of your samples prior to proceeding with the library preparation and sequencing process. RIN is determined by the result of microcapillary electrophoretic RNA separation and requires the use of a microcapillary electrophoresis machine, e.g., a Bioanalyzer system [20]. RIN or RQN values range from 10 (highly intact RNA) to 1 (completely degraded RNA). A high RIN (ideally, at least about 7–8) helps to guarantee the presence of full-length mRNA transcripts that can be incorporated into the sequencing library.

3.3 Experimental Design, Library Preparation, and Sequencing

After RNA extraction and quality assessment, library preparation and sequencing are performed. As there are a number of ways to generate data for RNA-seq experiments and as library preparation and sequencing are usually purchased as a service, we only provide a brief overview of aspects that should be considered. These include (a) the number of biological replicates, (b) the sequencing technology, (c) the library preparation method, and (d) read type, length, and output.

1. The number of biological replicates should be considered carefully, and an appropriate number should be chosen to capture the biological variability, which depends on the samples to be investigated [3, 23, 43]. A reasonable minimum would be four biological replicates, while the absolute minimum would be three biological replicates [21, 43]. However, a study by Schurch et al. highlights the importance of higher numbers of biological replicates [44].
2. Currently, most RNA-seq experiments are based on Illumina short-read sequencing [23], and the current protocol also focuses on this sequencing technology (*see* **Note 17**).
3. To eliminate highly abundant ribosomal RNA (rRNA), either poly(A) enrichment or rRNA depletion is used [45] (*see* **Note 18**). As the genome of *K. phaffii* exhibits a tight organization of genes and genes commonly overlap in opposite directions [5, 6, 8], a strand-specific library preparation is highly recommended (*see* **Note 19**).
4. The genome of *K. phaffii* is rather small with low sequence repetition, and splicing events are not common [8, 10, 21]. Hence, ten million single-end reads and a lower read length can be adequate for DGE analysis [21, 46] (*see* **Note 20**). However, many yeast RNA-seq experiments nowadays aim for higher read output and length and use paired-end reads [12, 15, 47, 48].

3.4 Data Analysis

Please note that this example workflow assumes strand-specific (fr-firststrand, stranded = reverse) and paired-end data. This protocol further requires command line use in a Linux/Unix-based environment (*see* Subheadings 3.4.1, 3.4.2, and 3.4.3) and the R [39] environment for statistical computing (*see* Subheading 3.4.4) (*see* **Note 21**).

3.4.1 Preprocessing of RNA-seq Data

The sequencing facility most commonly supplies the sequencing data as demultiplexed FASTQ files. Before proceeding with the RNA-seq data analysis, the data quality should be assessed.

1. Compute quality metrics for the sequencing data using FastQC [26] as follows (for each file separately, here exemplified for sample1 with the two FASTQ files `sample1_R1.fastq` and `sample1_R2.fastq`, as we assume paired-end data for this workflow):

```
fastqc sample1_R1.fastq
fastqc sample1_R2.fastq
```

   Repeat this step for all samples in your data set (*see* **Note 22**).

2. Summarize the FastQC output files to compile a quality report for the sequencing data using MultiQC [27] as follows (*see* **Notes 23** and **24**):

```
multiqc <path_to_all_FastQC_reports>
```

3. Optional: Depending on the quality of the sequencing reads, consider trimming the data (e.g., if the reads are contaminated with adapter sequences). A number of tools can be employed to achieve this step, including Trimmomatic [28] or Cutadapt [29]. After trimming, **steps 1** and **2** should be repeated.

3.4.2 Alignment of RNA-seq Data

Many RNA-seq data analysis workflows are based on read alignment to a reference genome followed by a quantification step to generate raw read counts, for which we present the necessary steps in this section and Subheading 3.4.3 (*see* **Note 25**):

1. Choose an appropriate aligner. The most commonly used ones are STAR [30] and HISAT2 [31]. We provide example commands for STAR in the following steps.

2. To be able to align the sequencing reads to a reference genome, the genome and annotation files need to be acquired and processed. Currently, the most up-to-date information for *K. phaffii* (CBS 7435) includes the reference genomes and annotations provided by Sturmberger et al. [8], Küberl et al. [5]/Valli et al. [9], and Love et al. [10] (*see* **Note 25**). Download the chosen reference (`pichia_genome.fa`) and the corresponding annotation file (`pichia_annotation.gtf`) (*see* **Notes 26** and **27**). Subsequently, build a genome index for STAR [30] alignments as follows (*see* **Notes 28–30**), which is generated in the specified folder (`pichia_index`):

```
STAR --runMode genomeGenerate \
--runThreadN <value> \
--genomeDir pichia_index \
--genomeFastaFiles pichia_genome.fa \
```

```
--sjdbGTFfile pichia_annotation.gtf \
--genomeSAindexNbases <value> \
--sjdbOverhang <value>
```

3. Align the sequencing data (for each sample separately, here exemplified for sample1 with the FASTQ files `sample1_R1.fastq` and `sample1_R2.fastq`) to the reference genome using STAR [30] as follows (*see* **Notes 30–32**):

```
STAR --genomeDir pichia_index \
--runThreadN <value> \
--readFilesIn sample1_R1.fastq sample1_R2.fastq \
--outFileNamePrefix sample1 \
--outSAMtype BAM Unsorted \
--twopassMode Basic \
--outSAMunmapped Within \
--alignIntronMin <value> \
--alignIntronMax <value> \
--alignMatesGapMax <value>
```

4. The alignment file (BAM file for sample1 generated in **step 3**) can be viewed using a genome viewer such as IGV [33] (*see* **Note 33**).
5. Finally, the quality of the alignment should be assessed (including alignment rate). To accomplish this, you can use the tool Qualimap [35] as follows (*see* **Notes 34–37**):

```
qualimap rnaseq --java-mem-size=<value> \
-pe \
-s \
-p strand-specific-reverse \
-outformat PDF:HTML \
-bam sample1Aligned.out.bam \
-gtf pichia_annotation.gtf
```

6. Execute **steps 3–5** for all samples in your data set (consider creating a loop for the STAR and Qualimap commands).
7. A summary report of all quality reports can be generated using MultiQC [27] as follows (*see* **Notes 23** and **24**):

```
multiqc <path_to_all_QC_reports>
```

3.4.3 Quantification of RNA-seq Data

Based on the alignment and annotation file, a quantification step is carried out to generate raw read counts for each feature (gene).

1. A number of tools exist to perform this step, including htseq-count [37] and featureCounts [38]. Use htseq-count for quantification as follows (again exemplified for sample1, *see* **Notes 38–40**):

```
htseq-count -r name \
--stranded=reverse sample1Aligned.out.bam \
pichia_annotation.gtf > sample1Counts.txt
```

   Repeat this step for all samples in your data set (*see* **Note 41**).

2. For further downstream analysis, the raw counts (`sample1-Counts.txt` for sample1 together with the count data for other samples in your data set) are usually read into R [39] (*see* **Notes 42** and **43**, *see* Subheading 3.4.4 for more details).

3.4.4 Downstream Analysis of RNA-seq Data

Commonly, the goal of an RNA-seq experiment is DGE analysis. Among the existing tools, DESeq2 [40], voom-limma [41], and edgeR [42] are widely used R packages for DGE analysis. As numerous protocols for the downstream analysis have been published (e.g., [49–52], *see* **Note 44**), we only provide a short example for DGE analysis in R based on the R package DESeq2 (see below) and a brief summary of a typical DGE analysis workflow: After importing the count data into R, filtering and normalization are typically carried out, providing the basis for the computation of differentially expressed genes [3, 49, 53]. In general, it is useful to perform dimension reduction to visualize similarities as well as dissimilarities between the samples [49]. Depending on the chosen R package, different statistical models/methods are used to compute a list of differentially expressed genes [22, 44, 53] (*see* **Note 45**). An essential step in DGE analysis is the definition of the design matrix for your experimental conditions as well as the selection of the appropriate comparisons of interest [54]. To obtain more insight into the results, a number of further analyses and visualizations can be carried out, including gene ontology (GO) enrichment analyses and heatmaps. (Strand-specific) reverse transcription polymerase chain reaction (RT-PCR) or reverse transcription quantitative PCR (RT-qPCR) can be used to validate the expression of certain genes of interest [55].

Here, we provide a short example for DGE analysis in R based on the R package DESeq2 [40]. More detailed information as well as elaborate examples can be found in the DESeq2 vignette.

1. Load the R package DESeq2 (*see* **Note 3**).

```
library(DESeq2)
```

2. The R package DESeq2 expects un-normalized counts and provides a function for data import (`DESeqDataSetFromHTSeqCount`) to build a `DESeqDataSet` object based on htseq-count data [52] (as generated in Subheading 3.4.3). To create an object of type `DESeqDataSet` (variable: `HTSeqCountData`), provide the following parameters: (a) a table listing the sample metadata (`SampleDataFrame`), (b) a directory containing the count data files, and (c) a custom design formula. The table (`SampleDataFrame`) provides the sample information (one row for each sample) and is of the following structure: (a) the first column lists the sample names, (b) the second column contains the file names of the count data files, and (c) the following columns hold the experimental conditions and further sample metadata. The design formula specifies which columns describe the experimental conditions and how to use this information for the DGE analysis [52] (in its simplest form: `~ group`, where `group` is the name of a column in the `SampleDataFrame`, and this column provides a grouping factor; more details are available in the DESeq2 vignette; *see* **Note 46**).

```
HTSeqCountData <- DESeqDataSetFromHTSeqCount(
  sampleTable = SampleDataFrame,
  directory = <path_to_working_directory>,
  design = <design_formula>)
```

3. Optional: Pre-filter the data to remove low-count genes, e.g., as follows (*see* **Note 47**):

```
FilteredGenes <- rowSums(counts(HTSeqCountData))>=<value>
FilteredHTSeqCountData <- HTSeqCountData[FilteredGenes,]
```

4. Perform DGE analysis and create a result table for a specific contrast containing log2 fold changes, p-values, and adjusted p-values as follows (if you are using unfiltered data, use `HTSeqCountData` instead of `FilteredHTSeqCountData`) (*see* **Notes 48–50**):

```
DESeq2Data <- DESeq(FilteredHTSeqCountData)
result <- results(DESeq2Data,
    contrast=<contrast_of_interest>,
    alpha=<value>)
summary(result)
```

5. Important visualizations, e.g., heatmaps, principal component analysis, or volcano plots, can be computed using functions provided by the R package DESeq2 (e.g., the `plotPCA` function) and by other R packages such as gplots [56], pheatmap [57], or EnhancedVolcano [58].

4 Notes

1. All solutions are diluted with and dissolved in sterile RNase-free "Fresenius" H_2O and sterile filtered before use.
2. When preparing the DNase I solution, do not vortex it. Mix it gently by pipetting.
3. Note that many of the stated tools can be installed by means of the conda package manager or through BiocManager (for Bioconductor R Packages).
4. Tool versions used for testing this protocol: FastQC v0.11.9, MultiQC v1.12, STAR v2.7.10a, Samtools v1.6, IGV 2.8.9, QualiMap v.2.2.1, htseq-count v2.0.1, R 4.0.5, and DESeq2 1.30.1.
5. Use identical baffled shake flasks for cultivation of *K. phaffii* cells, especially when comparing different strains or conditions, since *K. phaffii* is very sensitive to changes in oxygen supply.
6. Clean your working area and all needed materials and machines with RNase decontamination solution directly before starting the RNA extraction procedure. Wear fresh gloves at any time, and additionally, decontaminate your gloves with RNase decontaminant regularly.
7. It is recommended to perform the extraction procedure without breaks and in a workspace that is not shared with others to avoid potential contamination or degradation of RNA.
8. Cell amounts needed for efficient RNA isolation can vary from strain to strain, since this is mostly influenced by the effectiveness of the cell disruption procedure. Therefore, it is recommended to additionally harvest, e.g., 36 OD_{600} equivalents instead of 18 and store them at −80 °C as backups. This allows for a quick repetition of the RNA extraction protocol without having to cultivate cells again, which will save time.
9. For mechanical cell disruption, we suggest using 4× the amount suggested by the provider's RNA extraction manual ($4 \times 10^{\wedge 7}$ cells) to yield high RNA amounts (150 ng μL^{-1} – 250 ng μL^{-1}).
10. You can optionally freeze the cell pellets after this step with liquid nitrogen and store them at −80 °C until RNA extraction is performed.

11. In our hands, mechanical cell disruption worked better than enzymatic treatment, e.g., with Zymolyase, which resulted in RNA degradation.
12. Work continuously on ice while performing the RNA extraction.
13. After extraction, RNA samples are immediately frozen in liquid nitrogen and stored at −80 °C until RNA is further used.
14. Shipment of RNA samples has to be done on dry ice, and samples should not thaw during the shipping process.
15. The protocol as described in this book chapter allowed us to isolate 100 μL of 150–250 ng μL^{-1} RNA. Therefore, 5 μL of isolated RNA was prepared to be loaded onto the agarose gels.
16. The more RNA you load, the better the signal and the easier the assessment of whether any degradation occurred.
17. Long-read sequencing technologies such as Oxford Nanopore [59] can be useful depending on the goal of the experiment, e.g., to sequence full-length transcripts [3, 23].
18. Poly(A) selection captures transcripts with a poly(A) tail at the cost of excluding possible non-polyadenylated transcripts of interest that could be retained with rRNA depletion methods [23, 45]. Consider that there are trade-offs between both methodologies and choose the library preparation method that is a better fit for your experiment.
19. Strand-specific library preparation enables the retention of strand information and has been linked to increased gene quantification accuracy for genes that overlap and are transcribed from opposite strands [60].
20. Note these parameters need to be adapted depending on the goals of your experiment (e.g., if the assembly of new transcripts is of interest).
21. The data analysis can also be performed in Galaxy [61], which is an open-source, web-based platform that allows for bioinformatics analyses, including RNA-seq data analysis.
22. FastQC also accepts multiple FASTQ files as input, which will be processed sequentially:

    ```
    fastqc *.fastq
    ```

    See the FastQC user guide for help with the interpretation of results.

    Adapt the parameter `-t` to define the number of threads to be used, allowing for parallel processing of multiple FASTQ files (e.g., `fastqc -t 6 *.fastq`).
23. You can provide a list of report files (as a text file including paths to the report files) using the option `--file-list`.

24. Note that MultiQC searches through the specified directory and summarizes present reports of all supported tools (which include FastQC, STAR, and Qualimap) unless otherwise specified.
25. Please note that there are a number of ways to analyze RNA-seq datasets. A review by Stark et al. offers an overview of possible data analysis routes [3].
26. An overview of the publicly available genome assemblies and annotation information for *Pichia pastoris* strains can be found in the chapter '*Pichia pastoris* genomic resources: Accessibility, recommendations, and pitfalls' within this Pichia protocols edition. This chapter also provides information on downloading these resources. Please also consider the recommendations concerning the choice of the genome assembly and read the details provided in the overview table and in the notes on genome completeness and additional contigs!
27. In addition to GTF files, GFF3 files are also accepted by many tools. However, consult the corresponding user guide if and how the tool's input parameters need to be adjusted. Read more about the file formats in the chapter '*Pichia pastoris* genomic resources: Accessibility, recommendations, and pitfalls' within this Pichia protocols edition.
28. The parameter `--genomeSAindexNbases` should be min (14, log2(genome length) / 2 – 1) according to the STAR manual.
29. The parameter `--sjdbOverhang` should be set to the read length –1 (default: 100) according to the STAR manual.
30. Adapt the parameter `--runThreadN` to define the number of threads to be used.
31. While the default parameters are usually adequate and commonly used, especially for higher eukaryotes, tweaking of parameters might optimize the alignment process for *K. phaffii*. For example, we suggest that you consider adjusting the values for `--alignIntronMin`, `--alignIntronMax` and `--alignMatesGapMax`. Please consult the STAR manual for more details.
32. An interesting book chapter describes the optimization of parameters for mapping with STAR [62].
33. Note that you usually need a coordinate-sorted and indexed BAM file when using a genome viewer such as IGV [33]. To do so, sort the alignment BAM file by position using Samtools [32] as follows:

```
samtools sort -@ <value> \
-o sample1Aligned.sortedByCoord.out.bam \
sample1Aligned.out.bam
```

Furthermore, index a coordinate-sorted BAM file as follows:

```
samtools index -@ <value> sample1Aligned.sortedByCoord.out.bam
```

Adapt the parameter -@ to define the number of threads to be used.

Some additional information on IGV can be found in the chapter '*Pichia pastoris* genomic resources: Accessibility, recommendations, and pitfalls' within this Pichia protocols edition.

34. The option `--java-mem-size` may need adjustments to increase the memory size (e.g., `--java-mem-size = 8G`).
35. The -pe flag is used for paired-end data and allows counting fragments instead of reads. Adjust this according to the specifics of your sequencing data.
36. The `-p` flag is used to indicate the library preparation protocol (default: `non-strand-specific`) and can be: `strand-specific-forward`, `strand-specific-reverse` or `non-strand-specific`. Adjust this according to the specifics of your sequencing data. Note that Qualimap RNA-seq QC sorts BAM files by name for paired-end data prior to paired-end analysis. The -s flag is used to specify that the input file was already sorted by name and hence that no sorting by name is required. For information on how to sort a BAM file by name *see* **Note 40.**
37. The strand-specificity should also be assessed using, e.g., the RSeQC [36] tool box.
38. htseq-count [37] discards multi-mapping reads by default. Consult the user manual for more information.
39. Note that you need to set the `--stranded` parameter according to the specifics of your sequencing data. Depending on the library preparation protocol, set the `--stranded` parameter to no (for unstranded data), `reverse` or `yes` (default). Choosing this correctly is crucial and should be considered carefully!
40. Note that htseq-count expects a BAM file sorted by read name for paired-end data by default (default: `-r name`, highly recommended). You can sort the alignment by name using Samtools [32] as follows:

```
samtools sort -@ <value> -n \
-o sample1Aligned.sortedByName.out.bam \
sample1Aligned.out.bam
```

Adapt the parameter -@ to define the number of threads to be used.

Of further note, older versions of htseq-count may require you to set the parameter `--format/-f` to `sam/bam` depending on your input file format.

41. htseq-count also accepts multiple BAM files as input. You may use -n to define the number of parallel CPU processes to be used, which allows for parallel processing of the BAM files.
42. The user guides for edgeR [42], voom-limma [41], and DESeq2 [40] provide details on data import.
43. The quantification tool featureCounts can directly be used within the R environment [63].
44. Detailed user guides for edgeR [42], voom-limma [41], and DESeq2 [40] with examples are available.
45. Schurch et al. [44] compared DGE analysis tools (in relation to a given number of biological replicates) and can help you choose the right tool for your experiment.
46. Note that the grouping factor (and further sample metadata used in the design formula) should typically be provided as factors.
47. The pre-filtering retains only rows, which have at least as many reads in total as specified by the pre-filtering cutoff. Note that according to the DESeq2 vignette, pre-filtering increases calculation speed and can enhance visualizations.
48. Note that the DESeq function provides a wrapper for DGE analysis steps.
49. The `contrast` argument defines the comparison to be extracted from your DESeq2 results. For more details, visit the DESeq2 vignette or consult the help using `?results`.
50. DESeq2 carries out an automatic independent filtering step. The `alpha` argument sets the significance cutoff used for its optimization and is 0.1 by default. Note that `alpha` should be adjusted to the chosen FDR cutoff of your experiment. The DESeq2 vignette provides more information on independent filtering.

Acknowledgments

This work was supported by the BioTechMed-Graz Young Researcher Group Project 'StemP' (A.R. and A. E.-A.). The project was also supported by the COMET center acib (A. E.-A.), which is funded by BMK, BMDW, SFG, Standortagentur Tirol, the Government of Lower Austria, and Vienna Business Agency in the framework of COMET—Competence Centers for Excellent Technologies. The COMET-Funding Program is managed by the Austrian Research Promotion Agency FFG. V.S. was supported by the

Austrian Research Promotion Agency (FFG). J.F. and J.C.K. were supported by the Christian Doppler Research Association. The financial support by the Austrian Federal Ministry for Digital and Economic Affairs, the National Foundation for Research, Technology, and Development, and the Christian Doppler Research Association (Christian Doppler Laboratory for Innovative *Pichia pastoris* Host and Vector Systems) is gratefully acknowledged. The financial support from the participating industrial partners is gratefully acknowledged. Also, we want to thank Prof. Anton Glieder and Dr. Gerhard Thallinger for helpful discussions.

Competing Interest Statement The authors have no conflicts of interest to declare that are relevant to the content of this chapter.

References

1. Ahmad M, Hirz M, Pichler H et al (2014) Protein expression in *Pichia pastoris*: recent achievements and perspectives for heterologous protein production. Appl Microbiol Biotechnol 98:5301–5317
2. Bernauer L, Radkohl A, Lehmayer LGK et al (2021) *Komagataella phaffii* as emerging model organism in fundamental research. Front Microbiol 11:607028
3. Stark R, Grzelak M, Hadfield J (2019) RNA sequencing: the teenage years. Nat Rev Genet 20:631–656
4. Doughty T, Kerkhoven E (2020) Extracting novel hypotheses and findings from RNA-seq data. FEMS Yeast Res 20:foaa007
5. Küberl A, Schneider J, Thallinger GG et al (2011) High-quality genome sequence of *Pichia pastoris* CBS7435. J Biotechnol 154: 312–320
6. De Schutter K, Lin Y-C, Tiels P et al (2009) Genome sequence of the recombinant protein production host Pichia pastoris. Nat Biotechnol 27:561–566
7. Mattanovich D, Graf A, Stadlmann J et al (2009) Genome, secretome and glucose transport highlight unique features of the protein production host *Pichia pastoris*. Microb Cell Factories 8:29
8. Sturmberger L, Chappell T, Geier M et al (2016) Refined *Pichia pastoris* reference genome sequence. J Biotechnol 235:121–131
9. Valli M, Tatto NE, Peymann A et al (2016) Curation of the genome annotation of *Pichia pastoris* (*Komagataella phaffii*) CBS7435 from gene level to protein function. FEMS Yeast Res 16:fow051
10. Love KR, Shah KA, Whittaker CA et al (2016) Comparative genomics and transcriptomics of *Pichia pastoris*. BMC Genomics 17:550
11. Krappinger JC, Aguilar Gomez CM, Hoenikl A et al (2024) dMAD7 is a promising tool for targeted gene regulation in the methylotrophic yeast *Komagataella phaffii*. N Biotechnol 83: 110–120
12. Matthews CB, Kuo A, Love KR et al (2018) Development of a general defined medium for *Pichia pastoris*. Biotechnol Bioeng 115:103–113
13. Dou W, Zhu Q, Zhang M et al (2021) Screening and evaluation of the strong endogenous promoters in *Pichia pastoris*. Microb Cell Factories 20:1–12
14. Schwarzhans JP, Luttermann T, Geier M et al (2017) Towards systems metabolic engineering in *Pichia pastoris*. Biotechnol Adv 35: 681–710
15. Paes BG, Steindorff AS, Formighieri EF et al (2021) Physiological characterization and transcriptome analysis of *Pichia pastoris* reveals its response to lignocellulose-derived inhibitors. AMB Express 11:2
16. Mühlhofer M, Berchtold E, Stratil CG et al (2019) The heat shock response in yeast maintains protein homeostasis by chaperoning and replenishing proteins. Cell Rep 29:4593–4607.e8
17. Strassburg K, Walther D, Takahashi H et al (2010) Dynamic transcriptional and metabolic responses in yeast adapting to temperature stress. OMICS 14:249–259

18. Tan SC, Yiap BC (2009) DNA, RNA, and protein extraction: the past and the present. J Biomed Biotechnol 2009:574398
19. Stevenson HS, Wang Y, Muller R et al (2015) Long-term stability of total RNA in RNAstable® as evaluated by expression microarray. Biopreserv Biobank 13:114–122
20. Schroeder A, Mueller O, Stocker S et al (2006) The RIN: an RNA integrity number for assigning integrity values to RNA measurements. BMC Mol Biol 7:1–14
21. Hesketh AR (2019) RNA sequencing best practices: experimental protocol and data analysis. In: Oliver SG, Castrillo JI (eds) Yeast systems biology: methods and protocols. Springer New York, New York, pp 113–129
22. Costa-Silva J, Domingues D, Lopes FM (2017) RNA-Seq differential expression analysis: an extended review and a software tool. PLoS One 12:1–18
23. Krappinger JC, Bonstingl L, Pansy K et al (2021) Non-coding natural antisense transcripts: analysis and application. J Biotechnol 340:75–101
24. Radkohl A, Schusterbauer V, Bernauer L et al (2024) Human sterols are overproduced, stored and excreted in yeasts. Int J Mol Sci 25:781
25. Koch CM, Chiu SF, Akbarpour M et al (2018) A beginner's guide to analysis of RNA sequencing data. Am J Respir Cell Mol Biol 59:145–157
26. Andrews S (2010) Babraham bioinformatics—FastQC a quality control tool for high throughput sequence data. Babraham Bioinformatics—FastQC A Quality Control tool for High Throughput Sequence Data https://www.bioinformatics.babraham.ac.uk/projects/fastqc/
27. Ewels P, Magnusson M, Lundin S et al (2016) MultiQC: summarize analysis results for multiple tools and samples in a single report. Bioinformatics 32:3047–3048
28. Bolger AM, Lohse M, Usadel B (2014) Trimmomatic: a flexible trimmer for Illumina sequence data. Bioinformatics 30:2114–2120
29. Martin M (2011) Cutadapt removes adapter sequences from high-throughput sequencing reads. EMBnet J 17:10
30. Dobin A, Davis CA, Schlesinger F et al (2013) STAR: ultrafast universal RNA-seq aligner. Bioinformatics 29:15–21
31. Kim D, Paggi JM, Park C et al (2019) Graph-based genome alignment and genotyping with HISAT2 and HISAT-genotype. Nat Biotechnol 37:907–915
32. Danecek P, Bonfield JK, Liddle J et al (2021) Twelve years of SAMtools and BCFtools. Gigascience 10:giab008
33. Robinson JT, Thorvaldsdóttir H, Winckler W et al (2011) Integrative genomics viewer. Nat Biotechnol 29:24–26
34. García-Alcalde F, Okonechnikov K, Carbonell J et al (2012) Qualimap: evaluating next-generation sequencing alignment data. Bioinformatics 28:2678–2679
35. Okonechnikov K, Conesa A, García-Alcalde F (2016) Qualimap 2: advanced multi-sample quality control for high-throughput sequencing data. Bioinformatics 32:292–294
36. Wang L, Wang S, Li W (2012) RSeQC: quality control of RNA-seq experiments. Bioinformatics 28:2184–2185
37. Anders S, Pyl PT, Huber W (2015) HTSeq—a Python framework to work with high-throughput sequencing data. Bioinformatics 31:166–169
38. Liao Y, Smyth GK, Shi W (2014) featureCounts: an efficient general purpose program for assigning sequence reads to genomic features. Bioinformatics 30:923–930
39. R-Core Team (2022) R: A Language and Environment for Statistical Computing. https://www.r-project.org
40. Love MI, Huber W, Anders S (2014) Moderated estimation of fold change and dispersion for RNA-seq data with DESeq2. Genome Biol 15:550
41. Ritchie ME, Phipson B, Wu D et al (2015) Limma powers differential expression analyses for RNA-sequencing and microarray studies. Nucleic Acids Res 43:e47
42. Robinson MD, McCarthy DJ, Smyth GK (2009) edgeR: a Bioconductor package for differential expression analysis of digital gene expression data. Bioinformatics 26:139–140
43. Esteve-Codina A (2018) RNA-Seq data analysis, applications and challenges. In: Jaumot J, Bedia C, Tauler R (eds) Data analysis for omic sciences: methods and applications. Elsevier B. V, Amsterdam, pp 71–106
44. Schurch NJ, Schofield P, Gierliński M et al (2016) How many biological replicates are needed in an RNA-seq experiment and which differential expression tool should you use? RNA 22:839–851
45. Zhao W, He X, Hoadley KA et al (2014) Comparison of RNA-Seq by poly (A) capture, ribosomal RNA depletion, and DNA microarray for expression profiling. BMC Genomics 15:419
46. Bohn S (2021) Protocol for RNA-seq expression analysis in yeast. Bio-protocol 11:e4161

47. Wang X, Zhao X, Luo H et al (2022) Metabolic engineering of Komagataella phaffii for synergetic utilization of glucose and glycerol. Yeast 39:412–421
48. Zhang C, Ma Y, Miao H et al (2020) Transcriptomic analysis of *Pichia pastoris* (*Komagataella phaffii*) GS115 during heterologous protein production using a high-cell-density fed-batch cultivation strategy. Front Microbiol 11:463
49. Law CW, Alhamdoosh M, Su S et al (2016) RNA-seq analysis is easy as 1-2-3 with limma, Glimma and edgeR. F1000Research 5. ISCB Comm J-1408
50. Chen Y, Lun ATL, Smyth GK (2016) From reads to genes to pathways: differential expression analysis of RNA-Seq experiments using Rsubread and the edgeR quasi-likelihood pipeline. F1000Res 5:1438
51. Lun ATL, Chen Y, Smyth GK (2016) It's DE-licious: a recipe for differential expression analyses of RNA-seq experiments using quasi-likelihood methods in edgeR. In: Mathé E, Davis S (eds) Statistical genomics. Methods in molecular biology, vol 1418. Humana Press, New York, pp 391–416
52. Love MI, Anders S, Kim V et al (2016) RNA-Seq workflow: gene-level exploratory analysis and differential expression. F1000Res 4:1070
53. Van den Berge K, Hembach KM, Soneson C et al (2019) RNA sequencing data: hitchhiker's guide to expression analysis. Annu Rev Biomed Data Sci 2:139–173
54. Law CW, Zeglinski K, Dong X et al (2020) A guide to creating design matrices for gene expression experiments. F1000Res 9:1444
55. Yu X-W, Sun W-H, Wang Y-Z et al (2017) Identification of novel factors enhancing recombinant protein production in multi-copy *Komagataella phaffii* based on transcriptomic analysis of overexpression effects. Sci Rep 7:16249
56. Warnes G, Bolker B, Bonebakker L, et al (2015) Gplots: various R programming tools for plotting data. R package version 2.17.0. gplots: Various R Programming Tools for Plotting Data. R package version 2.17.0. http://cran.r-project.org/package=gplots
57. Kolde R (2019) Pheatmap. pheatmap https://cran.r-project.org/web/packages/pheatmap/index.html
58. Blighe K, Rana S, Lewis M (2022) Enhanced volcano: publication-ready volcano plots with enhanced colouring and labeling. https://github.com/kevinblighe/EnhancedVolcano
59. Jain M, Olsen HE, Paten B et al (2016) The Oxford nanopore minION: delivery of nanopore sequencing to the genomics community. Genome Biol 17:239
60. Zhao S, Zhang Y, Gordon W et al (2015) Comparison of stranded and non-stranded RNA-seq transcriptome profiling and investigation of gene overlap. BMC Genomics 16: 675
61. Afgan E, Baker D, Batut B et al (2018) The Galaxy platform for accessible, reproducible and collaborative biomedical analyses: 2018 update. Nucleic Acids Res 46:W537–W544
62. Dobin A, Gingeras TR (2016) Optimizing RNA-seq mapping with STAR. In: Carugo O, Eisenhaber F (eds) Methods in molecular biology, vol 1415. Humana Press, New York, pp 245–262
63. Liao Y, Smyth GK, Shi W (2019) The R package Rsubread is easier, faster, cheaper and better for alignment and quantification of RNA sequencing reads. Nucleic Acids Res 47:e47

Part III

Cell and Systems Biology

Chapter 5

Localization of Proteins in the Secretory Pathway of *Pichia pastoris* Using Fluorescence Microscopy

Verena Puxbaum, Brigitte Gasser, and Diethard Mattanovich

Abstract

Fluorescence microscopy is a widely used technique to localize and monitor intracellular proteins. Here we present different methods established for *Pichia pastoris* (syn *Komagataella* spp.), like immunofluorescence microscopy, the use of fluorescent fusion proteins, and various dyes to stain organelles. Furthermore, the new super-resolution technique STED is presented. The described methods mainly focus on studying the secretory pathway but are in principle applicable also to other research questions.

Key words *Pichia pastoris*, *Komagataella phaffii*, Fluorescence microscopy, Immunofluorescence, Fluorescent fusion protein, STED, Protein secretion

1 Introduction

Exploring details of the secretion process in *Komagataella phaffii* (commonly known as *Pichia pastoris*) is receiving increased attention in the last years. A key element, therefore, is visualizing recombinant proteins on their way through the secretory pathway. This helps to gain further knowledge of the secretory process and eventually can provide valuable information for strain improvement. We established imaging methods in our lab to detect specific proteins in a fixed state (immunofluorescence) or in living cells (fluorescent fusion proteins), concentrating on processes related to recombinant protein production, which will be presented here in further detail.

The major aim of the presented methods is to investigate heterologous proteins on their secretory pathway and to detect bottlenecks they might encounter in specific organelles. The challenges of secreting a heterologous protein involve translocation into the endoplasmic reticulum (ER), transport from the ER to the Golgi and further vesicular transport from the Golgi to the

Xavier Garcia-Ortega et al. (eds.), *Pichia pastoris: Methods and Protocols*, Methods in Molecular Biology, vol. 2697, https://doi.org/10.1007/978-1-0716-4779-0_5,

plasma membrane, as well as different redox conditions in different compartments.

In this chapter we will present protocols for microscopy with immunofluorescence and fluorescent fusion proteins in *P. pastoris* with emphasis on secreted proteins.

1.1 Fluorescent Fusion Proteins

The main application of fluorescent fusion proteins is live cell imaging. Due to the efforts of Martin Chalfie, Roger Tsien, and many others, the green fluorescent protein from the jellyfish *Aequorea victoria* can be fused to the protein of interest either C- or N-terminally and be tracked and analyzed in the living state [1, 2]. The most used GFP variant today is EGFP, which, compared to wtGFP, exhibits much brighter fluorescence when excited at 488 nm, folds much faster, and is codon-optimized for mammalian cells [3]. Furthermore, chromophore formation is similar at 28 and 37 °C, which is important for the use in yeast, fungi, and mammalian cultures [4]. Detailed information on the design of fluorescent fusion proteins can be found in [5]. By now a real fruit garden of fluorescent proteins is available (covering the spectrum from mBanana to mPlum [6, 7]). The well-known red fluorescent protein was originally derived from the coral *Discosoma* sp. as a tetramer (DsRed) [8]. Meanwhile, different red fluorescent proteins are available as monomers with improved characteristics due to several point mutations [9, 10]. Among the red fluorescent proteins, mCherry is the most widely used. It is a monomer with fast maturation time, high photostability, and reasonably bright fluorescence. Further information on different fluorescent proteins from the Tsien lab is summarized in [11]. Recently, a novel monomeric yellow-green fluorescent protein called mNeonGreen attracted attention in the field. It was originally derived as a tetrameric protein from *Branchiostoma lanceolatum*. Optimization due to several point mutations resulted in the brightest monomeric green or yellow fluorescent protein described so far, which is also easy to tag [12]. Therefore, mNeonGreen is a very promising fluorescent protein for live cell microscopy in *P. pastoris* as well as for all other species. Properties of different common fluorescent proteins can be found at, e.g., https://www.fpbase.org/table/

Concerning the use of fluorescent (fusion) proteins in the secretory pathway, it is important to note that the redox state of the ER is different than of the cytosol. As GFP is naturally localized in the cytosol of jellyfish cells, the more oxidizing redox state of the secretory organelles is a challenge to obtain correctly folded and active fluorophores. Hence, for fluorescent fusion proteins to be used within the secretory pathway, we recommend only truly monomeric FPs like superfolder GFP (sfGFP) [13], photostable monomeric superfolder GFP (msGFP2) [14], or the cysteine-free oxidation-resistant oxGFP [15] to avoid misfolding resulting in non-fluorescent proteins [13–17]. In addition to oxGFP, the

Snapp lab designed oxidation-resistant fluorescent protein variants (oxFPs) of several common FPs, which cover the spectrum from blue (oxBFP) via green (oxGFP, moxNeonGreen) to cyan (oxCerulean) and yellow (oxVenus) and are well suited for studying oxidizing cellular environments such as the secretory pathway [15].

Fluorescent proteins have also been optimized for use in the yeast *Saccharomyces cerevisiae*. Two blue, seven green, and seven red fluorescent proteins were codon-optimized and tested for brightness, photostability, and correct subcellular localization of a C-terminally tagged protein. Under these conditions, EGFP remained the best green fluorescent protein, Tag-RFP-T and mRuby2 performed best for the red fluorescent proteins, and mTagBFP was recommended as a blue fluorescent protein [18]. Initially, codon-optimized mNeonGreen exhibited lower fluorescence than expected in *S. cerevisiae* [19]; however, this was recently overcome by removing secondary structure in the RNA near the start codon [20].

No similar comparative studies have been published for *P. pastoris* yet. We tested fusions of (recombinant) secretory proteins with yeast codon-optimized sfGFP, mCherry, and oxGFP and obtained similar patterns as for immunofluorescence staining of the recombinant secretory protein ([21, 22] and Figs. 1, 2, 3, 4 and 5). Furthermore, msGFP and its photostable variant msGFP2 were shown to outperform EGFP for fusion proteins localized to the secretory pathway in *P. pastoris* [14, 23, 24]. For multicolor co-localization studies within the secretory pathway, proteins were routinely fused to a rapidly maturing dsRed variant developed by the Glick lab [25, 26] or mCherry [23, 27]. In the latter study, three-color fluorescent imaging in *P. pastoris* was achieved by tagging a lowly abundant secretory protein with HaloTag and subsequent labeling [23]. We also tested yeast-optimized mRuby2 and mKO2, which outperformed mCherry in terms of fluorescence intensity in *S. cerevisiae* [18], for their use within the secretory pathway in *P. pastoris* by fusing them to human serum albumin (HSA), but we observed mislocalization compared to the pattern we observed for sfGFP, oxGFP, and immunofluorescence staining. As expected, yeast-optimized sfGFP performed very well and showed correct localization of the tested fusion proteins.

In addition, specific fluorescent fusion proteins can be used as genetically encoded redox sensors [28, 29]. Prior to their use, their correct localization in the intended organelle (e.g., ER, cytosol) needs to be verified by fluorescence microscopy [30].

If no live cell monitoring is needed, fluorescent protein-producing cells can also be fixed with paraformaldehyde and viewed at a convenient time point. Usually the subcellular pattern remains unchanged; however, checking for the correct pattern compared to live cells when fixing for the first time is advisable.

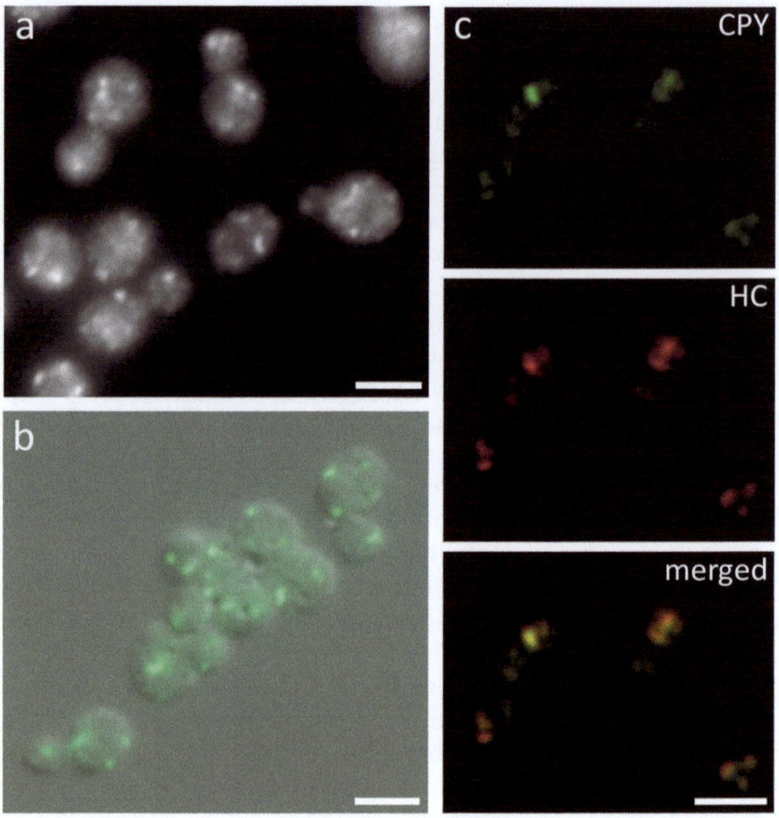

Fig. 1 Immunofluorescence microscopy visualizing *P. pastoris* organelles using self-made antibodies. (**a**) Secretory vesicles (marker protein Snc2) are visualized in CBS7435 using guinea pig anti-Snc2 antibodies followed by anti-guinea pig IgG Biotin and Streptavidin 488. (**b**) The late Golgi is stained with a monoclonal anti-Sec7 antibody followed by donkey anti-mouse IgG Alexa Fluor 488. The merged image of fluorescence and DIC is shown. (**c**) The vacuolar marker protein CPY is stained with a rabbit anti-CPY antibody followed by a donkey anti-rabbit IgG biotin and streptavidin Alexa Fluor 488 (CPY). A recombinant scFv-Fc fusion protein that shows a vacuolar pattern in the cells was detected with goat anti-human IgG (γ-chain specific) followed by donkey anti-goat IgG Alexa Fluor 555 (HC). The overlay of the fluorescence images is shown at the bottom (merged). Bar, 5 μm

1.2 Immunofluorescence

The challenge in the work with immunofluorescence microscopy is the lack of commercially available antibodies against *P. pastoris* proteins. The only commercially available antibody that worked fine in our hands (apart from the antibodies against different heterologous proteins) is anti-HDEL (2E7, Santa Cruz). Therefore, there was the need for self-made antibodies. To detect secretory vesicles, polyclonal antibodies against two peptides of the v-SNARE Snc2 were made. The antibodies were purified against the peptides (Fig. 1a). It proved valuable to first screen the sera of 10 animals and then go for the 2 animals with the least background signal. For

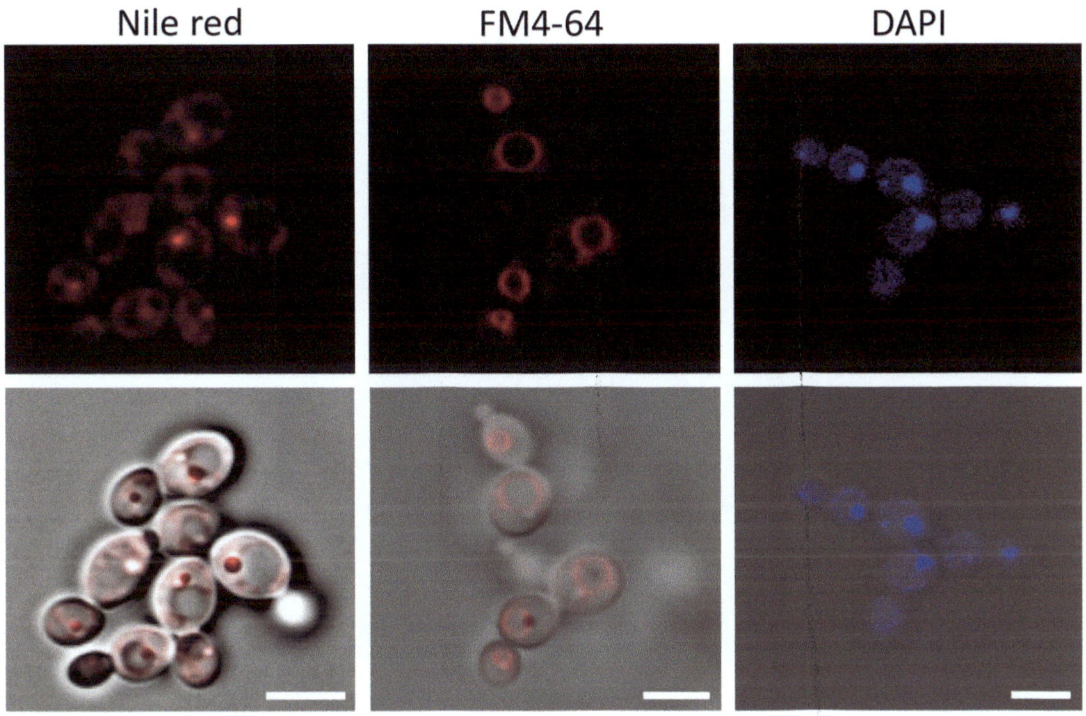

Fig. 2 Staining of lipid droplets, vacuole, and nuclei using the dyes Nile Red, FM4–64, and DAPI. The cells were stained according to the protocol and viewed in a confocal laser scanning microscope using a 561 nm laser for Nile Red and FM4–64 and a 405 nm laser for DAPI. The fluorescence images and the merged images with bright-field are shown. Bar, 5 μm

the Golgi marker Sec7, a monoclonal antibody was made (Fig. 1b). The first 310 amino acids of the protein produced in *E. coli* were injected into mice. An anti-carboxypeptidase Y (CPY) antibody was made against the mature *P. pastoris* CPY as a vacuolar marker (Fig. 1c). This antibody was kindly provided by Günther Daum and affinity-purified with protein A by Karl-Heinz Grillitsch, which made the antibody suitable for immunofluorescence microscopy. Purification was based on Roberts et al. [31]. As heterologous protein intended for secretion might be targeted to the vacuole [22, 24, 32, 33], a vacuolar antibody is necessary to achieve the entire intracellular localization pattern.

For detecting recombinant proteins, most of the commercial antibodies worked fine, like anti-HSA (Bethyl Laboratories), anti-human kappa light chain (bound and free, Sigma), anti-human IgG (gamma-chain specific, Sigma), and anti-human IgG (Fab-specific, Sigma). As secondary antibodies, Alexa Fluor antibodies from Life Technologies produced in donkeys proved to be valuable antibodies for yeast microscopy.

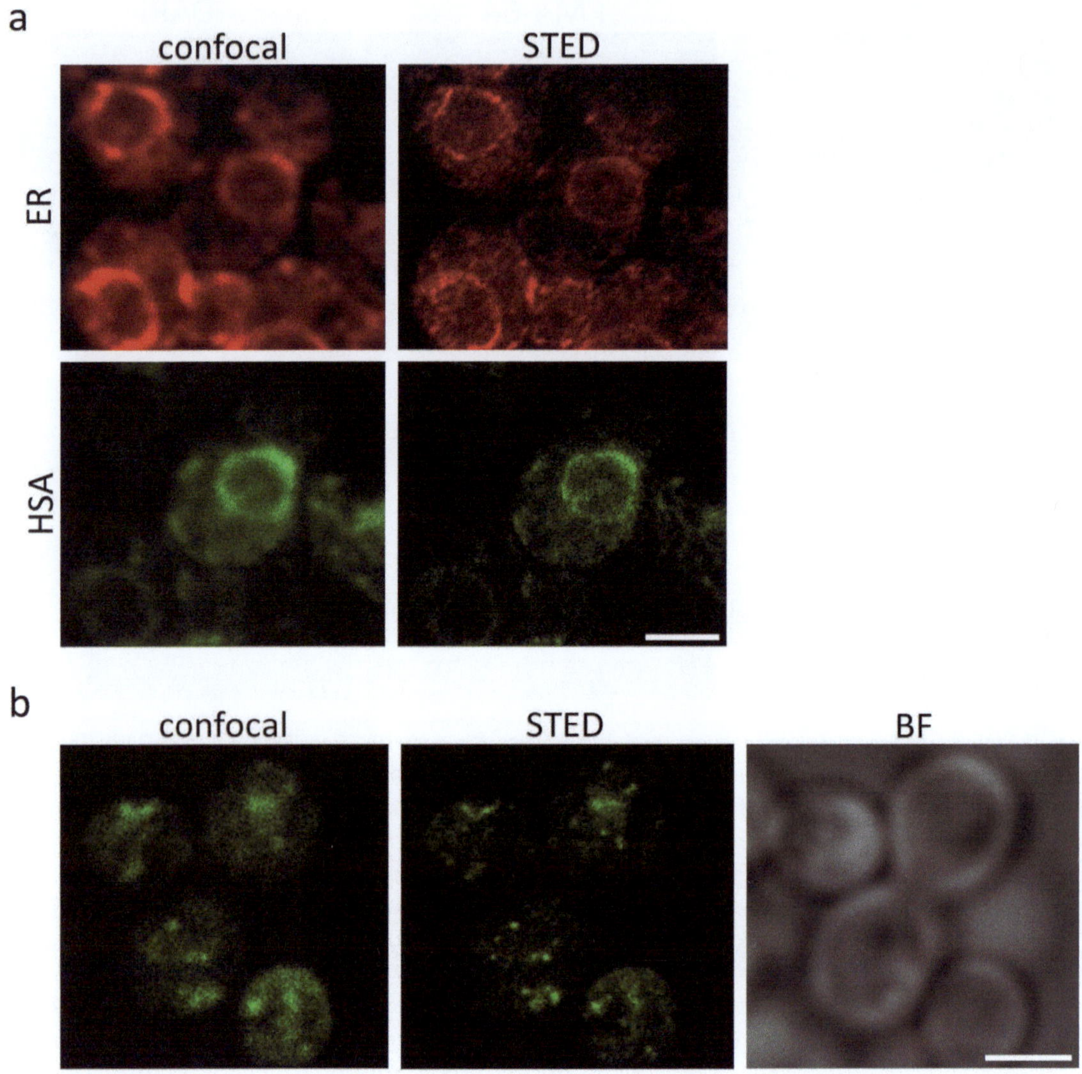

Fig. 3 Confocal and STED microscopy of HSA and Snc2-sfGFP in *P. pastoris.* (**a**) Confocal and STED images of an immunofluorescence double-staining of HSA and the ER. HSA was stained with goat anti-HSA IgG, bovine anti-goat IgG biotin, and streptavidin V500. The ER was stained with mouse anti-HDEL and goat anti-mouse IgG Oregon Green 488. As the HSA and the Oregon Green 488 antibodies are from the same species, the staining was performed sequentially, starting with the anti-HSA staining, followed by intense washing and the anti-HDEL staining. (**b**) *P. pastoris* expressing the fusion protein Snc2-sfGFP (marker for secretory vesicles) was paraformaldehyde-fixed and viewed in confocal and STED mode. Also, a bright-field image (BF) is shown. Bar, 3 μm

If no primary antibodies are available, the use of small epitope tags, like c-myc, HA, 6xHis, GluGlu, or FLAG, can be helpful and easy to use [34]. As they are short amino acid sequences, there is a lower risk of interference with the protein of interest, as in the case of fluorescent proteins.

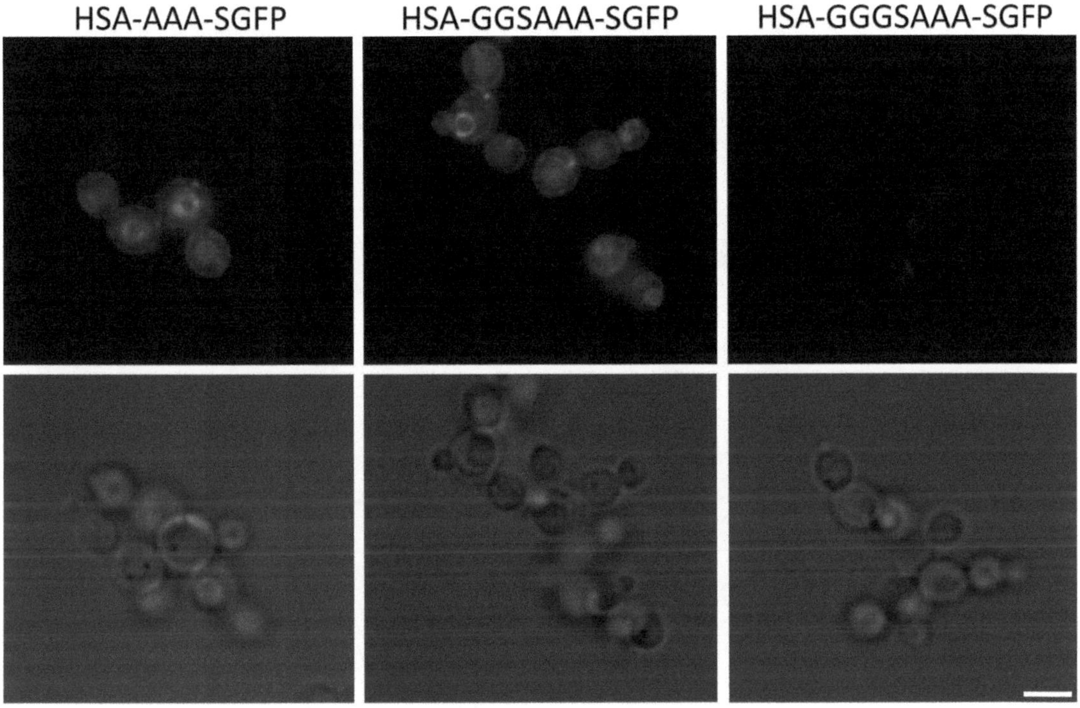

Fig. 4 Fluorescence microscopy of *P. pastoris* producing HSA-sfGFP fusion proteins with different linkers. The linkers with the amino acid sequences AAA, GGSAAA, and GGGSAAA were tested. The fluorescence images (top row) and the DIC images (bottom row) are shown. Fusion proteins with the first two linkers display intense fluorescence, whereas the fusion protein with the longest linker (GGGSAAA) showed only low fluorescence intensity. The images shown are representative images from ten different clones and were taken with identical microscope and camera settings. Also, even longer linkers were tested with a similar outcome as for GGGSAAA. Bar, 5 μm

1.3 Fluorescent Dyes

Here, fluorescent dyes for general applications are presented. Fluorescent dyes are excellent tools if they show specific staining like FM4-64. FM4-64 (*N*-(3-triethylammoniumpropyl)-4-(6-(4-(diethylamino) phenyl) hexatrienyl) pyridinium dibromide) is a lipophilic dye which stains the plasma membrane and, due to endocytosis, finally stains the vacuolar membrane [35]. A further lipophilic dye that worked nicely in *P. pastoris* is Nile red, which stains lipid droplets [36]. The cell nucleus staining with DAPI (4′,6-diamidino-2-phenylindole) worked only with fixed cells (FA/MeOH). There are protocols for DAPI staining in living cells, e.g., [37], but the live staining did not work in our hands. Images of *P. pastoris* stained with the mentioned dyes are presented in Fig. 2.

1.4 Super-Resolution

Super-resolution microscopy is a promising technique for more detailed cellular information. These new technologies are able to break or bypass Abbe's diffraction limit and shift the optical resolution down to macromolecular levels (see overview by Schermelleh

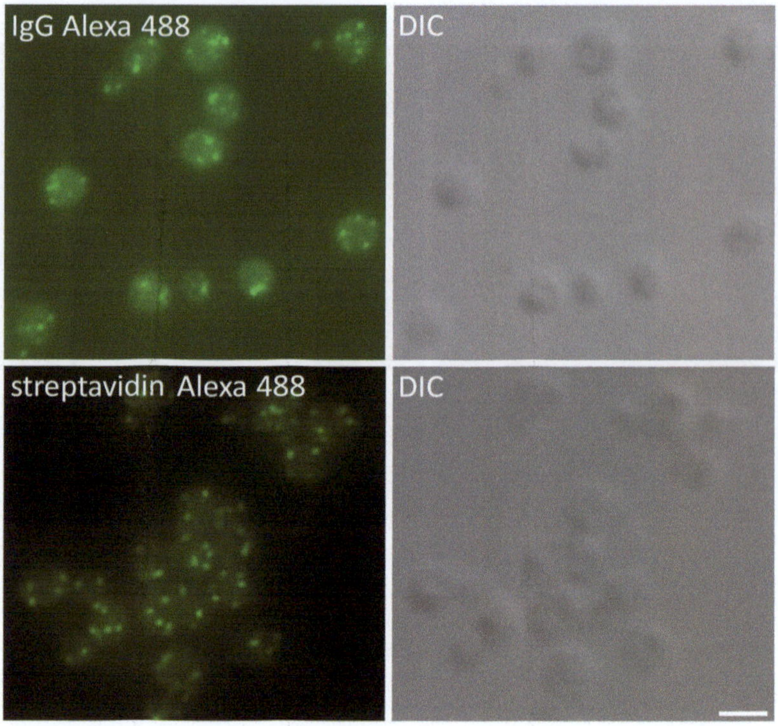

Fig. 5 Immunofluorescence microscopy of *P. pastoris* using self-made rabbit anti-Sec7 antiserum. In the upper row, the fluorescence and bright field images of the cells incubated with donkey anti-rabbit IgG Alexa Fluor 488 are shown. The lower row represents the images of the cells incubated with donkey anti-rabbit IgG biotin followed by streptavidin Alexa Fluor 488, resulting in an improved signal-to-noise ratio. Bar, 5 μm

et al. [38]). With these techniques it is possible to increase resolution two-fold using SIM (structured illumination microscopy) and up to ten-fold using STORM/PALM (stochastic optical reconstruction microscopy/photo-activated localization microscopy) or stimulated emission depletion (STED) compared to standard wide-field combined with deconvolution or confocal laser scanning microscopy (CLSM). An overview of resolution-enhancing microscopy techniques and recent advances can be found in [38–42]. STED, SIM, and STORM/PALM offer great spatial resolution, while better temporal resolution is achieved by advanced methods such as fluctuation-enhanced structured illumination microscopy (fFE-SIM) [43, 44] or by super-resolution confocal live imaging microscopy (SCLIM), which even allows for simultaneous three-color and four-dimensional live cell imaging with high spatiotemporal resolution [45].

Over the last decade, super-resolution microscopy in yeast has mainly been used to study molecular interactions (e.g., protein-protein) and to unravel the basic mechanistic processes in membrane and organelle organization and dynamics, but not for more applied research questions [46–56].

We used STED in immunofluorescence and co-stained the recombinant protein HSA and the ER. Compared to the confocal image, finer structures are visible in the STED image (Fig. 3a). For STED, the secondary antibody has to be selected carefully. Due to the depletion laser, only a narrow range of fluorophores can be applied. We used Oregon Green 488 (Thermo Fisher Scientific) and V500 (BD Horizon).

As already mentioned, fluorescent fusion proteins can also be fixed by paraformaldehyde solution. This is of particular interest for high-resolution images, which eliminates the risk of specimen movement during image acquisition. *P. pastoris* expressing an additional copy of the vesicular marker Snc2-sfGFP under control of the *GAP* promoter is shown in Fig. 3b. The cells were cultivated in minimal medium and paraformaldehyde fixed for microscopy. Compared to the confocal image, small single dots can be distinguished from each other in the STED image, which appear as one single dot in the confocal image. Therefore, for very small structures like vesicles, which could be only poorly resolved before, super-resolution is a powerful tool to obtain high-quality information on organelle patterns.

2 Materials

2.1 Cultivation Media (See Note 1)

1. YPD (per liter): 10 g yeast extract, 20 g peptone, 22 g glucose monohydrate. pH set to 7.4–7.6 with KOH.
2. M2 minimal medium: 20 g citric acid, 3.15 g $(NH_4)_2HPO_4$, 0.03 g $CaCl_2.2H_2O$, 0.8 g KCl, 0.5 g $MgSO_4.7H_2O$, 2 mL biotin stock solution, 1.5 mL PTM0 trace salts stock solution, and a carbon source. The pH is set to 5.0 with 5 M KOH solution. The final volume is one liter after the addition of the carbon source.
3. PTM0 trace salts stock solution: 6.0 g $CuSO_4.5H_2O$, 0.08 g NaI, 3.0 g $MnSO_4.H_2O$, 0.2 g $Na_2MoO_4.2H_2O$, 0.02 g H_3BO_3, 0.5 g $CoCl_2$, 20.0 g $ZnCl_2$, 5.0 g $FeSO_4.7H_2O$ and 5.0 mL H_2SO_4 (95–98% w/w)). The final volume is one liter.
4. Biotin stock solution: 0.2 g biotin per liter.
5. Carbon source: 2% (w/v) glucose, i.e., 22 g glucose monohydrate per liter (*see* **Note 2**).

2.2 Fluorescent Dyes

1. FM4-64: Stock solution: 100 μg (1 vial) FM4-64 (Molecular Probes, Thermo Fisher Scientific) in 10 μL DMSO; store at −20 °C for no longer than two weeks. Working solution: dilute stock solution 1:10 in 1x PBS and keep solution on ice in the dark until use.

2. Nile red: Stock solution: 1 mM Nile red (Molecular Probes, Thermo Fisher Scientific) in anhydrous DMSO, stored at −20 °C. Working solution: Dilute stock solution 1:1000 in 1x PBS; use immediately.
3. DAPI: Stock solution: 5 mg DAPI hydrochloride (fluoropure grade) in 1 mL DMF; store at −20 °C. Working solution: pre-dilute stock solution 1:1000 in H_2O. Dilute 1:100 in mounting medium (e.g., Vectashield) to yield a final concentration of 50 ng/mL.
4. 37% formaldehyde solution
5. 1x PBS: 0.2 g KH_2PO_4, 1.8 g $Na_2HPO_4.2\ H_2O$, 0.2 g KCl, 8.0 g NaCl. The final volume is one liter, and the pH should be 7.4.
6. 100 mM potassium phosphate (pH = 6.5): 3.3 mL 1 M K_2HPO_4, 6.8 mL 1 M KH_2PO_4, 90 mL H_2O, for storage: 200 μL 10% NaN_3-solution. The final volume is 100 mL.
7. 1 M K_2HPO_4: 17.4 g in 100 mL final volume
8. 1 M KH_2PO_4: 13.6 g in 100 mL final volume.

2.3 Microscopy of Fluorescence Fusion Proteins of Secretory Proteins

1. Low melt agarose.
2. Masking tape.
3. Glass slides and coverslips (18 × 18 mm, 24 × 50 mm).
4. Fixing reagent: Mix 10 mL 16% formaldehyde (methanol-free) with 30 mL 1x PBS and add $MgCl_2$ at a final concentration of 1 mM.
5. 1x PBS.
6. Polysine slides.

2.4 Immunofluorescence

1. IBIDI 15 μ-Chamber 12-well glass slide and coverslips (1.5H) or imaging chambers CG 8 wells Cover glass bottom (PAA) and glass slides.
2. 0.01% Poly-L-lysine solution: dilute 0.1% poly-L-lysine 1:10 in ddH_2O.
3. Fixing reagent: Mix 10 mL 16% formaldehyde (methanol-free) with 30 mL 1x PBS and add $MgCl_2$ at a final concentration of 1 mM.
4. 100 mM potassium phosphate buffer (pH = 6.5) containing 1 mM $MgCl_2$.
5. Yeast staining sorbitol buffer: 100 mM potassium phosphate containing 0.9 M sorbitol.
6. 5 U/μL Lyticase from *Arthrobacter luteus*: Dissolve 50 KU of lyticase in 5 mL sterile 10 mM Tris–HCl (pH = 7.5) and add 5 mL sterile 100% glycerol. Store in aliquots at −20 °C.

7. β-Mercaptoethanol.
8. Blocking solution: 1% BSA in 1x PBS.
9. Antibody dilution solution: 0.5% BSA + 0.1% Triton in 1x PBS.
10. Methanol or acetone (pure), −20 °C.
11. Vectashield (Vectorlabs) or similar mounting medium.

2.5 STED

1. ProLong Gold Antifade Mountant (Molecular Probes, Thermo Fisher Scientific).
2. Imaging chambers CG 8 wells Cover Glass Bottom (PAA).
3. Glass slides.

3 Methods

3.1 Staining with Fluorescent Dyes

3.1.1 Cultivation for All Stainings

1. 5 mL YPD containing the respective antibiotics should be inoculated for an overnight pre-culture and incubated at 25 ° C with rigorous shaking at 180 rpm.
2. For the main culture, 10 mL of M2 minimal medium with 2% glucose (pH = 5–6) shall be inoculated at OD_{600} = 0.1 (*see* **Note 2**).
3. *P. pastoris* cells are grown for approx. 24 h at 25 °C and shaking at 180 rpm to an OD_{600} of 0.5-1.
4. The cells are either viewed directly (fluorescent proteins) or fixed with formaldehyde and stained.

3.1.2 Staining of Vacuoles Using FM4-64

1. Harvest cells by centrifugation (2500 rpm for 3 min, room temperature) in the morning and resuspend the cells in M2 minimal medium to obtain 1 mL with an OD_{600} of 15.
2. Pipet 200 μL cell suspension in a 1.5 mL centrifuge tube.
3. Add 2 μL FM4–64 working solution to cell suspension.
4. Incubate in a thermoblock at 30 °C for 15 min with shaking at 600 rpm; cover with aluminum foil.
5. Harvest cells by centrifugation at 2500 rpm for 3 min in a microcentrifuge at room temperature; remove supernatant.
6. Resuspend cells in 200 μL fresh M2 minimal medium.
7. Incubate for 60 min at 30 °C with shaking at 600 rpm.
8. Harvest cells by centrifugation at 2500 rpm for 3 min in a microcentrifuge at room temperature; remove supernatant.
9. Resuspend cells in 100 μL fresh, pre-warmed M2 minimal medium.
10. Pipet 3-5 μL of cell suspension on a glass slide, cover with a coverslip, and view in a fluorescence microscope using an

appropriate filter cube like the Leica N2.1 filter cube (excitation: BP 515-560, emission: LP 590).

3.1.3 Staining of Lipid Particles Using Nile Red

1. Pipet an aliquot of cells that corresponds to $OD_{600} = 8/0.5$ mL into a 2 mL microcentrifuge tube.
2. Harvest cells by centrifugation at 2500 rpm for 3 min in a microcentrifuge at room temperature; remove supernatant.
3. Resuspend cells in 500 μL Nile red working solution.
4. Incubate at room temperature or 30 °C for 10 min and cover with aluminum foil.
5. Harvest cells by centrifugation at 2500 rpm for 3 min in a microcentrifuge at room temperature; remove supernatant.
6. Wash cells once with 1x PBS, and harvest them as in the previous step.
7. Resuspend cells in 500 μL pre-warmed 1x PBS.
8. Monitor in a fluorescence microscope using an appropriate filter cube like the Leica N2.1 filter cube (excitation: BP 515-560, emission: LP 590).

3.1.4 Staining of Yeast Nucleus Using DAPI for Fluorescent Microscopy

1. Mix 1 mL of culture (grown to OD_{600} of 0.5-1) with 0.1 mL of 37% formaldehyde solution. Incubate for 1 hour on an end-over-end rotator (6–7 rpm) at room temperature.
2. Harvest cells by centrifugation at 3500 rpm for 3 min in a microcentrifuge at room temperature; remove supernatant.
3. Wash cells twice with 1x PBS, and harvest them as in the previous step.
4. Resuspend cells in 300 μL 1x PBS.
5. Pipet 5 μL of cells into a 1.5 mL microcentrifuge tube. Mix with 5 μL mounting medium containing 50 ng/mL DAPI. Mix well (*see* **Note 3**).
6. Pipet 3 μL on a microscope slide, cover with a coverslip, and view in a fluorescence microscope using an appropriate filter cube like the Leica A filter cube (excitation: BP 340-380, emission: LP 425).

3.2 Fluorescence Fusion Proteins of Secretory Proteins

3.2.1 Generation of Strains Containing Fluorescence Fusion Proteins

1. Genetically tag proteins of the secretory pathway with suitable fluorescent proteins (FP) (*see* **Note 4**).
2. Check by immunoblotting for secretion ability of the fusion protein and the correct fusion protein band in the cells by using anti-FP antibodies.

3.2.2 Cultivation of Strains Containing Fluorescence Fusion Proteins

1. Inoculate 5 mL YPD with respective antibiotics for an overnight culture, and incubate at 25 °C with rigorous shaking at 180 rpm.
2. Inoculate 10 mL of M2 minimal medium with 2–5 μL of the pre-culture and let it grow for 20–24 h. Check OD_{600}. Use cells with an OD ~ 1.
3. For identification of positive clones, pipet 5 μL of suspension on a glass slide, cover with a coverslip, and view in a fluorescence microscope using an appropriate filter cube like the Leica L5 filter cube (excitation: BP 460-500, emission: BP 527/30) (*see* **Note 5**).
4. For monitoring of the fluorescent fusion protein, see below.

3.2.3 Live Cell Monitoring with Agarose Pad

1. Melt 4% (w/v) agarose in M2 minimal medium (e.g., 40 mg agarose in 1 mL medium) in a 2 mL microcentrifuge tube at 90 °C for approx. 15 min in a thermoblock.
2. Tape a microscope slide with one layer of masking band on the edges of the slide around the entire slide.
3. Pipet 100 μL of liquid agarose in the middle of the glass slide and cover the glass slide with a second slide.
4. Take off the second slide after 10 min. Usually, the agarose pad is located on this slide.
5. At the microscope, put 20 μL of a 24 h culture grown in minimal medium (OD_{600} = ~1) on the agarose pad and cover with a 24 × 50 mm microscope slide.
6. Monitor cells in a confocal laser scanning microscope with low laser intensity for several hours. Make a z-stack (0.3 μm) every 3–5 min if your protein is moving slowly, and decrease the timeframe if your protein is moving faster (*see* **Note 6**).

3.2.4 Fixation of Fluorescent Fusion Proteins

1. Use a culture grown in minimal medium up to an OD_{600} of about 1.
2. Harvest 8 mL of culture by centrifugation at 2000 rpm for 4 min in a microcentrifuge at room temperature, and remove supernatant.
3. Resuspend cells in 5 mL fixing solution.
4. Incubate at room temperature for 1 h on an end-over-end mixer at low speed (6–7 rpm).
5. Wash cells 3 times with 1x PBS containing 1 mM $MgCl_2$.
6. Resuspended cells in 1x PBS to a final OD_{600} of about 1.
7. Pipet 5 μL of the suspension on a glass slide, cover with a coverslip, and view in a fluorescence microscope using the appropriate filter cube (*see* **Note 7**).

3.3 Immunofluorescence Microscopy

Immunofluorescence (IF) microscopy is used for (co-)localization studies. We are using a combination of the protocols reported in Rossanese et al. and Kilmartin et al. [57, 58] with a main interest in localization of recombinant secretory proteins.

1. Coat the appropriate number of wells of a chamber slide (e.g., 15 μ-Chamber 12-well glass slide or Imaging chambers CG 8-well cover glass bottom) with poly-L-lysine (Sigma) by adding 150 μL 0.01% poly-L-lysine to each well and incubate at room temperature for 5 min (*see* **Note 8**).
2. Remove the liquid and wash wells 3 times with 300 μL H_2O.
3. Let the chamber slide dry.
4. Use a culture grown for approx. 24 h in minimal medium up to an OD_{600} of 0.5–1.
5. Pipette 8 mL of the culture onto a 50 mL Whatman Klari-Flex bottle-top filter (pore size 0.22 μm). Apply vacuum. As soon as the medium has passed through the filter, break the vacuum and add 5 mL of fixing solution. Alternatively, harvest 8 mL suspension by centrifugation at 2000 rpm for 4 min and add 5 mL fixing solution to the cells.
6. Softly resuspend the cells and transfer the suspension to a 15 mL centrifugation tube.
7. Incubate at room temperature for 2 h on an end-over-end mixer at low speed (6–7 rpm).
8. Centrifuge for 3 min at 2500 rpm in a low-speed centrifuge.
9. Remove the supernatant as much as possible.
10. Wash cells 2 times with 5 mL 100 mM potassium phosphate buffer (pH = 6.5) containing 1 mM $MgCl_2$ (pellet cells by centrifuging 3 min at 2500 rpm) and remove the supernatant.
11. Resuspend cells in yeast staining sorbitol buffer to a final OD_{600} of 10.
12. Transfer 100 μL of the cell suspension to a 1.5 mL microcentrifuge tube.
13. Add 0.6 μL β-mercaptoethanol and 1 μL lyticase (5 U).
14. Incubate suspension in a thermoblock for 30 min at 30 °C and mix occasionally (*see* **Note 9**).
15. Centrifuge for 3 min at 2500 rpm. From now on, handle spheroblasts on ice until they are pipetted onto the slide.
16. Wash spheroblasts 2 times with 0.5 mL yeast staining sorbitol buffer (pellet cells by centrifuging 3 min at 2500 rpm).
17. Gently resuspend pellet in 100 μL yeast staining sorbitol buffer (*see* **Note 10**).
18. Add 30 μL of spheroblast suspension to each well.

19. Incubate for 10 min at room temperature.
20. Remove the excess liquid with a cotton swab and let the chamber slide dry completely.
21. Add 100 μL pre-cooled methanol or acetone (−20 °C) to each well (*see* **Note 11**).
22. Incubate for 5 min, remove the liquid, and let the chamber slide dry.
23. Add 300 μL of blocking solution and incubate overnight at 4 °C in a humid chamber.
24. Centrifuge at full speed for 5 min at 4 °C, and remove supernatant.
25. Remove the excess of blocking solution.
26. Dilute primary antibody in antibody dilution solution (*see* **Note 12**). Add 100 μL of primary antibody solution to every well and incubate for 1 h at room temperature in a humid chamber (*see* **Note 13**).
27. Centrifuge at full speed for 5 min at 4 °C, and remove supernatant.
28. Wash cells six times with 300 μL antibody dilution solution (without antibody).
29. Dilute secondary antibody in antibody dilution solution (*see* **Note 14**). Add 100 μL of secondary antibody solution to every well.
30. Incubate 45 min at room temperature in the dark in a humid chamber.
31. Centrifuge at full speed for 5 min at 4 °C, and remove supernatant.
32. Wash six times with 300 μL antibody dilution solution (without antibody).
33. Wash once with 300 μL 1x PBS.
34. Treat cells with 200 μL fixing solution for 10 min at RT.
35. Wash three times with 300 μL 1x PBS.
36. Wash once with 300 μL ddH_2O.
37. Remove remaining liquid, remove the plastic grid, and add 5 μL Vectashield mounting media to each sample.
38. If co-staining of the nucleus is wanted, add DAPI diluted in mounting media at a concentration of 50 ng/mL. Alternatively, add 100 μL of 0.1 μg/mL DAPI diluted in 1x PBS to each well and incubate for 15 min before mounting. Afterwards, wash cells 3 times with 1x PBS.

39. Add a coverslip (24 × 60 mm, 1.5 H) or put a coverslip on a glass slide depending on the chamber used, press down gently, and seal with nail polish. Keep samples in the dark.
40. View in a fluorescence microscope using the appropriate filter cubes.

3.4 STED

Sample preparation is similar to immunofluorescence microscopy (*see* Subheading 3.3.) with small changes. The 8-well coverslip chamber from PAA worked better than IBIDI chamber slides with 1.5 high-precision coverslips. Due to the larger wells, use twice as much of the solutions as indicated above. It is pointed out by Leica that Vectashield is not usable as a mounting medium; instead, ProLong Gold shall be used (*see* **Note 15**). The first antibody solution is incubated over the weekend (while the Leica protocol suggests using the antibody at higher concentrations than for IF, we used the same concentration but incubated over the weekend at 4 °C). Due to the depletion laser, the variety of secondary antibodies is restricted. We used Oregon Green and BD Horizon V500. The samples were viewed in a Leica SP8 gated STED microscope.

4 Notes

1. For cultivation, any minimal medium that is buffered to a pH between 5.0 and 6.5 is suited. We used M2 minimal medium, but also YNB or related media can be used. Buffering is important as *P. pastoris* cells tend to form aggregates or flocs around pH 4.0.
2. In case strains containing methanol-inducible promoter systems shall be investigated, methanol (1%) has to be used as a carbon source instead of glucose, and the time to reach the required OD_{600} has to be adapted accordingly.
3. This step has to be done directly at the microscope; otherwise, the background might be too high. If you experience a very high background, the cells can be stained in the microcentrifuge tube by adding 300 μL of DAPI solution (c = 0.1 μg/mL) to the cell pellet, incubating for 15 min, and washing cells 3 times with 1x PBS before viewing in the microscope.
4. If you tag proteins that are dedicated to the secretory pathway, be sure to use fluorescent proteins suitable or optimized for oxidizing environments like sfGFP, oxGFP, or moxNeonGreen. Be careful when designing the linker region. We observed differences in fluorescence intensity depending on the length of the linker (Fig. 4).

5. If cells are heavily floating, use Polysine adhesion slides (Thermo Fisher Scientific). Due to the positive charge of the poly-L-lysine derivative and the negative charge of the cell surface, the cells adhere to the slides and can be easily viewed in the microscope.
6. Make as few images as possible as your sample is bleaching and the laser radiation is harming the cells. Mounting live cells on agar is especially important if you want to monitor cells for a longer period of time, e.g., to monitor cell proliferation.
7. Compare the pattern to the signal of the respective live cells to exclude artifacts due to fixation.
8. Using the cover glass bottom slides yields better imaging results due to the close proximity of the cells to the objective. If you use 8 well slides, you have to double the amount of solutions mentioned throughout the protocol due to larger wells.
9. This is a very crucial step. Be sure to flip the microcentrifuge tubes every 5–10 min. For us, 30 min at 30 °C always worked perfectly fine, but when you experience problems with the staining, vary the incubation time for better results.
10. To handle spheroblasts gently, cut off the narrow part of the tip.
11. Try out which solution works better for your antibody. If possible, the use of methanol is preferable, as the cells remain more structured and are therefore easier to view in the microscope. Usually acetone works better when antibodies are directed against the native protein, whereas methanol is of use if antibodies were made against a peptide [59].
12. If you have poorly working antisera, use biotin/streptavidin for signal enhancement. After incubation with the primary antibody and washing of the wells, incubate with a biotinylated secondary antibody for 1 h. After washing 6 times with 0.5% BSA in PBST, incubate with streptavidin Alexa Fluor 488 conjugate for 45 min–60 min. An example of the difference between using only a secondary antibody compared to the use of a streptavidin conjugate can be seen in Fig. 5.
13. If you are using poorly working antisera, incubate overnight at 4 °C.
14. Even if nowadays the use of FITC and Cy3 is not common anymore, we want to point out that, especially for small cells like yeast, it is important to work with bright and stable fluorophores, e.g., Alexa Fluor antibodies. Oregon Green 488 and Texas Red fluorophores are also working very well.

15. When using ProLong Gold, be sure to warm it up to room temperature before use, and after use, let it dry for 24 h. Sealing with nail polish is not necessary.

Acknowledgments

Research on *Pichia pastoris* microscopy was supported by the COMET center acib: Next Generation Bioproduction, funded by the Austrian BMIMI, BMWET, Standortagentur Tirol, Government of Lower Austria, Vienna Business Agency, and BOKU through the COMET Funding Program, managed by the FFG. Further support by the BMWET and Nationalstiftung FTE through the Christian Doppler Research Association, and the Austrian Science Fund (FWF) 10.55776/COE17 within the Cluster of Excellence Circular Bioengineering, is acknowledged. This article contributes to the Yeast4Bio COST Action 18229, supported by COST (European Cooperation in Science and Technology).

We thank the BOKU Core Facility Multiscale Imaging (formerly VIBT Imaging Center) for access to Leica fluorescence microscope devices.

Competing Interest Statement The authors have no conflicts of interest to declare that are relevant to the content of this chapter.

References

1. Chalfie M, Tu Y, Euskirchen G, Ward W, Prasher D (1994) Green fluorescent protein as a marker for gene expression. Science 263: 802–805. https://doi.org/10.1126/science.8303295
2. Tsien R (1998) The green fluorescent protein. Annu Rev Biochem 67:509–544. https://doi.org/10.1146/annurev.biochem.67.1.509
3. Yang T, Cheng L, Kain S (1996) Optimized codon usage and chromophore mutations provide enhanced sensitivity with the green fluorescent protein. Nucleic Acids Res 24:4592–4593. https://doi.org/10.1093/nar/24.22.4592
4. Patterson G, Knobel S, Sharif W, Kain S, Piston D (1997) Use of the green fluorescent protein and its mutants in quantitative fluorescence microscopy. Biophys J 73:2782–2790. https://doi.org/10.1016/S0006-3495(97)78307-3
5. Snapp E (2005) Design and use of fluorescent fusion proteins in cell biology. Curr Protoc Cell Biol Chapter 21:21.4.1. https://doi.org/10.1002/0471143030.cb2104s27
6. Tsien R (2009) Constructing and exploiting the fluorescent protein paintbox (Nobel lecture). Angew Chem Int Ed Engl 48:5612–5626. https://doi.org/10.1002/anie.200901916
7. Shaner N, Patterson G, Davidson M (2007) Advances in fluorescent protein technology. J Cell Sci 120:4247–4260. https://doi.org/10.1242/jcs.005801
8. Matz M, Fradkov A, Labas Y, Savitsky A, Zaraisky A, Markelov M, Lukyanov S (1999) Fluorescent proteins from nonbioluminescent Anthozoa species. Nat Biotechnol 17:969–973. https://doi.org/10.1038/13657
9. Shaner N, Lin M, McKeown M, Steinbach P, Hazelwood K, Davidson M, Tsien R (2008) Improving the photostability of bright monomeric orange and red fluorescent proteins. Nat Methods 5:545–551. https://doi.org/10.1038/nmeth.1209

10. Shaner N, Campbell R, Steinbach P, Giepmans B, Palmer A, Tsien R (2004) Improved monomeric red, orange and yellow fluorescent proteins derived from *Discosoma* sp. red fluorescent protein. Nat Biotechnol 22:1567–1572. https://doi.org/10.1038/nbt1037
11. Shaner N, Steinbach P, Tsien R (2005) A guide to choosing fluorescent proteins. Nat Methods 2:905–909. https://doi.org/10.1038/nmeth819
12. Shaner N, Lambert G, Chammas A, Ni Y, Cranfill P, Baird M, Sell B, Allen J, Day R, Israelsson M, Davidson M, Wang J (2013) A bright monomeric green fluorescent protein derived from *Branchiostoma lanceolatum*. Nat Methods 10:407–409. https://doi.org/10.1038/nmeth.2413
13. Pédelacq J, Cabantous S, Tran T, Terwilliger T, Waldo G (2006) Engineering and characterization of a superfolder green fluorescent protein. Nat Biotechnol 24:79–88. https://doi.org/10.1038/nbt1172
14. Valbuena F, Fitzgerald I, Strack R, Andruska N, Smith L, Glick B (2020) A photostable monomeric superfolder green fluorescent protein. Traffic 21:534–544. https://doi.org/10.1111/tra.12737
15. Costantini LM, Baloban M, Markwardt ML, Rizzo M, Guo F, Verkhusha VV, Snapp EL (2015) A palette of fluorescent proteins optimized for diverse cellular environments. Nat Commun 6:7670. https://doi.org/10.1038/ncomms8670
16. Aronson D, Costantini L, Snapp E (2011) Superfolder GFP is fluorescent in oxidizing environments when targeted via the Sec translocon. Traffic 12:543–548. https://doi.org/10.1111/j.1600-0854.2011.01168.x
17. Suzuki T, Arai S, Takeuchi M, Sakurai C, Ebana H, Higashi T, Hashimoto H, Hatsuzawa K, Wada I (2012) Development of cysteine-free fluorescent proteins for the oxidative environment. PLoS One 7:e37551. https://doi.org/10.1371/journal.pone.0037551
18. Lee S, Lim W, Thorn K (2013) Improved blue, green, and red fluorescent protein tagging vectors for *S. Cerevisiae*. PLoS One 8:e67902. https://doi.org/10.1371/journal.pone.0067902
19. Kaishima M, Ishii J, Matsuno T, Fukuda N, Kondo A (2016) Expression of varied GFPs in *Saccharomyces cerevisiae*: codon optimization yields stronger than expected expression and fluorescence intensity. Sci Rep 6:35932. https://doi.org/10.1038/srep35932
20. Hector R, Mertens J, Nichols N (2021) Increased expression of the fluorescent reporter protein ymNeonGreen in *Saccharomyces cerevisiae* by reducing RNA secondary structure near the start codon. Biotechnol Rep 33:e00697. https://doi.org/10.1016/j.btre.2021.e00697
21. Puxbaum V, Gasser B, Mattanovich D (2016) The bud tip is the cellular hot spot of protein secretion in yeasts. Appl Microbiol Biotechnol 100:8159–8168. https://doi.org/10.1007/s00253-016-7674-6
22. Marsalek L, Puxbaum V, Buchetics M, Mattanovich D, Gasser B (2019) Disruption of vacuolar protein sorting components of the HOPS complex leads to enhanced secretion of recombinant proteins in *Pichia pastoris*. Microb Cell Factories 18:119. https://doi.org/10.1186/s12934-019-1155-4
23. Roy Chowdhury S, Bhattacharjee C, Casler J, Jain B, Glick B, Bhattacharyya D (2020) ER arrival sites associate with ER exit sites to create bidirectional transport portals. J Cell Biol 219:e201902114. https://doi.org/10.1083/jcb.201902114
24. Fitzgerald I, Glick BS (2014) Secretion of a foreign protein from budding yeasts is enhanced by cotranslational translocation and by suppression of vacuolar targeting. Microb Cell Factories 13:125. https://doi.org/10.1186/s12934-014-0125-0
25. Bevis BJ, Hammond AT, Reinke CA, Glick BS (2002) *De novo* formation of transitional ER sites and Golgi structures in *Pichia pastoris*. Nat Cell Biol 4:750–756. https://doi.org/10.1038/ncb852
26. Bevis B, Glick B (2002) Rapidly maturing variants of the *Discosoma* red fluorescent protein (DsRed). Nat Biotechnol 20:83–87. https://doi.org/10.1038/nbt0102-83
27. Puxbaum V, Mattanovich D, Gasser B (2015) Quo vadis? The challenges of recombinant protein folding and secretion in *Pichia pastoris*. Appl Microbiol Biotechnol 99:2925–2938. https://doi.org/10.1007/s00253-015-6470-z
28. Kostyuk A, Panova A, Kokova A, Kotova D, Maltsev D, Podgorny O, Belousov V, Bilan D (2020) In vivo imaging with genetically encoded redox biosensors. Int J Mol Sci 21:8164. https://doi.org/10.3390/ijms21218164
29. Pang Y, Zhang H, Ai H (2021) Genetically encoded fluorescent redox indicators for unveiling redox signaling and oxidative toxicity. Chem Res Toxicol 34:1826–1845. https://doi.org/10.1021/acs.chemrestox.1c00149

30. Delic M, Rebnegger C, Wanka F, Puxbaum V, Haberhauer-Troyer C, Hann S, Kollensperger G, Mattanovich D, Gasser B (2012) Oxidative protein folding and unfolded protein response elicit differing redox regulation in endoplasmic reticulum and cytosol of yeast. Free Radic Biol Med 52:2000–2012. https://doi.org/10.1016/j.freeradbiomed.2012.02.048
31. Roberts C, Raymond C, Yamashiro C, Stevens T (1991) Methods for studying the yeast vacuole. Methods Enzymol 194:644–661. https://doi.org/10.1016/0076-6879(91)94047-g
32. Idiris A, Tohda H, Sasaki M, Okada K, Kumagai H, Giga-Hama Y, Takegawa K (2010) Enhanced protein secretion from multiprotease-deficient fission yeast by modification of its vacuolar protein sorting pathway. Appl Microbiol Biotechnol 85:667–677. https://doi.org/10.1007/s00253-009-2151-0
33. Heiss S, Puxbaum V, Gruber C, Altmann F, Mattanovich D, Gasser B (2015) Multistep processing of the secretion leader of the extracellular protein Epx1 in *Pichia pastoris* and implications for protein localization. Microbiology 161:1356–1368. https://doi.org/10.1099/mic.0.000105
34. Brizzard B (2008) Epitope tagging. BioTechniques 44:693–695. https://doi.org/10.2144/000112841
35. Vida TA, Emr SD (1995) A new vital stain for visualizing vacuolar membrane dynamics and endocytosis in yeast. J Cell Biol 128:779–792. https://doi.org/10.1083/jcb.128.5.779
36. Greenspan P, Mayer E, Fowler S (1985) Nile red: a selective fluorescent stain for intracellular lipid droplets. J Cell Biol 100:965–973. https://doi.org/10.1083/jcb.100.3.965
37. Farre J, Shirahama-Noda K, Zhang L, Booher K, Subramani S (2007) Localization of proteins and organelles using fluorescence microscopy. Methods Mol Biol 389:239–250. https://doi.org/10.1007/978-1-59745-456-8_17
38. Schermelleh L, Heintzmann R, Leonhardt H (2010) A guide to super-resolution fluorescence microscopy. J Cell Biol 190:165–175. https://doi.org/10.1083/jcb.201002018
39. Feng H, Wang X, Xu Z, Zhang X, Gao Y (2018) Super-resolution fluorescence microscopy for single cell imaging. Adv Exp Med Biol 1068:59–71. https://doi.org/10.1007/978-981-13-0502-3_6
40. Valli J, Garcia-Burgos A, Rooney L, Vale de Melo E, Oliveira B, Duncan R, Rickman C (2021) Seeing beyond the limit: a guide to choosing the right super-resolution microscopy technique. J Biol Chem 297:100791. https://doi.org/10.1016/j.jbc.2021.100791
41. Shashkova S, Leake M (2017) Single-molecule fluorescence microscopy review: shedding new light on old problems. Biosci Rep 37:BSR20170031. https://doi.org/10.1042/BSR20170031
42. Toomre D, Bewersdorf J (2010) A new wave of cellular imaging. Annu Rev Cell Dev Biol 26:285–314. https://doi.org/10.1146/annurev-cellbio-100109-104048
43. Yao L, Zhang L, Chen L, Gong X, Zhong J, Wang B, Fei Y, Mi L, Ma J (2021) Dynamic structure of yeast septin by fast fluctuation-enhanced structured illumination microscopy. Microorganisms 9:2255. https://doi.org/10.3390/microorganisms9112255
44. Unruh J, Slaughter B, Jaspersen S (2018) Functional analysis of the yeast LINC complex using fluctuation spectroscopy and super-resolution imaging. Methods Mol Biol 1840:137–161. https://doi.org/10.1007/978-1-4939-8691-0_12
45. Kurokawa K, Nakano A (2020) Live-cell imaging by super-resolution confocal live imaging microscopy (SCLIM): simultaneous three-color and four-dimensional live cell imaging with high space and time resolution. Bio Protoc 10(17):e3732. https://doi.org/10.21769/BioProtoc.3732
46. Sing C, Yang E, Higuchi-Sanabria R, Pon L, Boldogh I, Swayne T (2022) Imaging the actin cytoskeleton in fixed budding yeast cells. Methods Mol Biol 2364:81–100. https://doi.org/10.1007/978-1-0716-1661-1_4
47. Da Mota M, Cau J, Mateos-Langerak J, Lengronne A, Pasero P, Poli J (2021) 3D positioning of tagged DNA loci by widefield and super-resolution fluorescence imaging of fixed yeast nuclei. STAR Protoc 2(2):100525. https://doi.org/10.1016/j.xpro.2021.100525
48. Rodriguez-Gallardo S, Kurokawa K, Sabido-Bozo S, Cortes-Gomez A, Perez-Linero A, Aguilera-Romero A, Lopez S, Waga M, Nakano A, Muñiz M (2021) Assay for dual cargo sorting into endoplasmic reticulum exit sites imaged by 3D super-resolution confocal live imaging microscopy (SCLIM). PLoS One 16:e0258111. https://doi.org/10.1371/journal.pone.0258111
49. Black L, Tollis S, Fu G, Fiche J, Dorsey S, Cheng J, Ghazal G, Notley S, Crevier B, Bigness J, Nollmann M, Tyers M, Royer C (2020) G1/S transcription factors assemble in increasing numbers of discrete clusters through

G1 phase. J Cell Biol 219(9):e202003041. https://doi.org/10.1083/jcb.202003041

50. Adhikari S, Banerjee C, Moscatelli J, Puchner E (2020) Conventional BODIPY conjugates for live-cell super-resolution microscopy and single-molecule tracking. J Vis Exp 160: 60950. https://doi.org/10.3791/60950
51. Tojima T, Suda Y, Ishii M, Kurokawa K, Nakano A (2019) Spatiotemporal dissection of the trans-Golgi network in budding yeast. J Cell Sci 132:jcs231159. https://doi.org/10.1242/jcs.231159
52. Alexandrov A, Grosfeld E, Dergalev A, Kushnirov V, Chuprov-Netochin R, Tyurin-Kuzmin P, Kireev I, Ter-Avanesyan M, Leonov S, Agaphonov M (2019) Analysis of novel hyperosmotic shock response suggests 'beads in liquid' cytosol structure. Biol Open 8:bio044529. https://doi.org/10.1242/bio.044529
53. Stoldt S, Wenzel D, Kehrein K, Riedel D, Ott M, Jakobs S (2018) Spatial orchestration of mitochondrial translation and OXPHOS complex assembly. Nat Cell Biol 20(5): 528–534. https://doi.org/10.1038/s41556-018-0090-7
54. Stagge F, Mitronova G, Belov V, Wurm C, Jakobs S (2013) SNAP-, CLIP- and halo-tag labelling of budding yeast cells. PLoS One 8:1–9. https://doi.org/10.1371/journal.pone.0078745
55. Daboussi L, Costaguta G, Payne G (2012) Phosphoinositide-mediated clathrin adaptor progression at the trans-Golgi network. Nat Cell Biol 14:239–248. https://doi.org/10.1038/ncb2427
56. Rankin B, Moneron G, Wurm C, Nelson J, Walter A, Schwarzer D, Schroeder J, Colón-Ramos D, Hell S (2011) Nanoscopy in a living multicellular organism expressing GFP. Biophys J 100:63–65. https://doi.org/10.1016/j.bpj.2011.05.020
57. Rossanese O, Soderholm J, Bevis B, Sears I, O'Connor J, Williamson E, Glick B (1999) Golgi structure correlates with transitional endoplasmic reticulum organization in *Pichia pastoris* and *Saccharomyces cerevisiae*. J Cell Biol 145(1):69–81. https://doi.org/10.1083/jcb.145.1.69
58. Kilmartin J, Adams A (1984) Structural rearrangements of tubulin and actin during the cell cycle of the yeast *Saccharomyces*. J Cell Biol 98: 922–933. https://doi.org/10.1083/jcb.98.3.922
59. Glick B (2007) Fluorescence microscopy and thin-section electron microscopy. Methods Mol Biol 389:251–260. https://doi.org/10.1007/978-1-59745-456-8_18

Chapter 6

Auxin-Induced Degradation of Target Proteins in *Pichia pastoris*

Leonie Lehmayer, Lukas Bernauer, and Anita Emmerstorfer-Augustin

Abstract

The disruption of protein production is a major approach used for investigating protein function in a cell. In yeasts, this is mostly achieved by introducing an insertion or deletion ("indel") into a certain gene or by disrupting the whole reading frame. However, these methods have noteworthy limitations. For example, whole gene deletions may influence promoter or terminator regions of genes located up- and downstream of this particular gene; essential genes cannot simply be knocked out in order to study their function; and the overall absence of a protein can allow time for compensatory mechanisms that confound interpretation of results. To overcome these limitations, this chapter presents the application of the auxin-inducible degron (AID) technology for fast and efficient degradation of a target protein in *Komagataella phaffii*. The AID system is based on the formation of a chimeric SCF complex with the plant F-box protein TIR1, which, in the presence of auxin, leads to the polyubiquitination of the target sequence, the so-called AID tag. If the AID tag is fused to a protein of interest, this protein is immediately subjected to ubiquitin-mediated proteasome degradation upon incubation with auxin. In this work, the usability and efficiency of auxin-inducible degradation of Gut1, a well-studied glycerol kinase with an important role in glycerol metabolism, is demonstrated. Important parameters for experimental design for the use of this system in *K. phaffii* are being evaluated. The results prove that this toolkit can be expanded to all areas of *K. phaffii*-related research.

Key words *Komagataella phaffii*, *Pichia pastoris*, Auxin, Degron, Protein depletion, Conditional mutants

1 Introduction

A conditional mutant is an organism that resembles the wild type under normal environmental conditions but displays a characteristic mutant phenotype under restrictive or altered environmental conditions. The responsible mutation is called a conditional mutation and often involves a protein variant that is subjected to degradation under certain restrictive conditions. If a conditional mutant is incapable of vegetative growth under restrictive conditions, the mutation it harbors is a conditional lethal mutation, and the gene

Xavier Garcia-Ortega et al. (eds.), *Pichia pastoris: Methods and Protocols*, Methods in Molecular Biology, vol. 2697, https://doi.org/10.1007/978-1-0716-4779-0_6,

and function that are defective are referred to as essential. The most common type of conditional mutants are temperature-sensitive (ts) mutants, which are efficiently degraded in the host organism once shifted to higher temperatures [1]. While the use of ts mutants has been reported numerous times, their application has several drawbacks, e.g., their establishment can be very time-consuming, and a temperature shift is often not ideal since it may cause certain stress responses in the cell [2].

Hence, different methods have been developed that are easier to implement. One of these methods is the auxin-inducible degron (AID) technology [3, 4]. The AID technology allows rapid, targeted protein depletion induced by the natural plant hormone auxin. To apply the AID system in non-plant cells, an auxin-inducible destabilizing domain, also referred to as a "degron," is fused to the target protein. In addition, the gene of the plant auxin receptor F-box protein *TIR1* has to be co-expressed, forming a chimeric SKP1-CUL1-TIR1 (SCF^{TIR1}) ubiquitin E3 ligase with endogenous components in eukaryotic cells. Since the proteins of the SCF complex are highly conserved from plants to fungal cells and homologous genes of *CUL1* and *SKP1* are endogenously expressed in all eukaryotic cells [5, 6], the system has successfully been applied in diverse cell types, mostly mammalian cells [7–9], but also in *Saccharomyces cerevisiae* [4, 10] and *Schizosaccharomyces pombe* [11]. The chimeric E3 ligase recruits the degron in an auxin-dependent manner (Fig. 1a). The addition of small molecules of the auxin class, e.g., indole-3-acetic acid (IAA, the natural auxin) or 1-naphthalene acetic acid (1-NAA, a synthetic auxin), to the culture medium promotes the interaction of the chimeric E3 ligase with the degron-fused protein, which leads to polyubiquitination and rapid proteasomal degradation.

Using the AID technology for the generation of conditional mutants has several advantages over other methods. First, no temperature shift is needed to achieve protein degradation; second, it can be applied to a broad range of different eukaryotic host organisms; and third, the system is reversible, which means that after complete degradation of the target protein, auxin can be washed out and the target protein starts to reaccumulate. This allows for elegant analysis of protein function, which is especially interesting when looking at short-time influences of protein disruption in pulse-chase experiments or when studying an essential gene. In the presence of auxin, TIR1, which itself is located inside of the nucleus, readily translocates into the cytosol to form a functional SCF complex. A broad range of proteins can be suspected of degradation using this system, as the only requirement is that the AID degron needs to be accessible to the SCF^{TIR1}-complex either in the cytosol or the nucleus. The targeted protein can be monomeric or be part of a macromolecular complex, and even transmembrane proteins of the cell membrane or organellar membranes get

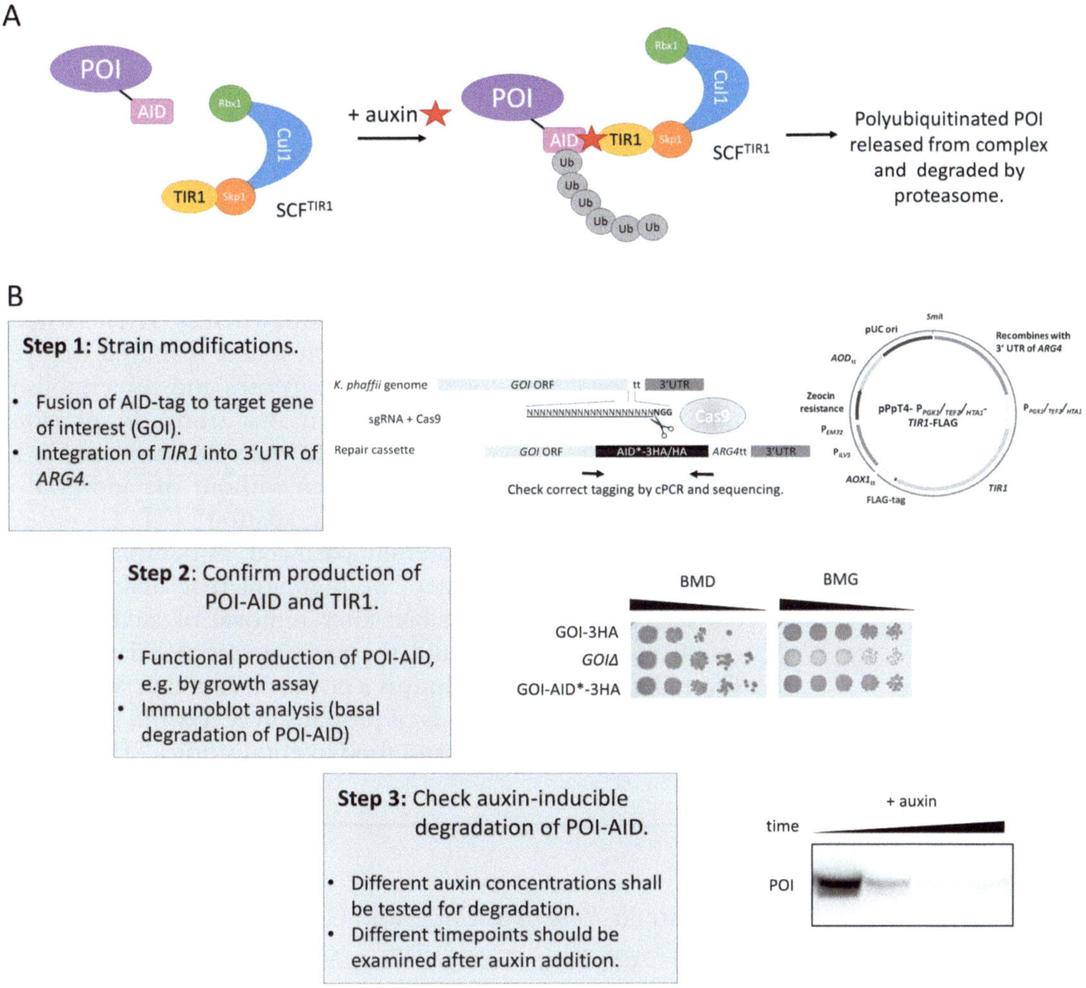

Fig. 1 The AID technology in *K. phaffii*. (**a**) *TIR1* is expressed heterologously and forms a complex with *K. phaffii* endogenous Skp1, Cul1, and Rbx1. Additionally, a gene of interest (GOI) is expressed from its original locus as an AID-tagged fusion. Upon incubation with auxin, the chimeric SCF^{TIR1}-complex interacts with the AID degron, leading to polyubiquitination and degradation of the protein of interest (POI) by the proteasome. (**b**) Overview of the workflow for introducing the AID system into *K. phaffii*. Step 1: Fusion of AID tag to target genes and introduction of *TIR1*. A CRISPR/Cas9 genome engineering approach was used for C-terminal tagging *GUT1* with an AID*-3HA-tag or HA-tag. Plasmids expressing *TIR1* were linearized with *SmiI* to be integrated into the 3′UTR of the *ARG4* locus of *K. phaffii*. Step 2: Determination of functionality of the AID-tagged protein and basal degradation in strains co-expressing *TIR1* by spot assays and immunoblot. Step 3: Testing of auxin-inducible degradation using different auxin concentrations and incubation times. (Figure adapted from [12])

degraded—provided that the tag is exposed to the cytosolic side, independent of N- or C-terminal tagging [8, 10].

Recently, our group developed a simple strategy for the implementation of the AID technology in *Komagataella phaffii* (previously *Pichia pastoris*) [12]. On the basis of this publication, we would like to provide a thorough and easy-to-follow method description of how the system can be applied and modified in order to target other proteins of interest in *K. phaffii*. All plasmids described in this work are readily available. As a model protein, we have chosen Gut1, a glycerol kinase that has often served as a target protein in *K. phaffii,* e.g., for testing targeted protein, disruption applying disruptive cassettes [13] or CRISPR/Cas9 [14]. Gut1 is an excellent target protein, since (I) it is non-essential for growth on glucose-containing media, and a gene knockout does not lead to cell death, which makes it easy to work with, whereas (II), it shows a clear growth phenotype when the carbon source is shifted to glycerol, which is beneficial for screening purposes and the establishment of new methods as presented in this study. In previous research, high expression levels of *TIR1* were shown to result in leaky degradation of target proteins even without the addition of auxin, which is referred to as "basal degradation" [7]. Therefore, *TIR1* expression has to be carefully balanced, as demonstrated in this book chapter. In a last step, we highlight that auxin-induced protein degradation is reversible; after removal of auxin, *GUT1* expression recovers quickly and the protein reaccumulates in the cell. Overall, the AID technology is a powerful tool that will extend the functional study of all kinds of proteins in *K. phaffii*, which will be highly useful in industrial and fundamental studies [15].

2 Materials

2.1 Strains, Vectors, and Reagents

1. Host strain *Komagataella phaffii* CBS7435 *his4Δ* (*see* **Note 1**).
2. Vectors: pPpHyg-Cas9-*GUT1* (pAEA452) [12] (*see* **Note 2**) containing a hygromycin selection marker (*see* **Note 3**); pPpKC2-*GUT1*-3HA (pAEA453) and pPpKC2-*GUT1*-AID*-3HA (pLL003) [12] containing an Ampicillin selection marker, pPpT4-P_{PGK1}-*TIR1*-FLAG (pAEA454), pPpT4-P_{TEF2}-*TIR1*-FLAG (pLL002), and pPpT4-P_{HTA1}-*TIR1*-FLAG (pLL001) [12] (*see* **Note 4**), containing a Zeocin™ selection marker.
3. *Gel* and PCR Clean-Up System (e.g., Promega GmbH, Mannheim, Germany).
4. Electroporator (e.g., Micro-Pulser™, Bio-Rad Laboratories, California, USA).
5. Electroporation cuvettes.
6. Thermomixer Comfort (Eppendorf AG, Hamburg, Germany).
7. Restriction endonuclease *Smi*I.
8. Electrocompetent *K. phaffii* cells prepared fresh according to Lin-Cereghino et al. [13].

9. 1 M sorbitol: Dissolve 18.17 g of sorbitol in 100 mL of ddH_2O. Autoclave and store at room temperature.
10. YPD: Dissolve 10 g of yeast extract and 20 g of peptone in 800 mL of ddH_2O and autoclave. Dissolve 20 g of dextrose in ddH_2O and autoclave. When both mixtures are cooled down to room temperature, mix and store at room temperature.
11. DreamTaq DNA Polymerase (Thermo Fisher Scientific Inc., St. Leon-Rot, Germany).

2.2 Cell Culture

1. 1 M potassium phosphate buffer, pH 6.0: Combine 132 mL of 1 M K_2HPO_4, 868 mL of 1 M KH_2PO_4. Confirm the pH and use phosphoric acid or KOH to adjust the pH, if necessary. Autoclave and store at room temperature.
2. Buffered minimal dextrose (BMD) medium: 2% glucose, 1.34% yeast nitrogen base (without amino acids), 4×10^{-5}% biotin, 0.4% histidine to permit growth of auxotrophs, and 100 mM phosphate buffer, pH 6.
3. Buffered minimal glycerol (BMG) medium: 1% glycerol, 1.34% Yeast Nitrogen Base (without amino acids), 4×10^{-5}% biotin, 0.4% histidine to permit growth of auxotrophs, and 100 mM phosphate buffer, pH 6.
4. Agar plates: Media for plates are solidified by the addition of agar to 2%. YPD plates: 1% yeast extract, 2% peptone, and 2% dextrose; buffered minimal dextrose (BMD) plates: 1.34% YNB (without amino acids), 4×10^{-5}% biotin, 0.4% histidine to permit growth of auxotrophs, 2% dextrose, and 100 mM phosphate buffer, pH 6; buffered minimal glycerol (BMG) plates: 1% glycerol, 13.4 g/l yeast nitrogen base (without amino acids), 4×10^{-5}% biotin, 0.4% histidine to permit growth of auxotrophs, and 100 mM phosphate buffer, pH 6. Antibiotics were used at the following concentrations: Hygromycin B (Formedium™, Norfolk, United Kingdom): yeast [200 mg/L], bacteria [50 mg/L], Zeocin™ (InvivoGen (Eubio), Vienna, Austria) [100 mg/L].
5. M 1-naphthaleneacetic acid (1-NAA) (Sigma-Aldrich®, Vienna, Austria): 186 mg of 1-NAA dissolved in 1 mL DMSO. Stored until use at −20 °C (*see* **Note 5**).
6. Glass culture tubes for overnight cultures (Sigma-Aldrich®, Vienna, Austria).
7. 250 mL shake flasks with baffles (Bartelt GmbH, Graz, Austria).
8. HT Infors Shaker (120 rpm, 28 °C, 80% humidity).
9. Plate Incubator (KB 115, 28 °C; BD115, 37 °C, Binder GmbH).

2.3 Analysis of Auxin-Induced Degradation of Target Proteins

2.3.1 Liquid Assays

1. 10 mL glass culture tubes (Sigma-Aldrich®, Vienna, Austria).
2. BMD or BMG medium.
3. 1 M 1-NAA (Sigma-Aldrich®, Vienna, Austria) in DMSO.

2.3.2 Spot Assays

1. 10 mL glass culture tubes.
2. 250 mL shake flasks with baffles.
3. Buffered minimal dextrose (BMD) medium or buffered minimal glycerol (BMG) medium.
4. Eppendorf Biophotometer plus.
5. Semi-micro cuvettes.
6. 1.5 mL reaction tubes with canonical bottoms.
7. 96-Well Flat-Bottom Microplate.
8. 48-pin replicator.
9. Glass petri dish for containment of ethanol.
10. Agar plates containing BMD or BMG with either DMSO as a vehicle or 1-NAA to initiate degradation of AID-tagged proteins (*see* **Notes 6** and 7).
11. 1 M 1-naphthaleneacetic acid (1-NAA) (Sigma-Aldrich®, Vienna, Austria): 186 mg of 1-NAA dissolved in 1 mL DMSO. Stored until use at −20 °C (*see* **Note 5**).
12. 28 °C Plate Incubator.

2.3.3 Immunoblotting

1. 10 mL glass culture tubes.
2. 250 mL shake flasks with baffles.
3. 1.5 mL reaction tubes with canonical bottoms.
4. Eppendorf Microcentrifuge 5415 R.
5. 1.85 M Sodium hydroxide (NaOH): Dissolve 14.8 g of NaOH in ddH_2O and store at 4 °C.
6. Lysis solution: Freshly add 75 μL of beta-mercaptoethanol to 925 μL of ice-cold 1.85 M NaOH. Keep on ice until use.
7. 50% Trichloroacetic acid (TCA): 50 g of TCA dissolved in 50 mL ddH_20. Store in the fridge until use.
8. TBST-buffer: 0.25 M Tris-buffered saline (TRIS), 1.5 M NaCl, 0.003% Tween-20, pH 7.5.
9. Blocking buffer: Freshly dissolve 5% skim milk powder in TBST. Store at 4 °C until use (but no longer than 24 h).
10. SDS-PAGE loading dye (e.g., NuPAGE® Loading Dye).

11. SDS-PAGE gel system (e.g., NuPAGE® system: Mini gel tank, 4–12% Bis Tris precast gels; Thermo Fisher Scientific Inc., St. Leon-Rot, Germany).
12. Power Supply (e.g., PowerEase® 300 W, Life Technologies, Carlsbad, CA, United States).
13. Nitrocellulose membrane (e.g., Hybond ECL, GE Healthcare, Buckinghamshire, United Kingdom).
14. MR-12 Rocker-Shaker (BioSan, Riga, Latvia).
15. Primary antibodies: mouse ANTI-FLAG® M2-Peroxidase (HRP) antibody (1:4000 in TBST with 2.5% milk powder) (Sigma-Aldrich®, Vienna, Austria); peroxidase-conjugated anti-HA 3F10 from rat (1:2500 in TBST with 2.5% milk powder) (Roche, Basel, Switzerland); and rabbit anti-GAPDH (1:5000 in TBST with 2.5% milk powder) (Institute of Biochemistry, Graz University of Technology, Austria) [16].
16. HRP-conjugated secondary antibody (goat anti-rabbit, A9169; 1:10,000 dilution in TBST with 2.5% milk powder) (Sigma-Aldrich®, Vienna, Austria).
17. Clarity Max Western ECL Substrate (Bio-Rad, Vienna, Austria).
18. Gel documentation and analysis system (e.g., G:Box, Syngene, Cambridge, UK).

3 Methods

The construction of the plasmids needed for AID-tagging of target genes and the plasmids for expression of *TIR1* has been described in detail by Lehmayer et al., 2022 [12]. Hence, we only give a brief overview (Fig. 1b) of the modifications needed to generate AID-tagged genes in *TIR1*-expressing strains but lay a special focus on the methods applied to test auxin-induced protein degradation.

3.1 Strain Constructions

To obtain optimal conditions, a suitable *TIR1* expression level and dose of auxin required for efficient degradation have to be determined experimentally. Based on our research, the cellular localization dictates both the necessary *TIR1* expression and auxin concentration and therefore varies between different target proteins [12]. In order to determine the ideal level of *TIR1* expression, promoters of different strengths controlling its expression can be tested (*see* **Note 8**). One has to consider that the higher the expression level of *TIR1*, the stronger the basal degradation of the AID-tagged protein will be, which is why a good balance between efficient degradation with minimal basal degradation has to be found. Cytosolic proteins are readily degraded even with low auxin concentration and low *TIR1* expression, whereas transmembrane proteins require a higher dose of both [12]. In case basal degradation is still a problem, it is recommended to test AID2, which employs an OsTIR1^{F74G} mutant and an auxin derivative, 5-Ph-IAA, shows no detectable leaky degradation, and requires a 1000-times lower ligand concentration [12, 17].

3.1.1 AID-Tagging of Genes in K. Phaffii *(See Notes 9–12)*

1. Cut 1 μg of the repair vectors pPpKC2-*GUT1*-3HA and pPpKC2-*GUT1*-AID*-3HA with *Smi*I (*see* **Notes 13** and **14**).
2. Purify linear plasmid DNA by agarose gel electrophoresis followed by gel clean-up of repair cassettes (*Wizard*® SV *Gel* and PCR Clean-Up System). Elute the linear fragment in 40 μL of ddH_2O.
3. Prepare fresh electrocompetent *K. phaffii* cells according to Lin-Cereghino et al. [13].
4. Add 1 μg of purified repair cassette (either pPpKC2-*GUT1*-3HA or pPpKC2-*GUT1*-AID*-3HA) and 100 ng of pPpHyg-Cas9-*GUT1*tt to 80 μL of electrocompetent *K. phaffii* cells and transfer to a 2 mm gap electroporation cuvette (keep on ice).
5. Pulse the cells in a Micro-Pulser™ Electroporator using the program for *Pichia* electroporation (2 kV, 1 pulse). Immediately add 500 μl of 1 M sorbitol and 500 μl of YPD to the pulsed cells.
6. Incubate the cells at 28 °C for 2 h at 500 rpm in a thermomixer.
7. Pellet the cells by brief centrifugation at 2500 rpm, remove the excess supernatant, and resuspend the cell pellet in the remaining liquid.
8. Directly plate the cell suspension on YPD-Hygromycin agar plates to select for transformants containing pPpHyg-Cas9-*GUT1*tt.
9. Incubate plates for 3–5 days at 28 °C.
10. Streak out ~20 colonies on fresh YPD-Hygromycin plates.
11. Verify correct integration of 3HA and AID*-3HA by colony PCR using DreamTaq DNA Polymerase and primers flanking the genomic regions outside of the repair cassette.
12. Cure verified strains from pPpHyg-Cas9-*GUT1*tt by single-colony streak out on YPD plates lacking hygromycin. Repeat twice (*see* **Note 15**).
13. Check for functional expression of tagged *GUT1* on BMG plates and in liquid BMG media (*see* **Note 18**).
14. For the preparation of the strains used in a spot assay, cultivate cells overnight to saturation in 2 mL of BMD.
15. Dilute cells in 50 mL of BMD to an OD_{600} of 0.04.

16. When the cells reach early stationary phase (OD_{600} of 7.0–10.0), dilute to $OD_{600} = 1.0$ with BMD in a 96-well microtiter plate.
17. Perform six times fivefold serial dilutions (6 × 1:5) in BMD.
18. Spot cells of each dilution using a multiprong inoculator on agar plates containing BMD or BMG medium (*see* **Notes 16** and **17**).
19. Let the cell spots dry before turning the plates upside down.
20. Incubate plates at 28 °C for 3–5 days and take images. In the case of functional AID-tagged Gut1, no growth defect should be observed on BMG.

3.1.2 Expression of *TIR1* in K. phaffii

1. Cut 1 μg of pPpT4-P_{PKG1}/P_{TEF2}/P_{HTA1}-*TIR1*-FLAG (*see* **Note 18**) with *Smi*I.
2. Purify linear plasmid DNA by agarose gel electrophoresis (single band on gel at ~5830 bp) followed by gel clean-up (*Gel* and PCR Clean-Up System). Elute the linear fragment in 40 μL of ddH_2O.
3. Prepare fresh electrocompetent *K. phaffii* cells expressing AID- and HA-tagged target genes according to Lin-Cereghino et al. [13].
4. Transform strains with 100 ng of purified, linearized *TIR1* plasmid (*see* **Note 19**) by adding the DNA to 80 μL of electrocompetent *K. phaffii* cells and transferring to a 2 mm gap electroporation cuvette (keep on ice).
5. Pulse the cells in a Micro-Pulser™ Electroporator using the program for *Pichia* electroporation (2 kV, 1 pulse). Immediately add 500 μl of 1 M sorbitol and 500 μL of YPD to the pulsed cells.
6. Incubate the cells at 28 °C at 500 rpm in a thermomixer for 2 h.
7. Pellet cells and plate on YPD-Zeocin agar plates to select for transformants expressing the *TIR1* cassette.
8. Incubate plates for 3–5 days at 28 °C.
9. Streak out ~4 colonies on fresh YPD-Zeocin plates (*see* **Note 20**).
10. To verify expression of *TIR1* by immunoblotting, start by pre-growing the cells overnight to saturation in 2 mL of BMD (*see* **Note 21**).
11. Dilute cells in 50 mL BMD to an OD_{600} of 0.04.
12. When the cells are growing in the middle exponential phase (OD_{600} – 4.0–5.0), harvest 4 OD_{600} equivalents of the culture by centrifugation at maximum speed for 1 min, and remove the supernatant (*see* **Note 22**).

13. For cell disruption, resuspend the cell pellet in 300 μL of lysis solution (*see* **Note 23**) and incubate on ice for 10 min.
14. Add 300 μL of ice-cold 50% TCA, mix by gently inverting 3 times, and incubate on ice for 1 h.
15. Centrifuge at maximum speed at 4 °C for 5 min.
16. Remove supernatant and carefully rinse the tube and the cell pellet with 1 mL ice-cold water (*see* **Notes 24–25**).
17. Resuspend cell pellet in 50 μL of SDS running buffer (*see* **Note 26**).
18. Optional step: Denature protein at 70 °C for 10 min (*see* **Note 27**).
19. Centrifuge at 10,000 rpm for 20 s to pellet any cell debris.
20. Load 12 μL of the soluble fraction onto a NuPAGE SDS gel. Perform the SDS-PAGE and immunoblotting according to the manual provided by the manufacturer.
21. Block the membrane for 4 h at room temperature with blocking buffer.
22. Pour off the blocking buffer and carefully overlay the membrane with TBST. Shake on a Rocker-Shaker at room temperature for 5 min. Pour off the TBST and repeat this **step 4** times.
23. Pour off the TBST and incubate the membrane with primary antibodies (anti-FLAG for detection of TIR1, anti-HA for detection of Gut1, or anti-GAPDH as loading control) overnight at 4 °C on a Rocker-Shaker.
24. Wash the membrane with TBST five times for 5 min (*see* **step 22**).
25. If required, incubate the membrane with secondary antibodies (e.g., anti-rabbit for anti-GAPDH). Otherwise, continue with **step 27**.
26. Wash the membrane with TBST five times for 5 min.
27. Incubate the membrane with chemiluminescence solution for a few seconds and detect proteins using a G:Box.
28. Determine optimal *TIR1* expression by investigating the basal degradation of the target protein, test degradation by incubation with auxin, and determine a suitable auxin concentration.

3.2 Spot Assays of Basal and Auxin-Induced Degradation of Target Proteins

Basal degradation of AID-tagged Gut1 by TIR1 can be assessed by comparing growth on glucose and glycerol (Fig. 2**a**) and by comparing signal intensities on immunoblots (Fig. 2**b**). Moderate to high expression levels of *TIR1* lead to significant basal degradation, even in the absence of auxin. Additionally, the minimal concentration of auxin that is required for rapid and efficient degradation varies between different target proteins and has to be determined

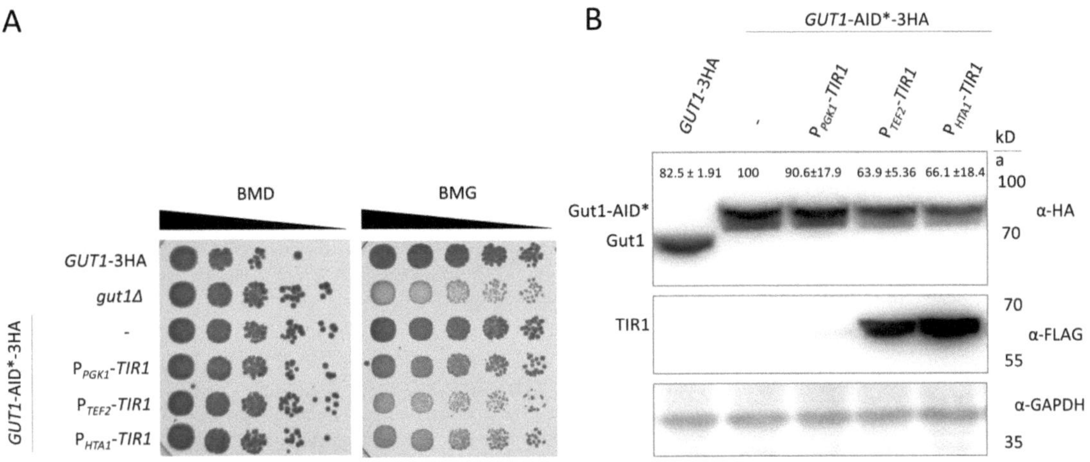

Fig. 2 Functionality and basal degradation of AID-tagged Gut1 and TIR1. (**a**) Spot assays of strains expressing AID*-3HA, 3HA-tagged *GUT1,* and *GUT1*-AID*-3HA strains co-expressing *TIR1* with low, moderate, and high expression levels compared to a *gut1*Δ strain on non-restrictive (BMD) and restrictive (BMG) medium. AID-tagging of Gut1 did not influence its functionality, as no growth phenotype could be observed when growing on BMG. Moderate to high expression of *TIR1* (in *GUT1*-AID*-3HA strains co-expressing P_{TEF2}-*TIR1* or P_{HTA1}-*TIR1*) leads to the same growth defect as a *gut1*Δ, indicating strong basal degradation, even in the absence of auxin. (**b**) Basal degradation of Gut1-AID*-3HA (α-HA) in strains with moderate (P_{TEF1}-TIR1) to high expression of TIR1 (P_{HTA1}-TIR1) is detected by a signal loss of 30–40% on immunoblots. *TIR1* expression from the three different promoters of various expression levels (α-FLAG). (Figure adapted from [12])

experimentally. In the case of Gut1, an auxin concentration of 0.1 mM 1-NAA was sufficient—in combination with the weakest *TIR1* expression—to obtain the same phenotype as a *gut1*Δ strain (Fig. 3a). Although yeast does not seem to be as affected by higher auxin concentrations as mammalian cells [10, 15], the stress caused by auxin should be kept to a minimum.

In order to test basal degradation of Gut1 and to assess the optimal auxin concentration for degradation, spot assays can be performed:

1. Cultivate and dilute strains as described in Subheading 3.1.1, **steps 15–18**.
2. Spot cells of each dilution using a multiprong inoculator on agar plates containing BMD, and BMG supplemented with various auxin concentrations (0.1—1.0 mM 1-NAA). As vehicle controls, also spot on plates supplemented with the same amount of DMSO as used in the respective auxin plates.
3. Incubate the plates at 28 °C for 3–5 days. Two things have to be monitored: (I) Cells co-expressing AID-tagged *GUT1* and *TIR1* should not exhibit any growth defect on BMG, since this would be an indication for basal degradation (Fig. 2a). (II) The perfect auxin concentration can be assessed by comparing the growth of strains that do not exhibit any basal degradation of

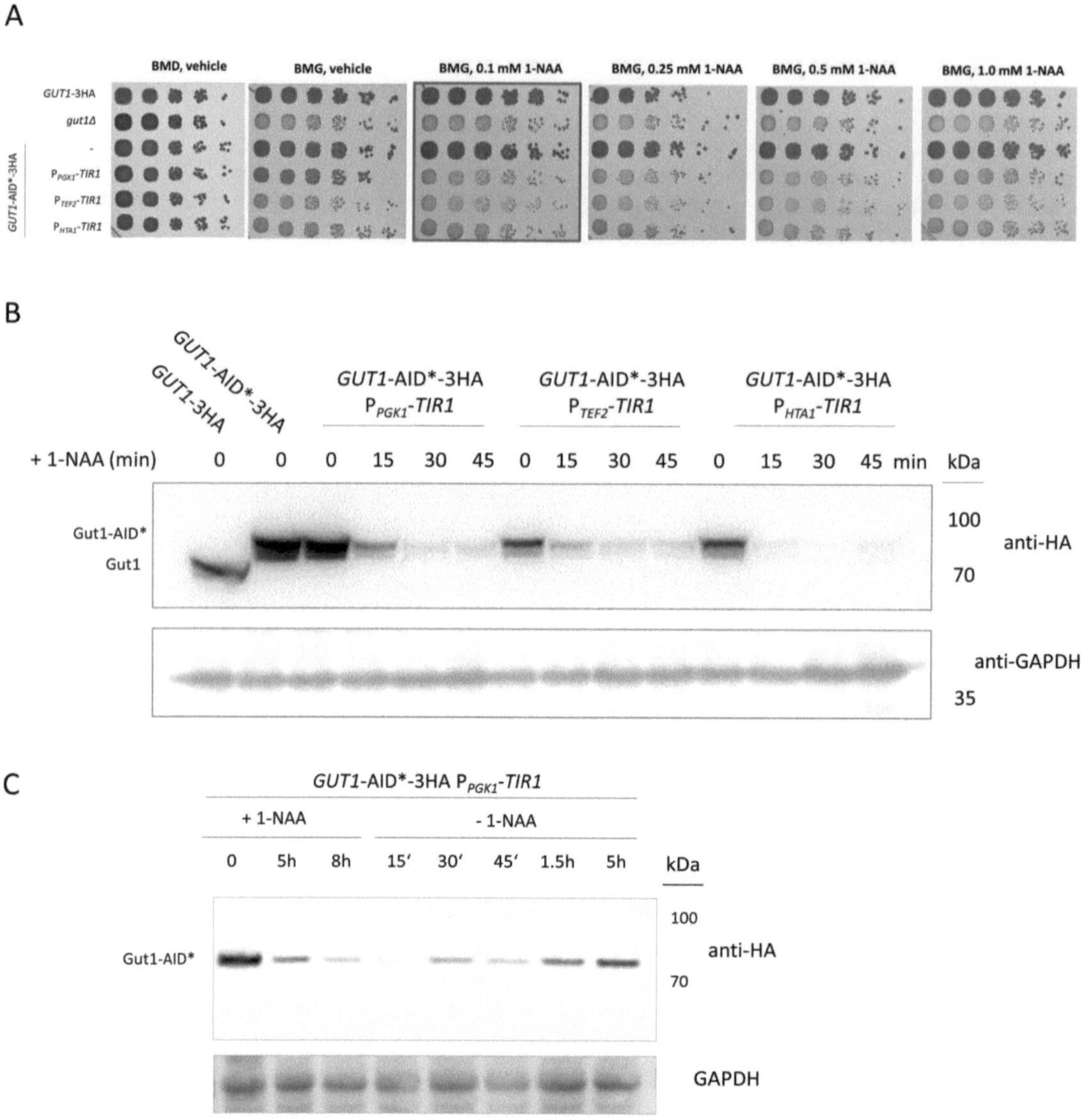

Fig. 3 Auxin-induced degradation of Gut1. (**a**) Spot assays to test different auxin concentrations (0.1 mM–1.0 mM 1-NAA) for degradation of Gut1-AID*-3HA. 0.1 mM 1-NAA supplementation (highlighted with a grey rectangle) is sufficient to result in the same growth defect of the *GUT1*-AID*-3HA P_{PGK1}-*TIR1* strain as in the *gut1*Δ control strain. (**b**) Immunoblot of auxin-induced degradation of Gut1-AID*-3HA strains co-expressing *TIR1* of variable expression strength. Upon incubation with auxin, Gut1-AID*-3HA is depleted within 15–30 min. GAPDH was used as a loading control. (**c**) Reversibility of the AID system. After incubation with auxin, followed by its removal, Gut1-AID*-3HA starts to re-accumulate in the cell. (Figure adapted from [12])

Gut1 on BMG and BMG containing 1-NAA. In the case of Gut1, 0.1 mM 1-NAA was already sufficient to achieve the desired phenotype.

3.3 Time Course of Auxin-Induced Degradation of Target Protein

To monitor the degradation of the tagged protein in strains co-expressing *TIR1*, incubate cells with 1-NAA in liquid medium. Dependent on the level of *TIR1* expression, the protein can be depleted within 15–45 min (Fig. 3b). The reversibility of auxin-inducible protein degradation can easily be tested by washing the cells after auxin incubation with fresh medium followed by immunoblot detection (Fig. 3c).

3.3.1 Monitoring of Degradation by Immunoblot Analysis

1. Prepare cells as described in Subheading 3.1.2, **steps 11–13**.
2. Harvest 4 OD_{600} equivalents of the culture by centrifugation at maximum speed for 1 min, and remove supernatant. This is timepoint 0. Immediately store the cell pellet until use at 4 °C.
3. Add 1-NAA (optimal final concentration can be determined by spot assays as described in Subheading 3.2, **step 1**) (*see* **Notes 28–29**) and incubate the cells at 28 °C and 120 rpm.
4. Harvest 4 OD_{600} equivalents of the culture after 15, 30, and 45 min by centrifugation at maximum speed for 1 min, and remove supernatant.
5. Discard the supernatant and either freeze the cell pellet at –20 °C or directly continue with step 13 of Subheading 3.1.2.

3.3.2 Reversibility of Auxin-Induced Protein Degradation

1. Perform the same **steps 1–3** as described in Subheading 3.3.1, incubate with auxin until the timepoint is reached where all protein is depleted (e.g., 30 min in case of Gut1-AID*-3HA).
2. Harvest 4 OD_{600} equivalents of the culture by centrifugation at maximum speed for 1 min, remove the supernatant, and freeze at −20 °C until further use (this is time point 0).
3. Centrifuge the rest of the cell culture (~ 50 mL) at 2500 rpm for 5 min.
4. Remove the supernatant and gently wash the cell pellet by resuspending it in 50 mL of BMD.
5. Centrifuge again at 2500 rpm for 5 min.
6. Remove supernatant and gently resuspend the cell pellet in 50 mL of BMD.
7. Incubate at 180 rpm and 28 °C and harvest 4 OD_{600} equivalents of the culture by centrifugation at maximum speed for 1 min after incubation for 30 min, 1 h, 2 h, 4 h (*see* **Note 30**).
8. Discard the supernatant and either freeze the cell pellet at –20 °C or directly continue with step 13 of Subheading 3.1.2.

4 Notes

1. *Komagataella phaffii* strain CBS7435 *his4*Δ [13] is derived from the *P. pastoris* wild-type strain NRRL-Y11430 (ATCC 76273). The data presented in this manuscript are exclusively based on *K. phaffii* strain CBS7435 *his4*Δ.
2. pPpHyg-Cas9-*GUT1*tt (pAEA452) is derived from pPpT4-pHTX-PARS1-*Hs*Cas9 [18]. The original Zeocin resistance gene in pPpT4-pHTX-PARS1-*Hs*Cas9 has been replaced by a hygromycin resistance gene by cutting the plasmid with restriction enzymes *Mfe*I and *Nco*I, yielding pPpHyg-Cas9-*GUT1*tt (pAEA452). All plasmid maps can be downloaded as supplementary information from [12].
3. *E. coli* cells replicating pPpHyg-Cas9-*GUT1*tt grow rather slowly on LB-Hygromycin plates because of the weak promoter for expression of the hygromycin resistance gene. This is especially important in case the guide RNA binding site should be replaced to target a different gene (a detailed overview of how to do so is given in the Supplementary Information of Lehmayer et al., 2022 [12]); after cloning and transformation of *E. coli*, it might take 1 ½ to 2 days for *E. coli* colonies to appear on LB-Hygromycin plates.
4. pPpT4-P_{PGK1}-TIR1-FLAG (pAEA454), pPpT4-P_{TEF2}-TIR1-FLAG (pLL002), and pPpT4-P_{HTA1}-TIR1-FLAG (pLL001) were constructed using pPpT4_eGFP_RFP_P_*TEF2/HTA1* [13] as both the vector backbone and the source of the promoters P_{TEF2} and P_{HTA1}. The P_{PGK1} sequence was derived from the *K. phaffii* genome. The *TIR1* sequence, originating from *Oryza sativa*, was derived from the plasmid pMK200 (yeast codon optimized, NBRP ID: BYP7569; Masato Kanemaki). All TIR1-expressing plasmids can be obtained from Addgene.
5. Stock solutions of 500 mM 1-NAA were prepared in DMSO and can be stored at −20 °C for up to 1 month.
6. Minimal media such as BMD or BMG supplemented with auxin have to be buffered (e.g., to pH 6) for cells to grow. YPD medium, on the contrary, does not need to be, but can be buffered [4].
7. The auxin stock solution (and DMSO as vehicle control) should only be added after autoclaving of the medium, since it is not stable under high temperatures and pressure. When auxin is added to the medium, a whitish precipitate appears, which quickly dissolves when mixed.
8. In our studies, we used the following promoters with increasing strength: P_{PGK1} (protein ID CCA37205.1), an

endogenous promoter sequence of low expression strength; P_{TEF2} (protein ID CCA37646.1), a homologue to P_{TEF1} in *S. cerevisiae* with moderate expression strength; and P_{HTA1} (protein ID SCV12077.1), an endogenous promoter sequence of strong expression strength.

9. Always use a strain that does not express *TIR1* for fusing an AID*-tag to your gene of interest in order to be able to check functional expression of the tagged gene without any basal degradation.
10. Make sure to add the AID tag to the cytosolic part of your protein of interest to ensure auxin-induced interaction with TIR1.
11. It is also wise to generate a protein variant that carries only HA-tag, applying the same CRISPR strategy to check whether the AID-tag increases/decreases protein expression, stability, or functionality.
12. When establishing a CRISPR/Cas9-based tagging strategy, look for a PAM and sgRNA targeting sequence in the transcription termination region of the gene of interest and then simply replace this sequence with a different one on your repair cassette (e.g., use the transcription termination sequence of *ARG4* instead). Alternatively, silent mutations can be introduced into the PAM sequence on your repair cassette in order to avoid recutting of the repaired locus.
13. Cutting of pPpKC2-*GUT1*-3HA and pPpKC2-*GUT1*-AID*-3HA with *Smi*I generates two fragments (4465 and 3567, or 3807 bp, respectively). The bigger one represents the vector backbone, and the smaller one represents the repair cassette, which should be excised for purification.
14. *Smi*I creates blunt ends that generate perfect complementary DNA strands, which facilitates integration into the *GUT1* locus.
15. *K. phaffii* cells will readily lose plasmids carrying a selection marker if there is no selection pressure applied [15]. After two single-colony streak-outs on YPD plates, the cells should lose the plasmid. This can be verified by streaking out the cells on YPD-Hygromycin plates, on which they are no longer able to grow.
16. Choose conditions where a phenotype can be observed in case the function of the tagged protein is reduced.
17. Take plates out of the fridge 30–60 min before spotting, let excess water evaporate in a sterile environment, and warm up to 28 °C in an incubator.

18. The plasmid contains regions homologous to the downstream *ARG4* sequence and is thereby efficiently integrated into that locus upon linearization.
19. Do not use too much of the linearized vector to avoid multiple, random integrations into the genome.
20. The transformation and integration of pPpT4-P_{PKG1}/P_{TEF2}/P_{HTA1}-*TIR1*-FLAG cassettes into *K. phaffii* is highly efficient, so only a small number of colonies (3–4) need to be tested for expression of *TIR1* and subsequent auxin-induced degradation of the target protein.
21. A preliminary test for positive transformants can be done by incubating the cells in medium supplemented with auxin, where a degraded target protein would show a phenotype (in the case of Gut1, BMG was used).
22. If not immediately used in further experiments, store harvested cell pellets at −20 °C.
23. Always add β-mercaptoethanol freshly to 1.85 M NaOH, because β-mercaptoethanol is not stable.
24. Make sure to completely remove all of the supernatant, as any liquid residues would lead to variations in total protein concentration.
25. In this step, do not resuspend the pellet. Only carefully rinse the tube to neutralize any residual TCA. Afterward, immediately remove the water using the same pipette tip.
26. The pellet containing cell debris and protein does not readily resuspend. Hence, use a pipette tip to carefully rub the pellet against the plastic tube for proper resuspension. Do not vigorously pipette the pellet up and down, since this causes protein unfolding and massive foam formation.
27. Heat denaturation shall not be applied if investigating transmembrane proteins. In this case, skip denaturation and continue with **step 19**.
28. Make sure to timely take the auxin stock solution out of the freezer, as DMSO takes some time to thaw.
29. Fill up the volume of the cell culture to 50 mL to compensate for potential loss of volume due to evaporation. To do this, the cells can be transferred into a sterile 50 mL plastic tube, and the required volume of BMD can be added before adding auxin (e.g., for a final concentration of 0.5 mM, 50 μL of 500 mM stock solution added to 50 mL cell culture).
30. The required time to reach the same protein expression level as pre-induction with auxin has to be determined experimentally; the named timepoints are just provided as an example.

However, when performing this experiment for the first time, a minimum of 4 h of regeneration should be considered.

Acknowledgments

This work was supported by the BioTechMed-Graz Young Researcher Group Project "StemP" (L. L., L. B., and A. E.-A.). The project was also supported by the COMET center ACIB (A. E.-A.), which is funded by BMK, BMDW, SFG, Standortagentur Tirol, the Government of Lower Austria, and the Vienna Business Agency in the framework of COMET—Competence Centers for Excellent Technologies. The COMET-Funding Program is managed by the Austrian Research Promotion Agency FFG. Also, we want to thank Prof. Anton Glieder and Prof. Harald Pichler for helpful discussions.

Competing Interest Statement The authors have no conflicts of interest to declare that are relevant to the content of this chapter.

References

1. Dohmen RJ, Wu P, Varshavsky A (1994) Heat-inducible degron: a method for constructing temperature-sensitive mutants. Science 263: 1273–1276. https://doi.org/10.1126/SCIENCE.8122109
2. Morano KA, Grant CM, Moye-Rowley WS (2012) The response to heat shock and oxidative stress in *Saccharomyces cerevisiae*. Genetics 190(4):1157–1195. https://doi.org/10.1534/genetics.111.128033
3. Gray WM, Kepinski S, Rouse D, Leyser O, Estelle M (2001) Auxin regulates SCFTIR1-dependent degradation of AUX/IAA proteins. Nature 414:271–276. https://doi.org/10.1038/35104500
4. Nishimura K, Fukagawa T, Takisawa H, Kakimoto T, Kanemaki M (2009) An auxin-based degron system for the rapid depletion of proteins in nonplant cells. Nat Methods 6: 917–922. https://doi.org/10.1038/nmeth.1401
5. Lyapina SA, Correll CC, Kipreos ET, Deshaies RJ (1998) Human CUL1 forms an evolutionarily conserved ubiquitin ligase complex (SCF) with SKP1 and an F-box protein. Proc Natl Acad Sci USA 95:7451–7456. https://doi.org/10.1073/pnas.95.13.7451
6. Willems AR, Schwab M, Tyers M (2004) A hitchhiker's guide to the cullin ubiquitin ligases: SCF and its kin. Biochim Biophys Acta, Mol Cell Res 1695:133–170. https://doi.org/10.1016/j.bbamcr.2004.09.027
7. Natsume T, Kiyomitsu T, Saga Y, Kanemaki MT (2016) Rapid protein depletion in human cells by auxin-inducible Degron tagging with short homology donors. Cell Rep 15:210–218. https://doi.org/10.1016/j.celrep.2016.03.001
8. Camlin NJ, Evans JP (2019) Auxin-inducible protein degradation as a novel approach for protein depletion and reverse genetic discoveries in mammalian oocytes. Biol Reprod 101: 704–718. https://doi.org/10.1093/biolre/ioz113
9. Holland AJ, Fachinetti D, Han JS, Cleveland DW (2012) Inducible, reversible system for the rapid and complete degradation of proteins in mammalian cells. Proc Natl Acad Sci USA 109: E3350. https://doi.org/10.1073/pnas.1216880109
10. Morawska M, Ulrich HD (2013) An expanded tool kit for the auxin-inducible degron system in budding yeast. Yeast 30:341–351. https://doi.org/10.1002/yea.2967
11. Kanke M, Nishimura K, Kanemaki M, Kakimoto T, Takahashi TS, Nakagawa T, Masukata H (2011) Auxin-inducible protein depletion system in fission yeast. BMC Cell

Biol 12. https://doi.org/10.1186/1471-2121-12-8

12. Lehmayer L, Bernauer L, Emmerstorfer-Augustin A (2022) Applying the auxin-based degron system for the inducible, reversible and complete protein degradation in *Komagataella phaffii*. iScience 25:104888. https://doi.org/10.1016/j.isci.2022.104888
13. Näätsaari L, Mistlberger B, Ruth C, Hajek T, Hartner FS, Glieder A (2012) Deletion of the *Pichia pastoris KU70* homologue facilitates platform strain generation for gene expression and synthetic biology. PLoS One 7:e39720. https://doi.org/10.1371/journal.pone.0039720
14. Weninger A, Hatzl AM, Schmid C, Vogl T, Glieder A (2016) Combinatorial optimization of CRISPR/Cas9 expression enables precision genome engineering in the methylotrophic yeast *Pichia pastoris*. J Biotechnol 235:139–149. https://doi.org/10.1016/j.jbiotec.2016.03.027
15. Bernauer L, Radkohl A, Lehmayer LGK, Emmerstorfer-Augustin A (2021) *Komagataella phaffii* as emerging model organism in fundamental research. Front Microbiol 11. https://doi.org/10.3389/fmicb.2020.607028
16. Grillitsch K, Tarazona P, Klug L, Wriessnegger T, Zellnig G, Leitner E, Feussner I, Daum G (2014) Isolation and characterization of the plasma membrane from the yeast *Pichia pastoris*. Biochim Biophys Acta Biomembr 1838:1889–1897. https://doi.org/10.1016/j.bbamem.2014.03.012
17. Yesbolatova A, Saito Y, Kitamoto N, Makino-Itou H, Ajima R, Nakano R, Nakaoka H, Fukui K, Gamo K, Tominari Y, Takeuchi H, Saga Y, Hayashi K, Ichiro, Kanemaki MT (2020) The auxin-inducible degron 2 technology provides sharp degradation control in yeast, mammalian cells, and mice. Nat Commun 11:5701. https://doi.org/10.1038/s41467-020-19532-z
18. Weninger A, Fischer JE, Raschmanová H, Kniely C, Vogl T, Glieder A (2018) Expanding the CRISPR/Cas9 toolkit for *Pichia pastoris* with efficient donor integration and alternative resistance markers. J Cell Biochem 119:3183–3198. https://doi.org/10.1002/jcb.26474

Chapter 7

Establishing Reference Genes for *Pichia pastoris* Quantitative Real-Time Polymerase Chain Reaction Analyses

Kai Büchner, Lana Vidal, Roland Kerpes, and Thomas Becker

Abstract

Quantitative real-time polymerase chain reaction (qPCR) can be used to investigate the relative expression of selected genes under different conditions; hence, it requires a robust endogenous reference consisting of various constitutive genes, also known as housekeeping genes. Thorough scrutiny of several genes under the specific conditions of the later experiments provides the most stably expressed candidate genes. The pairwise comparison and variance analysis of these candidate genes also yield the minimum number of reference genes for reliable RT-qPCR analysis. Therefore, this study selected three reference genes for the normalization and relative quantification of the gene expression of a *Komagataella phaffii* (previously classified as *Pichia pastoris*) wild-type strain using dextrose and methanol as different carbon sources.

Key words *Pichia pastoris*, *Komagataella phaffii*, Real-time quantitative PCR (RT-qPCR), Relative quantification, Reference genes

1 Introduction

Since its development by Higuchi et al. in 1993 [1], the quantitative real-time polymerase chain reaction (qPCR) has developed into a broadly adopted technique. PCR amplification combined with fluorescent signaling allows for the rapid and sensitive determination and quantification of nucleic acids, making it a valuable tool for both the quantification of gene copy numbers in genomic DNA [2] and of gene transcription [3]. Generally, fluorescence signals are detected at the end of the elongation step of each PCR cycle, showing signals increasing in strength with each round of DNA multiplication. The measured fluorescence intensities are plotted as a function of the PCR cycles. The theoretical cycle at which this curve passes a predetermined threshold value is the quantification cycle (Cq) value [4]. Higher Cq values show the later passing of the threshold and a lower abundance of template DNA. The two most

Xavier Garcia-Ortega et al. (eds.), *Pichia pastoris: Methods and Protocols*, Methods in Molecular Biology, vol. 2697,
https://doi.org/10.1007/978-1-0716-4779-0_7,

common detection techniques use either an intercalating fluorescent dye (non-specific detection) [1] or fluorescent probes (specific detection) [5]. Notably, the earliest established fluorescent probes, the hydrolysis probes of the popular TaqMan® system, consist of a fluorophore and a quencher bound to an oligonucleotide specifically binding to the amplified region. The probe is cleaved and displaced when the polymerase reaches this binding site. Through this cleavage, the quencher is removed from the proximity of the fluorophore, and a fluorescence signal becomes detectable. Since both the forward- and reverse primers and the fluorescence probe are distinctly designed for the target gene, this technique ensures high precision and minimizes false-positive signals.

Furthermore, fluorescent probes can be synthesized with fluorophores of different wavelengths, permitting multiplexing, i.e., the detection of multiple different target genes in the same reaction sample [6]. In contrast, intercalating dyes bind non-specifically to double-stranded DNA [7]. This low specificity offers high flexibility in the experimental setup since new targets can be added simply by designing new primers without additional probes; however, this simultaneously prohibits multiplexing. Furthermore, the need to design only two out of three specific oligonucleotides permits a broader range of possible amplicons, thus lowering qPCR specificity. This holds true especially for complementary DNA (cDNA) reverse transcribed from challenging RNA sequences.

In *Komagataella phaffii* research and its associated biotechnological applications, qPCR is predominantly adopted to quantify gene expression. This technique compares a target sample with a calibrator sample, either assessing different strains under the same growth conditions (i.e., wild type as calibrator and a mutant strain as target) or one strain under different growth conditions (i.e., growth phase as calibrator vs. induction phase as target). In these applications, qPCR offers distinct advantages over more global techniques like RNA-seq; next to lower costs per sample, qPCR instruments are more prevalent in labs than sequencing platforms [8]. Also, the fewer data generated enable a streamlined evaluation process of a set of known genes of interest (GOI). Furthermore, due to the technique's extensive application, it is of interest to the research community to report qPCR findings in an easily accessible way. Therefore, the MIQE (Minimum Information for Publication of Quantitative Real-Time PCR Experiments) guidelines were published to unify the reporting style for qPCR experiments, providing a checklist with essential and desirable information [4]. These guidelines also propose a consensus terminology that is used throughout this protocol.

Relative expression studies comprise RNA extraction, cDNA synthesis, and qPCR. They use cDNA synthesized from mRNA as the PCR template. In the first-strand cDNA synthesis, a reverse transcriptase elongates a primer annealed to the RNA in the $5' \rightarrow 3'$

direction, using deoxynucleotides. Random hexamer primers, oligo (dT) primers, and anchored oligo(dT) primers are the most widely applied. The latter two anneal to the 3′-poly(A) tail, selecting for mRNA in the process. Oligo(dT) primers consist of a string of deoxythymidylic acid (dT) residues. In anchored primers, these are followed by two additional nucleotides represented by VN, where V is dA, dC, or dG and N is dA, dC, dG, or dT. Due to this anchor, the primer anneals only at the 5′ end of the mRNA's poly(A) tail, leading to more efficient cDNA synthesis [9, 10].

To render qPCR analysis results free from influences such as different cell counts in the samples or varying efficiency and purity of RNA extractions, the expression data is normalized to reference genes as internal controls [11]. The expression of these genes is assumed to be unaffected by the specific growth rate and other experimental conditions, which are not necessarily provided under all circumstances. Especially with methanol-induced *K. phaffii* production strains, the carbon source switch vastly influences gene expression [12]. Furthermore, ectopically integrating expression cassettes affects gene expression. While gene copy number is the primary influence on expression levels [13], *K. phaffii's* propensity for integration through non-homologous end joining can also induce a high clonal variation [14]. Additionally, a microarray study in different human cancer cell lines revealed that among an array of 8000 genes, no target exhibited an expression ratio that varied by less than twofold [15]. This variety of gene expression highlights the importance of a robust reference gene system adapted to the strains of the specific experiment.

Whether normalization through one or more reference genes is chosen directly influences which model can be applied to assess relative expression experiments. The following clause illustrates the evolution of formulas to convert Cq values into normalized relative quantities (NRQ), depicting the change in gene of interest (GOI) expression between the target and calibrator sample. The development toward a robust way to calculate NRQs resulted in an approach including multiple reference genes. The earliest method, established by Livak and Schmittgen, assumes a similar efficiency of 100% for target and reference primers (formula (1)). This amplicon duplication with every PCR cycle is included through the value 2 as the function's base. The value ΔCq_{target} in the exponent considers the relative amount of the GOI in the target sample, normalized to a single endogenous reference gene. This normalized target sample is then put in relation to an equally normalized calibrator sample ($\Delta Cq_{calibrator}$). Due to the same base in the exponential functions of the target and calibrator samples, the power law allows summarizing the fraction to a difference, as shown in formula (1) as $\Delta\Delta Cq$ [16].

$$NRQ = \frac{2^{-\Delta Cq_{target}}}{2^{-\Delta Cq_{calibrator}}} = 2^{-(\Delta Cq_{target} - \Delta Cq_{calibrator})} = 2^{-\Delta\Delta Cq} \quad (1)$$

Pfaffl further developed this model to consider differences in primer efficiencies (E) of the GOI and the reference gene [11].

$$NRQ = \frac{E_{GOI}^{(Cq_{calibrator} - Cq_{target})}}{E_{ref}^{(Cq_{calibrator} - Cq_{target})}} = \frac{E_{GOI}^{\Delta Cq_{GOI}}}{E_{ref}^{\Delta Cq_{ref}}} \quad (2)$$

However, Vandesompele et al. showed that normalization to a single reference gene led to a 3.0- and 6.4-fold erroneous normalization in 25% and 10% of investigated cases, respectively [17]. Therefore, Hellemans et al. extended the model through the geometric mean of a reference gene system to cater for this instability of single endogenous controls [18]. This is shown as the denominator in formula (3), with f as the total number of reference genes.

$$NRQ = \frac{E_{GOI}^{\Delta Cq_{GOI}}}{\sqrt[f]{\prod_0^f E_{ref}^{\Delta Cq_{ref}}}} \quad (3)$$

This denominator includes the endogenous control genes expressed most stably under the given experimental conditions. The following protocol describes a screening process for detecting suitable genes.

Around ten candidate genes are typically assessed to identify robust reference genes in the experimental system at hand [19]. One of the most popular algorithms to calculate expression stability is geNorm [17]. In this algorithm, the M value is calculated, representing the expression stability of the candidate reference genes. In short, for every combination of a candidate gene (j) with any other candidate gene (k), the pairwise variation V_{jk} is determined. The stability measure for this gene (M_j) results from the arithmetic mean of all pairwise variations. Higher M_j values signify higher average variation and, therefore, lower expression stability of gene j compared to the other candidate genes. A calculation of the variation $V_{n/n+1}$ also determines the minimum number of reference genes needed: beginning with the most stable two candidate genes, the variation of n and $n + 1$ (i.e., the best two vs. the best three genes in the first iteration) is calculated. As a general guideline, using an extra ($n + 1$)th reference gene is unnecessary when the $V_{n/n+1}$ drops below a 0.15 cutoff value.

The literature provides an extensive catalog of established reference genes for prevalent model organisms, like the yeast *Saccharomyces cerevisiae* [20, 21], that can be used as a starting point for evaluating *K. phaffii* endogenous controls [22]. Table 1 shows the nine putative reference genes used in this protocol. A prudent

Table 1
Reference gene primer sequences with amplification efficiencies E_{ref} and geNorm stability rank

Name	Sequence	Amplicon / bp	E_{ref}	geNorm rank	Function of the target
ACT1-FWD **ACT1-REV**	AGGCTTCTGGTATCGACCAAAC AGTAGTACCACCGGACATAACG	104	2.016	3	ACTin, structural protein of the cytoskeleton
ALG9-FWD ALG-REV	TGGACCAGTTTGCTTCAGATACTC CGTAGACGCATAGAAGGAGGTAG	100	1.985	5	Asparagine Linked Glycosylation, mannosyltransferase, involved in N-linked glycosylation
TAF10-FWD **TAF10-REV**	ACAACCTCAAACGGGAGCTTC TAGTGCCCGTACTTGTTCCTC	89	2.161	1	TATA-binding protein-Associated Factor, involved in transcription initiation and in chromatin modification
TDH3-FWD TDH3-REV	CGACGCTTCTGCCGGTATTC TCGACGACTCTGGTGGAGTAAC	93	1.998	8	Triosephosphate DeHydrogenase, involved in glycolysis and gluconeogenesis
TEF1-FWD **TEF1-REV**	ACTGACAAGCCTCTGAGATTGC GCCTTGATGACACCGGTCTC	98	1.965	2	Translational Elongation Factor, involved elongation of nascent polypeptides at the A site of the ribosome
TFC1-FWD TFC1-REV	TGGTGATACCTTCCCATCAGTTG CCATAACGCTCAGAAGGCTTCTC	114	1.958	6	Subunit of RNA polymerase III Transcription initiation Factor Complex
TPI1-FWD TPI1-REV	TCACCGAGAACAAGCAGAAG TAGGCACCAGAAGCTTTGTC	70	1.902	4	Triose Phosphate Isomerase, an abundant glycolytic enzyme
TRS65-FWD TRS65-REV	TTTGGGCAGGAAGCAAGACC AGATGCTGAGTCCTGCGAATG	101	1.999	7	TRansport protein particle (TRAPP) complex II Subunit; involved in intra-Golgi traffic regulation
UBC6-FWD UBC6-REV	TACCGCCCATCCTAGTGAATC TCAGGAGGACCATCAATGAC	63	2.117	9	UBiquitin-Conjugating enzyme, involved in endoplasmic reticulum associated degradation

The three primer pairs making up a stable set of reference genes are marked in bold. The geNorm rank is only valid for the experiment described in this chapter and must be reevaluated for other experimental conditions. Letters in the protein or function names, from which the gene names are derived, are written in capitals and underlined measure is a wide distribution of candidate genes over diverse cellular structures and metabolic pathways, with preferably only one gene per pathway. This way, effectors influencing a single pathway affect a minimum number of reference genes. The gene *ACT1*, which codes for a structural protein of the cytoskeleton, was chosen as it is among the most popular reference genes [23]. Other chosen genes were e.g. *UBC6*, coding for a ubiquitin-conjugating enzyme involved in endoplasmic reticulum associated protein degradation; *TDH3*, coding for a glyceraldehyde-3-phosphate dehydrogenase; and *TAF10*, the product of which is involved in RNA polymerase II transcription initiation [20]. Another popular endogenous control, ribosomal RNA, was omitted as a reference due to its high abundance relative to mRNA [17] and due to it not being reverse transcribed with the anchored oligo(dT) primers chosen for this protocol.

2 Materials

2.1 Cultivation

1. Any *P. pastoris* strain of interest. This study was done using the *Komagataella phaffii* BSYBG11 (Δ*aox1,* mut^S) strain (bisy GmbH, Hofstätten an der Raab, Austria).
2. BMD1 medium: 1% dextrose, 1.34% YNB, 4×10^{-5}% biotin, 200 mM Potassium phosphate buffer pH 6.0.
3. BMM2 medium: 1% methanol, 1.34% YNB, 4×10^{-5}% biotin, 200 mM Potassium phosphate buffer pH 6.0.
4. Sterile 50 mL centrifugation tubes.
5. Sterile baffled 500 mL shake flasks, covered with cheesecloth.
6. Shaking incubator set to 28 °C, 130 rpm.

2.2 RNA Extraction and Reverse Transcription

1. 5% (v/v) phenol in ethanol (purity >95%; cooled on ice).
2. TRIzol™ Reagent (Thermo Fisher Scientific Inc., Waltham, USA).
3. DEPC-treated $_{dd}H_2O$: Prepare five 10 mL aliquots of $_{dd}H_2O$, add 1:1000 diethyl pyrocarbonate (DEPC), shake to disperse droplets evenly, and incubate overnight at RT. Autoclave twice in the liquid program.
4. Chloroform, U.S.P.
5. Dimethylformamide, MQ 200 (*see* **Note 1**).
6. Formaldehyde, 37%, U.S.P.
7. RNA precipitation solution: 1.2 M NaCl, 0.8 M disodium citrate. pH adjustment is not necessary.
8. 75% (v/v) ethanol: 11.70 mL 96% ethanol molecular biology grade, 3.3 mL DEPC-treated $_{dd}H_2O$.

9. 10× MOPS buffer: 8.38 g MOPS (3-(N-morpholino)propanesulphonic acid, 1.64 g sodium acetate trihydrate, 0.744 g EDTA (ethylenediaminetetraacetic acid, disodium salt). Add 160 mL $_{dd}H_2O$, adjust to pH 7.0 with 10 M NaOH, and fill up to 200 mL total volume. Sterile filter into an autoclaved flask using a 0.2 μm syringe filter.
10. Reducing 1.5% agarose gel: 0.6 g agarose, 28.8 mL DEPC-treated $_{dd}H_2O$, heat until molten; then add 4 mL 10× MOPS buffer, 7.2 mL formaldehyde, and 3 μL Midori green, and cast.
11. 6× RNA loading buffer (500 μL bidest. Glycerol, 1.46 mg EDTA dissolved in 500 μL DEPC-treated $_{dd}H_2O$ and adjusted to pH 8.0 with NaOH).
12. Oligo d(T)$_{23}$ VN (New England BioLabs Inc., Ipswich, USA).
13. ProtoScript® II M-MuLV reverse transcriptase (New England BioLabs Inc., Ipswich, USA).
14. Glass beads Ø 0.5 mm, covered in 5.8 M HCl for 1 h at room temperature, then rinsed 10 × 10 min with $_{dd}H_2O$, autoclaved, and dried.
15. TissueLyser II with Adapter Set 2 × 24 (Qiagen N.V., Venlo, Netherlands).
16. NanoDrop 2000c spectrophotometer (Thermo Fisher Scientific Inc., Waltham, USA).
17. Filter tips, RNase free.

2.3 RT-qPCR Primer Validation and Selection of Reference Genes

1. Oligonucleotide primers, 10 μM (Table 1).
2. Biozym Blue S'Green qPCR Kit Separate ROX (Biozym Scientific, Hessisch Oldendorf, Germany).
3. 96 well PCR plate, white, LC 480 (Biozym Scientific, Hessisch Oldendorf, Germany).
4. Adhesive clear qPCR seal (Biozym Scientific, Hessisch Oldendorf, Germany).
5. LightCycler 480 II (Roche, Basel, Switzerland).
6. Filter tips, RNase free.

2.4 Software

1. Primer3web, version 4.1.0 (Whitehead Institute for Biomedical Research, Cambridge, Massachusetts, USA. https://primer3.ut.ee/) [24].
2. qbase+ software, version 3.0 (Biogazelle, Zwijnaarde, Belgium. www.qbaseplus.com) [18].

3 Methods

3.1 Cell Cultivation and Sampling

Generally, strain cultivation should be carried out in biological triplicates under the conditions of the later experiment. Here, sampling times were 24 h post-inoculation and 20 h post-induction through methanol addition. All metabolic activity needs to be quenched instantly upon cell harvest to generate precise data displaying the cells' state at the sampling time point [25].

1. Prepare three independent seed cultures by inoculating 5 mL BMD1 medium in 50 mL centrifugation tubes with conical bottom with a freshly grown *K. phaffii* colony each.
2. Incubate the cultures overnight at 28 °C and 130 rpm in a slightly inclined position. The lid should not be screwed on fully to allow free gas diffusion and should be fixated with an adhesive strip.
3. Fill three sterile 500 mL baffled shake flasks with 25 mL BMD1 medium, inoculate to an OD_{600} of 1.0, and close the flasks with a layer of sterile cotton fabric.
4. Incubate the cultures in shake flasks for 48 h at 28 °C and 130 rpm.
5. About 10 min before each sampling time point, fill one 15 mL conical tube per shake flask with 2 mL of ice-cold 5% (v/v) phenol in ethanol solution and store on ice.
6. Take the first sample 24 h after inoculation under sterile conditions. Add 4 mL of the cell suspension to a prepared tube from the previous step; mix by inverting.
7. For induction conditions, add 25 mL BMM2 and incubate under the same conditions as before. Take the second sample 20 h after the addition of the BMM2 medium.
8. Transfer 2 mL aliquots of the solution to microcentrifuge tubes and spin down for 5 min at 4 °C and 4500× *g*.
9. Carefully remove the supernatant. The pellets are best used freshly but can also be stored at −80 °C until further use.

3.2 RNA Extraction and cDNA Synthesis

The use of gloves during this step is essential since a) RNA is prone to degradation (among others, through ribonucleases from human hair and skin) and b) the monophasic lysis reagent (MLR) TRIzol™, used for the extraction, is a potent skin irritant (*see* **Note 2**). Here, cells are disrupted mechanically with glass beads while suspended in MLR. [26].

1. Pre-cool the TissueLyser II adapter set overnight at −20 °C.
2. Add 1 mL TRIzol™ per 100 mg wet cell weight of yeast pellet. (*see* **Note 3**).

3. Resuspend the pellet by pipetting up and down slowly.
4. Add about 500 μL of acid-washed glass beads to the cell suspension, measured by liquid displacement. (*see* **Note 4**).
5. Disrupt cells in a Qiagen TissueLyser II for 5 min at 30 Hz.
6. Incubate the samples for 5 min at room temperature to allow the dissociation of nucleoprotein complexes.
7. Add 200 μL chloroform, vortex briefly, and incubate for approximately 5 min at room temperature.
8. Centrifuge at 4 °C and 13,000× *g* for 15 min to separate the suspension into three distinct phases: the upper, colorless, aqueous phase contains RNA, the white interphase contains DNA, and the bottom, red, organic phase contains proteins.
9. Prefill fresh RNase-free 1.5 mL microcentrifuge tubes with 250 μL isopropanol and 250 μL RNA precipitation solution [27]. Aspirate the upper phase of the samples from **step 8** carefully without disturbing the interphase, and add it to the prepared tubes. Mix by inverting the tubes several times.
10. Incubate the samples at 4 °C for 15 min.
11. Centrifuge the solution at 4 °C and 13,000× *g* for 15 min and aspirate the supernatant with a pipette. (*see* **Note 5**).
12. Wash the pellet by adding 1 mL of 75% ethanol and flick the tube a few times.
13. Centrifuge the solution at 4 °C and 13,000× *g* for 5 min and aspirate the supernatant with a pipette.
14. Repeat **steps 12** and **13**.
15. Briefly air-dry the RNA pellet for 5–10 min by placing the open tubes sideways on an RNase-free surface. (*see* **Note 6**).
16. Add 50 μL of DEPC-treated $_{dd}H_2O$ and dissolve the pellet by incubation at 65 °C for about 10 min.
17. Quantify the RNA in solution using a NanoDrop 2000c spectrophotometer.
18. Control RNA quality by running a denaturing 1.5% (w/v) agarose gel in 1× MOPS buffer. For denaturation of RNA before the gel electrophoresis, add 4 μL RNA solution, 2 μL 10x MOPS buffer, 4 μL 5× RNA loading buffer, and 10 μL dimethylformamide (*see* **Note 1**). Incubate the sample at 90 °C for 2 min and put it on ice directly afterward. Then load the gel with up to the total sample volume (depending on the size of the cavities) and run it for 90 min at 70 V. (*see* **Note 7**) Fig. 1 shows the ribosomal RNA bands visible in intact RNA.
19. Synthesize cDNA using NEB ProtoScript® II M-MuLV reverse transcriptase according to manufacturer's instructions [28] with 1 μg of RNA and anchored $d(T)_{23}$ VN primers.

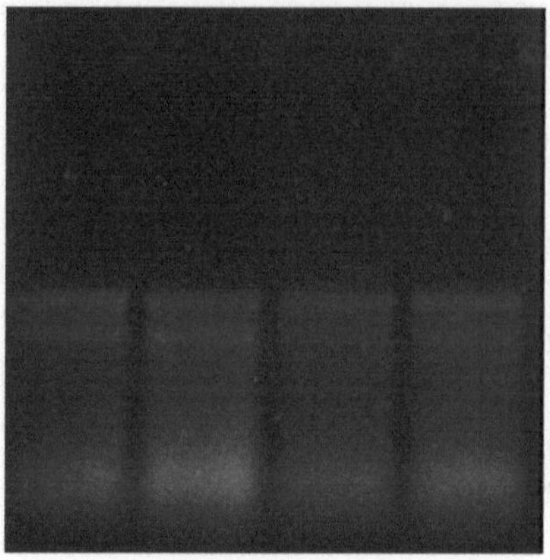

Fig. 1 Quality assessment of extracted RNA in a denaturing agarose gel. Clear 25S and 18S ribosomal RNA (rRNA) bands signify minimal sample degradation. Sharp separation of 5.8S and 5S bands is often influenced by the agarose gel itself

Additionally, prepare a no reverse transcription (No-RT, RT for "reverse transcription" in this context is not to be confused with RT for "real-time" in RT-qPCR) control in the later RT-qPCR run by adding water instead of the reverse transcriptase. The cDNA can be stored at −20 °C for several months; avoid repeated freeze-thaw cycles.

3.3 Primer Validation and Determination of Efficiency

The typical RT-qPCR experiment comprises three stages: i) primer design, validation, and determination of primer efficiency for the specified investigation; ii) establishment of a reference gene system; and iii) investigation of the target gene expression relative to a control population, both normalized to their reference genes. The validation of the correct amplification of the target is an important quality control step for all RT-qPCR primers. Individual amplification efficiencies E_{gene} can be used in later experiments for a more precise calculation of relative expression levels using the qbase + algorithm.

Although the practical approach to validation is the same for all RT-qPCR primers, this protocol focuses on establishing a reference gene system. This is due to the data analyses requiring plate setups different from those needed for investigating relative gene expression, i.e., dilution series. Inquiry of expression levels depends heavily on the targeted genes and the expression conditions. Therefore, the primer pairs presented in Table 1 must be reevaluated under the respective experimental conditions following the protocol below.

1. Select RT-qPCR targets from housekeeping genes in the literature [20, 22].

2. Design primer pairs based on the corresponding gene sequences and optimize for a melting temperature (Tm) of 60 °C (calculated in Primer 3 software according to the PCR master mix instructions). The amplicons should have a total length of 95–120 bp. Table 1 shows the primers corresponding to the selected putative reference genes (*see* **Note 8**). Since reverse transcription starts at the poly(A) tail, primers should be located close to the 3′ end of the target sequence. This design renders the amplification more robust against prematurely terminated reverse transcription.
3. Prepare a fivefold serial dilution of one cDNA sample, covering five points in total, with cDNA from the undiluted reverse transcription mix as the first step. Take up 7 μL of the RT mix and add it to 28 μL sterile $_{dd}H_2O$ in a fresh tube by slowly pipetting up and down. In the same way, mix 7 μL of the previous dilution with 28 μL sterile $_{dd}H_2O$. Especially abundant targets can also be measured using a tenfold dilution series.
4. For each primer pair, prepare mix 1 from 100 μL of 2× qPCR S'Green BlueMix and 8 μL of 10 μM forward and reverse primer. Prepare mix 2 from 18 μL cDNA for every dilution step and 133.2 μL autoclaved $_{dd}H_2O$.
5. Pipette 5.2 μL of mix 1 in triplicates for every primer pair in an RT-qPCR plate (*see* **Note 9**).
6. Add 4.2 μL of mix 2 to get triplicates of every combination of primers and cDNA dilution, as shown in Fig. 2 (*see* **Note 10**).
7. As a No-RT control, prepare mix 2 as before and add water instead of cDNA. Any signal will originate from residual gDNA in the RNA extraction.
8. Seal the plate and spin down briefly at 500× *g* and room temperature for 30 s.

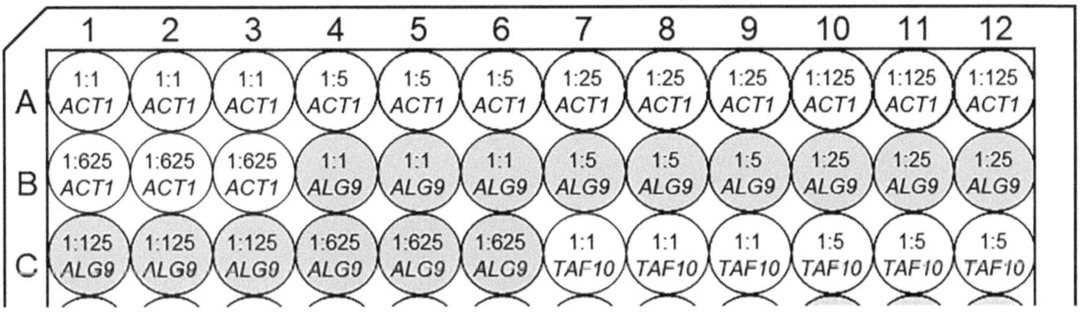

Fig. 2 Plate setup RT-qPCR of cDNA dilution rows to determine primer efficiencies

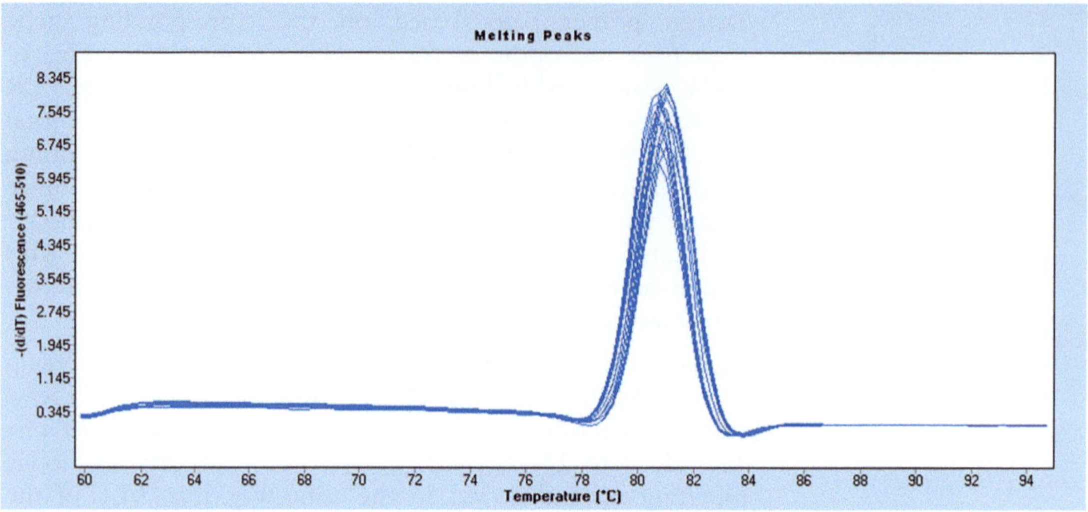

Fig. 3 Melting peaks of the TRS65 amplicons after a 5-point dilution series RT-qPCR. Pure amplification of the target leads to unimodal (single-peaked) curves with the peaks centered around the melting temperature. Undesired side products would show as additional peaks or shifts in the peak location on the X-axis, indicating unsuitable primers for the target

9. Place the plate in the LightCycler 480 II and run the RT-qPCR at the following conditions: 95 °C/2:00 min – (95 °C/5 s – 60 °C/30 s) × 55 – 60 °C to 95 °C/1 min (gradient).
10. Control for primer specificity via melting curve analysis. As shown in Fig. 3, the melting peaks of the amplicons should be unimodal and centered around their melting temperature. Additionally, after the PCR, run a sample of each primer pair reaction on a 2% TAE agarose gel (75 min at 85 V) with a ladder appropriate to the small amplicon size. Take samples from the wells with the highest template concentration; only a single band should be visible.
11. Calculate primer efficiency using Roche Light Cycler® II 480 Software release 1.5.0 SP4; the efficiency should be between 1,8 and 2,2 (90–110%). (*see* **Note 11**).

3.4 RT-qPCR Selection of Reference Genes

Use all sufficiently efficient and specific primer pairs for RT-qPCR reactions with cDNA templates from both sampling time points for reference gene selection. This experiment's set of nine primers leads to 120 reactions, including controls. To determine a set of reference genes, no less than eight primers should be analyzed.

1. For each primer pair, prepare mix 1 from 100 μL of 2× qPCR S'Green BlueMix and 8 μL of 10 μM forward and reverse primer. Prepare mix 2 from 18 μL cDNA generated from each sampling time point and biological replicate and 133.2 μL autoclaved $_{dd}H_2O$.

2. Pipette 4.2 μL of mix 2 in duplicates for every sampling time-point and biological replicate, with each of the six samples filling two columns.
3. Add 5.8 μL of mix 1 to the plate with one primer pair per row.
4. As No-RT control, prepare mix 2 as before and add water instead of cDNA. Include an No-RT control for every template, pipetted in duplicates. Any signal will originate from residual gDNA in the RNA extraction (*see* **Note 12**).
5. Seal the plate and spin down briefly at 500× *g* and room temperature for 1 min.
6. Place the plate in the LightCycler 480 II and run the RT-qPCR at the following conditions: 95 °C/2:00 min – (95 °C/5 s – 60 °C/30 s) × 50 – 60 °C to 95 °C/1 min (gradient).
7. Control for specific amplification via melting curve analysis.
8. Enter the detected Cq values into the qbase+ software and run the geNorm analysis using the program's wizard. geNorm will rank the stability of reference gene expression (Fig. 4) and calculate the minimum number of genes needed for a sufficiently stable internal control (Fig. 5). In these calculations, a lower geNorm M value signifies less variation in gene expression under the different experimental conditions and, therefore, a more stable gene.

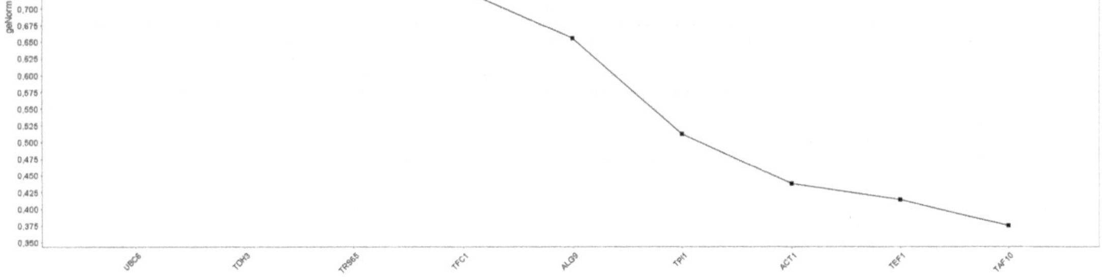

Fig. 4 geNorm analysis showing the expression stability of the target genes. A lower geNorm M indicates higher stability, i.e., more consistent expression under different experimental conditions

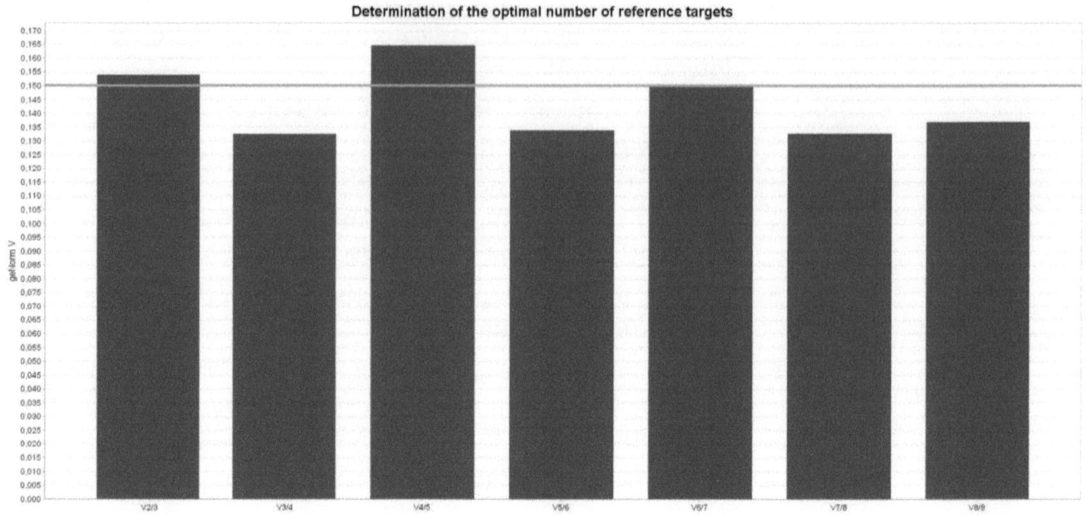

Fig. 5 geNorm analysis illustrating the optimal number of reference genes for stable internal control. The algorithm iterates through, comparing variations (geNormV) of the most stable n versus the most stable $n + 1$ genes, starting with two versus three (V2/3). Here, a minimal number of three genes is needed (from V3/4) for a geNorm V below the generally accepted 0.150 threshold.

4 Notes

1. Instead of dimethylformamide, formaldehyde can be used.
2. Use filter tips throughout this protocol's RNA extraction and RNA handling steps. This will protect pipettes against corroding reagents and the samples against contamination with RNases.
3. Sampling as described here will lead to ~25 mg/mL wet cells in quenching solution. Pool pellets from the same biological replicate to reach 80–100 mg pellet weight in total before extraction.
4. Seal the tubes with Parafilm and label them both on the lid and the sidewall. Be cautious of glass beads wedged between the inner rim of the lid and the tube, as they may lead to TRIzol™ leaking from the tubes during cell disruption.
5. Remove the bulk of the supernatant using a 1000 μL tip. Spin down any residual liquid and remove it using a 20 μL tip.
6. Overdried RNA pellets are hard to solve in the following steps. Storing the RNA pellet and solute overnight at −80 °C can aid resuspension.
7. The organic solvent content of the gel makes it softer than TAE or TBE gels of the same agarose concentration. Pre-cool the running buffer overnight to 4 °C to prevent overheating.

8. Sequences were retrieved from the NCBI nucleotide database [29] with the search term architecture "(gene product name) AND '*Komagataella phaffii*'[porgn:__txid644223]" or in the annotation data of strain CBS7435 at pichiagenome.org [30].
9. White wells are preferable to clear ones in RT-qPCR plates since they enhance the fluorescence signals.
10. Always follow good pipetting practices when pipetting RT-qPCR mixes and plates. This includes flushing pipette tips with the solution by pipetting up and down once before aspiration, allowing the pressure in the pipette to equalize by resting the tip shortly in the solution after aspirating it, and holding the pipette tip inclined with contact to the side of a well when ejecting the liquid.
11. Primer efficiency should lie between 90% and 110% for the experiment. From formula (5), a 100% efficiency equals a slope of −3.32, making −3.91 to −2.92 an acceptable range.

$$Efficiency\ (\%) = \left(10^{\frac{-1}{slope}} - 1\right) \times 100 \tag{5}$$

Convert primer efficiencies according to formula (6) to calculate the NRQ.

$$E_{gene} = \left(\frac{Efficiency\ (\%)}{100}\right) + 1 \tag{6}$$

12. The selection of reference genes generates a total of $n = 2 \times 3 \times 2 \times 9 = 108$ reactions from two time points, three biological and two technical replicates, and nine candidate genes without No-RT control. As the number of reactions surpasses 96, it is advisable to run one identical reaction (with the same primers and cDNA template, in triplicate, pipetted to the same well numbers) on every plate to account for variation between runs.

Competing Interest Statement

The authors have no conflicts of interest to declare that are relevant to the content of this chapter.

References

1. Higuchi R, Fockler C, Dollinger G et al (1993) Kinetic PCR analysis: real-time monitoring of DNA amplification reactions. Nat Biotechnol 11(9):1026–1030. https://doi.org/10.1038/nbt0993-1026
2. Abad S, Kitz K, Hörmann A et al (2010) Real-time PCR-based determination of gene copy numbers in Pichia pastoris. Biotechnol J 5(4): 413–420. https://doi.org/10.1002/biot.200900233

3. Gause WC, Adamovicz J (1994) The use of the PCR to quantitate gene expression. Genome Res 3:S123
4. Bustin SA, Benes V, Kubista M et al (2009) The MIQE guidelines: minimum information for publication of quantitative real-time PCR experiments. Clin Chem 55(4):611–622
5. Holland PM, Abramson RD, Watson R et al (1991) Detection of specific polymerase chain reaction product by utilizing the 5′- 3′ exonuclease activity of Thermus aquaticus DNA polymerase. Proc Natl Acad Sci 88(16):7276–7280. https://doi.org/10.1073/pnas.88.16.7276
6. Henegariu O, Heerema NA, Dlouhy SR et al (1997) Multiplex PCR: critical parameters and step-by-step protocol. BioTechniques 23(3): 504–511. https://doi.org/10.2144/97233rr01
7. Nafisi S, Saboury AA, Keramat N et al (2007) Stability and structural features of DNA intercalation with ethidium bromide, acridine orange and methylene blue. J Mol Struct 827(1–3):35–43. https://doi.org/10.1016/j.molstruc.2006.05.004
8. Nonis A, de Nardi B, Nonis A (2014) Choosing between RT-qPCR and RNA-seq: a back-of-the-envelope estimate towards the definition of the break-even-point. Anal Bioanal Chem 406(15):3533–3536. https://doi.org/10.1007/s00216-014-7687-x
9. Khan AS, Wilcox AS, Hopkins JA et al (1991) Efficient double stranded sequencing of cDNA clones containing long poly(A) tails using anchored poly(dT) primers. Nucleic Acids Res 19(7):1715
10. Liao J, Gong Z (1997) Sequencing of 3' cDNA clones using anchored oligo(dT) primers. BioTechniques 23(3):368–370. https://doi.org/10.2144/97233bm03
11. Pfaffl MW (2001) A new mathematical model for relative quantification in real-time RT-PCR. Nucleic Acids Res 29(9):e45. https://doi.org/10.1093/nar/29.9.e45
12. Vogl T, Sturmberger L, Kickenweiz T et al (2016) A toolbox of diverse promoters related to methanol utilization: functionally verified parts for heterologous pathway expression in *Pichia pastoris*. ACS Synth Biol 5(2): 172–186. https://doi.org/10.1021/acssynbio.5b00199
13. Vogl T, Gebbie L, Palfreyman RW et al (2018) Effect of plasmid design and type of integration event on recombinant protein expression in *Pichia pastoris*. Appl Environ Microbiol 84(6). https://doi.org/10.1128/AEM.02712-17
14. Schwarzhans J-P, Wibberg D, Winkler A et al (2016) Non-canonical integration events in *Pichia pastoris* encountered during standard transformation analysed with genome sequencing. Sci Rep 6(1):38952. https://doi.org/10.1038/srep38952
15. Ross DT, Scherf U, Eisen MB et al (2000) Systematic variation in gene expression patterns in human cancer cell lines. Nat Genet 24:227–235
16. Livak KJ, Schmittgen TD (2001) Analysis of relative gene expression data using real-time quantitative PCR and the $2^{(-\Delta\Delta C(T))}$ method. Methods 25(4):402–408. https://doi.org/10.1006/meth.2001.1262
17. Vandesompele J, de Preter K, Pattyn F et al (2002) Accurate normalization of real-time quantitative RT-PCR data by geometric averaging of multiple control genes. Genome Biol 3(7):RESEARCH0034
18. Hellemans J, Mortier G, de Paepe A et al (2007) qBase relative quantification framework and software for management and automated analysis of real-time quantitative PCR data. Genome Biol 8(2):R19. https://doi.org/10.1186/gb-2007-8-2-r19
19. Derveaux S, Vandesompele J, Hellemans J (2010) How to do successful gene expression analysis using real-time PCR. Methods 50(4): 227–230. https://doi.org/10.1016/j.ymeth.2009.11.001
20. Teste M-A, Duquenne M, François JM et al (2009) Validation of reference genes for quantitative expression analysis by real-time RT-PCR in Saccharomyces cerevisiae. BMC Molecular Biol 10(1):99. https://doi.org/10.1186/1471-2199-10-99
21. Nadai C, Campanaro S, Giacomini A et al (2015) Selection and validation of reference genes for quantitative real-time PCR studies during Saccharomyces cerevisiae alcoholic fermentation in the presence of sulfite. Int J Food Microbiol 215:49–56. https://doi.org/10.1016/j.ijfoodmicro.2015.08.012
22. Rebnegger C, Graf AB, Valli M et al (2014) In *Pichia pastoris*, growth rate regulates protein synthesis and secretion, mating and stress response. Biotechnol J 9(4):511–525. https://doi.org/10.1002/biot.201300334
23. Chapman JR, Waldenström J (2015) With reference to reference genes: a systematic review of endogenous controls in gene expression studies. PLoS One 10(11):e0141853. https://doi.org/10.1371/journal.pone.0141853

24. Untergasser A, Cutcutache I, Koressaar T et al (2012) Primer3—new capabilities and interfaces. Nucleic Acids Res 40(15):e115. https://doi.org/10.1093/nar/gks596
25. Dragosits M, Mattanovich D, Gasser B (2011) Induction and measurement of UPR and osmotic stress in the yeast *Pichia pastoris*. Meth Enzymol 489:165–188. https://doi.org/10.1016/B978-0-12-385116-1.00010-8
26. Öztürk S, Demir İ, Çalık P (2019) Isolation of high-quality RNA from *Pichia pastoris*. Curr Protoc Protein Sci 98(1):e101. https://doi.org/10.1002/cpps.101
27. Sambrook J, Russel DW (2001) Molecular cloning: a laboratory manual, vol 1, 3rd edn. Cold Spring Harbor Laboratory Press
28. New England BioLabs Inc. First Strand cDNA Synthesis (Standard Protocol) (NEB #M0368). https://international.neb.com/protocols/2016/04/26/first-strand-cdna-synthesis-standard-protocol-neb-m0368. Accessed 09 Feb 2022
29. National Center for Biotechnology Information National Library of Medicine (US), National Center for Biotechnology Information. https://www.ncbi.nlm.nih.gov/. Accessed 09 Feb 2022
30. Valli M, Tatto NE, Peymann A et al (2016) Curation of the genome annotation of *Pichia pastoris* (*Komagataella phaffii*) CBS7435 from gene level to protein function. FEMS Yeast Res 16(6):fow051. https://doi.org/10.1093/femsyr/fow051

Chapter 8

ID-MS-Based Quantitative Analysis of Metabolites in *Pichia pastoris*: A Step-by-Step Protocol

Marc Carnicer, S. Aljoscha Wahl, Reza Maleki Seifar, Joan Albiol, Walter van Gulik, and Pau Ferrer

Abstract

Quantitative metabolomics is based on a set of experimental approaches to accurately quantify intracellular metabolite concentrations. This allows us to characterize the response of a metabolic network (i.e., the metabolic phenotype) to an environmental or genetic perturbation. Here, we describe a four-step protocol adapted to the methylotrophic yeast *Komagataella phaffii:* (1) separation of the cells from the fermentation broth by cold filtration and addition of ^{13}C-labeled cell extract, (2) a metabolic quenching step based on aqueous cold methanol, (3) a metabolite extraction method based on boiling ethanol, and (4) quantification by isotope dilution mass spectrometry (LC-IDMS/MS and/or GC-IDMS). This method allows us to quantify most metabolites of central carbon metabolism, including glycolytic, tricarboxylic acid cycle, and pentose phosphate pathway intermediates, as well as cofactors and free amino acids. This method has been validated for *K. phaffii* grown on glucose, as well as on a mixture of carbon substrates such as methanol in combination with glucose or glycerol.

Key words *Pichia pastoris*, *Komagataella phaffii*, Metabolomics, Microbial metabolism, Intracellular metabolites, ^{13}C-labeled internal standard, Quenching, Extraction, Isotope dilution mass spectrometry

1 Introduction

In past years, metabolomics has developed methods for the quantitative and broad analysis of microbial metabolism, mainly due to improvements in MS-based analytical procedures [1–4]. Together with advances in the analysis of intracellular fluxes via ^{13}C isotopic labeling [5], a quantitative analysis of metabolic pathways is well within reach.

To study the function of metabolic reaction networks, it is crucial to obtain a representative and accurate snapshot of the quantitative state of metabolic pathways. Most metabolites, especially those of the central carbon metabolism, have turnover times in the order of seconds or less, which highlights the need for a rapid

Xavier Garcia-Ortega et al. (eds.), *Pichia pastoris: Methods and Protocols*, Methods in Molecular Biology, vol. 2697,
https://doi.org/10.1007/978-1-0716-4779-0_8,

sampling protocol to properly capture a representative cultivation state. Next to rapid sampling and inactivation, sampling protocols must ensure that there are no losses of metabolites during the sampling (leakage-free quenching) [6, 7] and the subsequent extraction procedure. Additionally, the intra- and extracellular space should be well separated, e.g., if substantial amounts of metabolites are present in the extracellular medium, those need to be efficiently removed. Therefore, a sample treatment procedure allowing separation of the intracellular and extracellular metabolite pools, as well as a degradation-free extraction method, is required [8].

Although many efforts have been directed toward the development of a universal method, no consensus solution has been found because of the vast diversity in cell properties [7, 9, 10].

Rapid sampling, quenching, separation, and extraction protocols for the intracellular metabolite quantification in *Komagataella phaffii* (also known as *Pichia pastoris*) have been presented and successfully applied over the past years, allowing accurate, reliable, and reproducible metabolite determinations [11, 12].

Here we provide detailed protocols for the quantitative analysis of the *K. phaffii* endo- and exometabolome, based on the work of Carnicer et al. [11]. They cultivated the cells under well-controlled conditions (growth rate, pH, oxygen supply) in steady-state chemostats and optimized/validated rapid sampling, cold methanol-based quenching, and hot ethanol-based extraction procedures for *K. phaffii.* Metabolite quantification was carried out by isotope dilution mass spectrometry (LC-IDMS/MS and GC-IDMS) [13]. Validation of the procedure was performed through quantification of metabolites in all different sample fractions (whole broth, quenched and washed cells, quenching and washing solutions, and culture filtrate) combined with a mass balance approach. The optimized protocols can be applied for quantitative analysis of intra- and extracellular metabolites in *K. phaffii* under dynamic conditions as well as ^{13}C-labeling experiments, including instationary metabolic flux analysis [14, 15].

Overall, the detailed procedure includes (1) fast sampling from a yeast culture and quenching using cold methanol, (2) separation and washing of the cells by cold filtration and addition of ^{13}C-labeled cell extract as an internal standard [16], (3) extraction of intracellular metabolites using boiling ethanol, and (4) a metabolite quantification and data analysis step. Specifically, the sampling process described here assumes that the cells are grown in well-controlled bioreactors (growth rate, pH, pO_2, etc.) and with online monitoring. Moreover, the bioreactors should allow fast sample collection, which can be best achieved by a sampling port in direct proximity to the culture broth, with a short tubing distance between the sample port and quenching solution, and overpressure inside the bioreactor. Notably, in this protocol cells are separated

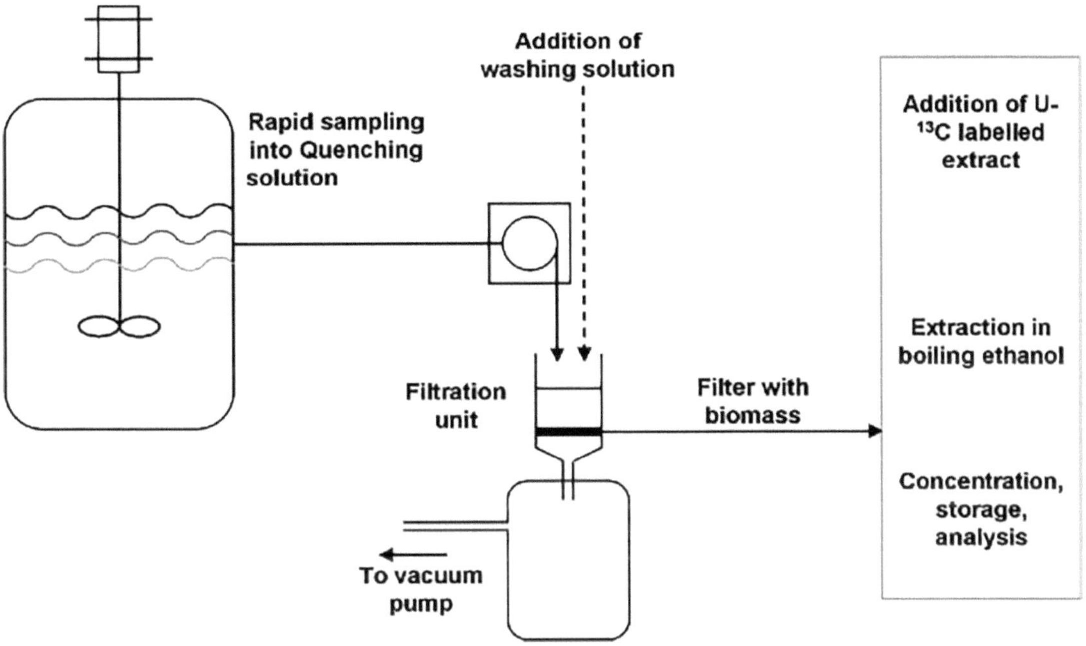

Fig. 1 A schematic overview of the cold filtration based sampling and processing procedure (taken from [17]). Rapid sampling is performed using the device of Lange et al. [18] using under pressure. Remark: The double valve to apply under pressure on the sample tube should be included

from the fermentation broth and washed by a cold filtration-based procedure, in contrast to other protocols based on cold centrifugation [13]. Figure 1 shows a schematic overview of the sampling procedure based on cold filtration performed in *K. phaffii*. Possible alternative sampling set-ups to facilitate the fast sampling and cell quenching should be evaluated before starting the experiment.

A comprehensive protocol for ID-MS-based quantitative metabolomics has been previously published by Wahl and co-workers [13], using *S. cerevisiae* as a reference case.

2 Materials

2.1 Cultivation of K. phaffii

The cultivation procedure and sampling scheme (number of samples, frequency, volume, cell amount) for metabolome profiling will vary depending on the purpose and needs of the study. The desired setup for the following metabolomics protocol is a fully instrumented bench-top bioreactor, to which a fast-sampling workflow procedure can be easily implemented (*see* **Note 1**).

2.2 Materials Needed for Sample Preparation

1. Quenching solution: 60% (v/v) methanol.
2. Extraction solution: 75% (v/v) ethanol.

3. ^{13}C internal standard solution, for example, ^{13}C-labeled cell extract [3, 16]. The internal ^{13}C must contain all relevant metabolites as U-^{13}C-labeled isotopes.
4. Membrane disk filters 47 mm in diameter, 0.45μm pore size (Pall Corporation, USA).
5. 0.2μm Durapore PVDF centrifuge filters.
6. 2-mL Eppendorf tubes.
7. Polypropylene test tubes (14 mL, 17 mm diameter).
8. 50-mL Falcon tubes.

2.3 Equipment for Sample Preparation

1. Bench-top bioreactor equipment, preferably equipped with CO_2 and O_2 off-gas analyzers.
2. Spectrophotometer equipped with a visible (400–700 nm) light source.
3. HPLC equipment, ion exchange column (e.g., Aminex HPX-87H, Bio-Rad Laboratories, Inc.), and HPLC vials (HPLC is used for extracellular metabolite analyses).
4. Rapid sampling setup for the bioreactor, e.g., as described in Fig. 1 [18]. The use of this device will allow controlling the sample volume taken from the bioreactor as well as reducing the residence time of the cells in the tube before being quenched.
5. Cryostat at −27 °C, the optimal temperature for quenching *K. phaffii* cells [11].
6. RapidVap (Labconco, USA) to evaporate/concentrate the samples. The amount of the metabolome samples that are going to be evaporated will be 30 mL. Therefore, the RapidVap should be adjusted to be able to cope with these volumes.
7. Refrigerated bench-top centrifuge for 2-, 15-, and 50-mL tubes.
8. 2× water baths. Operating conditions: 75 °C and 95 °C.
9. Vacuum filtration setup.

2.4 Materials and Equipment for LC-MS/MS Analyses

The LC-MS/MS and GC-MS analytical platforms used in this method have been recently described in Wahl et al. [13].

2.4.1 LC-MS/MS-Based Analysis for Glycolytic and Tricarboxylic Acid (TCA) Cycle Intermediates

1. HPLC: Alliance HT pump 2795 system (Waters, USA).
2. MS: Quattro LC™ triple quadrupole (Waters, UK).
3. 515 HPLC pump (Waters, USA).
4. Two-position ten-port stainless steel LabPro Selector valve (Rheodyne, USA).
5. Two-way LabPro Solvent select (Rheodyne, USA).

6. ASRS® 300 4 mm self-regenerating suppressor (Thermo Scientific, USA).
7. HPLC propylene screw top 300μl vials with cap and bonded pre-slit PTFE/silicon septa (Waters, USA).
8. Sodium hydroxide.
9. Methanol, HPLC grade.
10. LC column for glycolytic and TCA cycle metabolites: IonPac AS11 (Dionex, Sunnyvale, USA) 4 mm inner diameter column to 250 mm in length.
11. Guard column: An XTerra® MS C18 column with a 3 mm inner diameter and 20 mm length (Waters, Ireland).
12. Polypropylene sample vials (Waters).

2.4.2 LC-MS/MS-Based Analysis for Nucleotides and Coenzymes

1. HPLC: Alliance HT pump 2795 system (Waters, USA).
2. MS: Quattro Premier XE triple quadrupole mass spectrometer equipped with an electrospray ion source (Micromass MS Technologies—Waters, UK).
3. 515 HPLC pump (Waters, USA).
4. Ten-port Selector valve (Waters, USA).
5. LC column for analysis of nucleotides and coenzymes: AcQuity™ UPLC® BEH C18 (1.7μm, 100 mm × 2.1 mm i. d., Waters, Ireland).
6. Guard column: VanGuard™™ pre-column, AcQuity UPLC® BEH C18 (1.7μm, 5 mm × 2.1 mm, Waters, Ireland).
7. Polypropylene sample vials (Waters).

2.4.3 The GC-MS/MS-Based Analysis for Amino Acids Includes

1. GC: Agilent GC system 7890 A (Agilent Technologies, CA, USA).
2. MS: Agilent 5975C quadrupole MS equipped with an electron ionization (EI) ion source (Agilent Technologies, CA, USA). MassHunter software is used to control the MS and spectral processing.
3. Agilent automatic liquid sampler 7683B (Agilent Technologies, CA, USA).
4. Programmed temperature vaporizing (PTV) injector (Gerstel, Mühlheim, Germany), equipped with a glass liner containing glass wool.
5. GC column: ZB-50 column (30 m × 0.25 mm × 0.25μm, Phenomenex, CA, USA).
6. O-methoxyamine hydrochloride (MOX) stored in desiccators.
7. Anhydrous pyridine.

8. N-Methyl-N-(tert-butyldimethylsilyl)trifluoroacetamide (MTBSTFA) for amino acid derivatization.
9. Glass sample vials (Agilent Technologies).

3 Methods

3.1 K. phaffii Cultivation

Grow the desired *K. phaffii* strain in the required conditions (*see* **Notes 1** and **2**).

3.2 Before Sampling

1. For each intracellular metabolome sample prepare:
   (a) *1 sampling tube*: 5 mL of quenching solution in 14-mL polypropylene tubes. Number and weigh them. Store at – 40 °C.
   (b) *1 washing tube*: 10 mL of quenching solution in 14-mL polystyrene tubes. Store at −40 °C.
   (c) *1 extraction tube*: 30 mL of extraction solution in 50-mL Falcon tubes. Store at 4 °C. (*see* **Note 3**)
2. Turn on the cryostat at −27 °C, preferably the day before the sampling.
3. Connect the rapid sampling setup to the bioreactor and calibrate the valves for sampling the required broth volume (*see* **Note 4**).

3.3 Rapid Sampling

1. Take the ^{13}C-labeled extract from −80 °C and let it thaw at 4 ° C (*see* **Note 5**).
2. Place the sampling tube and washing tube in the cryostat.
3. Set each of the two water baths at 75 °C and 95 °C, respectively.
4. Place the extraction tube in the 75 °C water bath and wait 3–5 min to warm up the solution. Ethanol solution does not boil at that temperature.
5. Sample 600μL of broth, preferably using a rapid sampling setup, into a sampling vial (containing 5 mL of 60% methanol at −27 °C).
6. Weigh the sampling tube for calculation of the exact sample weight (*see* **Note 6**).
7. Filter the cell suspensions through 0.45μm pore size membrane disk filters (47 mm diameter) by using a vacuum pump.
8. Wash the filters to remove as much extracellular metabolites as possible by pouring 10 mL of quenching solution (washing tube) on the filter cake as soon as the cake starts to fall dry.

9. Pipette 120μL of ^{13}C-labeled extract on top of the dry filter cake (*see* **Note 6**).
10. Once the filter cake looks dry, stop the vacuum pump and transfer the filter cake into the extraction tube at 75 °C.
11. Place the extraction tube containing the filter in the 95 °C water bath for 3 min.
12. Remove the extraction tube from the water bath and place it back in the cryostat (*see* **Note 7**).
13. Perform dry cell weight (DCW) analysis to know the extract biomass concentration in the bioreactor.
14. After sampling: Metabolite extraction.
15. Turn on the RapidVap and wait until it is ready.
16. Remove the filter disk from the extraction solution.
17. Place the extraction tubes in the RapidVap and follow the standard protocol for drying.
18. Meanwhile, cool down the Eppendorf centrifuge at 1 °C.
19. When the drying is completed, take the samples from the RapidVap and resuspend the sediments with 600μL of ultra-pure water.
20. Transfer the resuspended solutions to Eppendorf tubes and centrifuge them at 12,000 × *g* for 5 min at 1 °C.
21. Transfer the supernatants to 0.2μm Durapore PVDF centrifuge filters.
22. Filter by centrifugation again at 12,000 × *g* for 5 min at 1 °C.
23. Transfer the filtrate into screw-cap sample vials.
24. Store the samples at −80 °C until analysis.

3.4 Preparation of Standards: Calibration Curves

The calibration curves for analysis of metabolites are constructed based on the ratio between the ^{12}C-metabolite/U-^{13}C-metabolite areas. The U-^{13}C-metabolites are used as internal standards to overcome potential metabolite losses or degradation during sample extraction, inefficient derivatization, matrix effects during analytical measurements, etc. [16]. Therefore, it is important to use the same ^{13}C-labeled cell extract used in the samples when preparing the standards.

The metabolite concentrations in the samples can vary depending on the metabolite being an intermediate or end product, the culture conditions, the cell density, the sample volume, etc. Therefore, the standard calibration ranges must be adjusted accordingly. Normally, 50–100μM is sufficient as the highest metabolite concentration in the calibration curves for the metabolites present in glycolysis (e.g., glucose 6-P, fructose bisphosphate, dihydroxyacetone, etc.), using this metabolome protocol for *K. phaffii* grown in glucose-limited chemostat cultures on glucose at a steady-state biomass concentration of about 4 g_{DCW}/L.

In addition, to facilitate the calculations it is preferred that the ratio between the total volume of each standard (e.g., 100μL) and the volume of ^{13}C-labeled extract added (e.g., 20μL) to the standards is equal to the ratio between the total volume of each sample after resuspension (e.g., 600μL) and the volume of ^{13}C-labeled extract added to the samples (e.g., 120μL). If this is not the case, a correction factor (CF) needs to be applied (*see* Subheading 3.7).

3.5 Metabolite Quantification with Mass Spectrometry

Quantitative analyses of metabolites are conducted using isotope dilution mass spectrometry-based analytical platforms (LC-IDMS/MS and GC-IDMS).

LC-IDMS/MS analysis of glycolytic and TCA metabolites is performed based on a modified method reported by van Dam et al. [1]. Briefly, a Waters Alliance HPLC 2795 system (Waters, USA), coupled to an ASRS® 300 4 mm self-regenerating suppressor (Thermo Scientific, USA) and a Waters Quattro LC triple quadrupole mass spectrometer (Waters, UK), is used. Metabolites are detected in negative ion mode with multiple reaction monitoring (MRM). Electrospray ionization is used as an interface between LC and MS. The capillary voltage is set at 2700 V; nitrogen gas is used as nebulizer gas at a flow rate of 88 L/h and as desolvation gas at a flow rate of 555 L/h with the temperature set at 250 °C. The ion block temperature is set at 90 °C. MRM transitions and corresponding instrument settings yielding the highest signal-to-noise ratio (S/N) are separately found for each individual metabolite by direct infusion of standard compounds into the MS.

GC-MS analysis of the pentose phosphate pathway metabolites was based on the method reported by [13, 19]. Briefly, a sample solution containing metabolites is lyophilized overnight, followed by derivatization reactions with 20 g/L MOX in anhydrous pyridine for 50 min at 70 °C. Then it is reacted for another 50 min at 70 °C with a mix of N-methyl-N-(trimethylsilyl) trifluoroacetamide (MSTFA) and trimethylchlorosilane (TMCS). The derivatized sample is centrifuged for 1 min at 10,000 × *g* and injected into the GC column using a PTV inlet program: initial temperature 70 °C, first ramp at 720 °C/min to 220 °C, hold for 5 min. A volume of 1μL is injected in splitless mode (for highly concentrated metabolites, in split mode 1/40 is used). The oven program is as follows: initial temperature 70 °C (hold for 1 min), then it is increased with a ramp of 10 °C/min to 320 °C with a final backflash of 5 column volumes at 320 °C. The MS setting is as follows: Transfer line 250 °C, source temperature 230 °C, and quadrupole temperature of 150 °C.

The MS/MS analyses of nucleotides and coenzymes are conducted based on methods reported by Seifar et al. [19, 21].

Briefly, a triple quadrupole mass spectrometer Waters Quattro Premier XE (Micromass MS Technologies-Waters) equipped with an electrospray ion source in negative ion mode is used for the determination of metabolites with MRM. The capillary voltage is set at 2800 V; nitrogen gas is used as cone gas with a flow rate of 50 L/h as well as desolvation gas with a flow rate of 700 L/h at 360 °C. The ion block temperature is set at 120 °C. MRM transitions and corresponding instrument settings yielding the highest S/N ratio are separately found for each individual metabolite by direct infusion of standard compounds into the MS.

Amino acids are quantified using a method reported by de Jonge et al. [22], with GC-MS on an Agilent GC system 7890A (Agilent Technologies, CA, USA) coupled to the EI ion source of an Agilent 5975C quadrupole MS (Agilent Technologies, CA, USA). MassHunter software is used to control the MS and for spectral processing. Metabolites are detected and monitored in selected ion monitoring mode. The operating temperature of the quadrupole is set by default to 150 °C, and the ion source temperature is set to 230 °C with EI at 70 eV. The transfer line from GC to MS is kept at a fixed temperature of 250 °C.

Prior to GC-MS analysis, the amino acids are derivatized with MTBSTFA to volatile tert-butyldimethyl silane derivatives.

3.6 Liquid Chromatography

3.6.1 Anion Exchange Chromatography for Separation of Central Carbon Metabolites

Liquid chromatographic separation of glycolytic and TCA cycle metabolites is performed by anion-exchange chromatography using a method adapted from the method reported by van Dam et al. [1]. The analytical column is equipped with a pre-column to trap hydrophobic contaminants. The analytical column is maintained at a temperature of 25 °C, while the auto sampler temperature is set to 4 °C. The injection volume is 10μL with partial loop injection.

1. A sample volume of a minimum of 40μL is required.
2. Prepare four solvents: Mobile phase A: 5% (v/v) methanol, B: 0.6 mM NaOH, C: 60 mM NaOH, and D: 300 mM NaOH.
3. Set the flow rate of the LC to 1 mL/min with A: 95% and B: 5%.
4. Set the flow rate of the HPLC pump for the methanol inflow (80% v/v) to the ion suppressor to 0.7 mL/min.
5. Set the pump for solvent selector to 0.9 mL/min. The solvent selector switches between solvents of 80% (v/v) methanol and 5% (v/v) methanol.
6. Switch the ion suppressor on (auto suppression external water mode, the ion suppressor removes the sodium ions from the effluent of the analytical column). The final flow to the MS is reduced to ~100μL/min (T union flow splitter).

Table 1
Gradient profile used for anion exchange chromatography of central carbon metabolites

Time (min)	A%	B%	C%	D%
0	95	5	0	0
1.5	95	5	0	0
1.51	95	0	5	0
17	75	0	25	0
22	0	0	100	0
24	0	0	100	
24.2	0	0	0	100
30	0	0	0	100
31	95	5	0	0
38	95	5	0	0

7. The chromatography is started with the solvent selector in position A (the guard column is switched in front of the analytical column).
8. Inject 10μL of the sample.
9. The NaOH gradient profile is generated using the four solvent mixtures, and the gradient is programmed as described in Table 1.

In the first 1.5 min, the pre-column and separation column are connected in series. Then the pre-column is switched to the waste and washed continuously with an 80% (v/v) methanol solution. The effluent from the separation column is switched to the waste after 26 min, until the start of the next run. The ASRS® 300 4 mm self-regenerating suppressor is washed with a continuous flow of 80% (v/v) methanol until the end of the run.

3.6.2 Ion Pair Reversed-Phase Liquid Chromatography for Separation of Nucleotides and Coenzymes

The chromatographic separation of adenosine mono-, di-, and tri-phosphates, NAD, NADH, NADP, NADPH, CoA, acetyl CoA, succinyl CoA, and FAD in cell extracts is achieved on a reversed-phase column. A pre-column is employed to protect the analytical column from particles present in the sample solutions. The analytical column is maintained at 25 °C while the autosampler temperature is set at 4 °C. The gradient profile and mobile phase compositions are the same as described in Seifar et al. [20, 21] with minor modifications. The injection volume is set to 10μL with partial loop injection.

Table 2
Gradient profile used for IP-UHPLC of nucleotides and coenzymes

Time (min)	B%
0	2
2	2
8	100
8.1	2
15	2

IP-UHPLC Ion pair reversed-phase liquid chromatography

1. Transfer 50 mL of sample to an HPLC vial.
2. Add 1 mL of 50 mM ion-pair reagent DBAA to the sample vial.
3. Prepare two solvents: A: 5% (v/v) acetonitrile, 2 mM DBAA, and mobile phase B: 84% (v/v) acetonitrile, 2 mM DBAA.
4. Set the flow rate of the LC to 1 mL/min with A: 90% and B: 10%.
5. Set the HPLC pump for 50% (v/v) acetonitrile to a flow of 0.1 mL/min (located between the analytical column and the electrospray of the MS).
6. Using a T union flow splitter the final flow of the MS is reduced to 0.1 mL/min.
7. Inject 5 mL of sample into the LC column.
8. The gradient profile used is shown in Table 2.

The effluent from the analytical column, with a flow of 0.2 mL/min, is mixed with a flow of 0.1 mL/min of 50% (v/v) acetonitrile solution, which is provided by a Waters 515 HPLC pump before entering the electrospray ionization of the MS system. The make-up flow is applied to enhance the MS signal at the beginning of the chromatographic separation, when the organic solvent concentration is rather low. During the first two minutes and the last five minutes of each run, the effluent from the analytical column is diverted to waste to prevent contamination of the MS system by salts in the sample matrix and ion pairs in the mobile phases.

3.7 Gas Chromatography for Separation of Free Amino Acids

The gas chromatography (GC) separation of amino acids is based on the method reported by de Jonge et al. [22]. For GC-MS analysis, chemical derivatization is required, for instance, as described in the following steps. Additional experimental tips are provided in [13]:

1. Lyophilization: Transfer 100μL of sample into a glass sample vial, close with a screw cap with a Teflon septum (punch holes with a syringe needle), and freeze-dry overnight. Make sure that the samples do not start melting by, for instance, using a cold (−80 °C) block.
2. Dissolve MOX at 20 g/L in (anhydrous) pyridine; shake until MOX is completely dissolved (prepare fresh).
3. Add 50μL of 20 g/L MOX solution and incubate for 50 min at 70 °C.
4. Open a fresh MSTFA ampule and add 50μL of TMCS.
5. Add 80μL of MSTFA-TMCS and incubate for 50 min at 70 °C.
6. Transfer 100μL of the derivatized sample into a glass vial and centrifuge for 1 min at 10,000 × *g*.
7. Transfer 70μL to a GC vial with an inlet and close tightly with a screw cap.

For GC-MS analysis, the following operating conditions are used:

1. Injection and PTV inlet program: Samples are kept in an automatic liquid sampler. A 1-μL sample is injected into the separation column using a PTV injector equipped with a glass liner containing glass wool. The PTV is used with a split of 1:15 for samples with a high concentration of amino acids. The initial temperature of PTV is set to 100 °C and held for 0.2 min, then raised at a rate of 720 °C/s to 220 °C and maintained at that temperature for 5 min, and is finally raised at a rate of 720 °C/s to 300 °C and held for another 22 min.
2. Oven program: The initial oven temperature is set to 100 °C and held for 1 min and then raised at a rate of 10 °C/s to 300 °C. A backflush technique is used for cleaning up the column from slow-eluting compounds with high boiling temperatures. The backflush is conducted by reversing the carrier gas flow into the inlet and venting through the split line for 3.7 min. Then the oven temperature is reduced to 100 °C within 4 min to start the new run.
3. MS settings: The helium carrier gas flow is set to 1.5 mL/min. The transfer line to MS is kept at a fixed temperature of 250 °C.

3.8 Calculation of Intracellular Metabolite Concentrations

In metabolomics, the metabolite pool sizes are usually expressed as amounts per gram DCW (μmol/g_{DCW}). These units present are close to primary data; no assumptions on cell volume or compartmentation are required. A possibility to obtain the biomass dry weight (DCW) concentration is to use an OD measurement when the OD/g_{DCW} relation is known. However, the researcher needs to validate that this relation is valid for the culture conditions applied.

The amount per gram of biomass values ($c_{Mi/x}$) can be easily calculated as μmol/g_{DCW} knowing the cell density of the cultures c_X (expressed as g_{DCW}/L), the concentration of the metabolites in the sample (c_{Mi}) expressed as μmol/l, the broth sample volume (V_{sample}; 0.6 mL in this protocol), and the calculated sample volume taken from the bioreactor (V_t):

$$c_{Mi/x} = c_{Mi} \cdot \frac{V_{\text{sample}}}{V_t} \cdot \frac{1}{cx}$$

Finally, the metabolite pool sizes need to be corrected by the following CF:

$$\text{CF} = \frac{V\,^{13}C_{\text{sample}}}{V_{\text{sample}}} \cdot \frac{V_{\text{standard}}}{V\,^{13}C_{\text{standard}}}$$

where $V_{13\text{Csample}}$ is the volume of ^{13}C extract added to the cell pellet (μL) and V_{standard} is the volume of ^{12}C standard (μL) with the respective addition of $V_{13\text{Cstandard}}$ of internal standard.

3.9 Preparation of ^{13}C Cell Extract

For the ^{13}C cell extract preparation, the content of ^{12}C must be minimized [3, 16]. Therefore, the cultivation is performed as fed-batch, starting with a low cell density (small inoculum) until a high cell density, fully grown on a ^{13}C carbon source, is reached. To balance the amount of intracellular metabolites, cells are harvested at different time points. Additionally, for the last harvest, a glucose pulse is applied to increase the amount of glycolytic intermediates in the extract. The different cell extracts are pooled, measured, and diluted. Aliquots of 1.2 mL are stored at −80 °C. For *K. phaffii*, a *S. cerevisiae* ^{13}C cell extract can also be used [11, 14].

4 Notes

1. It is desirable to grow cells in chemically defined media, i.e., where all the medium components are known, at moderate cell densities. Higher cell density cultures will need higher salt concentrations in the media, which may cause ion suppression effects in the electrospray ionization interface between the LC and the MS. The present protocol has been validated for cells growing on glucose [11] and glucose:methanol (80:20%, w/w) mixed feeds in chemostat cultures [14]. Since the characteristics of the cell membrane and cell wall of microorganisms may vary depending on the growth conditions, it is advisable to validate the quenching and extraction method for the new conditions, as described in [11]. Procedures for dynamic ^{13}C-labeling strategies for instationary ^{13}C-MFA can also be performed in chemostat cultures, as described in [14, 15]. Moreover, chemostat cultures are the basis for the

study of environmental perturbations, e.g., substrate pulses, where fast sampling under dynamic conditions allows characterizing the metabolic response to such perturbations [24]. A detailed protocol to perform continuous (i.e., chemostat) cultivations is presented in Chapter 33 of this book.

2. We advise to perform conventional analyses of cultivation parameters such as concentration of biomass and extracellular metabolites, as well as CO_2 and O_2 off-gas measurements. Protocols for conventional HPLC analyses of extracellular metabolites and determination of DCW (biomass concentration) are described in [23].
3. We advise to have several backup tubes in case some error/accident occurs during the fast sampling and quenching process and this needs to be repeated.
4. The sample volume is highly dependent on the biomass density, the applied analytical platform, and targeted metabolites. In this protocol, sampling 600μL of broth was sufficient. If some metabolite levels appear to be close to or below the detection limit, the sample volume and/or the concentration of the final extract could be increased. When increasing the sample volume, it is advised to increase the volume of quenching solution such that the ratio is kept the same.
5. The volume of the applied ^{13}C-labeled cell extract into the samples is highly dependent on its quality (i.e., the levels of the U-^{13}C-labelled metabolites). In this protocol, it is assumed that the addition of 120μL of ^{13}C-labeled cell extract to each sample results in comparable levels of ^{12}C and ^{13}C metabolites for most compounds in the final extract, which is beneficial for accurate quantification with IDMS. Typically, IDMS can operate over several orders of magnitude, but clearly, the noise increases when the reference (or sample) peak has only a low signal intensity.
6. Make sure to completely wipe off the sample tubes after removing them from the cryostat before weighing them to avoid an overestimation of the taken broth volume. Sample volume has to be dispensed accurately because all measurements are going to be scaled regarding this value and the amount of ^{13}C extract added to the cells (*see* Subheading 3.8).
7. After the extraction step, cell enzymes are inactivated, and the metabolites are mixed with the ^{13}C-labeled cell extract. Samples do not strictly need to be at −40 °C anymore, but they should be kept cold anyway to limit degradation.

Acknowledgments

This work has been supported by the Spanish program on Chemical Process Technologies (project CTQ2013-42391-R) of the Spanish Ministry of Science and Innovation, and the Agència de Gestió d'Ajuts Universitaris i de Recerca (AGAUR) of the Catalan Government (Grant No. 2014-SGR-452).

Competing Interest Statement The authors declare that they have no conflicts of interest that are relevant to the content of this chapter.

References

1. van Dam JC, Eman MR, Frank J, Lange HC, van Dedem GWK, Heijnen SJ (2002) Analysis of glycolytic intermediates in *Saccharomyces cerevisiae* using anion exchange chromatography and electrospray ionization with tandem mass spectrometric detection. Anal Chim Acta 460: 209–218. https://doi.org/10.1016/S0003-2670(02)00240-4
2. Oldiges M, Lutz S, Pflug S, Schroer K, Stein N, Wiendahl C (2007) Metabolomics: current state and evolving methodologies and tools. Appl Microbiol Biotechnol 76:495–511. https://doi.org/10.1007/s00253-007-1029-2
3. Mashego MR, Wu L, van Dam JC, Ras C, Vinke JL, van Winden WA, van Gulik WM, Heijnen JJ (2004) MIRACLE: mass isotopomer ratio analysis of U-^{13}C-labeled extracts. A new method for accurate quantification of changes in concentrations of intracellular metabolites. Biotechnol Bioeng 85:620–628. https://doi.org/10.1002/bit.10907
4. Dettmer K, Aronov PA, Hammock BD (2007) Mass spectrometry-based metabolomics. Mass Spectrom Rev 26:51–78. https://doi.org/10.1002/mas.20108
5. Tang YJ, Martin HG, Myers S, Rodriguez S, Baidoo EE, Keasling JD (2009) Advances in analysis of microbial metabolic fluxes via ^{13}C isotopic labeling. Mass Spectrom Rev 28:362–375. https://doi.org/10.1002/mas.20191
6. Canelas AB, Ras C, ten Pierick A, van Dam JC, Heijnen JJ, van Gulik WM (2008) Leakage-free rapid quenching technique for yeast metabolomics. Metabolomics 4:226–239. https://doi.org/10.1007/s11306-008-0116-4
7. van Gulik WM (2010) Fast sampling for quantitative microbial metabolomics. Curr Opin Biotechnol 21:27–34. https://doi.org/10.1016/j.copbio.2010.01.008
8. Canelas AB, ten Pierick A, Ras C, Seifar RM, van Dam JC, van Gulik WM, Heijnen JJ (2009) Quantitative evaluation of intracellular metabolite extraction techniques for yeast metabolomics. Anal Chem 81:7379–7389. https://doi.org/10.1021/ac900999t
9. Villas-Boas SG, Hojer-Pedersen J, Akesson M, Smedsgaard J, Nielsen J (2005) Global metabolite analysis of yeast: evaluation of sample preparation methods. Yeast 22:1155–1169. https://doi.org/10.1002/yea.1308
10. Bolten CJ, Kiefer P, Letisse F, Portais JC, Wittmann C (2007) Sampling for metabolome analysis of microorganisms. Anal Chem 79: 3843–3849. https://doi.org/10.1021/ac0623888
11. Carnicer M, Canelas AB, ten Pierick A, Zeng Z, van Dam J, Albiol J, Ferrer P, Heijnen JJ, van Gulik W (2012) Development of quantitative metabolomics for *Pichia pastoris*. Metabolomics 8:284–298. https://doi.org/10.1007/s11306-011-0308-1
12. Tredwell GD, Edwards-Jones B, Leak DJ, Bundy JG (2011) The development of metabolomic sampling procedures for *Pichia pastoris*, and baseline metabolome data. PLoS One 6:e16286. https://doi.org/10.1371/journal.pone.0016286
13. Wahl AS, Seifar RM, ten Pierick A, Ras C, van Dam J, Heijnen JJ, van Gulik WM (2014) Quantitative metabolomics using ID-MS. Meth Mol Biol 1191:91–105. https://doi.org/10.1007/978-1-4939-1170-7_6
14. Jordà J, Suarez C, Carnicer M, ten Pierick A, Heijnen JJ, van Gulik W, Ferrer P, Albiol J, Wahl A (2013) Glucose-methanol co-utilization in *Pichia pastoris* studied by metabolomics and instationary ^{13}C flux analysis. BMC Syst Biol 7:17. https://doi.org/10.1186/1752-0509-7-17

15. Jordà J, Rojas HC, Carnicer M, Wahl A, Ferrer P, Albiol J (2014) Quantitative metabolomics and instationary ^{13}C-metabolic flux analysis reveals impact of recombinant protein production on trehalose and energy metabolism in *Pichia pastoris*. Metab 4:281–299. https://doi.org/10.3390/metabo4020281
16. Wu L, Mashego MR, van Dam JC, Proell AM, Vinke JL, Ras C, van Winden WA, van Gulik WM, Heijnen JJ (2005) Quantitative analysis of the microbial metabolome by isotope dilution mass spectrometry using uniformly ^{13}C-labeled cell extracts as internal standards. Anal Biochem 336:164–171. https://doi.org/10.1016/j.ab.2004.09.001
17. Douma RD, de Jonge LP, Jonker CT, Seifar RM, Heijnen JJ, van Gulik WM (2010) Intracellular metabolite determination in the presence of extracellular abundance: application to the penicillin biosynthesis pathway in *Penicillium chrysogenum*. Biotechnol Bioeng 107: 105–115. https://doi.org/10.1002/bit.22786
18. Lange HC, Eman M, van ZG, Visser D, van Dam JC, Frank J, de Mattos MJ, Heijnen JJ (2001) Improved rapid sampling for *in vivo* kinetics of intracellular metabolites in *Saccharomyces cerevisiae*. Biotechnol Bioeng 75:406–415. https://doi.org/10.1002/bit.10048
19. Cipollina C, ten Pierick A, Canelas AB, Seifar RM, van Maris AJ, van Dam JC, Heijnen JJ (2009) A comprehensive method for the quantification of the non-oxidative pentose phosphate pathway intermediates in *Saccharomyces cerevisiae* by GC-IDMS. J Chromatogr B Analyt Technol Biomed Life Sci 877:3231–3236. https://doi.org/10.1016/j.jchromb.2009.07.019
20. Seifar RM, Ras C, van Dam JC, van Gulik WM, Heijnen JJ, van Winden WA (2009) Simultaneous quantification of free nucleotides in complex biological samples using ion pair reversed phase liquid chromatography isotope dilution tandem mass spectrometry. Anal Biochem 388:213–219. https://doi.org/10.1016/j.ab.2009.02.025
21. Seifar RM, Ras C, Deshmukh AT, Bekers KM, Suarez-Mendez CA, da Cruz AL, van Gulik WM, Heijnen JJ (2013) Quantitative analysis of intracellular coenzymes in *Saccharomyces cerevisiae* using ion pair reversed phase ultra high performance liquid chromatography tandem mass spectrometry. J Chromatogr A 1311: 115–120. https://doi.org/10.1016/j.chroma.2013.08.076
22. de Jonge LP, Buijs NA, ten Pierick A, Deshmukh A, Zhao Z, Kiel JA, Heijnen JJ, van Gulik WM (2011) Scale-down of penicillin production in *Penicillium chrysogenum*. Biotechnol J 6:944–958. https://doi.org/10.1002/biot.201000409
23. Carnicer M, Baumann K, Toplitz I, Sanchez-Ferrando F, Mattanovich D, Ferrer P, Albiol J (2009) Macromolecular and elemental composition analysis and extracellular metabolite balances of *Pichia pastoris* growing at different oxygen levels. Microb Cell Factories 8:65. https://doi.org/10.1186/1475-2859-8-65
24. Taymaz-Nikerel H, De MM, Baart G, Maertens J, Heijnen JJ, van Gulik W (2013) Changes in substrate availability in *Escherichia coli* lead to rapid metabolite, flux and growth rate responses. Metab Eng 16:115–129. https://doi.org/10.1016/j.ymben.2013.01.004

Chapter 9

High-Throughput Stationary ^{13}C-Based Metabolic Flux Analysis of *Pichia pastoris* Strains

Albert Fina, Pierre Millard, Cecilia Berges, Pau Ferrer, Joan Albiol, and Stéphanie Heux

Abstract

^{13}C-based Metabolic Flux Analysis (^{13}C-MFA) is an experimental approach that allows the quantification of the intracellular reaction rates (fluxes) that has become a valuable tool in the field of metabolic systems engineering. Rapid advances in the field of synthetic biology are leading to the generation of large numbers of engineered strains for the same research project. Thereby, the use of high-throughput cultivation and analytical platforms is required to test and characterize these sets of strains and extract meaningful information for further strain engineering cycles. Here, a detailed protocol for the characterization of the fluxome of a set of *Komagataella phaffii* strains using a high-throughput cultivation platform is described.

Key words *Pichia pastoris*, *Komagataella phaffii*, Fluxomics, ^{13}C-based Metabolic Flux Analysis, High-Throughput Bioreactor Cultures, Genome-Scale Model, Systems Biology, Quantitative Physiology

1 Introduction

Fluxomics can be defined as a set of methodologies for quantitative analysis of the distribution of metabolic fluxes (i.e., in vivo rates of metabolic reaction) in an organism, providing a tool for analysis of metabolic phenotypes. Determining the metabolic flux distribution of a microorganism can be used to assess the impact of a genetic modification or environmental perturbation at the metabolic level, as well as to identify genetic engineering targets for further strain improvements. However, besides the extracellular (input-output) exchange fluxes, it is generally not possible to directly measure intracellular metabolic fluxes in a metabolic network. Thus, several techniques combining experimental measurements with

Supplementary Information The online version contains supplementary material available at https://doi.org/10.1007/978-1-0716-4779-0_9.

Xavier Garcia-Ortega et al. (eds.), *Pichia pastoris: Methods and Protocols*, Methods in Molecular Biology, vol. 2697,
https://doi.org/10.1007/978-1-0716-4779-0_9,

mathematical modeling approaches have been developed for their determination. The most well-established of those techniques rely on the use of isotopically labeled substrates and on the measurement of label incorporation into metabolic intermediates and by-products. There exist multiple protocols for fluxomic analyses, which are usually grouped into two main sub-groups: stationary and instationary (also known as isotopic non-stationary) fluxomics [1]. Stationary ^{13}C-based Metabolic Flux Analysis (^{13}C-MFA) relies on the evaluation of the metabolic fluxes under metabolic and isotopic steady-state conditions, that is, defined as constant metabolic fluxes, metabolite concentration, and labeling pattern of the metabolites over time. Therefore, samples are withdrawn from the culture once the isotopic and metabolic steady states have been reached (i.e., from a chemostat culture or a mid-exponential phase of a batch culture). Afterwards, isotope incorporation into metabolites can be determined using mass spectrometry (MS) coupled with liquid chromatography (LC) or gas chromatography (GC), which provide access to their isotopologue distribution, and/or nuclear magnetic resonance (NMR), which provides positional labeling information [2]. In the case of instationary fluxomics, a switch from non-labeled to partially or fully labeled substrate (or vice versa) is performed in a culture under steady-state conditions. Afterward, multiple samples are withdrawn while labeling progresses toward its steady state. Again, the fluxes can be determined from the isotopologue distribution or the positional labeling of the metabolites. In this case, the metabolic fluxes and the concentration of the metabolites remain constant, but the labeling of the metabolites changes over time. Both stationary and instationary fluxomics of *Komagataella phaffii* (previously *Pichia pastoris)* strains have been described [3–5]. Compared to the instationary approach, stationary ^{13}C-MFA requires fewer samples, resulting in a less experimentally and computationally demanding task. In contrast, the instationary approach enhances the precision of flux measurements, improves the coverage of the metabolic network (e.g., to measure fluxes through parallel reactions), and allows measuring fluxes in microorganisms that metabolize C1 compounds such as CO_2 or methanol [6] (*see* **Note 1**).

To perform a fluxomics experiment, a model of the pathways of interest, typically central carbon metabolism, is required. Early metabolic models of *K. phaffii*'s core metabolism were obtained by combining literature data, specific biomass composition, and assuming similarity with *Saccharomyces cerevisiae* when no data were available [5, 7, 8]. Later, the use of a core model derived from the *K. phaffii* Genome-Scale Model (GSM) *iMT1026v3* was reported [9]. The GSM *iMT1026v3* contains the biomass composition of *K. phaffii* for different C-sources (glucose, glycerol, methanol, and combinations thereof), as well as the reactions for the synthesis of model heterologous proteins [10]. Moreover, not all enzymes are active in all culture conditions. Therefore, the compression of this GSM into a context-specific, substrate-dependent core model is preferred [11, 12].

The ease of genetic engineering of *K. phaffii* due to multiple recent advances in the synthetic biology field has led to the rapid generation of multiple strains in a short time [13–15]. To study the effect of genetic manipulations on strain metabolism, it is convenient to be able to analyze a large battery of strains using fluxomics. Performing high-throughput instationary fluxomics would require a significantly higher number of samples, together with fast sampling and quenching procedures. As not all the steps of the process can be fully automatized (sample treatment, peak integration, etc.), high-throughput instationary fluxomics is still beyond the current feasible automatization capabilities. Therefore, the current protocol will describe the workflow to perform high-throughput stationary ^{13}C-MFA in batch cultures of *K. phaffii*. A simplified representation of the workflow described in this chapter can be found in Fig. 1.

A high-throughput fluxomics approach for yeast has previously been described [16]. In that protocol, cells were cultivated on 96-deep well plates, where cell growth was manually monitored. A single sample was taken at the mid-exponential phase, and the fluxes were derived from the analysis of this sample. The determination of the extracellular exchange fluxes was obtained by fitting an exponential equation to the first and last points of the exponential phase. In the present case, the high-throughput stationary ^{13}C-MFA protocol will describe the use of a robotic platform with multiple mini bioreactors, which allows better control of the growth conditions (pH, O_2, etc.). Multiple samples can be collected to calculate the substrates and product exchange rates. Therefore, biases caused by a lag phase between strains can be avoided. Moreover, the automatization of absorbance density measurements reduces the noise derived from withdrawing the plate from the incubator shaker to measure the biomass concentration. In this protocol, a custom-made platform that was designed specifically for high-throughput fluxomics was used [17]. This platform contains 48 mini bioreactors, and samples are withdrawn and treated (quenching, centrifugation, and/or filtration) automatically. However, other high-throughput cultivation platforms could be adapted to the same use [18].

The methodology described herein has been effectively applied to analyze the fluxome of a series of ten 3-hydroxypropionic-producing *K. phaffii* strains, each harboring different genetic modifications, growing under two distinct pH conditions, 5 and 3.5. The findings detailed in this study allowed us to dissect changes in flux distribution resulting from genetic modifications in each strain, thereby providing a basis for the design of further metabolic engineering strategies for enhanced 3-hydroxypropionic acid

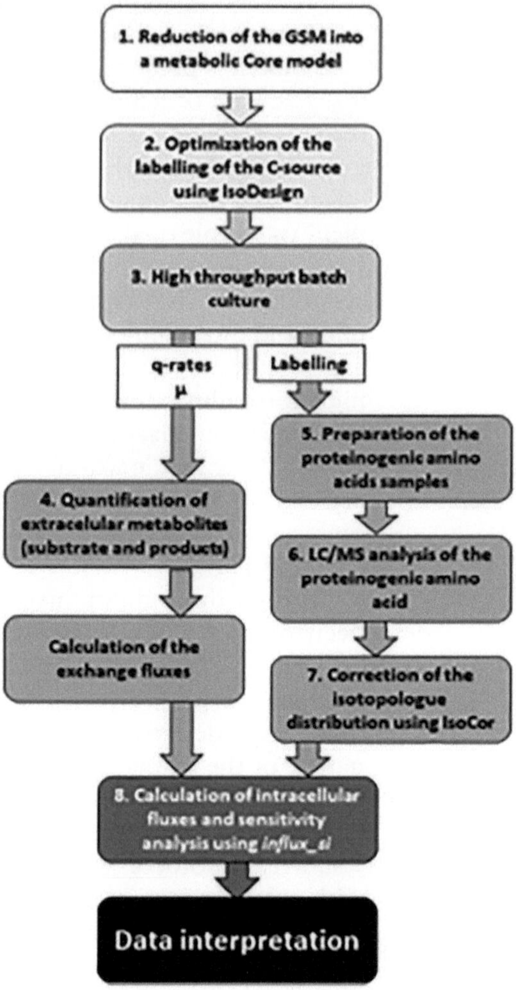

Fig. 1 Workflow for the high-throughput metabolic flux profiling of a set of *K. phaffii* strains. (Taken from [19])

production. Additionally, the results at pH 3.5 highlight the protocol's usefulness for examining fluxomic adaptations to different environmental conditions, beyond merely assessing the impact of genetic modifications [19].

In this work, the isotopic distribution of the proteinogenic amino acids analyzed using LC/MS is described. However, all the sections of the protocol besides Subheading 3.6 would remain identical if the positional labeling of the proteinogenic amino acids was determined using NMR. Similarly, Subheadings 3.1 and 3.2 of this protocol would remain without changes if the intracellular metabolite isotopologue distributions were measured. The protocols described elsewhere to quench the metabolism and analyze the intracellular metabolites from the cell culture could be used

[5, 20]. Additionally, Subheadings 3.4, 3.7, and 3.8 would still apply, using the data obtained from the intracellular metabolite measurements, instead of the proteinogenic amino acid results.

2 Materials

2.1 Software Required (See Note 2)

1. *MATLAB* (Matlab inc., Mathworks, MA, USA).
2. *Cobra Toolbox* [21] is a *MATLAB* toolbox for the constraint-based analysis of metabolic networks.
3. *CellNetAnalyzer* toolbox for *MATLAB* [22], which is a group of methods to analyze the structure and function of metabolic, signaling, and regulatory networks.
4. *influx_si* [23], which requires the installation of *Anaconda* with *Python3*. A detailed guideline for the installation of *influx_si* can be found at https://influx-si.readthedocs.io/en/latest/install.html#installation-with-conda. *Influx_si* is used to calculate the fluxes of metabolic reactions from a given model and the labeling data from fluxomics experiments.
5. *IsoDesign* [24]. This software is used to optimize the labeling of the C-source for fluxomics applications. All the required details are described in the cited article.
6. *PhysioFit* [25], an R-based tool to calculate the cell growth and the production and consumption rates. It can be downloaded from https://github.com/MetaSys-LISBP/PhysioFit
7. *IsoCor* [26], a *Python3*-based tool which is used to correct the MS labeling data for the presence of naturally occurring isotopes (^{2}H, ^{13}C, ^{15}N, etc.). Its installation details with Python 3.5 can be found in the following link: https://isocor.readthedocs.io/en/latest/quickstart.html

2.2 Starting Genome-Scale Metabolic (GSM) Model

1. The *iMT1026v3* (Tomàs-Gamisans et al., 2018; https://www.ebi.ac.uk/biomodels/MODEL1612130000) or any other *K. phaffii* GSM can be used.

2.3 High-Throughput Mini-Bioreactor Culture Components

1. YPG medium: 10% w/v yeast extract, 20% w/v peptone, and 1% v/v glycerol.
2. High-throughput culture platform: The use of a high-throughput fluxomics robotic platform, including an automated sampling, quenching, and processing of samples for fluxomic applications, is recommended [17]. However, other multiple mini-bioreactor systems, including inline, online, or atline OD_{600} measurements, such as the ones reviewed elsewhere [18], could also be used.

3. Batch medium: 2.5 g L^{-1} glycerol (conveniently partially ^{13}C-labeled), 1.8 g L^{-1} citric acid, 0.02 g L^{-1} $CaCl_2 \cdot 2\ H_2O$, 12.6 g L^{-1} $(NH_4)_2HPO_4$, 0.5 g L^{-1} $MgSO_4 \cdot 7\ H_2O$, 0.9 g L^{-1} KCl, 0.4 mg L^{-1} biotin, and 4.6 ml L^{-1} of PTM1 trace salts [28]. The pH was adjusted to 5.0 using HCl. The ^{13}C-labeled glycerol could be replaced with other labeled C-sources such as ^{13}C-labeled glucose, depending on the condition to investigate/question to address.
4. To sterilize the medium, mix all medium compounds except the biotin, the PTM1 salts, and the labeled substrate. Autoclave the medium. Filter-sterilize the concentrated C-source, the biotin, and the PTM1 salts. Add the filter-sterilized components to the autoclaved medium components in sterile conditions.

2.4 Equipment for Sample Preparation

1. Centrifuge.
2. 6 N HCl.
3. Rotavapor (Büchi, Switzerland).

2.5 Materials and Equipment for Proteinogenic Amino Acids Labeling Analysis Using LC/MS

1. Column: Discovery HS F5 HPLC column of 150 × 2.1 mm with particle size 5 μm (Supelco Bellefonte, PEN, USA).
2. Precolumn: Discovery HS F5 Supelguard Cartridge of 20 × 2.1 mm with particle size 5 μm (Supelco Bellefonte, PEN, USA).
3. Liquid chromatography system, such as UHPLC Vanquish (Thermo Fisher Scientific, MA, USA).
4. High resolution mass spectrometry detector, such as Orbitrap Q Exactive Plus, with a heated electrospray ionization (HESI) source (Thermo Fisher Scientific, MA, USA).
5. Acetonitrile and formic acid, LC-MS grade.
6. Ultrapure water (milli Q).
7. Solvent A: 0.1% v/v formic acid in ultrapure water.
8. Solvent B: 0.1% v/v formic acid in acetonitrile.

3 Methods

The following sections describe step by step the workflow to determine the fluxome of a *K. phaffii* strain. Starting with the experimental design (construction of the model and optimization), the execution of the experiments (cultures, sampling, and subsequent analyses), and finally, performing the calculation of the fluxes and the sensitivity analysis of the results.

3.1 Reduction of the GSM into a Metabolic Core Model

1. Install MATLAB, COBRA toolbox, and *CellNetAnalyzer* toolbox if not already done. Launch *MATLAB* and the *CobraToolbox* on your computer. Load the GSM *iMT1026v3*, which is provided in the Supplementary Materials (*iMT1026v3.mat*). Check that the installations and the model are running by constraining the uptake rates for all C-sources using the function *changeRxnBounds* to set the lower boundaries to 0. Set the lower bound of glycerol to -1 mmol $\cdot$ $g_{CDW}{}^{-1}$ $\cdot$ h^{-1} and change the upper boundaries of the biomass composition reactions for growth on glycerol (*LIPIDS_glyc*, *PROTEINS_glyc*, *STEROLS_glyc*, and *BIOMASS_glyc*) to 1000. The boundaries for the other biomass composition reactions, available for growth on alternative carbon sources, should be set to 0. Set the value of the ATP of maintenance (*ATPM*) to 2.51 mmol $\cdot$ $g_{CDW}{}^{-1}$ $\cdot$ h^{-1} [27]. Perform FBA using the maximization of the growth rate as the objective function. The maximal growth rate result should be 0.0587 h^{-1}.
2. The following model reduction steps are based on the *NetworkReducer* algorithm [29]. Once working installations of *MATLAB* and the *Cobra Toolbox* are ready, launch *CellNetAnalyzer* by running the function *startcna.m*. Afterwards, upload the *PpaGS.mat* model found in the Supplementary Materials of this chapter, which corresponds to the *iMT1026v3* model, in a format readable for *CellNetAnalyzer*. Alternatively, use the "*CNAcobra2cna*" function of the *CellNetAnalyzer Toolbox* (also included in the COBRA toolbox) to convert a COBRA model to *CellNetAnalyzer*. See *CellNetAnalyzer* manual for additional details. Use the *MATLAB* script *model_reduction.m* found in the Supplementary Materials to obtain a reduced model from the *K. phaffii* GSM. Change the reactions and metabolites that you want to be protected in the final model using the vectors *protect_reac* and *protect_met*, respectively (see **Note 3**). Specific phenotypes can also be protected using the vector *protect_func*. In the *model_reduction.m* file, mentioned above, reactions coupled to the measurements of glycerol uptake rate and growth rate were protected [9, 27]. As a result of this step, a Pruned Model is obtained from the GSM. The Pruned Model is also included in the supplementary materials for completeness (*PpaPruned.mat*).
3. Use the *MATLAB* script *model_compression.m* from the Supplementary Materials to compress the Pruned Model into a Core Model. Again, protect the desired reactions and metabolites using the vectors *protect_reac_pr* and *protect_spec*. In the present case, the amino acids should be protected in the *protect_spec* vector, as it is required that the biomass formation equation contain the individual stoichiometry of each amino

acid. As a result of this step, a Core Model is obtained from the Pruned Model. The resulting core model is included in the Supplementary Materials (*PpaCore.mat*).

4. Check the consistency of the reduced model by repeating **step 1** of this block using the Pruned and the Core models. Use the function *convertCNAModelToCbModel* of the *CellNetAnalyzer Toolbox* to convert the model from *CellNetAnalyzer* into *COBRA toolbox* format. Make sure the predicted maximal growth rate is the same for the GSM, the Pruned, and the Core models.
5. Before writing the model in Flux TaBuLar (FTBL) format [30, 31], the core model can be further manually curated. First, make sure the biosynthetic reactions use the precursors from the right cellular compartment, considering the knowledge from previous literature. For example, in the assayed conditions, alanine is derived from mitochondrial pyruvate [7], while the *PpaCore.mat* model considers that it is made from cytosolic pyruvate. Similarly, biosynthesis of lysine uses acetyl-CoA from the mitochondria and cytosolic α-ketoglutarate, which was also corrected in the *PpaCore.mat* model [32, 33], as the P*paCore.mat* model described the synthesis of lysine from cytosolic acetyl-CoA and mitochondrial α-ketoglutarate. Moreover, the biosynthetic pathway toward isoleucine uses mitochondrial pyruvate instead of the intermediary metabolite 2-(α-hydroxyethyl)thiamine diphosphate [7]. Finally, the compartment where biosynthesis of glutamate, glutamine, aspartate, and asparagine occurs is not known. Therefore, the biosynthetic reactions of these amino acids in both the cytoplasm and the mitochondria need to be considered. These corrections were applied to the model using the script *PpaCoreCorrection.m*, which generates the model *PpaCore2.mat*. Both documents can be found in the Supplementary Materials.
6. Lump the amino acid biosynthetic reactions of the *PpaCore2.mat* model that produce intermediary metabolites that only intervene in the synthesis of one amino acid. As some of these intermediary metabolites might also be present in the biomass formation equation, correct the biomass equation (*see* **Note 4**). To do so, you need to take the stoichiometric matrix and multiply the column of the reaction producing the component that you want to remove by the stoichiometric value of the component in the biomass reaction, and then subtract the result from the biomass reaction column, keeping the resulting column as the new biomass equation. This can be easily done in a spreadsheet (check the Supplementary Materials file *Model_-reactions.xlsx* (sheet *Biomass stoichiom. Modification*)). At the end, you will get a biomass stoichiometric equation like the one

shown in the Supplementary Materials file *Model_reactions.xlsx* (sheet *Biomass stoichiom. Simplification*), where the stoichiometric coefficient for each amino acid is equal to its reported abundance in the biomass [27].

7. Write the model in FTBL. This format only accounts for the molecules involved in the C-transitions of the reactions (excluding the molecules involved in the redox, energy, or nitrogen exchange). You can find the correspondence of the reactions of the PpaCore model with the reactions of the FTBL model in the "*PpaCore to FTBL model*" sheet of the *Model_reactions.xlsx* Supplementary Materials file. You will also find the template FTBL format of the *K. phaffii* Core model in the Supplementary Materials (*PpaCoreFTBL.ftbl*). The FTBL model can also be obtained from BIOMODELS [34], accession number: MODEL2404240001.Influx contains the functions *influx_s* and *influx_i* for stationary and instationary ^{13}C-MFA, respectively. As we are focusing on the use of stationary ^{13}C-MFA, test the function *influx_s* with the *PpaCoreFTBL.ftbl* model. Make sure the model is running (you can check the error message in the *.err* file, or the results provided in the *.kvh* file). Setting each reaction as constrained ©, dependent (D), or free (F) is crucial for the model. Constrained reactions correspond to the reactions with fluxes that have a known flux. For instance, the backward flux of reactions assumed to be irreversible in vivo can be constrained to 0. The free reactions correspond to the reactions not directly determined by the resulting linear system, thus corresponding to the degrees of freedom of the system. Therefore, they will have to be identified. The free reactions are typically found at the crossing nodes, where the flux will be split into several branches. The model will fit the fluxes of the free reactions to fit the measured data. Finally, the dependent reactions are the rest of the reactions. The fluxes of the dependent reactions are determined from the free and constrained fluxes to ensure all metabolites of the network are balanced (consistent with the metabolic steady-state condition). You can take advantage of the additional arguments "*–ffguess –noopt*" in *influx_si*. This tool will help you on the classification of the fluxes into C, D, or F.

3.2 Optimization of the Labeling of the C-source Using IsoDesign

1. Once the FTBL model is running on *influx_s*, use *IsoDesign* to test the optimal C-source combination to obtain the highest resolution from the measured data. A template file for *IsoDeisgn* is provided (*PpaCoreFTBL_IsoDesign.ftbl*). In that file, a combination of five alternative ^{13}C-labeled glycerol substrates is included (U-$^{12}C_3$-glycerol, 1-$^{13}C_1$-glycerol, 2-$^{13}C_1$-glycerol, 1,3-$^{13}C_2$-glycerol, and U-$^{13}C_3$-glycerol).

2. The output file obtained from *IsoDesign* contains the calculated standard deviation (SD) for each flux given each combination of labeled glycerol used as substrate. Using a spreadsheet or any other convenient computational tool (*Matlab, Python, R,* etc.), one can calculate the number of acceptably determined fluxes, given a certain criterion, such as *SD* below a given threshold value. The substrate combinations can also be ranked according to the maximal precision for a single flux or a set of fluxes. Therefore, the most appropriate substrate will primarily depend on the biological question addressed in the study.

3.3 High-Throughput Batch Culture

1. Inoculate appropriate samples of the *K. phaffii* clones to be analyzed in 50 mL Falcon tubes containing 5 mL YPG medium supplemented with the appropriate antibiotic (if needed). Cultivate the cells overnight in an incubator shaker at 30 °C and 200 rpm.
2. Use the saturated overnight cultures to inoculate a 250 mL shake flask culture containing 25 mL YPG medium at a starting OD_{600} of 1 (*see* **Note 5**). Cultivate the cells for 6–10 h in an incubator shaker at 30 °C and 200 rpm. The fastest-growing strains may be inoculated at a lower OD_{600}. The objective is to obtain cells at the exponential phase (OD_{600} in the range 5–10) in 6–10 h.
3. Add the batch sterile medium into each reactor of the high-throughput culture platform. The volume may be adjusted depending on the cultivation setup.
4. Take the shake flasks from the incubator. Use the cultures from step 2 to inoculate each reactor at a starting OD_{600} of 0.025. The temperature of the reactors is set to 28 °C.
5. Using 2.5 g L^{-1} of glycerol, the exponential phase ends when the cells reach an OD_{600} around 2 ~ 2.5. Thus, 500 μL samples for proteinogenic amino acid analysis are withdrawn from the cultures when the OD_{600} reaches 1 to ensure the cultures are at a mid-exponential phase (metabolic pseudo-steady state). Ideally, the samples should be automatically withdrawn from the culture to minimize manual interventions (see **Notes 6** and 7).
6. Immediately centrifuge the samples at 4 °C and discard the supernatant. Store the cell pellet at −20 °C until further use.
7. Throughout the exponential growth of the cell culture, 250 μL samples need to be withdrawn for the analysis of the culture supernatant. A total of ten 250 μL samples can be withdrawn from all cultures, each taken at 1-h intervals. The samples can be automatically withdrawn, placed on 96-well plates with a 0.45 μm filter bottom, and vacuum filtered immediately. Afterwards, the supernatants are automatically placed into a new

96-well plate, which can be stored at −20 °C. If an automated sampling and filtering system is not available, the process can be manually performed using syringe filters.

3.4 Quantification of Extracellular Metabolites (Substrates and Products) and Calculation of Exchange Fluxes

1. Quantification of the concentration of ^{13}C-labeled substrate (s) and by-products in culture supernatants using 1D-^{1}H NMR on a Bruker Advance III 800 MHz spectrometer (Bruker BioSpin, Germany) with a 5 mm CQPI cryoprobe is described, but any other relevant method (HPLC or LC-MS) may be used.
2. 180 μL supernatant samples were mixed with 20 μL of a D_2O stock solution containing 10 mM 3-(trimethylsilyl)-[2,2,3,3-^{2}-H_4]-propionic acid sodium salt (TSP). TSP was used as an internal standard and for spectra calibration.
3. The probe temperature of the NMR spectrometer was set at 280 K, and the NMR data was recorded after a 30° presaturation pulse (zgpr30), followed by a relaxation delay of 7 s.
4. The NMR spectrum was analyzed using TopSpin 4.3.1 (Bruker BioSpin, Germany).
5. The growth rate and the specific production/consumption rates need to be calculated. To do so, take the OD_{600} data of the cultures and the supernatant concentration of the substrate and products from the NMR analysis. The OD_{600} data needs to be converted into Cell Dry Weight (CDW) concentration. Do the appropriate experiments to determine such a parameter for the spectrophotometer, the *K. phaffii* strain, and the OD_{600} range you are using.
6. Using *PhysioFit* [25] or any other similar custom-made script or software, calculate the μ_{max} and the specific production/ consumption rates. Use only the measurements of the exponential phase of the culture ($OD_{600} < 2.5$ for the conditions described in this protocol).

3.5 Preparation of the Proteinogenic Amino Acid Samples

1. The frozen pellets were resuspended in 150 μL 6 N HCl and incubated overnight (16 h) at 110 °C. Use a glass tube with an acid-resistant cap sealing, so HCl does not leak from the tube. The HCl will hydrolyze the cells as well as the peptide bonds of the proteins.
2. Evaporate the HCl in the rotavapor. Make sure the rotavapor and the tubing connections can stand HCl, and connect the gas outlet to a condenser, so that HCl gas is not released into the environment. Use the pressure gradient described in Table 1.
3. Resuspend the pellet in 100 μL ultrapure water and evaporate the water using the rotavapor. This step is performed to remove all the HCl.

Table 1
Vacuum evaporator gradient protocol description

Starting pressure (mbar)	Ending pressure (mbar)	Time (min)
800	500	10
500	500	1
500	300	10
300	200	10
200	200	5
200	100	15
100	100	10
100	75	10
75	75	5
75	0	15
0	0	∞

4. Repeat the previous step.
5. Finally, resuspend the pellet in 200 μL ultrapure water. Centrifuge at 10,000 g for 5 min. The pellet contains the biomass debris, while the supernatant contains the free proteinogenic amino acids.
6. Take 40 μL of the supernatant and transfer to a tube with 460 μL of ultrapure water.

3.6 LC/MS Analysis of the Proteinogenic Amino Acid

1. A previously described protocol was used [35].
2. Liquid chromatography: The precolumn and column described in the materials section are used. Set the flow of the eluent at 0.25 mL · min^{-1}, the sampler temperature at 4 °C, the column temperature at 30 °C, and the volume of injection at 5 μL. Use the gradient described in Table 2.
3. Mass Spectrometry: Use the parameters described in Table 3. Set the equipment to measure all the *m/z* positive ratios described in the Supplementary Materials (Supplementary Tables, Table S1; *see* **Note 8**). The relative abundance of each isotopologue can be calculated from its relative area compared to the sum of the areas of all isotopologues for the same amino acid.
4. Place 100 μL of the samples containing the proteinogenic amino acids described in Subheading 3.5, **step 6** into HPLC glass vials with inserts and run the protocol to perform the analyses.

Table 2
Gradient description for the solvents A and B of the Liquid Chromatography

Time (min)	Solvent A (%)	Solvent B (%)
0	98	2
2	98	2
10	95	5
16	65	35
20	0	100
24	0	100
24.1	98	2
30	98	2

Table 3
Mass Spectrometry parameters

Variable	Setpoint
Mode	Positive
Acquisition mode	FTMS mode
Resolution	70,000 *(400 m/z)*
Capillary temperature	320 °C
Source heater temperature	300 °C
Sheath gas flow rate	40 arbitrary units
Auxiliary gas flow rate	10 arbitrary units
S-Lens RF level	40%
Source voltage	5 kV
Duration	30 min

3.7 Correction of the Isotopologue Distribution Using IsoCor

1. The isotopologue distribution needs to be corrected for the presence of the isotopes at natural abundance. To do so, provide the isotopologue distribution obtained from the LC/MS analysis to *IsoCor* and select the appropriate correction options (isotopic purity of the tracer and natural abundance of the tracer element).
2. *IsoCor* generates a results file that can be used in *influx_si* to calculate the fluxes. A detailed protocol for the use of *IsoCor* can be found elsewhere ([26]; https://isocor.readthedocs.io/).

3.8 Calculation of Intracellular Fluxes and Sensitivity Analysis Using influx_si

1. The measured μ_{max} and q-rates need to be normalized by the specific substrate uptake rate (qS), as the fluxes of all strains and cultures will be scaled to the qS for each strain. The quotient of the μ_{max} or q-rates with the qS is equivalent to the biomass and the product yields. Add the yield results (and associated standard deviations) to the *MEASUREMENTS* section of the *PpaCoreFTBL.ftbl* file (*see* **Note 9**).
2. Add the corrected isotopologue distribution data obtained with *IsoCor* to the *MASS_SPECTROMETRY* section of the *PpaCoreFTBL.ftbl* file containing the measurements for that strain. Run *influx_s* to obtain the calculated fluxes.
3. The flux results can be found in the *linear stats* section of the results "*.kvh*" file. Moreover, at the end of the "*.kvh*" file, the statistical goodness of fit obtained from a chi-squared test is shown. The results clearly indicate whether the measurements were accurately fitted (at a confidence level of 95%). If the fit is satisfactory, the calculated fluxes (and associated standard deviation) can be interpreted. An inconsistent fit may be due to (i) errors in measurements and/or (ii) errors in the model. For instance, it is possible that reactions which were not considered for the model carry a significant flux and thereby need to be included in the model to fit the data satisfactorily. Therefore, some iterations between the raw data and the model may be required until an acceptable fit is achieved.
4. In case *influx_si* encounters an error in the calculation process, the cause of the error will be provided in a *.err* file. For instance, this situation may occur if the definition of some fluxes as Constrained (C), Free (F), or Dependent (D) is not consistent (see Subheading 3.1, **step 8**).
5. Finally, you may use some additional features of *influx_s* to get the most information out of your data. For example, using the additional argument "*—sens mc = 100,*" you can perform a Monte Carlo sensitivity analysis with 100 simulations to assess the robustness of the results. You can also use the additional argument "*sln*" to minimize the sum of fluxes that are not identifiable from the available isotopic measurements, following the principle that cells would have a fitness advantage of using the least amount of enzyme [36]. Use of "*sln*" is therefore particularly interesting when some fluxes are poorly determined.

4 Notes

1. Using stationary or instationary ^{13}C-MFA, the fluxes are calculated from the changes in the labeling of the metabolites resulting from the flux, the stoichiometry, and the C-transitions of a

reaction. Fluxomics of microorganisms growing on a one-carbon source (i.e., methanol or CO_2) can only be performed using isotopically non-stationary fluxomics, as incorporation of the labeled or unlabeled substrate would happen randomly due to the absence of C-bonds in the substrate. Recently, isotopically non-stationary fluxomics using methanol as a sole C-source was reported in *Bacillus methanolicus* [37], but a similar approach has not been used in *K. phaffii*, yet. Nevertheless, both stationary and isotopically non-stationary fluxomics in *Pichia pastoris* using glycerol/methanol or glucose/methanol mixtures have been reported [38, 39]. *IsoDesign* can also be used to optimize the labeling of the substrates when two or more C-sources are co-consumed. However, the high-throughput approach reported here might not apply, as the system is performed in batch cultures. Under such conditions where catabolite repression is active, as there is an excess of glucose or glycerol, the cells would not co-consume methanol.

2. All the software and computational tools used for this study contain tutorials and/or test functions. Doing the tutorials for all the tools before starting the experimental design is highly recommended. An example of the code used can be found at "https://github.com/ENG4BIODEQBA/HT13CMFA".
3. If the objective of the project is to characterize the fluxome of a *K. phaffii* strain producing a heterologous chemical compound, make sure to protect the reactions producing the precursors of this product. Later, add the heterologous pathway to both the Core and the FTBL Core models.
4. Simplification of the model by lumping the reactions producing intermediary metabolites will greatly reduce the overall number of reactions. That will be useful, as this will reduce the number of reactions for the C-transitions model. If two reactions are lumped and an intermediary metabolite is removed, make sure that the intermediary is not involved in any other reactions. This can be easily done by checking the corresponding metabolite row in the stoichiometric matrix. In case it is involved in other reactions, make sure to correct the stoichiometry of the other reactions, too. Make sure also that the carbon positions from the substrate(s) are properly mapped to the product(s) of the lumped reaction.
5. The OD_{600} measures always refer to the absorbance at 600 nm using a 1 cm cuvette. If measured using a different system (i.e., microplate reader), the correlation between the two spectrophotometers should be tested.
6. Smaller volumes may be withdrawn from the culture (i.e., 50 μL), but this should be considered at Subheading 3.5, **step 6**. In case only 50 μL are withdrawn, the proteinogenic

amino acids from Subheading 3.5, **step 4** (the supernatant after the centrifugation step) can be directly placed into an HPLC vial with insert.

7. If the biomass samples cannot be withdrawn and processed immediately, it is recommended to store the samples in quenching solution at −20 °C for up to 2 h. The quenching solution consists of 40% v/v acetonitrile, 40% v/v methanol, 20% water, and 0.1 M formic acid. Use LC/MS-grade reagents. In that case, an additional step to evaporate all the quenching solution is required before the HCl is added to the cell pellet. The same rotavapor and the same protocol described in Table 1 can be used.
8. Methionine and tryptophan cannot be measured using the reported method. Moreover, asparagine and glutamine are converted into aspartate and glutamate during the acidic hydrolysis of the biomass with HCl. Therefore, the method provides the isotopologue distribution of aspartate plus asparagine and glutamate plus glutamine instead of the individual isotopologue distribution of each amino acid.
9. The results from the ^{13}C-MFA can be used as measurements of the *PpaCore2.mat* stoichiometric model to perform Flux Balance Analysis (FBA) using the *CobraToolbox* [21].

Acknowledgments

This work has been supported by grant 2021-SGR00143 from the Agència de Gestió d'Ajuts Universitaris i de Recerca (AGAUR) of the Catalan Government. The AF PhD scholarship was funded with an FPU grant (FPU17/05434) from the Spanish Ministry of Science.

Competing Interest Statement The authors have no conflicts of interest to declare that are relevant to the content of this chapter.

References

1. Wiechert W, Nöh K (2005) From stationary to instationary metabolic flux analysis. Adv Biochem Eng Biotechnol 92:145–172. https://doi.org/10.1007/b98921
2. Millard P, Sokol S, Kohlstedt M et al (2021) IsoSolve: an integrative framework to improve isotopic coverage and consolidate isotopic measurements by mass spectrometry and/or nuclear magnetic resonance. Anal Chem 93: 9428–9436. https://doi.org/10.1021/acs.analchem.1c01064
3. Ferrer P, Albiol J (2014) ^{13}C-based metabolic flux analysis in yeast: the *Pichia pastoris* case. Methods Mol Biol 1152:209–232. https://doi.org/10.1007/978-1-4939-0563-8_13
4. Ferrer P, Albiol J (2014) ^{13}C-based metabolic flux analysis of recombinant *Pichia pastoris*. Methods Mol Biol 1191:291–313. https://doi.org/10.1007/978-1-4939-1170-7_17

5. Jordà J, Rojas H, Carnicer M et al (2014) Quantitative metabolomics and instationary ^{13}C-metabolic flux analysis reveals impact of recombinant protein production on trehalose and energy metabolism in *Pichia pastoris*. Meta 4:281–299. https://doi.org/10.3390/metabo4020281
6. Nöh K, Wiechert W (2011) The benefits of being transient: isotope-based metabolic flux analysis at the short time scale. Appl Microbiol Biotechnol 91:1247–1265. https://doi.org/10.1007/s00253-011-3390-4
7. Solà A, Maaheimo H, Ylönen K et al (2004) Amino acid biosynthesis and metabolic flux profiling of *Pichia pastoris*. Eur J Biochem 271:2462–2470. https://doi.org/10.1111/j.1432-1033.2004.04176.x
8. Baumann K, Carnicer M, Dragosits M et al (2010) A multi-level study of recombinant *Pichia pastoris* in different oxygen conditions. BMC Syst Biol 4:141. https://doi.org/10.1186/1752-0509-4-141
9. Tomàs-Gamisans M, Ødum ASR, Workman M et al (2019) Glycerol metabolism of *Pichia pastoris* (*Komagataella* spp.) characterised by ^{13} C-based metabolic flux analysis. New Biotechnol 50:52–59. https://doi.org/10.1016/j.nbt.2019.01.005
10. Tomàs-Gamisans M, Ferrer P, Albiol J (2016) Integration and validation of the genome-scale metabolic models of *Pichia pastoris*: a comprehensive update of protein glycosylation pathways, lipid and energy metabolism. PLoS One 11:e0148031. https://doi.org/10.1371/journal.pone.0148031
11. Robaina Estévez S, Nikoloski Z (2014) Generalized framework for context-specific metabolic model extraction methods. Front Plant Sci 5:1–11. https://doi.org/10.3389/fpls.2014.00491
12. Richelle A, Chiang AWT, Kuo CC, Lewis NE (2019) Increasing consensus of context-specific metabolic models by integrating data-inferred cell functions. PLoS Comput Biol 15:1–19. https://doi.org/10.1371/journal.pcbi.1006867
13. De S, Mattanovich D, Ferrer P, Gasser B (2021) Established tools and emerging trends for the production of recombinant proteins and metabolites in *Pichia pastoris*. Essays Biochem 65:293–307. https://doi.org/10.1042/EBC20200138
14. Prielhofer R, Barrero JJ, Steuer S et al (2017) GoldenPiCS: a Golden Gate-derived modular cloning system for applied synthetic biology in the yeast *Pichia pastoris*. BMC Syst Biol 11:1–14. https://doi.org/10.1186/s12918-017-0492-3
15. Weninger A, Hatzl AM, Schmid C et al (2016) Combinatorial optimization of CRISPR/Cas9 expression enables precision genome engineering in the methylotrophic yeast *Pichia pastoris*. J Biotechnol 235:139–149. https://doi.org/10.1016/j.jbiotec.2016.03.027
16. Ebert BE, Blank LM (2014) Successful downsizing for high-throughput ^{13}C-MFA applications. Methods Mol Biol 1191:127–142. https://doi.org/10.1007/978-1-4939-1170-7_8
17. Heux S, Juliette P, Stéphane M et al (2014) A novel platform for automated high-throughput fluxome profiling of metabolic variants. Metab Eng 25:8–19. https://doi.org/10.1016/j.ymben.2014.06.001
18. Bareither R, Pollard D (2011) A review of advanced small-scale parallel bioreactor technology for accelerated process development: current state and future need. Biotechnol Prog 27:2–14. https://doi.org/10.1002/btpr.522
19. Fina A, Millard P, Albiol J et al (2023) High throughput ^{13}C-metabolic flux analysis of 3-hydroxypropionic acid producing *Pichia pastoris* reveals limited availability of acetyl-CoA and ATP due to tight control of the glycolytic flux. Microb Cell Factories 22:1–16. https://doi.org/10.1186/s12934-023-02123-0
20. Carnicer M, Canelas AB, ten Pierick A et al (2012) Development of quantitative metabolomics for *Pichia pastoris*. Metabolomics 8:284–298. https://doi.org/10.1007/s11306-011-0308-1
21. Becker SA, Feist AM, Mo ML et al (2007) Quantitative prediction of cellular metabolism with constraint-based models: the COBRA toolbox. Nat Protoc 2:727–738. https://doi.org/10.1038/nprot.2007.99
22. Klamt S, Saez-Rodriguez J, Gilles ED (2007) Structural and functional analysis of cellular networks with CellNetAnalyzer. BMC Syst Biol 1:2. https://doi.org/10.1186/1752-0509-1-2
23. Sokol S, Millard P, Portais JC (2012) Influx_s: increasing numerical stability and precision for metabolic flux analysis in isotope labelling experiments. Bioinformatics 28:687–693. https://doi.org/10.1093/bioinformatics/btr716
24. Millard P, Sokol S, Letisse F, Portais J-C (2014) IsoDesign: a software for optimizing the design of ^{13}C-metabolic flux analysis experiments. Biotechnol Bioeng 111:202–208. https://doi.org/10.1002/bit.24997
25. Peiro C, Millard P, de Simone A et al (2019) Chemical and metabolic controls on

dihydroxyacetone metabolism lead to suboptimal growth of *Escherichia coli*. Appl Environ Microbiol 85:e00768–e00719. https://doi.org/10.1128/AEM.00768-19

26. Millard P, Delépine B, Guionnet M et al (2019) IsoCor: isotope correction for high-resolution MS labeling experiments. Bioinformatics 35:4484–4487. https://doi.org/10.1093/bioinformatics/btz209
27. Tomàs-Gamisans M, Ferrer P, Albiol J (2018) Fine-tuning the *P. pastoris* iMT1026 genome-scale metabolic model for improved prediction of growth on methanol or glycerol as sole carbon sources. Microb Biotechnol 11:224–237. https://doi.org/10.1111/1751-7915.12871
28. Maurer M, Kühleitner M, Gasser B, Mattanovich D (2006) Versatile modeling and optimization of fed batch processes for the production of secreted heterologous proteins with *Pichia pastoris*. Microb Cell Factories 5:1–10. https://doi.org/10.1186/1475-2859-5-37
29. Erdrich P, Steuer R, Klamt S (2015) An algorithm for the reduction of genome-scale metabolic network models to meaningful core models. BMC Syst Biol 9:1–12. https://doi.org/10.1186/s12918-015-0191-x
30. Wiechert W, de Graaf AA (1997) Bidirectional reaction steps in metabolic networks. Biotechnol Bioeng 55:101–135. https://doi.org/10.1002/(SICI)1097-0290(19970705)55:1<118::AID-BIT13>3.0.CO;2-I
31. Wiechert W, Möllney M, Petersen S, de Graaf AA (2001) A universal framework for ^{13}C metabolic flux analysis. Metab Eng 3:265–283. https://doi.org/10.1006/mben.2001.0188
32. Gombert AK, dos Santos MM, Christensen B, Nielsen J (2001) Network identification and flux quantification in the central metabolism of *Saccharomyces cerevisiae* under different conditions of glucose repression. J Bacteriol 183:1441–1451. https://doi.org/10.1128/JB.183.4.1441-1451.2001
33. Lehnen M, Ebert BE, Blank LM (2017) A comprehensive evaluation of constraining amino acid biosynthesis in compartmented models for metabolic flux analysis. Metab Eng Commun 5:34–44. https://doi.org/10.1016/j.meteno.2017.07.001
34. Chelliah V, Juty N, Ajmera I et al (2015) BioModels: ten-year anniversary. Nucleic Acids Res 43:D542–D548. https://doi.org/10.1093/nar/gku1181
35. Heuillet M, Bellvert F, Cahoreau E et al (2018) Methodology for the validation of isotopic analyses by mass spectrometry in stable-isotope labeling experiments. Anal Chem 90:1852–1860. https://doi.org/10.1021/acs.analchem.7b03886
36. Holzhütter HG (2004) The principle of flux minimization and its application to estimate stationary fluxes in metabolic networks. Eur J Biochem 271:2905–2922. https://doi.org/10.1111/j.1432-1033.2004.04213.x
37. Delépine B, Gil López M, Carnicer M et al (2020) Charting the metabolic landscape of the facultative methylotroph *Bacillus methanolicus*. mSystems 5:e00745. https://doi.org/10.1128/mSystems00745-20
38. Solà A, Jouhten P, Maaheimo H et al (2007) Metabolic flux profiling of *Pichia pastoris* grown on glycerol/methanol mixtures in chemostat cultures at low and high dilution rates. Microbiology 153:281–290. https://doi.org/10.1099/mic.0.29263-0
39. Jordà J, Suarez C, Carnicer M et al (2013) Glucose-methanol co-utilization in *Pichia pastoris* studied by metabolomics and instationary ^{13}C flux analysis. BMC Syst Biol 7:17. https://doi.org/10.1186/1752-0509-7-17

Chapter 10

Determination of Biomass Elemental Composition of *Pichia pastoris*

Arnau Gasset, Francisco Valero, and José Luis Montesinos-Seguí

Abstract

A quantitative determination of a consistent biomass elemental composition is required for a wide range of applications, including metabolic engineering, bioprocess parameter calculation, and medium optimization, among others. Its determination relies on the use of different analytical techniques for the range of biochemical compounds considered. In this chapter, a robust methodology is proposed for its determination of yeast, specifically *Komagataella phaffii* (previously named *Pichia pastoris*).

Key words *Pichia pastoris*, *Komagataella phaffii*, Biomass elemental composition, Metabolic engineering, Metabolic flux analysis, Macromolecular components

1 Introduction

To analyze the metabolic performance of the cell growing in different conditions and/or with different genetic backgrounds, it is necessary to combine data from different sources to obtain consistent results. With this aim, macromolecular and biomass elemental composition must be well known [1]. Furthermore, the application of metabolic flux analysis (MFA) through numerical techniques requires an accurate determination of a consistent biomass composition, since this biomass elemental composition must match the combination of the different macromolecular components involved in the bioprocess.

On the other hand, bioprocess optimization needs an accurate calculation of all parameters to compare different cultivation strategies, which can range from substrate selection and oxygen conditions (among others) to the application of different operational modes. However, the bioreactor's operation involves a great number of measurements, which typically contain an associated error.

Xavier Garcia-Ortega et al. (eds.), *Pichia pastoris: Methods and Protocols*, Methods in Molecular Biology, vol. 2697,
https://doi.org/10.1007/978-1-0716-4779-0_10,

To minimize this set of associated errors and to assure that no gross errors have been made in any single parameter determination, as well as that the lack of one major metabolite or missed by-product has been overlooked, carbon and electron balances are often applied to the set of parameters calculated from the measurements, in a process called data reconciliation [2, 3]. These balances consist of modifying all specific rates (μ, q_S, q_{O2}, q_{CO2}, ...) to assure that all carbon atoms and electrons that enter the system (through substrate addition, aeration, etc.) are equal to those that leave the system (through sampling, off-gas, biomass generation, accumulation of substrate and products/by-products in the final culture broth, etc.) [2, 3]. In the same manner as with the MFA, to apply carbon and/or electron balances is mandatory to know accurately the elemental composition of biomass, since it affects not only biomass generation but also all specific rate determinations. Thus, a proper elemental composition determination must be conducted before data reconciliation.

Additionally, an elemental composition analysis should be done for each strain and culture condition (protein production, cell stress application, operational mode...) since the elemental composition may change slightly between different conditions and operational modes, even between different batch growth cycles. Moreover, growth medium composition and substrate utilization can also affect biomass elemental composition.

There are two main procedures to analyze biomass elemental composition, depending on the equipment available. The most common elemental analyzers can measure the carbon (C), hydrogen (H), nitrogen (N), and sulfur (S) content of any sample. In addition, other elemental analyzers can measure the oxygen (O) content. If the last one is available, the procedure basically consists of centrifuging a culture sample and washing it with Tris-HCl buffer (pH = 7.5). Then, the next stage is to immediately freeze it, followed by a lyophilization step, and introduce it into an elemental analyzer. Alternatively, if only the first one is available, the O content must be calculated by difference, determining first the ash content (all mineral elements that cannot be burned) by exposing the sample to temperatures over 500 °C. It is assumed that the water content in the samples after the lyophilization step and the mass fraction of hydrogen in the ash can be neglected. However, it is proposed an additional step of drying at 105 °C, but care must be taken that this does not remove volatile compounds and does not cause thermal degradation of cellular components.

2 Materials

2.1 Sample Preparation

1. 50 mL centrifugation tubes.
2. 20 mM Tris-HCl (tris(hydroxymethyl)aminomethane), pH = 7.5.
3. Phosphorus pentoxide (if ash determination is needed).
4. Centrifuge with adapter for 50 mL centrifugation tubes.

2.2 Component Analysis

1. For C, H, O, N, and S determination, an elemental analyzer must be used. One example for the measurement of C, H, N, and S composition is the elemental analyzer EA CE 1108 (Thermo Fisher Scientific) (*see* **Note 1**).
2. For O determination, one example is the elemental analyzer EA Flash 2000 (Thermo Fisher Scientific) (*see* **Note 2**).
3. Analytic balance (with milligram precision).
4. Centrifuge with adapter for 50 mL centrifugation tubes.
5. −80 °C freezer.
6. Freeze dryer.
7. Desiccator filled with silica gel (*see* **Note 3**).
8. Ceramic cups for ash determination.
9. Oven at 550 °C for ash determination.

3 Methods

3.1 Sample Preparation

1. Extract a sample from the fermenter/flask. To assure a minimum of 1 g of lyophilized sample with total homogeneity, the sample volume should be around 50 mL if the biomass concentration is higher than 50 g of dry cell weight (DCW) per liter or 100 mL if the biomass concentration is lower (*see* **Note 4**).
2. Centrifuge for 5 min at 4500 × *g* and 4 °C.
3. Discard the supernatant and resuspend the pellet with approximately the same initial volume with 20 mM Tris-HCl buffer (pH = 7.5) (*see* **Note 5**).
4. Repeat the washing procedure 3 more times.
5. Freeze the sample at −80 °C (*see* **Note 6**).
6. Lyophilize the frozen sample for 24 h. A crystalline powder should be obtained after all the water has evaporated.
7. Grind the sample manually with a lab spatula to a homogenous fine powder.

8. After lyophilization, optionally dry the sample in a desiccator for 24 h if there is any indication that it might still contain moisture.
9. Keep the dry sample hermetically sealed and store it at room temperature in the absence of light until elemental composition analysis for the minimum period.

3.2 Component Analysis

1. Weigh 1000 mg (*see* **Note 7**) of lyophilized (and dried if applied) yeast powder and introduce it into the elemental analyzer. Take at least 3 aliquots of 1000 mg of homogenized lyophilized yeast powder and introduce each of them separately in the analyzer for three independent combustions.
2. Until this step, the procedure is equal irrespective of the availability of oxygen determination. If oxygen composition can be measured, then the ash content can be considered rather equal to the remaining percentage other than C, H, O, N, and S.
3. Alternatively, if no oxygen composition measurement is available, it can be estimated based on the following rule:

$$\%O = 100 - \%C - \%H - \%N - \%S - \%Ashes$$

4. To calculate the ash fraction, weigh accurately 1000 mg of lyophilized yeast powder in a ceramic or glass container (no plastic, *see* **Note 8**) and put it into the oven at 550 °C. After a minimum of 6 h, introduce it in the desiccator (containing phosphorus pentoxide) under a water-free atmosphere until it cools down, and weigh the sample again. Assure that the sample weight is constant. If not, keep the sample in the oven until the measured weight is constant. The remaining weight is equivalent to the ash content.

3.3 Evaluation of Samples

1. Table 1 shows a standard elemental composition of *K. phaffii* growing on a chemostat (see Chapter 33 in this volume) with glucose as a carbon source. It may be used as an approximation if needed.

Table 1
The standard elemental composition of *K. phaffii* growing on a chemostat with glucose as a carbon source [1]

% C	% H	% O	% N	% S	% Ashes
40–45*	6–8*	32–36*	7–8*	0.1–0.2*	6–12*

*% w/w

2. The chemical formula of the biomass can be calculated by dividing the percentage of each component by its molecular weight. Then, it is usually normalized by the carbon atoms. For example, for a given composition:

$$43.0\%C; 7.50\%H; 35.4\%O; 7.60\%N; 0.15\%S$$

3. The chemical formula would be:

C: 43.0 / 12 = 3.58;	3.58 / 3.58 = 1.00
H: 7.50 / 1 = 7.50;	7.50 / 3.58 = 2.09
O: 35.4 / 16 = 2.21;	2.21 / 3.58 = 0.617
N: 7.60 / 14 = 0.543;	0.543 / 3.58 = 0.152
S: 0.15 / 32 = 0.005;	0.005 / 3.58 = 0.001
$CH_{2.09}O_{0.617}N_{0.152}S_{0.001}$	

4. For the calculation of the C-mol value (the grams of biomass per mole of carbon of the biomass), Formula 1 can be used:

$$C - \text{mol} = \frac{1}{\frac{\left(\frac{\%C}{100}\right)}{12}} \quad (1)$$

4 Notes

1. Its functioning is based on the combustion of the sample using vanadium pentoxide and tin powder as combustion helpers and the posterior analysis of oxides generated in the combustion. Sulfonamide can be used as a standard.
2. Its functioning is based on the pyrolysis of the sample and the posterior analysis of the generated compounds.
3. If ash determination is needed, the desiccator should contain phosphorus pentoxide.
4. If biomass concentration is lower than 10 gDCW/L, consider taking higher sampling volumes.
5. It is very important to completely wash the sample with 20 mM Tris-HCl. Thus, assure that all the pellet is resuspended. Incomplete sample washing could lead to errors in elemental composition, since substrate, salts, or other compounds could be introduced to the elemental analyzer if they settle together with the biomass in the centrifugation steps. However, resuspend the pellet carefully in this washing step to avoid cell lysis and the removal of soluble intracellular components.

6. In this step, the pellet layer must not exceed 0.5 cm thick. The sample can be frozen by submerging the 50 mL centrifugation tube in liquid nitrogen and then placed in the freeze-dryer container and stored in the freezer at −80 °C. Alternatively, it can be frozen by placing it in the freeze-dryer container and directly putting it in the freezer at −80 °C.
7. It is not necessary to weigh exactly 1000 mg. However, what is really necessary is that the weighing is very accurate and correctly noted (milligram precision).
8. If an oxygen analyzer is not available and ash content must be measured, make sure that the weight of the container does not change when introduced into the oven. For that reason, use ceramic containers like ceramic cups.

Acknowledgments

This work was funded by the Project PID2022-136936OB-I00 MCIN/AEI/10.13039/501100011033/FEDER, UE of the Spanish Ministry of Science and Innovation.

Competing Interest Statement The authors have no conflicts of interest to declare that are relevant to the content of this chapter.

References

1. Carnicer M, Baumann K, Töplitz I et al (2009) Macromolecular and elemental composition analysis and extracellular metabolite balances of *Pichia pastoris* growing at different oxygen levels. Microb Cell Factories 8:65. https://doi.org/10.1186/1475-2859-8-65
2. Ponte X, Montesinos-Seguí JL, Valero F (2016) Bioprocess efficiency in *Rhizopus oryzae* lipase production by *Pichia pastoris* under the control of P_{AOX1} is oxygen tension dependent. Process Biochem 51:1954–1963. https://doi.org/10.1016/j.procbio.2016.08.030
3. Jordà J, Jouhten P, Cámara E et al (2012) Metabolic flux profiling of recombinant protein secreting *Pichia pastoris* growing on glucose: methanol mixtures. Microb Cell Factories 11: 57. https://doi.org/10.1186/1475-2859-11-57

Part IV

DNA Parts and Expression Constructs

Chapter 11

Pichia pastoris Expression Plasmid Assembly Using Type IIS Restriction Enzymes

Alexandra Charlotte Ilse Grebe, Lia Jade Schallengruber, Carla Magdalena Aguilar Gomez, and Lukas Rieder

Abstract

A variety of cloning techniques are commonly used for the assembly of expression cassettes used for the transformation of *Komagataella phaffii* and, finally, the production of recombinant proteins. Restriction endonucleases are particularly useful due to their ability to recognize and cleave specific sequences in the DNA, allowing simple and reliable cloning strategies. However, the recognition sites of type II restriction enzymes leave a trace upon digestion since they cut within the palindromic recognition site. Hence, type IIS restriction enzymes that cleave outside of their recognition site became valuable tools for convenient and seamless single- and multi-gene cloning.

Key words *Pichia pastoris*, *Komagataella phaffii*, Type IIS restriction enzymes, Seamless cloning

1 Introduction

Molecular cloning allows the engineering of organisms with desired gene elements to imbue them with specific characteristics [1]. Since the discovery that restriction enzymes produce specific fragments of a DNA sequence, they have become an important tool for molecular cloning and the construction of vectors for recombinant DNA as they allow for targeted and specific DNA digestion based on specific recognition sites [2, 3].

Restriction enzymes are Endonucleases that recognize and cleave both strands of DNA at specific sequences. There are three different types of restriction enzymes, classified as I, II, and III, which differ regarding their cleavage method, subunit composition, and cofactor requirements. Type I enzymes normally have three different subunits and cleave DNA at seemingly random

Alexandra Charlotte Ilse Grebe and Lia Jade Schallengruber contributed equally with all other contributors.

Xavier Garcia-Ortega et al. (eds.), *Pichia pastoris: Methods and Protocols*, Methods in Molecular Biology, vol. 2697, https://doi.org/10.1007/978-1-0716-4779-0_11,

locations far from the recognition site [4]. Type II enzymes have two identical subunits and typically cleave DNA close to or within 4–8 bp of the recognition sequence. They require Mg^{2+} as a cofactor [5]. "Standard" type II enzymes recognize palindromic and partially palindromic sequences and cleave DNA around or within the recognition site, resulting in "sticky" (with a 5′ or 3′ overhang) or "blunt" (no overhang) ends. They have high specificity and are widely commercially available [2]. Such enzymes include popular restriction enzymes such as *Eco*RI, *Bam*HI, and *Hind*III [6]. Type III enzymes must interact with two different unmethylated DNA sequences in inversely repeated head-to-head orientations for effective cleavage at 25–27 bp downstream [6–8].

The ligation of DNA fragments produced by restriction enzyme digestion into predefined standard *Komagataella phaffii* (reclassified from *Pichia pastoris*) expression vectors facilitates the reliable and simple construction of expression vectors. Standard expression vectors for restriction/ligation cloning exist for intra- and extracellularly targeted proteins. Early vectors of Invitrogen, such as pHILD2, pPIC9K, and the pPICZ series as well as the PichiaPink™ vectors, are all based on available multiple cloning sites and the use of type II restriction enzymes. They are still frequently used, but scars are produced at the cloning sites due to the specific recognition sites. This might cause a significant influence on the efficiency of gene expression.

Alternatively, new vector designs avoid the use of type II restriction enzyme recognition sites, instead making use of type IIS restriction enzymes and allowing seamless cloning [4]. The subclass type IIS restriction enzymes typically have a recognition site of 4–7 bp. However, the defined cleavage site is 1–20 bp away from this recognition site [9]. Their recognition sites are non-palindromic, and the generated overhangs depend on the sequence neighboring the recognition site. Their particular characteristics (the recognition site is independent of the generated overhang) make them useful for seamless cloning [10]. Two digested fragments can be ligated to form a product that lacks the initial restriction site by properly locating the recognition and cleavage sites [11]. This enables the combination of restriction and ligation in a single one-pot reaction in which vector, insert, restriction enzyme, and ligase are mixed [12]. Thus, cloning/restriction procedures can be simplified and significantly accelerated. Furthermore, upon cleavage with a type IIS restriction enzyme, it is possible to assemble multiple DNA fragments in any desired order and direction in a single ligation reaction [13], especially desired for the cloning of metabolic pathways. There are more than 240 commonly known and used type II restriction endonucleases, of which 80 are known to be type IIS and over 50 are commercially offered by known suppliers of cloning reagents [12]. Typical type IIS restriction enzymes used for the construction of *K. phaffii* expression vectors are *Bsa*I, *Sap*I, and *Mly*I [9].

The restriction enzyme *Bsa*I (*Eco*31I) recognizes the sequence 5′-GGTCTC(N1)/(N5)-3′ and produces 4 bp sticky overhangs that can be used for in-frame cloning of the desired product (Fig. 1) [9]. It is the enzyme utilized in many vectors (e.g., pJAZ) of BioGrammatics, Inc. (Carlsbad, CA, USA), one of the early companies with a focus on *P. pastoris* and founded by the inventors of the early classic *Pichia* expression system for protein production by *K. phaffii*.

Another type IIS restriction enzyme, namely *Sap*I (*Lgu*I), recognizes the sequence 5′-GCTCTTC(N1)/(N4)-3′ and produces 3 bp sticky overhangs (Fig. 2) [9] (*see* **Note 1**). It is commonly used by ATUM, a company offering gene synthesis and gene expression services in the Bay Area, CA. Also, most vectors from the Bisy GmbH (Hofstätten an der Raab, Austria), such as the vectors pBSY3Z or pBSY3H, enable cloning with this enzyme for type IIS cloning (Fig. 3), in addition to commonly used cloning by recombination, such as Gibson cloning.

A type IIS restriction enzyme that creates blunt overhangs is *Mly*I (*Sch*I). It recognizes the sequence 5′-GAGTC(N5)/N(5)-3′, producing 5 bp long blunt overhangs (Fig. 4) [9] (*see* **Note 2**). The blunt-end cloning vectors such as pBSYAOXsec_blunt (Bisy GmbH, Hofstätten an der Raab, Austria), for example, also offer an *Mly*I site for type IIS cloning at the preferred site for the translation start between promoter and transcription terminator sites of the vector [14].

The fact that type IIS enzymes don't cleave the DNA at their recognition site also enables seamless molecular cloning via so-called Golden Gate Assembly [15]. As described previously, the sticky ends produced by enzymes such as *Sap*I and *Bsa*I allow orientated cloning, enabling for the application in a simple one-pot

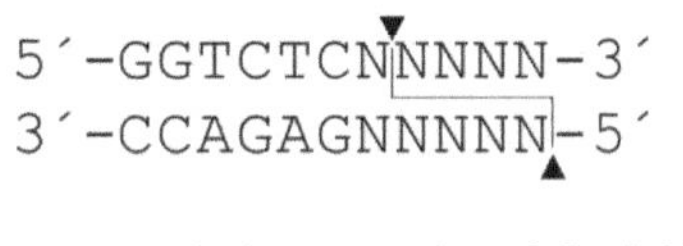

Fig. 1 Recognition sequence and cleavage site of *Bsa*I. Variable base pairs are denoted as "N," cleavage site indicated by arrows and lines

Fig. 2 Recognition sequence and cleavage site of *Sap*I. Variable base pairs are denoted as "N," cleavage site indicated by arrows and lines

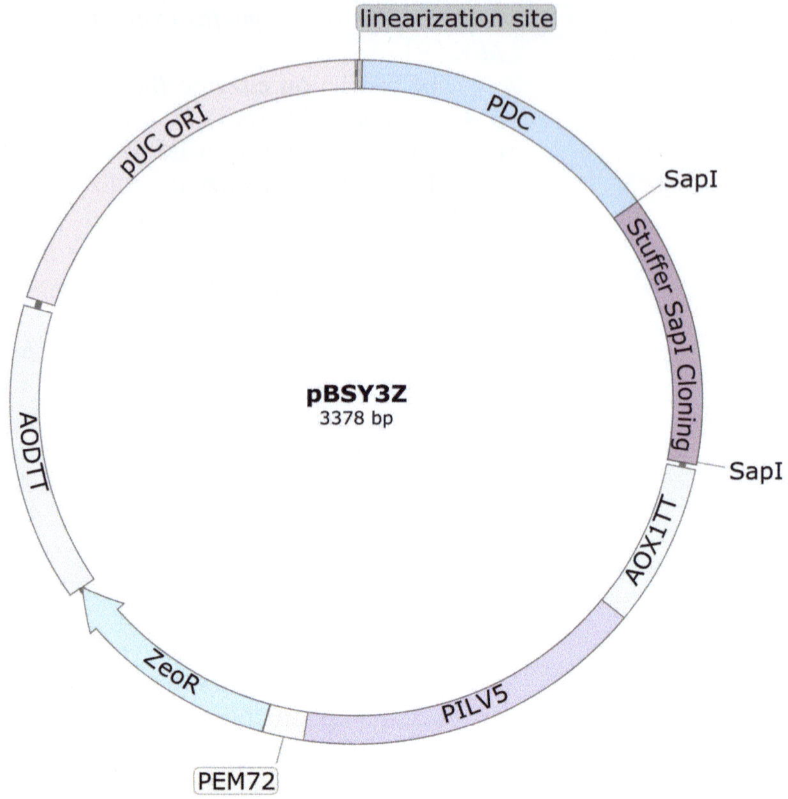

Fig. 3 Plasmid map of the pBSY3Z expression plasmid. Commercial plasmid from Bisy GmbH (Hofstätten an der Raab, Austria) carrying a stuffer fragment between the PDC promoter and the *AOX1* transcription terminator (*AOX1*TT) flanked by *Sap*I restriction sites, a Zeocin resistance gene (ZeoR) under the control of the eukaryotic promoter P_{ILV5} and the prokaryotic promoter P_{EM72} and the *AOD* transcription terminator (*AOD*TT) as well as a (*Smi*I) linearization site and the pUC origin of replication (pUC ORI)

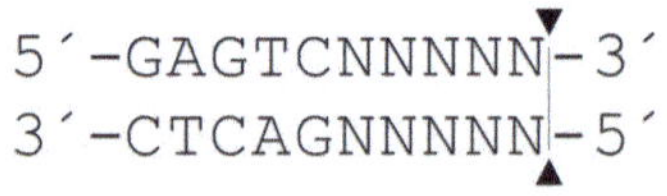

Fig. 4 Recognition sequence and cleavage site of *Mly*I. Variable base pairs are denoted as "N," cleavage site indicated by arrows and lines

reaction (Fig. 5) and a flexible design of the desired sticky-end desired overhangs. For example, for the enzyme *Sap*I, generating a 3-base overhang, those 3 bases are usually the universal ATG (start codon) at the 5′ end of the target gene and TGA (a stop codon) for the 3′ site.

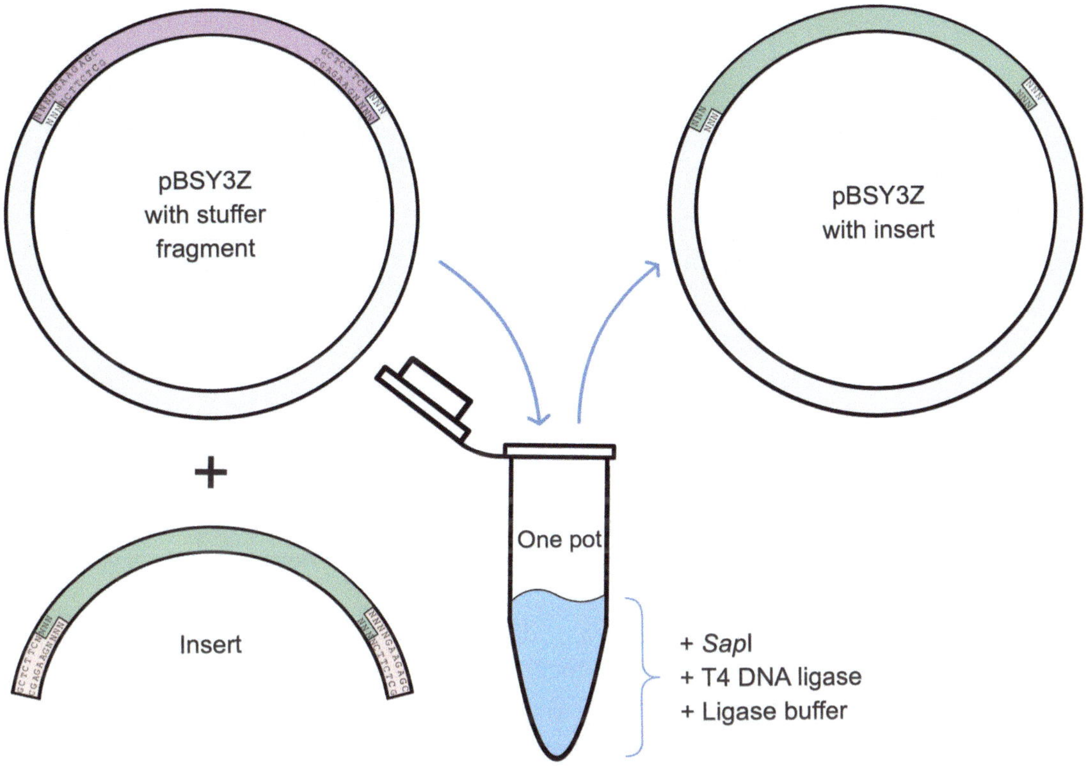

Fig. 5 Scheme of one-pot cloning with pBSY3Z using *Sap*I. Vector backbone shown in light blue, stuffer fragment in purple, insert (including GOI) in green, flanked by external *Sap*I recognition sequences shown in orange. *Sap*I recognition sequences as well as the associated cleavage sites are indicated via the corresponding nucleotide sequence and lines

2 Materials

In this chapter, cloning is exemplified based on the vector pBSY3Z (bisy GmbH, Hofstätten an der Raab, Austria), which is made for *Sap*I seamless restriction cloning and employs a 500 bp fragment of the CAT1 promoter of *K. phaffii* to regulate transcription of cloned target genes. This promoter is repressed by high concentrations of repressing carbon sources such as glucose or glycerol, derepressed to a middle-strength promoter upon depletion of such carbon sources, and can be further strongly induced with inducing carbon sources such as methanol or oleic acid.

2.1 Preparation of the Insert

1. Gene of interest (GOI), insert (synthetic DNA, PCR product, …).
2. Phusion™ Plus DNA Polymerase with appropriate buffer.
3. 10 μM of each primer pair is needed for amplification of the GOI that introduces the required restriction sites. For example, if using *Sap*I (e.g., using the vector pBSY3Z, Bisy GmbH,

Hofstätten an der Raab, Austria), the following primers are recommended: 5′-XXXXGCTCTTCTNNN-3′ (FWD) and 5′--XXXXGCTCTTCTNNN-3′ (REV). Variable base pairs enabling restriction enzyme binding are hereby denoted as "X" and base pairs that are dependent on the insert and will create specific overhangs for cloning are designated as "N" (*see* **Note 3**).

4. 10 mM dNTPs.
5. PCR tubes.
6. Microcentrifuge.
7. Thermocycler.
8. 1.5 % (w/v) agarose.
9. 1× TAE-buffer.
10. Gel electrophoresis chamber.
11. Power supply, (e.g., Bio-Rad Power Pac™ Basic from Bio-Rad).
12. Gel imager (e.g., UVP GelDoc-It™ Imager from Bio-Rad or other photo documentation systems).
13. PCR clean-up kit (e.g., Wizard® SV Gel and PCR Clean-Up System from Promega).
14. 1.5 mL reaction tube.
15. Thermomixer.
16. Nanodrop 1000 Spectrophotometer.

2.2 One-Pot Cloning

1. Plasmid that is designed for type IIS cloning and has *Sap*I restriction sites (e.g., pBSY3Z, Bisy GmbH, Hofstätten an der Raab, Austria).
2. GOI with *Sap*I restriction sites (*see* **Notes 4** and **5**).
3. *Sap*I (10 U/μL).
4. T4 DNA Ligase (400 U/μL) (*see* **Note 6**).
5. 10× CutSmart buffer.
6. ATP (10 mM) (*see* **Note 7**).
7. 1.5 mL reaction tubes or PCR tubes.
8. Thermocycler.
9. Ice box filled with ice.
10. Chemically competent *E. coli* cells of a cloning strain (e.g., XL1-blue).
11. Thermomixer.
12. SOC medium: 0.5% (w/v) yeast extract, 2% (w/v) tryptone, 20 mM (w/v) glucose, 10 mM NaCl, 2.5 mM KCl, 10 mM

$MgCl_2$, 10 mM $MgSO_4$. Autoclave and store at room temperature.

13. Plastic Petri dishes: 94 mm × 16 mm.
14. LB-Zeo plates: 0.5% (w/v) yeast extract, 1% (w/v) tryptone, 1% (w/v) NaCl and 1.5% (w/v) agar-agar. Autoclave, then add Zeocin™ to a final concentration of 25 μg/mL.
15. Plasmid isolation kit (e.g., Promega Wizard® Plus SV Minipreps DNA Purification Systems).
16. Possibility for plasmid sequencing.

3 Methods

3.1 Preparation of the Insert

In order to conduct a one-pot reaction, it is necessary to initially modify the GOI through PCR, adding requisite flanking cleavage sites. Subsequently, the PCR product is validated via gel electrophoresis and then purified.

1. Amplify GOI with primers that introduce the required restriction sites via PCR (Tables 1 and 2).
2. Mix 5 μL of the PCR reaction with 1 μL of loading dye and load on the 1.5% (w/v) agarose gel.
3. Separate DNA by gel electrophoresis at 120 V and 400 mA for 35 min in 1× TAE buffer.
4. If the amplicon has the expected size, purify the remaining 45 μL of the PCR reaction using the PCR clean-up kit and assess DNA concentration using the Nanodrop.

Table 1
PCR reaction mixture for amplification of the GOI to introduce the necessary overhangs for the One-Pot cloning approach

Component	Volume
Polymerase	0.5 μL
Reaction buffer	10 μL
Template DNA	0.5–1 ng (X μL)
FWD primer	2.5 μL
REV primer	2.5 μL
dNTPs	1 μL
Nuclease-free H_2O	33.5 − X μL
Total	50 μL

X is the volume of template DNA which is dependent on its concentration

Table 2
Thermal cycling profile used for the amplification of GOI

Time	Temperature	# of cycles
30 s	98 °C	1
10 s	98 °C	35
20 s	X °C (primer dependent)	
30 s/kb	72 °C	
10 min	72 °C	1
∞	4 °C	1

Table 3
Composition of the one-pot reaction mixture

Component	Amount
10× CutSmart buffer	2 μL
ATP (10 mM)	2 μL
*Sap*I (10 U/μL)	0.25 μL
T4 DNA ligase (400 U/μL)	0.25 μL
Vector	X μL (75–100 ng)
PCR product	Y μL (2 times molar excess relative to the vector)
Nuclease-free H_2O	20-X-Y μL
Total	20 μL

3.2 One-Pot Cloning

Once the insert is prepared, the one-pot reaction can be set up. Following a predetermined incubation period, *E. coli* is transformed with the reaction mixture, the plasmids are isolated, and the sequences of the obtained clones are confirmed by sequencing.

1. Set up the one-pot cloning reaction (Table 3) (*see* **Note 8**).
2. Incubate the reaction for 1 h at room temperature.
3. Add 10 μL of the reaction mix to 50 μL of the chemically competent *E. coli* cells.
4. Incubate on ice for 20 min.
5. Heat shock at 42 °C for 30 s.
6. Add 500 μL SOC medium and regenerate cells for 1.5 h at 37 ° C and 350 rpm.
7. Plate 100 μL of the regenerated cells on LB-Zeo plates (*see* **Note 9**).

8. Incubate overnight at 37 °C.
9. Cultivate transformants for plasmid isolation.
10. Isolate the plasmid DNA using a plasmid isolation kit.
11. Prepare samples for sequencing.
12. Plasmids with the correct sequence can now be used for *K. phaffii* transformation (*see* **Notes 10** and **11**).

4 Notes

1. If desired and designed properly, the 3 bp triplet overhang can be used to encode a start codon, amino acid, stop codon, etc.
2. Fragments cleaved with *Mly*I (or other type IIS restriction enzymes that produce blunt ends) can be used for homologous recombination or blunt-end ligation with any other blunt DNA, for example, if the cloning site does not offer universal sequence motifs such as for the start or stop codon.
3. If the GOI encodes an intracellular protein without tags, one may choose ATG on the forward primer and TTA on the reverse primer as the 3 bp overhangs (NNN), allowing the creation of standardized primers for different GOIs. For secreted and tagged proteins, NNN shall be modified to overlap with the signal sequence or tag, respectively. It must be ensured that the resulting overhangs are compatible with the overhangs created on the vector. Hence, in this case, one may need to modify the overhangs on the vector via proper design.

   One can further add longer sequence stretches that are homologous to the GOI to the 3′ end of the primers (after NNN) to ensure proper amplification via PCR; this will not change the overhang created via type IIS endonuclease restriction cut. Variable base pairs "X" can be varied in length and sequence to allow for optimized primer length, GC content, and annealing temperature, given that a minimum length required for restriction enzyme binding is achieved.
4. If the insert is designed in silico, it is advisable to directly design and order the synthetic gene flanked by the approriate recognition sites.
5. Ensure that the GOI does not encode the recognition sequence of the desired type IIS enzyme. Should such a scenario arise, remove the recognition site along the codon optimization process.
6. It is recommended to thaw and resuspend the T4 DNA ligase prior to usage. It should be the final component added to the mixture.

7. The ATP should be thawed once and mixed carefully to make sure the solution is homogeneous prior to use. If needed, the ATP can be portioned into aliquots for single use due to its limited stability.
8. The NEBioCalculator can be used to calculate the molar ratios according to the length and concentration of the desired insert and vector.
9. Alternatively, or additionally, centrifuge the tube at 16,000 rpm for 5 min to pellet the cells. Discard the supernatant, resuspend the cells in the remaining media, and plate the entire resulting suspension.
10. The plasmid must be linearized before transformation.
11. Alternatively, the whole reaction mix can be used for *K. phaffii* transformation.

Acknowledgments

A.C.I.G., L.J.S., and L.R. have received funding from the EIC Pathfinder project NewCat under grant No. 101046815. The financial support of C.M.A.G. by the Austrian Federal Ministry for Digital and Economic Affairs, the National Foundation for Research, Technology and Development and the Christian Doppler Research Association (Christian Doppler Laboratory for innovative *Pichia pastoris* host and vector systems) and by the industrial partners Novo Nordisk A/S, BioGrammatics, SeSam Biotech, ZETA and Solar Biotech is gratefully acknowledged.

Competing Interest Statement The authors declare that they have no conflicts of interest that are relevant to the content of this chapter.

References

1. Weber E, Engler C, Gruetzner R, Werner S, Marillonnet S (2011) A modular cloning system for standardized assembly of multigene constructs. PLoS One 6(2):e16765. https://doi.org/10.1371/journal.pone.0016765
2. Roberts RJ (2005) How restriction enzymes became the workhorses of molecular biology. Proc Natl Acad Sci USA 102(17):5905–5908. https://doi.org/10.1073/pnas.0500923102
3. Danna K, Nathans D (1971) Specific cleavage of simian virus 40 DNA by restriction endonuclease of *Hemophilus influenzae*. Proc Natl Acad Sci USA 68(12):2913–2917. https://doi.org/10.1073/pnas.68.12.2913
4. Pingoud A, Alves J, Geiger R (1993) Restriction enzymes. In: Enzymes of molecular biology, vol 16. Humana Press, New Jersey, pp 107–200. https://doi.org/10.1385/0-89603-234-5:107
5. Williams RJ (2003) Restriction endonucleases: classification, properties, and applications. MB 23(3):225–244. https://doi.org/10.1385/MB:23:3:225
6. Green MR, Sambrook J (2021) Restriction enzymes. Cold Spring Harb Protoc 2021(4): pdb.top101360. https://doi.org/10.1101/pdb.top101360

7. Boyer HW (1971) DNA restriction and modification mechanisms in bacteria. Ann Rev Microbiol 25(1):153–176. https://doi.org/10.1146/annurev.mi.25.100171.001101
8. Rao DN, Dryden DTF, Bheemanaik S (2014) Type III restriction-modification enzymes: a historical perspective. Nucleic Acids Res 42(1):45–55. https://doi.org/10.1093/nar/gkt616
9. Szybalski W, Kim SC, Hasan N, Podhajska AJ (1991) Class-IIS restriction enzymes — a review. Gene 100:13–26. https://doi.org/10.1016/0378-1119(91)90345-C
10. Tóth E et al (2014) Restriction enzyme body doubles and PCR cloning: on the general use of type IIS restriction enzymes for cloning. PLoS One 9(3):e90896. https://doi.org/10.1371/journal.pone.0090896
11. Fromme T, Klingenspor M (2007) Rapid single step subcloning procedure by combined action of type II and type IIs endonucleases with ligase. J Biol Eng 1(1):7. https://doi.org/10.1186/1754-1611-1-7
12. Zhu Z, Samuelson JC, Zhou J, Dore A, Xu S (2004) Engineering strand-specific DNA nicking enzymes from the type IIS restriction endonucleases BsaI, BsmBI, and BsmAI. J Mol Biol 337(3):573–583. https://doi.org/10.1016/j.jmb.2004.02.003
13. Beck R, Burtscher H (1994) Introduction of arbitrary sequences into genes by use of class IIs restriction enzymes. Nucl Acids Res 22(5):886–887. https://doi.org/10.1093/nar/22.5.886
14. Raschmanová H et al (2019) Single-cell approach to monitor the unfolded protein response during biotechnological processes with *Pichia pastoris*. Front Microbiol 10:335. https://doi.org/10.3389/fmicb.2019.00335
15. Engler C, Kandzia R, Marillonnet S (2008) A one pot, one step, precision cloning method with high throughput capability. PLoS ONE 3(11):e3647. https://doi.org/10.1371/journal.pone.0003647

Chapter 12

Autonomously Replicating Sequence-Bearing Plasmids Utilized in *Pichia pastoris*

Carsten Pichler, Florian Weiss, and Anton Glieder

Abstract

Plasmids are a common tool in biotechnology to deliver recombinant DNA into microbial cells for the production of enzymes, pharmaceutical proteins, chemicals, or metabolites. Therefore, a stable plasmid system that provides reliable gene expression over generations is essential for the successful utilization of single-cell organisms in research and production applications. Most *Komagataella phaffii* expression clones are generated by the integration of linear plasmids into the genome, as circular episomal plasmids are not stable under non-selective conditions. The low rate of homology-directed specific integration and the large variation among transformants of random integration limit the organism's application in enzyme engineering approaches or comparative studies where high transformation rates and uniform expression levels are desired. In the yeast *Saccharomyces cerevisiae*, the problem of circular plasmid stability and partition to the daughter cells during mitosis has been solved by combining centromeric sequences or elements of the 2-micron plasmid with an autonomously replicating sequence (ARS) that serves as an origin of replication. Similar attempts have not yet been successful or widely adapted in *K. phaffii*; hence, permanent selection pressure is required to maintain episomal plasmids in *K. phaffii*. There are no reports so far about functional 2-micron plasmids for *P. pastoris*, and CEN/ARS plasmids for *P. pastoris* are usually rather large and do not provide the high transformation rates as known for episomal plasmids of *S. cerevisiae* expression systems. However, the availability of a broad set of resistance, auxotrophic, and carbon source utilization markers facilitates reliable plasmid selection in small-scale screening applications and recently also proved to be successful for bioreactor-scale expression. This allows the combined advantages of high transformation rates and low clonal variability of ARS plasmids to be exploited. This article describes the successful utilization of ARS1-containing plasmids in *K. phaffii*, including antibiotic-free selection, complementation of knockout strains, or even for the application of CRISPR/Cas by transient gRNA and CAS9 gene expression in *K. phaffii*.

Key words *Pichia pastoris*, *Komagataella phaffii*, Autonomously replicating sequence, ARS plasmid, PARS1, Episomal expression

---

Carsten Pichler and Florian Weiss contributed equally with all other contributors.

Xavier Garcia-Ortega et al. (eds.), *Pichia pastoris: Methods and Protocols*, Methods in Molecular Biology, vol. 2697, https://doi.org/10.1007/978-1-0716-4779-0_12,

1 Introduction

Plasmids are a common tool to deliver genetic information for recombinant protein production or genetic engineering in single-cell microorganisms like bacteria or yeasts and therefore play a major role in biotechnology, like in the production of proteins or enzymes [1, 2]. To ensure extrachromosomal plasmid stability and consistent copy numbers over generations, specific DNA sequences are necessary to enable integration into the cell's endogenous replication process. In bacteria, these sequences contain a binding site (DnaA-boxes) for the initiator protein DnaA, DNA unwinding elements (DUE; AT-rich sequences), and binding sites for assisting proteins. All those sequences together are called the origin of replication (*ori*) [3].

In the budding yeast *Saccharomyces cerevisiae*, autonomously replicating sequences (ARSs) support the replication initiation and stable maintenance of episomal plasmids. Many of those ARSs were identified from the genome (roughly 400 sequences), where they exhibit function as the origin of replication. They are a few hundred base pairs long and mainly contain three to six domains (A, B1, B2, B3, B4, and C) that are essential for their function. These domains include the eleven base pairs consensus sequence (ACS; ARS consensus sequence: [A|T]TTTAT[A|G]TTT[A|T]), DUE, binding sites for the origin recognition complex, and other sequences for DNA-protein interactions [3, 4].

For the methylotrophic yeast *Komagataella phaffii* (reclassified from *Pichia pastoris*), multiple ARSs have been described recently as functional, including a mitochondrial ARS, panARS, and even an efficient new ARS from *Candida boidinii* [5–8]. The two most extensively studied ARSs of *K. phaffii*, which have been well-documented and described since 1985, are the *Pichia* autonomously replicating sequence 1 (PARS1; 164 bp) and 2 (PARS2; 385 bp) [9]. PARS1 and PARS2, as well as *S. cerevisiae* ARSs, display a high adenine and thymine content, and both exhibit homology to the *S. cerevisiae* ACS. They have at least ten out of eleven matching base pairs, with PARS1: ATT**A**ATATTTA and PARS2: TTT**C**ATATTTT/TT**A**TATATTTA. The boldly highlighted nucleotides in PARS1 and PARS2 represent the difference between the *S. cerevisiae* and the *K. phaffii* ACS). Additionally, the PARSs contain multiple eight to ten base pairs of direct and inverted repeats [9]. Despite these similarities, most *Pichia* ARSs do not function in *S. cerevisiae*, and *vice versa*. It seems that either a sequence within the ARS fragments inhibits the function in the respective organism or that host-specific sequences additional to the consensus sequences are necessary for ARS activity [7, 9]. In addition to PARS1 and PARS2, a vast number of replicators in *Pichia* have been identified. Approximately one-third of them contain G/C-rich sequences, some of which may harbor the potential to stabilize episomal plasmids [10].

Due to the known instability of episomal plasmids, the integration of plasmids into the *K. phaffii* genome either via unspecific non-homologous end-joining or site-specific homology-directed repair is the most common practice for production clone generation. While both approaches bring individual advantages for certain applications, the big variation in recombinant gene expression levels of the former and the low efficiency of the latter method are not well-suited for applications where uniform expression and high transformation rates are desired, e.g., for screening mutagenesis libraries or comparative analyses of genetic elements. Circular ARS-containing plasmids have been described to show transformation efficiencies of 10^6 CFU/μg compared to linear integrative plasmid efficiencies of 10^3 CFU/μg and, due to their episomal maintenance, lead to more homogenous recombinant protein production among transformants [11–13]. They can be present in multiple copies (up to 30 copies per cell) but are inherently unstable (up to 20% loss rate per cell division). Therefore, selective pressure is required to retain the plasmid over generations [9, 11, 14–16]. This drawback of low mitotic stability, especially of plasmids with an ARS1 of *K. phaffii*, can be used as an advantage when transient expression is beneficial, e.g., when employing CRISPR/CAS systems or FLP or CRE recombinases for genome engineering. In this case, the loss of the method's plasmid(s) is usually desired and can simply be reached by removing the selective pressure [17, 18].

ARS-containing plasmids were successfully stabilized in *S. cerevisiae* by the addition of a cell-derived centromeric sequence (CEN3), but with the drawback that this reduced the copy number in the direction of one to two plasmids per cell [19], and observed transformation rates were low compared to ARS-based episomal plasmids. Efforts to establish centromere-stabilized plasmids in *K. phaffii* have not been widely adapted so far. Partially, this might be due to a large plasmid size (centromeric sequences consist of large inverted repeats), low copy numbers, and a poor transformation rate [11, 20, 21].

As an alternative to the above-mentioned plasmid systems, there is still the 2-micron plasmid from *S. cerevisiae* available. The 2-micron plasmid is a stable circular DNA element that can also maintain high-copy numbers. Those features were used for the production of recombinant proteins in *S. cerevisiae*. Although this efficient episomal plasmid system is well established in *S. cerevisiae*, it was not possible to transfer this system to *K. phaffii* [22, 23]. Therefore, the best approaches for episomal plasmids in *K. phaffii* are still ARS-containing plasmids in combination with permanent selection pressure.

One of the commonly used methods is based on maintaining the selection pressure by utilizing an antibiotic marker system [8, 18, 20, 24–27]. Antibiotics have proven to be convenient additives for culturing and screening positive *K. phaffii* transformants in small-scale [8, 18]. There are several of these marker systems available, and all of them are well established for use in *K. phaffii* (e.g., the *ble* resistance gene against zeocin, Fig. 1a, or the *hph* resistance gene against hygromycin B) [8, 18, 20, 24–27]. Although most of these genes derive from bacterial organisms and play an essential role in research, *K. phaffii* strains bearing antibiotic-resistance genes are undesirable when it comes to recombinant protein production for the pharmaceutical or food and feed industries [28]. The cost, long-term instability, and necessity to remove them during downstream processing make antibiotics not feasible for large-scale production processes [28, 29]. Besides that, the antibiotic resistance genes can impose a potential metabolic

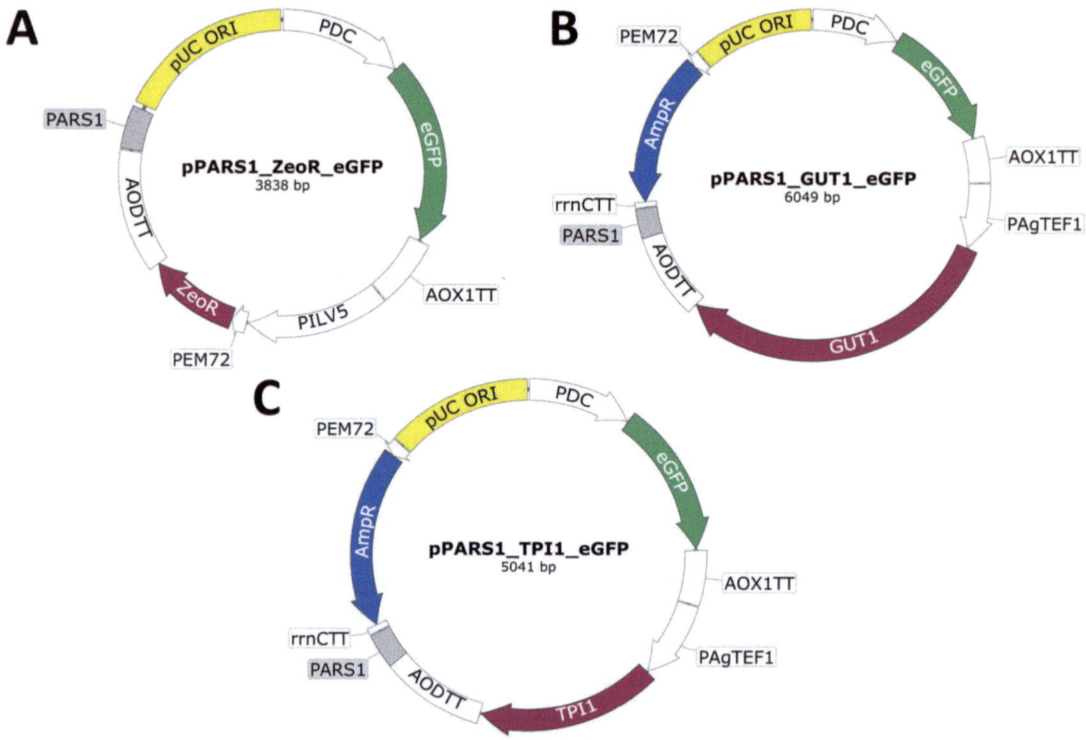

Fig. 1 Plasmid maps representing exemplary ARS plasmids with antibiotic/Zeocin- (**a**), *GUT1*- (**b**) and *TPI1*- (**c**) based selection. Besides the *K. phaffii* selection marker cassettes, the *Pichia* autonomously replicating sequence 1 (PARS1), an eGFP expression cassette controlled by the PDC promoter and the pUC Ori for replication in *E. coli* are present on the plasmids. Plasmids **b** and **c** additionally harbor an Ampicillin resistance cassette to make these vectors suitable as *E. coli*—*K. phaffii* shuttle vectors. For this purpose, the Zeocin resistance gene in vector **a** is additionally controlled by the synthetic bacterial promoter P_{EM72}. The plasmid sequences can be retrieved from the GenBank database with the accession numbers PP793737 (pPARS1_-ZeoR_eGFP), PP793738 (pPARS1_GUT1_eGFP) and PP793736 (pPARS1_TPI1_eGFP)

burden, affect the cells' growth rate, or interfere with cellular functions such as gene expression [29].

Alternatively, auxotrophic *K. phaffii* strains (e.g., Δ*arg4*, Δ*ura3*, Δ*his4*, Δ*met2*, Δ*ura5*, Δ*ade1*, Δ*pro3*, and Δ*pha2*), which are unable to synthesize certain essential substances, can be complemented with the respective genes presented on plasmids [5, 10, 30]. Contrary to the antibiotic marker selection, prototrophic markers are not dominant; hence, certain *K. phaffii* knockout strains deficient in the respective gene are necessary [13, 31]. While this allows for simple selection in defined media, cross-feeding events during cultivation might reduce the selective pressure and lead to marker-less subpopulations [32, 33].

Carbon source utilization-based (auto-) selection represents the third option that can be used to maintain selection pressure in *K. phaffii*. Therefore, genes that are essential for the glycerol utilization pathway (Δ*gut1*, Fig. 1b or Δ*tpi1*, Fig. 1c) were deleted to obtain the respective deficiency strains [13, 31]. By complementing these *K. phaffii* strains with the corresponding gene, positive transformants can be selected on defined medium with glycerol as the sole carbon source. This method demonstrates an inexpensive and reliable way for transformant selection of both integrative and episomal ARS plasmids and subsequent protein production in *K. phaffii*. It is presumed that the utilization of inexpensive carbon sources for cross-feeding is less problematic due to the larger quantities required for cell growth in comparison to the use of amino acids or nucleotides [13].

The described high transformation rates and uniform expression levels make ARS plasmids an excellent choice for screening gene or mutant libraries in protein engineering approaches. They address the limitations of genomic integration and ensure consistent titers of proteins produced by recombinant expression across various individual transformants, given that selective pressure is maintained.

In this article, a universal protocol for the handling of ARS plasmids in *K. phaffii* employing antibiotic or glycerol-utilization markers is illustrated. Based on the objective, the obtained transformants can undergo screening on a small scale using 96-deep well plates (96-DWPs) [34], enabling high-throughput screening of gene variants/libraries in protein engineering approaches. But they can also be cultivated on a larger scale to produce significant amounts of the protein of interest for further analysis. Note that selective pressure must be applied at all times to maintain the plasmid within the cell.

2 Materials

2.1 Host Strains and Plasmid

1. Desired *K. phaffii* strain as host for transformations (*see* **Note 1**).
2. A plasmid of choice containing an ARS, a selection marker cassette, and a gene of interest (GOI). Three exemplary plasmids are shown in Fig. 1a–c (*see* **Notes 2** and **3**). The plasmid sequences were deposited in the GenBank database and can be retrieved with the accession numbers PP793737 (pPARS1_-ZeoR_eGFP), PP793738 (pPARS1_GUT1_eGFP), and PP793736 (pPARS1_TPI1_eGFP).

2.2 Equipment and Kits

1. Microcentrifuge.
2. Centrifuge for 50 mL tubes.
3. Electroporation system with appropriate electroporation cuvettes (2 mm gap).
4. Heat block with shaking device.
5. 28 °C incubator.
6. 94/16 mm Petri dishes with ventilation cams.
7. Sterile spatula for plating cell suspension on selection agar plates.
8. Sterile reaction tubes (1.5 mL, 2 mL, 15 mL, and 50 mL).
9. Sterile pipette tips.
10. Spectrophotometer and cuvettes for optical density determination.

2.3 Media and Reagents

1. $_{dd}H_2O$: double deionized water, autoclaved.
2. 1 M dithiothreitol (DTT): Dissolve 15.4 g DTT in 100 mL $_{dd}H_2O$, sterile filter, and store at −20°C.
3. 1 M sorbitol: Dissolve 182.2 g sorbitol in 1.0 L $_{dd}H_2O$ and autoclave.
4. 20% (w/v) Glucose solution: dissolve 200 g D(+)-glucose in 1.0 L $_{dd}H_2O$ and autoclave.
5. Antibiotic stock solutions: For preparation of selective agar plates, e.g., 100 mg/mL Zeocin; sterile filter, and store at −20°C.
6. 4×10^{-5}% (w/v) Biotin stock solution: Dissolve 10 mg biotin in 50 mL $_{dd}H_2O$, sterile filter, and stored at 4°C.
7. BEDS: Dissolve 0.816 g bicine and 91 g sorbitol in 400 mL $_{dd}H_2O$. Adjust the pH to 8.3 using 3M NaOH solution. Add 15 mL ethylene glycol and 25 mL dimethyl sulfoxide, fill up to 500 mL with $_{dd}H_2O$, sterile filter, and store at −20°C.

8. 1M Potassium phosphate buffer (PP_i): 46.4 g K_2HPO_4 and 100 g KH_2PO_4 in 900 mL $_{dd}H_2O$. Adjust the pH to 6.0 with 3M KOH or 85% H_3PO_4 and fill up to 1.0 L with $_{dd}H_2O$, autoclave.
9. 13.4% (w/v) YNB: Autoclave 134 g YNB (yeast nitrogen base without amino acids and 0.5% (w/v) ammonium sulfate) dissolved in 1.0 L $_{dd}H_2O$ (*see* **Note 4**).
10. BMG1: Add 10 g of glycerol to 650 mL of $_{dd}H_2O$ and autoclave. Let it cool down and add 200 mL 1M PP_i-buffer (pH 6.0), 100 mL 13.4% (w/v) YNB, and 2 mL 4×10^{-5}% (w/v) biotin (*see* **Notes 4** and **5**).
11. BMG1 agar: Prepare BMG1 medium and add 15 g/L agar-agar before autoclaving. When lukewarm, pour the medium into Petri dishes (*see* **Note 6**).
12. YPD medium: Dissolve 20 g peptone and 10 g yeast extract in 900 mL $_{dd}H_2O$, then autoclave. Let the YPD medium cool down and add 100 mL sterile 20% glucose solution (the final concentration should be 2%) (*see* **Note 7**). If required, add the respective antibiotic in the desired concentration, e.g., 1 mL of 100 mg/mL Zeocin stock to reach a final concentration of 100 mg/L (*see* **Note 8**).
13. YPD agar: Prepare like YPD medium, but add 15 g/L agar-agar before autoclaving. When lukewarm, add 100 mL of the 20% glucose solution and the respective antibiotic in the desired concentration, e.g., 1 mL of 100 mg/mL Zeocin stock to reach a final concentration of 100 mg/L. Pour the YPD agar into Petri dishes (*see* **Notes 6** and **8**).
14. YPEtOH medium: Dissolve 20 g peptone and 10 g yeast extract in 975 mL $_{dd}H_2O$ and autoclave. Once it is cooled down, add 5 mL of 20% glucose and 20 mL of 100% ethanol.
15. YPD:Sorbitol: Mix equal volumes of YPD and 1M sorbitol.

3 Methods

3.1 Preparation of Competent *K. phaffii* Cells

For the preparation of electrocompetent *K. phaffii* cells, the "Condensed protocol for competent cell preparation and transformation of the methylotrophic yeast *Pichia pastoris*" was adapted from Lin-Cereghino et al. (2005) as follows [35]. All the following procedures must be carried out under sterile conditions:

1. Inoculate 5 mL YPD medium with a fresh *K. phaffii* colony for an overnight culture (ONC) and incubate in a 28 °C shaking incubator at 130 rpm. When growing Δ*tpi1* strains use YPEtOH medium.

2. The next day assess the cell density of the ONC at 600 nm using a spectrophotometer (OD_{600}).
3. Inoculate 50 mL YPD medium with the pre-culture to a final OD_{600} of 0.15–0.20 mL^{-1} (*see* **Note 9**). When growing Δ*tpi1* strains use YPEtOH medium.
4. Grow the *K. phaffii* culture to an OD_{600} of 0.8–1.0 mL^{-1} in a 28 °C shaking incubator at 130 rpm (*see* **Note 10**).
5. Transfer the *K. phaffii* culture to 50 mL sterile reaction tubes.
6. Centrifuge the culture at 500 × g for 5 min at 4 °C and pour off the supernatant completely.
7. Resuspend the pellet in 9.0 mL ice-cold BEDS solution supplemented with 1.0 mL 1M DTT.
8. Shake the cell suspension for 5–10 min at 100 rpm at 28 °C.
9. Centrifuge the culture again at 500 × g for 5 min at 4 °C and remove the supernatant completely.
10. Resuspend the cell pellet in 1.0 mL BEDS solution without DTT and store on ice until used.

3.2 *K. phaffii* Transformation

The following steps should be performed on ice to prevent loss of transformation efficiency.

1. Mix approximately 5–20 ng of circular plasmid DNA with 80 μL of freshly prepared electrocompetent *K. phaffii* cells in a 1.5 mL reaction tube (*see* **Note 11**).
2. Incubate the transformation mixture on ice for ~10 min (*see* **Note 12**).
3. Transfer the solution consisting of electrocompetent *K. phaffii* cells and the circular plasmid into pre-cooled electroporation cuvettes (gap 2 mm) shortly before electroporation (*see* **Note 13**).
4. Electroporate the competent *K. phaffii* cells using the following parameters: charging voltage, 2.0 kV, resistance, 200 Ω; and capacitance, 50 μF (*see* **Note 14**).
5. The *K. phaffii* cells are now ready for regeneration.

3.3 *K. phaffii* Regeneration and Selection

3.3.1 Antibiotic-Based Selection

1. Immediately after electroporation, add and resuspend the transformed *K. phaffii* cells in 1.0 mL YPD:Sorbitol.
2. Transfer resuspended *K. phaffii* cells to a sterile 2 mL reaction tube.
3. Incubate the freshly transformed *K. phaffii* cell suspension in a shaker at 28 °C and 600 rpm for two to three hours (*see* **Note 15**).

4. Take the antibiotic-based selection agar plates from the fridge (4 °C) and put them in the sterile workbench for warming up and drying (*see* **Note 16**).
5. Plate 100 μL of the regenerated, transformed *K. phaffii* cell suspension on a dried antibiotic-based selection agar plate and spread the cell suspension with a sterile spatula until the whole liquid is soaked by the agar plate.
6. Centrifuge the remaining cell suspension in the 2 mL reaction tube at 500 × g for 5 min and decant the supernatant.
7. Resuspend the cell pellet in the remaining media in the 2 mL reaction tube.
8. Plate the rest on another dried antibiotic-based selection agar plate and spread the cell suspension with a sterile spatula until the whole liquid is soaked by the agar plate.

3.3.2 Carbon Source Utilization- and Auxotrophy-Based Selection

1. Immediately after electroporation, add and resuspend the transformed *K. phaffii* cells in 1.0 mL BMG1 solution when using the *K. phaffii* Δ*gut1* or Δ*tpi1* strain (*see* **Note 17**).
2. Transfer resuspended *K. phaffii* cells to a sterile 2 mL reaction tube.
3. Incubate the freshly transformed *K. phaffii* cell suspension in an incubator at 28 °C and 600 rpm for ~30 min (*see* **Note 18**).
4. Take BMG1 agar plates from the fridge (4°C) and put them in the sterile workbench for warming up and drying (*see* **Notes 16** and **17**).
5. Plate 100 μL of the regenerated transformed *K. phaffii* cell suspension on a dried BMG1 agar plate and spread the cell suspension with a sterile spatula until the whole liquid is soaked by the agar plate.
6. Centrifuge the remaining cell suspension in the 2 mL reaction tube at 500 × g for 5 min and decant the supernatant.
7. Resuspend the cell pellet in the remaining media in the 2 mL reaction tube.
8. Plate the rest on another dried BMG1 agar plate and spread the cell suspension with a sterile spatula until the whole liquid is soaked by the agar plate.

3.4 Incubation of Transformation Plates

1. Incubate the agar plates with the transformed *K. phaffii* cells bottom-up in a 28 °C incubator.
2. After 48–72 hours, *K. phaffii* transformants should be visible on antibiotic selection agar plates.
3. Transformants growing on minimal media selection agar usually take up to four days to become visible.

4. Once *K. phaffii* transformants have been obtained, they can be used for cultivation and screening, e.g., in small-scale 96-DWPs or baffled shake flasks containing the respective selective medium.

4 Notes

1. The described methods can be carried out with common wild-type (-like) strains such as *K. phaffii* NRRL Y-11430, CBS7435, NRRL Y-7556, X-33 or (BSY)BG10, or Mut^S strains such as (BSY)BG11 (Δ*aox1*) for antibiotic-based selection or with respective platform strains for auxotrophic-based (Δ*arg4*, Δ*ura3*, Δ*his4*) or carbon source-based auto-selection (Δ*gut1*, Δ*tpi1*).
2. Any other ARS-containing plasmid other than those shown in Fig. 1a–c can also be exploited for this purpose. Although ARS plasmids can be used with several other sequences, they are most effectively studied with PARS1, where no specific restrictions regarding freedom to operate were seen so far. In contrast, some other ARSs are protected by patents. Different antibiotics and auxotrophic or carbon source-based markers are suitable for selection. For expression of the GOI or reporter gene, any promoter is applicable, e.g., constitutive, de-repressed, or inducible promoters.
3. The plasmids shown in Fig. 1a–c can be used as a PCR template to amplify the sequences ranging from the 5′-end of the PDC (500 bp promoter fragment of the *K. phaffii* CTA1 gene) sequence to the 3′-end of the AODTT sequence. The resulting product without the ARS sequence is suitable for the application as a linear plasmid for the integration into the *K. phaffii* genome, which would lead to a stable and antibiotic-resistance-free genotype. The described transformation, regeneration, and selection procedures are equally applicable to linear plasmids.
4. After autoclaving, precipitates formed in the 13.4% (w/v) YNB stock solution, which settled to the bottom of the flask after a while. This did not affect the minimal media (see Subheading 2.3, **items 9–11**) for cultivations. Alternatively, the 13.4% (w/v) YNB stock solution can be sterilized by filtration using a 0.22 μm pore-size filter. Always store the 13.4% (w/v) YNB stock solution at ambient temperature and in a dark place to avoid adverse photochemical reactions. The quality and composition of YNB seemed to vary between different suppliers. This can cause problems, especially when using auxotrophic or carbon source-based selection. BD Difco™ yeast nitrogen base w/o amino acids and 0.5% (w/v) ammonium sulfate (REF

291920; Becton, Dickinson and Company, NJ, USA) has been a reliable product for all applications.

5. Store the BMG1 medium at ambient temperature and in a dark place.
6. Store agar plates at 4°C. To ensure reliable results do not use agar plates older than four weeks due to reduced antibiotic effects or contamination of minimal media selection agar plates.
7. Store the YPD medium at ambient temperature as long as it does not contain antibiotic(s).
8. Ensure that antibiotics are added only to media with temperatures below 60 °C.
9. Starting volumes can be scaled up or down.
10. Based on a doubling time of 100–120 min, *K. phaffii* should reach OD_{600} 0.8–1.0 mL^{-1} in 4–5 h.
11. For comparison of plasmid transformation efficiencies, use equimolar amounts of plasmid DNA. 5 ng circular plasmid DNA is the bottom limit. With higher amounts, plating dilutions may become inevitable. The following formula can be used for calculating equimolar amounts of plasmid DNA:

$$\text{Equimolar amount [ng]} = \frac{\text{Smallest plasmid [ng]} \times \text{Size desired plasmid [bp]}}{\text{Size smallest plasmid [bp]}}$$

12. To increase the transformation efficiency, incubate the freshly prepared electrocompetent *K. phaffii* cells together with the added DNA on ice for better distribution and adhesion of the DNA to the cell surface. Normally, five to ten minutes of incubation time is enough, but it can be extended if necessary.
13. Put the electroporation cuvettes (gap 2 mm) on ice again before transformation. Wipe the cuvettes dry using a paper towel to prevent electrical shorts during electroporation.
14. Depending on the electroporator used, the parameters may differ.
15. Regeneration time for antibiotic selection can be shortened to a minimum of 60 min. Doing so may adversely affect transformation efficiency since the expression of the resistance gene is required to survive on the selective agar plates. However, do not exceed 3 h of regeneration to prevent needless cell division of the freshly transformed *K. phaffii* cells.
16. Slightly open the lid of the Petri dish in the laminar flow so that the agar plates can dry.

17. When using prototrophic markers, other carbon sources like glucose can be used in a defined minimal medium. It is important, however, that the respective amino acids or nucleotides are not present in the medium.
18. In theory, no regeneration time is required for this type of selection, but it might increase transformation efficiencies.

Acknowledgments

CP was funded by the doc.fund program CATALOX [Project DOC 46-B21 CATALOX] by the Austrian Science Fund (FWF). FW was funded by the Austrian Federal Ministry for Digital and Economic Affairs, the National Foundation for Research, Technology and Development. This work was performed as part of the Christian Doppler Laboratory. The financial support from the Austrian Federal Ministry for Digital and Economic Affairs and the National Foundation for Research, Technology and Development and from the contributing company partners Novo Nordisk, BioGrammatics, SeSam Biotech, Zeta and SolarBiotech is gratefully acknowledged. We also want to thank Lukas Rieder for the fruitful discussions as well as for editing the manuscript.

Competing Interest Statement The authors declare that they have no conflicts of interest that are relevant to the content of this chapter.

References

1. Friehs K (2004) Plasmid copy number and plasmid stability. In: Scheper T (ed) New trends and developments in biochemical engineering. Springer Berlin Heidelberg, Berlin, Heidelberg, pp 47–82
2. Püllmann P, Weissenborn MJ (2021) Improving the heterologous production of fungal peroxygenases through an episomal *Pichia pastoris* promoter and signal peptide shuffling system. ACS Synth Biol 10(6):1360–1372
3. Ekundayo B, Bleichert F (2019) Origins of DNA replication. PLoS Genet 15(9):1–21
4. Dhar MK, Sehgal S, Kaul S (2012) Structure, replication efficiency and fragility of yeast ARS elements. Res Microbiol 163(4):243–253
5. Schwarzhans JP et al (2017) A mitochondrial autonomously replicating sequence from *Pichia pastoris* for uniform high level recombinant protein production. Front Microbiol 8 (May):1–15
6. Vogl T, Glieder A, Wasmayer R (2017) Novel episomal plasmid vectors
7. Liachko I, Dunham MJ (2014) An autonomously replicating sequence for use in a wide range of budding yeasts. FEMS Yeast Res 14(2):364–367
8. Camattari A et al (2016) Characterization of a panARS-based episomal vector in the methylotrophic yeast *Pichia pastoris* for recombinant protein production and synthetic biology applications. Microb Cell Fact 15(1):1–11
9. Cregg JM, Barringer KJ, Hessler AY, Madden KR (1985) *Pichia pastoris* as a host system for transformations. 5(12):3376–3385
10. Liachko I et al (2014) GC-rich DNA elements enable replication origin activity in the methylotrophic yeast *Pichia pastoris*. PLoS Genet 10(3)
11. Piva LC, De Marco JL, De Moraes LMP, Reis VCB, Torres FAG (2020) Construction and

characterization of centromeric plasmids for *Komagataella phaffii* using a color-based plasmid stability assay. PLoS One 15(7):1–14

12. Schwarzhans JP, Wibberg D, Winkler A, Luttermann T, Kalinowski J, Friehs K (2016) Integration event induced changes in recombinant protein productivity in *Pichia pastoris* discovered by whole genome sequencing and derived vector optimization. Microb Cell Fact 15(1):1–15
13. Weiss F, Requena-Moreno G, Pichler C, Valero F, Glieder A, Garcia-Ortega X (2024) Scalable protein production by *Komagataella phaffii* enabled by ARS plasmids and carbon source-based selection. Microb Cell Fact 23: 1–16
14. Struhl K, Stinchcomb DT, Scherer S, Davis RW (1979) High-frequency transformation of yeast: autonomous replication of hybrid DNA molecules. Proc Natl Acad Sci U S A 76(3): 1035–1039
15. Murray AW, Szostak JW (1983) Pedigree analysis of plasmid segregation in yeast. Cell 34(3): 961–970
16. Romanos MA, Scorer CA, Clare JJ (2002) Foreign gene expression in yeast. Yeast 8:1–66
17. Weninger A, Hatzl AM, Schmid C, Vogl T, Glieder A (2016) Combinatorial optimization of CRISPR/Cas9 expression enables precision genome engineering in the methylotrophic yeast *Pichia pastoris*. J Biotechnol 235:139–149
18. Smirnov K et al (2024) Comparison of CRISPR-MAD7 and CRISPR-Cas9 for gene disruptions in *Komagataella phaffii*
19. Clarke L, Carbon J (1980) Isolation of a yeast centromere and construction of functional small circular chromosomes. Nature 287(5782):504–509
20. Nakamura Y et al (2018) A stable, autonomously replicating plasmid vector containing *Pichia pastoris* centromeric DNA. Appl Environ Microbiol 84(15):1–16
21. Sturmberger L et al (2016) Refined *Pichia pastoris* reference genome sequence. J Biotechnol
22. Rizvi SMA, Prajapati HK, Ghosh SK (2018) The 2 micron plasmid: a selfish genetic element with an optimized survival strategy within *Saccharomyces cerevisiae*. Curr Genet 64(1):25–42
23. Gasser B, Mattanovich D (2019) Recombinant protein production in yeast, vol 1. Humana Press
24. Lee CC, Williams TG, Wong DWS, Robertson GH (2005) An episomal expression vector for screening mutant gene libraries in *Pichia pastoris*. Plasmid 54(1):80–85
25. Hong IP, Anderson S, Choi SG (2006) Evaluation of a new episomal vector based on the GAP promoter for structural genomics in *Pichia pastoris*. J Microbiol Biotechnol 16(9): 1362–1368
26. Püllmann P et al (2021) A modular two yeast species secretion system for the production and preparative application of unspecific peroxygenases. Commun Biol 4(1)
27. Madsen CK, Vismans G, Brinch-Pedersen H (2016) The PARS sequence increase the efficiency of stable *Pichia pastoris* transformation. J Microbiol Methods 129:1–7
28. Mapelli V (2014) Yeast metabolic engineering: methods and protocols, 1152nd edn. Humana Press
29. Lu ZY, Ma YL, Zhang JT, Fan NS, Huang BC, Jin RC (2020) A critical review of antibiotic removal strategies: Performance and mechanisms. J Water Process Eng 38 (September):101681
30. Ahmad M, Winkler CM, Kolmbauer M, Pichler H, Schwab H, Emmerstorfer-Augustin A (2019) *Pichia pastoris* protease-deficient and auxotrophic strains generated by a novel, user-friendly vector toolbox for gene deletion. Yeast 36(9):557–570
31. Näätsaari L, Mistlberger B, Ruth C, Hajek T, Hartner FS, Glieder A (2012) Deletion of the *Pichia pastoris KU70* homologue facilitates platform strain generation for gene expression and synthetic biology. PLoS One 7(6)
32. Pande S et al (2014) Fitness and stability of obligate cross-feeding interactions that emerge upon gene loss in bacteria. ISME J 8(5): 953–962
33. Pronk JT (2002) Auxotrophic yeast strains in fundamental and applied research. Appl Environ Microbiol 68(5):2095–2100
34. Weis R, Luiten R, Skranc W, Schwab H, Wubbolts M, Glieder A (2004) Reliable high-throughput screening with *Pichia pastoris* by limiting yeast cell death phenomena. FEMS Yeast Res 5(2):179–189
35. Lin-Cereghino J et al (2005) Condensed protocol for competent cell preparation and transformation of the methylotrophic yeast *Pichia pastoris*. Biotechniques 38(1):44–48

Chapter 13

Overlap Extension Construction and Direct Cloning of Loop-Out Prone DNA Expression Cassettes in *Pichia pastoris*: 75 k-γ-Secalin from *Secale cereale* (rye) as a Case Study

Kai Büchner, Carsten Pichler, Roland Kerpes, and Thomas Becker

Abstract

Overlap extension PCR (OE-PCR) is a simple technique allowing for the construction of linear expression cassettes for direct integration into the *Komagataella phaffii* genome. This strategy enables the assembly and amplification of expression cassettes of interest that are genetically unstable or cause physiological problems in *Escherichia coli* and accelerates the path from gene to yeast transformation. As a typical example, here we describe the cloning of the *Secale cereale* (rye) gene *Sec2*, coding for the prolamin 75k γ-secalin. This gene proved to be unstable in circular plasmids in *E. coli*; in silico analysis suggested a loop-out-prone secondary structure of the DNA due to repetitive sequences. In three individual reactions, overlapping DNA fragments of the expression cassette and the gene of interest were amplified, followed by a two-step PCR protocol joining these three fragments to a linear expression cassette. After purification, the linear expression construct was directly used for the transformation of *K. phaffii*.

Key words *Pichia pastoris*, *Komagataella phaffii*, Overlap extension cloning, Direct cloning, Loop-out mutations, Secalin

1 Introduction

The simple overlap extension polymerase chain reaction (OE-PCR) technique is often used for site-directed mutagenesis or gene splicing [1]. It can also be applied to combine three independent fragments into one linear expression cassette for direct integration into the *Komagataella phaffii* (previously called *Pichia pastoris*) genome [2]. To this end, the gene of interest (GOI) is amplified from a suitable source (i.e., synthetic DNA, genomic DNA, or complementary DNA from mRNA). Parts of the expression plasmid are amplified separately to yield the later expression cassette's 5′ (upstream of the GOI) and 3′ (downstream of the GOI) fragments (Fig. 1). While the 5′ fragment harbors the promoter for GOI

Xavier Garcia-Ortega et al. (eds.), *Pichia pastoris: Methods and Protocols*, Methods in Molecular Biology, vol. 2697, https://doi.org/10.1007/978-1-0716-4779-0_13,

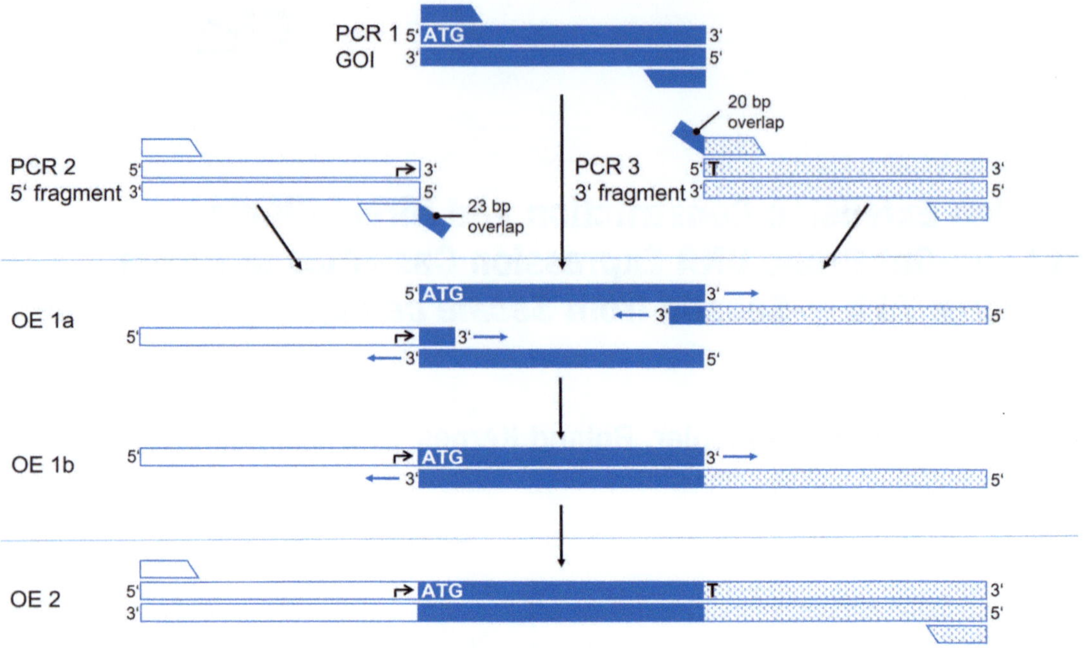

Fig. 1 Schema of the overlap extension PCR workflow. PCR 1: amplification of the gene of interest (GOI); PCR 2 and 3: amplification of the 5′ fragment containing the promoter and the 3′ fragment containing the terminator. OE 1a: fusion of either the 5′ or the 3′ fragment with the GOI. OE 1b: Joining the fragment intermediates from OE 1a to the whole linear expression cassette. OE 1a and OE 1b occur in the same 15-cycle PCR; blue arrows indicate replication direction. OE 2: PCR amplification of the entire linear cassette

expression, the 3′ fragment includes the terminator. Therefore, the fragments must conjoin in the correct orientation relative to each other to yield a functional expression cassette. In this protocol, the 5′ fragment forward primer contains a 23 bp overlap with the 5′ end of the GOI's coding strand. The 3′ fragment reverse primer has a 20 bp overlap with the 3′ end of the non-coding GOI strand. As for regular primer design, an 18–30 bp length is advisable for this overlap to minimize the risk of nonspecific amplification [3]. The OE-PCR itself is carried out in two-steps. At first, all three previously amplified fragments are combined in equimolar amounts in a PCR reaction without adding primers. Herein, the overlap sequences of the 5′ and 3′ fragments anneal to the GOI, acting as primers. After the first cycle, the amplicons comprise the 5′ fragment plus GOI as the sense strand and the 3′ fragment plus GOI as the antisense strand, respectively. These amplicons anneal from the second cycle on, and the GOI acts as a primer. At the end of the first PCR step, the three independent fragments are conjoined to a linear cassette. The second PCR step is carried out by adding forward and reverse primers to amplify the whole cassette, which can then be cleaned up through agarose gel electrophoresis for *K. phaffii* transformation and integration into its genome.

For the following protocol, we used OE-PCR for cloning the *Secale cereale* (*S. cereale*) gene *Sec2*, coding for the prolamin 75k γ-secalin [4]. Prolamins are the major family of storage proteins in most cereal grains [5]. They are of significant interest in food contaminant analysis and clinical studies, as they can induce celiac disease [6].

The prolamin encoding gene *Sec2* from *S. cereale* harbors multiple repetitive regions. In silico analysis [7] of the GOI shows several loops of single-stranded DNA (Fig. 2) that suggest a propensity for the loss of sequence parts through homologous recombination. Omitting the transformation of *Escherichia coli (E. coli)* with the circular plasmid containing the GOI lowers the risk of such

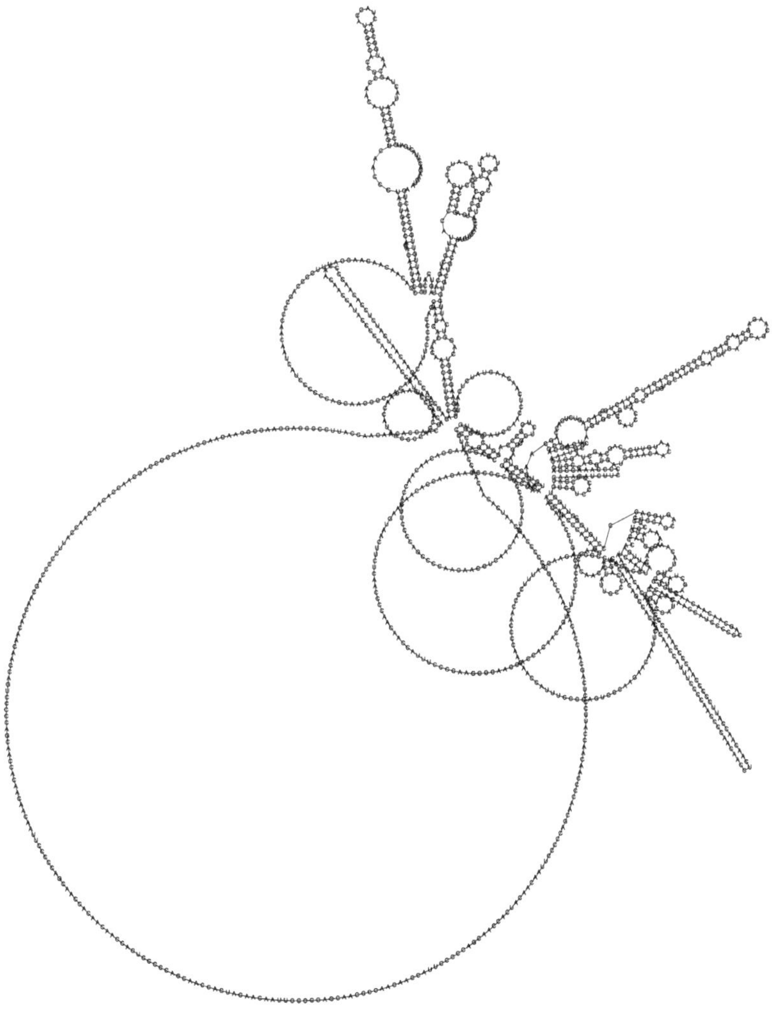

Fig. 2 Minimum Free Energy (MFE) plain structure drawing of the *Secale cereale Sec2* gene, coding for the prolamin 75k γ-secalin. The predicted loop sizes roughly correspond to the length by which the gene of interest was shortened after cloning in *E. coli* (data not shown)

loop-out events occurring. For this protocol, the *Sec2* gene was amplified from genomic rye DNA using primers without overlap sequences, reducing the risk of false priming in the complex genomic DNA. After separation on a 1% agarose TAE gel, the *Sec2* amplicon showed a band at approximately 1400 bp. Whereas Gibson cloning of the amplified *Sec2* ORF into the pBSY3Z expression plasmid and propagation in *E. coli* XL1 only yielded bands of around 700 bp on 1% agarose gels of colony PCRs, transformants with the correct integrated expression cassette and insert size were obtained via OE-PCR followed by *K. phaffii* electroporation and selection.

2 Materials

2.1 Host Strains and Vectors

1. *Komagataella phaffii* BSYBG11 mut^S strain, electrocompetent [8].
2. Plasmid pBSY3Z (Bisy GmbH, Hofstätten an der Raab, Austria) (*see* **Note 1**).

2.2 Kit, Reagents, and Equipment

1. gDNA from rye (*Secale cereale* cv. Elias), isolated using a CTAB-based DNA-extraction protocol [9].
2. $_{dd}H_2O$ deionized water, autoclaved.
3. Oligonucleotide primers (Table 1).
4. Phire Plant Direct PCR Master Mix (Thermo Fisher Scientific Inc., Waltham, USA).
5. Q5® High-Fidelity DNA Polymerase (New England Biolabs, Ipswich, USA) (*see* **Note 2**).

Table 1
Sequences of the oligonucleotide primers

Primer name	Sequence 5′-3′
pBSYleft-FWD	(P)-AAATTAATCGAACTCCGAATGCGGTTC
pBSYleft-REV	ATTGCAAGCATGAGTAAGGTCTT CATTTTAATTGTAAGTCTTGACTAGAGCAAG
pBSYright-FWD	TCACTGGTATTGTTGGCCACTAAGCGGCCGCTCAAGAG
pBSYright-REV	(P)-AAATACATGTGAGCAAAAGGCCAGCAAAAGG
gSEC-FWD	ATGAAGACCTTACTCATGCTTGC
gSEC-REV	TTAGTGGCCAACAATACCAGTG
pBSYseq-FWD	CGGTAAGTTTCATTATCCCAG

Overlap sequences with the *S. cereale Sec2* gene are underlined

6. *Dpn*I restriction endonuclease (New England Biolabs, Ipswich, USA).
7. Promega Wizard® SV Gel and PCR Clean-Up System (Promega Corporation, Fitchburg, USA).
8. 50x TAE buffer: 2 M Tris base [tris(hydroxymethyl)-aminomethane], 1 M acetic acid, 50 mM EDTA (pH 8.0; adjusted with 2 M NaOH).
9. 1x TAE buffer: diluted from 50x TAE buffer stock.
10. 1% Agarose gel: boil 0.4 g agarose in 40 mL 1x TAE buffer, cool down to approx. 60 °C, and add 2 μL MIDORI Green Advance (Nippon Genetics Europe GmbH, Düren, Germany) before casting the gel.
11. Low TE Buffer (10 mM Tris–HCl (pH 8.0), 0.1 mM EDTA).
12. Gene Pulser electroporation system (Bio-Rad, Hercules, USA).
13. PCR cycler (BioCenter, GE9612T) (*see* **Note 3**).
14. Agarose gel electrophoresis system (Bio-Rad PowerPac basic electrophoresis system).
15. Centrifuge for 1.5 mL and 50 mL tubes.
16. Heat block (Eppendorf Thermomixer comfort 5355 R).
17. NanoDrop 2000c spectrophotometer (Thermo Fisher Scientific Inc., Waltham, USA).

2.3 Culture Media

1. YPD-A: Autoclave 20 g peptone, 10 g yeast extract, and 15 g agar agar in 950 mL of deionized water with a magnetic stir bar. Add 50 mL of a separately autoclaved 200 g/L glucose solution. The final glucose concentration should be 1%. Stir until cooled to approx. 50 °C and pour into 93 mm diameter petri dishes with ventilation cams. For plates containing Zeocin™, add the antibiotic to a 100 μg/mL concentration directly before pouring.
2. YPD-L: Prepare liquid media the same as YPD-A without adding agar agar.
3. 1:1 YPD:1 M Sorbitol: Add an equal volume of autoclaved 1 M sorbitol to YPD-L.

3 Methods

All PCR reagents and kits are used according to the manufacturer's instruction manuals if not stated otherwise. For PCRs, autoclaved ddH2O is used. Annealing temperatures are determined using the manufacturer's calculators [10, 11]. Reactions are carried out on ice if not stated otherwise.

3.1 Construction of Expression Cassettes

1. Resuspend lyophilized primers in low TE buffer (*see* **Note 4**) according to manufacturer's instructions (*see* **Note 5**) to a final concentration of 100 μM stock. Dilute this primer stock further in a ratio of 1:10 with $_{dd}H_2O$ to reach the desired concentration of 10 μM.
2. Amplify the 1410 bp long *Sec2* gene from rye gDNA using the Phire Plant Polymerase (*see* **Note 6**). Prepare a reaction mix of 25 μL Phire Plant Direct PCR Master Mix (2X) containing 350 ng gDNA and primers gSEC-FWD and gSEC-REV (**Note 7**).
3. Set the PCR cycler's temperature program to 98 °C/5 min – (98 °C/12 s – 63.9 °C/15 s – 72 °C/45 s) × 36 – 72 °C/ 6 min.
4. Purify the amplified *Sec2* gene from a 1% agarose TAE gel using the Wizard® SV Gel and PCR Clean-Up System (*see* **Note 8**). A band of approximately 1400 bp should be visible with gDNA from rye cultivar Elias.
5. To amplify the 3′ fragment from pBSY3Z (*see* **Note 9**), prepare a Q5 HiFi polymerase reaction mix according to manufacturers' instructions, containing 10 ng plasmid DNA and primers pBSYright-FWD and pBSYright-REV.
6. Conduct the PCR of the 2.9 kb vector arm using the following temperature program: 98 °C/30 s – (98 °C/10 s – 69 °C/ 20 s – 72 °C/90 s) × 35 – 72 °C/2 min.
7. To amplify the 5′ fragment from pBSY3Z (*see* **Note 11**), prepare a Q5 HiFi polymerase reaction mix according to manufacturers' instructions, containing 10 ng plasmid DNA and primers pBSYleft-FWD and pBSYleft-REV.
8. For the amplification of this 0.5 kb vector arm, set the PCR cycler's temperature program to 98 °C/30 s – (98 °C/10 s – 64 °C/20 s – 72 °C/15 s) × 35 – 72 °C/2 min.
9. Add 1 μL of *Dpn*I restriction endonuclease to both PCR mixes and digest for at least 2 h at 37 °C.
10. Purify the amplification products from a 1% agarose TAE gel using the Wizard® SV Gel and PCR Clean-Up System. Measure the DNA concentration and purity using the NanoDrop 2000c photometer.
11. To assemble the whole expression cassette, add equimolar amounts of 5′ fragment, 75k γ-secalin gDNA, and 3′ fragment to a reaction containing 5 μL 5x Q5 reaction buffer, 1 μL 10 mM dNTPs, 0.5 μL Q5 polymerase, and $_{dd}H_2O$ to 25 μL. Calculate the molar amounts in relation to 10 ng of the 3′ fragment being the longest fragment. (*see* **Note 12**).

12. Run a 15-cycle PCR with the annealing temperature calculated from the fragment overlaps with the GOI. The elongation time is calculated from the combined length of the GOI and the longer 3′ fragment, resulting in the following temperature program: 98 °C/30 s – (98 °C/10 s – 67 °C/20 s – 72 °C/2 min) × 15 – 72 °C/2 min.
13. Since the following PCR yields unspecific side products, it is advisable to prepare several reactions and pool them for the cleanup. For a total of four reactions, prepare a 200 μL Q5 PCR master mix without template and with the primers pBSYleft-FWD and pBSYright-REV. Aliquot the master mix to 50 μL into four PCR tubes and add 1 μL of PCR mixture from step 9 as a template. Amplify the full linear cassette using the following temperature program: 98 °C/30 s – (98 °C/10 s – 68 °C/20 s – 72 °C/2 min 30 s) × 25 – 72 °C/2 min.
14. Perform electrophoresis using the total volume of all four reactions on a 1% agarose TAE gel and purify the DNA band corresponding to the size of the whole fragment using the Wizard® SV Gel and PCR Clean-Up System. This yields the linear expression cassette composed of the 5′ and 3′ fragments flanking the GOI. Measure the DNA concentration and purity using the NanoDrop 2000c photometer (*see* **Note 12**).

3.2 Transformation

1. Incubate sterile 2 mm gap electroporation cuvettes on ice for a minimum of 10 min.
2. Add 8 μL (300–1000 ng total DNA) of gel-purified linear cassette DNA to 80 μL of freshly prepared electrocompetent *K. phaffii* BSYBG11 mut^S cells in a PCR tube. Mix and incubate for at least 5 min on ice, then transfer the cell suspension to the precooled electroporation cuvettes. For a negative control sample, use sterile $_{dd}H_2O$ instead of DNA.
3. Pulse the cells in the Gene Pulser II system at 2.5 kV, 200 Ω resistance, and 25 μF capacitance (*see* **Note 12**). Add 1 mL of sterile 1:1 YPD:1 M sorbitol solution immediately after the pulse.
4. Incubate the mixture for 1–1.5 h at 28 °C, in a 50 mL centrifugation tube with a conical bottom, shaking at 130 rpm.
5. Plate 200–250 μL of the cell suspension on YPD-A plates containing 100 μg/mL Zeocin™.
6. Spin down the remaining cell suspension at 1500× *g* for 5 min, decant the supernatant, and resuspend the pellet in the residual liquid. Plate the cell suspension on YPD-A plates containing 100 μg/mL Zeocin™.
7. Incubate all plates at 28 °C for 2–3 days until colonies have formed.

3.3 Test for Genetic Integrity

The genetic integrity of the transformed strains can be tested via colony PCR (cPCR) and subsequent sequencing. It is best to use primers that anneal in the promoter and terminator regions controlling the GOI transcription since most sequencing results only get reliable about 15–20 bp downstream of the sequencing primers.

1. Prepare a master plate by drawing and labeling a grid on a YPD-A plate containing 100 μg/mL Zeocin™.
2. Prepare a Phire Plant PCR reaction mix (25 μL 2x Phire Plant Master Mix (2X), 2.5 μL 10 μM pBSYseq-FWD, 2.5 μL 10 μM gSEC-REV, 20 μL $_{dd}H_2O$). Aliquot 10 μL of the reaction mixture into PCR tubes labeled according to the master plate's grid.
3. Under sterile conditions, pick at least five colonies from the transformant plate using sterile filter tips. Transfer the colonies to the master plate with one streak of the filter tip. Then, resuspend the residual cells in the correspondingly labeled PCR tube by pipetting up and down.
4. Perform cPCRs (*see* **Note 17**) using the following temperature protocol: 98 °C/5 min – (98 °C/12 s – 59.3 °C/15 s – 72 °C/45 s) × 36 – 72 °C/6 min.
5. Perform electrophoresis using the cPCR products on a 1% agarose TAE gel and check for bands of the expected size (Fig. 3).

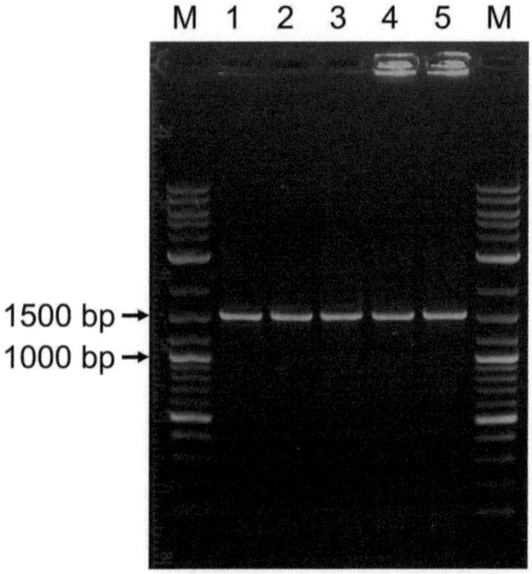

Fig. 3 Colony PCR of *K. phaffii* BSYBG11 mut^S, transformed with the linear expression cassette containing the gene of interest *Sec2* from rye cultivar Elias. The bands show part of the cassette, amplified with a forward primer that anneals in the promoter region of the cassette, and a reverse primer that anneals at the '3' end of the GOI. Marker (M): NEB 1 kb plus ladder

6. Pick colonies showing correct bands for sequencing. Purify the corresponding cPCR products using the Promega Wizard® SV Gel and PCR Clean-Up System, and prepare the samples for sequencing according to the sequencing service's instructions.
7. Check sequencing results for correct sequence identity using NCBI BLAST (https://blast.ncbi.nlm.nih.gov/Blast.cgi). If this is the case, prepare glycerol stocks and store them at − 80 °C (*see* **Note 8**).

4 Notes

1. The techniques described in this protocol are also suitable for other *K. phaffii* strains and expression plasmids; primers must be constructed accordingly.
2. Any other high-fidelity polymerase is suitable for this purpose.
3. Also, other suppliers' electroporation cuvettes might be suitable, e.g., 2 mm electroporation cuvettes supplied by VWR™ International.
4. Any PCR cycler will be suitable for amplifying the *Sec2* gene from genomic DNA.
5. Also, $_{dd}H_2O$ can be used to resolve the lyophilized primers.
6. This buffer's low EDTA concentrations and high pH will protect oligonucleotide stocks from DNases and acid hydrolysis, respectively. By further diluting the stock solutions in $_{dd}H_2O$, EDTA concentrations are lowered to not interfere with PCR reactions.
7. For the amplification of the *Sec2* gene from gDNA, different polymerases were tested using varying annealing temperatures. Phire Plant Polymerase Master Mix under the given conditions yielded the cleanest results. For other GOI's, testing an array of different polymerases and annealing temperatures is advisable.
8. These primers were designed using a consensus sequence rendered from the entries of the NCBI nucleotide database found under the search term "(75k gamma secalin) AND 'Secale cereale' [porgn:__txid4550]." Database searches in the NCBI database show that, depending on the origin variety, the *Sec2* coding sequence length may vary. However, the genes' 5′ and 3′ ends are conserved and specific for the respective prolamin protein variety. The native sequence is amplified here, including the start ATG with the predicted 19 amino acid secretion signal. The native secretion signal sequence should be omitted if other signal sequences are used.
9. The manufacturer's protocol was modified as follows: (a) After the first centrifugation of the dissolved gel mixture on the

column, the flowthrough is loaded on the column a second time, and the centrifugation is repeated. (b) After completing the washing steps, the column is placed in the heat block at 50 ° C for a minimum of 10 min. to evaporate residual ethanol. (c) The DNA is eluted in two steps with 25 μL $_{dd}H_2O$ at 50 °C each. The incubation time is lengthened to 10 min. For each step.

10. The reverse primer of the 5′ fragment and the forward primer of the 3′ fragment will introduce a ≥20 bp overlap with the 75k γ-secalin gene. The length of the overlap regions should be chosen to have matching annealing temperatures. Ideally, the overhangs include a 3' GC clamp to increase PCR efficiency.

11. The molecular weight of the DNA fragments was calculated using the vector map and the https://www.bioinformatics.org/sms2/dna_mw.html calculator [12]. The molecular weight of the *Sec2* gene was calculated for the estimated length of 1400 bp and an average of 650 g/(mol × bp) for double-stranded DNA. The molar amount (mol) of the 3′ fragment (right-side fragment; rsf) was calculated as

$$mol_{rsf} = \frac{10\ ng}{M_{rsf}}$$

with M_{rsf} as the molar mass of the 3′ fragment.

The volume V of each fragment f was calculated as

$$V_f = \frac{mol_{rsf} \cdot M_f}{c_f}$$

with M_f as the molar mass of the fragment and c_f as the concentration of the eluted fragment.

12. A DNA concentration of 60 ng/μL or more is desirable, and a minimum of 37.5 ng/μL is needed for the following transformation step. This way, the minimum total amount of DNA (300 ng) is attained by adding 8 μL of solution to 80 μL of cell suspension.

13. It is of utmost importance that the cuvettes are dry and the cell suspension is free of excessive salt content. A time constant of 4.3–4.8 msec is typical for successful electroporations.

14. Since both the promoter and terminator of the expression cassette are also present in the wild-type *K. phaffii* BSYBG11 mut^S genome, it is best to perform a cPCR with one primer binding in the GOI sequence and one primer binding in either the promoter or terminator sequence.

15. For the given *S. cereale* cultivar Elias, a sequence identity of >99% could be achieved with NCBI 75k γ-secalin entries.

Competing Interest Statement

The authors have no conflicts of interest to declare that are relevant to the content of this chapter.

References

1. Heckman KL, Pease LR (2007) Gene splicing and mutagenesis by PCR-driven overlap extension. Nat Protoc 2(4):924–932. https://doi.org/10.1038/nprot.2007.132
2. Liu Z, Pscheidt B, Avi M, Gaisberger R, Hartner FS, Schuster C, Skranc W, Gruber K, Glieder A (2008) Laboratory evolved biocatalysts for stereoselective syntheses of substituted benzaldehyde cyanohydrins. Chembiochem 9(1):58–61. https://doi.org/10.1002/cbic.200700514
3. Apte A, Daniel S (2009) PCR primer design. Cold Spring Harb Protoc 2009(3):pdb.ip65. https://doi.org/10.1101/pdb.ip65
4. Sybenga J, Parmar S, van Eden J et al (1991) Mapping of the seed storage protein locus Sec-2 in rye. Genetica 84(2):101–105. https://doi.org/10.1007/BF00116549
5. Shewry PR, Halford NG (2002) Cereal seed storage proteins: structures, properties and role in grain utilization. J Exp Bot 53(370):947–958. https://doi.org/10.1093/jexbot/53.370.947
6. Vader L, Stepniak DT, Bunnik EM et al (2003) Characterization of cereal toxicity for celiac disease patients based on protein homology in grains. Gastroenterology 125(4):1105–1113. https://doi.org/10.1016/S0016-5085(03)01204-6
7. Gruber AR, Lorenz R, Bernhart SH et al (2008) The Vienna RNA websuite. Nucleic Acids Res 36(2):W70–W74. https://doi.org/10.1093/nar/gkn188
8. Lin-Cereghino J, Wong WW, Xiong S et al (2005) Condensed protocol for competent cell preparation and transformation of the methylotrophic yeast *Pichia pastoris*. BioTechniques 38(1):44–48. https://doi.org/10.2144/05381BM04
9. Linkmeyer A, Götz M, Hu L, Asam S, Rychlik M, Hausladen H, Hess M, Hückelhoven R (2013) Assessment and introduction of quantitative resistance to *Fusarium* head blight in elite spring barley. Phytopathology 103(12):1252–1259. https://doi.org/10.1094/PHYTO-02-13-0056-R
10. New England Biolabs NEB Tm Calculator v1.13.0. Accessed 29 Apr 2024., https://tmcalculator.neb.com/#!/main
11. ThermoFisher Scientific Tm Calculator. Accessed 29 Apr 2024., https://www.thermofisher.com/at/en/home/brands/thermo-scientific/molecular-biology/molecular-biology-learning-center/molecular-biology-resource-library/thermo-scientific-web-tools/tm-calculator.html
12. Stothard P (2000) The sequence manipulation suite: JavaScript programs for analyzing and formatting protein and DNA sequences. BioTechniques 28(6):1102–1104. https://doi.org/10.2144/00286ir01

Chapter 14

In Vivo Plasmid Assembly in NHEJ-Deficient *Pichia pastoris* Strains

Leon Wolbank, Florian Weiss, and Anton Glieder

Abstract

The integration of linear plasmids into the genome is the most common method for genetic engineering of *Komagataella phaffii* (reclassified from *Pichia pastoris*). This, however, results in the drawbacks of lower transformation efficiencies and diverse clonal variations of expression levels compared to typical experiments with baker's yeast. This is due to the affected genomic loci or copy number effects, making mutant library screenings or comparative expression studies challenging. The use of autonomously replicating plasmids, which are maintained extrachromosomally, can overcome these limitations. To speed up the library generation process and reduce the bottleneck of plasmid cloning, the episomal plasmids can be directly assembled in vivo in *K. phaffii* via homology-directed repair in a non-homologous end-joining deficient Δ*ku70* strain. Time-consuming cloning in vitro followed by vector amplification can be avoided. At the same time, in principle, no specific sequences for selection and replication in *E. coli* are needed, enabling a vector design solely employing yeast DNA. Typically, libraries in the range of 10^5–10^6 CFU/μg DNA can be obtained by following the protocol described in this chapter.

Key words *Pichia pastoris*, *Komagataella phaffii*, Episomal plasmid, Homologous recombination, Carbon source selection, *GUT1*

1 Introduction

Linear plasmids designed for random or site-directed integration into the genome are the most common method for the transformation of *Komagataella phaffii* (reclassified from *Pichia pastoris*) [1, 2]. The former leads to variation of transgene expression among transformants caused by the affected genomic loci and plasmid copy numbers, which can be beneficial to identify a good producer strain [3, 4]. The latter shows low efficiency of correct integration due to non-homologous end-joining (NHEJ) being the preferred DNA repair mechanism over homology-directed repair (HDR) in

Authors Leon Wolbank and Florian Weiss have equally contributed to this chapter.

Xavier Garcia-Ortega et al. (eds.), *Pichia pastoris: Methods and Protocols*, Methods in Molecular Biology, vol. 2697,
https://doi.org/10.1007/978-1-0716-4779-0_14,

K. phaffii [5]. Hence, these approaches are not suitable for some applications such as large mutant library screenings, where high transformation efficiencies and uniform expression levels are desired [6, 7]. The required high transformation efficiencies can be reached using plasmids with an autonomous replicating sequence (ARS), which also lead to less heterogenous expression levels among transformants [8, 9].

To avoid additional cloning steps and loss of mutant library variability, the plasmid can be assembled in vivo via HDR [7]. In wild-type strains, it is necessary to use a split-marker system to keep the empty vector background to a minimum [7]. This, however, results in a reduction of overall transformation efficiency. An alternative to avoid split-marker systems would be the use of a NHEJ-deficient strain such as the *ku70* deletion strain of *K. phaffii* [5]. The Ku70p homologue is a key player in NHEJ and its deletion results in an almost complete loss of the cellular DNA repair mechanism [5]. Subsequently, the deletion strain almost exclusively repairs DNA double-strand breaks using HDR, allowing in vivo assembly of circular ARS plasmids from two or more fragments [5, 7]. This results in a higher transformation rate than observed with split-marker systems, and the size of the insert can be reduced [7].

This chapter describes the workflow of the in vivo plasmid assembly approach from in silico design to transformation and screening based on an exemplary ARS plasmid expressing eGFP using the *GUT1* selection marker.

2 Materials

2.1 Preparation of Cloning Fragments

1. 10 μM primers suitable to amplify the backbone and the insert with the desired length of overlapping regions (*see* Fig. 1 and Table 1).
2. Phusion™ Plus DNA polymerase (or another high-fidelity polymerase) with appropriate buffer.
3. 10 mM dNTPs.
4. PCR tubes.
5. Thermocycler.
6. PCR clean-up kit.

2.2 Transformation of K. phaffii

1. Electrocompetent *K. phaffii* cells of a strain of choice, e.g., a BSYBG11 *Δgut1Δku70* double deletion strain when the *GUT1* selection marker is used.
2. Icebox filled with ice.

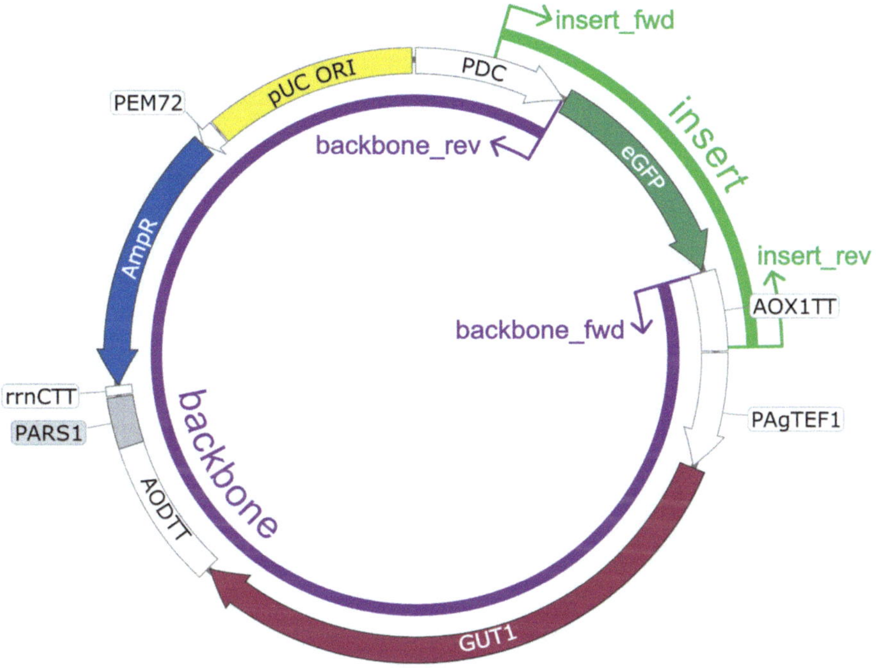

Fig. 1 Map of the plasmid with GenBank accession number PP793738 with indicated positions of designed primers and resulting backbone and insert fragments. The plasmid PP793738 contains a beta-lactamase gene (ampicillin resistance) and a pUC origin of replication, which are included for the potential amplification in *Escherichia coli*. They can also be omitted to reduce the overall size of the plasmid and to avoid bacterial or resistance genes in the experiments. With the *ku70* deletion strain, there is no need to use a split-marker system, which allows to decrease the size of the insert. Therefore, the insert comprises only the GOI (in this case, eGFP) and 250 bp of sequences homologous to the backbone. The promoter and terminator for the GOI are located on the backbone. Additionally, the backbone includes the selection marker *GUT1* with its respective promoter and terminator, as well as the PARS1

Table 1
List of primers designed to amplify the backbone and insert for in vivo cloning from the plasmid with the GenBank accession number PP793738

Primer name	Sequence 5′-3′
insert_fwd	CTTCTCGCACGAGACCGAATC
insert_rev	GGTCTCACTTAATCTTCTGTACTCTGAAG
backbone_fwd	TCAAGAGGATGTCAGAATGCCATTTG
backbone_rev	TTTAATTGTAAGTCTTGACTAGAGCAAGTG

3. Electroporation cuvettes with 2 mm gap.
4. 1.5 mL reaction tubes with conical bottom.
5. Electroporator.
6. Thermomixer.

7. Plastic petri dishes: 94 mm × 16 mm with vents.
8. 4×10^{-5}% (w/v) biotin stock solution: dissolve 10 mg biotin in 50 mL ddH_2O, sterile filtrate, and store at 4 °C.
9. 1 M potassium phosphate buffer: 46.4 g K_2HPO_4 and 100 g KH_2PO_4 in 900 mL ddH_2O. Adjust the pH to 6.0 with 3 M KOH or 85% H_3PO_4 and fill up to 1.0 L with ddH_2O, autoclave.
10. 13.4% (w/v) YNB: autoclave 134 g YNB (BD™ Difco™ Yeast Nitrogen Base w/o amino acids) dissolved in 1.0 L ddH_2O.
11. BMG1: Add 10 g of glycerol and fill up to 700 mL with ddH_2O and autoclave. Let it cool down and add 200 mL 1 M potassium phosphate buffer (pH 6.0), 100 mL 13.4% (w/v) YNB, and 2 mL 4×10^{-5}% (w/v) biotin (*see* **Note 1**).
12. BMG1 agar: Prepare like BMG1 medium but add 15 g/L agar-agar before autoclavation. When lukewarm pour medium into the petri dishes and let it solidify.

2.3 K. phaffii Cultivation and Activity Screening

1. Sterile toothpicks.
2. Sterile 96-deepwell plates (DWP) (e.g., from Bel-Art Products, USA).
3. BMG1.
4. BMG0.5: Add 5 g of glycerol and fill up to 700 mL with ddH_2O and autoclave. Let it cool down and add 200 mL 1 M potassium phosphate buffer (pH 6.0), 100 mL 13.4% (w/v) YNB, and 2 mL 4×10^{-5}% (w/v) biotin (*see* **Note 1**).
5. BMG2.5: Add 25 g of glycerol and fill up to 700 mL with ddH_2O and autoclave. Let it cool down and add 200 mL 1 M potassium phosphate buffer (pH 6.0), 100 mL 13.4% (w/v) YNB, and 2 mL 4×10^{-5}% (w/v) biotin (*see* **Note 1**).
6. 0.2 M potassium phosphate buffer: Dilute 200 mL of 1 M potassium phosphate buffer with 800 mL ddH_2O.
7. Multichannel pipette.
8. Thermo Scientific™ Nunc MicroWell 96-Well Optical-Bottom Plates with Polymer Base (Catalog number: 265301) for fluorescence measurements.
9. Plate reader capable of measuring fluorescence.
10. Desired screening equipment, when other protein of interest is used.

3 Methods

3.1 In Silico *Design and Preparation of Vector Backbone and Insert*

For the homology-directed-repair (HDR)-based in vivo assembly, overlapping homologous sequences are needed. Therefore, a set of PCR primers must be designed that creates two fragments with overlapping regions between the fragments. An optimal length of homologous sequence for highly efficient HDR is around 250 bp (*see* **Note 2**) [5, 10].

1. Design a set of primers that creates two fragments with around 250 bp overlapping regions between the fragments.
2. If the screening of a mutated protein of interest (POI) is desired, then error-prone PCR can be performed to create a mutant library or synthesized inserts with mutated gene variants of the POI can also be ordered and used for transformation.

   To give an example, a set of primers was designed for the plasmid with the GenBank accession number PP793738 (*see* Table 1 and Fig. 1).
3. The PCR reactions are prepared according to the manufacturer's instructions, and the reaction parameters are adjusted to the melting temperature of the designed primers and the fragment size. Use a maximum of 1 ng of template plasmid. The PCR parameters for the example reactions are shown in Table 2.
4. Subsequently, the product is purified using a PCR purification kit (*see* **Note 3**).
5. Determine the quantity of the DNA, e.g., using a NanoDrop spectrophotometer.

3.2 *Transformation of* K. phaffii

1. Pre-cool electroporation cuvettes on ice.
2. Mix 80 μL of competent cells with 50 ng of backbone (*see* **Note 4**) and a fivefold molar excess of insert to the vector backbone

Table 2
PCR profile of the example reactions using the primers listed in Table 1

	Backbone	Insert	
98 °C	30 s		
98 °C	10 s		35×
60 °C	15 s		
72 °C	120 s	30 s	
72 °C	5 min		
4 °C	∞		

in a pre-cooled cuvette [10]. For the example backbone and insert amplified from plasmid PP793738 using the primers listed in Table 1, this would be 58 ng of insert.

3. The cell and DNA mixtures are incubated on ice for 5 to 10 min prior to electroporation.
4. Electroporate at 2 KV; resistance, 200 Ω; capacitance, 25 μF.
5. Immediately after electroporation, the cells are resuspended in 1 mL of sterile BMG1.
6. Regenerate the cells for 30 min at 28 °C and 600 rpm in a thermomixer.
7. The regenerated transformations are then plated on BMG1 agar plates (*see* **Notes 5** and **6**).
8. Incubate the plates at 28 °C for 4–5 days until colonies are visible.
9. Following this procedure for the exemplary plasmid, transformation efficiencies in the range of 10^5 to 10^6 CFU/μg were obtained (*see* **Note 4**).

*3.3 **K. phaffii** High-Throughput Cultivation and Screening*

There are different ways to screen the obtained transformants. One way with a high throughput is to cultivate them in 96-deepwell plates (DWP) [9, 11]. Since episomal plasmids require permanent selection pressure in *K. phaffii*, it is absolutely essential that appropriate selective media are used for cultivation. In the case of the *GUT1* selection marker, buffered minimal media with glycerol as sole carbon source are used.

1. Add 250 μL of sterile BMG1 in each well of a DWP (*see* **Notes 6** and **7**).
2. Inoculate each well of the DWP with a distinct *K. phaffii* transformant using sterile toothpicks.
3. Cultivate the DWP at 28 °C, 320 rpm, and 80% humidity for 60 hours.
4. After 60 h, add 250 μL of BMG0.5 under sterile conditions.
5. Add 50 μL of BMG2.5 after 72 and 84 h.
6. After a total of 108 h of cultivation, the OD_{600} of the cultivations can be measured, and the desired screening can be carried out (*see* **Note 8**).
7. In case of the example plasmid, intracellular eGFP is used as reporter protein. To determine the fluorescence levels, prepare 200 μL of a 1:20 dilution in the microtiter plate using 0.2 M potassium phosphate buffer, excite at 488 nm, and measure the emission at 507 nm in the plate reader (*see* **Note 9**).

4 Notes

1. Alternatively, BMG media of different glycerol concentrations can be prepared from buffered minimal media without carbon source (BM_) by adding the respective amounts of 50% glycerol:

   Autoclave 700 mL ddH_2O. Let it cool down and add 200 mL 1 M potassium phosphate buffer (pH 6.0), 100 mL 13.4% (w/v) YNB, and 2 mL 4×10^{-5}% (w/v) biotin.

   Then add 0.5 mL, 1 mL, or 2.5 mL of sterile 50% glycerol (w/w) to 50 mL BM_ to obtain BMG0.5 (0.5% glycerol), BMG1 (1% glycerol), or BMG2.5 (2.5% glycerol), respectively.
2. The length of homologous regions can also be decreased to as low as 25 bp, but this leads to a decrease in recombination and subsequently transformation efficiency, and therefore a loss in library size. Homologous sequences of such short length can simply be attached to the cloning fragments by PCR primers carrying the additional sequences at the 5′ end. Increasing the homologous regions to more than 250 bp, however, does not increase the transformation efficiency any further.
3. When circular plasmids with an ARS sequence are used as PCR templates, transformational background caused by the template plasmid used for the PCRs may be observed during the screening procedures. In this case, *Dpn*I digestion following the PCR can be done to remove template DNA from the PCR product, which minimizes the risk of transformational background caused by the template DNA. *Dpn*I digestion is carried out according to the manufacturer's instructions.
4. The total number of transformants can be increased by using more DNA and competent cells. This, however, increases the chance of a cell being transformed with multiple copies of the plasmid, which might result in transformants carrying more than one gene variant and lower homogeneity of expression levels while screening a library [10].
5. To ensure that single colonies are obtained, plate multiple volumes, e.g., 10 μL, 100 μL, and 900 μL.
6. Minimal media containing glycerol as sole carbon source is necessary to select for the episomal plasmids due to the used *GUT1* selection marker.
7. When using a constitutive promoter, feeding is not necessary. Cultivation can be batch phase only with a higher initial volume or glycerol concentration.
8. When using the PDC or other promoters that are repressed when an excess of carbon source is present, it is important that the glycerol is added in de-repressing concentrations.

Additionally, all carbon sources should be completely consumed before harvesting the cultivations to maximize recombinant protein production.

9. When the POI is secreted to the culture broth, the cultivations are harvested by centrifugation at RT and 3000 g for 10 min and the supernatant is used for further analysis.

Acknowledgments

This work was performed as part of the Christian Doppler Laboratory. The financial support from the Austrian Federal Ministry for Digital and Economic Affairs and the National Foundation for Research, Technology and Development and from the contributing company partners Novo Nordisk, BioGrammatics, SeSam Biotech, Zeta and SolarBiotech is gratefully acknowledged.

Competing Interest Statement The authors have no conflicts of interest to declare that are relevant to the content of this chapter.

References

1. Ahmad M, Hirz M, Pichler H, Schwab H (2014) Protein expression in Pichia pastoris: recent achievements and perspectives for heterologous protein production. Appl Microbiol Biotechnol 98(12):5301–5317. https://doi.org/10.1007/s00253-014-5732-5
2. Cereghino JL, Cregg JM (2000) Heterologous protein expression in the methylotrophic yeast Pichia pastoris. FEMS Microbiol Rev 24(1):45–66. https://doi.org/10.1111/j.1574-6976.2000.tb00532.x
3. Vogl T, Gebbie L, Palfreyman RW, Speight R (2018) Effect of plasmid design and type of integration event on recombinant protein expression in Pichia pastoris. Appl Environ Microbiol 84(6):e02712–e02717. https://doi.org/10.1128/AEM.02712-17
4. Hohenblum H, Gasser B, Maurer M, Borth N, Mattanovich D (2004) Effects of gene dosage, promoters, and substrates on unfolded protein stress of recombinant Pichia pastoris. Biotechnol Bioeng 85(4):367–375. https://doi.org/10.1002/bit.10904
5. Näätsaari L, Mistlberger B, Ruth C, Hajek T, Hartner FS, Glieder A (2012) Deletion of the Pichia pastoris KU70 homologue facilitates platform strain generation for gene expression and synthetic biology. PLoS One 7(6):e39720. https://doi.org/10.1371/journal.pone.0039720
6. Lee CC, Williams TG, Wong DWS, Robertson GH (2005) An episomal expression vector for screening mutant gene libraries in *Pichia pastoris*. Plasmid 54(1):80–85. https://doi.org/10.1016/j.plasmid.2004.12.001
7. Camattari A et al (2016) Characterization of a panARS-based episomal vector in the methylotrophic yeast Pichia pastoris for recombinant protein production and synthetic biology applications. Microb Cell Factories 15(1):139. https://doi.org/10.1186/s12934-016-0540-5
8. Cregg JM, Barringer KJ, Hessler AY, Madden KR (1985) Pichia pastoris as a host system for transformations. Mol Cell Biol 5(12):3376–3385. https://doi.org/10.1128/mcb.5.12.3376-3385.1985

9. Weiss F, Requena-Moreno G, Pichler C, Valero F, Glieder A, Garcia-Ortega X (2024) Scalable protein production by Komagataella phaffii enabled by ARS plasmids and carbon source-based selection. Microb Cell Factories 23:116. https://doi.org/10.1186/s12934-024-02368-3
10. Vogl T, Glieder A, Wasmayer R (2020, July 08) Novel episomal plasmid vectors. EP3356534B1 [Online]. Available: https://patents.google.com/patent/EP3356534B1/en. Accessed 05 Jan 2024
11. Weis R, Luiten R, Skranc W, Schwab H, Wubbolts M, Glieder A (2004) Reliable high-throughput screening with Pichia pastoris by limiting yeast cell death phenomena. FEMS Yeast Res 5(2):179–189. https://doi.org/10.1016/j.femsyr.2004.06.016

Chapter 15

Protein Targeting Signals and Strategies for *Pichia pastoris*

Sarah Gangl and Katharina Ebner

Abstract

Besides the simplicity of genetic modification and convenient low-effort cultivation, *Pichia pastoris* (today mostly *Komagataella phaffii* species are used for biotechnology) comes with another great advantage: simple targeting of recombinant proteins to different compartments of the host cell. Indeed, the most prominent pathway for recombinant proteins is the secretory pathway, whereby the protein is guided through the endoplasmic reticulum (ER) and Golgi apparatus, resulting in secretion at the cell surface. Secreted proteins can be separated from the cells, forgoing expensive cell disruption and purification processes. However, there are further localization sites for recombinant proteins (e.g., mitochondria, nucleus, peroxisomes, vacuole, cytoplasmic membrane, cell wall, etc.), technologically relevant.

Key words *Pichia pastoris*, *Komagataella phaffii*, Signal peptide, Secretion, Targeting, Localization, Recombinant protein

1 Introduction

1.1 Signal Sequences: Similarity or Diversity?

Within the cell, there are many final destinations for proteins after translation. Each protein has a certain function and is demanded by a certain cell compartment. The protein's localization is determined by a signal or targeting sequence (i.e., signal peptides). There is no general consensus sequence described for signal sequences, nonetheless, common structural features exist, e.g., for proteins which are targeted to the ER: i.e., a hydrophobic core region flanked by basic amino acids at the N-terminal end, and a polar C-terminal end providing a signal peptidase cleavage site [1, 2]. Many signal sequences from different organisms have been described and lead proteins into mutual targeting pathways employing related mechanistic principles. This contributes to the fact that some signal sequences were reported as successfully transferrable between different host organisms (e.g., yeast to mammals) [3]. On the other hand, e.g., Bird *et al.* reported that the vacuolar enzyme carboxypeptidase from *Saccharomyces cerevisiae* (CPY), which is known to carry an N-terminal signal sequence mediating

Xavier Garcia-Ortega et al. (eds.), *Pichia pastoris: Methods and Protocols*, Methods in Molecular Biology, vol. 2697, https://doi.org/10.1007/978-1-0716-4779-0_15,

the protein's transfer into the yeast ER, cannot be transferred through a mammalian ER membrane. Surprisingly, marginal mutations within the ER-directing signal sequence, as well as the exchange of the signal sequence by a mammalian one, enabled protein translocation into the mammalian ER. Summarizing the complexity of signal sequences, it can be said that although sharing common features, there are different mechanisms of how they are recognized and how translocation is facilitated [1, 2], and a small subset of sequences is experimentally preferred for heterologous protein targeting.

1.2 The Yeast Secretory Pathway

The yeast secretory pathway is a complex cellular transport system encompassing different membrane-enclosed compartments, like the ER and the Golgi apparatus. Although some proteins are secreted by non-conventional secretory pathways [39], for the entrance into the secretory pathway, the N-terminal signal sequence is in charge of guiding the protein into the ER (co- or post-translationally). Once located within the ER, the protein can undergo folding and conformational changes, e.g., disulfide bridge formation and glycosylation. Before the protein is transported to the cell surface, it is often further processed within the Golgi apparatus [5].

In eukaryotic cells, two initial steps can mediate the translocation through the ER membrane, whereby the interaction between the protein and the translocation machinery is essential: For co-translational translocation, the synthesis of the protein is believed to slow down or even freezes as soon as the signal sequence binds the signal recognition particle (SRP). The synthesis continues when SRP interacts with the corresponding receptor. This endoplasmic membrane receptor is located on the surface of the ER, enabling protein translocation. Other signal sequences, being incapable of SRP interaction, facilitate protein translocation after the protein is completely or almost translated, resulting in a post-translational translocation [1, 6]. Whether a certain signal sequence mediates co- or post-translational translocation can vary in dependence on the host organism: Pre-pro-alpha-MF signal sequence (from *S. cerevisiae*) can interact with SRP when expressed in mammalian cells, although in yeast cells, this signal sequence typically mediates post-translational translocation [6].

1.3 Intracellular Targeting

Intracellular targeting of recombinant proteins presents an important option when it comes to the expression of specialized recombinant proteins in *P. pastoris*. A variety of different targeting sites exist within the yeast cell.

1.3.1 Nucleus Targeting

The "classic" translocation to the nucleus refers to short nuclear localization signals (NLSs), which are described as mono- or bipartite, according to the number of basic amino acid stretches.

Proteins providing NLSs are bound by a specific receptor, which mediates translocation through nuclear pores. After the translocation, the GTPase (Ran) interacts with the cooperating receptor protein, and the NLS-protein is liberated into the nucleoplasm. This function of NLSs can also be applied to nonnucleic proteins: Transfer of NLS motifs to cytoplasmatic proteins can promote their entry into the nucleoplasm, highlighting NLSs as interesting tools to facilitate genome engineering [13, 14]. In *P. pastoris*, foreign NLS from SV40 T antigen (Simian Virus 40) [15], *Xenopus laevis* nucleoplasmin protein (*Xl*Nuc, [13]), human proto-oncogene (*Hs*Myc, [13]), as well as transcription factor Swi5p from *S. cerevisiae* (*Sc*SWI5, [16]) mediated translocation of eGFP into the nucleus [14]. Additionally, the NLS from SV40 T antigen was C-terminally fused to a codon-optimized Cas9 to target the enzyme into the nucleus of *P. pastoris* for genome editing [17]. In addition, Weniger *et al.* described NLSs natively occurring in the genome of *P. pastoris* [18]. Some of these NLSs were identified as homologues to *S. cerevisiae* nuclear proteins [19]. NLSs such as *Pp*Set7 and *Pp*Uba1 from *P. pastoris* were able to mediate eGFP translocation into the nucleus [14].

1.3.2 Vacuolar Targeting

There are two possible ways that proteins can be transported into the vacuoles of *S. cerevisiae* or *P. pastoris*: utilization of the secretory pathway compartments (via a N-terminal signal peptide) or transport through the cytoplasm-to-vacuole pathway (Cvt-pathway) [2, 20]. The latter refers to vacuolar proteins which lack a signal sequence for the transport via the secretory pathway. API (aminopeptidase I) in *S. cerevisiae*, for instance, was described to form a complex of dodecamers within the cytoplasm. A double membrane vesicle is then formed around the protein, directing it to the vacuole. A single membrane vesicle is set free into the vacuole and decomposed in the vacuolar lumen. This pathway can be used as protective mechanism of the cell, as described for Ams1p (a mannose glycosidase) in *P. pastoris*. Whereby not only protein size but also function could potentially interfere with cellular components during transport via the secretory pathway: Since Ams1p is considered constitutively active, the mannose glycosidase function might affect other proteins during transport through the secretory pathway [20, 21]. Besides the Cyt-pathway, proteins can utilize compartments of the secretory pathway to reach the vacuole. For instance, the amino-terminal signal sequence of the vacuolar protein CPY (carboxypeptidase Y) mediates translocation into the ER, followed by translocation into the Golgi apparatus and transport to the vacuole [2]. For *P. pastoris*, a successfully utilized signal sequence for vacuolar targeting of recombinant proteins was derived from proteinase A (a vacuolar enzyme): THCAS ($\Delta 9$-tetrahydrocannabinolic acid) synthase from *Cannabis sativa* was targeted into the vacuoles of different *P. pastoris* strains – wild type and vacuolar protease knockouts. The combination of a proteinase A-deficient strain and a high-copy vector revealed the highest intracellular volumetric activity (98+-5 pkat/mL) [22].

1.3.3 Peroxisome Targeting

The first peroxisomal targeting signal (PTS1) was described for firefly luciferase, consisting of C-terminal serine, lysine, and leucine (SKL) [28]. In addition, various PTSs are described for *Hansenula polymorpha*, such as ARF for the enzyme AOXp, as well as SKI for catalase [29]. In general, PTSs can be considered conserved from yeasts to humans. After recognized by a receptor, translocation into peroxisomes is mediated by peroxisomal translocation machineries [30]. Targeting recombinant proteins into peroxisomes can have different intentions: in a scientific context or for product engineering. Especially in methylotrophic yeasts, such as *P. pastoris*, peroxisomes are in the scientific focus of interest: Targeted marker proteins can help observe peroxisomal development and maturation during metabolic changes. SKL can be used for heterologous protein targeting in yeast when fused at the C-terminus of the recombinant protein. Monosov *et al.* used GFP-SKL fusion as a marker to track peroxisomes in living *P. pastoris* cells using fluorescence microscopes [31]. Furthermore, Sakai *et al.* used GFP-SKL to investigate microautophagy in *P. pastoris* [32].

Beside the scientific purpose, peroxisomal targeting can also be beneficial regarding product engineering or storage. Accumulation of heterologous proteins within peroxisomes was described to prevent undesired protein modification in *H. polymorpha*. In addition, methanol-dependent growth leads to enlargement of peroxisomes in methylotrophic yeasts (i.e., *P. pastoris*, *H. polymorpha*), stimulating the formation of large compartments for the accumulation of (perhaps toxic) recombinant proteins. Not only *H. polymorpha*, but also *P. pastoris* seems to be an interesting candidate for peroxisomal targeting. In 2009, Wang *et al.* utilized peroxisomal targeting for the expression of the catalytic domain of KDR kinase insert domain receptor (i.e., vascular endothelial growth factor receptor-2, VEGFR2) in *P. pastoris*. This protein is of special interest regarding cancer treatment due to its therapeutic potential in tumor angiogenesis. By the addition of the SKL signal sequence at the C-terminus, Wang *et al.* directed the heterologous protein to peroxisomes. This special protein target location was meant to combat the potentially negative effects of phosphorylation, which can lead to inhibited cell growth [29]. *In vivo* compartmentalization of biocatalytic reactions and metabolic pathways by peroxisomal targeting of enzymes meanwhile became a key technology for whole cell biocatalysis and metabolic engineering employing *P. pastoris* [38, 42]. Similar to the addition of C-terminal signal sequences for peroxisomal targeting, N-terminal mitochondrial sequences were successfully added to proteins for the localization in mitochondria.

1.3.4 ER Retention

Beside its key role as a transit stop during the secretory pathway – first place to fold disulfide bridge-containing proteins and initiator place for typical eukaryotic posttranslational modifications such as N-glycosylation – the ER can also function as final location for recombinant proteins, e.g., for chaperones or helper proteins. For example, Summpunn *et al.* utilized an ER-retention signal (HDEL) to target the chaperones GroEL and GroES into the ER in a manner to mediate adequate folding of recombinant phytase. However, to ensure functional chaperones, GroEL and GroES have to previously undergo the processing during the passage of the secretory pathway, for instance, guided by pre-pro-alpha-MF signal sequence as described by Summpunn et al. [23]. After the proteins have been processed within the Golgi apparatus, the HDEL sequence at the C-terminus facilitates the proteins' return to the ER [23]. In *P. pastoris*, adequate folding of heterologous proteins containing disulfide bridges can be promoted by over-expression of ER-directed protein disulfide isomerase (PDI) using a HDEL-retention signal, as reported by Warsame et al. [24]. Further in this context, Callewaert *et al.* used HDEL retention signal to target mannosidase to the ER engineering yeast N-glycosylation toward mammalian-like glycosylation [25].

However, not only chaperones and helper proteins have potential benefits when targeted to the ER, but also heterologous protein production of the plant protein proamarantin (from *Amaranthus hypochondriacus*) was increased using an ER-retention signal in *P. pastoris*. In plants, proamarantin trimerizes within the ER and is targeted to storage vacuoles for further processing [26, 27. However, accumulation of recombinant proamarantin in *P. pastoris* vacuoles can have a negative effect on protein yield since vacuolar proteolytic degradation occurs. In order to circumvent potential vacuolar accumulation, recombinant proamarantin can be retained in the ER using a plant-based ER-retention signal sequence (KDEL). In this study, KDEL successfully increased protein production and avoided degradation when fused to proamarantin [26].

1.4 Extracellular Targeting

P. pastoris is known for its efficient secretion competencies for heterologous proteins despite naturally secreting endogenous proteins in low quantities. This enables the accumulation of the heterologous proteins within the culture medium and usually results in efficient downstream processing and pure products [4].

1.4.1 Alpha-MF Secretion Signal Sequence, Ost1 or a Combination?

The alpha mating factor (alpha-MF, α-MF) signal sequence is one of the most frequently described signal sequences used for heterologous protein expression in *P. pastoris*. The sequence originates from *S. cerevisiae*, where it is responsible for the secretion of α-mating pheromones, having a key role in the yeast's mating process. Alpha-MF signal sequence is produced as a precursor, characterized as pre-pro-alpha-MF, whereby the pre-signal contains 19 amino acids (aa) and is cleaved by a signal peptidase within the ER. The second part of the signal sequence, described as pro-signal, contains 67 aa including three N-glycosylation motifs and a KR (lysine, arginine) motif near the C-terminal end of the signal peptide. In the Golgi, the signal peptide is cleaved by endopeptidase Kex2p at the KR motif. The most C-terminal part of this signal peptide is the cleavage site for another endopeptidase, i.e., Ste13p (acting in the Golgi apparatus) which finalizes the signal peptide processing by cleaving the C-terminal EA (glutamic acid, alanine) repeats [7–9]. In 2014, Fitzgerald and Glick revealed studies regarding the functions of pre- and pro-signal of alpha-MF sequence by expressing monomeric superfolder GFP (msGFP) as pre-pro-alpha-msGFP and pre-alpha-msGFP in *S. cerevisiae.* They observed vast amounts of pre-pro-alpha-msGFP translocating into the ER, while pre-alpha-msGFP was accumulating within the cytosol. They suggest that the pro-signal plays a key role in posttranslational translocation [5]. Further in this context, the signal sequence of dolichyl-diphosphooligosaccharide-protein glycosyltransferase subunit 1 (OST1, pre-OST1) was published by Willa and Forte *et al.*, which originates from *S. cerevisiae* and is described as hydrophobic and capable of SRP interaction [10]. Fitzgerald and Glick expressed pre-OST1-msGFP to examine whether co-translational translocation is able to prevent msGFP accumulation within the cytosol. The function of the missing pro-signal was de facto compensated by the pre-OST1 signal sequence, and efficient secretion was described for pre-OST1-msGFP. Co-translational translocation most probably prevents msGFP from blocking translocation due to folding within the cytosol. However, pre-OST1-msGFP tended to remain within the ER, while pre-pro-alpha-msGFP was more efficiently secreted to the supernatant. Accordingly, the combination of pre-OST1-pro-alpha-msGFP led to the most efficient secretion, suggesting a key role of the pro-sequence for ER export [5]. The fusion of pre--OST1-pro-alpha was also described by Barrero *et al.* They investigated the secretion of the oligomeric model protein E2-Crimson. This protein is capable of folding rapidly within the cytosol as soon as it is translated, causing problems for secretion employing the pre-pro-alpha mating factor signal peptide of *S. cerevisiae*, known to support posttranslational translocation. Therefore, co-translational translocation seemed to be an adequate choice and was confirmed by their experiments. However, to prevent accumulation within the ER, the ER export signal of pro-alpha was required. ER export is mediated by a transmembrane receptor (Erv29), which plays a key role in the efficiency of the alpha-factor signal sequence. Finally, a combination of pre-OST1 and the ser42 variant of the *S. cerevisiae* mat alpha-pro region was successfully used to increase the titer of secreted E2-Crimson [4].

However, neither the native pre-pro-alpha-MF nor the pre--OST1-alpha-MF-hybrid signal sequence represents a universal remedy for heterologous protein expression in *P. pastoris*, since both of them are prone to leaving C-terminal EA extensions on the protein of interest due to incomplete processing by Ste13p in *P. pastoris*. For many protein products, the EA repeat at the C-terminal end of the signal peptide is not needed for efficient secretion and can simply be deleted. Alternatively, the pre-OST1 signal peptide alone has also been reported to mediate efficient secretion of heterologous proteins [11]. For example, the pre--pro-alpha-MF signal sequence was described as nonsuitable for the heterologous expression of LPMOs (lytic polysaccharide monooxygenases) in *P. pastoris* due to incomplete signal peptide processing interfering with enzyme activity. Therefore, the utilization of native LPMO signal peptides was mandatory [12] until the pre-OST1 sequence was described as an efficient alternative in 2021 [11].

1.4.2 Alpha-MF Secretion Signal Variants

Various variants of pre-pro-alpha-MF signal sequence have been investigated. Among others, e.g., Rakestraw et al. [40] published a comprehensive study on alpha-MF signal sequence mutations and improved secretion in *S. cerevisiae*. However, such benefits are not always transferable to related hosts such as *P. pastoris* or characteristic for specific product molecules. For *P. pastoris*, for instance, an allelic variant of pro-signal was described to be able to prevent aggregation of the recombinant protein within the ER. Leu42 instead of Ser42 decreases aggregation, most probably through the mediation of rapid ER exit of the recombinant protein [4]. Interestingly, Ser42 seems to be a variant of some *S. cerevisiae* strains while Leu42 is the natural amino acid in *S. cerevisiae* S288C and many other sequenced strains. Additional mutations were described as beneficial for nanobody secretion by *P. pastoris* [41]. Further in this context, Lin-Cereghino *et al.* revealed a study including the systematic deletion of amino acid stretches utilizing site-directed mutagenesis [9].

2 Methods

2.1 Customizing a Signal Sequence

By now, there are many commercially available vectors providing suitable, codon-optimized signal sequences (e.g., Biogrammatics Inc., bisy GmbH, Thermo Fischer) that can be cloned in frame with the gene of interest. Thermo Fisher Scientific Inc., for instance, offers a collection of eight different secretion signal sequences as

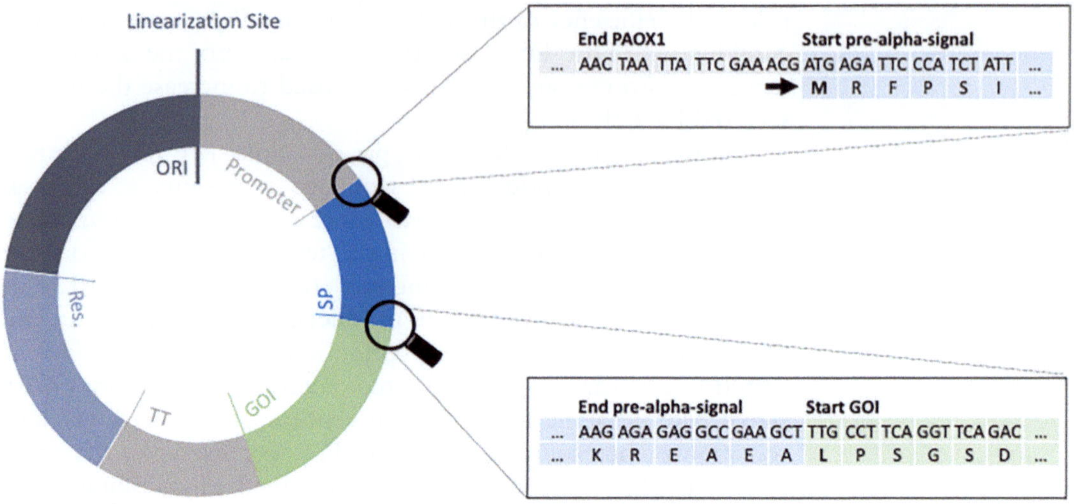

Fig. 1 Example for a standard expression vector showing in-frame cloning of alpha-MF signal sequence and a gene of interest (GOI). Promoter ... constitutive (e.g., P_{GAP}) or inducible (e.g., P_{AOX1}); SP ... signal peptide (pre-pro-alpha-MF signal sequence shown); GOI ... gene of interest; TT ... terminator (e.g., *AOX1*TT); Res. ... resistance cassette (e.g., promoter for bacterial and yeast expression, Zeocin resistance gene, terminator), ORI ... origin of replication. Pre-pro-alpha-MF signal sequence starting with methionine allows the first amino acid of the GOI to differ from the initial methionine

part of the PichiaPink™ yeast expression system. To identify the most suitable signal sequence for a recombinant protein, literature research and comparative experiments are indispensable [33] and also codon usage adaption plays a key role regarding heterologous signal sequences in *P. pastoris* [14, 33]. In addition, special requirements on the protein product have to be considered, such as the potentially negative influence of signal peptide remnants (produced by incorrect processing). In this context, a signal sequence should be chosen which is processed adequately. The C-terminus of the alpha-mating factor signal sequence, for instance, tends to be processed inefficiently by Ste13 protease. As a consequence, nonhomogeneous N-terminal ends might remain on the target protein [34, 35]. For many proteins, this EA repeat coding sequence can simply be removed without reducing secretion efficiency and often increases the fraction of correctly processed mature protein products. In general, the start codon (i.e., methionine for eukaryotes) is provided by an N-terminal secretion signal sequence. In-frame cloning to an N-terminal signal sequence allows the GOI to start with all possible amino acids and can be facilitated using type IIS restriction enzymes (e.g., BsaI enabling "seamless" cloning [33]) or Gibson isothermal assembly [36]. Figure 1 shows an example of a standard expression vector with a GOI cloned in-frame with a signal sequence facilitating secretion. Using seamless cloning technologies or synthetic genes enables the use of the full diversity of available secretion signals, in contrast to the few available standard vectors for cloning using standard type II restriction enzymes and ligase to obtain in-frame fusions.

2.2 Guidance for Directing a Recombinant Protein to a Specific Target

For targeting any protein to the desired cell compartment, the encoding gene of interest has to be retrieved from a database and prepared. A selection of signal sequences evaluated for the use in *P. pastoris* is shown in Table 1. Many companies are provide expression vectors already including specific signal sequences (*see* **Note 1**).

2.2.1 Preparing the Gene of Interest

1. Retrieve a desired protein sequence from databank (e.g., NCBI; https://www.ncbi.nlm.nih.gov).
2. Prediction of native signal sequences of the retrieved sequence using online tools like PrediSi (http://www.predisi.de), SignalP (https://services.healthtech.dtu.dk/services/SignalP-6.0/), PSORT (https://www.psort.org), or Signal-BLAST (http://sigpep.services.came.sbg.ac.at/signalblast.html).
3. Remove the predicted signal sequenced if not suitable for targeting to the intended target location (*see* **Note 2**).
4. Perform a back translation of your GOI to adapt to *P. pastoris* codon usage (e.g., GeneDesigner 2.0 by ATUM) (*see* **Note 3**).

2.2.2 In Silico Cloning

1. Choose a plasmid backbone with or without the chosen signal sequence.
2. For in silico cloning, load the plasmid in a genetic engineering software (e.g., SnapGene®).
3. Virtually clone your GOI, depending on the selected plasmid with or without the selected signal sequence (Table 1), downstream of the promoter in your target backbone (*see* **Note 4**).
4. Make sure the reading frame, starting at the 5′-ATG, covers the signal sequence and the GOI.
5. When an N-terminal signal sequence is used, the 5′-ATG must be removed from the protein of interest, as it has to be provided by the N-terminal signal sequence.

3 Notes

1. For extracellular targeting, evaluation of multiple signal sequences (if possible, with complementary processing modes) is recommended to increase the possibility of correct processing and high-yield secretion.
2. To get the most precise targeting prediction possible, we recommend evaluating the sequence of interest with multiple tools in parallel.

Table 1
A selection of protein targeting sequences, which were successfully used in *P. pastoris*

Mechanism/ target	Signal sequence	Sequence (N-terminal → C-terminal)	Origin	Ref.
Secretion (post-translational)	Alpha-mating factor	MRFPSIFTAVLFAASSALAAPVN TTTEDETAQIPAEAVIG YSDLEGDFDVAVLPFSN SNNGLLFINTTIASIAAKEEG VSLEKREAEA	*S. cerevisiae*	[12]
Secretion (co-translational)	Pre-OST1	MRQVWFSWIVGLFLCFFNVSSA	*S. cerevisiae*	[11]
Secretion (co-translational)	Pre-OST1- pro-alpha -MF Hybrid	MRQVWFSWIVGLFLCFFNVSSAAP VNTTTEDETAQIPAEAVIGYS DLEGDFDVAVLPFSNSNNGLLFIN TTIASIAAKEEGVSLEKREAEA	*S. cerevisiae*	[5]
Peroxisomes	PTS1 for firefly luciferase, consensus for peroxisomal targeting	SKL	*Photinus pyralis*	[28] [31] [38]
Vacuoles	signal sequence of the vacuolar protein CPY	MKAFTSLLCGLGLSTTLAKAISL QRPL	*S. cerevisiae*	[1]
Nuclear targeting	NLS from SV40 T antigen	PKKKRKV	*Simian Virus 40*	[13] [14]
	*Xl*Nuc (NLS from *Xenopus laevis* nucleoplasmin protein	KRPAAATKKAGQAKKKK	*Xenopus laevis*	[13] [14]
	*Hs*Myc (NLS from human proto-oncogene)	PAAKRVKLD	*Homo sapiens*	[13] [14]
	*Sc*SWI5 (NLS from transcription factor Swi5p)	KYENVVIKRSPRKRGRPRK	*S. cerevisiae*	[14]
	*Pp*Set7	KRKLEEEEGSKRNKRIKG	*P. pastoris*	[14]
	*Pp*Uba1	KRPLEIEQEETYSKRKKSTI	*P. pastoris*	[14]
ER retention	K/HDEL	K/HDEL	Plants/ yeasts	[37]

3. We recommend the *P. pastoris* high methanol codon usage following Abad et al. [38].
4. If yeast signal sequences are added for protein targeting in *P. pastoris*, native signal peptides of heterologous proteins need to be removed from those sequences to avoid duplicated targeting sequences.

Acknowledgments

This work received funding from the European Union in the frame of the ROBUSTOO project (grant agreement nr. 101135119) under the Horizon Europe Framework Programme.

The COMET center acib: Next Generation Bioproduction is funded by BMK, BMDW, SFG, Standortagentur Tirol, Government of Lower Austria und Vienna Business Agency in the framework of COMET—Competence Centers for Excellent Technologies. The COMET-Funding Program is managed by the Austrian Research Promotion Agency FFG.

Competing Interest Statement The authors have no conflicts of interest to declare that are relevant to the content of this chapter.

References

1. Bird P, Gething MJ, Sambrook J (1987) Translocation in yeast and mammalian cells: Not all signal sequences are functionally equivalent. J Cell Biol. https://doi.org/10.1083/jcb.105.6.2905
2. Blachly-Dyson E, Stevens TH (1987) Yeast carboxypeptidase Y can be translocated and glycosylated without its amino-terminal signal sequence. J Cell Biol. https://doi.org/10.1083/jcb.104.5.1183
3. Zheng N, Gierasch LM (1996) Signal sequences: the same yet different. Cell. https://doi.org/10.1016/S0092-8674(00)80159-2
4. Barrero JJ, Casler JC, Valero F, Ferrer P, Glick BS (2018) An improved secretion signal enhances the secretion of model proteins from Pichia pastoris. Microb Cell Fact. https://doi.org/10.1186/s12934-018-1009-5
5. Fitzgerald I, Glick BS (2014) Secretion of a foreign protein from budding yeasts is enhanced by cotranslational translocation and by suppression of vacuolar targeting. Microb Cell Fact. https://doi.org/10.1186/s12934-014-0125-0
6. Ng DTW, Brown JD, Walter P (1996) Signal sequences specify the targeting route to the endoplasmic reticulum membrane. J Cell Biol. https://doi.org/10.1083/jcb.134.2.269
7. Waters MG, Evans EA, Blobel G (1988) Prepro-α-factor has a cleavable signal sequence. J Biol Chem. https://doi.org/10.1016/s0021-9258(18)68773-3
8. Ahmad M, Hirz M, Pichler H, Schwab H (2014) Protein expression in Pichia pastoris: Recent achievements and perspectives for heterologous protein production. Appl Microbiol Biotechnol. https://doi.org/10.1007/s00253-014-5732-5
9. Lin-Cereghino GP et al (2013) The effect of α-mating factor secretion signal mutations on recombinant protein expression in Pichia pastoris. Gene. https://doi.org/10.1016/j.gene.2013.01.062
10. Willer M, Forte GMA, Stirling CJ (2008) Sec61p is required for ERAD-L: Genetic dissection of the translocation and ERAD-L functions of Sec61P using novel derivatives of CPY. J Biol Chem. https://doi.org/10.1074/jbc.M803054200
11. Rieder L, Ebner K, Glieder A, Sørlie M (2021) Novel molecular biological tools for the efficient expression of fungal lytic polysaccharide monooxygenases in Pichia pastoris. Biotechnol Biofuels. https://doi.org/10.1186/s13068-021-01971-5
12. Tanghe M et al (2015) Recombinant expression of Trichoderma reesei Cel61A in Pichia pastoris: optimizing yield and N-terminal processing. Mol Biotechnol. https://doi.org/10.1007/s12033-015-9887-9
13. Conti E, Kuriyan J (2000) Crystallographic analysis of the specific yet versatile recognition of distinct nuclear localization signals by karyopherin α. Structure. https://doi.org/10.1016/S0969-2126(00)00107-6
14. Weninger A, Glieder A, Vogl T (2015) A toolbox of endogenous and heterologous nuclear localization sequences for the methylotrophic

yeastpichia pastoris. FEMS Yeast Res. https://doi.org/10.1093/femsyr/fov082

15. Hobl B, Hock B, Schneck S, Fischer R, Mack M (2013) Bacteriophage T7 RNA polymerase-based expression in Pichia pastoris. Protein Expr Purif. https://doi.org/10.1016/J.PEP.2013.09.004
16. Jans DA (1995) The regulation of protein transport to the nucleus by phosphorylation. Biochem J. https://doi.org/10.1042/bj3110705
17. Weninger A, Fischer JE, Raschmanová H, Kniely C, Vogl T, Glieder A (2018) Expanding the CRISPR/Cas9 toolkit for Pichia pastoris with efficient donor integration and alternative resistance markers. J Cell Biochem. https://doi.org/10.1002/jcb.26474
18. De Schutter K et al (2009) Genome sequence of the recombinant protein production host Pichia pastoris. Nat Biotechnol. https://doi.org/10.1038/nbt.1544
19. Cherry JM et al (2012) Saccharomyces Genome Database: the genomics resource of budding yeast. Nucleic Acids Res. https://doi.org/10.1093/nar/gkr1029
20. Farré JC, Vidal J, Subramani S (2007) A cytoplasm to vacuole targeting pathway in P. pastoris. Autophagy. https://doi.org/10.4161/AUTO.3905
21. Hutchins MU, Klionsky DJ (2001) Vacuolar localization of oligomeric α-Mannosidase requires the cytoplasm to vacuole targeting and autophagy pathway components in Saccharomyces cerevisiae. J Biol Chem. 276(23):20491. https://doi.org/10.1074/JBC.M101150200
22. Zirpel B, Stehle F, Kayser O (2015) Production of Δ9-tetrahydrocannabinolic acid from cannabigerolic acid by whole cells of Pichia (Komagataella) pastoris expressing Δ9-tetrahydrocannabinolic acid synthase from Cannabis sativa L. Biotechnol Lett 37(9):1869–1875. https://doi.org/10.1007/S10529-015-1853-X
23. Summpunn P, Jomrit J, Panbangred W (2018) Improvement of extracellular bacterial protein production in Pichia pastoris by co-expression of endoplasmic reticulum residing GroEL-GroES. J Biosci Bioeng 125(3):268–274. https://doi.org/10.1016/J.JBIOSC.2017.09.007
24. Warsame A, Vad R, Kristensen T, Øyen TB (2001) Characterization of a gene encoding a Pichia pastoris protein disulfide isomerase. Biochem Biophys Res Commun 281(5):1176–1182. https://doi.org/10.1006/BBRC.2001.4479
25. Callewaert N et al (2001) Use of HDEL-tagged Trichoderma reesei mannosyl oligosaccharide 1,2-alpha-D-mannosidase for N-glycan engineering in Pichia pastoris. FEBS Lett 503(2–3):173–178. https://doi.org/10.1016/S0014-5793(01)02676-X
26. Medina-Godoy S, Valdez-Ortiz A, Valverde ME, Paredes-López O (2006) Endoplasmic reticulum-retention C-terminal sequence enhances production of an 11S seed globulin from Amaranthus hypochondriacus in Pichia pastoris. Biotechnol J 1(10):1085–1092. https://doi.org/10.1002/BIOT.200600126
27. Müntz K (1998) Deposition of storage proteins. Plant Mol Biol 38(1–2):77–99. https://doi.org/10.1007/978-94-011-5298-3_4
28. Gould SJ, Keller GA, Subramani S (1987) Identification of a peroxisomal targeting signal at the carboxy terminus of firefly luciferase. J Cell Biol 105(6 Pt 2):2923–2931. https://doi.org/10.1083/JCB.105.6.2923
29. Van Dijk R, Faber KN, Kiel JAKW, Veenhuis M, Van Der Klei I (2000) The methylotrophic yeast Hansenula polymorpha: a versatile cell factory. Enzyme Microb Technol 26(9–10):793–800. https://doi.org/10.1016/S0141-0229(00)00173-3
30. Rachubinski RA, Subramani S (1995) How proteins penetrate peroxisomes. Cell 83(4):525–528. https://doi.org/10.1016/0092-8674(95)90091-8
31. Monosov E, Wenzel TJ, Luers GH, Heyman JA, Subramani S (1996) I labeling of peroxisomes with green fluorescent protein in living I! pastoris Cells'. J Histochem Cytochem 44(6):181–189
32. Sakai Y, Koller A, Rangell LK, Keller GA, Subramani S (1998) Peroxisome degradation by microautophagy in Pichia pastoris: identification of specific steps and morphological intermediates. J Cell Biol 141(3):625–636. https://doi.org/10.1083/jcb.141.3.625
33. Cregg JM, Tolstorukov I, Kusari A, Sunga J, Madden K, Chappell T (2009) Chapter 13 Expression in the yeast Pichia pastoris. In: Methods in Enzymology
34. Almeida MS, Cabral KS, De Medeiros LN, Valente AP, Almeida FCL, Kurtenbach E (2001) cDNA cloning and heterologous expression of functional Cysteine-Rich antifungal protein Psd1 in the yeast Pichia pastoris. Arch Biochem Biophys 395(2):199–207. https://doi.org/10.1006/ABBI.2001.2564
35. Cabral KMS, Almeida MS, Valente AP, Almeida FCL, Kurtenbach E (2003) Production of the active antifungal Pisum sativum defensin 1 (Psd1) in Pichia pastoris:

overcoming the inefficiency of the STE13 protease. Protein Exp Purif 31(1):115–122. https://doi.org/10.1016/S1046-5928(03)00136-0

36. Gibson DG, Young L, Chuang RY, Venter JC, Hutchison CA, Smith HO (2009) Enzymatic assembly of DNA molecules up to several hundred kilobases. Nat Methods. https://doi.org/10.1038/nmeth.1318
37. Rossanese OW et al (2001) A role for Actin, Cdc1p, and Myo2p in the inheritance of Late Golgi elements in Saccharomyces cerevisiae. J Cell Biol 153(1):47. https://doi.org/10.1083/JCB.153.1.47
38. Abad S et al (2010) Stepwise engineering of a Pichia pastoris D-amino acid oxidase whole cell catalyst. Microb Cell Fact 9:24. https://doi.org/10.1186/1475-2859-9-24
39. Miura N, Ueda M (2018) Evaluation of unconventional protein secretion by Saccharomyces cerevisiae and other Fungi. Cells 7(9):128. https://doi.org/10.3390/cells7090128
40. Rakestraw JA, Sazinsky SL, Piatesi A, Antipov E, Wittrup KD (2009) Directed evolution of a secretory leader for the improved expression of heterologous proteins and full-length antibodies in Saccharomyces cerevisiae. Biotechnol Bioeng 103(6):1192–1201. https://doi.org/10.1002/bit.22338
41. Brigé A, Schotte P, Walcarius B (2012) Method for the production of immunoglobulin single variable domains. WO 2012/152823 A1, November 15, 2012
42. DeLoache W, Russ Z, Dueber J (2016) Towards repurposing the yeast peroxisome for compartmentalizing heterologous metabolic pathways. Nat Commun 7:11152. https://doi.org/10.1038/ncomms11152

Chapter 16

Cloning of MFα Secretion Signal Variants for Protein Secretion by *Pichia pastoris*

Magdalena Merkaš, Odyssefs-Ioannis Pantelakis, and Anita Emmerstorfer-Augustin

Abstract

Protein secretion is a complex, multistep process that relies on signal sequences to guide recombinant proteins through various stages of the secretion pathway, ultimately targeting these proteins to the extracellular matrix. The mating factor alpha (MFα) signal peptide, initially identified for its role in the mating factor protein export in *Saccharomyces cerevisiae*, has emerged as the most effective tool for the secretory production of recombinant proteins in yeast expression systems, including *Komagataella phaffii* (reclassified from *Pichia pastoris*). Over the years, numerous studies have aimed at enhancing the efficiency and processability of the MFα signal sequence, resulting in several optimized variants. In this chapter, we highlight the most frequently used and highly effective MFα signal leader variants. We present a streamlined and universal cloning strategy using pPpT4_MFα as a vector backbone and the gene encoding eGFP as the gene of interest to facilitate the use of these variants for recombinant protein secretion. This step-by-step approach also provides the precise (codon optimized) nucleotide and amino acid sequences of these MFα signal leader variants.

Key words *Pichia pastoris*, *Komagataella phaffii*, MFα, Protein secretion, Mutational analysis

1 Introduction

Haploid yeast can fuse to form a diploid when cells of opposite mating types (termed **a** and α in *Saccharomyces cerevisiae*) are mixed. **a**-type cells secrete the **a**-factor mating pheromone, while α cells secrete α-mating factor (MFα) [1]. Secretion of the MFα peptide is facilitated by the MFα signal sequence, which directs the peptide into the secretory pathway, where it is further processed [2]. Once outside the cell, the MFα peptide acts as a pheromone, prompting a haploid **a**-type cell to initiate the mating response. This response leads to various cellular changes, including cell cycle arrest and morphological alterations, ultimately resulting in the formation of mating projections that facilitate cell fusion [3, 4]. Soon after the

Xavier Garcia-Ortega et al. (eds.), *Pichia pastoris: Methods and Protocols*, Methods in Molecular Biology, vol. 2697,
https://doi.org/10.1007/978-1-0716-4779-0_16,

discovery of these processes in the early 1980s, researchers found that other proteins can be targeted into the yeast secretory pathway through N-terminal fusion of the MFα signal sequence. This was first demonstrated with epidermal growth factor in *S. cerevisiae* [5]. Since then, numerous studies have followed, revealing that this complex process involves multiple steps across various cellular compartments before the desired protein ultimately reaches the cell exterior [6, 7]. Additionally, it has been shown that the *S. cerevisiae* MFα signal sequence can be applied for efficient protein secretion in other yeasts, such as *Komagataella phaffii* (reclassified from *Pichia pastoris*) [8].

Protein secretion begins on ribosomes. Signal recognition particles (SRP) bind secretion signal sequences of freshly synthesized proteins and guide them to the endoplasmic reticulum (ER), where proteins are processed. Vesicular transport then carries proteins to the Golgi for further modification and sorting before secretion via secretory vesicles to the cell membrane [9, 10]. This process can occur co-translationally or post-translationally, utilizing the Sec61 complex with different channel partners [9]. The main distinction lies in the timing of targeting and translocation: post-translationally synthesized proteins are selectively targeted for translocation into the secretory pathway, while co-translationally synthesized proteins begin translocation into the ER during translation [9]. The MFα signal peptide is classified as a post-translational or signal recognition particle signal sequence [11]. Within this pathway, proteins undergo post-translational translocation into the endoplasmic reticulum lumen, where various processes occur, including folding, disulfide bond formation, and initiation of post-translational modifications like glycosylation [12].

Sequence analysis of the prepro-α-factor (UniProt: P01149) revealed that it consists of multiple functional segments [2]. The primary, full-length structure (165 amino acids) starts with a 19-amino-acid N-terminal region (pre-region), followed by a 66-amino-acid middle segment (pro-region), and a segment with four tandem repeats of MFα peptides, interspersed with repeating spacer peptides "Lys-Arg-Glu-Ala-Asp-Ala-Glu-Ala" [2]. Additionally, the pro-region contains three N-linked glycosylation sites and a dibasic (KR) Kex2 endopeptidase cleavage site [13]. The prepro-α-factor is a fully inactive precursor protein and must be processed along the secretory pathway in three consecutive steps. In the first step, the signal recognition particle recognizes the leader sequence [13], which grants translocation of the cargo protein across the endoplasmic reticulum [14]. At the same time, cleavage of the pre-peptide and N-glycosylation at the three Asn residues in the pro-region are facilitated [15]. In the second step, the yeast Kex2 Ca^{2+}-dependent serine endopeptidase [16] binds the Kex2 cleavage site and cleaves off the signal peptide after the dibasic Lys and Arg residues in the late Golgi compartment [15]. Finally,

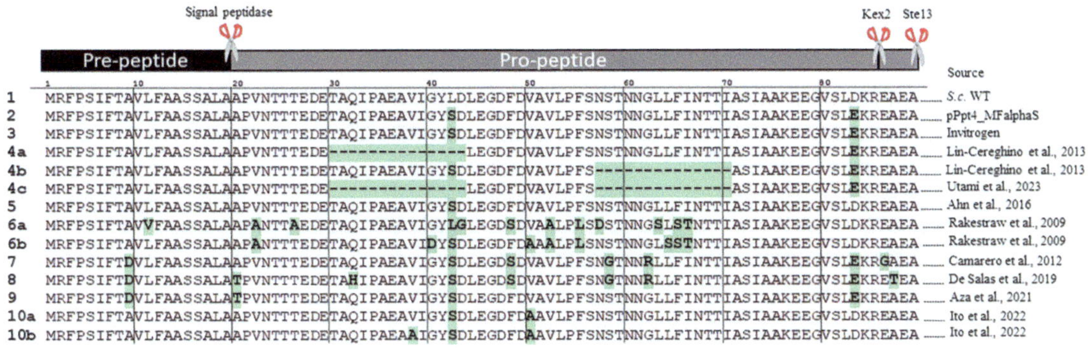

Fig. 1 Multiple sequence alignment of MFα variants from Table 1 using Clustal Omega and pyBoxshade for graphic visualization. The pre-peptide is processed by a signal peptidase in the ER, the pro-peptide by Kex2, and the EAEA repeat by Ste13 in the late Golgi. Green background and font indicate sequence alterations

Ste13, a membrane-bound dipeptidyl aminopeptidase, recognizes and processes the Glu-Ala-Glu-Ala sequences between tandem MFα peptides in the Golgi, and the cargo protein enters secretory vesicles [13, 17].

Although *K. phaffii* has successfully been applied for the secretory production of many different recombinant proteins [18], not every protein can be secreted, even when combined with proper secretion leader sequences. When utilizing MFα for secretion, several factors are important to consider for efficient and successful recombinant protein production. Since the MFα signal sequence facilitates secretion through post-translational translocation, it allows proteins to fold prematurely. Consequently, proteins may accumulate in the cytoplasm, unable to reach the endoplasmic reticulum, or may even block the translocon pores [19]. Incorrect signal sequence processing can also influence the ability of protein secretion. The endogenous *K. phaffii* Ste13 protein was shown to be less efficient than *S. cerevisiae* Ste13 for processing of the MFα leader sequence, leading to the presence of Glu-Ala-Glu-Ala overhangs on the N-terminus of the secreted product, while inefficient processing by protease Kex2 is rather rare [12]. Therefore, a common strategy is to simply omit the Glu-Ala-Glu-Ala sequence, which was shown not to affect proteolytic processing of the pro signal peptide by Kex2 [20–22]. Despite these variations in N-terminal processing, the *S. cerevisiae* MFα signal peptide remains the most used leader sequence in *K. phaffii*. This is because few alternative signal peptides have consistently achieved higher product quality or higher secreted titers of recombinant protein [22, 23]. Also, over the years, diverse modifications have led to improved variants of the MFα signal peptide, like deletion of specific amino acids [24], codon context (CC) optimization [25, 26] or mutational variations [27, 28] (Fig. 1 and Table 1). In this chapter, we provide an overview of the currently most frequently

Table 1
Overview of native and improved MFα signal sequence variants

No.	Sequence (nucleotides/amino acids)		Source
1	ATGAGATTTCCTTCAATTTTTACTGCAGTTTTATTCGCAGCATCCTCCGCATTAGCTGCTCCAG TCAACACTACAACAGAAGATGAAACGGCACAAATTCCGGCTGAAGCTGTCATCGGTTAC TTAGATTTAGAAGGGGATTTCGATGTTGCTGTTTTGCCATTTTCCAACAGCACAAATAACGGG TTATTGTTTATAAATACTACTATTGCCAGCATTGCTGCTAAAGAAGAAGGGGTATCTTTGGA TAAAAGAGAGGCTGAAGCT MRFPSIFTAVLFAASSALAAPVNTTTEDETAQIPAEAVIGYLDLEGDFDVAVLPFSNSTNNGLLFIN TTIASIAAKEEGVSLDKREAEA	(−) WT	[2]
2	ATGAGATTCCCATCTATTTTCACCGCTGTCTTGTTCGCTGCCTCCTCTGCATTGGCTGCCCCTG TTAACACTACCACTGAAGACGAGACTGCTCAAATTCCAGCTGAAGCAGTTATCGGTTACTC TGACCTTGAGGGTGATTTCGACGTCGCTGTTTTGCCTTTCTCTAACTCCACTAACAACGG TTTGTTGTTCATTAACACCACTATCGCTTCCATTGCTGCTAAGGAAGAGGGTGTCTCTC TCGAGAAGAGAGAGGCCGAAGCT MRFPSIFTAVLFAASSALAAPVNTTTEDETAQIPAEAVIGYSDLEGDFDVAVLPFSNSTNNGLLFIN TTIASIAAKEEGVSLEKREAEA	CO	pPpT4_MFalphaS, codon optimized: Leto, Entelechon®
3	ATGAGATTTCCTTCAATTTTTACTGCTGTTTTATTCGCAGCATCCTCCGCATTAGCTGCTCCAG TCAACACTACAACAGAAGATGAAACGGCACAAATTCCGGCTGAAGCTGTCATCGGTTAC TCAGATTTAGAAGGGGATTTCGATGTTGCTGTTTTGCCATTTTCCAACAGCACAAATAACGGG TTATTGTTTATAAATACTACTATTGCCAGCATTGCTGCTAAAGAAGAAGGGGTATCTC TCGAGAAAAGAGAGGCTGAAGCT MRFPSIFTAVLFAASSALAAPVNTTTEDETAQIPAEAVIGYSDLEGDFDVAVLPFSNSTNNGLLFIN TTIASIAAKEEGVSLEKREAEA	(−)	Invitrogen (verified by Barrero et al. [29]

4a	ATGAGATTTCCTTCAATTTTTACTGCTGTTTTATTCGCAGCATCCTCCGCATTAGCTGCTCCAG TCAACACTACAACAGAAGATGAATTAGAAGGGGATTTCGATGTTGCTGTTTTGCCA TTTTCCAACAGCACAAATAACGGGTTATTGTTTATAAATACTACTATTGCCAGCATTGCTGC TAAAGAAGAAGGGGTATCTCTCGAGAAAAGAGAGGCTGAAGCT MRFPSIFTAVLFAASSALAAPVNTTTEDELEGDFDVAVLPFSNSTNNGLLFINTTIASIAAKEEG VSLEKREAEA	Δ	[24]
4b	ATGAGATTTCCTTCAATTTTTACTGCTGTTTTATTCGCAGCATCCTCCGCATTAGCTGCTCCAG TCAACACTACAACAGAAGATGAAACGGCACAAATTCCGGCTGAAGCTGTCATCGGTTAC TCAGATTTAGAAGGGGATTTCGATGTTGCTGTTTTGCCATTTTCCGCCAGCATTGCTGC TAAAGAAGAAGGGGTATCTCTCGAGAAAAGAGAGGCTGAAGCT MRFPSIFTAVLFAASSALAAPVNTTTEDETAQIPAEAVIGYSDLEGDFDVAVLPFSASIAAKEEGVSLEK REAEA	Δ	
4c	ATGAGATTTCCTTCAATTTTTACTGCTGTTTTATTCGCAGCATCCTCCGCATTAGCTGCTCCAG TCAACACTACAACAGAAGATGAATTAGAAGGGGATTTCGATGTTGCTGTTTTGCCA TTTTCCGCCAGCATTGCTGCTAAAGAAGAAGGGGTATCTCTCGAGAAAAGAGAGGCTGAAGC T MRFPSIFTAVLFAASSALAAPVNTTTEDELEGDFDVAVLPFSASIAAKEEGVSLEKREAEA	Δ	[30]
5	ATGAGATTCCCATCTATCTTTACTGCCGTTTTGTTTGCTGCTTCCTCTGCTTTGGCAGCTCCAG TTAACACAACTACTGAGGATGAAACTGCTCAAATTCCTGCTGAGGCAGTTATTGGTTACTC TGATTTGGAAGGTGACTTTGATGTTGCTGTCTTGCCTTTCTCTAACTCCACCAACAACGG TTTGTTGTTCATCAACACTACCATTGCCTCTATTGCTGCCAAGGAAGAGGGTGTTTC TTTGGACAAGAGAGAGGCTGAAGCT MRFPSIFTAVLFAASSALAAPVNTTTEDETAQIPAEAVIGYSDLEGDFDVAVLPFSNSTNNGLLFIN TTIASIAAKEEGVSLDKREAEA	CO	[26]
6a	ATGAGATTCCCATCCATCTTCACTGCTGTTGTTTTCGCTGCTTCTTCCGCTTTGGCTGC TCCAGCTAACACTACTGCTGAAGATGAGACTGCTCAGATTCCAGCTGAGGCTGTTATTGG	M	[28]

(continued)

Table 1
(continued)

No.	Sequence (nucleotides/amino acids)		Source
	TTACTTGGGTTTGGAAGGTGACTCCGACGTTGCTGCTTTGCCATTGTCTGATTCCAC TAACAACGGTTCCTTGTCCACAAACACTACTATCGCTTCCATTGCTGCTAAAGAAGAGGGAG TTTCCTTGGACAAGAGAGAGGCTGAAGCT MRFPSIFTAVVFAASSALAAPANTTAEDETAQIPAEAVIGYLGLEGDSDVAALPLSDSTNNGSLSTN TTIASIAAKEEGVSLDKREAEA		
6b	ATGAGATTCCCATCCATCTTCACTGCTGTTTTGTTCGCTGCTTCTTCCGCTTTGGCTGC TCCAGCTAACACTACTACTGAAGATGAGACTGCTCAGATCCCAGCTGAGGCTGTTATTGAC TACTCTGACTTGGAAGGTGACTTCGACGCTGCTGCTTTGCCATTGTCTAACTCCAC TAACAACGGTTTGTCCTCCACAAACACTACAATCGCTTCCATTGCTGCTAAAGAAGAGGGAG TTTCCTTGGACAAGAGAGAGGCTGAAGCT MRFPSIFTAVLFAASSALAAPANTTTEDETAQIPAEAVIDYSDLEGDFDAAALPLSNSTNNGLSSTN TTIASIAAKEEGVSLDKREAEA	M	
7	ATGAGATTTCCTTCAATTTTTACTGATGTTTTATTCGCAGCATCCTCCGCATTAGCTGCTCCAG TCAACACTACAACAGAAGATGAAACGGCACAAATTCCGGCTGAAGCTGTCATCGGTTAC TCAGATTTAGAAGGGGATTCCGATGTTGCTGTTTTGCCATTTTCCAACGGCACAAATAACAGG TTATTGTTTATAAATACTACTATTGCCAGCATTGCTGCTAAAGAAGAAGGGGTATCTC TCGAGAAAAGAGGGGCTGAAGCT MRFPSIFTDVLFAASSALAAPVNTTTEDETAQIPAEAVIGYSDLEGDSDVAVLPFSNGTNNRLLFIN TTIASIAAKEEGVSLEKRGAEA	M	[31]
8	ATGAGATTTCCTTCAATTTTTACTGATGTTTTATTCGCAGCATCCTCCGCATTAGCTACTCCAG TCAACACTACAACAGAAGATGAAACGGCACATATTCCGGCTGAAGCTGTCATCGGTTAC TCAGATTTAGAAGGGGATTCCGATGTTGCTGTTTTGCCATTTTCCAACGGCACAAATAACAGA TTATTGTTTATAAATACTACTATTGCCAGCATTGCTGCTAAAGAAGAAGGGGTATCTC TCGAGAAAAGAGAGACTGAAGCT MRFPSIFTDVLFAASSALATPVNTTTEDETAHIPAEAVIGYSDLEGDSDVAVLPFSNGTNNRLLFIN TTIASIAAKEEGVSLEKRETEA	M	[27]

9	ATGAGATTTCCTTCAATTTTTACTGATGTTTTATTCGCAGCATCCTCCGCATTAGCTACTCCAG TCAACACTACAACAGAAGATGAAACGGCACAAATTCCGGCTGAAGCTGTCATCGGTTAC TCAGATTTAGAAGGGGATTTCGATGTTGCTGTTTTGCCATTTTCCAACAGCACAAATAACGGG TTATTGTTTATAAATACTACTATTGCCAGCATTGCTGCTAAAGAAGAAGGGGTATCTC TCGAGAAAAGAGAGGCTGAAGCT MRFPSIFTDVLFAASSALATPVNTTTEDETAQIPAEAVIGYSDLEGDFDVAVLPFSNSTNNGLLFIN TTIASIAAKEEGVSLEKREAEA	M	[32]
10a	ATGAGATTTCCTTCAATTTTTACTGCTGTTTTATTCGCAGCATCCTCCGCATTAGCTGCTCCAG TCAACACTACAACAGAAGATGAAACGGCACAAATTCCGGCTGAAGCTGTCATCGGTTAC TCAGATTTAGAAGGGGATTTCGATGCTGCTGTTTTGCCATTTTCCAACAGCACAAATAACGGG TTATTGTTTATAAATACTACTATTGCCAGCATTGCTGCTAAAGAAGAAGGGGTATCTCTCGA TAAAAGAGAGGCTGAAGCT MRFPSIFTAVLFAASSALAAPVNTTTEDETAQIPAEAVIGYSDLEGDFDAAVLPFSNSTNNGLLFIN TTIASIAAKEEGVSLDKREAEA	M	[33]
10b	ATGAGATTTCCTTCAATTTTTACTGCTGTTTTATTCGCAGCATCCTCCGCATTAGCTGCTCCAG TCAACACTACAACAGAAGATGAAACGGCACAAATTCCGGCTGAAGCTGCCATCGGTTAC TCAGATTTAGAAGGGGATTTCGATGCTGCTGTTTTGCCATTTTCCAACAGCACAAATAACGGG TTATTGTTTATAAATACTACTATTGCCAGCATTGCTGCTAAAGAAGAAGGGGTATCTCTCGA TAAAAGAGAGGCTGAAGCT MRFPSIFTAVLFAASSALAAPVNTTTEDETAQIPAEAAIGYSDLEGDFDAAVLPFSNSTNNGLLFIN TTIASIAAKEEGVSLDKREAEA	M	

Alterations: *CO* codon optimization, *Δ* deletion, *M* mutation used MFα signal sequence variants and provide a simple, straightforward step-by-step description of how to apply them for secretory production of recombinant proteins in *K. phaffii*, using eGFP as a model protein.

2 Materials

2.1 DNA Sequence Design

1. DNA sequences (ideally in GenBank format, *see* **Note 1**) of the vector backbone pPpT4_MFα (originally named pPpT4_Alpha_S, [29]) (*see* **Note 2**).
2. DNA sequence of the gene of interest (e.g., eGFP) (ideally in GenBank format) (*see* **Note 3**).
3. DNA sequence of the MFα signal sequence variant (ideally in GenBank format) (Table 1).
4. Software to manipulate DNA sequences in silico (*see* **Note 4**).
5. Primers (5′ - 3′), underlined parts bind the respective DNA template:

   First fragment (vector backbone pPpT4_MFα):

   fw(pPpT4): *TACGTACCTAGGGCGGCCGC*,
   rv(pPpT4): *CGTTTCGGAATTCTTTCAATAATTAG*,

   Second fragment (gene of interest (GOI), e.g., eGFP):

   fw(eGFP):*GCTAGCAAAGGAGAAGAACTTTTC*,
   rv(eGFP): GCATTCTGACATCCTCTTGAGCGGCCGCCC TAGGTACGTA*TTACTTGTACAATTCATCCA*,

   Third fragment (MFα signal sequence variant: example 3, Invitrogen):

   fw(MFα): AGATCAAAAAACAACTAATTATTGAAA GAATTCCGAAACG*ATGAGATTCCCATCTATTTT*,
   rv(MFα): TTGGGACAACTCCAGT GAAAAGTTCTTCTCCTTTGCTAGC *AGCTTCGGCCTCTCTCTTCT*.

2.2 Cloning

1. High-fidelity DNA polymerase (e.g., Phusion® High-Fidelity DNA polymerase, New England Biolabs) (*see* **Note 5**).
2. 2.5 mM dNTPs each.
3. Buffer for high-fidelity DNA polymerase (*see* **Note 6**).
4. DNA templates (pPpT4_MFα vector, GOI—here eGFP, MFα signal sequence).
5. 10 μM forward and 10 μM reverse primers.
6. 6× Thermo Scientific™ DNA Gel Loading Dye 6×.
7. 50XTAE solution: Dissolve 18.61 g of EDTA disodium salt and 242.28 g of Tris base in 750 mL of distilled water. Add

60.05 mL of 100% (v/v) acetic acid and adjust pH to 8.0 using NaOH. Fill up to a total volume of 1 L and store at room temperature.

8. 1XTAE electrophoresis buffer: Take 20 mL of 50XTAE solution and dilute it with distilled water to 1 L. Store at room temperature.
9. 1% (w/v) agarose gel: Weight 1% (w/v) agarose in an adequate amount of 1XTAE electrophoresis buffer. Boil the mixture until the agarose is completely melted. Cool down gel solution to about 50 °C and add ethidium bromide to a final concentration of ~0.3–0.6 μg/mL (*see* **Note 7**).
10. Agarose gel electrophoresis equipment.
11. Gel and PCR Clean-Up System (e.g., Promega GmbH, Mannheim, Germany).
12. Gibson Assembly® Cloning Kit (e.g., New England Biolabs, Frankfurt, Germany) (*see* **Note 8**).
13. Electroporator (e.g., MicroPulser™, Bio-Rad Laboratories, California, USA).
14. Electroporation cuvettes.
15. Thermomixer Comfort (Eppendorf AG, Hamburg, Germany).
16. Plasmid Miniprep Kit.
17. Restriction endonucleases *Smi*I and *Xba*I.
18. LB medium: 1% w/v tryptone, 1% w/v NaCl, and 0.5% w/v yeast extract. Antibiotics are used at the following concentrations: Zeocin™ (InvivoGen (Eubio), Vienna, Austria) [100 mg/L].
19. LB agar medium: 1% w/v tryptone, 1% w/v NaCl, and 0.5% w/v yeast extract, and 2% w/v agar. Antibiotics are used at the following concentrations: Zeocin™ (InvivoGen (Eubio), Vienna, Austria) [100 mg/L].
20. SOC medium: 20 mM glucose, 2% w/v tryptone, 0.05% w/v NaCl, 0.5% w/v yeast extract, 2.5 mM KCl, 5 mM $MgCl_2$, 5 mM $MgSO_4$, and pH 7.0.
21. 10 mL glass culture tubes for overnight cultures (Sigma-Aldrich®, Vienna, Austria).
22. HT Infors Shaker.
23. Plate Incubator (BD115, 37 °C, Binder GmbH).

3 Methods

The following approach describes the step-by-step cloning of a desired gene of interest (e.g., eGFP) along with different variants of the secretion signal MFα into the shuttle vector pPpT4_MFα for secretory expression in the yeast *Komagataella phaffii*. The preparation of the plasmid is performed via Gibson assembly, using three DNA fragments: the vector backbone, the MFα variant, and the gene of interest (Fig. 2).

3.1 Preparation of the Three Fragments Required for Gibson Assembly

3.1.1 Preparation of the First Fragment: pPpT4_MFα Vector Backbone

The backbone fragment is obtained via PCR, using two primers that exclude the original MFα of the plasmid template (pPpT4_MFα). The forward primer fw(pPpT4) binds to the region downstream of MFα, while the reverse primer rv(pPpT4) binds to the region upstream of MFα (*see* **Notes 9** and **10**). This fragment is designed to be compatible at its 3′ and 5′ ends with any GOI or MFα signal sequence variant.

3.1.2 Preparation of the Second Fragment: GOI (e.g., eGFP)

For amplification of the second fragment, the GOI (in this case, eGFP) is used as a template. The forward primer should bind to base pairs encoding the N-terminus of the GOI, and the reverse primer should bind to base pairs encoding the C-terminus of the GOI (eGFP) (green), adding a 20–40 bp overhang homologous to the pPpT4_MFα vector backbone for Gibson assembly (*see* **Note 11**).

3.1.3 Preparation of the Third Fragment: MFα Signal Sequence Variant

In a third PCR, the desired MFα sequence variants (either obtained from the original source or as synthetic genes or G-blocks based on the nucleotide sequence listed in Table 1) serve as templates. The forward primer should bind to the base pairs that encode the N-terminus of the MFα variant, with 20–40 bp overhangs homologous to the vector backbone. The reverse primer should bind to base pairs that encode the C-terminus of MFα variant, with appropriate 20–40 bp overhangs homologous to the upstream region of the GOI for Gibson assembly.

1. Generate first, second, and third fragments using a high-fidelity polymerase according to the manufacturer's protocol with the templates and primers as explained above (*see* **Note 11**).
2. Separate all DNA fragments on 1% (w/v) agarose gel by electrophoresis.
3. Excise the DNA fragments from the gel and purify them using a Gel and PCR Clean-Up System according to the manufacturer's protocol (*see* **Note 12**).

3.2 Gibson Cloning and *E. coli* Transformation

1. According to the manufacturer's protocol, all fragments are added in equimolar amounts, in a total volume of 5–15 μL of Gibson assembly master mix on ice (*see* **Note 13**).
2. Mix the content by carefully pipetting on ice.

3. Incubate reactions at 50 °C for 60 min.
4. Transform NEB 5-alpha Competent *E. coli* (High Efficiency) (provided with the Gibson Assembly® Cloning Kit) with 2 μL of the assembly reaction, following the transformation protocol (*see* **Note 14**).
5. Regenerate the transformed *E. coli* cells in 1 mL of SOC medium at 600 rpm for 40–60 min at 37 °C (in a Thermomixer).
6. Plate regenerated clones on LB-Zeocin plates and incubate at 37 °C overnight.

3.3 Confirmation of Correct Cloning

1. Cultivate single transformants (2–4) obtained from Gibson assembly in 2 mL of LB-Zeocin in 10 mL glass culture tubes at 37 °C overnight.
2. Isolate plasmids using a plasmid miniprep kit following the manufacturer's instructions.
3. For restriction analysis, use two single-cutting restriction enzymes. In the case of pPpT4_MFα, *Smi*I (also called *Swa*I) and *Xba*I are recommended. Use 200 ng of the isolated plasmid, add 1 μL of 10× restriction enzyme buffer, 0.5 μL of *Smi*I, 0.5 μL of *Xba*I, and fill up to 10 μL with ddH_2O. Incubate for 1 h at 37 °C.
4. Analyze samples by electrophoresis on 1% (w/v) agarose gels. In case of correct assembly, bands should be visible at 2774 and 1763 bp (*see* **Note 15**).
5. Send plasmids that give correct bands after restriction analysis to sequencing.

4 Notes

1. The GenBank file format includes detailed annotations such as the origin of replication, antibiotic resistance genes, and other genetic elements. It provides a comprehensive representation of the vector's features and is widely supported by various bioinformatics tools, including those used for in silico DNA manipulation and databases.
2. The pPpT4_MFα (pPpT4_Alpha_S, GenBank accession # JQ519691.1) vector contains the following elements: P_*AOX1*Syn (*AOX1* promoter for methanol-induced expression of the target gene in *K. phaffi*), MFα, *AOX1*TTSyn (*AOX1* terminator for target gene transcription termination in *K. phaffi*), pUC origin (pUC replication origin for *E. coli*), *AOD*TT (*AOD* terminator for target gene transcription termination in *K. phaffi*), Zeocin Syn (Zeocin resistance, selection

marker in *E.coli* and *K. phaffi*), and P_*ILV5* (*ILV5* promoter for expression of respective antibiotic resistance gene in *K. phaffi*).

3. The DNA sequence of eGFP is available on NCBI (sequence ID: AGT98535.1).
4. Recommended tools for DNA manipulation are SnapGene, Benchling, ape, or VectorNTI.
5. Any high-fidelity DNA polymerase can be used instead (e.g., Q5® High-Fidelity DNA polymerase).
6. When preparing PCR reactions, make sure to follow the manufacturer's recommendations regarding the concentrations of the different reagents and PCR conditions.
7. For DNA staining, any other DNA staining method can be used by following the manufacturer's instructions.
8. The Gibson assembly master mix can also be homemade, as described, for example, here: DOI: dx.doi.org/10.17504/protocols.io.ba2yigfw.
9. The plasmid sequence for pPpT4_MFα is available at https://www.novoprolabs.com/plus/novobuilder/?vid=V003935 and pPpT4_MFα-eGFP at https://www.novoprolabs.com/vector/Vge3tiojq.
10. The primers should bind directly after (fw(pPpT4)) and before (rv(pPpT4)) the MFα signal sequence. This is especially important for rv(pPpT4), since it amplifies the region immediately upstream of the ATG start codon of the MFα signal sequence and contains the Kozak consensus sequence that is mandatory for ribosome binding during translation.
11. Make sure to use the correct primer annealing temperature and elongation time. Annealing temperatures of the primer regions that bind to the corresponding templates can be calculated via online tools (e.g., Tm calculator, New England BioLabs).
12. Ideally, the purified PCR product should have a high concentration (>50 ng/μL) and purity ($A_{260/280}$ of 1.8–2.0, measured spectrophotometrically, e.g., by Nanodrop) for the Gibson assembly to work well. Otherwise, the total volume of DNA required for the assembly may exceed the 5 μL that can be added to the Gibson assembly master mix.
13. Use 50–100 ng for each of the three fragments. When assembling a vector backbone and smaller inserts, it may be recommended to use a 3 times molar excess of each insert.
14. In case a homemade Gibson assembly mix is used, competent *E. coli* cells would also have to be prepared. If an electroporator is available, it is highly recommended to prepare electrocompetent *E. coli* cells, for example, as described here: dx.doi.org/10.17504/protocols.io.vzfe73n. Otherwise, chemically

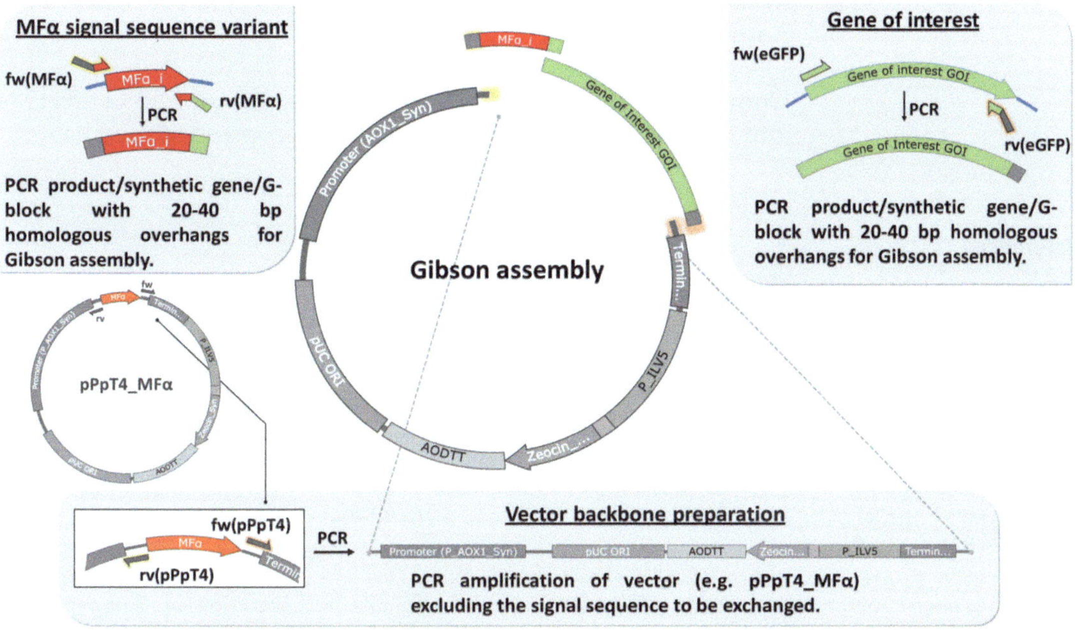

Fig. 2 Cloning of the expression plasmid. The following three DNA fragments are required for Gibson assembly: the vector backbone, the MFα signal sequence variant, and the gene of interest. All fragments must contain 20–40 bp homologous regions to each other. The vector backbone is prepared by PCR amplification of the vector pPpT4_MFα, excluding the MFα sequence, using primers fw(pPpT4) and rv (pPpT4). The gene of interest (in this case, eGFP) was amplified using primers fw(eGFP) and rv(eGFP). The MFα signal sequence variant (MFα_i) is prepared by PCR, applying primers fw(MFα) and rv(MFα)

competent cells also work (dx.doi.org/10.17504/protocols.io.gtcbwiw), but usually yield lower transformation efficiencies.

15. In case a different GOI is cloned, the band size at 2774 bp will differ. The eGFP gene has a size of 717 bp.

Acknowledgments

This work was supported by the FFG-Bridge project "ProSek" (project number FO999898875) (M.M) in collaboration with the bisy GmbH, the Horizon Europe Marie Skłodowska-Curie Actions—Doctoral Network BiocatCodeExpander (grant agreement No 101072686) (O.P.), and the BioTechMed-Graz Young Researcher Group Project "StemP" (A. E.-A.). The project was also supported by the COMET center acib (A. E.-A.), which is funded by BMK, BMDW, SFG, Standortagentur Tirol, Government of Lower Austria, and the Vienna Business Agency in the framework of COMET—Competence Centers for Excellent Technologies. The COMET-Funding Program is managed by the Austrian

Research Promotion Agency, FFG. Also, we want to thank Prof. Anton Glieder for helpful discussions.

Competing Interest Statement The authors have no conflicts of interest to declare that are relevant to the content of this chapter.

References

1. Haber JE (2012) Mating-type genes and MAT switching in *Saccharomyces cerevisiae*. Genetics 191(1):33. https://doi.org/10.1534/GENETICS.111.134577
2. Kurjan J, Herskowitz I (1982) Structure of a yeast pheromone gene (MF alpha): a putative alpha-factor precursor contains four tandem copies of mature alpha-factor. Cell 30(3): 933–943. https://doi.org/10.1016/0092-8674(82)90298-7
3. Bardwell L (2004) A walk-through of the yeast mating pheromone response pathway. Elsevier Inc. https://doi.org/10.1016/j.peptides.2003.10.022
4. Bender A, Sprague GF (1989) Pheromones and pheromone receptors are the primary determinants of mating specificity in the yeast *Saccharomyces cererrisiae*. Genetics 121(3): 463–476. https://doi.org/10.1093/genetics/121.3.463
5. Brake AJ et al (1984) Alpha-factor-directed synthesis and secretion of mature foreign proteins in *Saccharomyces cerevisiae*. Proc Natl Acad Sci USA 81(15):4642–4646. https://doi.org/10.1073/PNAS.81.15.4642
6. Fuller RS, Brake AJ, Thorner J (1989) Intracellular targeting and structural conservation of a prohormone-processing endoprotease. Science 246(4929):482–486. https://doi.org/10.1126/SCIENCE.2683070
7. Wilcox CA, Fuller RS (1991) Posttranslational processing of the prohormone-cleaving Kex2 protease in the *Saccharomyces cerevisiae* secretory pathway. J Cell Biol 115(2):297–307. https://doi.org/10.1083/JCB.115.2.297
8. Cereghino JL, Cregg JM (2000) Heterologous protein expression in the methylotrophic yeast *Pichia pastoris*. FEMS Microbiol Rev 24(1): 45–66. https://doi.org/10.1111/j.1574-6976.2000.tb00532.x
9. Delic M, Valli M, Graf AB, Pfeffer M, Mattanovich D, Gasser B (2013) The secretory pathway: exploring yeast diversity. FEMS Microbiol Rev 37(6):872–914. https://doi.org/10.1111/1574-6976.12020
10. Sun Z, Brodsky JL (2019) Protein quality control in the secretory pathway. Rockefeller University Press. https://doi.org/10.1083/jcb.201906047
11. Ng DTW, Brown JD, Walter P (1996) Signal sequences specify the targeting route to the endoplasmic reticulum membrane. J Cell Biol 134(2):269–278. https://doi.org/10.1083/jcb.201906047
12. Puxbaum V, Mattanovich D, Gasser B (2015) Quo vadis? The challenges of recombinant protein folding and secretion in *Pichia pastoris*. Appl Microbiol Biotechnol 99(7):2925–2938. https://doi.org/10.1007/s00253-015-6470-z. Springer
13. Aggarwal S, Mishra S (2020) Differential role of segments of α-mating factor secretion signal in *Pichia pastoris* towards granulocyte colony-stimulating factor emerging from a wild type or codon optimized copy of the gene. Microb Cell Factories 19(1). https://doi.org/10.1186/s12934-020-01460-8
14. Johnson N, Powis K, High S (2013) Post-translational translocation into the endoplasmic reticulum. Biochim Biophys Acta, Mol Cell Res 1833(11):2403–2409. https://doi.org/10.1016/j.bbamcr.2012.12.008
15. Aggarwal S, Mishra S (2021) Modifications in the Kex2 P1' cleavage site in the α-MAT secretion signal lead to higher production of human granulocyte colony-stimulating factor in *Pichia pastoris*. World J Microbiol Biotechnol 37(11). https://doi.org/10.1007/s11274-021-03167-3
16. Fuller RS, Brakes A, Thorner J (1989) Yeast prohormone processing enzyme (*KEX2* gene product) is a Ca2'-dependent serine protease (prohormone convertase/membrane-bound protease/precursor maturation/secretory pathway/posttranslational modification). Proc Natl Acad Sci U S A 86(5):1434–1438. https://doi.org/10.1073/pnas.86.5.1434
17. Julius D, Blair L, Brake A, Sprague G, Thorner J (1983) Yeast α factor is processed from a larger precursor polypeptide: the essential role of a membrane-bound dipeptidyl

aminopeptidase. Cell 32(3):839–852. https://doi.org/10.1016/0092-8674(83)90070-3
18. Barone GD et al (2023) Industrial production of proteins with *Pichia pastoris—Komagataella phaffii*. Biomolecules 13(3). https://doi.org/10.3390/BIOM13030441
19. Ast T, Michaelis S, Schuldiner M (2016) The protease Ste24 clears clogged translocons. Cell 164(1–2):103–114. https://doi.org/10.1016/j.cell.2015.11.053
20. Joshi KK, Sahni G (2010) Molecular cloning, expression, purification and characterization of truncated forms of human plasminogen in *Pichia pastoris* expression system. Process Biochem 45(8):1251–1260. https://doi.org/10.1016/j.procbio.2010.04.014
21. Neiers F, Belloir C, Poirier N, Naumer C, Krohn M, Briand L (2021) Comparison of different signal peptides for the efficient secretion of the sweet-tasting plant protein Brazzein in *Pichia pastoris*. Life 11(1):46. https://doi.org/10.3390/LIFE11010046
22. Kurniatin PA, Masduki FF, Nurainy N, Ihsanawati, Natalia D (2019) The use of α-MF signal peptide without spacer for producing insulin aspart precursor in *Pichia pastoris* KM71. Am J Biochem Biotechnol 15(4):198–207. https://doi.org/10.3844/AJBBSP.2019.198.207
23. Liang S, Li C, Ye Y, Lin Y (2013) Endogenous signal peptides efficiently mediate the secretion of recombinant proteins in *Pichia pastoris*. Biotechnol Lett 35(1):97–105. https://doi.org/10.1007/S10529-012-1055-8/FIGURES/3
24. Lin-Cereghino GP et al (2013) The effect of α-mating factor secretion signal mutations on recombinant protein expression in *Pichia pastoris*. Gene 519(2):311–317. https://doi.org/10.1016/j.gene.2013.01.062
25. Xiong AS, Yao QH, Peng RH, Han PL, Cheng ZM, Li Y (2005) High level expression of a recombinant acid phytase gene in *Pichia pastoris*. J Appl Microbiol 98(2):418–428. https://doi.org/10.1111/j.1365-2672.2004.02476.x
26. Ahn J, Jang MJ, Ang KS, Lee H, Choi ES, Lee DY (2016) Codon optimization of *Saccharomyces cerevisiae* mating factor alpha prepro-leader to improve recombinant protein production in *Pichia pastoris*. Biotechnol Lett 38(12):2137–2143. https://doi.org/10.1007/s10529-016-2203-3
27. Rakestraw JA, Sazinsky SL, Piatesi A, Antipov E, Wittrup KD (2009) Directed evolution of a secretory leader for the improved expression of heterologous proteins and full-length antibodies in *S. cerevisiae*. Biotechnol Bioeng 103(6):1192. https://doi.org/10.1002/BIT.22338
28. De Salas F et al (2019) Engineering of a fungal laccase to develop a robust, versatile and highly-expressed biocatalyst for sustainable chemistry. Green Chem 21(19):5374–5385. https://doi.org/10.1039/C9GC02475A
29. Näätsaari L, Mistlberger B, Ruth C, Hajek T, Hartner FS, Glieder A (2012) Deletion of the Pichia pastoris KU70 homologue facilitates platform strain generation for gene expression and synthetic biology. PLoS One 7(6):e39720. https://doi.org/10.1371/journal.pone.0039720
30. Barrero JJ, Casler JC, Valero F, Ferrer P, Glick BS (2018) An improved secretion signal enhances the secretion of model proteins from *Pichia pastoris*. Microb Cell Factories 17(1). https://doi.org/10.1186/s12934-018-1009-5

Chapter 17

Lytic Polysaccharide Monooxygenases as Reporters for Signal Peptide Evaluation in *Pichia pastoris*

Andrea K. Hönikl and Lukas Rieder

Abstract

Signal peptides are essential for the secretion of recombinant proteins using *Komagataella phaffii*. The N-terminal signal peptide is removed by proteases during the translocation process. However, this processing is not always complete, resulting in an N-terminal extension which is problematic for some proteins. In this chapter, we present a straightforward approach to evaluate signal peptide processing by a high-throughput solid-phase agar plate assay.

Key words *Pichia pastoris*, *Komagataella phaffii*, Signal peptides, Processing, Enzyme assay, LPMO

1 Introduction

One key advantage of using *Komagataella phaffii* (reclassified from *Pichia pastoris*) as the host for protein production is the organism's ability to secrete high amounts of recombinant protein while hardly secreting endogenous proteins. To export proteins to the extracellular environment of a *K. phaffii* culture, a signal peptide (SP) is typically fused to the N-terminus of the to-be-secreted target protein or replacing a native SP. For protein production using *K. phaffii*, the secretion signal peptide of the α-factor mating pheromone (MFα-SP) from *S. cerevisiae* is the gold standard. In many cases, this is still the most efficient SP for secreted recombinant protein production at high titers. The MFα-SP consists of a pre- and pro-region which are processed during the translocation through the ER and Golgi apparatus, respectively. Unfortunately, the processing of the pro-region by the STE13 protease is often incomplete, resulting in a Glu-Ala overhang at the N-terminus of the secreted protein [1]. This circumstance is problematic for various proteins, e.g., antibodies, that require a perfectly processed N-terminus for being functional. Hence, SPs combining good secretion capacities and correct processing are highly desirable.

Xavier Garcia-Ortega et al. (eds.), *Pichia pastoris: Methods and Protocols*, Methods in Molecular Biology, vol. 2697,
https://doi.org/10.1007/978-1-0716-4779-0_17,

A perfect model for evaluating alternative SPs concerning N-terminal processing in addition to its efficiency for secretion in *K. phaffii* is *Nc*AA9C, a cellulose-degrading AA9-type lytic polysaccharide monooxygenase (LPMO) [2]. Its eligibility stems from the active site architecture of LPMOs in which the N-terminal histidine is needed to coordinate the catalytic copper ion [3]. Thus, incomplete processing of the SP will result in secreted enzymes that are catalytically inactive [4, 5].

Here, we present protocols for (a) exchanging signal peptides in already existing expression plasmids, (b) *K. phaffii* cultivation, and (c) how the cellulose degradation activity of *Nc*AA9C can be assessed by using an agar plate-based high-throughput assay resulting in blue halo formation upon enzyme activity.

2 Materials

2.1 Expression Plasmid Assembly

1. *K. phaffii* expression plasmid to produce *Nc*AA9C (e.g., pBSY3Z-DAS2-NcAA9C_nat., GenBank accession number: PP793778.1, Fig. 1, *see* **Note 1**).
2. 10 μM stocks of FWD (5′-CATACAATTTTTCAGAAGGTGT CAGTCAAC-3′) and REV (5′-TTTTGATGTTTGATAGTTT GATAAGAGTGAAC -3′) primers for the amplification of pBSY3Z-DAS2-NcAA9C_nat. from the N-terminal histidine to the promoter (Figs. 1 and 2).
3. Phusion™ Plus DNA Polymerase (or another high-fidelity polymerase) with appropriate buffer.
4. 10 mM dNTPs.
5. Synthetic DNA elements encoding the alternative signal peptides including homologous regions to the promoter (5′-TC ACTCTTATCAAACTATCAAACATCAAAA -3) and the N-terminal LPMO region (5′- CATACAATTTTTCA GAAGGTGTCAGT-3) for Gibson assembly (Fig. 2).
6. Gibson assembly mix (e.g., New England Biolabs E2611).
7. Chemically competent *E. coli* cells of a cloning strain (e.g., XL1-blue).
8. Thermomixer.
9. Table centrifuge.
10. SOC medium: 0.5% (w/v) yeast extract, 2% (w/v) tryptone, 20 mM (w/v) glucose, 10 mM NaCl, 2.5 mM KCl, 10 mM $MgCl_2$, 10 mM $MgSO_4$. Autoclave and store at room temperature.
11. Plastic Petri dishes: 94 × 16 mm with vents (e.g., Greiner Bio-One, Frickenhausen, Germany; PS, Article-nr. 633,180).

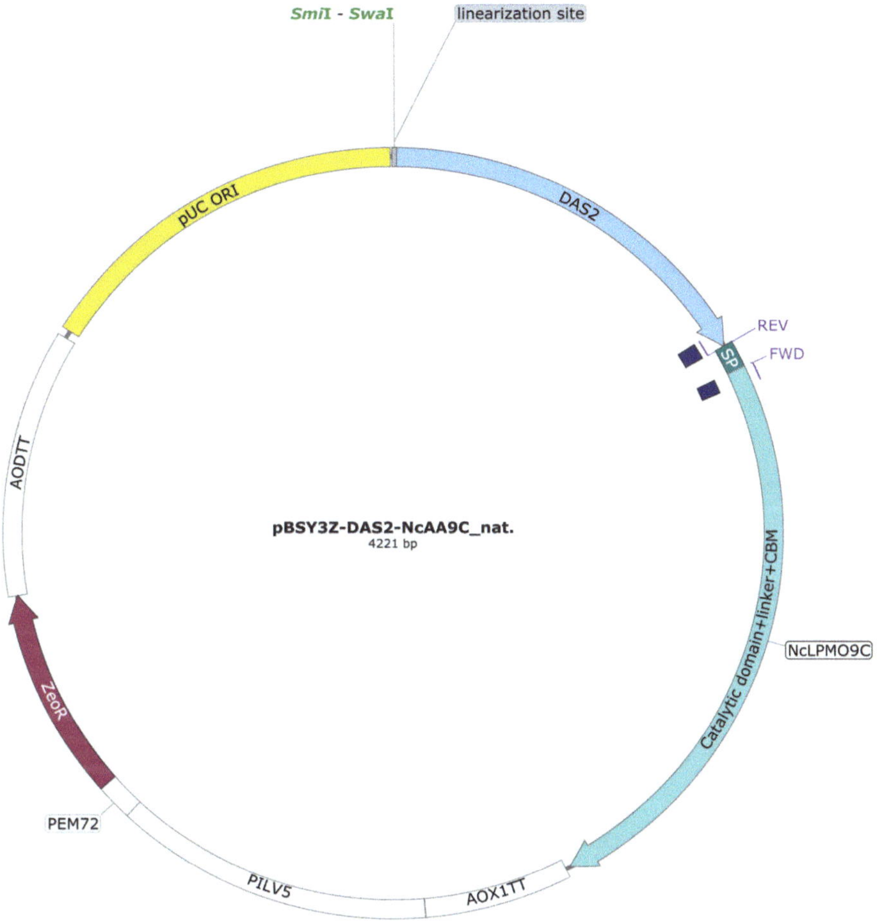

Fig. 1 Plasmid map of pBSY3Z-DAS2-NcAA9C_nat. (GenBank accession number: PP793778.1) to produce *Nc*AA9C with the native signal peptide in *K. phaffii*. Expression of the LPMO gene is initiated by the strong and methanol-inducible *DAS2* promoter and terminated by the *AOX1* terminator. The FWD and REV primers for amplification of the plasmid before the insertion of the alternative SP are shown in purple. The homologous regions suggested for Gibson assembly are shown in dark blue

12. LB-Zeo plates: 0.5% (w/v) yeast extract, 1% (w/v) tryptone, 1% (w/v) NaCl, and 1.5% (w/v) agar-agar. Autoclave. When lukewarm add Zeocin to a final concentration of 25 μg/mL, pour into Petri dishes, and let solidify.
13. Plasmid isolation kit (e.g., Promega Wizard® Plus SV Minipreps DNA Purification Systems).
14. Possibility for plasmid sequencing.

2.2 Transformation of K. phaffii

1. *Smi*I for linearization of pBSY3Z-DAS2-NcAA9C_nat (Fig. 1).
2. 13 mm 0.025 μm MCE membranes (Merck Millipore Ltd., Cork, Ireland).

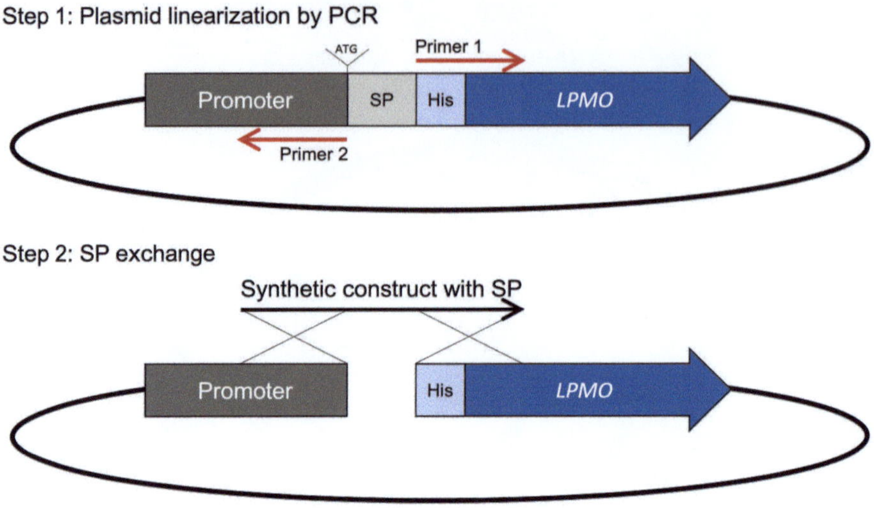

Fig. 2 Schematic illustration of the two steps needed for the exchange of SPs in expression plasmids. In Step 1, the plasmid is amplified by PCR to remove the native SP. In Step 2, the Gibson isothermal assembly is used to insert a synthetic DNA fragment with the to-be-tested SP

3. Thermomixer.
4. Electrocompetent *K. phaffii* cells of a strain of choice (e.g., BSYBG11).
5. Icebox filled with ice.
6. Electroporation cuvettes with 0.2 cm gap (e.g., BioRad 1,652,082).
7. 1.5 mL reaction tubes with conical bottom.
8. Electroporator (e.g., BioRad Gene Pulser).
9. YPD: 1% yeast extract, 2% peptone, 2% glucose. Autoclave and store at room temperature.
10. 1 M sorbitol: 182 g/L of sorbitol in ddH_2O. Autoclave and store at room temperature.
11. YPD/sorbitol: Mix equal volumes of YPD and 1 M sorbitol.
12. Plastic Petri dishes: 94 × 16 mm with vents (e.g., Greiner Bio-One, Frickenhausen, Germany; PS, Article-nr. 633,180).
13. YPD-Zeo plates: YPD with 1.5% (w/v) of agar-agar. Autoclave. When lukewarm add Zeocin to a final concentration of 100 μg/mL, pour the medium into Petri dishes, and let solidify.

2.3 K. phaffii Cultivation and Screening for LPMO Activity

1. Sterile toothpicks.
2. Sterile 96-deepwell plates.
3. 1 M potassium phosphate buffer: 615 mL 1 M K_2HPO_4 and 385 mL 1 M KH_2PO_4. Adjust pH to 6 using KOH. Autoclave and store at room temperature.

4. YNB: 13.4% (w/v) yeast nitrogen base w/o amino acids in ddH_2O. Autoclave and store at room temperature in the dark.
5. Biotin: 200 mg/L D-Biotin in ddH_2O. Filter sterilize using a 0.2-μm filter.
6. Glucose: 20% (w/v) glucose in ddH_2O. Autoclave and store at room temperature.
7. 100% methanol.
8. BMD1: 650 mL ddH_2O, 200 mL potassium phosphate buffer, 100 mL YNB, 2 mL biotin, 50 mL glucose.
9. Multichannel pipette.
10. 96-well replicator.
11. PTM1 trace salts w/o $CuSO_4$: 5 mL/L H_2SO_4, 65 g/L $FeSO_4{\cdot}7H_2O$, 20 g/L $ZnCl_2$, 0.6 g/L $CuSO_4$, 3.36 g/L $MnSO_4{\cdot}H_2O$, 0.82 g/L $CoCl_2{\cdot}7H_2O$, 0.2 g/L $Na_2MoO_4{\cdot}2H_2O$, 0.08 g/L NaI, 0.02 g/L H_3BO_3 in ddH_2O. Filter sterilize and store at +4 °C in the dark.
12. 1 M ascorbic acid.
13. Square Petri dish 120 × 120 × 17 mm.
14. Screening plates: 12.6 g $(NH_4)_2HPO_4$, 1.8 g citric acid anhydrous, 0.9 g KCl, 0.02 g $CaCl_2{\cdot}2\ H_2O$, 0.5 g $MgSO_4{\cdot}7H_2O$, and 15 g agar-agar in 967 mL ddH_2O. Autoclave. Once lukewarm add 2 mL biotin, 25 mL methanol, 5 mL PTM1 trace salts, 1 mL ascorbic acid, and 1 g AZCL-HEC substrate (*see* **Note 2**). Pour medium into a square Petri dish and let solidify (*see* **Note 3**).
15. Induction solution: 1 mM ascorbic acid and 5% (v/v) methanol in ddH_2O (sterile).

3 Methods

3.1 Expression Plasmid Assembly

In the first section of the protocol, we describe how SPs are efficiently exchanged in the pBSY3Z-DAS2-NcAA9C_nat. For this, the natural SP is removed by PCR and subsequently, the alternative SP is inserted using Gibson isothermal assembly (Fig. 2), employing double-stranded DNA fragments coding for the alternative SP and overlapping with the vector/reporter backbone DNA [6]. After verification of the DNA sequence, the plasmid can be used for transforming *K. phaffii*. Alternatively, similar new SP fusion proteins can be constructed, e.g., by overlap extension PCR or other seamless cloning technologies in addition to the use of fully synthetic hybrid genes.

1. Amplify the pBSY3Z-DAS2-NcAA9C_nat. using the given primers (Tables 1 and 2).
2. Prepare the Gibson isothermal assembly (Table 3).

Table 1
PCR reaction for the amplification of the pBSY3Z-DAS2-NcAA9C_nat. using the listed primers

Component	Volume
Polymerase	0.5 μL
Reaction buffer	10 μL
Template DNA	0.5–1 ng (X μL)
FWD primer	2.5 μL
REV primer	2.5 μL
dNTPs	1 μL
H_2O	33.5 – X μL
Total	50 μL

Table 2
PCR cycling profile used for the amplification of the pBSY3Z-DAS2-NcAA9C_nat. using the given primers

# of cycles	Time	Temperature (°C)
1	1 min	98
35	10 s	98
	15 s	58
	150 s	72
1	10 min	72
1	∞	10

3. Incubate the assembly mixture at 50 °C for 60 min.
4. Spin down the assembly mixture using a table centrifuge.
5. Thaw one aliquot of chemically competent *E. coli* cells per Gibson assembly reaction.
6. Preheat the thermomixer to 42 °C.
7. Mix 2.5 μL of the Gibson assembly reaction with the competent cells and flick the tube.
8. Place the tube on ice for 5–10 min to increase transformation efficiency.
9. Incubate the transformation mixture at 42 °C for 30 s.
10. Add 500 μL of SOC medium, turn down temperature to 37 ° C, and regenerate cells for at least 60 min.

Table 3
Gibson assembly composition to insert alternative signal peptides (*see* Note 4)

Ingredient	Volume
Gibson assembly mixture	15 μL
Plasmid backbone	100 ng (X μL)
Insert with signal peptide	Three-fold molar excess (Y μL)
H_2O	5-X-Y
Total	20 μL

11. Plate 100 μL and 900 μL of the regenerated cells on LB-Zeo plates.
12. Incubate the plates at 37 °C overnight.
13. Cultivate transformants for plasmid isolation.
14. Perform plasmid isolation and assess DNA concentration.
15. Prepare samples for sequencing.
16. Evaluate sequencing results and select plasmids with the correct sequence for the transformation of *K. phaffii*.

3.2 Transformation and Cultivation of K. phaffii

Once the *Nc*AA9C expression plasmids with the alternative signal peptides are sequence-verified, they can be used for the transformation of *K. phaffii*. Alternatively, signal peptide libraries can be screened, and sequencing is performed after the identification of the most efficient sequence variant by screening. In this protocol, an ectopic genome integration approach is described which is not suitable if the secretion capacity of the SP should be evaluated in parallel.

1. Linearize 2000 ng of the sequence-verified expression plasmids using *Smi*I according to the manufacturer's instructions.
2. Desalt the restriction digestion using the MCE membranes.
3. Pre-cool electroporation cuvettes on ice.
4. Mix ~1000 ng of the desalted and linearized DNA with one aliquot of competent *K. phaffii* cells in a 1.5-mL reaction tube.
5. Place the tube(s) on ice for 5–10 min.
6. Transfer the transformation mixture into electroporation cuvettes.
7. Electroporate (2 kV; resistance, 200 Ω; capacitance, 25 μF).
8. Immediately after the electroporation, add 1 mL of the YPD/-sorbitol mixture.
9. Regenerate the electroporated cells for at least 2 h at 28 °C and 350 rpm in a thermomixer.

10. Plate 100 and 900 μL of the regenerated cells on two separate YPD-Zeo plates.
11. Incubate the agar plates at 28 °C until colonies are visible.

3.3 *K. phaffii* Cultivation and Screening for LPMO Activity

To produce the LPMOs for assessing the processing of the alternative SPs, the transformants are cultivated in a defined medium for 24 h for biomass accumulation. After the cultivation, the biomass is replicated onto screening plates that induce protein production, contain reductant for the priming step in the LPMO reaction, and the substrate for the colorimetric activity detection. If active LPMO is produced, a blue halo is formed in the surrounding of the *K. phaffii* colony due to degradation products of the AZCL-HEC substrate, suggesting correct processing of the signal peptide (Fig. 3).

3.3.1 *K. phaffii* Cultivation

1. Add 500 μL of BMD1 into each well of a 96-deepwell plate using the multichannel pipette.
2. Inoculate each well of the DWP with a distinct *K. phaffii* transformant using sterile toothpicks (*see* **Note 5**).
3. Cultivate the cells in the DWP at 28 °C and 350 rpm for ~24 h.

3.3.2 LPMO Activity Screen

1. Transfer *K. phaffii* colonies from the DWP onto screening plates using a 96-well replicator.
2. Incubate the plates at 28 °C for 24 h.
3. Add 10 μL of the induction solution on top of each colony.
4. Incubate the plates at 28 °C for 24 h.
5. Optional: Repeat the induction process and incubate for another 24 h at 28 °C.
6. Colonies showing a blue halo (Fig. 3) are secreting functional LPMO suggesting correct processing of the signal peptide (*see* **Note 6**).

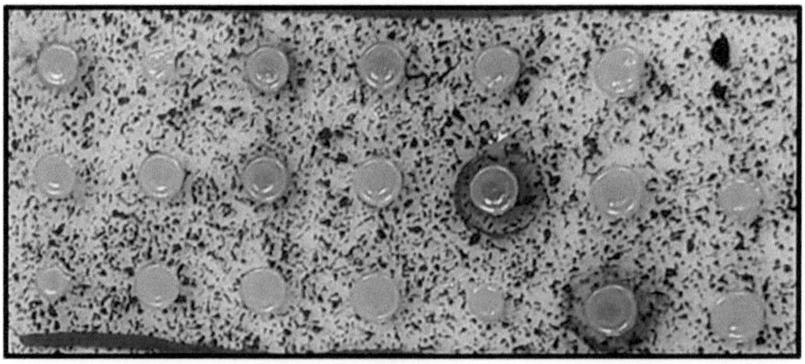

Fig. 3 Picture showing *K. phaffii* colonies growing on screening plates. The blue halo surrounding two of the colonies indicates the production of active *Nc*AA9C suggesting flawless SP processing

4 Notes

1. We recommend using an inducible (e.g., P_{DAS2}) or derepressed promoter (e.g., PDC) for controlling the LPMO gene expression.
2. AZCL-HEC is statically charged. Use anti-static measuring tools and pour the powder into the middle of the agarose mixture to avoid it sticking to all surfaces. Furthermore, mix the medium vigorously on a shaker to make sure the AZCL-HEC is dispersed evenly prior to pouring the screening plates.
3. Quickly pour thin AZCL-HEC square plates to avoid excessive sedimentation of AZCL-HEC.
4. The exact reaction conditions will depend on the Gibson assembly mixture used.
5. If the signal peptides should also be evaluated for secretion strength, the expression plasmid needs to be integrated into a specific locus to avoid clonal variation or one must look at ~30 individual transformants due to clonal variation effects and accept false positives which can be reanalyzed in rescreening procedures.
6. If incubated too long, unspecific blue color formation will occur potentially due to endogenously secreted enzymes [7].

Acknowledgments

This work was performed as part of the EIC Pathfinder project NewCat under the Horizon Europe program, grant number 101046815.

Competing Interest Statement The authors have no conflicts of interest to declare that are relevant to the content of this chapter.

References

1. Cereghino JL, Cregg JM (2000) Heterologous protein expression in the methylotrophic yeast *Pichia pastoris*. FEMS Microbiol Rev 24(1): 45–66. https://doi.org/10.1016/S0168-6445(99)00029-7
2. Isaksen T et al (2014) A C4-oxidizing lytic polysaccharide monooxygenase cleaving both cellulose and cello-oligosaccharides. J Biol Chem 289(5):2632–2642. https://doi.org/10.1074/jbc.M113.530196
3. Ciano L, Davies GJ, Tolman WB, Walton PH (2018) Bracing copper for the catalytic oxidation of C–H bonds. Nat Catal 1(8):571–577. https://doi.org/10.1038/s41929-018-0110-9
4. Tanghe M et al (2015) Recombinant expression of *Trichoderma reesei* Cel61A in Pichia pastoris: optimizing yield and N-terminal processing. Mol Biotechnol 57:1010–1017. https://doi.org/10.1007/s12033-015-9887-9

5. Rieder L, Ebner K, Glieder A, Sørlie M (2021) Novel molecular biological tools for the efficient expression of fungal lytic polysaccharide monooxygenases in *Pichia pastoris*. Biotechnol Biofuels 14:122. https://doi.org/10.1186/s13068-021-01971-5
6. Gibson DG, Young L, Chuang RY, Venter JC, Hutchison CA, Smith HO (2009) Enzymatic assembly of DNA molecules up to several hundred kilobases. Nat Methods 6(5): 343–345. https://doi.org/10.1038/nmeth.1318
7. Burgard J et al (2020) The secretome of *Pichia pastoris* in fed-batch cultivations is largely independent of the carbon source but changes quantitatively over cultivation time. Microb Biotechnol 13(2):479–494. https://doi.org/10.1111/1751-7915.13499

Chapter 18

Exploiting Self-Processing 2A Peptides for Multigene Expression in *Pichia pastoris*

Katharina Ebner and Martina Geier

Abstract

Metabolic engineering in *Komagataella phaffii* (reclassified from *Pichia pastoris*) often requires the implementation and expression of more than one (heterologous) gene. Instead of placing each gene under the separate control of a promoter, we present in this chapter a strategy for polycistronic pathway expression based on self-processing 2A peptides. This approach bears several advantages for multienzyme pathway expression such as compact pathway design, fast and simple pathway assembly, the generation of stable production strains, as well as the possibility for fine-tuning the expression of pathway genes.

Key words *Pichia pastoris*, *Komagataella phaffii*, Polycistronic expression, 2A peptides, Multienzyme pathways

1 Introduction

To produce fine chemicals or other valuable compounds with *Komagataella phaffii* (reclassified from *Pichia pastoris*), introducing a single enzyme activity—especially if the desired product should be derived from simple carbon sources—is often not sufficient. Thus, the challenge is to implement enzyme cascades or even whole pathways.

Typically, multigene expression relies on co-expression constructs that house the pathway genes, each controlled separately by the (same) promoter and terminator (i.e., one transcription unit). With increasing the number of pathway genes, the design and especially the assembly of these co-expression constructs get more complex. More importantly, the repeated use of homologous sequences can lead to recombination ("looping-out") events, causing genetic instability of the resulting strains. This is particularly critical when high levels of physiologically problematic proteins or metabolites are produced. Maintaining genetic stability in production strains, however, is essential for economically viable industrial

Xavier Garcia-Ortega et al. (eds.), *Pichia pastoris: Methods and Protocols*, Methods in Molecular Biology, vol. 2697,
https://doi.org/10.1007/978-1-0716-4779-0_18,

processes and consistent product quality over extended production times.

A completely different approach is to use only one promoter and express all the required genes from a single, polycistronic transcript. While this strategy is straightforward in prokaryotes, eukaryotes do not simply express polycistronic operons. However, polycistronic expression can be achieved using self-processing 2A peptides. These short peptides (up to 20 amino acids) originate from viral polyproteins and act cotranslationally, causing a "ribosomal skip" that impairs the formation of a peptide bond between the glycine and proline residue of the conserved "DxExNPGP" sequence. Translation then reinitiates, thus producing distinct proteins from a single transcript. It should be noted that the resulting proteins are modified in this process, i.e., proteins located upstream of the 2A peptide retain this short amino acid sequence (except the last proline) as a C-terminal extension, while proteins located downstream contain an N-terminal proline (Fig. 1) [1, 2]. Theoretically, the proteins of the 2A-based polycistron are produced in stoichiometric amounts; however, errors can occur during the skipping process such as the ribosome falls off the mRNA and fails to reinitiate translation or ribosome "read-trough" (unsuccessful skipping), resulting in the aborted translation of the downstream located proteins or the production of protein fusions, respectively [3].

2A peptides from various sources have successfully been used for the co-expression of two genes in a variety of eukaryotic systems, including *K. phaffii*, Table 1 summarizes the peptides evaluated for use in the latter [4]. Additionally, in *K. phaffii*, their potential for multigene expression was illustrated by the coordinate expression of nine genes from a single polycistronic transcript using the 2A peptide from the *Thosea asigna* virus (T2A) [3, 5]. This study highlighted the advantages of exploiting 2A peptides for whole pathway expression: (a) compact design and facilitated assembly of the expression construct due to the reduction of transcription units and the small size of 2A peptides; (b) generation of stable production strains due to the avoidance of repetitive

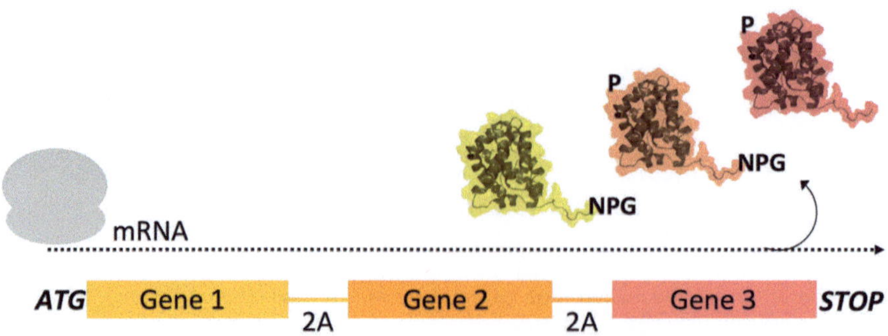

Fig. 1 Scheme illustrating the mechanism of 2A sequence-based polycistronic expression

Table 1
Viral 2A peptides shown to be functional in *K. phaffii*

2A peptide	Viral origin	Sequence
FMDV2A	Foot-and-mouth disease virus	QLLNFDLLKLAG**DVESNPG↑P**
T2A	*Thosea asigna* virus	RAEGRGSLLTCG**DVEENPG↑P**
P2A	Porcine teschovirus-1	ATNFSLLKQAG**DVEENPG↑P**

The consensus motif is highlighted in bold; the cleavage site is indicated by ↑

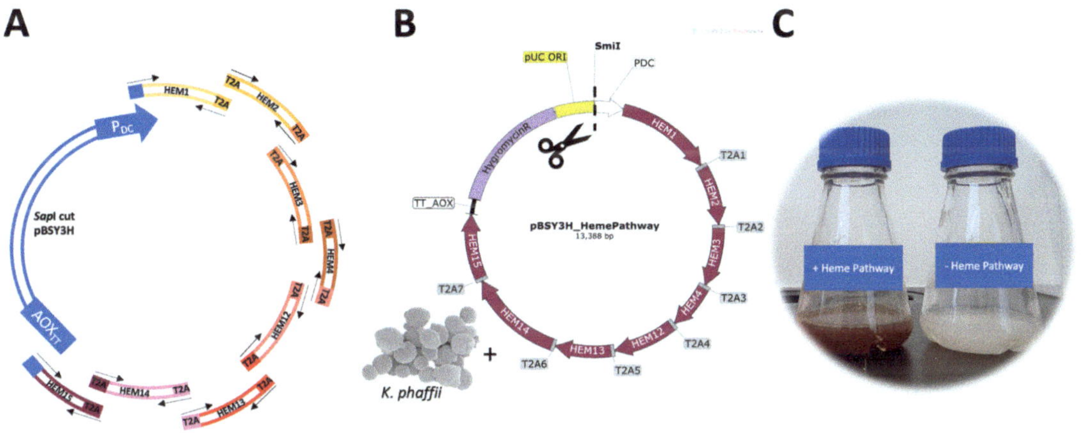

Fig. 2 Overview of the workflow for polycistronic expression of the heme pathway genes in *K. phaffii* via self-processing 2A peptides. **(a)** Schematic illustration of the cloning strategy employed to create a heme overexpression construct carrying all eight heme pathway genes (HEME 1-15) linked via 2A sequences using Gibson assembly. **(b)** Map of the resulting expression vector pBSY3H_HemePathway, which is SmiI linearized prior to transformation of *K. phaffii*. **(c)** Culture of a *K. phaffii* strain carrying the pBSY3H_HemePathway expression construct (+ Heme Pathway) and showing a characteristic red phenotype, which indicates the successful overproduction of heme, in comparison to the untransformed wildtype strain (− Heme Pathway)

sequences (e.g., promoter); and (c) tool to optimize the performance of a biosynthetic pathway by changing the order of 2A peptide sequence-linked genes within the polycistron.

Here, we present an easy guide illustrating how to implement this tool in a *K. phaffii* strain engineering workflow to facilitate the polycistronic expression of multiple genes and whole biosynthetic pathways. As a proof of concept, we describe the homologous overexpression of the *K. phaffii* heme biosynthesis pathway with a polycistronic construct containing all eight pathway genes fused via 2A sequences to boost cellular heme supply (Fig. 2). The overexpression of selected heme pathway genes (alone and in combination) has previously been shown to raise intracellular heme levels and enhance the production of hemoproteins. Therefore, the polycistronic expression of the whole pathway from a single construct presents an excellent target to showcase the benefits of using 2A sequences for multigene expression in *K. phaffii* [6, 7].

2 Materials

2.1 Expression Plasmid Design

1. Appropriate software for visualization and in silico design of DNA parts (plasmids, primer, etc.) (e.g., SnapGene, CLC Main Workbench).
2. The sequences of *K. phaffii* heme biosynthesis pathway genes (Table 2).
3. Differently codon-optimized 2A sequences encoding 2A peptide from the *Thosea asigna* virus (T2A) (Table 3) (*see* **Note 1**).

2.2 Plasmid Assembly

1. Genomic *K. phaffii* DNA to be used as a template to amplify heme pathway genes by PCR.
2. *E. coli–K. phaffii* shuttle plasmid to use as backbone for expression of the heme biosynthesis pathway (*see* **Note 2**). In the herein described case study, the bisy vector pBSY3H is used.
3. Phusion™ High-Fidelity DNA Polymerase (F530, Thermo Fisher Scientific Inc., Waltham, MA, United States) (or alternative high-fidelity polymerase) with appropriate buffer.
4. 10 mM dNTPs.
5. 10 μM stocks of single-stranded DNA oligonucleotides (primers) binding to the genes of interest, introducing 2A sequences to the respective pathway genes, and adding overlaps to adjacent DNA fragments for Gibson assembly (Table 4).
6. Gibson assembly mix (e.g., E2611, New England Biolabs, Frankfurt am Main, Germany).

Table 2
***K. phaffii* heme biosynthesis pathway enzymes and genes**

Pathway enzyme	GenBank accession No. (amino acid sequence)	Genomic locus tag	Short gene ID
5-aminolevulinate (ALA) synthase	CAH2448104.1	BQ9382_C2-3060	HEM1
Aminolevulinate (ALA) dehydratase	CAH2450948.1	BQ9382_C4-2970	HEM2
Porphobilinogen deaminase	CAH2446963.1	BQ9382_C1-5720	HEM3
Uroporphyrinogen III synthase	CAH2448309.1	BQ9382_C2-4085	HEM4
Uroporphyrinogen III decarboxylase	CAH2449163.1	BQ9382_C3-1210	HEM12
Coproporphyrinogen III oxidase	CAH2449176.1	BQ9382_C3-1280	HEM13
Protoporphyrinogen oxidase	CAH2451376.1	BQ9382_C4-5025	HEM14
Ferrochelatase	CAH2449954.1	BQ9382_C3-5040	HEM15

For the protein and corresponding nucleotide sequences, the unique identifiers, GenBank accession number, and genomic locus tag, respectively, are given. Additionally, for all genes a short gene ID is indicated

Table 3
Different nucleotide sequences encoding the *Thosea asigna* virus peptide (T2A) for the use in *K. phaffii*

Identifier	Sequence
T2A1	AGAGCTGAGGGTAGAGGTTCTTTGCTTACTTGCGGTGACGTTGAGGAAAAC CCAGGTCCA
T2A2	CGTGCCGAAGGACGTGGATCCCTTTTGACCTGCGGAGATGTCGAAGAGAAT CCTGGACCT
T2A3	AGAGCAGAAGGTCGTGGCTCATTGCTGACTTGTGGCGACGTGGAGGAAAAT CCCGGACCA
T2A4	CGTGCAGAGGGCCGTGGTTCCTTACTTACCTGCGGTGATGTGGAAGAAAAT CCAGGACCC
T2A5	CGTGCCGAGGGTAGGGGATCACTTCTTACATGTGGAGACGTCGAGGAGAAC CCTGGTCCA
T2A6	AGAGCTGAAGGAAGGGGTTCCCTGTTAACGTGTGGCGATGTTGAAGAGAAC CCCGGTCCT
T2A7	AGGGCAGAAGGCAGAGGATCTCTGTTGACTTGTGGTGATGTAGAGGAGAAT CCCGGCCCA

Sequences were derived from varied codon optimizations for *K. phaffii* to avoid repetitive homologous sequence stretches

7. Chemical-competent *E. coli* cells of a cloning strain (e.g., XL1-blue).
8. Thermomixer.
9. Table centrifuge.
10. SOC medium: 0.5% (w/v) yeast extract, 2% (w/v) tryptone, 20 mM (w/v) glucose, 10 mM NaCl, 2.5 mM KCl, 10 mM $MgCl_2$, 10 mM $MgSO_4$. Autoclave and store at room temperature.
11. Plastic Petri dishes round (PS) 94 × 16 mm with vents (633,180, Greiner Bio-One, Frickenhausen, Germany).
12. LB-hygromycin agar round plates: 0.5% (w/v) yeast extract, 1% (w/v) tryptone, 1% (w/v) NaCl, and 1.5% (w/v) agar-agar. Autoclave and add hygromycin when lukewarm to a final concentration of 100 μg/mL. Pour into Petri dishes and let solidify.
13. Plasmid isolation kit (e.g., A1460, Wizard® Plus SV Minipreps DNA Purification Systems, Promega GmbH, Mannheim, Germany).
14. Possibility to sequence-verify plasmids (services by, e.g., Microsynth AG, LGC Genomics).

Table 4
Forward (FW) and reverse (REV) primer for the amplification of the *K. phaffii* heme biosynthesis pathway genes carrying overlapping sequences of adjoining fragments (*italic underline*)

Gene	Primer
HEM1	FW: 5′-*CTCTAGTCAAGACTTACAATTAAA*ATGGAGTTTGTCGCCCGTCAGT-3′ REV: 5′-*AGTAAGCAAAGAACCTCTACCCTCAGCTCT*CAATCTGACTCCTGATGAGGTTTCGATCTG-3′
HEM2	FW: 5′-*GAGCTGAGGGTAGAGGTTCTTTGCTTACT*TGCGGTGACGTTGAGGAAAACCCAGGTCCAGTGCATAAGGCTGAATACTTGGACGACC-3′ REV: 5′-*AGGTCCAGGATTCTCTTCGACATCTCCGC*AGGTCAAAAGGGATCCACGTCCTTCGGCACGTTCAGATAACCACTCCAGGAATTCAGGGG-3′
HEM3	FW: 5′-*GCGGAGATGTCGAAGAGAATCCTGGACCT*AACCAAATCGAACAGAGCGGACCC-3′ REV: 5′-*GTCAGCAATGAGCCACGACCTTC*TGCTCTATGTAAACCCTCTTCCTTAATCTCTTTGATCTTGTTGAAG-3′
HEM4	FW: 5′-*GAAGGTCGTGGCTCATTGCTGAC*TTGTGGCGACGTGGAGGAAAATCCCGGACCACCAAAAGCCATTCTTCTGAAGAATAAAACTACACCG-3′ REV: 5′-*CACCGCAGGTAAGTAAGGAACCACGG*CCCTCTGCACGGTGCACTTTTTGTATAGACAATAGTAAAGAGATAGGCTCTGG-3′
HEM12	FW: 5′-*CCGTGGTTCCTTACTTACCTGCGGTG*ATGTGGAAGAAAATCCAGGACCCAGTAGATTTCCAGAACTGAAGAATGACCTTATTTTAAGGGC-3′ REV: 5′-*AGAAGTGATCCCCTACCCTCGG*CACGTTGAGATCCAATGCGATGACATTCTTGTAAGAAC-3′
HEM13	FW: 5′-*CCGAGGGTAGGGGATCACTTCT*TACATGTGGAGACGTCGAGGAGAACCCTGGTCCAGCCATCGACTCTGATATCAATCTAAGCTCTCC-3′ REV: 5′-*CTCTTCAACATCGCCACACGTTAACAGGGAA*CCCCTTCCTTCAGCTCTTACCCATTCAATAGGATTTTGTAGTACCTGCAATAACTCC-3′
HEM14	FW: 5′-*TTCCCTGTTAACGTGTGGCGATGTTGAAGAG*AACCCCGGTCCTCTGAAAAGTCTTGCACCAAATTCCTCAATTGCC-3′ REV: 5′-*CACCACAAGTCAACAGAGATCCTCTGCCTTC*TGCCCTAATGCCACTGAGGGTAGCCGAAT-3′
HEM15	FW: 5′-*GAAGGCAGAGGATCTCTGTTGACTTGTGGTG*ATGTAGAGGAGAATCCCGGCCCACTTAACCGTCGTTTCCAATCTACCGTGTC-3′ REV: 5′-*ATTCTGACATCCTCTTGAGCGGCCGC*TCAATGGGCTCTGAAAAACTCTTTTGGATCTGTG-3′

2.3 Transformation of *K. phaffii*

1. *Smi*I for linearization of pBSY3H_HemePathway (Fig. 2b).
2. Kit for the purification of DNA from agarose gel or PCR mixtures (e.g., A9281, Wizard® SV Gel and PCR Clean-Up System, Promega GmbH, Mannheim, Germany).
3. Thermomixer.
4. *K. phaffii* strain of choice (e.g., BG10 or BG11).
5. 50 mL reaction tubes with conical bottom.
6. Shake flasks with baffles (300–500 mL).

7. BEDS: 10 mM bicine-NaOH, pH 8.3, 3% (v/v) ethylene glycol, 5% (v/v) DMSO, 1 M sorbitol. Prepare in the following order: 1.6317 g bicine in 900 mL ddH_2O, adjust pH to 8.3 with NaOH, add 30 mL ethylene glycol, add 50 mL DMSO, and add 182.18 g sorbitol. Fill up to 1 L with ddH_2O and check pH, filter sterilize, and store in 10 mL aliquots at −20 ° C.
8. 1 M DTT: 1.54 g DTT in 10 mL ddH_2O. Filter sterilize and store aliquots at −20 °C.
9. Icebox filled with ice for cooling.
10. Electroporation cuvettes with 0.2 cm gap fitting 40–400 μL volume (e.g., 732–1136, VWR, Vienna, Austria).
11. 1.5 mL reaction tubes with conical bottom.
12. Electroporator (e.g., Eppendorf Eporator®, EP4309000027 Merck Millipore Ltd., Cork, Ireland).
13. YPD: 1% (w/v) yeast extract, 2% (w/v) peptone, 2% (w/v) glucose. Autoclave and store at room temperature.
14. 1 M sorbitol: 182 g/L sorbitol in 1 L ddH_2O. Autoclave and store at room temperature.
15. YPD/sorbitol: Mix equal volumes of YPD and 1 M sorbitol.
16. Plastic Petri dishes square (PS) 120 × 120 × 17 mm with vents (e.g., 688,102, Greiner Bio-One, Frickenhausen, Germany).
17. YPD-hygromycin agar round plates: 1% (w/v) yeast extract, 2% (w/v) peptone, 2% (w/v) glucose. Autoclave and add hygromycin when lukewarm to a final concentration of 300 μg/mL. Pour into petri dishes and let solidify.

2.4 Cultivation of K. phaffii

1. Sterile toothpicks.
2. Sterile 96-well deep-well cultivation plates (F37860–0000 Bel-Art Products, Wayne, NJ, United States).
3. 1 M potassium phosphate buffer: Mix 132 mL 1 M K_2HPO_4 and 868 mL 1 M KH_2PO_4 to reach a pH of 6.0 (if the pH needs to be adjusted use H_3PO_4 or KOH). Autoclave and store at room temperature.
4. 10× YNB, 13.4% (w/v): 134 g yeast nitrogen base w/o amino acids in 1 L ddH_2O. Autoclave and store at room temperature in the dark.
5. 500×biotin, 0.02%: 200 mg/L D-biotin in ddH_2O. Filter sterilize and store at 4 °C.
6. 10× dextrose: 110 g glucose monohydrate in 1 L ddH_2O. Autoclave and store at room temperature.
7. 100% methanol (MeOH).

8. BMD1, buffered minimal medium containing 1% (w/v) glucose: 100 mL 10× YNB, 2 mL 500× biotin, 200 mL 1 M potassium phosphate buffer, 100 mL 10× dextrose, add sterile ddH_2O to a volume of 1 L. Mix and store at 4 °C.
9. BMM2, buffered minimal medium containing 1% (v/v) methanol: 100 mL 10× YNB, 2 mL 500× biotin, 200 mL 1 M potassium phosphate buffer, 10 mL 100% MeOH, add sterile ddH_2O to a volume of 1 L. Mix and store at 4 °C.
10. BMM10, buffered minimal medium containing 5% (v/v) methanol: 100 mL 10× YNB, 2 mL 500× biotin, 200 mL 1 M potassium phosphate buffer, 50 mL 100% MeOH, add sterile ddH_2O to a volume of 1 L. Mix and store at 4 °C.
11. Multichannel pipette.
12. 96-pin microplate replicator (96-pin stamp).
13. Sterile plastic tray.
14. YPD-hygromycin agar square plates.

3 Methods

3.1 *In Silico* Construct and Primer Design

As a first crucial step for polycistronic expression using 2A sequences, we describe the in silico design of the overexpression construct pBSY3H_HemePathway, including primers needed for the amplification of the pathway genes that facilitate cloning via Gibson isothermal assembly. Alternatively, parts needed for the assembly of pBSY3H_HemePathway can be ordered as synthetic DNA parts. To facilitate polycistronic expression, it is important that all pathway genes share an open reading frame; therefore, only the first and the last gene of the pathway carry a start and stop codon, respectively.

1. Retrieve nucleotide sequence of heme pathway genes from *K. phaffii* genome using Table 2 and Chapter 2 in this book.
2. In silico cloning of heme pathway genes in the appropriate order (*see* **Note 3**) into the vector backbone pBSY3H (*see* **Note 2**).
3. To facilitate polycistronic expression, add 2A sequences between the genes of the pathway (*see* **Note 3**).
4. Remove start codons from all pathway genes but the first gene of the multi-ORF construct (i.e., HEM1).
5. Remove stop codons from all pathway genes but the last gene of the multi-ORF construct (i.e., HEM15).
6. Design primers for the amplification of the pathway genes introducing the 2A sequences and carrying overlapping sequences (at 3′ and 5′ ends) to adjoining parts of the construct

to facilitate cloning (*see* **Note 4**). Overlapping regions are designed in the 2A sequences linking genes together (for HEM2, HEM3, HEM4, HEM12, HEM13, HEM14), the P_{DC} (for HEM1), or the AOX_TT (for HEM15, Fig. 2a).

3.2 Expression Plasmid Assembly

After amplification of the pathway genes, the construct pBSY3H_-HemePathway can be assembled via Gibson isothermal assembly (*see* **Note 5**).

1. Amplify the heme pathway genes from the genome of *K. phaffii* using appropriate primers (Tables 4, 5, and 6).
2. Prepare the vector backbone for cloning by linearization using *Sap*I.
3. Prepare the Gibson isothermal assembly (Table 7).
4. Incubate the mixture at 50 °C for 60 min.
5. Spin down using a table centrifuge.
6. Mix 2.5 μL of the Gibson assembly reaction with one aliquot of chemically competent *E. coli* cells and incubate on ice for 5–10 min.
7. Heat shock (incubate at 42 °C) for 30 s.
8. Add 1 mL of SOC medium and regenerate cells at 37 °C for 45–60 min.
9. Plate 100 μL and 900 μL cells on LB-hygromycin agar plates.
10. Incubate at 37 °C for 16–24 h.
11. Cultivate 4–8 transformants for plasmid isolation.

Table 5
PCR mixture for the amplification of heme pathway genes from the genome of *K. phaffii*

Component	Volume
Polymerase	0.5 μL
Reaction buffer	10 μL
Template DNA	100 ng (X μL)
FWD primer	2.5 μL
REV primer	2.5 μL
dNTPs (10 mM)	1 μL
H_2O	33.5 – X μL
Total	50 μL

Table 6
PCR cycling profile for the amplification of heme pathway genes from the genome of *K. phaffii*

# of cycles	Time	Temperature (°C)
1	5 min	98
35	10 s	98
	20 s	58
	20 s per 1 kb	72
1	7 min	72
1	∞	10

Table 7
Composition of the mix needed for cloning via Gibson isothermal assembly

Ingredient	Volume
Gibson assembly mixture	15 μL
Plasmid backbone	100 ng (X μL)
Insert with signal peptide	Threefold molar excess (Y μL)
H_2O	5-X-Y (μL)
Total	20 μL

12. Perform plasmid isolation with appropriate kit and assess DNA quality and concentration.
13. Verify sequence of plasmid DNA using an appropriate provider.

3.3 Transformation and Cultivation of K. phaffii

Once the pBSY3H_HemePathway expression plasmid is sequence verified, it can be introduced into *K. phaffii*. The vector pBSY3H is designed for an ectopic integration into the genome, hence copy number or locus effects can occur and influence the expression positively or negatively; therefore, we recommend screening at least 21 transformants. Electrocompetent *K. phaffii* cells are, e.g., sold commercially (e.g., bisy GmbH or Biogrammatics) or can be self-made according to literature (e.g., Lin-Cereghino et al. [8]). Finally, initial screening of transformants for a high-level expression phenotype is done in microscale, using a two-phase batch and methanol induction strategy as described by Weis et al. [9]. For highly producing transformants, a red color phenotype of the culture broth (and the resulting cell pellet after centrifugation) can be seen (Fig. 2c).

3.3.1 K. phaffii Transformation

1. Inoculate a 5-mL YPD preculture of the desired *K. phaffii* strain the day before the transformation should happen and incubate for 12–16 h at 28 °C, 150 rpm.
2. Measure the optical density at 600 nm (OD_{600}) of the culture.
3. Use the preculture to inoculate a main culture of 50 mL YPD to an OD_{600} of 0.2 and incubate the culture at 28 °C, 150 rpm until an OD_{600} of 1 is reached, which typically takes about 4–5 h.
4. Meanwhile, digest 3000 ng of the sequence-verified pBSY3H_-HemePathway using *Smi*I according to the manufacturer's instructions.
5. Purify the DNA using a silica column-based Kit system and determine DNA quality and concentration.
6. When the main culture has reached an OD_{600} of 1, prepare competent cells using BEDS and DTT according to the "Condensed protocol" of Lin-Cereghino et al.
7. Precool electroporation cuvettes on ice.
8. Mix approximately 1000 ng of linearized vector DNA with an aliquot of competent *K. phaffii* cells in a 1.5-mL reaction tube and transfer the mixture into the electroporation cuvette under sterile conditions.
9. Incubate on ice for 5–10 min.
10. Electroporate (2 KV; resistance, 200 Ω; capacitance, 25 μF).
11. Immediately after the electroporation, add 1 mL of the YPD/-sorbitol mixture.
12. Regenerate the cells for 2–3 h at 28 °C and 350 rpm in a thermomixer.
13. Plate 100 and 900 μL of the cells on YPD-hygromycin agar plates.
14. Incubate the agar plates at 28 °C until colonies are visible (usually 48–60 h).

3.3.2 K. phaffii Cultivation and Screening

1. Fill 250 μL of BMD1 into each well of a 96-deep-well plate under sterile conditions.
2. Inoculate each well with a distinct single *K. phaffii* transformant using sterile toothpicks.
3. Cultivate for 60 h at 28 °C, 320 rpm under 80% humidity.
4. Prepare induction media (BMM2 and BMM10).
5. After 60 h of growth, save the clones by stamping them to YPD-hygromycin agar square plates using a heat-sterilized 96-pin stamp under sterile conditions.

6. Afterwards, induce cultures by the addition of 250 μL induction medium (BMM2).
7. Resume cultivation at 28 °C, 320 rpm under 80% humidity.
8. Feed cultures again by addition of 50 μL BMM10 approximately 68 h, 84 h, and 92 h after inoculation.
9. Stop cultivation after approximately 108 h total cultivation time.
10. Search for clones showing high-level expression by visually assessing the color of the cultivation broth. Alternatively, centrifuge the whole cultivation plate at 4 °C, 4000 rpm for 10 min, and visually assess phenotype of the resulting cell pellets. In case of successful overexpression of the heme pathway, red color formation can be detected (Fig. 2c).

4 Notes

1. To avoid repetitive stretches of homologous sequences within the polycistronic expression construct, different nucleotide sequences for the T2A peptide should be used, which optimally show at most 80% sequence identity.
2. We recommend using a derepressed (e.g., P_{DC}) promoter controlling pathway expression to lower the metabolic burden of the overproduction. However, vector backbones carrying inducible (e.g., P_{AOX1}) or constitutive (e.g., P_{GAP}) promoter sequences controlling pathway expression can also be used.
3. The order of the pathway genes within the multi-ORF construct might influence the expression level of the individual genes and hence the yield of the desired pathway product. To get the highest possible production yield, different orders of the pathway genes should be tested and bottleneck enzymes placed at the beginning.
4. Recommended specifications for primers: Tm of approximately 60 °C for binding to the respective gene, Tm of approximately 60 °C for overlapping sequence, overall GC content of approximately 40–60%.
5. In case the Gibson assembly is not successful, reduce the number of fragments by pre-assembling genes pairwise by oe-PCR (Chapter 13 in this book).

Acknowledgments

This work was performed in collaboration with the COMET center acib: Next Generation Bioproduction is funded by BMK, BMDW, SFG, Standortagentur Tirol, Government of Lower Austria und Vienna Business Agency in the framework of COMET—Competence Centers for Excellent Technologies. The COMET-Funding Program is managed by the Austrian Research Promotion Agency FFG.

Competing Interest Statement The authors have no conflicts of interest to declare that are relevant to the content of this chapter.

References

1. Ryan MD, King AMQ, Thomas GP (1991) Cleavage of foot-and-mouth disease virus polyprotein is mediated by residues located within a 19 amino acid sequence. J Gen Virol 72(11): 2727–2732. https://doi.org/10.1099/0022-1317-72-11-2727
2. de Felipe P, Luke GA, Hughes LE, Gani D, Halpin C, Ryan MD (2006) Eunum pluribus: multiple proteins from a self-processing polyprotein. Trends Biotechnol 24(2):68–75. https://doi.org/10.1016/j.tibtech.2005.12.006
3. Geier M, Fauland P, Vogl T, Glieder A (2015) Compact multi-enzyme pathways in P. pastoris. Chem Commun (Camb) 51(9):1643–1646. https://doi.org/10.1039/C4CC08502G
4. Wang X, Marchisio MA (2021) Synthetic polycistronic sequences in eukaryotes. Synth Syst Biotechnol 6(4):254–261. https://doi.org/10.1016/j.synbio.2021.09.003. KeAi Communications Co.
5. Geier M, Fauland P, Glieder A (2019) Compact and optimized metabolic pathway design in Pichia pastoris. EP 3212786 B1
6. Yu F et al (2023) Biosynthesis of high-active hemoproteins by the efficient heme-supply Pichia pastoris Chassis. Adv Sci 10(30). https://doi.org/10.1002/advs.202302826
7. Shao Y et al (2022) High-level secretory production of leghemoglobin in Pichia pastoris through enhanced globin expression and heme biosynthesis. Bioresour Technol 363:127884. https://doi.org/10.1016/j.biortech.2022.127884
8. Lin-Cereghino J et al (2005) Condensed protocol for competent cell preparation and transformation of the methylotrophic yeast Pichia pastoris. BioTechniques 38(1):44–48. https://doi.org/10.2144/05381BM04
9. Weis R, Luiten R, Skranc W, Schwab H, Wubbolts M, Glieder A (2004) Reliable high-throughput screening with by limiting yeast cell death phenomena. FEMS Yeast Res 5(2): 179–189. https://doi.org/10.1016/j.femsyr.2004.06.016

Chapter 19

Retrofitting of *AOX1* Promoter-Based *Pichia pastoris* Expression Strains for Methanol-Independent Protein Production

Tanja Michelle Schleichert, Lisa Wasserer, and Anton Glieder

Abstract

The classical *Pichia pastoris* expression system is built on derivative strains of *Komagataella phaffii* NRLL Y-11430 and relies on strong methanol-inducible promoters that can be tightly regulated like the alcohol oxidase 1 promoter (P_{AOX1}). The P_{AOX1} is tightly repressed by carbon sources such as glucose or glycerol and is strongly induced upon methanol induction in absence of repressing carbon sources. However, induction with toxic and flammable methanol is a considerable safety risk during storage and handling, and was also discussed to cause metabolic stress and increased cell lysis. Furthermore, methanol is not compatible with many existing production plants; therefore, prompting the search for innovative processes that allow alternative inducible, derepressed, or constitutive high-level protein production without the use of methanol. Alternatively, methanol-free protein production has been achieved by the overexpression of single transcription factors, which among others regulate diverse methanol utilization genes like the *AOX1* gene. This approach also allows to convert (retrofit) existing methanol-dependent production strains to efficient methanol-independent systems or even to increase the productivity of such strains by simple overexpression of intrinsic *Pichia pastoris* genes. In this chapter, we present an approach to retrofit P_{AOX1}-based expression strains for methanol-independent expression based on a simple example using the intracellular reporter protein eGFP.

Keywords *Pichia pastoris*, *Komagataella phaffii*, Transcription factors, *AOX1* promoter, Methanol independent, Retrofit

1 Introduction

Komagataella phaffii (previously named *Pichia pastoris*) is widely used as a microbial expression host for the production of a wide variety of recombinant proteins. This methylotrophic yeast demonstrated favorable technological properties such as simple and fast growth to high cell densities and efficient secretion of heterologous proteins, while showing a low background by its own secretome. Strong and tightly carbon source-regulated promoters like P_{AOX1} demonstrated their efficiency for gene expression [1, 2]. However,

Xavier Garcia-Ortega et al. (eds.), *Pichia pastoris: Methods and Protocols*, Methods in Molecular Biology, vol. 2697,
https://doi.org/10.1007/978-1-0716-4779-0_19,

since using highly flammable and toxic methanol for the induction includes safety risks, it is undesirable for many large-scale processes. Thus, there is the desire to avoid methanol, if possible [3], but it is still widely used in specific cases where methanol induction results in clearly higher productivity or product titers the process efficiency and product titers are higher with methanol-induced processes. Due to the known disadvantages, methanol-independent protein expression should be taken into consideration as an alternative method for the production of proteins employing *K. phaffii* [4, 5]. Some of the successful efforts in this direction delivered derepressible P_{AOX1} variants or promoters like the *K. phaffii* catalase gene (P_{CAT1}) and the *Hansenula polymorpha* (*Ogataea angusta*) formate dehydrogenase gene (P_{FMD}), which demonstrated strong transcriptional strength based on simple derepression without the need of methanol for induction. As an alternative and well-proven approach, rewiring the induction process of methanol-regulated gene expression can be applied. In those approaches, the induction by methanol is replaced by the overexpression of single transcriptional regulators, known as activators of methanol-inducible genes. That can lead to methanol-free strains [3, 6–9]. Although the knowledge about transcription factors is still limited, several studies investigated the impact of transcription factors on the regulation of the MUT pathway in *K. phaffii* [2, 3, 10]. In general, such transcription factors are mostly regulated on the transcriptional level in addition to a regulated subcellular localization by protein phosphorylation.

Already identified positive regulators for the MUT genes in *K. phaffii* (*Kp*) are the *KpADR*1 (called Mxr1 in early literature, [9, 11]), the *KpTRM*1 (often named *PRM*1 [6, 12]), and the homolog to the *Hansenula polymorpha MPP*1 protein *KpMPP*1 (most commonly named *MIT*1 (methanol-induced transcription factor 1) [8, 13]). These three most frequently studied transcriptional regulators were chosen as an example to retrofit the *AOX1* promoter-based eGFP expression strains (Table 1) for methanol-independent induction.

For example, Wang et al. have identified *KpMPP*1, in their publication named *MIT*1 (methanol-induced transcription factor), as a critical transcription factor for the induction of P_{AOX1}. However, its structure and function differ from those of its *Hansenula polymorpha* (*Hp*) homolog *HpMPP*1. The authors also reported that *KpADR*1 mainly functioned as a positive regulator during carbon derepression, whereas *KpMPP*1 and *KpTRM*1 functioned during methanol induction [8]. In general, similar effects have been observed in our studies leading to this protocol, although some differences were observed in respect to regulatory details. In this context, it is important to note that the *Komagataella phaffii MIT*1 gene, as described by most *K. phaffii* genome annotations in databases such as GenBank (NCBI accession # CAH2449970), is a

Table 1
Gene coding sequences of *K. phaffii* native transcription factors *KpADR1* (CCA40655.1), *KpMPP1* (CCA39317.1), and *KpTRM1* (CCA40959.1)

TF	Sequence 5′ → 3′
KpADR1	**CACTTGCTCTAGTCAAGACTTACAATTAAA**ATGAGCAATCTACCCCCAA CTTTTGGTTCCACTAGACAATCTCCAGAAGACCAATCACCTCCCGTG CCCAAGGAGCTGTCATTCAATGGGACCACACCCTCAGGAAAGCTACGCTTA TTTGTCTGTCAGACATGTACTCGAGCATTTGCTCGTCAGGAACACTTGAAA CGACACGAAAGGTCTCACACCAAGGAGAAACCTTTCAGCTGCGGCATTTG TTCTCGTAAATTCAGCCGTCGAGATCTGTTATTGAGACATGCCCAAAAACTG CACAGCAACTGCTCTGATGCGGCCATAACAAGACTAAGGCGCAAGGCAACT CGTCGGTCTTCTAATGCCGCGGTTCCATATCTGGTTCTACTCCGGTGACAA CGCCAAATACTATGGGTACGCCCGAAGATGGCGAGAAACGAAAAGTT CAGAAACTGGCCGGCCGCCGGGACTCAAATGAACAGAAACTGCAACTGCAAC AACAACATCTACAGCAACAACCACAGTTGCAATACCAACAATCTCTTAAGCA GCATGAAAATCAAGTCCAGCAGCCTGATCAAGATCCATTGATATCCCC GAGAATGCAATTATTCAATGATTCCAACCATCACGTAAACAATTTGTTTGAT CTTGGACTAAGAAGAGCTTCCTTCTCCGCCGTTAGTGGAAATAATTATGCCCA TTATGTGAATAATTTTCAACAAGATGCCTCTTCTACCAATCCAAATCAAGATT CAAATAATGCCGAATTTGAGAATATTGAATTTTCTACCCCACAAATGA TGCCCGTTGAAGATGCTGAAACTTGGATGAACAACATGGGTCCAATTCCG AACTTCTCTCTCGATGTGAACAGGAACATTGGTGATAGCTTTACAGATATACAA CACAAGAATTCAGAGCCTATTATATCCGAACCGCCCAAGGACACCGCTCC AAACGACAAGAAGTTGAATGGCTACTCTTTTTACGAAGCCCCCATCAAGCC ATTAGAATCCCTATTTTCTGTCAGGAATACAAAGAGAAACAAG TATAAAACAAATGACGACTCTCCAGACACCGTGGATAATAACTCCGCACCGG CTGCTAATACCATTCAAGAACTTGAGTCTTCTTTGAATGCATCCAAGAATTTTT GCTTGCCAACTGGTTATTCCTTCTATGGTAATTTGGACCAACAGACTTTCTC TAACACGTTATCATGCACTTCTTCTAATGCCACAATTTCGCCCATTCTACTCGA TAACTCCATTAATAATAACTCCACTAGTGACGTGAGACCAGAATTTAGAACAC AAAGTGTCACCTCTGAAATGAGTCAAGCCCCTCCCCCTCCTCAAAAAAACAAC TCGAAATATTCCACCGAAGTTCTTTTTACCAGCAACATGCGGTCGTTTATTCA CTACGCTCTTTCCAAGTATCCTTTTATTGGTGTGCCCACTCCAA CTCTTCCGGAGAACGAAAGACTAAATGAATATGCTGATTCATTCACCAA CCGTTTCTTAAATCATTATCCTTTCATACATGTCACGATTCTCAAAGAATA CTCCCTTTTCAAGGCAATTTTAGATGAGAATGAGTCGACTAAGAA CTGGGAAAATAATCAGTTTTACTTAGAGAACCAACGAATATCAA TTGTTTGTCTTCCTCTTTTGGTGGCTACGATAGGTGCAGTACTATCAAACAA CAAAAAGGATGCTTCGAATTTATACGAAGCTTCAAGGCGTTGTATTCATGTTTA CTTAGATTCCAGGAAAAAGATACCCACTTCCTTGTCCGCAAATAACAATGA CTCTCCACTTTGGCTAATTCAATCCCTGACGTTATCTGTTATGTATGGGTTA TTTGCGGACAATGACATTAGTTTGAATGTCGTGATCAGACAAGTTAACGCA CTTAATTCTCTGGTCAAGACTTCGGGCCTGAATAGGACCTCAATTATAGA TCTTTTCAACATCAACAAACCTTTGGATAATGAACTCTGGAATCAA TTCGTGAAAATAGAGTCCACCGTAAGGACAATCCACACGATTTTTCAAATCA GTTCCAACTTAAGCGCCTTGTACAATATTATTCCATCGTTGAAAATTGATGA CCTAATGATTACTCTACCAGTTCCCACAACACTTTGGCAAGCTGA TTCTTTTGTGAAATTCAAAAGTCTAAGTTACGGAAATCAGATCCCTTTTCAATA TACAAGAGTACTACAGAATTTGATTGATTACAATCAGCCATTGAGCGA TGGAAAATTTTTGTATGAAAACCATGTAAGTGAGTTTGGACTCATATGCCTACA GAATGGTCTACACCAATACAGCTATTTCCAAAAATTGACTGCTGTCAATAACA

(continued)

Table 1
(continued)

TF	Sequence 5′ → 3’
	GAGAAGATGCGCTATTCACAAAGGTTGTTAATTCACTTCACAGTTGGGATA GGATGATTTCGAATTCTGATTTGTTTCCAAAGAAGATATATCAGCAGA GTTGCTTGATTTTGGACTCAAAGTTGCTTAATAATTTCCTGATTGTCAAGA GCTCATTGAAAGTTTCGACCGGAGACGTTAGTTCTTTGAATAAGTTAAAA GAAAACGTGTGGCTTAAAAACTGGAATCAAGTGTGTGCTATCTATTATAACA GCTTCATGAACATTCCTGCTCCCAGTATTCAAAAGAAGTACAATGACATAGA GTTTGTGGATGACATGATTAATTTGAGTCTAATCATCATCAAGATTATGAAA CTCATTTTCTATAACAATGTCAAAGACAATTATGAGGATGAAAATGACTTCAAA TTGCAAGAGTTAAATTTAACATTTGACAATTTTGATGAGAAAATATCCTTGAA TTTGACAATATTATTCGATATATTTTTGATGATCTACAAGATAATTACCAATTA CGAGAAGTTTATGAAGATCAAACACAAGTTTAATTACTACAATTCTAATTCGAA TATAAGCTTCTTGCATCATTTCGAACTCTCCTCGGTTATCAATAACACCCAAA TGAACCAGAATGATTATATGAAAACAGATATTGATGAAAAGCTTGATCA GCTTTTCCACATCTATCAAACATTTTTCCGGCTGTATCTGGATTTAGAAAA GTTTATGAAGTTCAAATTCAACTATCATGACTTTGAGACAGAGTTTTCAA GTCTCTCAATATCCAATATACTGAACACTCATGCTGCTTCTAACAATGACA CAAATGCTGCTGATGCTATGAATGCCAAGGATGAAAAAATATCTCCCACAA CTTTGAATAGCGTATTACTTGCTGATGAAGGAAATGAAAATTCCGGTCGTAA TAACGATTCAGACCGCCTGTTCATGCTGAACGAGCTAATTAATTTTGAAGTA GGTTTGAAATTTCTCAAGATAGGTGAGTCATTTTTTGATTTCTTGTATGAGAA TAACTACAAGTTCATCCACTTCAAAAACTTAAATGACGGAATGTTCCACATCA GGATATACCTAGAAAACCGACTAGATGGTGGTGTCTAG**GCGGCCGCTCAAGA** **GGATGTCAGAATGCCATTTGCCTG**
KpMPP1	**CACTTGCTCTAGTCAAGACTTACAATTAAA**ATGAGTACCGCAGCCCCAATCAA GGAAGAAAGCCAATTTGCCCATTTGACCCTAATGAACAAGGATATA CCTTCGAACGCAAAACAGGCAAAGTCGAAAGTTTCAGCGGCCCCTGCTAAGA CGGGCTCCAAATCTGCTGGTGGATCTGGCAACAACAACGCTGCA CCTGTGAAAAAAAGAGTCCGCACGGGCTGTTTGACCTGCCGAAAGAAGCA CAAGAAATGTGACGAGAACAGAAACCCAAAATGTGACTTTTGCACTTTGAAA GGCTTGGAATGTGTCTGGCCAGAGAACAATAAGAAGAATATCTTCGTTAACAA CTCCATGAAGGATTTCTTAGGCAAGAAAACGGTGGATGGAGCTGATA GTCTCAATTTGGCCGTGAATCTGCAACAACAGCAGAGTTCAAACACAA TTGCCAATCAATCGCTTTCCTCAATTGGATTGGAAAGTTTTGGTTA CGGCTCTGGTATCAAAAACGAGTTTAACTTCCAAGACTTGATAGGTTCAAA CTCTGGCAGTTCAGATCCGACATTTTCAGTAGACGCTGACGAGGCCCAAAAA CTCGACATTTCCAACAAGAACAGTCGTAAGAGACAGAAACTA GGTTTGCTGCCGGTCAGCAATGCAACTTCCCATTTGAACGGTTTCAATGGAA TGTCCAATGGAAAGTCACACTCTTTCTCTTCACCGTCTGGGACTAATGACGA TGAACTAAGTGGCTTGATGTTCAACTCACCAAGCTTCAACCCCCTCACA GTTAACGATTCTACCAACAACAGCAACCACAATATAGGTTTGTCTCCGATGTCA TGCTTATTTTCTACAGTTCAAGAAGCATCTCAAAAAAAGCATGGAAATTCCA GTAGACACTTTTCATACCCATCTGGGCCGGAGGACCTTTGGTTCAATGA GTTCCAAAAACAGGCCCTCACAGCCAATGGAGAAAATGCTGTCCAACAGGGA GATGATGCTTCTAAGAACAACACAGCCATTCCTAAGGACCAGTCTTCGAA CTCATCGATTTTCAGTTCACGTTCTAGTGCAGCTTCTAGCAACTCAGGAGA CGATATTGGAAGGATGGGCCCATTCTCCAAAGGACCAGAGATTGAGTTCAA CTACGATTCTTTTTTGGAATCGTTGAAGGCAGAGTCACCCTCTTCTTCAAA GTACAATCTGCCGGAAACTTTGAAAGAGTACATGACCCTTAGTTCGTCTCA

(continued)

Table 1 (continued)

TF	Sequence 5′ → 3′
	TCTGAATAGTCAACACTCCGACACTTTGGCAAATGGCACTAACGGTAACTA TTCTAGCACCGTTTCCAACAACTTGAGCTTAAGTTTGAA CTCCTTCTCTTTCTCTGACAAGTTCTCATTGAGTCCACCAACAATCACTGA CGCCGAAAAGTTTTCATTGATGAGAAACTTCATTGACAACATCTCGCCA TGGTTTGACACTTTTGACAATACCAAACAGTTTGGAACAAAAATTCCA GTTCTGGCCAAAAAATGTTCTTCATTGTACTATGCCATTCTGGCTATA TCTTCTCGTCAAAGAGAAAGGATAAAGAAAGAGCACAATGAAAAAACATTGCAA TGCTACCAATACTCACTACAACAGCTCATCCCTACTGTTCAAAGCTCAAATAA TATTGAGTACATTATCACATGTATTCTCCTGAGTGTGTTCCACATCATGTCTA GTGAACCTTCAACCCAGAGGGACATCATTGTGTCATTGGCAAAATACATTCAA GCATGCAACATAAACGGATTTACATCTAATGACAAACTGGAAAAGAGTA TTTTCTGGAACTATGTCAATTTGGATTTGGCTACTTGTGCAATCGGTGAAGA GTCAATGGTCATTCCTTTTAGCTACTGGGTTAAAGAGACAACTGACTACAAGA CCATTCAAGATGTGAAGCCATTTTTCACCAAGAAGACTAGCACGACAACTGA CGATGACTTGGACGATATGTATGCCATCTACATGCTGTACATTAGTGGTAGAA TCATTAACCTGTTGAACTGCAGAGATGCGAAGCTCAATTTTGAGCCCAA GTGGGAGTTTTTGTGGAATGAACTCAATGAATGGGAATTGAACAAACCCTTGA CCTTTCAAAGTATTGTTCAGTTCAAGGCCAATGACGAATCGCAGGGCGGA TCAACTTTTCCAACTGTTCTATTCTCCAACTCTCGAAGCTGTTACAGTAACCA GCTGTATCATATGAGCTACATCATCTTAGTGCAGAATAAACCACGATTATA CAAAATCCCCTTTACTACAGTTTCTGCTTCAATGTCATCTCCATCGGACAA CAAAGCTGGGATGTCTGCTTCCAGCACACCTGCTTCAGACCACCA CGCTTCTGGTGATCATTTGTCTCCAAGAAGTGTAGAGCCCTCTCTTTCGACAA CGTTGAGCCCTCCGCCTAATGCAAACGGTGCAGGTAACAAGTTCCGCTCTA CGCTCTGGCATGCCAAGCAGATCTGTGGGATTTCTATCAACAACAACCACAA CAGCAATCTAGCAGCCAAAGTGAACTCATTGCAACCATTGTGGCA CGCTGGAAAGCTAATTAGTTCCAAGTCTGAACATACACAGTTGCTGAAA CTGTTGAACAACCTTGAGTGTGCAACAGGCTGGCCTATGAACTGGAA GGGCAAGGAGTTAATTGACTACTGGAATGTTGAAGAATAG**GCGGCCGCTCAA GAGGATGTCAG**
KpTRM1	**CACCATAACACTTGCTCTAGTCAAGACTTACAATTAAA**ATGCCTCCTAAACA TCGGCTGGAGCAGAGTATACAGCCCATGGCTTCTCAACAAATAGTA CCCGGTAATAAGGTTATTCTGCCGAATCCAAAAGTAGATGCAAAATCTA CCCCAAACATTTCAGTTCAGAAGAGAAGAAGAGTCACCAGAGCTTGTGATGAA TGTCGGAAAAAGAAGGTCAAATGTGATGGTCAACAACCATGCATTCATTGTA CCGTTTATTCCTATGAGTGCACTTACAGCCAACCTTCCAGTAAGAAGAGACA GGGACAATCTCTGAGTCTGAGTGCTCCGTCAAACATTAATGCAACAA GTTCCGTACAAAAATCTGTAAAACCTCCTGAAATCGATTTCCAAAGGATGAGA GACGCACTCAAATATTACGAAGATCTTTTAAACCAGTTGATATACCCCAACA GTGCTCCAACTGTTCGAGTTAATCCGATTCGTCTAGCATCGATCTTAAAACAA TTGAGAGCCGATAAATCAAGTGATGAATTAATTTCAGTCAAGGCTCTTTCTGA CAATTACATTGAGATGCTTCACAAAACGATGCAACAACCTGTACAGCAGCCA GCTCCTCCTTCATTGGGGCAAGGAGGGTCCTTCTCTAATCACAGTCCCAATCA TAATAATGCTTCTATTGATGGTTCCATAGAATCTAATCTAGGGAGGGAAATA CGTATCATATTACCTCCGAGAGATATTGCGCTGAAGCTTATCTACAAGA CTTGGGACAACGCGTGTGTACTTTTCCGCTTTTATCACAGACCCGCATTTA TTGAGGACCTGAATGAGTTATATGAAACAGATTTGGCAAACTACACCAATAAA CAACAAAGGTTTTTACCTCTTGTATATTCGGTGA

(continued)

Table 1
(continued)

TF	Sequence 5′ → 3’
	TGGCTTGTGGTGCTCTTTTTTGCAAGACTGATGGGATTAATCACGGCCAAAA GAGCTCCAAGCCCAAAGACTCTTCTGATGAAAGTCTCATAGACGATGA GGGTTACAAGTATTTTATTGCCGCAAGAAAACTAATAGATATCACGGATACCA GGGATACCTACGGAATTCAGACTATTGTTATGCTGATCATTTTTTTACAA TGTTCGGCTCGTCTTTCAACATGCTATTCTTATATTGGCATTGCTCTAAGA GCTGCATTGAGAGAAGGTTTGCATCGTCAGTTGAACTATCCTTTCAATCCAA TTGAGTTAGAAACAAGAAAGCGTCTTTTTTGGACTATCTATAAAATGGACA TCTATGTCAATACAATGCTGGGGCTTCCAAGAACCATTTCTGAAGAGGA TTTCGACCAGGAAATGCCTATCGAACTTGATGATGAGAACATTAGTGAAA CCGGATATAGGTTCGATTTACAAGGTACAAAGTTATCCAGTTCAGGAATA GCCAATGCTCACACTAGATTGATATTCATAATGAAGAAAATTGTGAAAAAATTA TATCCTGTCAAACTACAGAAACCAACCTCAAACAGTGGCGATACCCCACTTGA GAACAATGATTTATTGGCTCATGAAATCGTTCATGAACTTGAGATGGA TCTCCAAAATTGGGTCAATAGTCTACCTGCAGAACTAAAACCGGGGATAGAA CCACCGACCGAGTATTTTAAAGCTAACAGATTGCTTCATTTGGCATACCTGCA TGTCAAGATTATTCTCTACAGGCCATTTATTCATTACATCTCAGAAAAGGATAA GGTTGGAAATAGTTCTATCCCTCCGTCGCCCGAAGAGATCACTTCTATCGA GAAAGCCAAGAATTGTGTCAATGTTGCCAGAATTGTTGTTAAACTAGCCGAA GACATGATTAATAGGAAAATGTTAAGTGGTTCATATTGGTTTTCCATTTATA CCATTTTTTTTTCCGTGGCATGTCTGGTGTACTATGTTCATTTCGCTCCA CCGAAGAAAGACAATGGAGAACTGGATCCCCAATACATGGAAATCAAGAAAGA TACAGAGAGTGGAAGAGAGGTCTTAAATATCCTCAAAGATAGTAGTA TGGCGGCAAGAAGAACGTATAATATTCTCAACTCTTTGTTTGAGCAGTTAAACA GAAGAACTGCAAAGGTCAACCTAGCAAAGGCACAGCAACCACCATCA GGGTTGAATAACCCAGCTGCTACCCAGTATCAGAAACAGGGTGAACACAGGCA GTTACAACCAAGTAACTATTCTGGAACTGTGAAATCTGTGGACCCAGAGAATA TCGATTACTCTTCCTTTGGTTCTCAGTTTGAAAACACTAACATCGAAGA TGGTTCCTCAAATACAAAGATTGATCAGAAAGTGAATGGGGTGAACTACA TCGATGGTGTGTTTACAGGGATCAACCTAAATATGCCTAATCTCTCAGAAA CTTCTAACACTCAAGGTATCGATAATCCAGCATTTCAAAGTATAAACAA TTCTAATTTGAACAATAATTTTGTACAAACAAAGTACATTCCCGGCATGATGGA CCAGCTAGATATGAAAATTTTCGGAAGATTCCTTCCACCTTACATGCTGAA CTCCAACAAGGTTGAACAGGGACAAAATGAAAGGAACCTATCAGGCCAACCA TCCTCGTCGAATACTCCTGATGGATCACAACCTGTGACAGTTCTGGATGGA TTATACCCGTTGCAGAATGATAATAATAATAACCACGACCCAGGAAATTCAAA GTCTGTTGTAAATAACAGTAACTCGGTAGAAAACTTACTACAGAACTTTACAA TGGTGCCCTCGGGGTTGTCATCAACAGTGCAAAATCCTGAA GCGGCCCAAAAATTCAATAATCATATGTCAAACATATCGAATATGAATGA TCCAAGAAGAGCTAGCGTAGCTACATCAGATGGATCCAATGACATGGATCA TCATAGCCAAGGCCCGATAAACAAAGATTTGAAACCGTTGAGCAACTACGA GTTTGACGATCTCTTCTTTAATGATTGGACCACTGCGCCAGATACAATAAA TTTTGACAGTTAA**GCGGCCGCTCAAGAGGATGTCAGAATGC**

The homolog overhangs to the expression plasmid backbone are highlighted in bold homolog of the *S. cerevisiae MIT*1 gene (GenBank # NP_010909.1) and has no structural or functional relationship to the *Hansenula polymorpha MPP*1 gene and the *K. phaffii* homolog named *MPP*1 by Wang et al.

Chang et al. successfully demonstrated a synthetic positive feedback circuit due to the overexpression of *KpADR*1 driven by a weak *AOX2* promoter (P_{AOX2}) to allow P_{AOX1} to be induced under glycerol starvation or carbon-free conditions. They enhanced the recombinant protein production efficiency in *K. phaffii* and created a methanol-free induction system to eliminate the potential risks of methanol [7].

Takagi et al. constructed a methanol-free expression system of *K. phaffii* by engineering the strong methanol-inducible *DAS1* promoter (P_{DAS1}). The constitutive expression of *KpTRM*1 was demonstrated to produce a model protein without addition of methanol, especially when P_{DAS1} was used and not P_{AOX1}. *KpADR*1 did not have the same effect on P_{DAS1}, while it was more effective than *KpTRM*1 on P_{AOX1} [6].

Vogl et al. altered the regulation of P_{AOX1} by co-overexpressing single transcriptional regulators involved in methanol-dependent carbon source regulation. The derepressed overexpression of three transcription factors (*KpADR*1 [there named *MXR*1], *KpTRM*1 [there named *PRM*1], *KpMPP*1 [there named *MIT*1]), among others, was used to retrofit existing P_{AOX1}-based expression [3] strains to induce production of the target protein upon depletion of the repressing carbon source. Under derepressed conditions, using the P_{CAT1}, the overexpression of the three transcription factors led to differing activation of the P_{AOX1}. In small-scale experiments, the overexpressed transcription factors *KpMPP*1 and *KpADR*1 reached levels of the methanol-induced control construct up to 44% and 25%, respectively, while *KpTRM*1 reached 7%. Detrimental growth effects occurred with the *KpADR*1 construct in the presence of methanol, which coincides with the results from Lin-Cereghino et al. [3, 9].

The data from experiments performed for this protocol supported the findings of Vogl et al. [3] and even exceeded it in some aspects (Fig. 1). As a control experiment, the reporter strain was transformed with an empty expression vector lacking the transcription factor gene (EVC). The transcription factor overexpression (TF-OE) strain with derepressed overexpression of the *KpMPP*1 transcription factor gene showed the highest derepressed expression of eGFP (37.2-fold improvement, 44% of the methanol-induced expression level of the EVC) followed closely by *KpADR*1 (29.1-fold improvement, 35% of the methanol-induced expression level of the EVC) and then *KpTRM*1 (9% of the methanol-induced expression level of the EVC), which also shows a small increase (7.5-fold improvement) compared to the EVC. *KpMPP*1 (2.1-fold improvement) and *KpADR*1 (1.7-fold

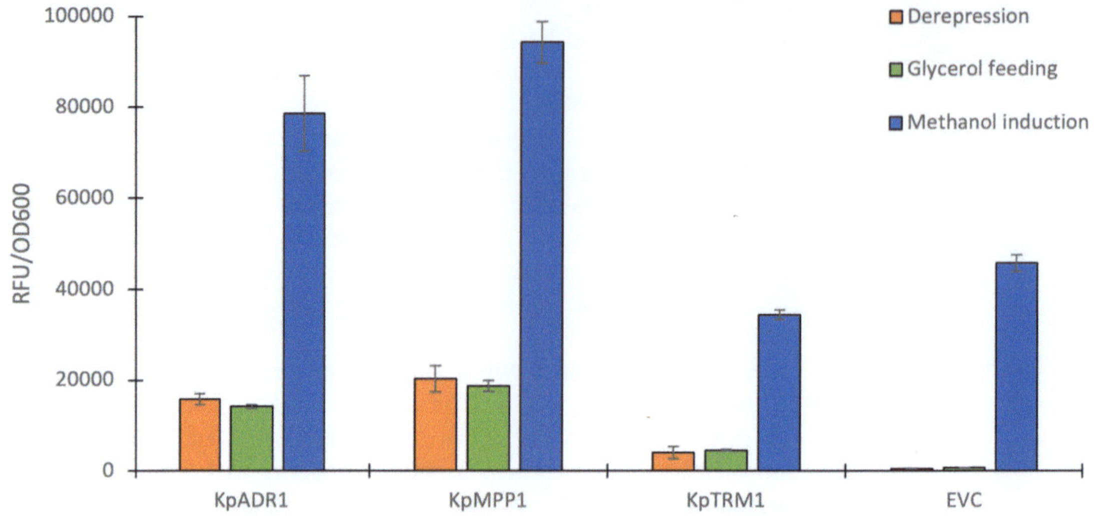

Fig. 1 Representative eGFP screening evaluation of the three *AOX1* promoter-based TF-OE strains with an EVC as a reference. Screening was performed under three cultivation conditions: derepression (~60 h), after glycerol feeding, and after methanol induction (~100 h). The TF-OE strains with their corresponding overexpression of a single transcription factor (*Kp*ADR1, *Kp*MPP1, *Kp*TRM1) are on the x-axis, and the normalized RFU/OD_{600} values are on the y-axis

improvement) could even increase the expression of the reporter under methanol induction in comparison to the EVC under the same conditions. In the performed small-scale experiments, the additional glycerol feeding to prolong the derepressed state did not show any improvement over the derepression phase after 60 h (*see* **Note 1**).

The derepressed overexpression of these three transcription factors (*KpADR*1, *KpTRM*1, and *KpMPP*1) can be used to retrofit *AOX1* promoter-based expression strains into glucose/glycerol-regulated, methanol-independent expression systems with higher space-time yields (STYs), and in addition also showed potential to increase titers of proteins in methanol-induced processes [3].

2 Materials

2.1 Assembly of Transcription Factor Overexpression (TF-OE) Plasmids

1. *K. phaffii* expression plasmid to use as the backbone for Gibson assembly (Fig. 2). For example, in this chapter, the commercial bisy plasmid pBSY3H) was used.
2. 10 μM primers suitable to amplify the backbone needed for Gibson assembly (Table 5).
3. Phusion™ High-Fidelity DNA Polymerase (or another high-fidelity polymerase) with appropriate buffer.

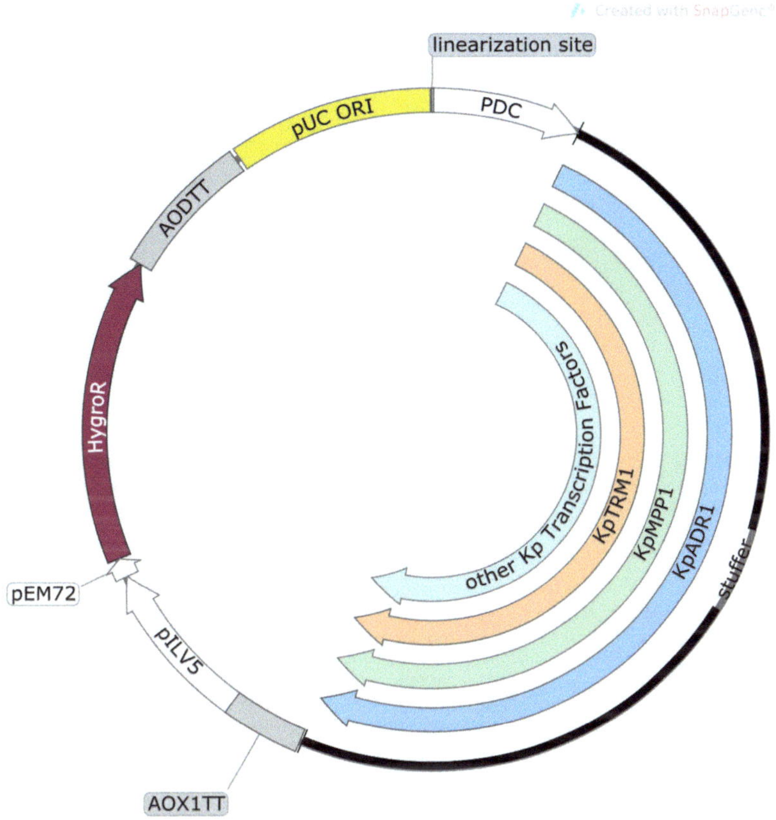

Fig. 2 Representative TF-OE plasmid with either the transcription factors (*ADR1*, *MPP1*, *TRM1*) or the stuffer fragment cloned downstream of the PDC promoter. Any other transcription factor can also be cloned instead of the stuffer fragment

4. 2 mM dNTP stock: 20 μL/mL of dATP (100 mM), dTTP (100 mM), dCTP (100 mM), and dGTP (100 mM) were mixed with nuclease-free water and stored at −4 °C.
5. The gene sequences of the *K. phaffii* transcription factors *ADR1*, *MPP1*, and *TRM1* including their homolog regions (highlighted in bold) to the promoter and the terminator (of the backbone) suitable for Gibson assembly are shown in Table 1.
6. Thermocycler.
7. Gibson assembly mix (e.g., New England Biolabs E2611).
8. Chemical-competent *E. coli* cells for cloning procedures (e.g., TOP10F′).
9. Thermomixer.
10. Table centrifuge.

11. 50× TAE buffer: 242 g/L of TRIS–HCl and 14.6 g/L of EDTA were dissolved in 57.1 g/l acetic acid (conc.), and the volume was adjusted to 1 L with dH_2O.
12. SOC medium: 0.5% (w/v) yeast extract, 2% (w/v) tryptone, 20 mM (w/v) glucose, 10 mM NaCl, 2.5 mM KCl, 10 mM $MgCl_2$, 10 mM $MgSO_4$. Autoclave and store at room temperature.
13. Plastic Petri dishes: 94 × 16 mm with vents (e.g., Greiner Bio-One, Frickenhausen, Germany; PS, Article-nr. 633,180).
14. LB-agar plates with appropriate ingredients for selection (in the experiments for this protocol, the antibiotic hygromycin was used): 0.5% (w/v) yeast extract, 1% (w/v) tryptone, 1% (w/v) NaCl, and 1.5% (w/v) agar-agar. Autoclaved and when lukewarm, hygromycin was added and media poured into Petri dishes to solidify. The standard end concentration for hygromycin in *E. coli* was 100 μg/mL.
15. DNA purification kit (e.g., Promega Wizard® SV Gel and PCR Clean up System).
16. Plasmid isolation kit (e.g., Promega Wizard® Plus SV Minipreps DNA Purification System).
17. DNA quantification (e.g., Thermo Scientific™ NanoDrop™ One).
18. Sequencing primer for plasmid sequencing (Table 5).

2.2 Transformation of K. phaffii AOX1 Promoter-Based Expression Strains with the TF-OE Plasmids

1. *Smi*I restriction enzyme for linearization of the TF-OE plasmid.
2. 13 mm, 0.025 μm MCE membranes (Merck Millipore Ltd., Cork, Ireland) for desalting.
3. Thermomixer.
4. Electrocompetent *K. phaffii* cells of *AOX1* promoter-based expression strain.
5. Icebox filled with ice.
6. Electroporation cuvettes with 0.2 cm gap (e.g., BioRad 1,652,082).
7. 1.5 mL reaction tubes with conical bottom.
8. Electroporator (e.g., Eppendorf Eporator®).
9. YPD: 1% yeast extract, 2% peptone, 2% glucose. Autoclaved and stored at room temperature.
10. 1 M sorbitol: 182 g/L of sorbitol in ddH_2O. Autoclaved and stored at room temperature.
11. YPD/sorbitol: Mix equal volumes of YPD and 1 M sorbitol.

12. Plastic Petri dishes: 94 × 16 mm with vents (e.g., Greiner Bio-One, Frickenhausen, Germany; PS, Article-nr. 633,180).
13. YPD agar plates with appropriate selection pressure (hygromycin and Zeocin): YPD with 1.5% (w/v) of agar-agar. Autoclaved and when lukewarm, hygromycin and Zeocin were added and media poured into Petri dishes to solidify. The standard end concentration of hygromycin for *K. phaffii* selection was 300 μg/mL and for Zeocin 100 μg/mL.
14. Antibiotics (100 mg/mL stock): Antibiotics had been dissolved in ddH_2O, filter sterilized, and aliquoted.

2.3 Cultivation and Screening of K. phaffii TF-OE Strains

1. Sterile toothpicks.
2. Sterile 96-deep-well plates (DWP).
3. 96-well replicator stamp.
4. Square Petri dish 120 × 120 × 17 mm for stamping.
5. Multichannel pipette.
6. Sterile plastic tray.
7. 1 M potassium phosphate buffer: 46.4 g of K_2HPO_4 and 100 g of KH_2PO_4 were dissolved in 1 L ddH_2O. pH adjusted, if necessary, by addition of 1 M KOH to a pH of 6. Autoclaved and stored at room temperature.
8. 13.4% (w/v) YNB: Dissolved yeast nitrogen base w/o amino acids in ddH_2O. Autoclaved and stored at room temperature in the dark.
9. 200 mg/L biotin: D-biotin in ddH_2O. Filter sterilized using a 0.2-μm filter.
10. 20% (w/v) glucose: Dissolved in ddH_2O. Autoclaved and stored at room temperature.
11. 50% (w/v) glycerol: Dissolved 250 g glycerol in 0.5 L ddH_2O and sterilized by autoclaving.
12. 100% methanol.
13. BMD1: 650 mL ddH_2O, 200 mL potassium phosphate buffer, 100 mL YNB, 2 mL biotin, 50 mL glucose.
14. BMM2: 690 mL ddH_2O, 200 mL potassium phosphate buffer, 100 mL YNB, 2 mL biotin, 10 mL 100% methanol.
15. BMM10: 650 mL ddH_2O, 200 mL potassium phosphate buffer, 100 mL YNB, 2 mL biotin, 50 mL 100% methanol.
16. BMG0.5: 690 mL ddH_2O, 200 mL potassium phosphate buffer, 100 mL YNB, 2 mL biotin, 10 mL 50% glycerol.
17. BMG2.5: 650 mL ddH_2O, 200 mL potassium phosphate buffer, 100 mL YNB, 2 mL biotin, 50 mL 50% glycerol.

2.4 Evaluation of the Screening Data

1. *Microplate reader*, e.g., *CLARIOstar*® (BMG LABTECH) for OD_{600} and eGFP fluorescence measurement.
2. Microtiter plates for screening, e.g., Nunc™ MicroWell™ 96-Well Optical-Bottom Plates with Polymer Base (Thermo Fisher Scientific Inc.).
3. Multichannel pipette or 96-well pipetting robot, e.g., *MINI 96 (INTEGRA Biosciences Corp.)*.

3 Methods

In this method, we demonstrated how to assemble *K. phaffii* expression plasmids based on the pBSY3H for the intracellular production of transcription factors that enabled the retrofitting of the *AOX1* promoter to avoid methanol for induction of protein production (Fig. 2). Subsequently, this protocol describes the procedure to transform *K. phaffii* strains that can already produce a target protein under the control of the P_{AOX1}. In the experiments performed for this protocol, we used eGFP as the reporter protein. To conclude, we provided instructions for the cultivation of the strains co-expressing the transcription factors and eGFP using a derepressing cultivation strategy with methanol or glycerol feeding, and how we evaluate their performance by assessing fluorescence.

3.1 Assembly of Transcription Factor Overexpression (TF-OE) Plasmids

1. Select (*see* **Note 2**) and order a transcription factor (e.g., Table 1) as synthetic DNA (*see* **Note 3**) including the homolog regions required for Gibson cloning.
2. Prepare PCR using the pBSY3H according to the scheme in Table 2 (*see* **Note 4**).

Table 2
PCR formulation for the amplification of the backbone

Component	Volume [μL]
Template (~10 ng DNA)	X
5× Phusion buffer (HF or GC)	10
Primer fwd (10 μM stock)	2.5
Primer rev (10 μM stock)	2.5
dNTP mix (2 mM stock)	5
Nuclease-free water	29-X-Y
DMSO (100%)	Y
Phusion High-Fidelity polymerase	0.5
Total	50

3. Perform a PCR according to the temperature profile in Table 3 to amplify the pBSY3H from the promoter (PDC) to the terminator (*AOX1*TT) (*see* **Notes 5**, **6**, and 7) with the appropriate primers (Table 4, *see* **Note 8**).

Table 3
PCR temperature and cycling profile used for the amplification of the backbone

Cycle step	Temperature [°C]	Time [s]	No. of cycles
Initial denaturation	98	30	1
Denaturation	98	20	30–35
Annealing	Dependent on primers	20	
Extension	72	20 s/1kbp + 30 s	
Final extension	72	420	1
Hold	4	∞	∞

Table 4
Primers for PCR of backbone as well as sequencing primers to check for correctly cloned TF-OE plasmids

Primer name	Sequence 5′ → 3′
AOX1TT_fw (backbone)	tcaagaggatgtcagaatgccatttg
PDC_rev (backbone)	tttaattgtaagtcttgactagagcaagtgttatg
PDC_fw_seq	accctctactaatggaattgaacaaag
AOX1TT_rv_seq	tcccaaacccctaccacaag
MPP1_seq1	atacatatcgtccaagtcatcgtc
MPP1_seq2	tcgaagactggtccttaggaatg
ADR1_seq1	cctcataattgtctttgacattgttatag
ADR1_seq2	attgtccgcaaataacccatac
ADR1_seq3	gagtagccattcaacttcttgtc
TRM1_seq1	ccatcgatgtagttcacccc
TRM1_seq2	ttgaggttggtttctgtagtttg
TRM1_seq3	cccaatgaaggaggagctg

Table 5
Gibson assembly composition to insert the various transcription factors

Ingredient	Volume [μL]
Gibson assembly mixture	15
pBSY3H backbone (100 ng)	X
Insert (threefold molar excess)	Y
H_2O	5-X-Y
Total	20 μL

4. Prepare 1% (w/v) agarose gel with enough wells for your samples and Midori Green for visualization of the DNA.
5. Mix 5 μL of the PCR with 1 μL of the 6× loading dye and load on the agarose gel together with a standard, e.g., O'GeneRuler Express.
6. Perform agarose gel electrophoresis at suitable conditions (e.g., 120 V, 400 mA for 30 min).
7. Inspect if an amplicon of ~3500 bp was obtained.
8. If the fragment is obtained, recover the DNA of the remaining 45 μL of the PCR using a purification kit (according to the manufacturer's instructions).
9. Assess DNA concentration of the recovered DNA using a Nanodrop.
10. Dilute the synthetic transcription factor DNA to a concentration of 50 ng/μL.
11. Prepare the Gibson assembly reaction according to Table 5 to clone the transcription factor inserts (*ADR1*, *MPP1*, *TRM1*) with the pBSY3H backbone.
12. Incubate the Gibson reaction at 50 °C for 60 min.
13. Spin down the reaction at full speed for 10 s.
14. Thaw one aliquot of chemically competent *E. coli* cells per Gibson assembly reaction and preheat the thermomixer to 42 °C.
15. Add 2.5 μL of the Gibson reaction to an aliquot of competent *E. coli* cells and flick the tube a few times.
16. Place the tube on ice for 5–20 min to increase transformation efficiency.
17. Incubate the transformation mixture at 42 °C for 45 s followed by 3 min on ice.

18. Add 1000 μL of SOC and recover the cells for 60 min at 37 °C and 350 rpm.
19. Plate 100 and 900 μL of the regenerated cells on LB-agar plates with appropriate selection pressure (hygromycin).
20. Incubate the plates at 37 °C overnight.
21. Cultivate transformants for plasmid isolation.
22. Perform plasmid isolation (according to the manufacturer's instructions) of the potential TF-OE plasmids.
23. Assess DNA concentration using a Nanodrop.
24. Prepare samples for sequencing with appropriate sequencing primers (Table 4).
25. Evaluate sequencing results and select TF-OE plasmids with the correct sequence for the transformation of *K. phaffii*.

3.2 Transformation of *K. phaffii AOX1* Promoter-Based Expression Strains with the TF-OE Plasmids

1. Select a *K. phaffii* expression strain that produces the desired protein under control of the P_{AOX1}.
2. Prepare competent cells of the *AOX1* promoter-based expression strain (e.g., condensed protocol from Lin-Cereghino et al. [14]).
3. Linearize the 2000 ng of the sequence-verified TF-OE plasmids with *Smi*I according to the manufacturer's instructions.
4. Desalt the restriction digestion using the MCE membranes for 30 min.
5. Precool electroporation cuvettes on ice.
6. Mix ~1000 ng of the desalted and linearized DNA (TF-OE plasmids) with one aliquot of competent *K. phaffii* cells in a 1.5-mL reaction tube (*see* **Note 9**).
7. Place the tubes on ice for 5–10 min.
8. Transfer the transformation mixture into the precooled electroporation cuvettes.
9. Electroporate at 2 KV; resistance, 200 Ω; capacitance, 25 μF.
10. Immediately after the electroporation, add 1 mL of the YPD/-sorbitol mixture.
11. Regenerate the electroporated cells for at least 2 h at 28 °C and 350 rpm in a thermomixer.
12. Plate 100 and 900 μL of the regenerated cells on YPD agar plates with appropriate selection conditions (both hygromycin and Zeocin).
13. Incubate the agar plates at 28 °C until colonies are visible (approximately 2–3 days).

3.3 Cultivation Strategy for Coproduction of eGFP and Transcription Factors

1. Add 250 μL of BMD1 into each well of a 96-DWP using the multichannel pipette.
2. Inoculate each well of the DWP with a distinct *K. phaffii* transformant (single colony) from transformation plates using sterile toothpicks (*see* **Note 10**).
3. Cultivate the cells in the DWP at 28 °C for ~60 h and 320 rpm (*see* **Note 11**).
4. Use a heat-sterilized 96-pin stamp to save the clones by replica plating onto a square YPD agar plate with appropriate selection pressures for both expression constructs.
5. After ~60 h of growth on glucose, the derepression state is usually reached in all wells.
   (a) The first screening under derepressed conditions (before induction) can be performed (e.g., measuring samples of the culture for eGFP fluorescence in a microtiter plate).
   (b) In case of a cultivation under derepressed conditions, 250 μL of BMG0.5 is added to each well resulting in a glycerol concentration of 0.5% OR in case of methanol induction, 250 μL of BMM2 is added to each well resulting in a methanol concentration of 0.5% (*see* **Note 12**).
   (c) The DWP is put again at 28 °C and 320 rpm.
6. Optional after ~68 h: Addition of 50 μL BMM10 (in case of methanol induction) or BMG2.5 (glycerol feeding in case of derepressed cultivation) (was not done in the experiments for this protocol).
7. After ~80 h: Addition of 50 μL BMM10 (in case of methanol induction) or BMG2.5 (glycerol feeding in case of derepressed cultivation).
8. After ~88 h: Addition of 50 μL BMM10 (in case of methanol induction) or BMG2.5 (glycerol feeding in case of derepressed cultivation).
9. After ~100 h: End of cultivation, methanol induction, or glycerol feeding.
   (a) The second analysis, i.e., after methanol induction or glycerol feeding, needs to be performed (in a microtiter plate).

3.4 Screening and Evaluation of the eGFP Fluorescence Within the Retrofitted Strains

For the measurement of OD_{600} and eGFP fluorescence, the *CLARIOstar*® (BMG LABTECH) in combination with the Nunc™ MicroWell™ 96-Well Optical-Bottom Plates with Polymer Base (Thermo Fisher Scientific Inc.) was used in the experiments for this protocol (*see* **Note 13**):

1. After ~60 h and ~100 h of cultivation, mix 10 μL of the cultivation with 190 μL ddH_2O in a Nunc™ MicroWell™ 96-Well Optical-Bottom Plates with Polymer Base under sterile conditions.
2. Place plate in a 96-well spectrophotometer.
3. Assess cell density at 600 nm.
4. Assess eGFP fluorescence at 488 nm excitation and 535 nm emission wavelength with the gain constantly set to 1100 and focal height at 5.6 nm (*see* **Note 14**).
5. Since eGFP is produced intracellularly, the obtained fluorescence values are normalized on the cell density. By this, relative fluorescence units (RFU) are obtained which enables to compare different transcription factors or strains with each other.
6. The screening was performed after approximately 60 h (Derepression) and at the end of the cultivation after ~100 h (after methanol induction or glycerol feeding).
7. Optionally, the screening can already be performed after approximately 24–48 h as well to obtain data under glucose-repressed conditions, since in most cases the carbon source is already depleted after that time (at least if the glucose or glycerol concentration in the batch medium is not higher than 1%) (*see* **Note 15**).

4 Notes

1. Since the glycerol feeding did not show any improvement over the derepression phase after 60 h, it is not a generally recommended procedure. However, this might give some additional effects in special cases due to an extended supply of energy for gene expression.
2. Suitable regulatory elements for overexpression targets can for example be identified from the literature mentioned in the Subheading 1. It can be trial and error to see which regulatory elements positively impact the derepressed expression of a specific gene of interest, so it is recommended to choose several at first and then narrow it down based on the obtained data. Due to possible titration effects, the results might especially depend on the copy number of the integrated target gene expression cassettes.
3. The transcription factors can be ordered as synthetic DNA, where they can also be codon optimized beforehand with, e.g., Gene Designer. In case of a synthetic codon-optimized gene, the chance of a homologous recombination in the host is lower and is therefore recommended.

4. The PCR reagent was prepared as a master mix which is shown in Table 2 and performed according to a certain PCR temperature program (Table 3). Dimethyl sulfoxide (DMSO) in combination with 5× GC Phusion buffer was only used in case of a GC-rich template where the PCR with the 5× HF Phusion buffer yielded a subpar result. The annealing temperature of the PCR depends on the melting temperature of the primers used, and the extension time varies according to the size of the amplified fragment.
5. For the TF-OE plasmid, an initial expression plasmid (pBSY3H) with the elements for cloning in *E. coli* and expression in *K. phaffii* is needed. pBSY3H includes the following elements:
   - PDC (a 500-bp fragment of the *K. phaffii* 1 promoter) Promoter.
   - Stuffer DNA fragment that is later exchanged by the selected transcription factors (e.g., *ADR1*, *MPP1*, *TRM1*).
   - *AOX1*TT, terminator at the 3′ site of the stuffer.
   - Promoters (P_{EM72} for *E. coli*, P_{ILV5} for *K. phaffii*) for the selection marker.
   - Selection marker gene (e.g., HygroR).
   - *AOD*TT, terminator for the yeast selection marker gene.
   - Origin of replication in *E. coli* (pUC ORI).
   - *Smi*I linearization site to linearize the plasmid for the subsequent transformation of *K. phaffii*.
6. For the overexpression of the transcription factors, differently regulated promoters are possible. For example, the PDC is a 500-bp fragment of the peroxisomal catalase 1 promoter P*CTA1* of *K. phaffii* and is repressed by the presence of glucose at higher concentration [15]. The expression is started when glucose is depleted, under derepressed conditions and can be further induced by methanol. Another option is a strong constitutive overexpression using the promoters P*GAP* [16] or of the core histone gene H2A P_{HTA1}, which are active in the presence of glucose and can be downregulated using methanol. P_{HTA1} is a truncated promoter of P_{HTX1}, where the 86-bp-long *HTB1*-core promoter was removed [17]. The constitutive expression of transcription factors may allow inducing promoters already on glucose; however, there might also be negative effects since strong methanol-inducible transcription factor overexpression may overburden the cellular machinery [5] and be detrimental to cells [9]. We recommend using a derepressed promoter (e.g., PDC) for controlling the transcription factor overexpression or an alternative inducible promoter, although a constitutive promoter (e.g., P_{GAP} or P_{HTA}) is also

possible but is rather recommended for a multicopy strain since there a more binding sites available for transcription factors (titration effect). In the experiments for this protocol, the PDC promoter was selected for a derepressed overexpression.

7. For the TF-OE plasmid, different selection markers are possible. It only needs to differ from the selection marker of the expression strain. For example, several markers based on antibiotic (e.g., hygromycin B or G418) resistance genes can be used for the selection of *K. phaffii* transformants. The aminoglycoside phosphotransferase conferring resistance to hygromycin was used in these experiments as the selection marker.
8. The TF-OE plasmid backbone should not have any overhangs for the transcription factor inserts, so it can be universally used for any inserts. It is sufficient for the various inserts to have overhangs to the plasmid backbone.
9. Optionally, the expression strain can also be transformed with an empty vector control (EVC) (just the backbone of our TF-OE plasmid without any transcription factor inserts) to later have a reference strain for evaluation. This is just to account for any integration effects and can also just be your initial *AOX1*-based expression strain.
10. In the initial screening, picking ~40 transformants per construct for expression analyses in 96-well plates would be optimal to avoid clonal variation observed in *K. phaffii* and to make sure that it is the transcription factor and not the integration locus responsible for the desired effect of derepressed expression of our gene of interest. For the rescreening, a few representative clones are chosen depending on the screening data analysis and streaked out for single colonies to avoid mixed populations. The chosen clones are then cultivated in at least triplicates to be able to verify the previous results.
11. Optionally, the inoculated DWP can be duplicated after a few hours to obtain identical plates that can be treated differently (e.g., compare effects of glycerol feeding and methanol induction). This was done in the experiments performed for this protocol.
12. Glycerol can be added to prolong the derepression effect, and methanol can be added to be able to compare the methanol-dependent with the methanol-free system.
13. Cultivation and screening of the *K. phaffii* TF-OE strains will vary depending on the protein of interest expressed under the P_{AOX1}. The protein of interest can be screened, e.g., for its activity, fluorescence, or amount of protein expressed.
14. Measuring optical density and eGFP fluorescence can be done simultaneously in the same microtiter plate if done correctly. A

suitable microtiter plate as well as spectrophotometer with the correct program is necessary. Additional shaking (10 s, 500 rpm) was performed before every plate reading. The gain was consistently set at 1100 and the focal height at 5.6 mm (depends on the microtiter plate used).

15. The full effect of derepressed overexpression of transcription activators needs to be evaluated in controlled bioreactor experiments where growth rates in carbon source-limited cultivations can be well controlled and optimized. Therefore, benefits for expression are usually bigger in bioreactor experiments compared to the small-scale experiments.

Acknowledgments

TS received funding by the FFG Femtech program.

Competing Interest Statement The authors have no conflicts of interest to declare that are relevant to the content of this chapter.

References

1. Vogl T, Hartner FS, Glieder A (2013) New opportunities by synthetic biology for biopharmaceutical production in Pichia pastoris. Curr Opin Biotechnol 24(6):1094–1101. https://doi.org/10.1016/j.copbio.2013.02.024
2. Vogl T, Glieder A (2013) Regulation of Pichia pastoris promoters and its consequences for protein production. New Biotechnol 30(4): 385–404. https://doi.org/10.1016/j.nbt.2012.11.010
3. Vogl T et al (2018) Methanol independent induction in Pichia pastoris by simple derepressed overexpression of single transcription factors. Biotechnol Bioeng 115(4): 1037–1050. https://doi.org/10.1002/bit.26529
4. Schotte P, Dewerte I, Groeve M, Keyser S, Brabandere V, Stanssens P (2016) Pichia pastoris MutS strains are prone to misincorporation of O-methyl-l-homoserine at methionine residues when methanol is used as the sole carbon source. Microb Cell Factories 15(1): 98. https://doi.org/10.1186/s12934-016-0499-2
5. Ruth C et al (2010) Variable production windows for porcine trypsinogen employing synthetic inducible promoter variants in Pichia pastoris. Syst Synth Biol 4(3):181–191. https://doi.org/10.1007/s11693-010-9057-0
6. Takagi S, Tsutsumi N, Terui Y, Kong XY, Yurimoto H, Sakai Y (2019) Engineering the expression system for Komagataella phaffii (Pichia pastoris): an attempt to develop a methanol-free expression system. FEMS Yeast Res 19(6). https://doi.org/10.1093/femsyr/foz059
7. Chang C-H, Hsiung H-A, Hong K-L, Huang C-T (2018) Enhancing the efficiency of the Pichia pastoris AOX1 promoter via the synthetic positive feedback circuit of transcription factor Mxr1. BMC Biotechnol 18(1):81. https://doi.org/10.1186/s12896-018-0492-4
8. Wang X et al (2016) Mit1 transcription factor mediates methanol signaling and regulates the alcohol oxidase 1 (AOX1) promoter in Pichia pastoris. J Biol Chem 291(12):6245–6261. https://doi.org/10.1074/jbc.M115.692053
9. Lin-Cereghino GP et al (2006) Mxr1p, a key regulator of the methanol utilization pathway and peroxisomal genes in Pichia pastoris. Mol Cell Biol 26(3):883–897. https://doi.org/10.1128/mcb.26.3.883-897.2006
10. Yurimoto H, Oku M, Sakai Y (2011) Yeast methylotrophy: metabolism, gene regulation and peroxisome homeostasis. Int J Microbiol

2011:1. https://doi.org/10.1155/2011/101298

11. Young ET, Dombek KM, Tachibana C, Ideker T (2003) Multiple pathways are co-regulated by the protein kinase Snf1 and the transcription factors Adr1 and Cat8. J Biol Chem 278(28): 26146–26158. https://doi.org/10.1074/jbc.M301981200
12. Sasano Y, Yurimoto H, Yanaka M, Sakai Y (2008) Trm1p, a Zn(II)2Cys6-type transcription factor, is a master regulator of methanol-specific gene activation in the methylotrophic yeast Candida boidinii. Eukaryot Cell 7(3): 527–536. https://doi.org/10.1128/EC.00403-07
13. Zutphen TV et al (2010) Adaptation of Hansenula polymorpha to methanol: a transcriptome analysis. [Online]. Available: http://www.biomedcentral.com/1471-2164/11/1
14. Lin-Cereghino J et al (2005) Condensed protocol for competent cell preparation and transformation of the methylotrophic yeast Pichia pastoris. BioTechniques 38(1):44–48
15. Fischer JE, Hatzl AM, Weninger A, Schmid C, Glieder A (2019) Methanol independent expression by Pichia pastoris employing derepression technologies. J Vis Exp 2019(143). https://doi.org/10.3791/58589
16. Waterham HR, Digan ME, Koutz PJ, Lair SV, Cregg JM, Bernardi A (1997) Isolation of the Pichia pastoris glyceraldehyde-3-phosphate dehydrogenase gene and regulation and use of its promoter. Gene 186:37–44
17. Vogl T et al (2018) Engineered bidirectional promoters enable rapid multi-gene co-expression optimization. Nat Commun 9:3589. https://doi.org/10.1038/s41467-018-05915-w

Part V

Pichia pastoris Expression Strains

Chapter 20

Selective Cultivation of Methylotrophic *Pichia pastoris* Laboratory and Natural Strains

Sarah Gangl, Marina Schmid, Maik Hilgarth, and Katharina Ebner

Abstract

During routine laboratory work, *Komagataella phaffii* (previously *Pichia pastoris*) is usually cultivated on rich medium agar plates or in rich or minimal salt-based liquid media, allowing optimal growth within a short time. However, to screen samples containing a diversity of microorganisms (e.g., wastewater or natural environmental samples) for the presence of viable *K. phaffii* or to aim for the isolation of novel strains, enhanced selective pressure is indispensable. In this chapter, we present a selective medium based on a chemically defined medium composition for *K. phaffii* in combination with methanol as the sole carbon source and a combination of antibiotics, which together suppress the growth of diverse bacteria. This selective medium is intended to enable an enhanced selective isolation and cultivation of the methylotrophic yeast *K. phaffii*, especially from samples containing a high microbiological background.

Key words *Pichia pastoris*, *Komagataella phaffii*, Selective growth, Microbial background, Methanol

1 Introduction

In the context of routine laboratory work, *Komagataella phaffii* (previously *Pichia pastoris*) isolates are commonly cultivated on YPD (yeast extract, peptone, dextrose) agar plates. For research and production strains carrying specific resistance markers, antibiotics can be added to maintain the selective pressure (e.g., for episomal vectors). Alternatively, minimal salt-based media are used, allowing auto-selection or selection based on complemented auxotrophies. Rich cultivation medium, e.g., YPD, is not applicable for selective growth of *K. phaffii* cells in samples taken from complex natural environments containing high levels of microbial background, since a great variety of microorganisms is capable of forming colonies on YPD agar plates. To promote the selective growth of *K. phaffii*, the media composition needs to be restrictive, for instance, by providing a less common carbon source. Besides dextrose, *K. phaffii* cells are capable of utilizing a variety of

Xavier Garcia-Ortega et al. (eds.), *Pichia pastoris: Methods and Protocols*, Methods in Molecular Biology, vol. 2697,
https://doi.org/10.1007/978-1-0716-4779-0_20,

additional carbon sources for cellular growth (e.g., glycerol, sorbitol, methanol, etc.) [1]. The use of methanol as the sole carbon source was once the main aspect for this methylotrophic yeast to become famous during the last century when *K. phaffii* was used to produce single-cell protein for animal feed by Philips Petroleum [2]. Later, the use of the inducible *AOX1* (alcohol oxidase 1) promoter for recombinant protein production caused the formation of three (artificial) methanol utilization (Mut) phenotypes in *K. phaffii* laboratory strains: Mut^+ (intact *AOX1* and *AOX2*, growth on methanol), Mut^s (intact *AOX2*, slow growth on methanol), and Mut^- (no growth on methanol) [3]. Nowadays, the majority of production strains exhibit Mut^+ and Mut^s phenotypes, and even Mut^- is known to show some residual utilization of methanol. The utilization of methanol as the sole carbon source is generally limited to methylotrophic microorganisms, resulting in methanol as the ideal carbon source for the promotion of the selective growth of *K. phaffii*. For many other organisms, the presence of an elevated methanol concentration is either lethal or suppresses their growth significantly. Nevertheless, to detect traces of *K. phaffii* cells in samples associated with a high microbial background, the selective medium not only needs to favor the growth of *K. phaffii* over the growth of other microorganisms, but it also has to prevent the latter. Several antibiotics, which are frequently used in microbiological experiments, are known to have no detectable effect on *K. phaffii* cells [4]: Ampicillin is a beta-lactam antibiotic inhibiting bacterial peptidoglycan structure formation [5]; kanamycin is an aminoglycoside antibiotic mainly targeting protein synthesis mechanisms by binding the 16S rRNA of the 30S ribosomal unit [6], and chloramphenicol is able to inhibit bacterial peptidyl transferase activity by binding the 50S ribosomal unit [7]. Furthermore, Gentamicin (Gentamycin) is an aminoglycoside antibiotic interacting with ribosomal subunits leading to misreading of mRNA and inhibition of protein biosynthesis [8]. The combination of chemically defined minimal medium (ddH_2O, yeast nitrogen base, potassium phosphate buffer, biotin, and agarose) with methanol as the sole carbon source and antibiotics, which together suppress the growth of diverse bacteria, reduces the growth of other microorganisms than *K. phaffii* drastically compared to YPD (Fig. 1) and therefore supports the detection of viable *K.phaffii* cells in samples with high microbial background.

2 Materials

2.1 Media Composition

1. ddH_2O.
2. Agarose.

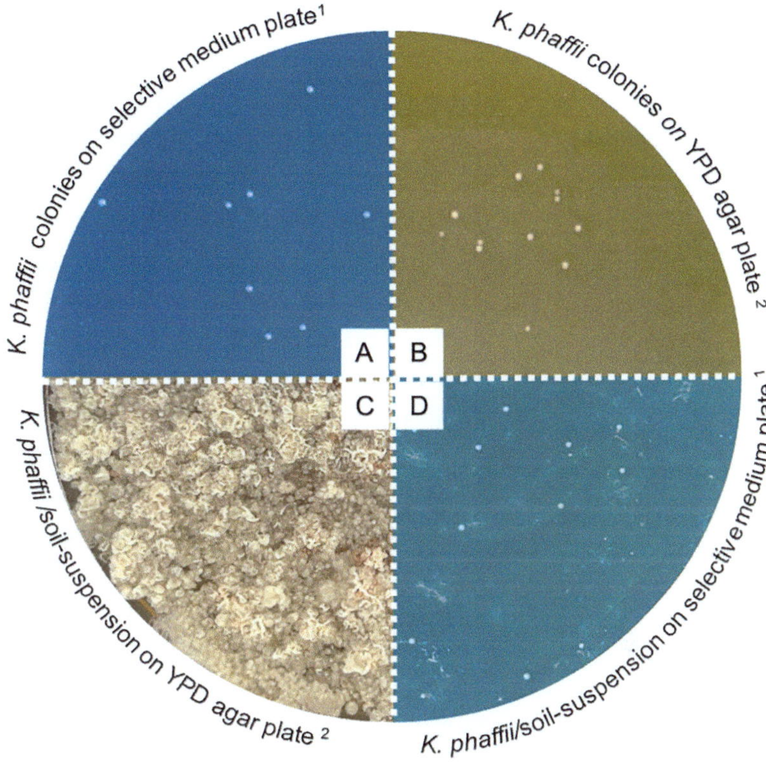

Fig. 1 Detection of *K. phaffii* in samples containing a high microbial background. (**a**) K. *phaffii* NRRL Y-11430 plated on a selective medium plate. (**b**) *K. phaffii* NRRL Y-11430 plated on a YPD agar plate. (**c**) *K. phaffii* NRRL Y-11430 in a mixture with 170 mg/mL soil plated on a YPD agar plate. (**d**) *K. phaffii* NRRL Y-11430 in a mixture with 170 mg/mL soil plated on a selective medium agar plate. 1 selective medium plates were incubated for 120 h at 28 °C. 2 YPD agar plates were incubated for 48 h at 28 °C

3. 13.4% w/v yeast nitrogen base stock solution.
4. 1 M potassium phosphate buffer, pH 6.
5. 0.02% (w/v) biotin stock solution.
6. 100 mg/mL water-based ampicillin stock solution.
7. 35 mg/mL methanol-based chloramphenicol stock solution.
8. 50 mg/mL water-based gentamicin stock solution.
9. 100 mg/mL water-based kanamycin stock solution.

2.2 Additional Equipment

1. Steam sterilizer.
2. Incubator.
3. Petri dishes.
4. Drigalski spatula.

3 Methods

3.1 Selective Medium Preparation

The selective medium is prepared as follows. We recommend preparing the stock solutions separately (autoclaved or sterile filtered). The medium composition is summarized in Table 1.

1. Autoclave 674 mL of ddH_2O containing 10 g of agarose.
2. When cooled down to a processable temperature, add 200 mL of sterile 1 M potassium phosphate buffer (pH 6).
3. Add 100 mL of the sterile yeast nitrogen base solution.
4. Add 2 mL of the sterile biotin stock solution.
5. Add 1 mL of each antibiotic stock solution (ampicillin, chloramphenicol, gentamicin, and kanamycin) (*see* **Note 1**).
6. After the medium has cooled down below 60 °C, methanol can be added under constant stirring (*see* **Notes 1** and **2**).
7. Colorants might be added to the medium (*see* **Note 3**).
8. Pour the medium into Petri dishes and wait until the medium is solidified (*see* **Note 4**).

Table 1
Preparation of 1 L selective medium

For 1 L selective medium heat sterilize		
674 mL	ddH_2O including	final concentration
10 g	Agarose	1% w/v
Cool down and add (sterile)		
200 mL	1 M potassium phosphate buffer (pH 6)	200 mM
100 mL	10× yeast nitrogen base stock solution	1.34% w/v
Make sure temperature is below 60 °C and add (sterile)		
2 mL	500× biotin stock solution	0.00004% w/v
1 mL	100 mg/mL ampicillin stock solution	0.1 mg/mL
1 mL	35 mg/mL chloramphenicol stock solution	0.035 mg/mL
1 mL	100 mg/mL kanamycin stock solution	0.1 mg/mL
1 mL	50 mg/mL gentamicin stock solution	0.05 mg/mL
20 mL	100% MeOH	2% w/v
*	*Colorant* (e.g., *Brilliant Blue FCF (E 133)*	*

3.2 Detection of Viable K. phaffii *in High Background Samples*

The described medium can be used to favor the growth of methylotrophic yeasts over other microorganisms and can therefore support the estimation of the number of viable *K. phaffii* cells in samples with high microbial background (e.g., wastewater) or intensify the selective growth conditions for critical samples. The medium was verified regarding its growth-promoting properties, i.e., yielding similar cell counts (not significantly different) of selected *K. phaffii* strains compared to a non-selective medium. To screen a sample for potential *K. phaffii* strains, follow this procedure.

1. Mix the sample with water and/or dilute it appropriately (adapted to the expected number of *K. phaffii* cells).
2. Plate the dilutions on the selective medium plates using a sterile Drigalski spatula.
3. If possible, a positive control (the *K. phaffii* strain expected in the sample of interest) can be used to determine the required incubation time of the plates, since it might alter due to the *Mut* phenotype or special growth requirements of a certain strain. Alternatively, the required incubation time necessary to count colony-forming units for individual production strains can be determined preliminarily (*see* **Note 5**).
4. The plates are incubated at 28 °C for at least 120 h (depending on the *Mut* phenotype or other growth-related deficiencies of the *K. phaffii* strain) (*see* **Note 6**).
5. After the incubation, colonies might be verified as the expected strain by using specific primers for colony PCR (*see* **Note 7**).

4 Notes

1. The addition of heat-labile ingredients to hot solutions could reduce their availability (e.g., due to the evaporation of methanol) or their mode of action (e.g., antibiotics) [9].
2. During media preparation, a high local concentration of methanol might occur during the addition process, resulting in precipitation with potassium phosphate buffer [10]. It is recommended to add methanol slowly and under constant stirring to minimize this effect. However, precipitation within the agar might be observed.
3. Since the selective medium is colorless, the addition of colorants (e.g., Brilliant Blue FCF (E 133), *see* Fig. 1) might be useful for illustrated documentation.
4. We recommend preparing the selective media directly before the intended use to avoid evaporation of methanol and reduced function of the antibiotics due to storage.

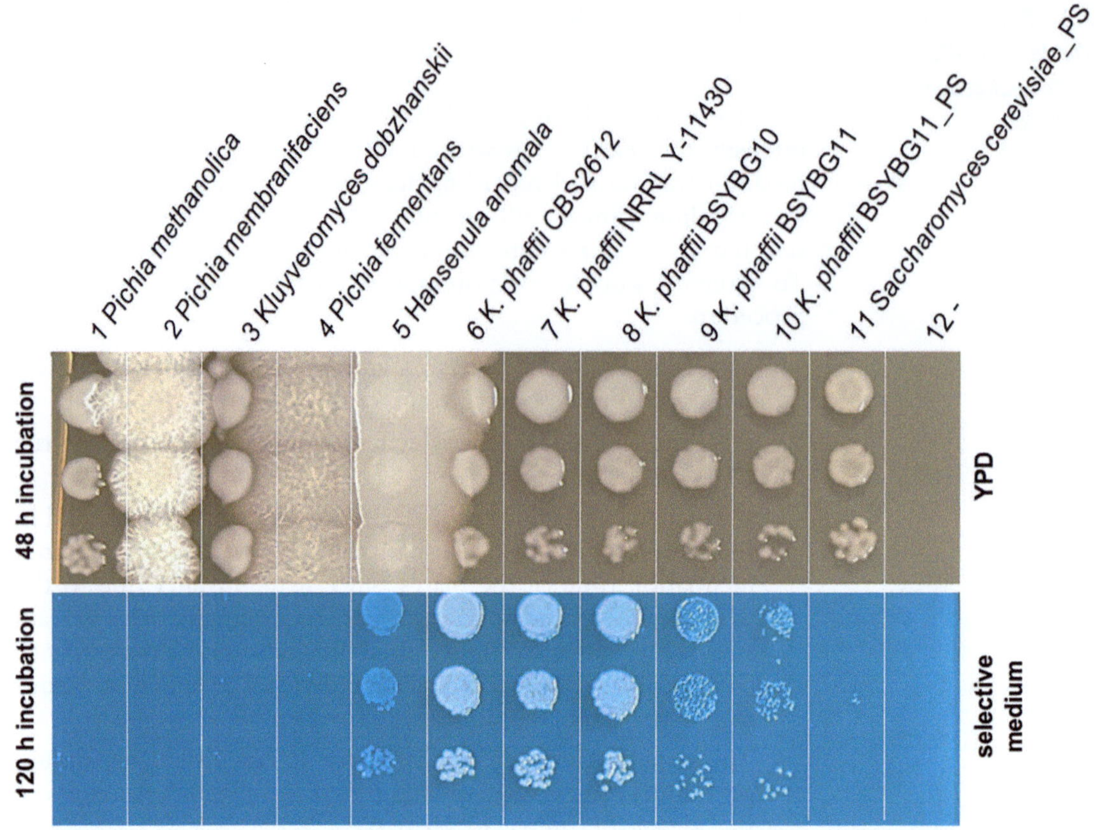

Fig. 2 Growth of different yeast species on YPD and selective medium. (1–5) yeast strains isolated from apple must (methanol-based isolation); (6) and (7) natural isolates of *K. phaffii*; (8) platform strain BSYBG10 (Mut+) derived from NRRL Y-11430; (9) platform strain BSYBG11 (MutS) derived from BSYBG10; (10) recombinant production strain based on BSYBG11; (11) recombinant *S. cerevisiae* production strain; (12) sterile control. (*see* **Note 8**)

5. We recommend testing the required incubation time for the selective cultivation of each individual strain, since deviations can occur due to special requirements or the *Mut* phenotype of a strain.
6. The agar plates should be stored and incubated in well-ventilated areas, since parts of the methanol might evaporate during incubation.
7. The selective growth of *K. phaffii* strains was verified for the here described media using soil and water samples. Nevertheless, contamination might occur due to fungi or other methylotrophic microbes (Fig. 2). If possible, a negative control (a sample not containing *K. phaffii*) is recommended to detect potential contamination. To verify a colony as *K. phaffii* on a DNA level, colony PCR using species- or even strain-specific primers might be useful after selective cultivation.

8. BSYBG10 and BSYBG11 are deposits of the BioGrammatics strains *K. phaffii* BG10 and BG11, respectively, at bisy.

Acknowledgments

Studies leading to this project were performed by bisy GmbH (Hofstätten, AUSTRIA) in collaboration with the acib COMET center and Roche Diagnostics GmbH, Penzberg, Germany. The COMET center, Next Generation Bioproduction, is funded by BMK, BMDW, SFG, Standortagentur Tirol, the Government of Lower Austria, and the Vienna Business Agency in the framework of COMET—Competence Centers for Excellent Technologies. The COMET-Funding Program is managed by the Austrian Research Promotion Agency FFG.

Competing Interest Statement The authors have no conflicts of interest to declare that are relevant to the content of this chapter.

References

1. Bernauer L, Radkohl A, Lehmayer LGK, Emmerstorfer-Augustin A (2021) *Komagataella phaffii* as emerging model organism in fundamental research. Front Microbiol 11. https://doi.org/10.3389/FMICB.2020.607028/FULL
2. Wegner GH (1990) Emerging applications of the methylotrophie yeasts. FEMS Microbiol Rev 87:279–284. https://doi.org/10.1111/j.1574-6968.1990.tb04925.x
3. Hartner FS, Glieder A (2006) Regulation of methanol utilisation pathway genes in yeasts. Microb Cell Fact 5. https://doi.org/10.1186/1475-2859-5-39
4. Haon M, Grisel S, Navarro D, Gruet A, Berrin JG, Bignon C (2015) Recombinant protein production facility for fungal biomass-degrading enzymes using the yeast Pichia pastoris. Front Microbiol 6(SEP). https://doi.org/10.3389/FMICB.2015.01002/ABSTRACT
5. Peechakara BV, Basit H, Gupta M (2023) Ampicillin. In: xPharm: the comprehensive pharmacology reference, pp 1–6. https://doi.org/10.1016/B978-008055232-3.61227-9
6. Faraji R, Parsa A, Torabi B, Withrow T (2006) Effects of kanamycin on the macromolecular composition of kanamycin sensitive Escherichia coli DH5! strain. J Exp Microbiol Immunol 9: 31–38
7. Xaplanteri MA, Andreou A, Dinos GP, Kalpaxis DL (2003) Effect of polyamines on the inhibition of peptidyltransferase by antibiotics: revisiting the mechanism of chloramphenicol action. Nucleic Acids Res 31(17):5074. https://doi.org/10.1093/NAR/GKG686
8. Tangy F, Moukkadem M, Vindimian E, Capmau M-L, Le Goffic F (1985) Mechanism of action of gentamicin components. Eur J Biochem 147(2):381–386. https://doi.org/10.1111/J.1432-1033.1985.TB08761.X
9. Traub WH, Leonhard B (1995) Heat stability of the antimicrobial activity of sixty-two antibacterial agents. J Antimicrob Chemother 35(1):149–154. https://doi.org/10.1093/JAC/35.1.149
10. Schellinger AP, Carr PW, Carr P (2004) Solubility of Buffers in Aqueous-Organic Eluents for Reversed-Phase Liquid Chromatography. LC GC N Am 22. Available: www.chromatographyonline.com

Chapter 21

Transformation of *Pichia pastoris* by Electroporation

Kirill Smirnov and Anton Glieder

Abstract

Transformation is one of the cornerstone techniques of molecular biotechnology, and it is an almost obligatory part of any study related to the investigation of gene functions, the production of recombinant proteins, or the alterations of the phenotypic features in *Komagataella phaffii*. One of the most commonly used transformation methods for delivering foreign DNA to *K. phaffii* cells is electroporation.

This chapter provides a simplified and fast protocol for preparing *K. phaffii* electrocompetent cells and for their transformation by electroporation. In addition, several parameters affecting the transformation efficiency are briefly described.

Key words *Pichia pastoris*, *Komagataella phaffii*, Electroporation, Transformation, Competent cells

1 Introduction

Transformation is the laboratory process of transferring external genetic material to the inner cell space, followed by the inclusion of the material in the genotype of the cell. The laboratory methods build on the natural mechanism used by many bacterial species to increase their genetic variability and/or to acquire metabolic, virulence, and antibiotic resistance genes [1]. In nature, such "horizontal gene transfer" is also common among eukaryotes, and it played and probably is still playing an important role in eukaryotic evolution [2, 3]. The knowledge about natural transformation phenomena was the theoretical basis for the development of laboratory transformation methods for a wide range of organisms, including yeasts like *Komagataella phaffii* (reclassified from *Pichia pastoris*).

Transformation as a laboratory technique consists of two parts—(a) making cells competent, i.e. increasing the cells' ability to uptake foreign genetic material from the environment and (b) transformation itself, triggered by certain conditions, created on purpose by a researcher, i.e. transformation method. Although both parts are interconnected and important, each type of transformation method requires a specific procedure for making cells

Xavier Garcia-Ortega et al. (eds.), *Pichia pastoris: Methods and Protocols*, Methods in Molecular Biology, vol. 2697,
https://doi.org/10.1007/978-1-0716-4779-0_21,

competent. There are three different and widely used transformation methods for *K. phaffii*: (a) a spheroplast method, (b) a chemical method such as the LiCl/LiAc-based method, and (c) electroporation [4–6]. Due to the robustness, simplicity of preparation of competent cells, and the high transformation rate, electroporation is the most widely used *K. phaffii* transformation method [6–11]. For electroporation, a high-intensity external electric field is applied to the cells for a short time (μs). This results in a transient increased permeability of the bilayer cell membrane through pore generation [12, 13].

The below-described procedures are based on the condensed protocol developed for electroporation by Lin-Cereghino et al. [14]; however, several adaptations have been made. One of the key steps in the protocol is treating cells with dithiothreitol (DTT), which reduces disulfide bonds and thereby provokes the partial loss of cell wall mannoproteins. As a result, the cell wall permeability increases [15, 16]. By adjusting some parameters as the amount of added DNA for transformation and/or the number of competent cells, with this protocol, a transformation efficiency of up to 3×10^6 CFU/μg can be achieved [17]. In addition, much time can be saved by replacing the time-consuming fresh growth of cells to a defined cell density with the use of cells from simple overnight cultures or cells grown on agar plates. Although transformation rates might be lower, for most applications, enough transformants will be obtained. For further simplification of the experimental procedure and in comparison to most widely used methods for the preparation of competent cells, most cooling steps can be skipped without a major negative influence on the transformation efficiency.

2 Materials

2.1 Competent Cells' Preparation

1. A suitable *K. phaffii* strain (*see* **Note 1**).
2. A thermomixer that can keep 28 °C and shake 2 mL tubes at 350 rpm.
3. Tabletop centrifuge.
4. 1 M DTT solution, filter sterilized (*see* **Note 2**).
5. BEDS solution (pH 8.3), filter sterilized (*see* **Note 2**).
6. Sterile 2 mL or 1.5 mL reaction tubes.
7. Single-channel microliter pipettes: 100–1000 μL and 20–200 μL volumes.
8. Sterile pipette tips matching the used microliter pipettes.
9. YPD medium: 1% Yeast extract, 2% Peptone, and 2% glucose. Autoclave and store at room temperature.

10. Orbital shaker-incubator.
11. Sterile 250 mL baffled flask(s).
12. Sterile 50 mL conical centrifuge tube(s).
13. Laboratory centrifuge with adapters for 50 mL tubes.

2.2 Transformation of *K. phaffii* by Electroporation

1. A desalted, linearized, or circular DNA fragment (*see* **Note 3**) containing a selectable marker (*see* **Note 4**) and a DNA element with the gene of interest.
2. Electrocompetent *K. phaffii* cells.
3. Thermomixer that can keep 28 °C and shake 2 mL tubes at 350 rpm (*see* **Note 5**).
4. Electroporation cuvettes with 0.2 cm gap (e.g. BioRad 1,652,082).
5. Sterile 2 mL and 1.5 mL reaction tubes.
6. Single-channel microliter pipettes: 100–1000 μL, 20–200 μL, and 1–10 μL volumes.
7. Sterile pipette tips matching the used microliter pipettes.
8. Electroporation device (e.g. BioRad Gene Pulser).
9. YPD: 1% Yeast extract, 2% Peptone, 2% glucose. Autoclave and store at room temperature.
10. 1 M sorbitol: 182 g/L of sorbitol dissolved in deionized water. Autoclave and store at room temperature.
11. YPDS: mix equal volumes of YPD and 1 M sorbitol (*see* **Note 6**).
12. Plastic Petri dishes: 94 mm × 16 mm with vents.
13. YPD agar plates (*see* **Note** 7) containing a selection agent: YPD with 1.5% (w/v) of agar-agar. Autoclave. When lukewarm, add a selection agent, usually an antibiotic, to a final concentration (*see* **Note 8**) and pour the medium into Petri dishes, and let solidify.

3 Methods

The present method describes the preparation of competent cells from liquid and agar plate *K. phaffii* cultures and their subsequent transformation. Due to the larger amount of biomass, the liquid culture as a starting point is recommended in scenarios with one strain that is to be transformed with multiple genetic constructs (>13). As a general rule, **steps 1**, **3**, **5**, and **9–11** should be performed within the sterile area. The protocol that uses agar plate cultures (Fig. 1) as the starting point is recommended when one or several strains are transformed with a few genetic constructs

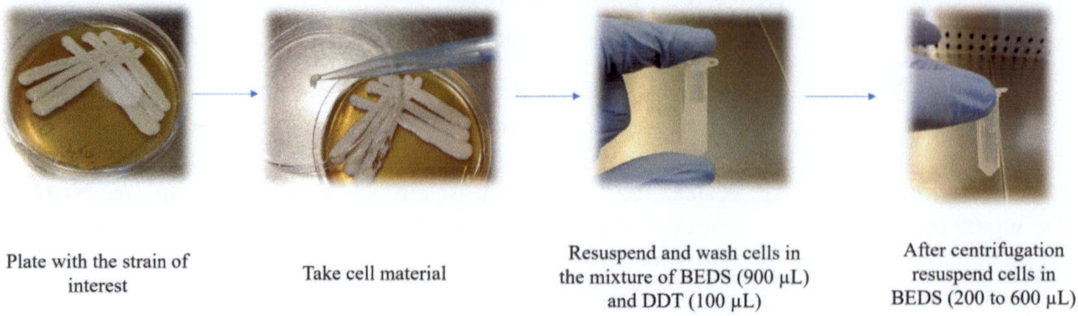

Fig. 1 The illustration of the main steps in the preparation of competent cells from plates

(e.g. five constructs per strain). Also in this case, as a general rule, **steps 1** and **4–6** should be performed within the sterile area (*see* **Note 9**). The transformation protocol is identical for all different options of this protocol for preparing the electrocompetent *K. phaffii* cells.

3.1 Preparing K. phaffii *Electrocompetent Cells from Liquid Cultures*

1. Inoculate a suitable *K. phaffii* strain in a 250 mL baffled flask with 50 mL of YPD.
2. Cultivate cells overnight at 28 °C, 130 rpm to obtain the cell culture with OD_{600} between 0.8 and 1.5 (*see* **Notes 10** and **11**).
3. Transfer the cell culture into a 50 mL tube.
4. Centrifuge the cells at 500 × g for 5 min.
5. Discard supernatant.
6. Resuspend the cell pellet in a mixture of 9 mL of BEDS and 1 mL of DDT.
7. Incubate the suspension at 28 °C, 130 rpm for 10 min in an orbital shaker.
8. Centrifuge the suspension at 500 × g for 5 min.
9. Discard the supernatant.
10. Resuspend the cell pellet in 1 mL of BEDS.
11. Prepare aliquots of 80 μL in 2 mL tubes.
12. Cells are ready for electroporation (*see* **Note 12**).

3.2 Preparing K. phaffii *Electrocompetent Cells from Agar Plate Cultures*

1. Take an agar plate culture of a suitable *K. phaffii* strain (*see* **Note 13**).
2. Use a 1 mL volume pipette tip (or inoculation loop) to harvest cell biomass (Fig. 1).
3. Resuspend cells in 1 mL of a mixture of BEDS (900 μL) and DTT (100 μL) in a 2 mL tube. The obtained resuspension usually has the desired OD_{600} between 10 and 16.

4. Incubate the resuspension in a thermomixer at 28 °C, 350 rpm for 10 min (*see* **Note 5**).
5. Centrifuge the resuspension at 500 × g for 5 min.
6. Discard the supernatant.
7. Resuspend the cell pellet in 200 mL of BEDS. If more aliquots for transformation are needed, the cell pellet can be resuspended in 640 mL of BEDS. Prepare aliquots of 80 μL in 2 mL tubes.
8. Cells are ready for electroporation.

3.3 Transformation of K. phaffii by Electroporation

1. Mix DNA fragments (*see* **Note 3**) dissolved in deionized water (*see* **Note 14**) with 80 μL of *K. phaffii* competent cells.
2. Incubate the mixture for 2–3 min at room temperature.
3. Transfer the cell-DNA mixture into an electroporation cuvette.
4. Electroporate (2 KV; resistance 200 Ω; capacitance, 25 μF, or follow the instructions of your device manufacturer).
5. Immediately after the electroporation, add 1 mL of the YPDS on top of the electroporated cells in the cuvette.
6. Transfer the cell mixture from the cuvette into a 2 mL tube.
7. Regenerate the transformed cells for at least 2 h at 28 °C, 350 rpm in a thermomixer (*see* **Note 15**).
8. Plate 100 μL of the regenerated cells on YPD agar plates (or other appropriate plates) containing a suitable selection agent (*see* **Note 16**).
9. Incubate plates at 28 °C until colonies are visible. It usually requires 2–3 days.

4 Notes

1. The described protocols have been tested with *K. phaffii* (BSY)BG10, (BSY)BG11, X-33, and (BSY)BG10 Δ*ku70* strains, but should equally be applicable for most other *K. phaffii* and *K. pastoris* strains.
2. It is recommended to aliquot BEDS and DTT solutions in 15 mL and 1.5 mL tubes, respectively, and store them at − 20 °C (the minimal shelf life under those conditions is 6 months). DTT and BEDS compositions are shown in Tables 1 and 2, respectively.
3. The amount of DNA used depends on the experiment design. According to our data, around 80 ng of linearized (or circular) plasmid is sufficient to obtain several hundred transformants. In case more transformants will be required, more vector DNA

Table 1
DTT composition

Component	Quantity	Manufacturer
1,4-Dithiothreitol (DTT)	154.2 g/L	Carl Roth GmbH, Germany, Article-nr. 6908.2
Deionized water	~1000 mL[a]	–

[a]This volume is given only as an example. Typically, 50 1-mL aliquots of 1.0 M DTT solution are enough for several months

Table 2
BEDS composition

Component	Quantity	Manufacturer
Bicine	1.632 g/L	Carl Roth GmbH, Germany, Article-nr. 9162.1
D-Sorbitol	182 g/L	Carl Roth GmbH, Germany, Article-nr. 6213.2
Deionized water	~920 mL[a]	–
Ethylene glycol	30 mL/L	Carl Roth GmbH, Germany, Article-nr. 6881.1
Dimethylsulfoxide (DMS)	50 mL/L	Carl Roth GmbH, Germany, Article-nr. 7029.2

[a]Add 900 mL of water, adjust pH to 8.3 with 2 M NaOH. After adding ethylene glycol and DMSO bring up the volume to 1 L, and check the pH of the solution again

(1–5 μg) should be used, although the relative number of transformants per μg of DNA will most likely be lower.

4. Typical selectable markers for *K. phaffii* are genes providing resistance to antibiotics or restoring the ability of a strain to effectively grow on media with certain composition for antibiotic free selection (e.g. the *HIS4* gene can be a selectable marker for *K. phaffii* Δ*his4* strain on mineral medium, and the *GUT1* gene can be used as a selectable marker for *K. phaffii* Δ*gut1* strain on medium containing glycerol as a sole carbon source).
5. If a thermomixer is not accessible, it is possible to use a shaker-incubator.
6. Although *K. phaffii* can utilize sorbitol, in the YPDS mixture, sorbitol serves not as a carbon source but as an osmoprotector.
7. Complex media (e.g. YPD) is normally used for the selection of transformants on antibiotics because some of the antibiotics (e.g. Zeocin) are less effective on minimal mineral media. Normally, minimal mineral media are used for the selection of transformants whose prototrophy is restored due to transformation with recombinant DNA.

Table 3
Antibiotic concentrations for the selection of *K. phaffii* transformants

Antibiotic	Recommended concentration in YPD medium
Zeocin	100 μg/mL[a]
Hygromycin	300 μg/mL
G418 (Geneticin)	300 μg/mL

[a]Concentration can be increased to stimulate multicopy integration

8. The recommended concentrations of the widely used antibiotic for the selection of *K. phaffii* transformants are shown in Table 3.
9. In many protocols describing the transformation of *K. phaffii* by electroporation, there are recommendations to keep cells on ice and use ice-cold solutions. However, according to our experience, performing all transformation steps at room temperature does not dramatically affect transformation efficiency. Thus, the use of ice can be considered as optional.
10. These values are approximate. It was observed that competent cells prepared from the cell culture with OD_{600} 5.0 did not have noticeably decreased transformation efficiency, and this observation is in accordance with data from Wu et al. [17].
11. According to our experience, the (BSY)BG10 cell culture with OD_{600} around 0.04 needs around 16 h of cultivation at 28 °C, 130 rpm to reach the optical density of 1. Nevertheless, it depends on the individual characteristics of the used strain.
12. Competent cells can be stored at −80 °C. However, freezing might decrease transformation efficiency.
13. In general, it is recommended to use a plate with fresh cell culture (from 2 to 5 days). However, the drastic difference in transformation efficiency from cells that had been stored in a fridge for 1 month has not been observed.
14. The excess of salts in the transformation mixture can cause an arc. Also, it is recommended that the volume of the DNA mixture should not exceed 10% of the volume of competent cells.
15. Regeneration provides time for transformants to express antibiotic resistance genes. In case of using auxotrophic selectable markers, cells are preferably plated immediately after the transformation. It is also possible to incubate the transformed cells at 28 °C without shaking.
16. If more transformants are needed, the regenerated cells can be centrifuged at 7000 × g for 2 min. After that, decant the

supernatant and resuspend the cell pellet in the rest of the liquid. Plate the resuspension on a suitable agar plate.

Acknowledgments

This work was performed as part of the Christian Doppler Laboratory. The financial support from the Austrian Federal Ministry for Digital and Economic Affairs and the National Foundation for Research, Technology and Development and from the contributing company partners Novo Nordisk, BioGrammatics, SeSam Biotech, Zeta, and SolarBiotech is gratefully acknowledged.

Competing Interest Statement The authors have no conflicts of interest to declare that are relevant to the content of this chapter.

References

1. Winter M, Harms K, Johnsen PJ, Buckling A, Vos M (2023) Testing for the fitness benefits of natural transformation during community-embedded evolution. Microbiology (Reading) 169(8):1–10. https://doi.org/10.1099/mic.0.001375
2. Keeling PJ, Palmer JD (2008) Horizontal gene transfer in eukaryotic evolution. Nat Rev Genet 9(8):605–618. https://doi.org/10.1038/nrg2386
3. Gladyshev EA, Meselson M, Arkhipova IR (2008) Massive horizontal gene transfer in bdelloid rotifers. Science 320(5880):1210–1213. https://doi.org/10.1126/science.1156407
4. Invitrogen Corporation (2010) User manual – EasySelect™ Pichia expression kit. Cat K1740-01(25):95
5. Cregg JM, Barringer KJ, Hessler AY, Madden KR (1985) Pichia pastoris as a host system for transformations. Mol Cell Biol 5(12):3376–3385
6. Kumar S, Mannil A, Mutturi S (2020) Modified chemical method for efficient transformation and diagnosis in Pichia pastoris. Protein Expr Purif 174(March):105685. https://doi.org/10.1016/j.pep.2020.105685
7. Zavec D et al (2021) Beyond alcohol oxidase: the methylotrophic yeast Komagataella phaffii utilizes methanol also with its native alcohol dehydrogenase Adh2. FEMS Yeast Res 21(2):1–13. https://doi.org/10.1093/femsyr/foab009
8. Miele R, Barra D, Bonaccorsi di Patti MC (2007) A GATA-type transcription factor regulates expression of the high-affinity iron uptake system in the methylotrophic yeast Pichia pastoris. Arch Biochem Biophys 465(1):172–179. https://doi.org/10.1016/j.abb.2007.05.020
9. Heistinger L, Gasser B, Mattanovich D (2018) Creation of stable heterothallic strains of Komagataella phaffii. Mol Cell Biol 38(2):1–16
10. Weiss F, Requena-Moreno G, Pichler C, Valero F, Glieder A, Garcia-Ortega X (2024) Scalable protein production by Komagataella phaffii enabled by ARS plasmids and carbon source-based selection. Microb Cell Factories 23(1):1–16. https://doi.org/10.1186/s12934-024-02368-3
11. Kumar R (2019) Simplified protocol for faster transformation of (a large number of) Pichia pastoris strains. Yeast 36(6):399–410. https://doi.org/10.1002/yea.3383
12. Weaver JC, Chizmadzhev YA (1996) Theory of electroporation: a review. Bioelectrochem Bioenerg 41(2):135–160. https://doi.org/10.1016/S0302-4598(96)05062-3
13. Prasanna GL, Panda T (1997) Electroporation: basic principles, practical considerations and applications in molecular biology. Bioprocess Eng 16(5):261–264. https://doi.org/10.1007/s004490050319
14. Lin-Cereghino J et al (2005) Condensed protocol for competent cell preparation and transformation of the methylotrophic yeast Pichia

pastoris. BioTechniques 38(1):44–48. https://doi.org/10.2144/05381BM04

15. Zlotnik H, Pilar Fernandez M, Bowers B, Cabib E (1984) Saccharomyces cerevisiae mannoproteins form an external cell wall layer that determines wall porosity. J Bacteriol 159(3): 1018–1026. https://doi.org/10.1128/jb.159.3.1018-1026.1984
16. Meilhoc E, Masson JM, Teissié J (1990) High efficiency transformation of intact yeast cells by electric field pulses. Bio/Technology 8(3): 223–227. https://doi.org/10.1038/nbt0390-223
17. Wu S, Letchworth GJ (2004) High efficiency transformation by electroporation of Pichia pastoris pretreated with lithium acetate and dithiothreitol. BioTechniques 36(1):152–154. https://doi.org/10.2144/04361dd02

Chapter 22

High-Throughput Transformation of Chemical Competent *Pichia pastoris* Cells Using Lithium Acetate/Chloride and Carrier DNA

Kirill Smirnov, Amina Hasanagic, Lukas Rieder, and Anton Glieder

Abstract

One of the most commonly used *Komagataella phaffii* transformation protocols is electroporation. Electroporation, i.e., applying an electric field to cells with the consequence that the cell membrane becomes permeable, is considered a fast and efficient method for introducing recombinant DNA into *K. phaffii* cells. The transformation method provides, depending on the quality of the electrocompetent cells, transformation efficiency between 2.1×10^4 and 1.1×10^5 CFU/μg of linear DNA (Kumar, *Yeast* 36(6):399–410, 2019). However, electroporation as a transformation method is hardly adaptable to a 96-well format and thus inconvenient for performing several transformations in parallel.

In this chapter, we describe protocols to produce chemically competent *K. phaffii* cells using lithium acetate/chloride and the subsequent transformation of the cells supported by the use of carrier DNA. Although this transformation procedure is more laborious, time-consuming, and potentially suffers from a lower transformation efficiency (1×10^2 and 1×10^3 CFU/μg DNA) than the conventional electroporation method, the lithium acetate/chloride transformation protocol is beneficial when very high transformation rates such as in the case of library generation are not essential.

Key words *Pichia pastoris*, *Komagataella phaffii*, High-throughput transformation, Lithium acetate, Lithium chloride, Transformation

1 Introduction

The term "transformation" was first used by Griffith to describe heritable changes in certain characteristics of *Pneumococcus* [1, 2]. These days, "transformation" describes the general process by which an organism takes up exogenous DNA. Although the process of transformation was discovered as a natural phenomenon, it was soon adapted to laboratory usage and is today actively used by every molecular biological laboratory as a standard procedure for introducing recombinant DNA into a variety of hosts.

Xavier Garcia-Ortega et al. (eds.), *Pichia pastoris: Methods and Protocols*, Methods in Molecular Biology, vol. 2697,
https://doi.org/10.1007/978-1-0716-4779-0_22,

There are several protocols to introduce DNA of interest into *Komagataella phaffii* (reclassified from *Pichia pastoris*), including electroporation, spheroplast generation, and alkali cation method. Each of these methods has its own advantages and drawbacks determined by the procedures themselves [3]. Thus, the preferred transformation method is often based on the purpose of the research, availability of the laboratory equipment, and personal background of the researcher. For example, mutant library creation requires high transformation efficiency, and therefore electroporation, which has transformation efficiency between 2.1×10^4 and 1.1×10^6 CFU/μg DNA, is recommended for usage [3, 4]. On the other hand, the electroporation protocol is difficult to adjust to a high-throughput format, and it is a limiting factor of that method. 96-well electroporators are available, but in our hands resulted in reduced transformation rates (unpublished personal observations). In contrast to electroporation, chemical transformation by lithium chloride (LiCl) and lithium acetate (LiAc) can be easily performed in 96-well format and fully automatized using liquid handling robotics. Nevertheless, chemical transformation methods suffer from lower transformation efficiencies, as based on our experience, the number of transformants is in the range between 1×10^2 and 1×10^3 CFU/μg DNA and thus about 2 orders of magnitude lower than the transformation efficiency achieved by electroporation. However, for many applications, only a few transformants for each genetic construct are needed.

The first work showing that plasmid DNA uptake by yeast cells can be stimulated by treating the cells with specific monovalent alkali cations (N^+, K^+, Rb^+, Cs^+, and Li^+) and polyethylene glycol (PEG) was published in 1983 [5] and successfully adapted to generate and screen *K. phaffii* libraries expressing cytochrome P450 saturation mutagenesis libraries [6]. Inspired by this work, the transformation procedure described in this chapter is based on treating the yeast cells with LiAc or LiCl to make them competent. According to our experience, the usage of LiCl or LiAc as treating agents results in a similar rate of transformation efficiency and can thus be considered as equally good. Regarding how the transformations of LiCl or LiAc competent cells occur, it is presumed that DNA enters the cell through holes and pores generated by treatment with metal ions (e.g. Li^+) and PEG [5]. In addition, the uptake of the recombinant DNA by the cells is assisted by carrier DNA such as degraded Salmon sperm DNA. It is proposed that the yeast cell wall binds DNA molecules and an excessive amount of DNA carrier, which is bound to the cell wall, facilitates the uptake of recombinant DNA possible [1].

To sum up, the chemical transformation of *K. phaffii* was known for a long time, but not as popular as electroporation, which is most likely caused by the lower transformation efficiency and laborious procedure for competent cells preparation. Today, there is a high need for high-throughput technologies, including *K. phaffii* transformation. To overcome the current bottlenecks, this chapter provides an adapted, robust, and efficient protocol for the LiAc/LiCl-mediated high-throughput transformation of *K. phaffii* cells for the 96-well format.

2 Materials

1. YPD Medium:1% yeast extract, 2% peptone, 2% dextrose.
2. LiCl-TE-/LiAc-TE-transformation buffer: 40.0% (w/v) PEG 4000, 100 mM LiCl/LiAc, 1 mM EDTA, 10.0 mM DTT, 5% (v/v) DMSO, 20 mM Tris-HCl, pH 8.0 (*see* **Note 1**).
3. 96 deep-well plate (DWP).
4. "Transformation ready" DNA (*see* **Note 2**).
5. Carrier DNA (e.g. UltraPure™ Salmon Sperm DNA Solution from ThermoFisher Scientific, catalog number 15632011).
6. Benchtop centrifuge with rotors and adaptors for centrifugation of microcentrifuge tubes and deep-well plates.
7. Temperature-controlled shaking incubator.
8. Selective agar plates.
9. Deionized sterile H_2O (ddH_2O).

3 Methods

3.1 Preparation of LiAc/LiCl Chemical Competent Cells

1. Use a single colony of any *K. phaffii* laboratory strain of interest to inoculate a preculture in a 50 mL centrifugation tube filled with 5 mL YPD medium and incubate at 28 °C for 4–5 h under moderate shaking at 130 rpm.
2. Use the preculture from **step 1** to inoculate 50 mL YPD in a sterile 50 mL baffled Erlenmeyer flask sealed with a cotton cloth to a starting cell density (OD_{600}) of ca. 0.015. Grow the main culture overnight (14–16 h) at 28 °C at 130 rpm. The final OD_{600} should be between 1.5 and 5.
3. Transfer the main culture into a new, sterile 50 mL centrifuge tube and harvest the cells by centrifugation at 500 g for 5 min at room temperature.
4. Pour off the supernatant and resuspend the pellet in the same amount of ddH_2O (i.e., if the volume of the main culture was 50 mL, use 50 mL ddH_2O).
5. Centrifuge the cells at 500 g for 5 min at room temperature.
6. Repeat **steps 4** and **5**.

7. Decant the supernatant and resuspend the cell pellet, depending on the transformation procedure, in either 5 mL of LiCl-TE or LiAc-TE-transformation buffer.

3.2 Transformation Procedure

1. Prepare a 96 deep-well plate containing 10 μg (can be increased to up to 100 μg to improve transformation efficiency) of denatured (by heating in thermocycler 5 min at 95 °C) salmon sperm DNA and ~200 ng or 10–50 ng of linearized/circular plasmid DNA, respectively, per well (*see* **Note 3**).
2. Add 200 μL of LiAc/LiCl competent cells to each well containing plasmid and carrier DNA.
3. Incubate the deep-well plate for 2 h at 28 °C and 320 rpm.
4. For regeneration of the transformed cells, add 800 μL of YPD per well to the DWP.
5. Incubate the DWP again for 2 h at 28 °C and 320 rpm to initiate cell growth.
6. Centrifuge the DWP at room temperature for 5–7 min at 500 g and remove the supernatant by turning the plate upside down (*see* **Note 4**) followed by quick shaking (*see* **Note 5**).
7. Resuspend the cell pellet in the remaining media by pipetting up and down.
8. Plate the whole cell suspension from one well (~50 μL) on one agar plate containing the appropriate selection pressure and incubate according to your requirements.
9. Alternatively, an aliquot of 100–300 μL of the regeneration mix can be directly used for further growth and induction in deep-well plates, employing the whole transformation mix instead of the more time-consuming and laborious use of single colony transformants for detailed analysis and screening of individual transformants.

4 Notes

1. We recommend using stock solutions of 1 M LiCl/LiAc in deionized water, 1 M DTT in deionized water (this stock should be stored at −20 °C), 0.5 M EDTA (pH 8.0) and 1 M Tris-Cl (pH 8.0) for making transformation buffer. DTT and DMSO have to be added directly before usage (very important). LiCl and LiAc are interchangeable, and the difference in transformation efficiency was not observed. The transformation buffer has to be filter-sterilized.
2. It is possible to use either a linearized plasmid (at least 200 ng or higher, if a high number of transformants is required) or a circular episomal plasmid (at least 50 ng is recommended for

the usage). In general, transformation rates employing circular autonomously replicating plasmid DNAs are about 10–100 times higher than for integrative expression cassettes.

3. We recommend using multichannel pipettes for repetitive pipetting.
4. Make sure to check that the cells are pelleted before turning the plate upside down. A small droplet of medium will remain in each well, which is required for the next step.
5. Avoid cross-contamination between the wells.

Acknowledgments

This work was performed as part of the Christian Doppler Laboratory. The financial support from the Austrian Federal Ministry for Digital and Economic Affairs and the National Foundation for Research, Technology and Development and from the contributing company partners Novo Nordisk, BioGrammatics, SeSam Biotech, Zeta and SolarBiotech is gratefully acknowledged.

Competing Interest Statement The authors have no conflicts of interest to declare that are relevant to the content of this chapter.

References

1. Gietz RD, Woods RA (2001) Genetic transformation of yeast. BioTechniques 30(4):816–831. https://doi.org/10.2144/01304rv02
2. Griffith F (1966) The significance of penumococcal types. J Hyg (Lond) 64(2):129–175. https://doi.org/10.1017/S0022172400040420
3. Kumar R (2019) Simplified protocol for faster transformation of (a large number of) *Pichia pastoris* strains. Yeast 36(6):399–410. https://doi.org/10.1002/yea.3383
4. Weiss F, Requena-Moreno G, Pichler C, Valero F, Glieder A, Garcia-Ortega X (2024) Scalable protein production by *Komagataella phaffii* enabled by ARS plasmids and carbon source-based selection. Microb Cell Factories 23(1):1–16. https://doi.org/10.1186/s12934-024-02368-3
5. Ito H, Fukuda Y, Murata K, Kimura A (1983) Transformation of intact yeast cells treated with alkali cations. J Bacteriol 153(1):163–168. https://doi.org/10.1128/jb.153.1.163-168.1983
6. Geier M et al (2013) Double site saturation mutagenesis of the human cytochrome P450 2D6 results in regioselective steroid hydroxylation. FEBS J 280(13):3094–3108. https://doi.org/10.1111/febs.12270

Chapter 23

High-Efficiency Transformation of Integrative Libraries Under Zeocin Selection in *Pichia pastoris*

Morgane Boone, Hannah Eeckhaut, and Nico Callewaert

Abstract

High-efficiency transformation is often a major bottleneck in the generation of large and diverse libraries. This is especially the case for the yeast *Komagataella phaffii*, where integration in the genome and selection via Zeocin are often desired but negatively affect the transformation efficiency.

This chapter describes an improved transformation protocol for integrative plasmids in *K. phaffii* using Zeocin selection that achieves a high efficiency (10^5 integrative clones per μg DNA) in a time-saving manner. Additionally, we show that by fine-tuning Zeocin concentration, the number of multi-insert transformants can be controlled. This method should be valuable for the construction of large-pooled libraries in *K. phaffii* and can be applied in multiple fields of study.

Key words *Pichia pastoris*, *Komagataella phaffii*, Transformation, Library, Integrative, Zeocin, Liquid selection, Single-insert, Multi-insert

1 Introduction

The ability to perform and assess a very large number of reactions in a single experiment has revolutionized biology, and along with it, biotechnology. The pursuit of large-scale screening efforts in yeast has culminated in specific, high-affinity binders from variant libraries of naïve natural or synthetic antibodies, lectins, and other scaffolds [1–4]; enzymes with improved stability or altered ligand affinities through directed evolution [5]; and strains with enhanced heterologous protein expression potential [6]—to name a few. With high-throughput screening platforms like FACS, microfluidics, and microengraving enabling continuously faster and more efficient screening, the limitation mostly lies in the generation of large and diverse libraries, which is typically constrained by limits in cloning and transformation efficiencies. As the capability to

Morgane Boone and Hannah Eeckhaut contributed equally with all other contributors.

Xavier Garcia-Ortega et al. (eds.), *Pichia pastoris: Methods and Protocols*, Methods in Molecular Biology, vol. 2697,
https://doi.org/10.1007/978-1-0716-4779-0_23,

generate larger libraries, with more diverse and thus more informative search spaces, frequently correlates with a higher likelihood of finding a successful variant [7, 8], high-efficiency transformation is key.

For *Saccharomyces cerevisiae,* multiple high-efficiency transformation protocols exist. The popular Gietz and Schiestl heat shock method with lithium acetate, single-stranded carrier DNA, and PEG reaches transformation efficiencies of up to 10^6 transformants per µg of DNA [9, 10]. Combining electroporation with in vivo recombination cloning has achieved an efficiency of up to 10^8 transformants per µg of DNA [11]. These high-efficiency protocols have been instrumental in the generation of large libraries in many research disciplines, for example, leading to the generation of a 10^{10} naive antibody library [11].

In contrast, the potential for large library screening in *Komagataella* phaffii (previously named *Pichia pastoris*) has yet to be fully explored. One contributing factor is that transformation efficiencies in *K. phaffii* are typically lower than in *S. cerevisiae.* The reasons for this are twofold. First, the vast majority of expression vectors for *K. phaffii* still rely on integration into the genomic DNA. Compared to the episomal CEN/ARS plasmids commonly used in *S. cerevisiae,* this imposes an additional event that has to occur in order to obtain a transformant. Second, the use of the *Sh ble* gene (Zeocin) as a dominant antibiotic selection marker is popular in *K. phaffii* because the same antibiotic resistance cassette can be used in bacteria and *K. phaffii,* and a wild-type yeast host strain can be used. However, the use of Zeocin selection results in at least a ten-fold lower transformation efficiency than with auxotrophic markers [12].

As such, current transformation methods for *K. phaffii* using integrative plasmids with Zeocin selection only achieve an efficiency of 10^3–10^4 colony-forming units (CFU) per µg of DNA [12, 13], while for many applications, libraries of 10^7 clones or more are desired. This low efficiency makes the generation of (ultra)large libraries in *K. phaffii* very time-consuming and often not feasible without the use of automation. To our knowledge, only three papers reported the generation of a *K. phaffii* library of at least 10^6 clones [14–16] (excluding our own paper using the protocol described here [17]), of which one used mating of antibody-heavy and light chain libraries as a tool to increase library size [16].

In many protocols, *K. phaffii* cells are typically only allowed to recover for 1 h at 30 °C in 1 M sorbitol post-electroporation. We reasoned that the Sh ble protein (the *Sh ble* gene product of the Zeocin resistance cassette) may not reach adequate levels during this short recovery period, impeding the growth of transformed cells once put under Zeocin selection. This is especially important since the Sh ble protein is not an enzyme but instead inhibits Zeocin-induced cell death by binding Zeocin stoichiometrically.

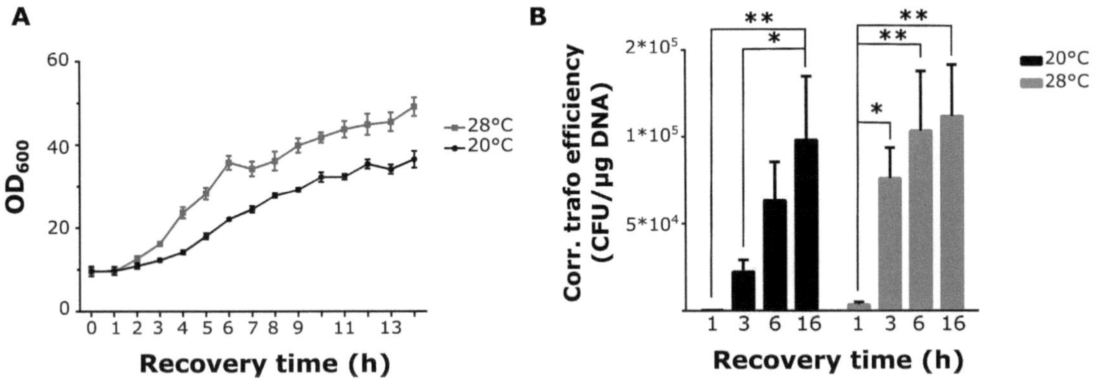

Fig. 1 Extending the recovery time of transformed *K. phaffii* cells in YPD improves transformation efficiency. **a** Recovery of *K. phaffii* transformant libraries in YPD medium at 20 °C or 28 °C. Electroporation was performed with 100 ng of linearized plasmid DNA. Mean OD_{600} +/− SEM ($n = 3$). **b** Comparison of transformation efficiencies between different recovery periods in YPD. Electroporation was performed with 100 ng of linearized plasmid DNA and selected on YPD pH 8.0 + 20 µg/mL zeocin agar plates at 30 °C after the recovery phase. From these plates, the colony-forming units (CFU) were determined. Two-way repeated measures ANOVA with Tukey's post-hoc test for multiple comparisons, mean corrected transformation efficiency +/− SEM ($n = 4$). *$p < 0.05$, **$p < 0.01$. (This figure has been reproduced from Boone et al. [17], under the Creative Commons Attribution 4.0 International License. http:/creativecommons.org/licenses/by/4.0/)

We found that, prior to Zeocin selection, the use of YPD medium instead of 1 M sorbitol during recovery of the transformants, combined with an extended recovery time of 6 h, significantly boosts the number of transformants, more than expected purely from cell division (Fig. 1a and b). Cell recovery at lower temperatures does not give a clear advantage in our hands (Fig. 1a and b). We also determined that increasing the YPD pH to 8.0 allows for efficient selection with lower Zeocin concentrations (20 µg/mL), which both reduces cost and simultaneously minimizes the fraction of multi-insert clones (Figs. 2 and 3). Finally, for pooled screens, we cut down hands-on time by introducing transformant selection via liquid culture rather than on solid agar plates.

The modified *K. phaffii* electroporation protocol allows for improved efficiency, cost, and ease of use by taking advantage of a 96-well electroporation set up, optimizing the conditions for recovery and Zeocin-based selection, and adapting the library selection format. Our improved transformation protocol for *K. phaffii*, presented in this chapter, reliably reaches an efficiency of 10^5 CFU per µg of DNA using integrative plasmids with Zeocin-based selection. We have used this method to create a 10^7 clone *K. phaffii* library without automation or the requirement for special strains or plasmids [17]. Since these simple modifications cumulatively improve and simplify the generation and manipulation of large *K. phaffii* expression libraries under Zeocin selection, this will pave the way toward more straightforward pooled screening in this yeast.

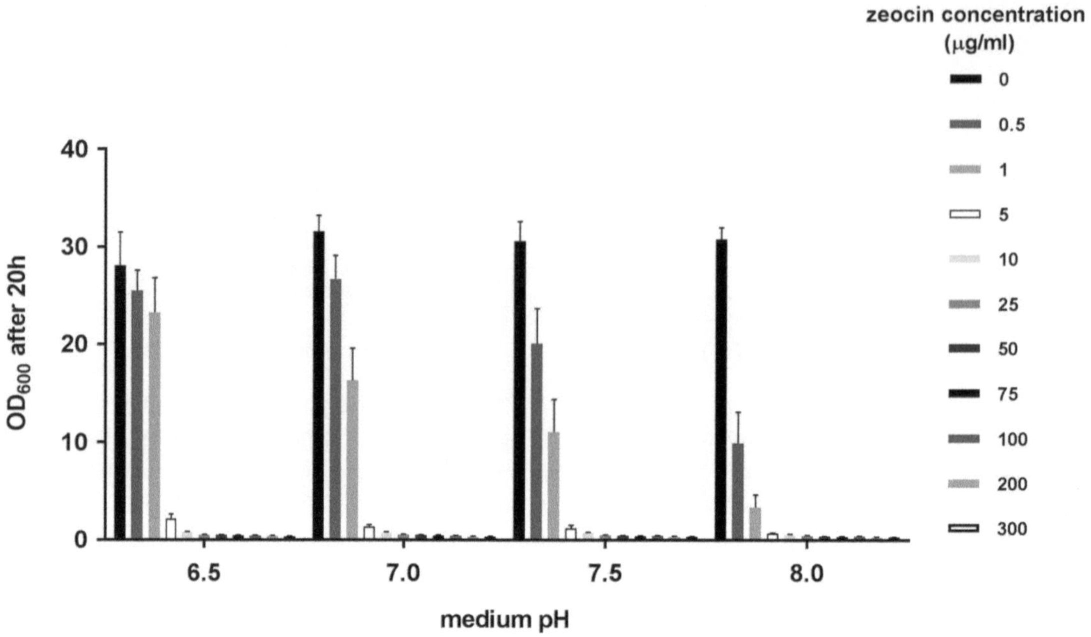

[zeocin] (μg/ml)	fraction transformed (%)	fraction transformants with multi-copy plasmid (%)
0	0.00	0.00
10	100.00	0.00
20	96.76	3.41
35	100.00	15.96
50	98.94	56.74
100	100.00	57.45
150	100.00	100.00
200	100.00	100.00
50 + spike	100.00	80.85
100 + spike	100.00	74.47
150 + spike	100.00	100.00

Fig. 2 a Minimal Zeocin concentration and optimal pH for growth inhibition of non-transformed cells in liquid medium. Growth of *K. phaffii* strain GS115 at 28 °C in liquid YPD supplemented with different concentrations of Zeocin and with different medium pH. Mean OD_{600} +/− SEM ($n = 4$–7). **b** Effect of Zeocin concentration on multi-copy insertion in liquid YPD medium. Electroporation was performed with 100 ng linearized plasmid DNA. After recovery, cells were selected in a liquid YPD medium with differing Zeocin concentrations. A yeast colony PCR was performed on at least 45 colonies per condition with primers spanning the library insert. The fraction of transformed cells is calculated from the number of colonies with one or more detectable bands after colony PCR (*see* Subheading 3.3). A gel pattern with multiple bands suggests multi-copy insertion of the plasmid. Spike = an additional 50 μg of Zeocin added every 12 h

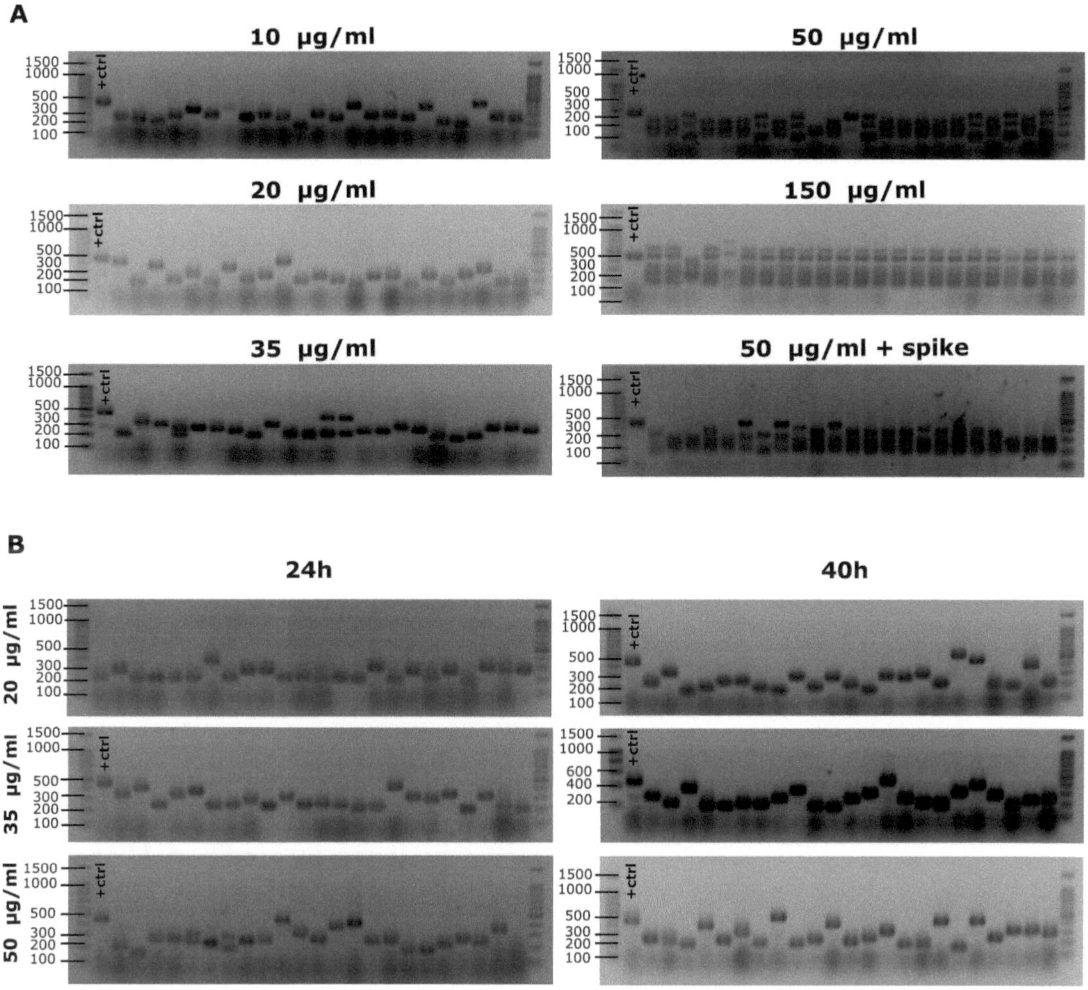

Fig. 3 Multi-copy integration increases with increasing Zeocin concentration. After electroporation and a 48-h selection in liquid YPD with differing Zeocin concentrations, library pools were diluted and plated on non-selective YPD plates and assessed for the presence of a library insert via colony PCR (*see* Subheading 3.3). The positive control is a colony from the GS115 strain with plasmid pPSDzeo-FLAG-hLYZ-V5-AOX1 [17]. Spike = an additional 50 μg of Zeocin added every 12 h

2 Materials

2.1 Library Size Considerations

Before starting, estimate the number of transformants (T) required to obtain a yeast library diversity (D) that is as feasibly close to the original plasmid library's diversity (D_{max}). We use the formula for library degeneracy by Bosley and Ostermeier [18].

1. $D = D_{max}(1 - e^{-T/D_{max}})$.
   - D_{max} = the diversity of the plasmid library that will be transformed.

This can be estimated by the number of unique clones in the *E. coli* starting library from which plasmid DNA was prepared. This number can be measured directly by deep sequencing of the plasmid pool extracted from *E. coli*. Alternatively, it can be estimated using the same formula, with T being the number of *E. coli* transformants and D_{max} the diversity of your input DNA library prior to cloning.

- D = the intended diversity of the transformed yeast library.

  In essence, this is the number of distinct plasmid variants targeted for representation in the yeast library. This number needs to be lower than D_{max}.

- T = the number of *K. phaffii* transformants obtained or to obtain.

2. Example: for a library of 2.5 million plasmid variants ($D_{max} = 2.5 \times 10^6$), you obtained 2.8 million *K. phaffii* transformants ($T = 2.8 \times 10^6$). This means your yeast library diversity is estimated at $2.5 \times 10^6 \times (1 - e^{-2.8/2.5}) = 1.68 \times 10^6$. To boost the yeast library diversity to 2.37 million, you would need 2.5× more transformants (seven million, since $2.5 \times 10^6 \times (1 - e^{-7/2.5}) = 2.37 \times 10^6$).

2.2 Linearization of Plasmid

1. Plasmid DNA: amount dependent on the number of transformants desired (1.5 μg per 10^5 transformants).
2. Restriction enzyme and respective buffer for linearization of plasmid (protocol tested for integration in the *AOX1* locus).
3. Milli-Q water.
4. Thermocycler.
5. Loading dye.
6. Agarose gel with DNA dye.
7. DNA ladder.
8. Agarose gel imager.
9. SPRI paramagnetic beads (we used CleanNGS beads from CleanNA).
10. Nanodrop.
11. PCR tubes.
12. Magnetic rack.
13. 70% ethanol.

2.3 Electroporation and Recovery

1. *K. phaffii* strain of choice (we used GS115).
2. Liquid YPD medium at pH 8.0, set pH with 1 M NaOH.
3. Sterile glass tubes, baffled shake flasks, and non-baffled shake flasks.

4. Spectrophotometer.
5. Sterile LiAc/DTT solution: 100 mM LiAc, 10 mM dithiothreitol (DTT), 0.6 M sorbitol, 10 mM Tris–HCl pH 7.5, sterilized via filtration.
6. Orbital shaker.
7. Incubator.
8. Centrifuge and sterile centrifuge tubes.
9. Sterile 1 M sorbitol (sterilized via filtration).
10. Gene Pulser electroporation system (BioRad, Hercules, USA) with an added HT-200 plate handler (BTX, Holliston, USA), or equivalent system.
11. 96-well format electroporation plate, 2 mm (BTX).
12. Zeocin (supplied as a sterile filtered blue solution at 100 mg/mL in HEPES buffer).
13. YPD pH 8.0 plates +20 μg/mL Zeocin, set pH with 1 M NaOH. Store plates for max. 1 week at 4 °C protected from light.
14. Spectrophotometer and cuvettes.
15. Pure glycerol for preparing solutions.
16. Cryotubes.
17. Glass beads for plating.

2.4 Yeast Colony PCR to Quantify Transformant Fraction and Multi-insert Transformant Fraction

1. 20 mM NaOH.
2. Milli-Q water.
3. GoTaq Green Master Mix (Promega).
4. PCR primers specific to the transformed plasmid.
5. Thermocycler.
6. Agarose gel with DNA dye.
7. Agarose gel imager.
8. DNA ladder.

3 Methods

3.1 Linearization of Plasmid Library

Generating a double-strand break in the plasmid part that is identical to a genomic locus targeted for integration enhances plasmid integration by homologous recombination. We commonly linearize in the *AOX1* promoter to specifically target integration at the *AOX1* promoter in the *K. phaffii* genome. We generally use either the *Mss*I or *Pme*I restriction enzyme (isoschizomers), as these cut only once in the *AOX1* promoter and they have relatively long recognition sites (8 bp), which reduces the frequency of occurrence in the variable elements of the library. Before proceeding with this protocol, it is important to check that these enzymes do not cut anywhere else in your plasmid.

The following protocol produces sufficient amounts (at least 100 μg) of linearized plasmid DNA to generate a 10^7 degeneracy *K. phaffii* library. Adjust these amounts proportionally to the library size.

1. Prepare the following mix on ice. Mix gently. If adapting these volumes, ensure that the volume of the restriction enzyme reagent does not exceed 1/10th of the total reaction volume.

Plasmid DNA (*see* Note 1)	**X μL (150 μg)**
Thermo Scientific 10× FastDigest buffer	150 μL
Thermo Scientific FastDigest MssI	75 μL
Milli-Q water	1275 – X μL (volume of plasmid DNA to be digested)
Total	1500 μL

2. Aliquot the mix into 50 μL reactions in PCR tubes on ice. Spin the tubes down and incubate for 30 min at 37 °C in a thermocycler with a heated lid.
3. Pool the reactions back together.
4. Take out 10 μL for a checkup on the gel. Run on an agarose gel to check for completeness of digestion.
5. Purify the remainder of the restriction digest pool with SPRI paramagnetic beads (we use CleanPCR beads from the brand CleanNA) (*see* **Note 2**). To summarize:
   (a) Vortex the CleanPCR reagent to fully resuspend the magnetic beads prior to usage.
   (b) Add 1.4× of CleanPCR reagent to the restriction digest sample. Incubate at room temperature for 5 min.
   (c) Place the tube(s) on a magnetic rack. Incubate at room temperature until the solution clears and all magnetic beads are pulled to the side of the tube. Remove and discard the cleared supernatant.
   (d) Add 70% ethanol to each tube (volume should be equal to the starting volume of the sample) without removing the tube from the rack. Make sure the beads are covered by ethanol. Incubate at room temperature for 30 s. Remove and discard the cleared ethanol. Repeat this step a second time.

(e) After the removal of the second ethanol wash, leave the tube on the magnetic rack to air dry the tubes.

(f) Once dry, remove the tube(s) from the magnetic rack and resuspend the beads in Milli-Q. Incubate at room temperature for 5 min.

(g) Place the tube(s) on a magnetic rack. Incubate at room temperature until the solution clears and all magnetic beads are pulled to the side of the tube. Transfer the cleared supernatant (containing the DNA) to a new tube.

6. Determine the DNA concentration via Nanodrop. Expect to retain approximately 105 μg (~70% yield).

3.2 Electroporation, Cell Recovery, and Transformant Selection

In this section the optimized electroporation protocol for *K. phaffii* initially presented in the introduction is described.

3.2.1 Preparation of Electrocompetent Cells

1. On day 1, inoculate the desired *K. phaffii* strain (we use GS115) in 5 mL of YPD in a sterile glass culture tube and leave to shake overnight at 28 °C.
2. On the morning of day 2, expand this preculture in 50 mL of YPD in a baffled 500 mL shake flask, i.e., a 1/10 dilution, and continue shaking at 28 °C.
3. On the evening of day 2, determine the OD_{600} of the 50 mL culture using a spectrophotometer. To obtain an OD_{600} of 1.5 the following morning, dilute the preculture in YPD according to the formula below, and continue growth at 28 °C.

(a) A 250 mL culture grown in a 2 L baffled shake flask gives enough cells for 23 transformation reactions. Constructing a 10^7 library at a transformation efficiency of 10^5 CFU/ug DNA, for example, would require five 250 mL cultures (in five 2 L baffled shake flasks).

(b) Calculate how many cells you need to add with the following formula:

- $$\frac{(V * 1.5)/2^n}{OD_{600}}$$

- where V is the volume of the culture, 1.5 is the desired OD_{600} of the culture, OD_{600} is the measured OD_{600} of the preculture, and n is the number of generations minus 1 (to allow the cells to recover from the stationary phase). For GS115 growing in YPD at 28 °C, we determined a log-phase generation time of 2 h.

4. On the morning of day 3, prepare electrocompetent cells. We use the method described in Wu and Letchworth [13]. To summarize:
   (a) Monitor the OD_{600} of your culture. When the culture reaches an OD_{600} of 1.5, harvest the cells at 1500 × g for 5 min at 4 °C using sterile centrifuge tubes. Discard the supernatant.
   (b) Resuspend the cells in LiAc/DTT solution at approximately 10^8 cells/mL in a non-baffled shake flask; this is about 200 mL of LiAc/DTT solution per 250 mL culture. Incubate for 30 min at 28 °C while gently shaking at 100 rpm on an orbital shaker.
   (c) Harvest the cells at 1500 × g for 5 min at 4 °C and discard the supernatant. Wash the cells in ice-cold 1 M sorbitol at approximately 3.3×10^8 cells/mL; this is about 37.5 mL of 1 M sorbitol per 250 mL culture. Repeat this washing step two more times.
   (d) Resuspend the cells in 1 M ice-cold sorbitol at approximately 10^{10} cells/mL; this is about 1.875 mL of 1 M sorbitol per 250 mL culture. Store the cells on ice until electroporation.

3.2.2 Electroporation

1. Pre-chill the 96-well format electroporation plate on ice (*see* **Note 3**). Aliquot 80 μL of competent cell suspension in every well. Afterwards, add 100 ng to 1 μg of purified linearized vector per well (*see* **Note 4**). Make sure to include a negative control (no DNA added) as well.
2. Incubate the electroporation plate with the competent cells and DNA on ice for 5 min.
3. Prepare a baffled shake flask with ice-cold YPD pH 8.0 medium (count on 1 mL per electroporation well, so 96 mL per 96-well electroporation plate). Use a separate flask for the negative control electroporation.
4. Perform the electroporation at 200 Ω, a capacitance of 25 μF, a capacitance extension of 125 μF, and a voltage of 1.5 kV using the Gene Pulser electroporation system (suggested equipment: BioRad, Hercules, USA, *see* **Note 5**), with an added HT-200 plate handler (suggested equipment: BTX, Holliston, USA). Be mindful of the arcing of the wells.
5. Immediately after electroporation, use a Pasteur pipette to gently but swiftly transfer and pool the electroporated cells from the electroporation plate to the ice-cold YPD pH 8.0 medium (*see* **Note 6**).

3.2.3 Recovery of Transformants

1. Determine the OD_{600} of each transformant culture (library and negative control) prior to recovery.
2. Grow the cultures for 6 h at 28 °C in a shaking incubator. In our hands, 6 h of recovery at 28 °C maximizes transformant recovery but avoids overgrowth of non-transformed cells (Fig. 1; *see* **Note 7**).
3. After 6 h, determine the OD_{600} of each culture again. The growth factor is calculated by dividing the OD_{600} after 6 h by the OD_{600} at the beginning of the recovery period. We expect the growth factor to be around 2.5–3.0.
4. Proceed to Subheadings 3.2.3.1 and 3.2.3.2 at the same time.

Assessing Transformation Efficiency and Library Diversity

1. Plate 10 and 20 μL, as well as 100 μL of a 1/100 dilution, of each recovered culture on fresh YPD pH 8.0 agar plates containing 20 μg/mL Zeocin. We plate cells using glass beads to ensure uniform dispersion. Do this in duplicate for every culture, including for the negative control. Inoculate the plates for 2–3 days at 30 °C to allow for colony growth.
2. Assess the transformation efficiencies by counting the colony-forming units (CFU) on the plates. No colonies should have grown on the negative control plates. Divide the CFU by the growth factor to correct for the growth that has happened during the electroporation recovery phase and thus obtain a growth-corrected number of *K. phaffii* transformants. Divide by the μg of DNA transformed to express the transformation efficiency in CFU per μg DNA.
3. Calculate the estimated library degeneracy, or the expected number of distinct sequences in the *K. phaffii* library, using the formula by Bosley and Ostermeier [18] as described in Subheading 2.1.

Liquid Selection of Transformants

1. Dilute each recovered culture 1/25 in YPD medium pH 8.0 containing 20 μg/mL Zeocin in baffled shake flasks (*see* **Note 8**).
2. Grow the cultures for 24 h at 28 °C to select for positive transformants.
3. Afterward, measure the OD_{600} of the cultures in a spectrophotometer.
4. Optional: Plate out a serial dilution of each culture on non-selective YPD agar plates and grow for 2 days at 28 °C for the yeast colony PCR (*see* Subheading 3.3).
5. Make 15% (w/w) glycerol stocks of the library. We recommend freezing at least 10× more cells than the number of transformants to avoid undersampling of the library.

3.3 Yeast Colony PCR to Quantify (Multi-insert) Transformant Fraction

In this section, we describe our protocol to quantify the fraction of cells without, with single-copy, or with multi-copy plasmid insertion of the library plasmids using a yeast colony PCR. Note that the single vs. multi-copy distinction can only be done reliably when the library consists of fragments of differing sizes. This design has the added advantage that the colony PCR will also give a quick indication of library diversity. If the library design does not allow separation based on size (for example, a library of point mutants of a protein), a Sanger or other long-read sequencing-based approach to assessing multi-copy integration could be used.

1. Design primers for the colony PCR. These should target the common flanks of the library inserts.
2. Pick a sufficiently large number of clones (we routinely screen roughly 50 colonies) from the non-selective plates of each library transformation (*see* Subheading 3.2.3.1) with a sterile pipette tip or toothpick and resuspend each in 20 μL of 20 mM NaOH in a PCR tube or plate. Do the same for:
   - Negative control: one colony of a *K. phaffii* strain that has not been transformed (in our case GS115).
   - Positive control: if possible, one colony of a *K. phaffii* strain that has been successfully transformed with a plasmid on which the colony PCR primers can anneal. If this is not available, a tiny quantity of the library plasmid prep used for transformation can be used.
3. Cover and boil the samples at 98 °C for 15 min in a PCR machine with a heated lid.
4. Add 80 μL of Milli-Q water to each PCR tube to dilute the NaOH and mix well.
5. Prepare the following master mix (quantities given for 50 reactions).

	Volume for 1 reaction (μL)	Volume for 50 reactions (μL)
Sterile Milli-Q water	5	250
2× GoTaq Green Master Mix (*see* **Note 9**)	12.5	625
Forward primer 10 μM	1.25	62.5
Reverse primer 10 μM	1.25	62.5
Yeast lysate mix	5	
Total	25	

6. Aliquot 20 μL of the master mix in PCR tubes or plates and add 5 μL of the respective yeast lysate mix to each PCR tube.

7. Perform PCR cycling in a thermocycler with a heated lid according to the following program.

	95 °C	**2 min**
	95 °C	45 s
30 cycles	X°C	30 s
	72 °C	Y
	72 °C	5 min
	12 °C	Hold

- Optimize the annealing temperature X for every primer pair.
- Allow approximately 1 min extension time Y for every 1 kbp of DNA to be amplified.

8. Perform agarose gel electrophoresis to assess the success of the PCR reaction. The GoTaq Green Master Mix conveniently already contains loading dye, so PCR reactions can be loaded directly.
9. Image the gel (for example, *see* Fig. 3). The lane of the negative control should show no amplification, while the positive control should give a band of the expected length. Divide the number of lanes where amplification is observed by the total number of lanes to obtain the fraction of transformants. If your library has a broad size distribution, you can estimate the fraction of multi-insert transformants by dividing the number of lanes showing multiple bands by the total number of lanes showing at least a single band.

4 Notes

1. Plasmid DNA should be prepared via plasmid purification kits according to yield requirements (miniprep, midiprep, maxiprep, megaprep, gigaprep).
2. It is essential that the digested plasmid is desalted, as salt causes arcing during electroporation. Arching leads to cell death.
3. In our experience, the disposable 96-well electroporation plates can be reused several times to cut down costs, provided that the plates are thoroughly cleaned immediately after use to remove any remaining lysed cells, DNA, and salts. We incubate the plates in a detergent solution and thoroughly wash them with distilled water to remove any traces of detergent. Plates can be sterilized by placing them under a UV light for a few hours and then covering them to keep them sterile until use.

4. We usually transform 1 μg of DNA per reaction to reduce time and cost. However, when small libraries are sufficient ($<10^6$ transformants), we do recommend using 100 ng, as this somewhat reduces the fraction of multi-insert clones (data not shown).
5. Optimal electroporation conditions vary from instrument to instrument. It is advisable to optimize this to ensure optimal transformation efficiency.
6. We use YPD medium for recovery instead of sorbitol, since it increases the number of transformants (data not shown).
7. When generating large libraries, it is key to maintain library diversity at every step. A critical moment is the liquid selection step, since at this step, clones with a higher differential growth rate could overgrow the culture. Therefore, we advise keeping the selection time short (we use 24 h), working with low concentrations of Zeocin (10–20 μg/mL) to avoid the selection of multi-copy or multi-insert fragments, and working with inducible promoters in repressive conditions to avoid differential metabolic burden/toxicity of expression of the protein variants.
8. Sufficient dilution is crucial, as active growth is required to allow for effective selection by Zeocin.
9. Other polymerase solutions should also work, given proper testing and optimization.

Competing Interest Statement

The authors have no conflicts of interest to declare that are relevant to the content of this chapter.

References

1. Feldhaus MJ, Siegel RW, Opresko LK, Coleman JR, Feldhaus JMW, Yeung YA, Cochran JR, Heinzelman P, Colby D, Swers J, Graff C, Wiley HS, Wittrup KD (2003) Flow-cytometric isolation of human antibodies from a nonimmune Saccharomyces cerevisiae surface display library. Nat Biotechnol 21: 163–170. https://doi.org/10.1038/nbt785
2. Boder ET, Midelfort KS, Wittrup KD (2000) Directed evolution of antibody fragments with monovalent femtomolar antigen-binding affinity. Proc Natl Acad Sci USA 97:10701–10705. https://doi.org/10.1073/pnas.170297297
3. Ryckaert S, Callewaert N, Jacobs PP, Dewaele S, Dewerte I, Contreras R (2008) Fishing for lectins from diverse sequence libraries by yeast surface display – an exploratory study. Glycobiology 18:137–144. https://doi.org/10.1093/glycob/cwm131
4. Woldring DR, Holec PV, Zhou H, Hackel BJ (2015) High-throughput ligand discovery reveals a sitewise gradient of diversity in broadly evolved hydrophilic fibronectin domains. PLoS One 10:e0138956. https://doi.org/10.1371/journal.pone.0138956
5. Ostafe R, Prodanovic R, Nazor J, Fischer R (2014) Ultra-high-throughput screening method for the directed evolution of glucose oxidase. Chem Biol 21:414–421. https://doi.org/10.1016/j.chembiol.2014.01.010
6. Wentz AE, Shusta EV (2007) A novel high-throughput screen reveals yeast genes that

increase secretion of heterologous proteins. Appl Environ Microbiol 73:1189–1198. https://doi.org/10.1128/AEM.02427-06

7. Agresti JJ, Antipov E, Abate AR, Ahn K, Rowat AC, Baret J-C, Marquez M, Klibanov AM, Griffiths AD, Weitz DA (2010) Ultrahigh-throughput screening in drop-based microfluidics for directed evolution. Proc Natl Acad Sci USA 107:4004–4009. https://doi.org/10.1073/pnas.0910781107
8. Ling XB (2008) High throughput screening informatics. Comb Chem High Throughput Screen 11:249–257. https://doi.org/10.2174/138620708783877726
9. Gietz RD, Schiestl RH (2007) Large-scale high-efficiency yeast transformation using the LiAc/SS carrier DNA/PEG method. Nat Protoc 2:38–41. https://doi.org/10.1038/nprot.2007.15
10. Gietz RD, Schiestl RH (2007) High-efficiency yeast transformation using the LiAc/SS carrier DNA/PEG method. Nat Protoc 2:31–34. https://doi.org/10.1038/nprot.2007.13
11. Benatuil L, Perez JM, Belk J, Hsieh C-M (2010) An improved yeast transformation method for the generation of very large human antibody libraries. Protein Eng Des Sel 23:155–159. https://doi.org/10.1093/protein/gzq002
12. Lin-Cereghino J, Wong WW, Xiong S, Giang W, Luong LT, Vu J, Johnson SD, Lin-Cereghino GP (2005) Condensed protocol for competent cell preparation and transformation of the methylotrophic yeast Pichia pastoris. BioTechniques 38:44–48. https://doi.org/10.2144/05381BM04
13. Wu S, Letchworth GJ (2004) High efficiency transformation by electroporation of Pichia pastoris pretreated with lithium acetate and dithiothreitol. BioTechniques 36:152–154. https://doi.org/10.2144/04361DD02
14. Ryckaert S, Pardon E, Steyaert J, Callewaert N (2010) Isolation of antigen-binding camelid heavy chain antibody fragments (nanobodies) from an immune library displayed on the surface of Pichia pastoris. J Biotechnol 145:93–98. https://doi.org/10.1016/j.jbiotec.2009.10.010
15. Stadlmayr G, Benakovitsch K, Gasser B, Mattanovich D, Sauer M (2010) Genome-scale analysis of library sorting (GALibSo): isolation of secretion enhancing factors for recombinant protein production in Pichia pastoris. Biotechnol Bioeng 105:543–555. https://doi.org/10.1002/bit.22573
16. Shaheen HH, Prinz B, Chen M-T, Pavoor T, Lin S, Houston-Cummings NR, Moore R, Stadheim TA, Zha D (2013) A dual-mode surface display system for the maturation and production of monoclonal antibodies in glyco-engineered Pichia pastoris. PLoS One 8: e70190. https://doi.org/10.1371/journal.pone.0070190
17. Boone M, Ramasamy P, Zuallaert J, Bouwmeester R, Van Moer B, Maddelein D, Turan D, Hulstaert N, Eeckhaut H, Vandermarliere E, Martens L, Degroeve S, De Neve W, Vranken W, Callewaert N (2021) Massively parallel interrogation of protein fragment secretability using SECRiFY reveals features influencing secretory system transit. Nat Commun 12:6414. https://doi.org/10.1038/s41467-021-26720-y
18. Bosley AD, Ostermeier M (2005) Mathematical expressions useful in the construction, description and evaluation of protein libraries. Biomol Eng 22:57–61. https://doi.org/10.1016/j.bioeng.2004.11.002

Chapter 24

High-Throughput Generation of *Pichia pastoris* Knock-Out Strains by Using CRISPR/Cas9

Kirill Smirnov, Lukas Rieder, and Anton Glieder

Abstract

The CRISPR/Cas9 (CRISPR is an acronym for clustered regularly interspaced short palindromic repeats) system is a powerful molecular biological tool simplifying the process of genome engineering. Originally demonstrated to function in human and mouse cells, the portfolio of organisms that can be engineered by the new and groundbreaking technology was soon expanded. In the past years, CRISPR/Cas9 tools for use in *Komagataella phaffii* were reported to allow the generation of *K. phaffii* mutant strains in less than 2 weeks. In addition, the *K. phaffii* tailored system uses episomal vectors for the expression of the CRISPR/Cas9 elements, which allows the recycling of the plasmid after the CRISPR editing to obtain empty mutant strains. This means that the engineered strains do not carry the expression cassette of the resistance marker and CRISPR/Cas9 plasmid in their genome and are therefore a superb starting point for further investigations.

In this chapter, we describe a pipeline for the high-throughput generation of *K. phaffii* mutant strains with interrupted open reading frames of genes, by using the CRISPR/Cas9 system in combination with error-prone repair of the double-strand break by NHEJ. The pipeline we developed consists of four steps: (a) CRISPR/Cas9 plasmids assembly, (b) transformation of *K. phaffii*, (c) screening for mutant strains, and (d) plasmid elimination and is due to the detailed description of every step being easily reproducible. To intensify and simplify the research work, most of the described procedures can be performed in a 96-well format.

Key words *Pichia pastoris*, *Komagataella phaffii*, CRISPR/Cas9, Genome engineering, Gene knock-out, Synthetic biology

1 Introduction

Several definitions of the term "genome editing" can be found in scientific literature. One of the definitions describes genome editing as the process by which the genome sequence is changed through the intervention of a DNA break or other DNA modifications [1]. A quite similar definition is given by Khan et al., who claim that genome-wide editing techniques can be interpreted as methods where DNA sequences are changed by deletions, mRNA

Xavier Garcia-Ortega et al. (eds.), *Pichia pastoris: Methods and Protocols*, Methods in Molecular Biology, vol. 2697,
https://doi.org/10.1007/978-1-0716-4779-0_24,

processing, and post-transcriptional modifications to result in altered gene expression, leading to functional behavior of proteins [2]. By using these definitions, a broad spectrum of methods, such as UV-based, chemical mutagenesis, and even physics-based methods, are suitable for altering genomic DNA and can therefore be considered as genome editing methods.

Another meaning of genome editing is suggested by Yang et al. [3], who define genome editing as site-specific genome modification inclusive of gene knockout, integration, and intentional point mutations. Dunn et al. point out that genome editing represents a group of newly developed capabilities that allow nucleotide-level targeting of genomic regions in a wide range of species [4]. These two definitions limit the number of possible methods that can be referred to as genome editing techniques. One of the crucial criteria that is included in both definitions is that genome editing methods are supposed to provide a relatively precise alteration of the genome (so-called nucleotide-level targeting). Following these definitions, we decided to focus only on modern genome editing techniques such as zinc-finger nucleases (ZFNs), transcription activator-like effector nucleases (TALENs), and CRISPR/Cas and to exclude conventional genome editing techniques such as homologous recombination. Nevertheless, classical approaches for yeast genome manipulation, which are based on site-specific recombinases (SSRs), remain a crucial tool for genome editing in *Komagataella phaffii* (reclassified from *Pichia pastoris*) and are therefore reviewed in brief as well [5–9].

1.1 Classical Site-Specific Recombination Systems

Site-specific recombination systems consist of SSRs, a class of specialized enzymes that catalyze strand-switching reactions, and two specific DNA sequences termed recombinase binding sites (RBS) [10]. Already in the early 1990s, three site-specific recombination systems, namely the Cre/lox (Cyclization recombinase/locus of crossing-over), FLP/FRT (flippase recombinase/flippase recognition target), and R recombinase, were used to manipulate genomic DNA of yeast and mammalian cells [11]. Today, the Cre/lox and FLP/FRT systems, in which the SSRs belong to the family of tyrosine recombinases [12], are still widely used tools for the genetic manipulation of eukaryotic genomes [13].

1.1.1 Cre-Lox

The Cre/lox system originates from the *Escherichia coli* (*E. coli*) bacteriophage P1 and was the first site-specific recombination system that was adapted for use in eukaryotes by Sauer and Henderson [12, 14]. The Cre recombinase is a 38 kDa tyrosine recombinase and is responsible for intra- and inter-molecular recombination at loxP recognition sites [15]. This SSR is, depending on the orientation of the target sequences, able to either excise or invert DNA elements located between the two short (34 bp) loxP sequences [14]. The canonical loxP sequence is 5′-*ATAACTTCGTATA*-**GCATACAT**-*TATACGAAGTTAT*-3′ and forms an 8 base pair (bp) core and element (bold) and two flanking inverted repeats (italic) [15]. Fascinatingly, the loxP sequence is in nature exclusively found in P1 phages and allows therefore very specific DNA manipulations via introducing loxP sites into a specific region in the genomic DNA and subsequent excision of the loxP-flanked DNA region by the Cre recombinase.

1.1.2 FLP-FRT

The FLP/FRT system originates from the autonomously replicating 2-μm plasmids of *Saccharomyces cerevisiae* (*S. cerevisiae*) [16] and has been proven to be an effective tool for cell engineering in the past [17]. The system is commonly used today for the site-specific excision of selectable marker genes from the genome of engineered organisms [13].

Similarly, to the Cre/lox system explained earlier, the FLP/FRT system consists of the two elements. In the FLP/FRT system, the SSR is the 48.6 kDa FLP [16] that targets the FRT site, which marks the RBS. The minimal sequence of the FRT site is 5′-*GAAGTTCCTATTC*-**TCTAGAAA**-*GTATAGGAACTTC*-3′ [18]. The sequence consists, identical to the loxP site, of an 8 bp core element (bold) which is flanked by inverted repeats (italic) [16]. However, in contrast to the loxP sites, the inverted repeats of the FRT site have to have a single bp mismatch (underlined) in their sequence [18].

One of the earliest applications of the FLP/FRT system in *K. phaffii* aimed to regenerate a selectable marker phenotype [19]. Later, the FLP/FRT system was adapted for generating knock-out strains [4, 8, 20].

1.2 Modern Genome Engineering Tools

1.2.1 ZFNs and TALENs

Zinc finger nucleases (ZFNs) are guided DNA cleavage reagents, which have separated DNA-binding and DNA cleavage domains [21]. ZFNs originate from *Fok*I, a natural type IIS restriction enzyme, and mark the first truly steerable DNA cleaving reagents as a specific DNA sequence within the genome of an organism could be targeted by protein engineering [22]. Each ZFN consists of a *Fok*I nuclease and a series of zinc fingers that are designed to bind the specific genome locus. An alternative to the classical ZNFs are the paired ZFNs, which can target two DNA sites simultaneously and introduce a double-strand break (DSB) [15]. Hence, homology directed repair (HDR) mechanisms are triggered by this genome editing tool, resulting in the introduction of insertions, point mutations, deletions, and translocations [23]. However, the synthesis and assembly of ZFNs is a complicated process, limiting their applicability [15].

Transcription activator-like effector nucleases (TALENs) are, like the ZFNs, formed from a nonspecific *Fok*I nuclease domain for DNA cleavage and an adaptable and specific DNA-binding domain. The DNA-binding domain consists of highly conserved repeats originating from transcription activator-like effectors (TALEs), which are secreted proteins originating from plant pathogens of the genus *Xanthomonas*. The altering of DNA-binding TALE repeats allows targeting a broad range of DNA sequences, and a DSB is initiated by *Fok*I nuclease [23].

Summing up, both technologies are based on the specific recognition of a target DNA sequence by customizable proteins with subsequent nuclease-induced DNA cleavage. Despite the active usage of ZFNs and TALENs for alterations of genes in *S. cerevisiae*, plants, and mammalian cells [4, 22, 23], there is, to the best of our knowledge, no report about the genome editing of *K. phaffii* using either of the two techniques.

1.2.2 CRISPR/Cas Systems

In nature, the CRISPR/Cas system is used by bacteria and archaea as a mechanism to protect themselves from foreign nucleic acid [24]. Whereas the acronym "CRISPR" stands for clustered regularly interspaced short palindromic repeats, the term "Cas" describes a nonspecific endonuclease. Explained at a basic level, the CRISPR/Cas system uses a small RNA to guide the Cas enzyme to a defined DNA/RNA sequence to introduce a double- or single-strand break, and then the DSB can be repaired by non-homologous end-joining (NHEJ) or by homology directed repair (HDR). Normally, the repair by NHEJ results in non-specific nucleotide insertions, deletions, or other mutations [25]. Soon after discovery, adaptations of the CRISPR/Cas technology for eukaryotic genome editing were made, enabling the accelerated discovery of novel gene functions and the development of new cell and animal models [26].

Nature provides a plethora of Cas proteins, and the most frequently used one is the Cas9 from *Streptococcus pyogenes* (*Sp*Cas9), because it was discovered early and adapted as a tool for target DNA editing [24]. As outlined earlier, the *Sp*Cas9 requires the presence of small RNA, leading the Cas protein to the target location. Originally, two RNAs are involved in the guiding of Cas9: the CRISPR RNA (crRNA), recognizing and pairing with a sequence within the target region, and the transacting CRISPR RNA (tracrRNA), linking the crRNA to Cas9. However, most of the modern CRISPR/Cas9 systems employ a single guide RNA (sgRNA) as a combination of crRNA and tracrRNA [27]. sgRNA has two functional elements. The first one is a guide RNA (gRNA or spacer), which is an 18–25-bp length sequence complementary to a target dsDNA sequence. The second element is a scaffold sequence (gRNA scaffold) that is necessary for the binding of the sgRNA molecule with Cas protein [27, 28]. An additional requirement for the proper functioning of the CRISPR/Cas9 system is that the genomic target locus is flanked at the 3′ by a 3-base pair (5′-NGG-3′) protospacer adjacent motif (PAM) that serves as a binding signal to the Cas9. When the system is designed correctly, the *Sp*Cas9 and gRNA form a ribonucleoprotein (RNP) complex that introduces a DSB three to four nucleotides upstream of the target site [25].

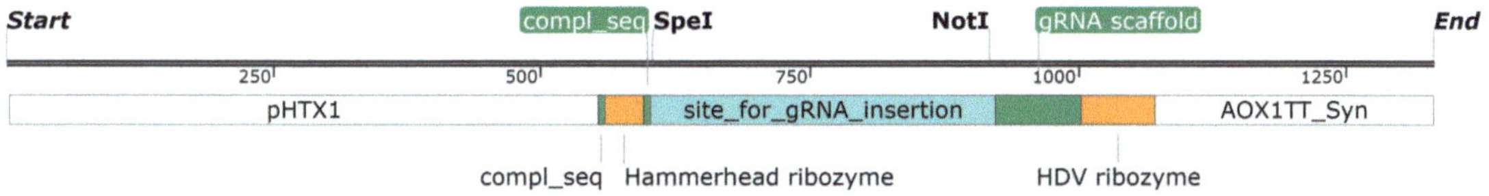

Fig. 1 Schematic representation of the sgRNA expression cassette found in the modified pCRISPR-Zeo plasmid. The transcription of the expression cassette is controlled by one side of the bidirectional RNA pol II promotor P_{HTX1}. The other side of the promoter controls the transcription of the Cas9. Downstream of the promoter, the hammerhead ribozyme flanked with six adenines marked as compl_seq is located, which is followed by a stuffer fragment that can be removed with the unique *Spe*I and *Not*I restriction sites and exchanged with the target-specific gRNA. The expression cassette is completed by the gRNA scaffold that is required to bind the Cas9 and the hepatitis delta virus (HDV). To terminate the transcription, the *AOX1* terminator (AOX1TT) is employed. For the exact design of the expression cassette, please read Ref. [30]. The figure was prepared using SnapGene, based on the pCRISPR-Zeo plasmid with NCBI accession number MW604246

One of the earliest reports about the usage of a CRISPR/Cas system in yeast (*S. cerevisiae*) was published by DiCarlo et al. in 2013 [29]. Three years later, Weninger et al. reported the first successful application of the CRISPR/Cas9 system in *K. phaffii* [30]. Interestingly, they discovered that the RNA Pol III promoters, which are used for the expression of sgRNA in *S. cerevisiae*, did not promote efficient genome editing in *K. phaffii*. To overcome this issue, they used RNA Pol II promoters for the transcription of the gRNA. To avoid the export of the gRNA from the nucleolus to the cytoplasm, they flanked the sgRNA with a hammerhead ribozyme at the 5′-end and a hepatitis delta virus (HDV) ribozyme sequence at the 3′-end (Fig. 1). These modifications ultimately resulted in a functional CRISPR/Cas9 system with almost 100% targeting efficiency for the use in *K. phaffii* [30]. Nevertheless, a recent publication demonstrated the RNA Pol-III-mediated expression of sgRNA in *K. phaffii* [31], indicating room for improving the CRISPR/Cas9 system as an editing tool for *K. phaffii*.

To sum up, we briefly described the most common tools for genome editing in *K. phaffii* such as site-specific recombination and CRISPR/Cas systems. In general, each system is composed of two functional elements, of which one is responsible for the recognition of a specific target sequence and the other one for the subsequent cleavage of it by an effector protein.

2 Materials

1. ddH_2O.
2. Yeast/*E.coli* shuttle vector that carries both a sgRNA expression cassette and *S. pyogenes* Cas9: e.g., pCRISPR-Zeo, GenBank accession # MW604246.

3. Gel/PCR clean-up kit (e.g. Wizard® SV Gel and PCR Clean-Up System, Promega, cat. no. A9282).
4. Restriction enzymes *Spe*I (e.g. NEB, cat. no. R3133S) and *Not*I (e.g. NEB, cat. no. R3189S) with appropriate reaction buffer.
5. 1% Agarose gel for DNA separation.
6. Plasmid isolation kit (e.g. Wizard® Plus SV Minipreps DNA Purification System, Promega, cat. no. A1460).
7. Proofreading DNA polymerase (e.g. NEB Q5 High-Fidelity DNA Polymerase, cat. no. M0491L).
8. 100 pmol/μL dNTP mix (e.g. NEB N0447S, cat. no. N0447S).
9. 100 pmol/μL Primers specific for user-designed guide RNA.
10. Gibson Isothermal Assembly mixture (e.g. Gibson Assembly® HiFi, Codex DNA, cat. no. GA1100).
11. Competent *E. coli* cells (e.g. XL1).
12. Zeocin (e.g. Thermo Fisher Scientific, cat. no. R25001).
13. LB Zeocin plates: 10 g/L tryptone, 10 g/L NaCl, 5 g/L yeast extract, and 15 g/L Agar-Agar for solidification with a final Zeocin concentration of 50 μg/mL.
14. PCR thermal cycler.
15. Gel electrophoresis equipment.
16. +37 °C incubator/shaker.
17. YPD medium: 10 g/L yeast extract, 20 g/L peptone and 20 g/L glucose).
18. YPD plates: YPD medium with 15 g/L Agar-Agar for solidification).
19. YPD Zeocin plates: YPD plates with a final Zeocin concentration of 100 μg/mL.
20. YPDS medium: 1:1 Mixture of YPD and 1 M Sorbitol.
21. BEDS solution: 10 mM Bicine-NaOH, pH 8.3, 3% (v/v) ethylene glycol, 5% (v/v) DMSO (molecular biology grade), 1 M Sorbitol.
22. 1.0 M dithiothreitol (DTT).
23. Phire Plant Direct PCR Master Mix (ThermoFischer Scientific, cat. no. F160S).
24. 96-deep-well cultivation plates (e.g. Bel-Art Products, Wayne, NJ, USA).
25. 96-well reaction plate (e.g. ThermoFisher Scientific, cat. no. N8010560).

3 Methods

3.1 Construction of the CRISPR/Cas9 Plasmid

The following section describes the essential steps for the assembly of the CRISPR/Cas9 plasmids (Fig. 2) carrying a desired gRNA sequence. This includes protocols for the design of the gRNA, the preparation of the backbone vector, and a single-oligonucleotide-based recombination cloning step, which allows fast and cheap cloning of the user-designed gRNA sequences into a CRISPR/Cas9 shuttle vector.

3.1.1 gRNA Design

1. Select the sequence(s) of the desired gene (or region) where you would like to introduce a frameshift mutation by insertion/deletion. Sources for the most updated *K. phaffii* genome sequences are e.g. NCBI (National Center for Biotechnology Information) with the accession numbers LT962476.2, LT962477.2, LT962478.2, and LT962479, for chromosomes 1, 2, 3, and 4, respectively [32], EMBL-EBI (European Bioinformatics Institute) [33], or Pichiagenome.org (University of Natural Resources and Life Sciences Vienna).
2. Create a multifasta file that contains target sequences of 500 bp length. If there is only one target sequence, copy it to the clipboard.
3. Go to the webpage CCTop-CRISPR/Cas9 target online predictor (https://cctop.cos.uni-heidelberg.de:8043/) of the University of Heidelberg [34]. Upload a multifasta file with your target sequences into the batch mode (or paste a sequence in a search line) and type your e-mail address into the corresponding field. Select as species *Komagataella phaffii* CBS 7435, make sure that the selected PAM motif is "NGG," the recommended length of the target site is 23 bases, and submit the job. Please remember that the PAM is dependent on the Cas protein used in the application.
4. Once the job is done, a notification is sent to your e-mail address. Follow the link and download the results. The results are organized in several folders; each folder represents one sequence that was uploaded in the multifasta file and contains files with gRNA sequences for each of the target sequences.
5. Select gRNA sequence(s) from the file(s) and remove the PAM sequence (5' NGG) from the 3′ end of the strand.
6. Incorporate the selected gRNA sequence(s) into oligonucleotides which contain 5′ and 3′ overhangs to hammerhead ribozyme and gRNA scaffold regions of the CRISPR/Cas9 vector backbone (black) for homologous recombination cloning (Fig. 1). The recommended oligonucleotide sequence is presented below. "$(N)_{23}$" (bold) marks the position that should be replaced with the newly designed gRNAs.

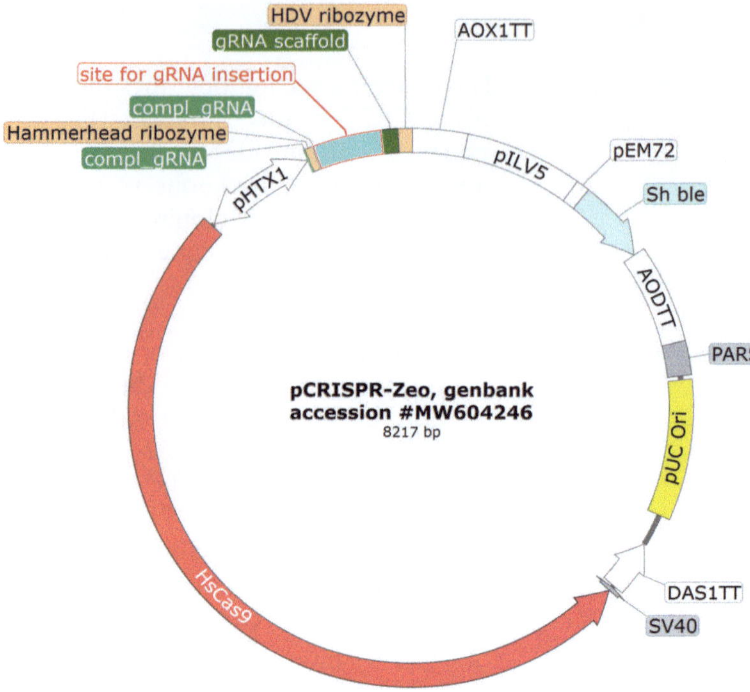

Fig. 2 Plasmid map of the CRISPR/Cas9 vector. Central to the plasmid are the Cas9 and sgRNA expression cassettes, which are both controlled by the constitutive bidirectional *HTX1* promoter (P_{HTX1}) and terminated by the *DAS1* and *AOX1* terminators, respectively. For plasmid propagation in *E. coli* during the cloning procedures, an origin of replication (pUC ori) is on the plasmid. For selection during plasmid assembly, and in the final eukaryotic host, a Zeocin expression cassette with promoters functioning in *E. coli* (P_{EM72}) and *K. phaffii* (P_{ILV5}) is encoded on the plasmid. Additionally, the plasmid harbors the PARS1 for episomal plasmid amplification and protein expression in *K. phaffii* and to remove the CRISPR/Cas9 plasmid after successful editing of the target gene. For the exact design of the plasmid, please read Ref. [30]. The "site for gRNA insertion" containing *Spe*I and *Not*I restriction sites was added additionally to the plasmid described by Weninger et al. [30] to facilitate cloning procedures. The figure was prepared using SnapGene

5′- AGTAAGCTCGTCAAAAAA$(\mathbf{N})_{23}$GTTTTAGAGCTAGAAATAG- 3'

7. Order the newly designed fragments containing the gRNAs as synthetic DNA oligomers (*see* **Note 1**) for subsequent cloning procedures.

3.1.2 CRISPR/Cas9 Plasmids Assembly

After the successful design and ordering of the gRNAs, the following protocols present two high-throughput methods for the insertion of the gRNAs into the CRISPR backbone vector pCRISPR-Zeo. The first protocol (from **step 1** to **step 11**) is based on a whole plasmid amplification PCR reaction, and the second protocol (from

step 12 to **step 18**) describes a single-based oligonucleotide Gibson assembly.

1. Restriction digestion of the pCRISPR-Zeo vector using *Not*I and *Spe*I according to the manual provided by the manufacturer. We typically use a reaction mix with a total volume of 100 μL that contains in total 5 μg of vector DNA. *Not*I and *Spe*I are added to a final concentration of 1 unit/μL per enzyme. To ensure complete digestion, it is recommended to incubate the reaction for 2 to 3 h.
2. The quality of the linearization has to be checked by agarose gel electrophoresis (*see* **Note 2**).
3. Recover the linearized vector from the gel by using either a PCR Clean-Up System kit or by ethanol precipitation (*see* **Note 3**).
4. Prepare a stock solution of the purified and linearized backbone vector with a final concentration of 1.5 ng/μL.
5. To assemble the CRISPR/Cas9 plasmids with the newly designed gRNAs, *E. coli*-mediated in vivo cloning is performed. For this, the plasmid is PCR amplified using primers with overlapping regions at the 5′-end. The forward primer consists of gRNA and overhanging sequences (see section 3.1.1 gRNA design). The reverse primer has the sequence: 5′-TGACGAGCTTACTCGTTTCGTCCTCACGGACTCAT CAGTTTTTTTTTGATTTG-3′.
6. The correct primer design for amplification of the pCRISPR-Zeo vector is exemplified in Fig. 3.
7. PCRs are performed according to the manual provided by the manufacturer. Typically, the PCR had a volume of 20 μL containing 1 μL linearized template backbone vector (1.5 ng/μL), 1 mM dNTPs, 0.05 μM form each primer, and 0.02 U/μL of a

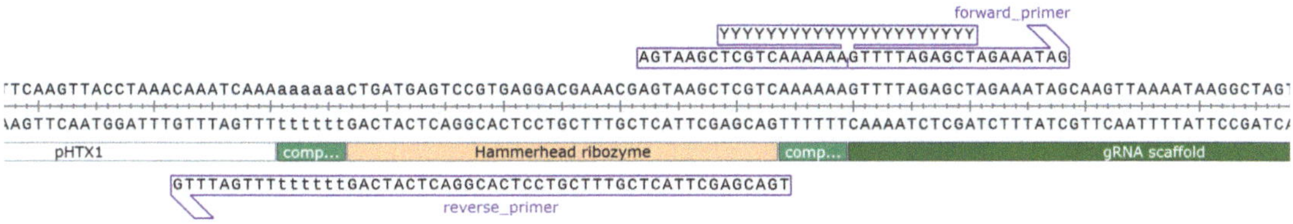

Fig. 3 Illustration of the sgRNA expression cassette with the recommended primer binding sites for amplification of the CRISPR/Cas9 plasmid. The forward primer binds to the hammerhead ribozyme and gRNA scaffold and corresponds to the designed gRNA oligonucleotides. The non-binding "YYYYYYYYYYYYYYYYYYYYYYY" marks to position of the target-specific gRNA sequence that will be inserted into the CRISPR/Cas9 plasmid during amplification. The reverse primer binds to the hammerhead ribozyme and the promoter region and can be directly copied as no target gene-specific modifications have to be made. Please note that the "site for gRNA insertion" was removed after restriction digestion using *Spe*I and *Not*I enzymes, and the forward and reverse primers must have an overlapping region for the subsequent in vivo cloning

high-fidelity DNA polymerase. The thermocycling conditions were: 95 °C/300 s—(95 °C/20 s—50–60 °C/20 s – 72 °C/510 s) × 25–72 °C/420 s – 4 °C/∞ (*see* **Note 4**). Please remember that the annealing temperature is dependent on the primer design.

8. Optional: assess the successful amplification of the plasmid by agarose gel electrophoresis. The length of the PCR product should be around 7 kb (the same size as the original backbone vector).
9. Use 5 μL of PCR reaction (*see* **Note 5**) for transformation of competent *E. coli* XL1 cells. Please follow the transformation procedure recommended by the manufacturer of the competent cells.
10. Pick at least two transformants and cultivate for isolation of the plasmid DNA.
11. Isolate plasmid DNA by using, e.g., a Miniprep DNA Purification System according to the manufacturer's instructions. To verify the successful insertion of the gRNA sequence, we recommend Sanger sequencing. When using the pCRISPR-Zeo vector, the recommended sequencing primer sequence is: 5′-TCTTCTTTATGCTGTGGCG-3′.
12. Perform a restriction digest of the pCRISPR-Zeo backbone vector using *Spe*I and *Not*I (follow the protocol outlined earlier).
13. Assess the quality of the linearization of the plasmid by agarose gel electrophoresis. The expected fragment size on the gel is ~7700 base pairs.
14. Purify the linearized plasmid with a PCR Clean-Up System according to the manufacturer's instructions or use ethanol precipitation (*see* **Note 6**) to remove the remaining, unwanted components of the restriction reaction.
15. For plasmid assembly, we recommend enzymatic Gibson assembly [35] with 50 ng of the linearized pCRISPR-Zeo backbone vector and 2 μL of a 10 μM 60-base oligonucleotide, which contains overhangs to the backbone vector and a gRNA sequence. If using a different cloning kit, follow the instructions of the manufacturer concerning DNA amounts and reaction conditions.
16. Transform any standard *E. coli* cloning strain with 5 μL of the plasmid assembly reaction mix according to the manufacturer's instructions, but ensure that the cells can be regenerated for 45–60 min prior to plating on LB agar plates supplemented with Zeocin. Incubate the plates overnight at 37 °C.
17. Pick at least two transformants and cultivate them in liquid LB medium or on LB agar plates for plasmid propagation.

18. Isolate plasmid DNA by using e.g., a Miniprep DNA Purification System according to the manufacturer's instructions. To verify the successful insertion of the gRNA sequence, we recommend Sanger sequencing. When using the pCRISPR-Zeo vector, the recommended sequencing primer sequence is: 5′--TCTTCTTTATGCTGTGGCG-3′.

3.2 Creating *K. phaffii* Mutant Strains

After the successful assembly of the CRISPR/Cas9 plasmid with the DNA coding for the gRNA, this section provides a pipeline for the transformation of *K. phaffii*. The standard method for the transformation of *K. phaffii* is electroporation. For confirmation of the CRISPR/Cas9-induced point mutation, PCR amplification of the targeted locus with subsequent sequencing is necessary. In addition, a protocol for episomal plasmid elimination to obtain an empty deletion strain is presented.

3.2.1 Transformation of *K. phaffii*

1. Use the cell material of any *K. phaffii* laboratory strain of interest to inoculate a preculture in a 50 mL centrifugation tube filled with 5 mL YPD medium and incubate at 28 °C for 4–5 h under moderate shaking (130 rpm).
2. After 4–5 h of cultivation, measure the cell density (OD_{600}) of the preculture and inoculate 50 mL YPD in a 250 mL baffled shake flask to a final OD_{600} of 0.002 and cultivate at 28 °C and 130 rpm for 20 h.
3. After 20 h of cultivation, assess the OD_{600} and if the obtained value is in the range from 0.8 to 5.0 (*see* **Note 7**), proceed with preparing competent cells.
4. To harvest the cells, transfer the culture into 50 mL centrifugation tubes and centrifuge at 25 °C and 500 g for 5 min (*see* **Note 8**).
5. Discard the supernatant and resuspend the cells in 9 mL BEDS and 1 mL 1 M DTT solution. Incubate under moderate shaking (130 rpm) at 28 °C for 5–10 min.
6. Repeat the centrifugation step explained in 4.
7. Discard the supernatant and resuspend cells in 2.5 mL BEDS. Divide the cell suspension into aliquots of 80 μL. In total, this protocol will provide competent cells for 30 transformations.
8. Take one aliquot of the competent *K. phaffii* cells and mix it with 200 ng of circular plasmid DNA.
9. Transfer the cell+DNA suspension into an electroporation cuvette and electroporate using a charging voltage of 1500 V, a resistance of 200 Ω, and a capacitance of 25–50 μF.
10. Immediately after the electroporation, add 1 mL YPDS medium to the electroporation cuvette. Transfer the whole transformation suspension from the cuvette into a 2 mL

reaction tube and regenerate the cells for 2–3 h at 28 °C and 120 rpm.

11. Plate 100 μL of the transformation suspension (*see* **Note 9**) on YPD agar plates supplemented with 100 μg/mL Zeocin and incubate the plates for 2 or 3 days at 28 °C.

3.2.2 Screening for Edited Strains

After the successful transformation of *K. phaffii*, this section describes the screening process to detect strains with edited target sequences. The screening process involves the cultivation of transformants in 96-deep-well plates (DWPs), amplification of target regions by PCR, and sending the amplicons for sequencing to examine whether the gene editing event took place.

1. Randomly select up to 96 transformants (*see* **Note 10**) and use them to inoculate a DWP filled with 250 μL of YPD medium containing Zeocin. One transformant per well.
2. Cultivate the transformants in the DWP for 48 h (*see* **Note 11**) at 28 °C, 80% humidity, and 320 rpm tilted in an orbital shaker (2.5 cm shaking diameter).
3. Replicate the deep-well plate on YPD square agar plates with and without supplemented Zeocin, respectively (*see* **Note 12**).
4. To identify if a gene editing event occurred, the target gene (s) are amplified using a colony-PCR and sent for sequencing. We recommend to use the Phire Plant Direct PCR Master Mix (Thermofisher Scientific). For each transformant, take 5 μL of the liquid cell culture from the DWP cultivation and mix it with 45 μL of the dilution buffer provided in the Phire Plant Direct PCR Master Mix (*see* **Note 13**).
5. Prepare a 96-well PCR plate and add 0.75 μL of forward and reverse primers, which specifically bind to the target gene, to each well. Subsequently, add 43.5 μL of 1× Phire Plant PCR Master Mix (following the manufacturer's protocol) to each well.
6. To finalize the PCR reaction mix, add 5 μL of the cell culture dilution, which was prepared in **step 4**, as a template for PCR. For amplification of the targeted gens the thermocycling conditions were chosen: 95 °C/300 s—(95 °C/20 s – 50–60 °C/20 s – 72 °C/90 s) × 35–72 °C/420 s – 4 °C/∞, Please note that the annealing temperature will be dependent on the primer design and the elongation time on the length of the target gene.
7. Confirm the success of the colony-PCR by gel electrophoresis.
8. Send the PCR fragments showing the correct size on the agarose gel for sequencing (follow the instructions provided by a sequencing service).

9. To analyze whether an indel is introduced into a target gene, align the sequencing result of the PCR fragments with the reference sequence of the target gene (*see* **Note 14**).

3.2.3 CRISPR/Cas9 Plasmid Elimination

1. Based on the sequencing results, pick a colony of the desired mutant strain from the antibiotic-free YPD square plate (*see* **Note 15**) and resuspend the cell material in 90 μL of sterile ddH_2O.
2. Homogenize the cell suspension by pipetting or vortexing and perform a serial dilution in microtiter plates (in that case, you can use a multichannel pipette to boost the dilution process) using ddH_2O until a dilution factor of 10^{-6} is reached. Make sure the cell suspension is properly mixed prior to each dilution step.
3. Plate 20 μL of the highest dilution on YPD and YPD-Zeo plates. Incubate the plates at 28 °C for 3 days (*see* **Note 16**).
4. After the incubation, it is expected that there are no colonies on the YPD plates with Zeocin and 10 to 30 clones on a YPD plate without the selection pressure. Finally, pick up a colony from the YPD plate to inoculate another ONC for the preparation of a glycerol stock (*see* **Note 17**).

4 Notes

1. If more than 40 oligonucleotides are needed, it is recommended to order them in a 96-well format.
2. The linearization is a critical step for the CRISPR/Cas9 plasmids. If traces of uncut vector are detected by gel electrophoresis, a small amount of additional restriction enzyme (e.g., 2 μL) can be added to the restriction reaction. Additionally, the restriction reaction can be incubated overnight.
3. Ethanol precipitation is the preferred method for purification because it results in a higher concentration of the linearized vector.
4. We recommend preparing a master mix containing all PCR components except the forward primers (with gRNAs) and aliquoting the master mix into PCR tubes. Add the forward primers as the last step since this is the only difference between the individual PCR reactions. The recommended volume of the PCR reaction is in the range from 10 to 20 μL.
5. Because of the low concentration of the template DNA, it is not necessary to treat PCR with *Dpn*I.

6. Ethanol precipitation is the recommended method because it results in a high concentration of the linearized backbone vector.
7. Although many protocols recommend preparing the *K. phaffii* competent cells from a cell culture with an OD_{600} between 0.8 and 1.2, we did not find any dramatic changes regarding the transformation efficiency when competent cells were made from a cell culture with an OD_{600} ranging between 2.0 and 5.0.
8. In our experience, there is no necessity to keep the cells on ice during the preparation of the competent cells or after aliquotation.
9. Picking colonies after the transformation is an essential step that affects the results. We recommend selecting the colonies that are most common in size and morphology on the transformation plate. Still, it might also be worth selecting a few colonies that are smaller size or show a strange morphology.
10. It is recommended to screen at least four clones of each knockout target. Sometimes it is better to select not only colonies of normal size but also small colonies for inoculation. Using YPD-Zeo allows for the elimination of cells which, for some reasons, do not carry a plasmid.
11. A shorter cultivation time is possible, but please take into consideration that some mutant strains might have a slow-growing phenotype.
12. Cells that grow on YPD agar without antibiotic tend to lose the ARS based episomal CRISPR plasmid due to the missing selection pressure. Thus, cultivation in/on media without Zeocin is required to obtain mutant strains without the CRISPR plasmid.
13. Preparation of the dilution can be performed in a 96-well-plate format.
14. Alignment of the sequences can be done in SnapGene, Clone-Manager, Vector NTI, or other software.
15. Growing the successfully edited clones in/on nonselective medium is crucial so that the isolates lose the CRISPR/Cas9 vector. The process can be performed in 96 DWP.
16. The amount of cell material that you take for the preparation of the serial dilutions will directly affect the number of colonies that you will obtain on YPD plates. Therefore, it is recommended to plate 20 μL not only from the 10^{-6} dilution but also from the 10^{-5} dilution.
17. It is recommended to examine the presence of indels in the target gene by PCR amplification and following the sequencing of the PCR fragment.

Acknowledgments

This work was performed as part of the Christian Doppler Laboratory. The financial support from the Austrian Federal Ministry for Digital and Economic Affairs and the National Foundation for Research, Technology and Development and from the contributing company partners Novo Nordisk, BioGrammatics, SeSam Biotech, Zeta and SolarBiotech is gratefully acknowledged.

Competing Interest Statement The authors have no conflicts of interest to declare that are relevant to the content of this chapter.

References

1. National T (2017) Somatic genome editing
2. Khan SH (2019) Genome-Editing Technologies: concept, pros, and cons of various genome-editing techniques and bioethical concerns for clinical application. Mol Ther Nucleic Acids 16(June):326–334. https://doi.org/10.1016/j.omtn.2019.02.027
3. Yang Z, Blenner M (2020) Genome editing systems across yeast species. Curr Opin Biotechnol 66:255–266. https://doi.org/10.1016/j.copbio.2020.08.011
4. Dunn DA, Pinkert CA (2014) Gene editing. Elsevier Inc.
5. Pan R, Zhang J, Shen WL, Tao ZQ, Li SP, Yan X (2011) Sequential deletion of Pichia pastoris genes by a self-excisable cassette. FEMS Yeast Res 11(3):292–298. https://doi.org/10.1111/j.1567-1364.2011.00716.x
6. Li C et al (2017) Recycling of a selectable marker with a self-excisable plasmid in Pichia pastoris. Sci Rep 7(1):1–10. https://doi.org/10.1038/s41598-017-11494-5
7. Ahmad M, Winkler CM, Kolmbauer M, Pichler H, Schwab H, Emmerstorfer-Augustin A (2019) Pichia pastoris protease-deficient and auxotrophic strains generated by a novel, user-friendly vector toolbox for gene deletion. Yeast 36(9):557–570. https://doi.org/10.1002/yea.3426
8. Krainer FW et al (2013) Knockout of an endogenous mannosyltransferase increases the homogeneity of glycoproteins produced in Pichia pastoris. Sci Rep 3. https://doi.org/10.1038/srep03279
9. Kang Z, Huang H, Zhang Y, Du G, Chen J (2017) Recent advances of molecular toolbox construction expand Pichia pastoris in synthetic biology applications. World J Microbiol Biotechnol 33(1):1–8. https://doi.org/10.1007/s11274-016-2185-2
10. Cody JP, Graham ND, Zhao C, Swyers NC, Birchler JA (2020) Site-specific recombinase genome engineering toolkit in maize. Plant Direct 4(3):1–9. https://doi.org/10.1002/pld3.209
11. Kilby NJ, Snaith MR, Murray JAH (1993) Site-specific recombinases: tools for genome engineering. Trends Genet 9(12):413–421. https://doi.org/10.1016/0168-9525(93)90104-P
12. Olorunniji FJ, Rosser SJ, Stark WM (2016) Site-specific recombinases: molecular machines for the genetic revolution. Biochem J 473(6):673–684. https://doi.org/10.1042/BJ20151112
13. Long DP et al (2012) FLP recombinase-mediated site-specific recombination in silkworm, Bombyx mori. PLoS One 7(6). https://doi.org/10.1371/journal.pone.0040150
14. Hubbard EJA (2014) FLP/FRT and Cre/lox recombination technology in C. elegans. Methods 68(3):417–424. https://doi.org/10.1016/j.ymeth.2014.05.007
15. Addgene (2011) Addgene's Plasmids 101, vol 2017, 3rd edn, p 193
16. Schweizer HP (2003) Applications of the Saccharomyces cerevisiae Flp-FRT system in bacterial genetics. J Mol Microbiol Biotechnol 5(2):67–77. https://doi.org/10.1159/000069976
17. Phan QV, Contzen J, Seemann P, Gossen M (2017) Site-specific chromosomal gene insertion: Flp recombinase versus Cas9 nuclease. Sci Rep 7(1):1–12. https://doi.org/10.1038/s41598-017-17651-0

18. Branda CS, Dymecki SM (2004) Talking about a revolution: the impact of site-specific recombinases on genetic analyses in mice. Dev Cell 6(1):7–28. https://doi.org/10.1016/S1534-5807(03)00399-X
19. Cregg JM, Madden KR (1989) Use of site-specific recombination to regenerate selectable markers. MGG Mol Gen Genet 219(1–2): 320–323. https://doi.org/10.1007/BF00261194
20. Guo F et al (2021) Metabolic engineering of Pichia pastoris for malic acid production from methanol. Biotechnol Bioeng 118(1): 357–371. https://doi.org/10.1002/bit.27575
21. Urnov FD, Rebar EJ, Holmes MC, Zhang HS, Gregory PD (2010) Genome editing with engineered zinc finger nucleases. Nat Rev Genet 11(9):636–646. https://doi.org/10.1038/nrg2842
22. Chandrasegaran S, Carroll D (2016) Origins of programmable nucleases for genome engineering. J Mol Biol 428(5):963–989. https://doi.org/10.1016/j.jmb.2015.10.014
23. Joung JK, Sander JD (2013) TALENs: a widely applicable technology for targeted genome editing. Nat Rev Mol Cell Biol 14(1):49–55. https://doi.org/10.1038/nrm3486
24. Swartjes T, Staals RHJ, Van Der Oost J (2020) Editor's cut: DNA cleavage by CRISPR RNA-guided nucleases Cas9 and Cas12a. Biochem Soc Trans 48(1):207–219. https://doi.org/10.1042/BST20190563
25. Resource AD (2021) CRISPR 101 a desktop resource introduction to addgene's CRISPR eBook. 2021(January)
26. Komor AC, Badran AH, Liu DR (2017) CRISPR-based technologies for the manipulation of eukaryotic genomes. Cell 168(1–2): 20–36. https://doi.org/10.1016/j.cell.2016.10.044
27. Jinek M, Chylinski K, Fonfara I, Hauer M, Doudna JA, Charpentier E (2012) A programmable dual-RNA-guided DNA endonuclease in adaptive bacterial immunity. Science (80-.) 337(6096):816–821. https://doi.org/10.1126/science.1225829
28. Goyon A et al (2021) Full sequencing of CRISPR/Cas9 single guide RNA (sgRNA) via parallel ribonuclease digestions and hydrophilic interaction liquid chromatography-high-resolution mass spectrometry analysis. Anal Chem 93(44):14792–14801. https://doi.org/10.1021/acs.analchem.1c03533
29. Dicarlo JE, Norville JE, Mali P, Rios X, Aach J, Church GM (2013) Genome engineering in Saccharomyces cerevisiae using CRISPR-Cas systems. Nucleic Acids Res 41(7):4336–4343. https://doi.org/10.1093/nar/gkt135
30. Weninger A, Hatzl AM, Schmid C, Vogl T, Glieder A (2016) Combinatorial optimization of CRISPR/Cas9 expression enables precision genome engineering in the methylotrophic yeast Pichia pastoris. J Biotechnol 235:139–149. https://doi.org/10.1016/j.jbiotec.2016.03.027
31. Dalvie NC et al (2021) Host-informed expression of CRISPR guide RNA for genomic engineering in Komagataella phaffii. ACS Synth Biol 9(1):26–35. https://doi.org/10.1021/acssynbio.9b00372.Host-informed
32. Bethesda (MD): National Library of Medicine (US), "National Center for Biotechnology Information," (1988)
33. H. (UK) Wellcome Genome Campus, "EMBL-EBI," (1992)
34. Stemmer M, Thumberger T, Del Sol Keyer M, Wittbrodt J, Mateo JL (2015) CCTop: an intuitive, flexible and reliable CRISPR/Cas9 target prediction tool. PLoS One 10(4):1–11. https://doi.org/10.1371/journal.pone.0124633
35. Gibson DG, Young L, Chuang R-Y, Venter JC, Hutchison CA, Smith HO (2009) Enzymatic assembly of DNA molecules up to several hundred kilobases. Nat Methods 6(5):343–345. https://doi.org/10.1038/nmeth.1318

Chapter 25

Engineering *Pichia pastoris* Strains Using CRISPR/Cas9 Technologies: The Basic Protocol

Hana Raschmanová, Astrid Weninger, and Karin Kovar

Abstract

The CRISPR/Cas9 (clustered regularly interspaced short palindromic repeats, CRISPR-associated protein 9) system has become a commonly used tool for genome editing and metabolic engineering. For *Komagataella phaffii*, commercialized as *Pichia pastor*is, the CRISPR/Cas9 protocol for genome editing was established in 2016 and since then has been employed to facilitate genetic modifications such as markerless gene disruptions and deletions as well as to enhance the efficiency of homologous recombination.

In this chapter, we describe a robust basic protocol for CRISPR-based genome editing, demonstrating near 100% targeting efficiency for gene inactivation via a frameshift mutation. As described in other chapters of this volume, CRISPR/Cas9 technologies for use in *P. pastoris* have been further optimized for various specific purposes.

Key words *Pichia pastoris*, *Komagataella phaffii*, CRISPR/Cas9, Guide RNA, Genome editing, Genome engineering, Homologous recombination, Gene knockout, Synthetic biology

1 Introduction

CRISPR/Cas9 is a naturally occurring defense mechanism against invading DNA in bacteria and *Archaea* [1] and has been adapted to function as a versatile tool for genome editing [2, 3], as well as metabolic engineering, transcriptional regulation, etc. [4]. This tool was also optimized for use in *P. pastoris* [5, 6]. CRISPR/Cas9 technology employs the following two components (Fig. 1a):

- Single guide RNA (sgRNA), commonly termed as guide RNA (gRNA).
- DNA endonuclease Cas9.

The sgRNA consists of a gRNA scaffold sequence, i.e., a structural element, enabling it to associate with Cas9, and a gRNA targeting sequence, i.e., a variable 20 bp element that is complementary to the targeted genome sequence. Through this variable

Xavier Garcia-Ortega et al. (eds.), *Pichia pastoris: Methods and Protocols*, Methods in Molecular Biology, vol. 2697,
https://doi.org/10.1007/978-1-0716-4779-0_25,

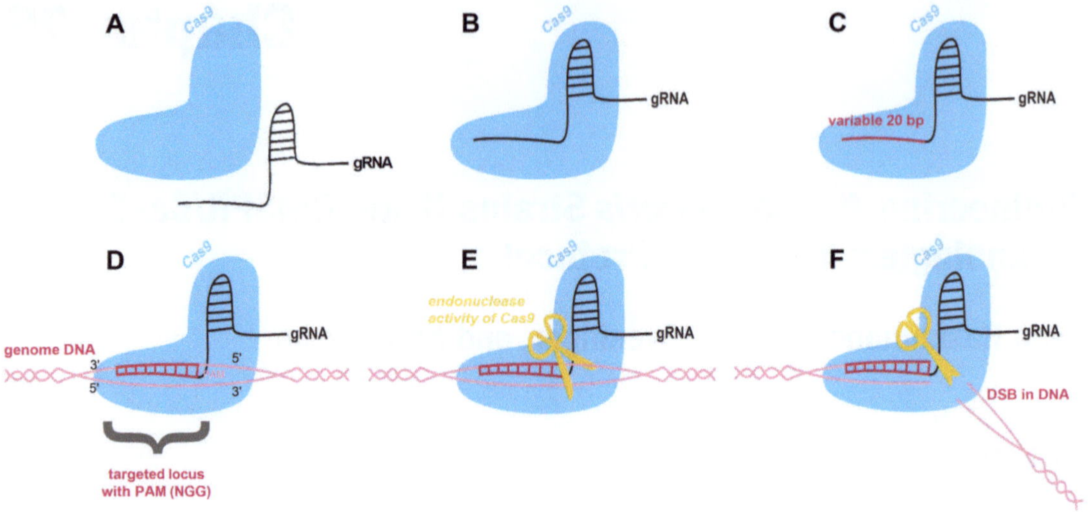

Fig. 1 Molecular mechanism of CRISPR/Cas9. gRNA acts as a navigator and Cas9 as scissors. (**a**) CRISPR/Cas9 consists of DNA endonuclease Cas9 and guide RNA (gRNA). (**b**) The Cas9 and gRNA form complex. (**c**) The gRNA contains a variable 20 bp part, which is complementary to a targeted genome sequence. (**d**) The endonuclease Cas9 is guided to the targeted genome sequence, located next to the protospacer adjacent motif (PAM) sequence, where the unwinding of DNA starts. (**e**) and (**f**) Cleavage of the DNA by Cas9, introduction of a double-strand break (DSB)

element of gRNA, the endonuclease Cas9 is guided to its target on a specific genome sequence (Fig. 1b). When referring to gRNA in this text further on, the complete sgRNA is meant, i.e., the gRNA consisting of the scaffold and the targeting sequence.

Besides its complementarity to the gRNA sequence, the target site is also determined by the presence of a protospacer adjacent motif (PAM). This is a short sequence in the genome, adjacent to the target site for Cas9, specifically NGG, where the unwinding of DNA is initiated. After the DNA is unwound, Cas9 cleaves the DNA by introducing a double-strand break (DSB, Fig. 1c). So in principle, gRNA acts as a navigator and Cas9 as scissors. A big advantage of CRISPR/Cas9 is that it can be retargeted simply by changing the 20 bp variable part of gRNA, which makes it a practicable tool for fast genome editing.

In *P. pastoris*, nonhomologous end joining (NHEJ) based on random insertions/deletions (indels) is a preferred DNA-repair mechanism compared with homologous recombination (HR) (Fig. 2). This makes targeted integration of expression cassettes in *P. pastoris* less effective than in *Saccharomyces cerevisiae*, where genetic modifications are naturally facilitated by efficient HR. The lack of simple and efficient methods for gene knockout, and especially integration of expression cassettes, has hindered genome engineering in *P. pastoris*. Employing CRISPR/Cas9, gene knockout can be performed simply: Cas9 introduces a DSB at the target

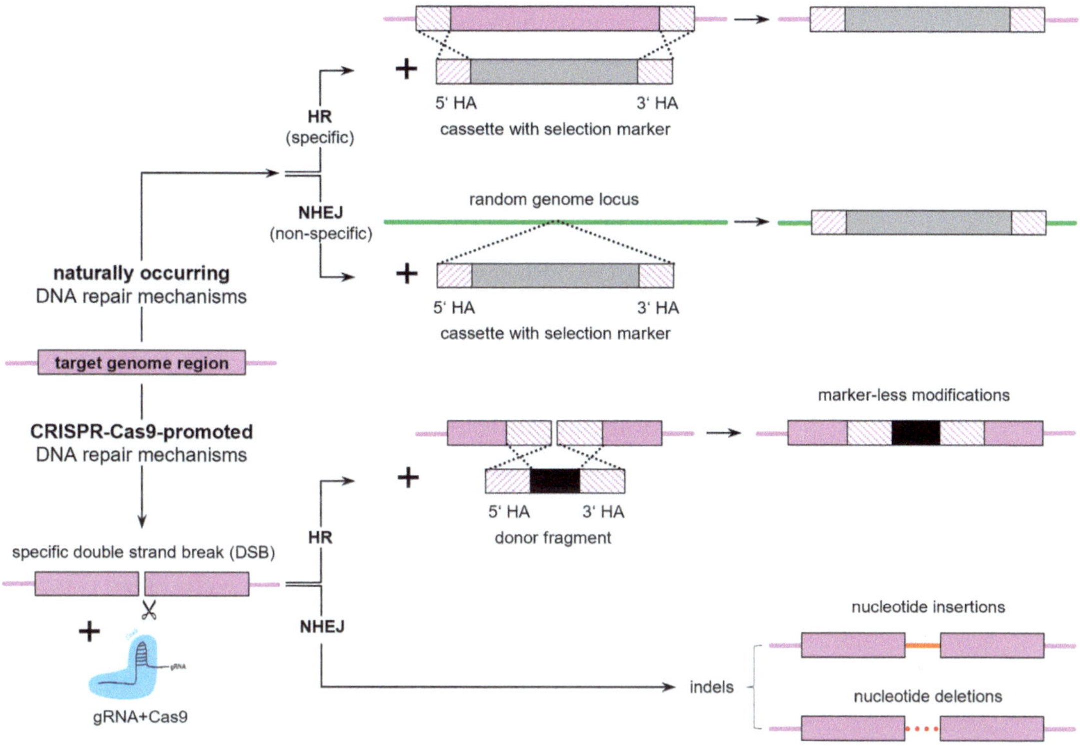

Fig. 2 Genetic modifications via naturally occurring and CRISPR/Cas-promoted DNA repair mechanisms. (Adapted from Ref. [4]). The genetic modifications based on naturally occurring DNA repair mechanisms only involve the delivery of an exogenous targeting cassette with 5′ and 3′ homology arms (HAs) identical to the genome. The cassette can be integrated via two mechanisms: homologous recombination (HR) or non-homologous end joining (NHEJ). HR leads to specific integration via the HAs, while NHEJ leads to integration in an arbitrary spot in the genome. Genetic modifications involving CRISPR/Cas9 systems are based on the introduction of a double-strand break (DSB) by the gRNA-Cas9 complex. If no exogenous DNA is provided, the DSB is repaired via NHEJ, resulting in insertions and deletions (indels) of varying lengths. Providing an exogenous donor DNA with HAs allows markerless HR modifications at generally increased frequencies

site, and this DSB is then repaired by NHEJ, whereby a frameshift can be introduced by an indel mutation (Fig. 2). Integration of the donor cassette with homologous arms via HR can then be facilitated in *P. pastoris* due to the targeted DSB introduced by CRISPR (Fig. 2). CRISPR/Cas9, therefore, allows for rapid and markerless genome engineering of *P. pastoris*.

2 Materials

Prepare all solutions at room temperature, using demineralized water for cultivation media and buffers, and molecular-biology-grade water for PCR and cloning reactions. Store heat or filter-sterilized cultivation media (liquid or solid) and stock solutions at

4 °C. Use a laminar flow box for sterile and aseptic work such as mixing cultivation media, handling the *Escherichia coli* and *P. pastoris* cultures, etc. Diligently follow all waste disposal regulations when disposing of waste materials.

Procurement details are given only when a very specific material that cannot be ordered elsewhere or a device is required and may be difficult to locate.

1. Plasmid pCRISPR-Zeo, carrying the Cas9 gene under the control of the bidirectional promoter P_{HTX1} and Zeocin resistance marker gene [5, 7].
2. Synthetic double-stranded DNA carrying gRNA.
3. Restriction endonuclease *NotI.*
4. Gibson assembly master mix.
5. Electrocompetent *E. coli* cells (e.g., DH5α or TOP10F').
6. Electrocompetent *P. pastoris* cells to be CRISPR modified.

   The strains tested with this protocol were killer plasmid-free BSYBG10, a single colony isolated of the BioGrammatics strain *Komagataella phaffii* BG10 and deposited in Europe (bisy GmbH, Hofstaetten a. d. Raab, Austria), and strains based on the *P. pastoris* parental strain CBS 435 (identical to NRRL Y-11430 and ATCC 76273) but can be applied with all other closely related strains such as strains derived from the Type strain NRLL Y-4290 (Y-7556, CBS2612, Pichia Open) or strains derived from the Thermo Fisher Scientific strains X33, GS115, or KM71(H) and related strains.
7. LB medium with Zeocin™ for *E. coli* clone selection: 5 g yeast extract, 10 g tryptone, 5 g NaCl, 20 g agar (for plates only), and 25 mg Zeocin™ per 1 litre. Dissolve 5 g yeast extract, 10 g tryptone, and 5 g NaCl in 800 mL of demineralized water. Adjust the pH to 7.5. When preparing plates, add 20 g agar. Fill with water to 1 litre and sterilize by autoclaving (121 °C for 20 min). Cool to approximately 60 °C and add Zeocin™ to a final concentration of 25 μg/mL.
8. YPD medium with Zeocin™ for *P. pastoris* clone selection: 20 g peptone, 10 g yeast extract, 20 g glucose, 20 g agar (for plates only), 100 mg Zeocin™ per 1 litre. Dissolve 20 g of peptone and 10 g of yeast extract in demineralized water. When preparing plates, add 20 g of agar. Fill to 900 mL and sterilize by autoclaving (121 °C for 20 min). Cool to approximately 60 °C, add 100 mL of glucose in the form of a separately autoclaved 20% glucose solution, and Zeocin™ to a final concentration of 100 μg/mL.
9. Primers.
10. Electroporator.

11. PCR thermocycler.
12. Incubator.
13. Orbital shaker.
14. 96-deep-well plates
15. Plasmid isolation kit.
16. DNA/plasmid purification kit.

3 Methods

Here, we describe the design of the CRISPR/Cas9 plasmid and its application for genome editing of *P. pastoris*, as based on the original protocol established by Weninger et al. [5], employing self-splicing for gRNA processing. As a model, we describe the knockout of the gene *HAC1* by CRISPR/Cas9 in *P. pastoris* BG10. However, the protocol below is generally applicable to other genetic manipulations.

3.1 Designing the CRISPR Plasmid

1. Search for the nucleotide sequence for the target region of interest in the *P. pastoris* genome, e.g., in the NCBI database, https://www.ncbi.nlm.nih.gov/

   For gene knockout, use the gene's coding sequence starting with ATG and ending with a stop codon.

   For example, for *HAC1*, the nucleotide sequence was:

   ATGCCCGTAGATTCTTCTCATAAGACAGCTAGCCCACTTC-
   CACCTCGTAAAAGAGCAAAGACGGAAGAAGAAAAGGAGCAGCGTC-
   GAGTGGAACGTATCCTACGTAATAGGAGAGCGGCCCATGCTTCCAGA-
   GAGAAGAAACGAAGACACGTTGAATTTCTGGAAAACCACGTCGTC-
   GACCTGGAATCTGCACTTCAAGAATCAGCCAAAGCCACTAACAAGTT-
   GAAAGAAATACAAGATATCATTGTTTCAAGGTTGGAAGCCTTAGGTGG-
   TACCGTCTCAGATTTGGATTTAACAGTTCCGGAAGTCGATTTTCC-
   CAAATCTTCTGATTTGGAACCCATGTCTGATCTCTCAACTTCTTC-
   GAAATCGGAGAAAGCATCTACATCCACTCGCAGATCTTTGACTGAG-
   GATCTGGACGAAGATGACGTCGCTGAATATGACGACGAAGAAGAGGAC-
   GAAGAGTTACCCAGGAAAATGAAAGTCTTAAACGACAAAAACAAGAG-
   CACATCTATCAAGCAGGAGAAGTTGAATGAACTTCCATCTCCTTTGT-
   CATCCGATTTTTCAGACGTAGATGAAGAAAAGTCAACTCTCACACATT-
   TAAAGTTGCAACAGCAACAACAACAACCAGTAGACAATTATGTTTC-
   TACTCCTTTGAGTCTTCCGGAGGATTCAGTTGATTTTATTAACCCAGG-
   TAACTTAAAAATAGAGTCCGATGAGAACTTCTTGTTGAGTTCAAA-
   TACTTTACAAATAAAACACGAAAATGACACCGACTACATTACTA-
   CAGCTCCATCAGGTTCCATCAATGATTTTTTTAATTCTTATGACAT-
   TAGCGACTCGAATCGGTTGCATCATCCAGCAGTGATGACGGATT-
   CATCTTTACACATTACAGCAGGCTCCATCGGCTTTTTCTCTT-
   GATTGGGGGGGGGGAAAGTTCTGTAGCAGGGAGGCGCAGTTCAGTTGG-

CACATATCAGTTGACATGCATAGCGATCAGGTGATTGCTCTTGT-
TAACTGGTATGCCAAGTCAACCGTTTTCGGCGTACCTACTACTC-
TACTTTAACTCTTGAATTGAGATGAAACAATCGCCCTTGTTCGA-
GACGCGTAGTTAATTTAAGATGAATCAATTTGCAACAAAAGAACAAAC-
GAAAAAGATAACAGCAAACTTAAGTTTCGTGGTCTGCAGCACCATT-
TACCGCTAATGCATTTGATTTAAATGACTTTGTATTCTTCCAGGAATAG

2. Determine the possible gRNA sequences (20 bp variable element) for your target. A PAM site sequence has to be adjacent to the target site in the genome. Various online tools are available for the prediction of gRNA sequences, e.g., CCTop-CRISPR/Cas9 target online predictor, https://crispr.cos.uni-heidelberg.de/.

   Copy and paste the nucleotide sequence of the target gene into the field "single query" in the predictor tool. In the field "PAM type", select "NGG". In the field "species", select "*Komagataella phaffii* CBS 7435". Press "Submit". A list of possible gRNAs is generated. If different prediction software is used or the gRNA is selected manually, make sure that the gRNA is not located in a *P. pastoris* intronic sequence.

3. Consider several different gRNAs from the generated list, since various gRNAs might have various efficiencies.

   For the model case of *HAC1* disruption, the following three predicted variants of gRNA were considered for further evaluation:

   gRNA1: AGAAAAGGAGCAGCGTCGAG
   gRNA2: GGAAAACCACGTCGTCGACC
   gRNA3: CAGATCTTTGACTGAGGATC

4. Avoid off-targeting (i.e., non-specific cleavage sites) of Cas9. The off-targeting is commonly predicted by the tool used for gRNA prediction. Nevertheless, you can always check for the off-target regions by blasting the sequence of your gRNA + PAM sequence (AGG, TGG, CGG, GGG, and all combinations) against the *P. pastoris* genome, using the NCBI Nucleotide Blast Tool, https://blast.ncbi.nlm.nih.gov/Blast.cgi

   In the field "Organism", choose "*Komagataella phaffii* CBS 7435".

   For example, to check the suitability of the gRNA1 from the previous step, you have to blast the following sequences against the genome of *Komagataella phaffii* CBS 7435:

   - AGAAAAGGAGCAGCGTCGAG<u>AGG.</u>
   - AGAAAAGGAGCAGCGTCGAG<u>TGG.</u>
   - AGAAAAGGAGCAGCGTCGAG<u>CGG.</u>
   - AGAAAAGGAGCAGCGTCGAG<u>GGG.</u>

     If there is more than one identical sequence in the result list of the blasting, do not use the gRNA, as this sequence

a

b

CCAGTTCAAGTTACCTAAACAAATCAAATTTTCTCTGATGAGTCCGTGAGGAC

GAAACGAGTAAGCTCGTCAGAAAAGGAGCAGCGTCGAGGTTTTAGAGCTAGAA

ATAGCAAGTTAAAATAAGGCTAGTCCGTTATCAACTTGAAAAAGTGGCACCGA

GTCGGTGCTTTTGGCCGGCATGGTCCCAGCCTCCTCGCTGGCGCCGGCTGGGC

AACATGCTTCGGCATGGCGAATGGGACTCAAGAGGATGTCAGAATGCC

Fig. 3 Exchangeable part of the CRISPR plasmid designed for *P. pastoris* by Weninger et al. (2016) [5]. (**a**) Schematic overview of the exchangeable cassette carrying the gRNA sequence to be inserted into the *NotI*-linearized backbone of plasmid pCRISPR-Zeo: Overhang to the P_{HTX1} promoter (orange), variable part of the hammerhead ribozyme (HH) complementary to the first six nucleotides of the gRNA (green), hammerhead ribozyme (green), 20 bp variable element of the gRNA (black), structural element of gRNA (black), hepatitis delta virus (HDV) ribozyme (blue), and overhang to the *AOX1* terminator (*AOX1TT*) (red). (**b**) Nucleotide sequence of the exchangeable cassette with an example of a gRNA sequence for targeting the *HAC1* gene for knockout: The variable part of the hammerhead ribozyme, with reverse-complementarity to the first six nucleotides of the gRNA, is underlined (TTTTCT). The 20 bp variable part of gRNA is framed (AGAAAAGGAG CAGCGTCGAG)

will certainly be a 100% off-target region in the genome. However, off-targeting can also take place at similar sites to the gRNA sequence and can hardly be predicted.

5. Once you have selected the right gRNA, design and order the synthetic DNA cassette carrying the overhang to the P_{HTX1} promoter, the variable part of the hammerhead ribozyme (*see* **Note 1**), the hammerhead ribozyme, the 20 bp variable element of the gRNA (very importantly: without PAM sequence), the structural element of gRNA, the HDV ribozyme and the overhang to the terminator [5]. *See* Fig. 3 for the nucleotide sequence.

   The synthetic ds DNA can be ordered, for example, from Integrated DNA Technologies, Inc., https://eu.idtdna.com/

3.2 Preparing the CRISPR Plasmid

1. Linearize the plasmid pCRISPR-Zeo [7] with *NotI*. Follow the manufacturer's instructions for the enzyme reaction conditions. Purify after linearization.
2. Insert the ordered synthetic DNA cassette (*see* **step 5** in Subheading 3.1) into the *NotI*-linearized and purified backbone of plasmid pCRISPR-Zeo [7]. This cloning can be done, e.g., by Gibson assembly, by mixing 50 ng of the plasmid backbone with an equimolar amount of the insert DNA and with the

Gibson assembly master mix. Follow the manufacturer's instructions for the assembly reaction conditions.

3. Transform the electrocompetent *E. coli* cells with the assembled plasmid (3 μL of the Gibson assembly reaction mixture) by electroporation, using the following settings at an electroporator: 2.5 kV/25 μF/200 Ω. Plate the transformed culture on the LB medium with Zeocin™ (agar plates) for selection. Insert the plates into an incubator set at 37 °C and incubate until the colonies appear (typically overnight).
4. Screen several random clones for the presence of the complete plasmid (with inserted gRNA cassette). As a screening method, you can use (colony) PCR for the detection of the inserted cassette. Alternatively, you can isolate the plasmid from the clones with a plasmid isolation kit and then cleave the plasmid with specific restriction endonucleases to check the size of the plasmid, i.e., to see if the cassette with the gRNA was inserted.
5. In selected positive clone(s), verify the insert, i.e., the gRNA cassette, by sequencing, e.g., by Microsynth (Vienna, Austria). For example, this primer sequence can be used for sequencing of the insert: TCCCAAACCCCTACCACAAG.
6. To obtain higher amounts of the plasmid (optional), culture the verified clone in liquid LB medium with Zeocin™. Insert the culture into an orbital shaker set at 37 °C and incubate. After reaching a desired optical density of the culture (indicated by the manufacturer of a plasmid isolation kit), isolate the plasmid from the bacterial cells. Purify the isolated plasmid using a DNA/plasmid purification kit.

3.3 Genome Editing Using the CRISPR Plasmid

1. (Co-)transform electrocompetent *P. pastoris* cells with 200 ng of the circular CRISPR-plasmid DNA, and optionally with 100 ng to 1 μg of donor DNA, by electroporation, using the following settings at an electroporator: 2 kV/25 μF/200 Ω. Plate the transformed culture on YPD medium with Zeocin™ (agar plates) for selection. Insert the plates into an incubator set at 30 °C and incubate until the colonies appear (typically 48 to 72 h).

   In the model case of *HAC1* knockout, the *P. pastoris* cells were only transformed with the CRISPR plasmid carrying the selected gRNA1. It was expected that the DSB introduced by Cas9 would cause a frameshift mutation in the *HAC1* coding sequence, resulting in a nonfunctional protein product after expression of the gene.
2. Confirm randomly selected clone(s) by sequencing (*see* **Note 2**). The targeting efficiency upon CRISPR transformation is very high (typically app. 80% and > 50% for difficult loci such as *OCH1*), and positive transformants can be easily identified by sequencing.

In our model experiment, the mutation introduced by gRNA1 and Cas9 in the *HAC1* locus of *P. pastoris* BG10 was shown to be the insertion of a single nucleotide "C" at position 86 downstream from ATG (... AGAAAAGGAGCAGCGTCCGAGTGG...).

3. Depending on the target locus, positive clones can be recognized from the parental strain, for instance, by a different phenotype (*see* **Note 3**). More commonly, however, the mutation does not result in a changed phenotype.

   In the case of targeting *HAC1*, the knockout mutants were expected to be sensitive to a reducing agent, dithiothreitol (DTT). Therefore, randomly chosen transformants were cultured in 96-deep-well plates with YPD and Zeocin™ for 48 h at 28 °C. Then, the transformants were plated on YPD agar plates with dithiothreitol and incubated at 28 °C for 3 days (Fig. 4). The sensitive clones, i.e., those not growing in the presence of DTT, were the positive ones.

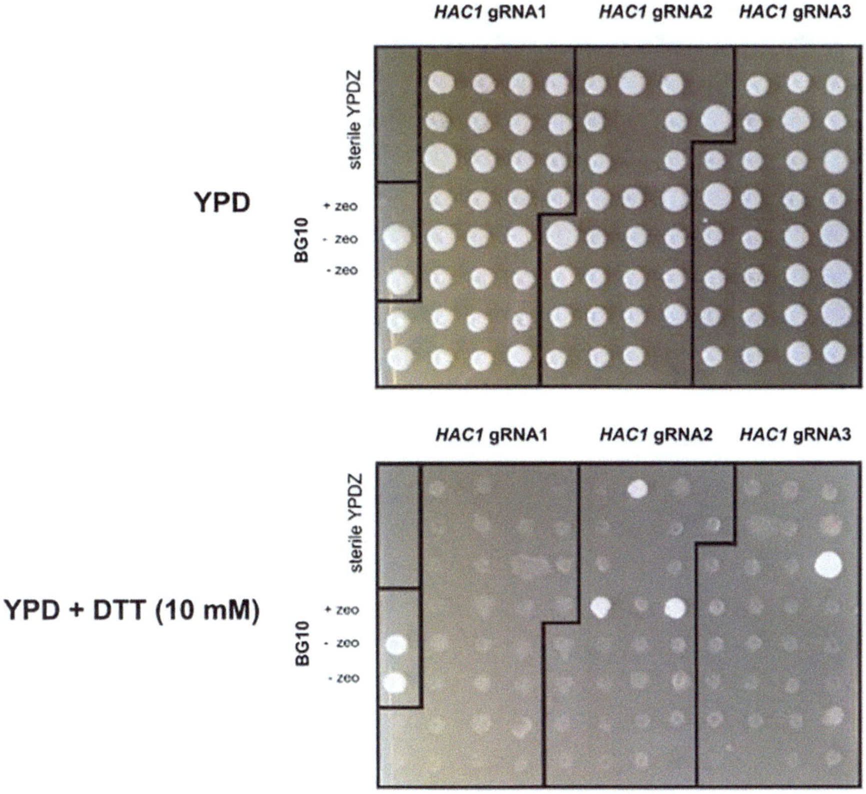

Fig. 4 Screening of clones by cultivation after transformation of *P. pastoris* BSYBG10 with a plasmid carrying gRNA and *Hs*Cas9 targeting the *HAC1* locus. Three different gRNAs (1, 2, 3) were designed to target the *HAC1* locus. Randomly chosen clones after transformation were incubated in 96-deep well plates containing YPD and Zeocin™ at 28 °C, 320 rpm for 48 h. The clones were then stamped on square agar plates with YPD and YPD plus 10 mM dithiothreitol (DTT). The plates were incubated for 2 days at 28 °C. Sterile YPDZ medium was used as a negative growth control, and *P. pastoris* BSYBG10 strain was used as a positive growth control

4. Remove the CRISPR plasmid by eliminating the selection stress, i.e., by cultivating the obtained strain in cultivation medium in the absence of Zeocin™ (*see* **Note 4**).

4 Notes

1. The variable part of the hammerhead ribozyme (HH) needs to be exchanged according to the gRNA sequence (Fig. 3). Namely, the variable part of the hammerhead ribozyme (HH) must be complementary to the first six nucleotides of the gRNA.
2. Positive clone(s) can be confirmed by colony PCR (e.g., Phire Plant Direct PCR Master Mix, Thermo Fisher Scientific), using primers against the mutated locus and subsequent sequencing of this locus.
3. High-throughput screening of the positive clones can be performed by cultivating the clones in Bel-Art (Sigma-Aldrich) deep-well plates in liquid YPD with Zeocin™, with subsequent transfer of the clones on selection agar plates (selection marker depending on the character of the introduced mutation). Using specifically the Bel-Art (Sigma-Aldrich) deep-well plates is considered to be critical, since other types of deep-well plates do not provide good oxygen transfer for the cultivation of *Pichia pastoris*.
4. The CRISPR plasmid contains an autonomously replicating sequence PARS1, which enables a transient expression and is responsible for the low stability of the plasmid. A single colony of the strain can be cultured in selective and nonselective media. The loss of the CRISPR plasmid is indicated by the fact that the colony does not grow in the selective media.

Competing Interest Statement The authors have no conflicts of interest to declare that are relevant to the content of this chapter.

References

1. Mojica FJM, Díez-Villaseñor C, García-Martínez J, Soria E (2005) Intervening sequences of regularly spaced prokaryotic repeats derive from foreign genetic elements. J Mol Evol 60:174–182. https://doi.org/10.1007/s00239-004-0046-3
2. Cong L, Ran FA, Cox D, Lin S, Barretto R, Habib N, Hsu PD, Wu X, Jiang W, Marraffini LA, Zhang F (2013) Multiplex genome engineering using CRISPR/Cas systems. Science 339:819–823. https://doi.org/10.1126/science.1231143
3. Jinek M, Chylinski K, Fonfara I, Hauer M, Doudna JA, Charpentier E (2012) A programmable dual-RNA-guided DNA endonuclease in adaptive bacterial immunity. Science 337:816–821. https://doi.org/10.1126/science.1225829
4. Raschmanová H, Weninger A, Glieder A, Kovar K, Vogl T (2018) Implementing

CRISPR-Cas technologies in conventional and non-conventional yeasts: current state and future prospects. Biotechnol Adv 36:641–665. https://doi.org/10.1016/j.biotechadv.2018.01.006

5. Weninger A, Hatzl AM, Schmid C, Vogl T, Glieder A (2015) Combinatorial optimization of CRISPR/Cas9 expression enables precision genome engineering in the methylotrophic yeast *Pichia pastoris*. J Biotechnol 235:139–149. https://doi.org/10.1016/j.jbiotec.2016.03.027
6. Weninger A, Fischer JE, Raschmanová H, Kniely C, Vogl T, Glieder A (2018) Expanding the CRISPR/Cas9 toolkit for *Pichia pastoris* with efficient donor integration and alternative resistance markers. J Cell Biochem 119:3183–3198. https://doi.org/10.1002/jcb.26474
7. Fischer JE, Glieder A (2022) CRISPR/Cas9 tool kit for efficient and targeted insertion/deletion mutagenesis of the *Komagataella phaffii* (*Pichia pastoris*). Genome Methods Mol Biol 2513:121–133. https://doi.org/10.1007/978-1-0716-2399-2_8

Chapter 26

Adaptive Laboratory Evolution of *Pichia pastoris* by Serial Batch Cultivation

Charles Moritz, Thomas Gassler, Martin Altvater, and Diethard Mattanovich

Abstract

Adaptive laboratory evolution (ALE) is a research tool widely applied for the improvement and investigation of microorganisms. Multiple yeast species have been effectively targeted with this method, and *Komagataella phaffii* specifically is now well-suited for ALE experimentation due to advancements in its characterization and genetic manipulation. Here is presented a simple and generalizable protocol for the adaptation of *K. phaffii* towards growth on minimal media with methanol as the sole carbon source, along with suggestions on how this protocol may be adapted for other purposes. We give detailed steps for the repeated subculturing as well as recommended pre-experimental testing and post-experimental analysis. We also include notes and tips from hands-on experience with this methodology as applied specifically to *K. phaffii*.

Key words *Komagataella phaffii*, *Pichia pastoris*, Adaptive laboratory evolution, Serial batch cultivation, Strain engineering, Reverse engineering

1 Introduction

Adaptive laboratory evolution (ALE) leverages our understanding of evolution by natural selection to investigate and improve microorganisms in a controlled setting [1, 2]. In addition to the discovery of underlying evolutionary principles [3–5], ALE is also increasingly applied in biotechnology for strain engineering [6, 7]. The two main approaches to performing ALE in the lab setting are continuous cultivation and serial batch cultivation. Continuous cultivation strategies are commonly used for heterologous production but are less common for performing ALE with *Komagataella phaffii* (reclassified from *Pichia pastoris*). For ALE purposes, however, chemostat or turbidostat cultivations can be performed in parallel operating bioreactors (e.g., DASGIP® or Infors® HT systems) or in more downscaled solutions allowing

Xavier Garcia-Ortega et al. (eds.), *Pichia pastoris: Methods and Protocols*, Methods in Molecular Biology, vol. 2697,
https://doi.org/10.1007/978-1-0716-4779-0_26,

for higher parallelization (e.g., Ambr® or eVOLVER systems). In general, these solutions are advantageous when a constant physiological state is required during ALE, as conditions such as pH, cell density, or media composition can be easily maintained [8, 9]. A particular advantage of chemostat systems is the ability to precisely control population size during evolution by setting a fixed dilution rate, presumably leading to reduced evolutionary bottlenecks [6, 8]. However, the high initial costs in terms of labor and investment are a major drawback for continuous cultivation during ALE of *K. phaffii,* preventing it from becoming a standard tool so far. For this reason, we focus on serial batch cultivation, which is considered the simplest, most accessible, and widely used approach for ALE, in which the evolving population is repeatedly subcultured to fresh media, usually attempting to maintain the cells in exponential phase [10]. Selective pressure may be applied and adjusted by altering the media composition and/or environmental conditions over the course of the experiment, which typically lasts for a period of weeks to months. Over time, individual clones which have acquired beneficial mutations will proliferate in the population, allowing for eventual isolation and genetic analysis of the improved phenotype [11]. ALE by serial batch cultivation in yeasts has historically focused on the model yeast *Saccharomyces cerevisiae*, but studies evolving "non-conventional" yeasts have been increasing [7, 10]. Phenotypes typically targeted for improvement include resistance to environmental stresses [12–16], substrate or nutrient utilization [17–23], and yield of bioproducts [24–26]. Investigations using ALE have also helped elucidate fundamental principles of yeast evolution, including the rate and form of mutations [27–29], response to environmental change [30–32], and clonal interference [33].

While examples of ALE applied in *Komagataella phaffii (Pichia pastoris)* in the literature are so far limited, there have been recent successful efforts to adapt strains to both alternative carbon sources and environmental stress [34–36], and to investigate the adaptive landscape for engineered and wild-type strains [37]. For biotechnological applications, *K. phaffii* is best known for the production of recombinant proteins, which can be difficult to directly improve through ALE due to growth trade-offs [38]. However, *K. phaffii* has also been engineered to produce other industrially relevant compounds [39–42], and novel methods for linking production to fitness during ALE are emerging [43–46]. With the advances in the availability of genetic parts [47–50], modular cloning and CRISPR technologies [51–53], metabolic models [54, 55], and curated genomic and proteomic data [56, 57], we believe ALE is well-poised to be a useful tool for applied and basic research with *K. phaffii* in the coming years.

Here we present an adaptable protocol for the evolution of *K. phaffii* by serial batch cultivation. We have modeled the experiment from the work of Moser et al. [35], wherein *K. phaffii* strain X-33 was evolved for improved growth on minimal media with methanol as the sole carbon source. We give suggestions for preliminary testing before starting the experiment (Subheading 3.1), followed by a detailed protocol for serial batch cultivation using both shake flasks (Subheading 3.1.1) and 24 deep well plates (Subheading 3.1.2). Finally, we provide information on the suggested analysis to be performed after the serial batch cultivation has concluded (Subheading 3.2). References to notes may be found throughout Subheadings 3.1 and 3.2, containing additional tips and tricks from our experience with this method.

2 Materials

2.1 Cell Cultivations and Transfer

1. Sterilized preculture medium: yeast peptone liquid medium (20 g/L hy soy peptone, 10 g/L yeast extract, pH 7.4–7.6) with 2% (*w* / *v*) glucose (YPD).
2. Sterilized main culture medium: buffered minimal medium (yeast nitrogen base, 0.1 M potassium phosphate, pH 6.0) (BM medium) and 1% (*w* / *v*) methanol as carbon source (BMM medium).
3. Sterile 100 mL flasks or sterile 24 deep well plates with a working volume of 10–20 mL or 2 mL, respectively.
4. Benchtop centrifuge.
5. Incubation chamber or room with shaker maintained between 25 and 30 °C.
6. Sterile 50 mL conical tubes or wide-bottom 15 mL conical tubes.
7. Optical density meter to measure optical density at 600 nm (OD_{600}) in cuvettes (shake flasks) or microtiter plates (24-well plates).

2.2 Culture Storage and Strain Isolation

1. Yeast isolation media: yeast peptone agar (20 g/L hy-soy peptone, 10 g/L yeast extract, pH 7.4–7.6, 20 g/L agar) with 2% (*w* / *v*) glucose (YPD).
2. Sterile 2 mL cryo-tubes.
3. 100% glycerol.
4. −70–80 °C culture storage freezer.

3 Methods

3.1 Pre-experimental Testing

To increase the chances of a successful ALE experiment, it is important to perform thorough preliminary testing to determine the optimal media, cultivation conditions, and selective pressure for the selected *K. phaffii* strain (Fig. 1). Once these conditions have been identified, we suggest growing at least one batch culture from inoculation through late stationary phase with regular sampling. This serves to identify the target OD_{600} range to avoid stationary phase sampling during early stages of the experiment (*see* **Note 1**). A number of *K. phaffii* strains have now been well characterized and would be highly suitable as ALE starting points, in particular strains CBS7435, CBS2612, X-33, and GS115. Engineered strains with defined or randomly introduced genetic alterations are also good candidates to be improved through ALE, but it should be noted that introduced genetic constructs not under positive selective pressure may be lost over the course of a long ALE experiment (*see* **Note 2**). Rich media for *K. phaffii*, such as YPD, can allow for a high growth rate and therefore a rapid ALE experiment; however, defined media such as buffered minimal media (BM media) would be preferable for ALE experiments where strains are being evolved for improved utilization of specific nutrients or substrates. The choice of whether to evolve your strain in shake flasks or 24 deep well plates may depend primarily on the equipment available, but there are also specific advantages for each. The use of 24 deep well plates increases the number of populations that can be evolved in parallel, while shake flasks can be more practical when longer

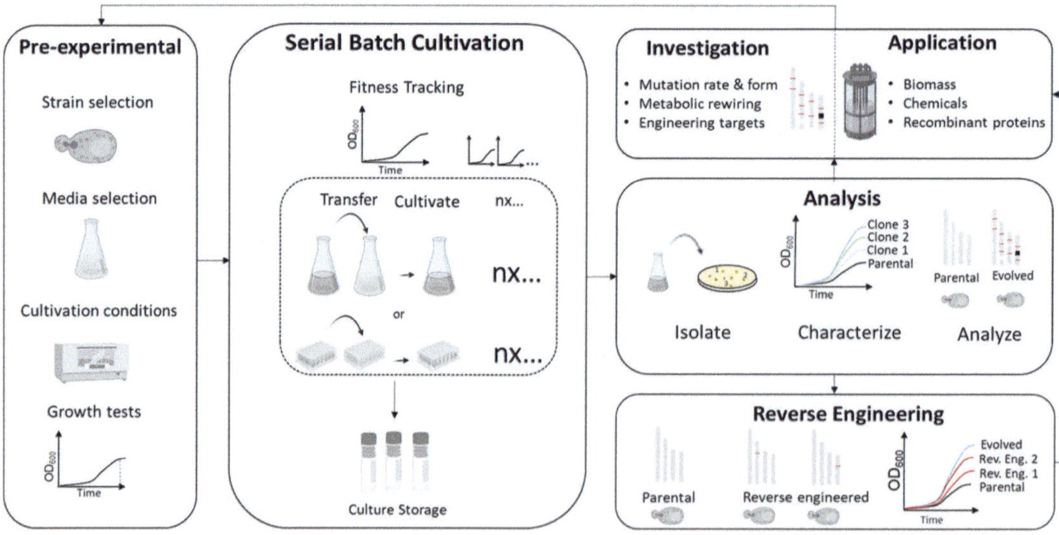

Fig. 1 Graphical overview of methodology for adaptive laboratory evolution of *P. pastoris* by serial batch cultivation

cultivation times or elevated temperatures are required, as the percentage of the culture volume lost to evaporation in shake flasks is less (*see* **Note 3**). Both approaches can be effectively used to perform ALE with *K. phaffii*; therefore, we have provided protocols for serial batch cultivation using both shake flasks and 24 deep well plates below.

3.1.1 Experimental Protocol—Shake Flasks (see ***Note 4****)*

1. Use 3–4 isolated colonies of the selected strain to inoculate one 10 mL YPD preculture in a 100 mL flask.
2. Grow the cultures at 25–30 °C and 180 rpm overnight.
3. Collect the entire sample in a sterile 50 mL conical tube and centrifuge at 2000 g for 5–10 min (*see* **Note 5**).
4. Add 2–5 mL BM medium and resuspend the pellet by pipetting or shaking.
5. Repeat the entire step two more times to ensure the preculture medium is removed.
6. Measure the optical density of the washed cell culture and calculate the volume needed to achieve a starting OD_{600} of 0.1 in a volume of 10 mL (*see* **Note 6**).
7. Transfer 10 mL of sterile main culture medium with carbon source (BMM medium) to the appropriate number of new 100 mL sterile flasks, then for each flask remove the calculated volume and replace it with well-mixed preculture (*see* **Note** 7).
8. Take a small sample for a starting OD_{600} measurement and transfer flasks to cultivation chamber for growth at 28 °C and 180 rpm (*see* **Note 8)**.
9. Use a sterile serological pipette to sample 1–3 mL of culture (*see* **Note 9**).
10. Place the culture flask back into the cultivation chamber.
11. Measure the OD_{600} of the sample and use this value to calculate the volume needed to inoculate a new 10 mL culture at OD_{600} of 0.1 (*see* **Note 10**).
12. Bring the culture flask back to the sterile hood and transfer the calculated volume from **step 3** to a sterile 50 mL tube.
13. Centrifuge at 2000 g for 5–10 min and remove supernatant.
14. Resuspend the cell pellet with 10–20 mL fresh BMM medium and transfer to a new labeled 100 mL flask.
15. Take a small sample for OD_{600} measurement and later calculation of growth rate.
16. Place the new shake flask into the cultivation chamber for growth at 28 °C and 180 rpm.
17. Repeat this process at regular intervals until satisfactory improvement in growth rate is observed.

*3.1.2 Experiment Protocol—24 Deep Well Plates (see **Note 4**)*

1. Use 3–4 isolated colonies of the selected strain to inoculate one 2 mL YPD preculture in a 24-deep well plate.
2. Grow the cultures at 25–30 °C and 280 rpm overnight.
3. Remove the plate from cultivation chamber and centrifuge at 2000 g for 5 min.
4. Remove the supernatant and add 2 mL BM medium per well and place on a shaking table for 5 min to resuspend.
5. Repeat this entire step two more times to ensure preculture media is removed.
6. Using a multichannel pipette, transfer 30–300 μL to a microtiter plate for OD_{600} measurement, and calculate the volume needed to achieve a starting OD_{600} of 0.1 in a volume of 2 mL (*see* **Note 11**).
7. Transfer 2 mL of main culture media with carbon source (BMM medium) to individual wells of a sterile 24-deep well plate (*see* **Note 7**).
8. For each well, remove the calculated volume and replace it with well-mixed preculture.
9. Place the new plate into the cultivation chamber for growth at 28 °C and 280 rpm (*see* **Note 8**).
10. Use a multichannel pipette to transfer 30–300 μL from the culture well to a 96-well microtiter plate.
11. Place the culture plate back into the cultivation chamber.
12. Measure the OD_{600} value and calculate the volume for inoculation of a new 2 mL culture at an OD_{600} of 0.1 (*see* **Note 10**).
13. Transfer the calculated volume from the culture to a new sterile 24-deep well plate (*see* **Note 12**).
14. Centrifuge the new plate at 2000 g for 5–10 min and remove the supernatant.
15. Add 2 mL of fresh main culture medium with carbon source (BMM medium) to each sample-containing well and place it into the cultivation chamber for growth at 28 °C and 280 rpm.
16. Repeat this process at regular intervals until a satisfactory improvement in growth rate is observed.

3.1.3 Additional Considerations or Actions During the Experiment

1. *Tracking fitness improvements:* Calculate the growth rate using the data obtained from regular OD_{600} measurements during serial batch cultivation. These data should be logged and monitored for improvement over time to identify when the conditions should be adjusted or when clones should be isolated for analysis. We also recommend screening for improvements in fitness throughout the ALE experiments by comparing the

evolved population to the clonal level at given intervals (e.g., after 50–100 generations) (*see* **Note 13**).

2. *Selective pressure increase:* If a satisfactory increase in growth rate is observed at one level of selective pressure, then the pressure may be increased in order to select for further improved phenotypes or avoid the stationary phase sampling. For example, if evolving for temperature tolerance, then the cultivation temperature may be increased slightly (+1–2 °C), or if evolving for organic acid tolerance, then the concentration of the organic acid may be increased slightly (+1–2 g/L) (*see* **Note 14**).
3. *Transfer volume reduction:* In the absence of either adjustable selective pressure or the ability to increase sampling frequency, stationary phase sampling can also be avoided by reducing the starting OD_{600}. It should be noted, however, that reducing the transfer volume can lead to genetic bottlenecks, meaning that clones containing potential beneficial mutations might not be transferred to the next batch culture and so would be purged from the experiment. Therefore, this should be performed only as necessary to avoid cells reaching the stationary phase before the next serial batch transfer can be performed.
4. *Culture storage:* At regular intervals—we suggest once per week or every 5–7 batch changes—prepare multiple glycerol stocks for each population to be stored at −80 °C. This should also be done prior to any selective pressure increase or transfer volume reduction to ensure the most recent population is not lost.
5. *Contamination detection:* Approximately every 50 generations, each population should be examined for contamination by microscopy and plating on YPD and LB agar. Additional differential and/or selective media may also be used to increase the visibility of possible contamination events.

3.2 Post-experimental Analysis

From each evolved population with improved fitness, at least three individual strains are isolated by plating and compared to the parental strain for fitness in the ALE condition. The best-performing clones are then characterized in detail. Typically, the genotype is revealed by whole genome sequencing (WGS) (*see* **Note 15**). All genetic changes between the evolved clones and the parental strain should be identified, including single nucleotide polymorphisms (SNPs), small insertions/deletions (InDels), and large-scale amplifications or deletions. Mutations should be examined in detail to determine those most likely to be linked to improved fitness, particularly those that are in coding sequences of a protein or associated regulator regions, and mutations that have arisen in multiple lineages. To demonstrate a causal effect on phenotype, these mutations can be reverse-engineered individually into the parental strain and screened against both the parental strain and the evolved strains. In addition to providing more direct evidence for the effect of the individual mutations, construction of the reverse-engineered strain is desirable because it may yield an improved variant of the parental strain without the potentially non-beneficial or deleterious mutations accumulated through the ALE experiment (*see* **Note 16**).

4 Notes

1. Stationary phase sampling is also possible for microbial ALE, as for instance in the foundational experiments by Lenski et al. [3, 4]. However, for most purposes, it will be beneficial to maintain growth conditions that are not limiting in substrate nutrients other than those which are specifically intended.
2. The use of antibiotic resistance markers is possible during ALE with yeast, but it should be done so at low concentrations to avoid selection for antibiotic resistance instead of for the intended phenotype.
3. In these cases where the potential for evaporation is high, it is suggested to adjust the volume with sterile water or main culture medium to normalize to the expected volume before OD_{600} measurement. In flasks, this can be performed by weighing the flask prior to the incubation and using this mass as the target to reach by adjusting prior to sampling.
4. Perform the following steps at the beginning of the ALE experiment for each strain selected for evolution.
5. Using a tube with a flat conical bottom to pour off the supernatant after centrifugation when diluting shake flask cultures will reduce the remaining liquid in the tube (e.g., KABE tubes N 95 S PPn or 50 mL Falcon tubes, not standard 15 mL Falcon tubes). The use of standard conical 15 mL tubes is possible but generally requires additional pipetting to remove the supernatant.
6. The starting OD of 0.1 was suitable for the example experiment [35], but this is by no means a requirement for ALE. For many experiments, a higher starting OD may be appropriate based on the growth rate or other factors. Having a higher OD also leads to a greater number of cumulative cell divisions (CCDs), providing more potential for beneficial mutations to occur. However, it should be avoided to start so dense that substrate or nutrient limitations would occur prior to

sampling. The best starting OD for the specific experiment should be identified during pre-experimental testing.

7. During this step, multiple ALE main cultures may be started from the single preculture. We recommend beginning by evolving at least 2–4 individual populations for each selected strain or different condition. This increases the potential for unique adaptive strategies to be identified and provides redundancy in the case of populations being lost from contamination or mishandling.
8. Perform the following steps at regular time intervals (e.g., 1–2 days) over the course of the experiment for each population (individual shake flasks or plate wells).
9. While sterile, the sampled volume may be used to create a glycerol stock if necessary, and the rest of this volume can be used for additional non-sterile analysis such as pH, HPLC, or GCMS.
10. The OD_{600} measurement in this step may indicate that not enough growth has taken place since the previous sampling (e.g., fewer than 3–4 cell doublings). In this case, continue the cultivation and start again with **step 1** after an appropriate amount of time based on the growth rate.
11. A standard 12-channel pipette can be used to efficiently sample from 24 deep well plates by interspacing tips on every other channel (e.g., only 6 of the 12 channels will have tips for sampling a full row). A serial dilution in water can be performed with two adjacent wells of the microtiter plate to reach a range compatible with the OD meter. It is most efficient to prepare a large number of tip boxes with 6 rows of tips before sterilization, which can be used over the course of the ALE experiment for plate sampling.
12. After the calculated volume is removed, an additional 1–2 mL sample can be taken at this step for a glycerol stock and/or pH, HPLC, or GCMS analysis.
13. From the evolved population, streaks are made on suitable agar plates (e.g., minimal medium and/or YPD agar plates), and singularized clones in technical triplicates are screened for changes in maximal growth rate. Alternatively, for pooled analysis, at least 12 individual colonies are used as inoculum for a preculture and are compared to the performance of the evolved population (i.e., preculture inoculated directly from ALE experiments) and to the ancestral strain.
14. Before changing the selective pressure, it is best to test the new condition, or a small set of new potential conditions, in a parallel experiment. This is done by taking biomass from the actual evolution cultures and testing the new conditions for

2–3 dilution rounds so that you can observe how the cells behave in the new environment. For example, if selecting for stress resistance, the new condition should decrease the growth rate compared to the current conditions; however, it should be confirmed beforehand that this is not reduced to zero or to such a low level that sufficient biomass will not be available for the next batch transfer.

15. We recommend using the fast and accurate method of Illumina next-generation sequencing (e.g., MiSeq platform) for SNPs. It is also suggested to perform sequencing on the single clone level after initial strain characterization (i.e., identification of desired phenotype). For accurate SNP detection, we recommend a sequencing coverage of above 60. If sequencing is performed on the population level (i.e., identification of fixed SNPs in the population by analysis of the frequencies), the coverage must be increased in accordance with the expected population diversity. We use the CLC Genomic Workbench 12.0 for genomic analysis, but other tools should also be considered (e.g., breseq v0.33.1). We recommend re-sequencing the parental strain in each round. Sequences are then mapped against the parental strain. Detected variants of the evolved strains are then filtered against the detected variants of the unevolved parental strains. For the filtering, we recommend a minimal threshold frequency setting of 65%.

16. If an evolved strain is chosen as a starting point for a certain biotechnological application, it is recommended to test it in multiple relevant conditions (e.g., different carbon sources, seeding density). Adverse trade-offs which could impede the successful implementation of processes based on ALE-derived strains can be identified using this procedure.

Acknowledgments

This work was supported by the Austrian Center of Industrial Biotechnology (ACIB) and the Austrian Science Fund (FWF W1224, Doctoral Program on Biomolecular Technology of Proteins (BioToP)). Objects in the figures were created using Biorender.com.

Competing Interest Statement The authors have no conflicts of interest to declare that are relevant to the content of this chapter.

References

1. Dettman JR, Rodrigue N, Melnyk AH, Wong A, Bailey SF, Kassen R (2012) Evolutionary insight from whole-genome sequencing of experimentally evolved microbes. Mol Ecol 21:2058–2077. https://doi.org/10.1111/j.1365-294X.2012.05484.x
2. Bennett AF, Hughes BS (2009) Microbial experimental evolution. Am J Physiol Regul Intergr Comp Physiol. https://doi.org/10.1152/ajpregu.90562.2008
3. Cooper VS, Lenski RE (2000) The population genetics of ecological specialization in evolving *Escherichia coli* populations. Nature 407:736–739
4. Woods RJ, Barrick JE, Cooper TF, Shrestha U, Kauth MR, Lenski RE (2011) Second-order selection for evolvability in a large *Escherichia coli* population. Science. https://doi.org/10.1126/science.1198914
5. Remigi P, Masson-Boivin C, Rocha EPC (2019) Experimental evolution as a tool to investigate natural processes and molecular functions. Trends Microbiol 27:623–634. https://doi.org/10.1111/j.1365-294X.2012.05484.x
6. Dragosits M, Mattanovich D (2013) Adaptive laboratory evolution – principles and applications for biotechnology. Microb Cell Factories. https://doi.org/10.1186/1475-2859-12-64
7. Mavrommati M, Daskalaki A, Papanikolaou S, Aggelis G (2022) Adaptive laboratory evolution principles and applications in industrial biotechnology. Biotechnol Adv. https://doi.org/10.1016/j.biotechadv.2021.107795
8. Gresham D, Dunham MJ (2014) The enduring utility of continuous culturing in experimental evolution. Genomics 104:399–405. https://doi.org/10.1016/j.ygeno.2014.09.015
9. Gresham D, Hong J (2015) The functional basis of adaptive evolution in chemostats. FEMS Microbiol Rev 39:2–16. https://doi.org/10.1111/1574-6976.12082
10. Sandberg TE, Salazar MJ, Weng LL, Palsson BO, Feist AM (2019) The emergence of adaptive laboratory evolution as an efficient tool for biological discovery and industrial biotechnology. Metab Eng 56:1–16. https://doi.org/10.1016/j.ymben.2019.08.004
11. Mozhayskiy V, Tagkopoulos I (2013) Microbial evolution in vivo and in silico: methods and applications. Integr Biol. https://doi.org/10.1039/c2ib20095c
12. Dhar R, Sägesser R, Weikert C, Yuan J, Wagner A (2011) Adaptation of *Saccharomyces cerevisiae* to saline stress through laboratory evolution. J Evol Biol 24:1135–1153. https://doi.org/10.1111/j.1420-9101.2011.02249.x
13. Fletcher E, Feizi A, Bisschops MMM, Hallström BM, Khoomrung S, Siewers V (2017) Evolutionary engineering reveals divergent paths when yeast is adapted to different acidic environments. Metab Eng 39:19–28. https://doi.org/10.1016/j.ymben.2016.10.010
14. Fiedurek J, Skowronek M, Gromada A (2011) Selection and adaptation of *Saccharomyces cerevisae* to increased ethanol tolerance and production. Pol J Microbiol 60:51–58
15. Caspeta L, Chen Y, Ghiaci P, Feizi A, Buskov S, Hallström BM et al (2014) Biofuels. Altered sterol composition renders yeast thermotolerant. Science 346:75–78. https://doi.org/10.1126/science.1258137
16. Totaro D, Rothbauer M, Steiger MG, Mayr T, Wang H, Lin Y et al (2020) Downscaling screening cultures in a multifunctional bioreactor array-on-a-chip for speeding up optimization of yeast-based lactic acid bioproduction. Biotechnol Bioeng. https://doi.org/10.1002/bit.27338
17. Li Y-C, Zeng W-Y, Gou M, Sun Z-Y, Xia Z-Y, Tang Y-Q (2017) Transcriptome changes in adaptive evolution of xylose-fermenting industrial *Saccharomyces cerevisiae* strains with δ-integration of different xylA genes. Appl Microbiol Biotechnol 101:7741–7753. https://doi.org/10.1007/s00253-017-8494-z
18. Wisselink HW, Toirkens MJ, del Rosario Franco Berriel M, Winkler AA, van Dijken JP, Pronk JT et al (2007) Engineering of *Saccharomyces cerevisiae* for efficient anaerobic alcoholic fermentation of L-arabinose. Appl Environ Microbiol 73:4881–4891. https://doi.org/10.1128/AEM.00177-07
19. Guimarães PMR, François J, Parrou JL, Teixeira JA, Domingues L (2008) Adaptive evolution of a lactose-consuming *Saccharomyces cerevisiae* recombinant. Appl Environ Microbiol 74:1748–1756. https://doi.org/10.1128/AEM.00186-08
20. Espinosa MI, Gonzalez-Garcia RA, Valgepea K, Plan MR, Scott C, Pretorius IS et al (2020) Adaptive laboratory evolution of native methanol assimilation in *Saccharomyces cerevisiae*. Nat Commun 11:5564. https://doi.org/10.1038/s41467-020-19390-9
21. Gassler T, Sauer M, Gasser B, Egermeier M, Troyer C, Causon T et al (2020) The industrial

yeast *Pichia pastoris* is converted from a heterotroph into an autotroph capable of growth on CO_2. Nat Biotechnol 38:210–216. https://doi.org/10.1038/s41587-019-0363-0

22. Perli T, Moonen DPI, van den Broek M, Pronk JT, Daran J-M (2020) Adaptive laboratory evolution and reverse engineering of single-vitamin prototrophies in *Saccharomyces cerevisiae*. Appl Environ Microbiol. https://doi.org/10.1128/AEM.00388-20
23. Hong K-K, Vongsangnak W, Vemuri GN, Nielsen J (2011) Unravelling evolutionary strategies of yeast for improving galactose utilization through integrated systems level analysis. Proc Natl Acad Sci USA 108:12179–12184. https://doi.org/10.1073/pnas.1103219108
24. Baek S-H, Kwon EY, Bae S-J, Cho B-R, Kim S-Y, Hahn J-S (2017) Improvement of d-lactic acid production in *Saccharomyces cerevisiae* under acidic conditions by evolutionary and rational metabolic engineering. Biotechnol J. https://doi.org/10.1002/biot.201700015
25. Liu L, Pan A, Spofford C, Zhou N, Alper HS (2015) An evolutionary metabolic engineering approach for enhancing lipogenesis in *Yarrowia lipolytica*. Metab Eng 29:36–45. https://doi.org/10.1016/j.ymben.2015.02.003
26. Reyes LH, Gomez JM, Kao KC (2014) Improving carotenoids production in yeast via adaptive laboratory evolution. Metab Eng 21:26–33. https://doi.org/10.1016/j.ymben.2013.11.002
27. Chang S-L, Lai H-Y, Tung S-Y, Leu J-Y (2013) Dynamic large-scale chromosomal rearrangements fuel rapid adaptation in yeast populations. PLoS Genet 9:e1003232. https://doi.org/10.1371/journal.pgen.1003232
28. Ene IV, Farrer RA, Hirakawa MP, Agwamba K, Cuomo CA, Bennett RJ (2018) Global analysis of mutations driving microevolution of a heterozygous diploid fungal pathogen. Proc Natl Acad Sci USA 115:E8688–E8697. https://doi.org/10.1073/pnas.1806002115
29. Oud B, Guadalupe-Medina V, Nijkamp JF, de Ridder D, Pronk JT, van Maris AJA et al (2013) Genome duplication and mutations in ACE2 cause multicellular, fast-sedimenting phenotypes in evolved *Saccharomyces cerevisiae*. Proc Natl Acad Sci USA 110:E4223–E4231. https://doi.org/10.1073/pnas.1305949110
30. Yona AH, Manor YS, Herbst RH, Romano GH, Mitchell A, Kupiec M et al (2012) Chromosomal duplication is a transient evolutionary solution to stress. Proc Natl Acad Sci USA 109:21010–21015. https://doi.org/10.1073/pnas.1211150109
31. Adamo GM, Lotti M, Tamás MJ, Brocca S (2012) Amplification of the CUP1 gene is associated with evolution of copper tolerance in *Saccharomyces cerevisiae*. Microbiology 158:2325–2335. https://doi.org/10.1099/mic.0.058024-0
32. Wenger JW, Piotrowski J, Nagarajan S, Chiotti K, Sherlock G, Rosenzweig F (2011) Hunger artists: yeast adapted to carbon limitation show trade-offs under carbon sufficiency. PLoS Genet 7:e1002202. https://doi.org/10.1371/journal.pgen.1002202
33. Kao KC, Sherlock G (2008) Molecular characterization of clonal interference during adaptive evolution in asexual populations of *Saccharomyces cerevisiae*. Nat Genet 40:1499–1504. https://doi.org/10.1038/ng.280
34. Gassler T, Baumschabl M, Sallaberger J, Egermeier M, Mattanovich D (2022) Adaptive laboratory evolution and reverse engineering enhances autotrophic growth in *Pichia pastoris*. Metab Eng 69:112–121. https://doi.org/10.1016/j.ymben.2021.11.007
35. Moser JW, Prielhofer R, Gerner SM, Graf AB, Wilson IBH, Mattanovich D et al (2017) Implications of evolutionary engineering for growth and recombinant protein production in methanol-based growth media in the yeast *Pichia pastoris*. Microb Cell Factories 16:49. https://doi.org/10.1186/s12934-017-0661-5
36. Lin N-X, He R-Z, Xu Y, Yu X-W (2021) Augmented peroxisomal ROS buffering capacity renders oxidative and thermal stress cross-tolerance in yeast. Microb Cell Fact 20:131. https://doi.org/10.1186/s12934-021-01623-1
37. Moser JW, Wilson IBH, Dragosits M (2017) The adaptive landscape of wildtype and glycosylation-deficient populations of the industrial yeast *Pichia pastoris*. BMC Genomics 18:1–16. https://doi.org/10.1186/s12864-017-3952-7
38. Mattanovich D, Branduardi P, Dato L, Gasser B, Sauer M, Porro D (2012) Recombinant protein production in yeasts. In: Recombinant gene expression. https://doi.org/10.1007/978-1-61779-433-9_17
39. de Lima PBA, Mulder KCL, Melo NTM, Carvalho LS, Menino GS, Mulinari E et al (2016) Novel homologous lactate transporter improves L-lactic acid production from glyc-

erol in recombinant strains of *Pichia pastoris*. Microb Cell Factories 15:158. https://doi.org/10.1186/s12934-016-0557-9
40. Guo F, Dai Z, Peng W, Zhang S, Zhou J, Ma J et al (2021) Metabolic engineering *of Pichia pastoris* for malic acid production from methanol. Biotechnol Bioeng 118:357–371. https://doi.org/10.1002/bit.27575
41. Siripong W, Wolf P, Kusumoputri TP, Downes JJ, Kocharin K, Tanapongpipat S et al (2018) Metabolic engineering of *Pichia pastoris* for production of isobutanol and isobutyl acetate. Biotechnol Biofuels. https://doi.org/10.1186/s13068-017-1003-x
42. Peña D, Gasser B, Zanghellini J, Steiger M, Mattanovich D (2018) Metabolic engineering of *Pichia pastoris*. Metab Eng 50:2–15. https://doi.org/10.1016/j.ymben.2018.04.017
43. Mahr R, Gätgens C, Gätgens J, Polen T, Kalinowski J, Frunzke J (2015) Biosensor-driven adaptive laboratory evolution of l-valine production in *Corynebacterium glutamicum*. Metab Eng 32:184–194. https://doi.org/10.1016/j.ymben.2015.09.017
44. Godara A, Kao KC (2020) Adaptive laboratory evolution for growth coupled microbial production. World J Microbiol Biotechnol 36:175. https://doi.org/10.1007/s11274-020-02946-8
45. Mans R, Daran J-MG, Pronk JT (2018) Under pressure: evolutionary engineering of yeast strains for improved performance in fuels and chemicals production. Curr Opin Biotechnol 50:47–56. https://doi.org/10.1016/j.copbio.2017.10.011
46. Konstantinidis D, Pereira F, Geissen E, Grkovska K, Kafkia E, Jouhten P et al (2021) Adaptive laboratory evolution of microbial co-cultures for improved metabolite secretion. Mol Syst Biol. https://doi.org/10.15252/msb.202010189
47. Vogl T, Sturmberger L, Kickenweiz T, Wasmayer R, Schmid C, Hatzl A-M et al (2016) A toolbox of diverse promoters related to methanol utilization: functionally verified parts for heterologous pathway expression in Pichia pastoris. ACS Synth Biol 5:172–186. https://doi.org/10.1021/acssynbio.5b00199
48. Garrigós-Martínez J, Vuoristo K, Nieto-Taype MA, Tähtiharju J, Uusitalo J, Tukiainen P et al (2021) Bioprocess performance analysis of novel methanol-independent promoters for recombinant protein production with *Pichia pastoris*. Microb Cell Factories 20:74. https://doi.org/10.1186/s12934-021-01564-9
49. Hartner FS, Ruth C, Langenegger D, Johnson SN, Hyka P, Lin-Cereghino GP et al (2008) Promoter library designed for fine-tuned gene expression in *Pichia pastoris*. Nucleic Acids Res 36:e76. https://doi.org/10.1093/nar/gkn369
50. Prielhofer R, Maurer M, Klein J, Wenger J, Kiziak C, Gasser B et al (2013) Induction without methanol: novel regulated promoters enable high-level expression in *Pichia pastoris*. Microb Cell Factories 12:5. https://doi.org/10.1186/1475-2859-12-5
51. Weninger A, Fischer JE, Raschmanová H, Kniely C, Vogl T, Glieder A (2018) Expanding the CRISPR/Cas9 toolkit for *Pichia pastoris* with efficient donor integration and alternative resistance markers. J Cell Biochem 119:3183–3198. https://doi.org/10.1002/jcb.26474
52. Prielhofer R, Barrero JJ, Steuer S, Gassler T, Zahrl R, Baumann K et al (2017) GoldenPiCS: a Golden Gate-derived modular cloning system for applied synthetic biology in the yeast *Pichia pastoris*. BMC Syst Biol 11:123. https://doi.org/10.1186/s12918-017-0492-3
53. Gassler T, Heistinger L, Mattanovich D, Gasser B, Prielhofer R (2019) CRISPR/Cas9-mediated homology-directed genome editing in *Pichia pastoris*. Methods Mol Biol 1923:211–225. https://doi.org/10.1007/978-1-4939-9024-5_9
54. Tomàs-Gamisans M, Ferrer P, Albiol J (2018) Fine-tuning the *P. pastoris* iMT1026 genome-scale metabolic model for improved prediction of growth on methanol or glycerol as sole carbon sources. Microb Biotechnol 11:224–237. https://doi.org/10.1111/1751-7915.12871
55. Sohn SB, Graf AB, Kim TY, Gasser B, Maurer M, Ferrer P et al (2010) Genome-scale metabolic model of methylotrophic yeast *Pichia pastoris* and its use for in silico analysis of heterologous protein production. Biotechnol J 5:705–715. https://doi.org/10.1002/biot.201000078
56. Valli M, Tatto NE, Peymann A, Gruber C, Landes N, Ekker H et al (2016) Curation of the genome annotation of *Pichia pastoris* (*Komagataella phaffii*) CBS7435 from gene level to protein function. FEMS Yeast Res 16. https://doi.org/10.1093/femsyr/fow051
57. Valli M, Grillitsch K, Grünwald-Gruber C, Tatto NE, Hrobath B, Klug L et al (2020) A subcellular proteome atlas of the yeast *Komagataella phaffii*. FEMS Yeast Res. https://doi.org/10.1093/femsyr/foaa001

Chapter 27

Classical Genetics of *Pichia pastoris*

Ilya Tolstorukov, Thomas G. Chappell, and James M. Cregg

Abstract

A significant advantage of *P. pastoris* as an experimental system and an eukaryotic expression platform is the ability to readily bring to bear both classical and molecular genetic approaches to a research problem and to develop technologies for manufacturing important proteins. Although the advent of yeast molecular biology has introduced new and exciting capabilities, classical genetics remains the approach of choice in many instances. These include the generation of random mutations by induced mutagenesis, alteration of any gene in the genome by targeted gene disruption or tagging, the removal of unwanted genetic background by backcrossing, the analysis of gene and protein interactions using mating and complementation, as well as the construction of strains with new combinations of mutant alleles and expressed proteins or polypeptides. This chapter describes the basis of *P. pastoris* classical genetics. Various methods of introducing of mutations into a strain genome are described. How classical genetic approaches can be applied to modern molecular studies and how they can be instrumental in the development of industrial strains to produce complex proteins will be discussed. Some basic information about available wild-type and genetically modified strains, as well as a method for their long-term storage, is provided.

An understanding of the *P. pastoris* life cycle and its manipulation is the key to genetic analysis and strain construction. In this chapter, a procedure for knocking out an essential gene is described, together with how to determine if the genomic locus involved is essential. The colony screening method to detect the secretion of a specific recombinant protein via the "Yeastern blot" is outlined. An application of *P. pastoris* classical genetics to the creation and production of therapeutic antibodies and other complex proteins is also discussed.

Key words *Pichia pastoris*, *Komagataella phaffii*, Yeast mating, Life cycle, Meiosis, Sporulation, Random spore analysis, Gene knock-outs and tagging, "Yeastern" blot

1 Introduction

To understand the classical genetic strategies employed with *Pichia pastoris* [1, 2], it is essential to understand the basics of its life cycle, which is characterized as homothallic and predominantly haploid [1, 3–5]. *P. pastoris* is an ascomycetous budding yeast that most commonly proliferates in the haploid state (Fig. 1). According to the experimentally proven model of homothallic behavior suggested by Tolstorukov and co-workers for another

Xavier Garcia-Ortega et al. (eds.), *Pichia pastoris: Methods and Protocols*, Methods in Molecular Biology, vol. 2697,
https://doi.org/10.1007/978-1-0716-4779-0_27,

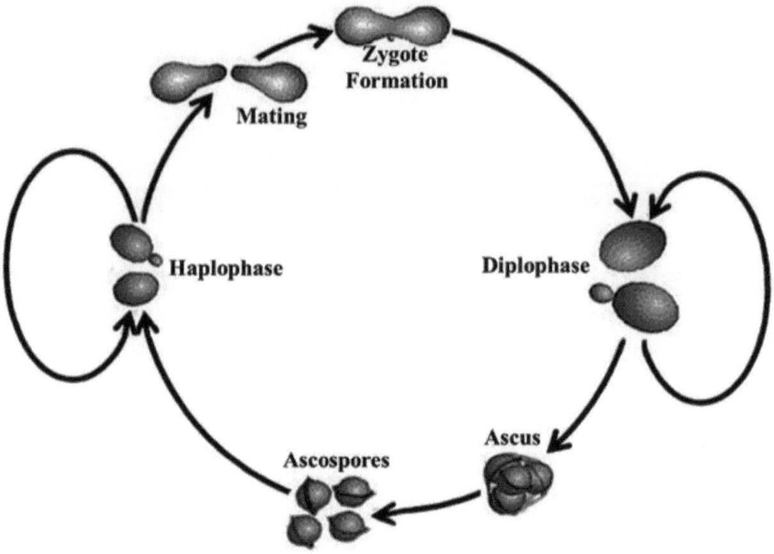

Fig. 1 Diagram of the *P. pastoris* life cycle. Adapted with modifications from [6]

methanol-assimilating species, *P. methanolica* [6, 7], and later confirmed for *P. pastoris* and *Hansenula polymorpha* (*P. angusta*, *Ogataea polymorpha*) [5, 8] another methanol-assimilating species of the same clade, each haploid cell of these yeasts can manifest one of two mating types, *a* or *α*. Only cells with opposite mating types can mate with each other, *a* x *α*, with the formation of an *a*/*α* diploid cell. Under certain culture conditions, the budding of a haploid cell leads to the formation of two cells, mother and daughter, with opposite mating types (e.g., *a*→*a* and *α*). This process is referred to as "mating type switching" [5, 9–13], which in methylotrophic yeasts is controlled by a flip-flop rearrangement of chromosomal DNA [5, 12, 13]. Neither mating nor mating type switching takes place under most growth conditions. However, under nitrogen starvation conditions, a significant proportion of the haploid cells switch mating type and can mate with the unswitched cells of the opposite type [5, 9–13]. If the resulting *a*/*α* diploid cells remain under nitrogen limitation, they will undergo meiosis and ascosporogenesis, resulting in asci containing four haploid ascospores. Alternatively, returning diploid cells to nitrogen-excess culture conditions results in stable diploid strains that can be isolated and propagated (Fig. 1). To selectively grow these diploid strains, it is convenient to begin mating with two phenotypically distinguishable haploid strains (e.g., *his4* and *arg4* mutant strains). Diploid strains from these crosses will grow on a minimal medium without supplementation, whereas the parental haploid strains and any self-mated diploids will not (Fig. 2). In the absence of auxotrophic mutations, orthogonal dominant drug-selection markers can also

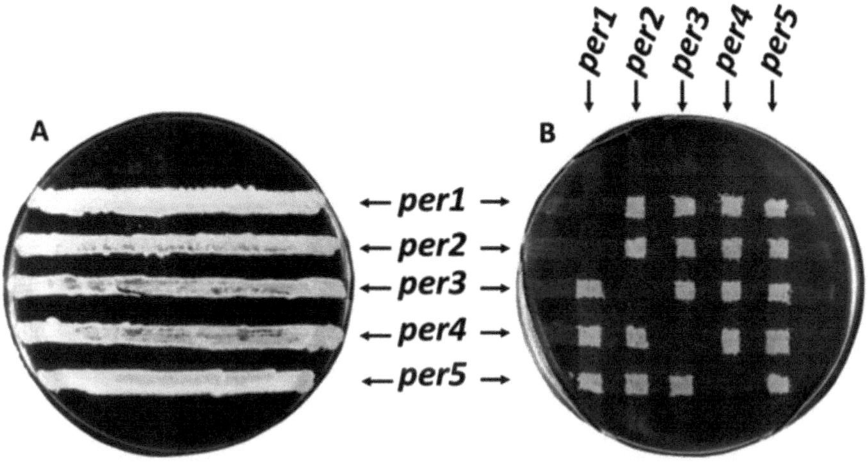

Fig. 2 Complementation analysis plates: a YPD medium plate in which five peroxisome biogenesis defective (*per1-per5*) methanol-utilization-negative strains are streaked for crossing (plate A); a YNM medium plate in which complementing diploid strains are selectively grown (plate B)

be used to isolate diploids resulting from the conjugation of the two independent strains.

Research published in 2014 [5] sheds light on the molecular mechanism of the *P. pastoris* mating type interconversion, which is controlled by the inversion of a 138-kb genomic fragment flanked by inverted repeats near one of the telomeres of chromosome 4. *MATa*1 and *MATa*2 are located at one end of the 138-kb region with *MATα*1 and *MATα*2 at the other end. Depending on their orientation, one of the two *MAT* loci is silenced by the adjacent telomeric region whereas the other, not near the telomere, is expressed, allowing the cell to manifest the corresponding mating type. This flip-flop chromosomal rearrangement model of mating type switching [5, 13] was elucidated from analysis of the genome sequences of *P. pastoris* [14–16] and two other taxonomically distinguished methanol-assimilating yeast species *Ogataea polymorpha* NCYC495 and *O. parapolymorpha* DL-1 [17], which have initially been identified as one species *H. polymorpha*.

The two-*MAT*-locus flip-flop chromosomal inversion model for methylotrophs differs from the three-*MAT*-locus "cassette" model of mating type interconversion described for *S. cerevisiae* [13, 18–24]. In *S. cerevisiae*, only one *MAT* locus located in the middle of the right arm of chromosome III is active and determines either *a* or *α* cell type of a haploid cell whereas two other *MAT* genes, *HML* and *HMR* located near the left (L) and right (R) telomers of the same chromosome III, are silent, playing the roles of information storage depots of *MATa* and *MATα* genes [18, 19, 23]. Interconversion or switching of mating types in *S. cerevisiae* is caused by a DNA conversion mechanism initiated by the formation of a double-strand DNA break (DSB) in the active *MAT* locus

[24]. The DSB occurs as a result of the activity of a site-specific endonuclease (*HO*) and is repaired by the transfer of information from one of the silent *MAT* genes to the active *MAT* gene. Historically, many *S. cerevisiae* strains used for genetic crosses have mutations that inactivate this switching mechanism, locking the mating type of a particular strain as either *a* or *α* and making it heterothallic.

A practical advantage of *P. pastoris* strains maintaining their wild-type mating mechanism and homothallism is that it is possible to cross any two haploid strains without first having to determine if they are of opposite mating types. One just needs two strains with complementing auxotrophic or dominant drug-selection markers. It should be noted, however, that mating *P. pastoris* strains without a strong selection to isolate the desired diploids from the background of the two input haploid strains is a challenge because the efficiency of mating is much lower than in *S. cerevisiae*. This difference in efficiency can be attributed in the most part to the different underlying strategies used for the two yeasts. In a mating between two heterothallic strains of *S. cerevisiae* of opposite mating types, every cell is capable of engaging in a productive conjugation event. In a mating of two homothallic *P. pastoris* strains, nonproductive self-conjugation events and the imbalance in the number of *a* and *α* cells result in significantly fewer productive cell fusions.

A haploid mating strategy is of practical use in synthesizing a variety of recombinant heteromultimeric proteins [25, 26]. The example we present in this chapter was utilized by Alder Biopharmaceuticals, Inc. for engineering of a *P. pastoris* strain secreting a full-length IgG1 molecule as migraine-preventing monoclonal antibody Eptinezumab targeted the CGRP receptor. Alder Biopharmaceuticals (now Lundbeck Seattle Biopharmaceuticals, a part of Lundbeck USA) was acquired by the Denmark-based H. Lundbeck [27]. In 2020, the U.S. FDA approved Eptinezumab as antimigraine medication, allowing Lundbeck to launch it on the pharmaceutical market under the trade name Vyepti.

The foundation of the mating strategy is the natural fusion of two haploid strains separately expressing heavy (HC) and light (LC) chains of an antibody. The assembled full-length human antibodies are heteromultimers consisting of two HC and two LC polypeptides covalently linked in tetrameric structures. The hybrid diploid strains obtained after mating of haploid cells expressing HC and LC molecules express both polypeptides and secrete them into the culture medium as assembled full-length antibody molecules, rMAb [25, 26, 28]. A mating strategy has also been applied for the discovery of novel MAbs using cell surface display libraries composed of heavy and light chain genes [29].

2 Materials

2.1 Strains

Most *P. pastoris* strains in common and commercial use are derivatives of the *Komagataella phaffii* strain NRRL Y-11430 (CBS 7435/ATCC 76273/CECT 11047), which was itself a Phillips Petroleum evolved isolate of the wild-type strain UCDFST 54–11.239 [30], which had originally been isolated by Eliza Knapp in 1954 from black oak exudate near Mather, California. This environmental isolate identified as *Pichia pastoris* was deposited at the ARS collection by Herman Phaff in 1957 (ID NRRL YB-4290). Later, the original Knapp's *P. pastoris* strain had also been deposited at other collections: CBS2612, NCYC 2543. In 1973, this strain was redeposited at the ARS collection as *Pichia pastoris* type strain NRRL Y-7556 by D. Yarrow, CBS, Delft.

According to the Pulsed Field Gel Electrophoresis [7] and genome sequencing data [14–16, 31], *K. phaffii* (*P. pastoris*) genome contains four chromosomes. In addition, the strain, NRRL Y-11430, like the parental type strain NRRL Y-7556, carries two cytoplasmic, autonomously replicating, multicopy double-stranded linear DNA molecules similar to the killer plasmids of other yeast species [32]. These linear plasmids were fortuitously lost in the mutagenized strain GS115 and in the BioGrammatics strain BG10 (*see* Table 1), a progeny of the patented strain NRRL Y-11430, which was purposefully cured of the plasmids. Thus, GS115, BG10, and their direct (non-crossed with other cell lines) derivatives and subsequent engineered strains lack the plasmids. Most other *P. pastoris* cell lines derived from the strain NRRL Y-11430 contain these cryptic plasmids, and back-crosses a plasmid-negative strain with these strains might restore the plasmids in the meiotic progenies of the hybrids.

Auxotrophic haploid strains are convenient for the selection of diploid strains or for the introduction of recombinant plasmids. A representative selection of such strains is listed in Table 1. The availability of the genome sequence and its annotation [14–16, 31, 33, 34] allows users to identify precisely the proteins and genes altered in mutant strains. While no obvious phenotypic effects other than histidine auxotrophy have been observed, it should be noted that GS115 contains background mutations relative to NRRL Y-11430, indicating a history of multiple rounds of mutagenesis.

In addition, it should be noted that phenotypically Arg^+ strains derived from GS190 still have an *arg4* mutation, which is covered by a heterologous *S. cerevisiae ARG4* gene inserted close to the *P. pastoris AOX1* locus [35]. Genetic crosses with these strains can unexpectedly generate progeny that are arginine auxotrophs.

Table 1
Some of the *P. pastoris* wild-type and mutant strains available for research

Strain	Genotype	Phenotype	Reference
NRRL Y-7556 (CBS2612[4], UCDFST 54–11.239[5], NCYC 2543)	Wild-type	Wild-type (Type strain)	[1, 2], & personal communication, UCD Phaff collection
NRRL Y-11430[1] (ATCC 76273, CBS 7435, CECT 11047)	Natural mutant (non-GMO)	Wild-type	[30]
BG10[2]	Natural mutant (non-GMO)	Wild-type	([39], *see also* [53])
BG21[2]	*sub2Δ*	Protease-deficient	([39], *see also* [53])
X-33[3]	GS115 + *HIS4*	Wild-type	[40]
GS115[3] (*see* **Note 1**)	*his4*	His	[54]
PichiaPink™[(3)]	*ade2*	Ade pink	[41]
SMD1163[2]	*his4 pep4 prb1*	His Protease-deficient	[55]
GS190[2]	*arg4*	Arg	*See* [3]
JC200[2]	*arg4 his4*	Arg His	[54]
JC220[2]	*ade1*	Ade (Pink, slow growing)	*See* [3]
JC227[2]	*ade1 arg4*	Ade Arg (Pink, slow growing)	[56]
JC233[2]	*his1*	His	*See* [3]
JC235[2]	*lis1 his4*	Lys His	*See* [3]
JC236[2]	*lis2 his4*	Lys His	*See* [3]
JC237[2]	*lys3*	Lys	*See* [3]
JC240[2]	*met2*	Met or Cys	*See* [3]
JC241[2]	*met3*	Met or Cys	*See* [3]
JC242[2]	*met4*	Met or Cys	*See* [3]
JC254[2]	*ura3*	Ura (Slow growing)	*See* [3]
JC255[3]	*ura5*	Ura (Slow growing)	*See* [3]
JC300[2]	*ade1 arg4 his4*	Ade Arg His (Pink, slow growing)	[56]
JC301[2]	*ade1 his4 ura3*	Ade His Ura (Pink, slow growing)	[56]
JC302[2]	*ade1 arg4 ura3*	Ade Arg Ura (Pink, slow growing)	[56]

(continued)

Table 1 (continued)

Strain	Genotype	Phenotype	Reference
JC303[2]	*arg4 his4 ura3*	Arg His Ura (Slow growing)	[56]
JC308[2]	*ade1 arg4 his4 ura3*	Ade Arg His Ura (Pink, slow growing)	[56]
SuperMan5 Pep Prb[2]	SuperMan5 *pep4Δ prb1Δ*	Blasticidn^R SuperMan5 Protease-deficient	[38]

[1]Available from ARS Culture (NRRL) patent Collection (http://nrrl.ncaur.usda.gov/)
[2]Available from "Biogrammatics, Inc." (http://biogrammatics.com)
[3]Available from "Thermo Fisher, Inc." (http://www.thermofisher.com)
[4]Available from CBS culture collection, (https://wi.knaw.nl/Collection)
[5]Available from UCD Phaff collection, (https://phaffcollection.ucdavis.edu/)

2.2 Media and Solutions

1. YPD medium: 1% (*w/v*) Yeast Extract, 2% (*w/v*) Bactopeptone, 0.4% (*w/v*) Dextrose (glucose).
2. YPD Zeo: YPD medium with 100 mg/L Zeocin (Zeocin solution must be added to the sterilized medium before pouring plates, pH of the medium must be adjusted to 7.4 before sterilization).
3. G418, Nourseothricin (Nat), and Blasticidin are prepared as filter-sterilized solutions and added to the sterile YPD medium before pouring plates to final concentrations 1000, 50, and 100 mg/L, respectively.
4. YNB medium: 0.67% (*w/v*) Yeast Nitrogen Base without amino acids and ammonium sulfate, 0.5% (*w/v*) Ammonium Sulfate. YNB can be supplemented with selected carbon sources such as 0.4% (*w/v*) dextrose for YND, or with 0.5% (*w/v*) methanol (YNM), ethanol (YNE), or glycerol (YNG). Amino acids and nucleotides are added to 50 mg/L as required for auxotrophic strains.
5. Mating (sporulation) medium: 0.5% (*w/v*) sodium acetate, 1% (*w/v*) potassium chloride, 1% (*w/v*) dextrose.
6. 5-FOA medium (for selection of Ura$^-$ mutants): YNG supplemented with 50 mg/L uracil and 750 mg/L 5-fluoroorotic acid.
7. For plates, agar is added to 2% (*w/v*) final concentration.
8. 75% (*w/v*) glycerol.
9. Diethyl ether.

2.3 Immunoblotting Reagents

1. 1 L transfer buffer: 25 m*M* Tris Base, 200 m*M* glycine, 20% (*w/v*) methanol (pH 8.5).
2. 1 L TBS (10×): 24.2 g/L Tris Base, 80 g/L NaCl (pH 7.6).
3. 1 L wash buffer (TBST): 100 mL 10× TBS, 0.05% (*w/v*) tween 20, 900 mL H_2O.
4. Blocking buffer: 30 mL 10× TBS, 0.05% (*w/v*) tween 20, 270 mL H_2O.
5. 15 mL blocking buffer 1: 15 mL blocking buffer, 1% (*w/v*) BSA.
6. 15 mL blocking buffer 5: 15 mL blocking buffer, 5% (w/v) BSA.

2.4 Other Materials

1. Velvet cloth for replica-plating (sterile).
2. Round paper filters (85–90 mm diameter, sterile).
3. Nitrocellulose immunoblotting membrane.
4. Extra-thick filter paper for protein transfer during immunoblotting.

3 Methods

3.1 Long-Term Strain Storage

Viable *P. pastoris* strains are readily stored frozen as ~30% glycerol stocks for long periods (>20 yr., *see* **Note 2**). For each strain to be stored, pick a single fresh colony from an appropriate plate and inoculate it into a sterile tube containing 2 mL of liquid YPD. After overnight incubation with vigorous shaking, transfer 1 mL of the culture to a sterile 2.0-mL cryovial containing 0.67 mL of 75% glycerol. Mix thoroughly and freeze at −70° to −80 °C.

To resurrect a stored strain, remove the cryovial from the freezer, immediately plunge a sterile inoculation loop or toothpick into the frozen material, transfer a small portion of the material to an agar plate containing an appropriate medium for recovery, and immediately return the culture to the freezer (*see* **Note 2**).

3.2 Selection for Uracil Auxotrophs Using 5-Fluoroorotic Acid

5-Fluoroorotic acid (5-FOA), an uracil biosynthetic pathway analogue, is metabolized to yield a toxic compound by certain enzymes in the pathway [36, 37]. As a result, organisms that are prototrophic for uracil synthesis (Ura^+) are sensitive to 5-FOA, whereas certain Ura^- auxotrophs cannot metabolize the drug and thus are resistant to it. Selection for 5-FOA resistant strains is a highly effective means of isolating Ura^- mutants affected in either of two Ura biosynthetic pathway genes, *URA3* and *URA5*, encoding orotidine-5′-phosphate decarboxylase and orotidine-5′-phosphate pyrophosphorylase, respectively. Preliminary mutagenic treatment increases the frequency of Ura^- mutants. However, Ura^- cells often exist at a low but significant frequency within the untreated cell population as well and can be readily selected by simply plating YPD-grown cells onto a 5-FOA plate.

1. To select for Ura^- mutants, spread ~2 OD_{600} units (~5e7 cells) on a 5-FOA plate. Resistant colonies appear after ~1 week of incubation at 30 °C.
2. Test the 5-FOA-resistant colonies for Ura phenotype by streaking them onto each of two YND plates, one with and another without uracil (*see* **Note 3**).
3. Isolate single-cell colonies of the 5-FOA-resistant mutants by spreading the cells of each strain onto either a YPD or a YND (+Ura) plate.
4. If it is necessary to determine which *URA* gene is affected in a new Ura^- strain, the mutant can be subjected to complementation testing (*see* Subheading 3.3) against known Ura^- mutants (Table 1) or analyzed by sequencing of DNA fragments amplified from *URA3* and *URA5* loci of each mutant. The mutated gene can also be identified by transformation of the mutant strain with vectors containing either a *P. pastoris* *URA3* or *URA5* gene.

3.3 Standard Genetic Crossing Procedure (Complementation and Recombination Analyses)

3.3.1 Mating and Selection of Diploids

The mating procedure constitutes the core of complementation analysis. The selection of diploids is the first step in strain construction associated with genetic recombination and backcrossing, but also can be the final step in the construction of protein production strains. Because *P. pastoris* is functionally homothallic, there is no need to consider mating type prior to crossing any pair of strains. However, because the mating efficiency of *P. pastoris* cells is rather low, it is essential that strains to be crossed contain complementary markers to allow for the selective growth of diploids and against the growth of parental strains and self-mated diploid cells. Auxotrophic markers are generally most convenient for this purpose (Table 1), but mutations in any gene that affect the growth phenotype such as genes controlling utilization of a specific carbon source (e.g., methanol or ethanol) or nitrogen source (e.g., methylamine) can be used as well as dominant drug-selection markers introduced during transformation.

1. To begin a mating experiment, select a fresh colony (no more than 1 week old) of each strain to be mated from a YPD plate and streak across the length of each of two YPD plates (Fig. 2a).
2. After overnight incubation (*see* **Note 4**), transfer the cell streaks from both plates onto a sterile velvet such that the streaks from one plate are perpendicular to those on the other.
3. Transfer the cross streaks from the velvet to a mating medium plate and incubate for 2 to 3 days at room temperature to

initiate mating. (For a single mating, cells of two strains can be mixed on a mating medium using sterile toothpicks. However, it might be advisable to have negative controls by plating both parental strains on the same plate separately from the mixed culture.)

4. After incubation, replica plate on an appropriate agar medium for the selection of complementing diploids. Colonies of diploids will arise at the junctions of the streaks after approximately 3 days of incubation at 30 °C (Fig. 2b). (The products of a single mating and controls can be streaked out onto the selective agar by toothpicks.) Diploid cells of *P. pastoris* are approximately twice as large as haploid cells and can be distinguished by examination under a light microscope. However, the microscopic examination of the diploid cultures can be simpler by comparison of the efficiency of sporulation after incubation of the hybrids and parental haploid strains on mating (sporulation) medium (*see* below) because a high portion of the cells in the diploid is meiotically competent and develop easily distinguishable asci with 2–4 ascospores.

3.3.2 Sporulation and Spore Analysis

Diploid strains of *P. pastoris* efficiently undergo meiosis and sporulation in response to nitrogen limitation.

1. To initiate sporulation, transfer grown diploid colonies from a YPD or YND plate to a mating (sporulation) plate either by replica plating or with an inoculation loop or toothpick. Incubate the plate for 3 to 4 days at room temperature. The sporulation of this species correlates with the accumulation of a red pigment in the ascus. The higher efficiency of sporulation, a more intense color of the colony. A sporulated diploid culture is easily distinguished by its tan color relative to the white color of a haploid culture or by a high number of asci in the cell culture as observed by light or phase-contrast microscopy. Because *P. pastoris* spores are small and adhere to one another, tetrad dissection via micromanipulator, as typically done in *S. cerevisiae*, is difficult. Moreover, the spore germination (ascospore viability rate) in *P. pastoris* cultures is not as high as needed for tetrad analysis. Therefore, spore products and the isolation of recombinant haploid cultures are performed using a random spore analysis (RSA) procedure.
2. For RSA, transfer an inoculation loop full of sporulated material to a 1.5-mL microcentrifuge tube containing 0.7 mL of sterile water and vortex the mixture.
3. In a fume hood, add 0.7 mL of diethyl ether to the spore preparation, and vortex thoroughly for approximately 5 min at room temperature. The ether treatment selectively kills vegetative cells remaining in spore preparations.

4. In the hood, centrifuge the cells for 2 min, remove the ether (the upper phase), and resuspend the pellet in the remaining water. Serially dilute the sample and spread 0.01 and 0.1 mL of the suspensions onto a nonselective medium (e.g., YPD).
5. After incubation for 3 to 5 d at 30 °C, replica plate from the plates that contain 100 to 1000 colonies onto a set of master YPD plates and then replicated onto a series of plates containing suitable diagnostic media. For example, to analyze the spore products resulting from a cross of GS115 (*his4*) and GS190 (*arg4*), appropriate diagnostic media would be YND supplemented with:
   (a) No amino acids;
   (b) Arginine;
   (c) Histidine;
   (d) Arginine and histidine.
6. Compare or score the phenotype of individual colonies on each of the diagnostic plates and identify ones with the desired recombinant phenotype(s). This can also be done at the DNA level by sequencing the corresponding PCR products.
7. Select several colonies that appear to have the appropriate recombinant phenotype and streak for single colonies onto a nonselective medium plate. Retest single colonies from each selected recombinant colony on the same set of diagnostic plates (*see* **Note 5**).

Mating can be used not only for the recombination of mutant alleles but also for the regeneration of selectable markers by ectopic gene conversion. This low-frequency event stimulated by meiosis is a result of the interaction of an allele located at the normal chromosomal locus with an altered copy of an identical DNA fragment (gene) introduced into another locus of the *P. pastoris* genome using gene-engineering manipulations. The ectopic gene conversion leads to a transfer of information between similar but not allelic genes in both directions; the result of such events depends on the availability of selective methods for isolation of the corresponding products of the conversion. An example of an ectopic recombination of a *his4* mutation in *P. pastoris* in which both structural genes for alcohol oxidase were disrupted by auxotrophic markers (genotype: *arg4 his4 aox1Δ::SARG4 aox2Δ::PHIS4*) was described by [35]. This prototrophic strain, which was unable to grow on methanol, was crossed with an *arg4 his4* double mutant by selecting for a prototrophic diploid capable of utilizing methanol. A portion of the segregants was unable to grow on methanol but were now auxotrophic for histidine as the result of a gene conversion event between the *his4* gene at its normal locus recombining with the

HIS4 gene at the *AOX2* locus (*arg4 his4 aox1Δ::SARG4 aox2Δ::Phis4*).

3.3.3 Mating as a Method to Construct Diploid Strains Expressing Full-Length Tetrameric Antibodies

Production of complex protein therapeutics such as recombinant monoclonal antibodies (rMAbs) requires specific approaches for strain development. Immunoglobulin G (IgG) is a heterotetrameric molecule formed by the assembly of two heavy (HC) and two light (LC) polypeptide chains connected by interchain disulfide bonds. Initially, to engineer mammalian cell lines expressing IgG molecules, researchers introduced sequentially two plasmids expressing HC and LC polypeptides into the host cell. Later, a cotransfection technique allowing for the simultaneous transfection of two constructs into the cell line, as well as vectors carrying both HC and LC expression cassettes, was developed and introduced into the rMAb manufacturing industry based on the mammalian expression platform.

Yeast mating provides an efficient approach to bring two or more recombinant constructs into one production strain. There are at least two additional advantages of this mating strategy. One is a potential ability to generate a variety of different polypeptide combinations, like that required for the construction of a combinatorial library. The second is the possibility of optimizing the expression of each polypeptide separately prior to construction of the final diploid strain so that equimolar concentrations of HC and LC are available for assembly [25, 26].

1. Construct two recombinant plasmids carrying genes for HC and LC polypeptides of a chosen monoclonal antibody, under control of strong promoters. Two different dominant drug-selection markers are preferable for the HC and LC plasmid constructs. This will allow us to do both: (1) Select more active haploid transformants carrying a range of the plasmid copy numbers, which can readily be identified by relative resistance to the antibiotic or by the Yeastern (see below) (2). Cross the selected haploid transformants to obtain diploid hybrids expressing both polypeptides, HC and LC.
2. Transform competent cells of a *P. pastoris* strain with the constructed plasmids to obtain transformants.
3. Select for positive transformants by plating on appropriate media (e.g., YPD Zeo or YPD G418) for 3–4 days at 30 °C and collect single colonies of transformants by streaking them out onto master plates with the same drug.
4. Analyze the relative resistance of the collected colonies to the antibiotics by plating them onto agar with increased concentrations of the drug. Cultures with higher relative resistance might express a higher amount of the product.

5. Use colony PCR to analyze the presence of the plasmid DNA integrated into the genome of the transformants.
6. Use the Yeastern blot method (Subheading 3.4) with appropriate primary diagnostic antibodies to identify colonies secreting the highest amounts of HC or LC polypeptide on plates. For constitutive promoters, replica plating from master YPD/antibiotic plates incubated for 1–2 days at 30 °C is sufficient. For methanol inducible expression, prior incubation for 1–2 days on plates with methanol (YNM) for induction of the transcription is required. The broth samples of the selected transformants grown in liquid media can also be analyzed by dot-blotting with the same diagnostic antibodies.
7. Cross the HC and LC secreting strains using the protocol described in Subheading 3.3.1. For the selection of diploid cultures obtained by mating of auxotrophic mutants, use YND medium. Selection of diploid hybrids of drug-resistant transformants can be done on YPD plates containing both drugs.
8. Streak out positive diploid colonies onto appropriate selective plates to obtain single colonies and incubate for 2–3 days at 30 °C. Collect a few single colonies from each cross as patches to a fresh selection plate.
9. Use the Yeastern blot (Subheading 3.4.) or dot-blotting to identify colonies secreting the highest amount of antibody.

3.3.4 Mating and Random Spore Analysis as a Method to Generate Diversity in Haploid Heterologous Protein Expression Strains

Mating and random spore analysis can be used as a method to test a variety of genetic backgrounds for the expression of a heterologous protein. Rather than transforming the heterologous protein expression cassette into a panel of genetic backgrounds and risk having potentially different integration events in each of the strains, the cassette can be inserted into the genome of a starting strain and then crossed with other strains to generate an expression panel. As an example, transformation of an expression cassette at the *AOX1* locus using Zeocin selection in a *sub2Δ* protease deletion strain (BG21, *see* Table 1) can be crossed to a SuperMan5 GlycoSwitch® strain [38] with *pep4Δ* and *prb1Δ* protease deletions to generate a panel of 16 different genetic backgrounds to test the expression of a particular heterologous protein. Since the SuperMan5 strains are tagged with a Blasticidin resistance marker at the mutated *OCH1* locus, diploids from the cross can be selectively grown on YPD Blasticidin/Zeocin plates. The *AOX1*, *OCH1*, and three protease loci are separated on the *P. pastoris* chromosomes by enough distance that they do not show genetic linkage (Fig. 3). The 5 loci segregate independently, giving 16 different combinations of *OCH1* and proteases in a background containing the heterologous protein expression cassette.

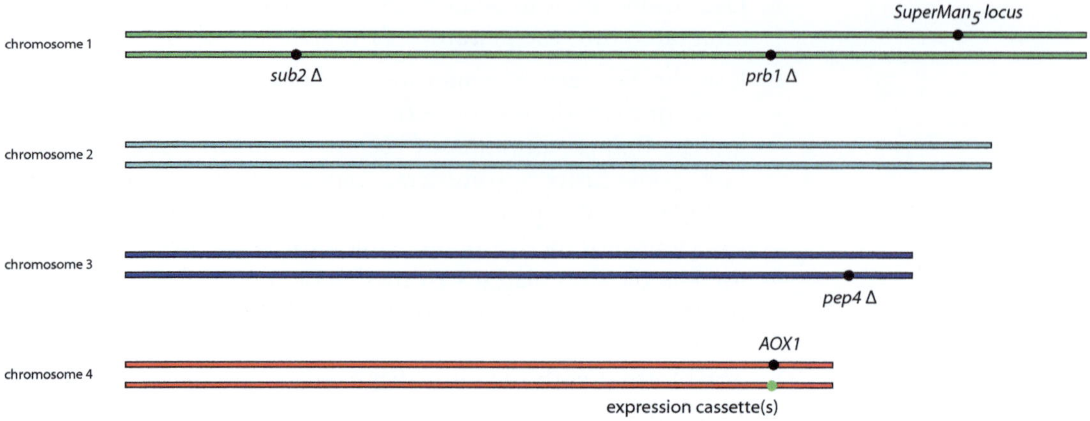

Fig. 3 Distribution of protease, *OCH1*, and *AOX1* loci on *P. pastoris* chromosomes. Meiotic crossover frequency during the sporulation of a diploid strain results in the random distribution of the five loci in the resulting haploid progeny

1. Construct a plasmid for the expression of the heterologous protein. For this specific example, a pJAZ or pPICZ [39–41] vector allows for Zeocin selection.
2. Transform competent BG21 (*sub2Δ* protease deletion, Table 1) cells with the expression plasmid, selecting for Zeocin resistance.
3. Similar to the method described in the Subheading 3.3.3, select for positive transformants and verify genomic integration at the *AOX1* locus by colony PCR analysis.
4. Using an appropriate screening method, determine protein expression levels in the initial population of isolates.
5. Grow both the BG21 expression strain and the SuperMan5 double protease deletion strain (SuperMan5 pep4 prb1) as patches on YPD.
6. Cross two strains (Subheading 3.3.1) and isolate single colonies of the diploid hybrid on a YPD Blasticidin/Zeocin plate.
7. Follow the method (Subheading 3.3.2) to sporulate and isolate haploids that contain the expression cassette (50% of isolates should grow on YPD Zeo). Half of these should grow on Blasticidin, containing the mutant *OCH1* allele, while the unmarked protease deletions will have segregated randomly.
8. At this point, haploid isolates can either be tested using a protein expression assay or scored for protease deletions by colony PCR.

3.4 The Yeastern Blot

The Yeastern blot is a convenient means of qualitatively screening large numbers of yeast colonies directly on plates for expression of a recombinant protein (*see* **Note 6**). However, readers should be aware that the correlation between the size of the "halo" surrounding a colony and the amount of recombinant product is not always linear, and all results with this method should be confirmed with a standard Western blot or other assays.

1. Transfer freshly grown yeast colonies from the surface of an agar plate onto a sterile piece of Whatman No. 1 filter paper by a standard replica-plating method. The filter paper should be cut to a size that exactly fits within the plate.
2. Place the filter with yeast cells onto the surface of a fresh plate containing an appropriate induction medium and incubate the plate for 1–2 days.
3. Prepare a piece of nitrocellulose membrane of the same size and dimensions as the filter paper by soaking the membrane for 5 min or more in 15 mL of transfer buffer. Soak two additional pieces of cut filter paper in transfer buffer.
4. Prepare a sandwich of the papers as described below:
   (a) Place one piece of the filter paper on the anode platform of a Western blotter. (It is essential to remove all bubbles between membrane layers. This can be done by rolling a pipette over their surface.)
   (b) Place the soaked nitrocellulose filter on top of the filter paper.
   (c) Place the filter paper with replica-plated yeast cells on top of the nitrocellulose paper.
   (d) Place the second piece of soaked filter paper on top of the filter with cells.
   (e) Finally, place the cathode plate on top of the sandwich.
5. Transfer proteins to the nitrocellulose membrane with a constant current (1–4 mA/cm^2) for 1 h.
6. Remove the nitrocellulose membrane and wash it for 5 min in 15 mL TBS buffer.
7. Replace the TBS buffer with 15 mL of TBST. Place the filter in blocking buffer and rock gently at room temperature for 1 h.
8. Transfer filter to 15 mL of TBST buffer and wash by rocking at room temperature for 5 min.
9. Prepare primary antibody dilution in 15 mL of blocking buffer according to the vendor's recommendations (typically, 0.5 μg/mL).
10. Incubate membrane overnight at 2–8 °C with rocking action (can be as short as 1–3 h at room temperature depending on antibody).

11. Wash membrane in at least four changes of TBST buffer (15 mL each, 5 min/wash), then briefly rinse in fresh TBST (*see* **Note 7**).
12. Prepare an enzyme-conjugated secondary antibody dilution in blocking buffer according to the vendor's recommendations (typically 3 μL in 15 mL of blocking buffer).
13. Incubate membrane in secondary antibody solution for 1 h, rocking at room temperature.
14. Repeat **step 11**.
15. Visualize the signals with appropriate reagents and available methods. The intensity and size of the signal around each colony approximately reflects the secretion level of the recombinant product by the different colonies/strains.

3.5 Knockout or Tagging of Essential and Nonessential Genes

Methods for chromosomal gene disruption and deletion by homologous recombination in *P. pastoris* are derived from the earlier pioneering work in *S. cerevisiae*. In general, linear DNA molecules containing either a dominant drug or auxotrophic complementing marker, along with chromosomal homology at both ends, can be integrated into the genome by homologous reciprocal crossover events. Off-target, nonhomologous integration events also occur and are more likely in *P. pastoris* than in *S. cerevisiae*.

In both yeasts, transformation with exogenous linear DNA molecules leads to the activation of DSB repair pathways. Eukaryotic cells contain two highly conserved molecular "machines" that repair DSBs by either homologous recombination (HR) or nonhomologous end joining (NHEJ) (reviewed in [42, 43]). The interplay between these two mechanisms of repair determines the ratio of homologous to nonhomologous integration events of a transformed DNA molecule. NHEJ repair is promoted by Ku70/80 association with DNA ends, while HR repair requires 5′ to 3′ exonuclease activity in conjunction with the RAD52 epistasis group of proteins. In practice, *S. cerevisiae* HR repair can be accomplished with relatively short homology arms (~50 bp) for interaction with the RAD proteins, while most other eukaryotic organisms need more homology to overcome an NHEJ bias in DSB repair [44].

Methods of gene disruption by insertion of an altered DNA fragment into a chromosomal locus were developed for *P. pastoris* during the early stages of the development of the yeast as an expression system [35, 45, 46]. Such methods were further improved by the development of new constructs, by increasing the efficiency of homologous (targeting) recombination in strains with mutations of genes controlling the DNA recombination/reparation pathway [47], and by improvement of DNA transformation methods.

In a previous edition of this book, we provided an example of disruption and analysis of the gene Pp*GAL83* whose product is involved in the control of all gluconeogenetic pathways in *P. pastoris* [4]. Inability to generate the haploid Pp*Gal83::ARG4* or Pp*Gal83::Zeo* disruption strains led us to the idea that a serious alteration of the Pp*GAL83* locus was lethal for the cells. The disruption constructs Pp*Gal83::ARG4* or Pp*Gal83::Zeo* were transformed into the cells of a His$^-$ Arg$^-$ diploid strain obtained by mating of two polyauxotrophic haploid strains, *ade1 arg4 his4* X *met1 arg4 his4,* and a few transformants were proven to carry both the wild-type and disrupted alleles in the *GAL83* locus. However, all attempts to rescue a meiotic haploid product with a disrupted *GAL83* gene were unsuccessful, meaning that the obtained disruption is lethal for the cells.

The approach within the Pichia community of attempting gene deletions/inactivations in haploid rather than diploid cells has led to anomalous results in several cases. A prominent example is the confusion that has arisen over the *OCH1* locus, which encodes the mannosyltransferase responsible for the initiation of long-chain mannosylation of N-linked oligosaccharides. Deletion (or strong repression) of this activity is a critical first step in N-linked glycosylation engineering of secreted *P. pastoris* proteins. In two published reports, strains disrupted in the *OCH1* locus showed little if any alteration in their growth phenotype [48, 49]. In two other publications, disruption or deletion of the *OCH1* ORF resulted in a severe growth defect [50, 51]. All these results were generated by putative homologous recombination in haploid strains. Here, we provide an example of the gene disruption/gene deletion/gene tagging technique by tagging *P. pastoris OCH1* gene product with a V5 epitope tag in a diploid strain. The V5 epitope is inserted into the nucleotide sugar binding site of the encoded mannosyltransferase, tagging and inactivating the protein simultaneously. For the Och1p tagging, a BG10/SuperMan5 diploid was constructed after mating of BG10 and SuperMan5 strains that were preliminarily marked with Zeocin and Nourseothricin resistance at the *AOX1* locus (*see* Fig. 3). These drug-resistant strains were obtained by introducing the *PmeI* linearized Biogrammtics' vectors, pJAZ, and pJAN, respectively, into each strain.

3.5.1 Using Synthetic DNA to Create Homologous Integration Vectors

We have tried various approaches for constructing linear DNA molecules for homologous genomic targeting in *P. pastoris*. Our criteria for a generally applicable process include the ability to target the mutation or deletion of any codon in the genome (including deletions up to and including full open reading frames), sequence verification after integration, and marker recycling. Pan et al. [52] published an approach with a self-excising *cre/lox* marker cassette where homologous recombination arms are attached by assembly PCR. In our hands, their method resulted in too much effort in the

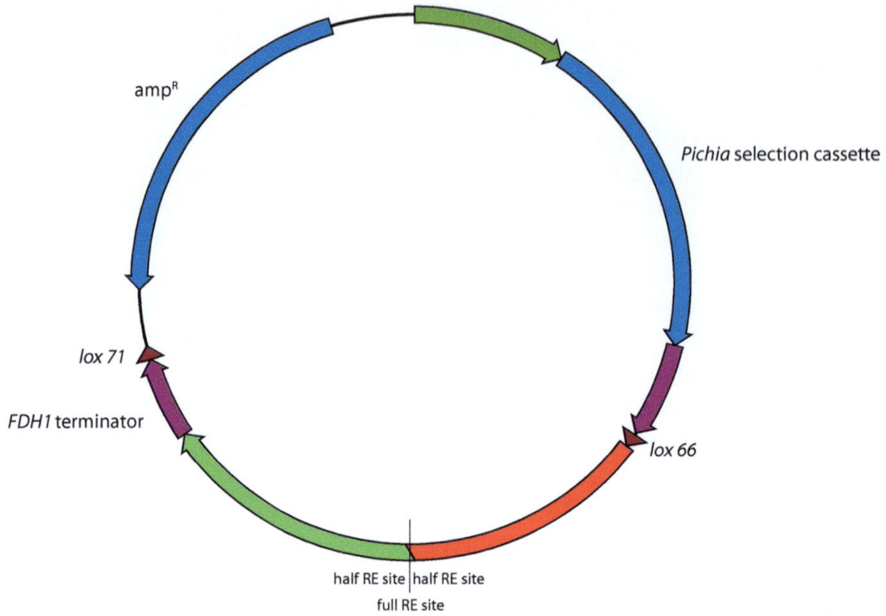

Fig. 4 Construction of a *P. pastoris* targeting vector using restriction enzyme half-sites to create a linearization site with perfect homology to the genome

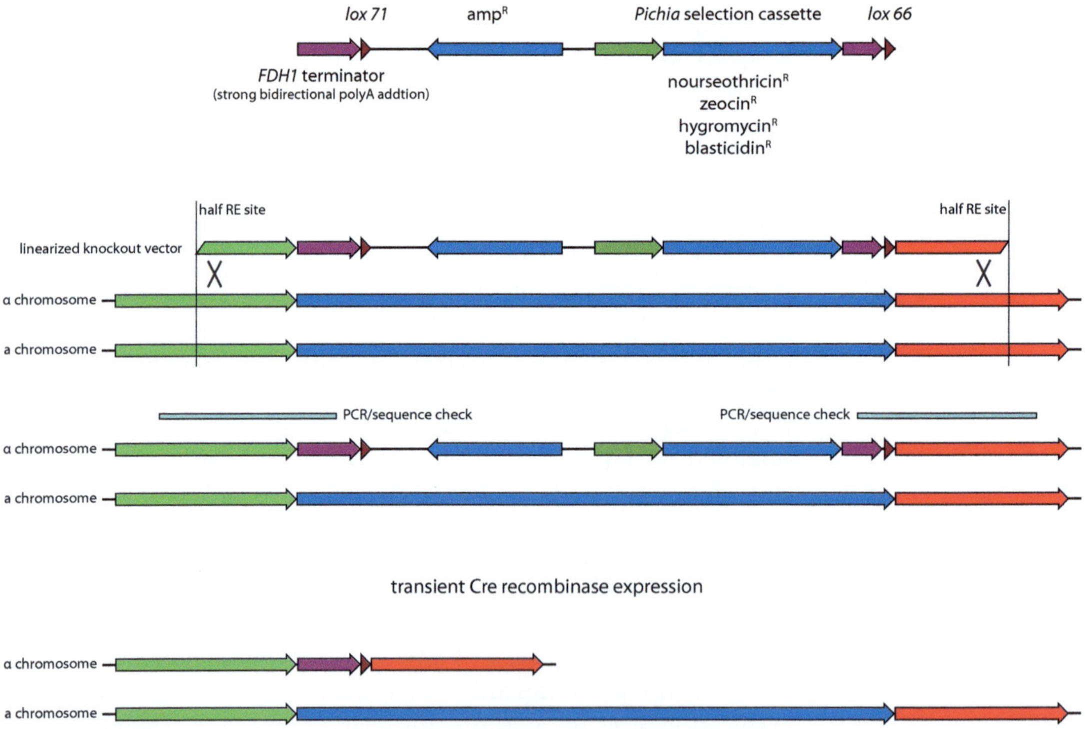

Fig. 5 Outline of an ORF knockout in a diploid strain. Based on experiments in *S. cerevisiae*, less than 0.1% of genes show haploinsufficiency and deletion of one copy in a diploid strain grows normally due to the presence of the second copy

Table 2
Restriction Enzyme Half Sites for Homology Arm Termination

Restriction Enzyme	5′ half site	3′ half site
*Eco*RV	ATC---	---GAT
*Pml*I	GTG---	---CAC
*Sma*I	GGG---	---CCC
*Pme*I	AAAC---	---GTTT
*Afe*I	GCT---	---AGC

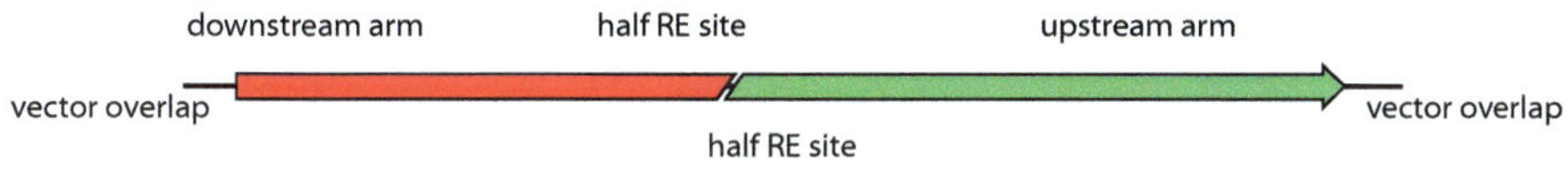

Fig. 6 Design of synthetic DNA fragment(s) for isothermal assembly. See Table 3 for 5′ and 3′ overlaps into BioGrammatics' targeting vectors

Table 3
Extensions for Isothermal Assembly into BioGrammatics' Targeting Backbones

Restriction Enzyme	5′ extension	3′ extension
*Nhe*I	ATAGCATACATTATACGAACGG TAG	CATGCTCGAGCTTACTTAAGA TGCA
*Xho*I	ATTATACGAACGGTAGCTAGCA TGC	GCTTACTTAAGATGCACCGGTGC TA
*Age*I	CATGCTCGAGCTTACTTAAGA TGCA	TGCTAGGATCCATGC TTGGCGGCCG
*Bam*HI	CTTACTTAAGATGCACCGGTGC TAG	CATGCTTGGCGGCCGCGGTCTCG TA
*Not*I	ACCGGTGCTAGGATCCATGC TTGGC	GCGGTCTCGTAATTGAAATGTA TTT
*Bsa*I	TCCATGCTTGGCGGCCGCGGTC TCG	TGAAATGTATTTAATTTGATATTAA

Note: *Nhe*I/*Bsa*I double digest removes all of the polylinker. For double digests, use the 5′ extension of the uppermost RE in the list and the 3′ extension of the lowermost RE in the list

initial PCR assembly step, where the presence of the *lox* sites on the cassette often necessitated extensive PCR optimization. In addition, carbon starvation during colony growth seems to lead to early, unwanted *cre* expression from the *AOX1* promoter with premature excision of the marker cassette. Instead, we have found that the use of synthetic double-stranded DNA in combination with isothermal assembly allows us to easily create vectors that can be linearized for targeting to any location in the genome. By introducing the *cre* recombinase later on an autonomously replicating plasmid, excision can be more precisely controlled.

Figures 4 and 5 briefly outline our general approach. One key feature is the use of restriction enzyme (RE) "half sites" present at high density in the *P. pastoris* genome to create circular vectors that can be linearized without introducing any nonhomologous bases at either end of the transforming DNA. Create a double-crossover gene knockout vector using the following steps.

1. Scan 400 to 700 bp upstream and downstream from the region of interest for RE "half sites" detailed in Table 2 (*see* **Note 8**). The region of interest for V5 epitope tagging of Och1p is from the active site of the protein to the termination codon. For gene deletions, the region of interest would be from the start to the stop of the ORF.
2. Confirm that the upstream and downstream homology arms (from the region of interest to the enzyme "half sites") do not contain any restriction sites for the RE planned for the final construct.
3. As shown in Fig. 6, use in silico DNA construction to join the proposed downstream arm (with a "half site" at its 3′ end) to the 5′ end of the upstream arm (with a "half site" at its 5'end). This in silico construction will contain the proposed RE site at the junction of the two homology arms. For BioGrammatics' targeting vector backbones, add isothermal assembly overlaps as detailed in Table 3.
4. Design either a single synthetic DNA fragment containing both arms and vector overlaps or a combination of synthetic DNA fragments that contain internal overlaps for isothermal assembly. Overlaps do not need to span the RE site at the junction of the arms.
5. Linearize the targeting vector to generate overlapping ends compatible with the synthetic DNA fragment(s).
6. Assemble and sequence-verify final homologous integration vector.

3.5.2 Creating a WT/SuperMan5 Diploid

1. Grow individual patches of the two haploid strains on YPD plates.
2. Cross the Zeocin-resistant (Zeo^R) and Nourseothricin-resistant (Nat^R) strains using the protocol described in Subheading 3.3.1 and isolate diploids that can grow on YPD Zeo/Nat plates.
3. Linearize the plasmid using the RE site present at the downstream/upstream arm junction.

4. Gel purify and isolate the linear plasmid. For electroporation, elute DNA from purification resin with unbuffered water.
5. Make electrocompetent cells of the selected Zeo^R Nat^R diploid (*see* **Note 9**) and transform them with the obtained DNA.
6. Plate transformants on selective medium and incubate at 30 °C for 2–3 days.
7. Pick individual colonies and test for retention of diploid markers.
8. Streak on YPD plates for single colony isolation. Retest diploid state.

3.5.3 Sporulation and Germination of WT/ SuperMan5 Diploid

1. Follow the methods described in Subheading 3.3.2 to generate and germinate random haploid spores of the V5-tagged diploid strain.
2. Pick individual colonies and screen for the presence or absence of the tagging vector, based on drug resistance/sensitivity.
3. Verify that drug-resistant isolates retain the tagging vector integrated at the *OCH1* locus, with the V5 tag still present in the *OCH1* open reading frame.

Figure 7 shows a section of a YPD plate containing colonies derived from random spore germination of an Och1p tagging integration. Colony growth and morphology fall into two classes (circled in red and green, respectively), with one class (red) growing much more slowly and showing "ragged" colony edges (several colonies in this microscopic field have been picked with sterile toothpicks and show disrupted morphology). Drug resistance and PCR verification showed that slow-growing colonies contained tagged/inactivated Och1p, while normal colonies contained a

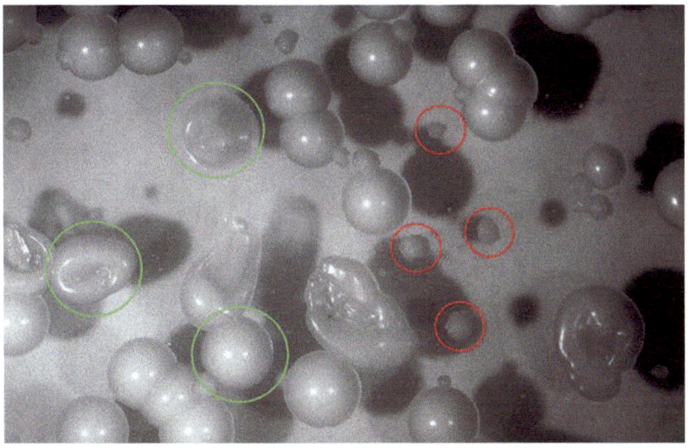

Fig. 7 Microscopic image of colonies grown from random spores generated from an Och1p tagged/inactivated diploid strain. Random spore analysis showed approximately 50/50 distribution of normal vs. slow-growing isolates

copy of the wild-type *OCH1* gene. Performing this integration in a diploid strain demonstrated conclusively that the disruption of *OCH1* results in abnormal growth. In the case where *OCH1* disruption had shown normal growth [49], Gehlsen and Chappell [38] subsequently demonstrated that the transforming vector had rearranged after transformation to reestablish expression of an altered form of Och1p.

4 Notes

1. Genome sequence of the strain GS115 is available [14, 15, 33].
2. Freezing in 30% glycerol solution kills ~90% of the vegetative cells. However, once frozen, the remaining cells maintain their viability at low temperature for a long time. Thus, it is critical not to allow the frozen culture to thaw because the next round of freezing might kill ~90% of the remaining viable cells.
3. *P. pastoris* Ura^- grows much slower than most of the other auxotrophic mutants and the wild-type strain. It might take several days before the evident growth of the Ura^- mutants can be detected on plates with uracil.
4. The incubation time of slow-growing *P. pastoris* mutants, like *ura3*, *ade1*, and such, can be longer before the streaks are ready for transferring to a mating medium. It can be wise not to streak strains having different growth rates on the same master plate.
5. Isolation of single colonies of the recombinant cultures is an important step because *P. pastoris* spores adhere tightly to one another, and initial colonies resulting from spore germination in the RSA procedure frequently contain cells derived from more than one spore. Another consequence of spore clumping is that markers appear not to segregate 1:1 as one would expect for monogenic determinants but to be biased toward the dominant or wild-type phenotype. For example, in the GS115 x GS190 cross described in the Subheading 3.3.2, more His^+ Arg^+ products will be apparent than the 25% expected in the random population, and His^- Arg^- spore products will appear to be underrepresented.
6. The Yeastern blot technique can be applied for the detection of any protein to which a good antibody is available or can be developed. Universal antitag antibodies can be used for the detection of epitope-tagged proteins.
7. The good wash from the primary antibody removes all nonspecific signals.

8. Approximately 95% of the genome can be targeted using EcoRV, PmeI, or SmaI restriction sites. Additional restriction enzymes in Table 2 allow targeting of the entire genome.
9. *P. pastoris* diploid cells can be transformed using the same electroporation protocol used for haploid cells. Cells will maintain their diploid state at high frequency during the transformation and selection protocol, so the initial selection plates only need to contain the drug for the marker present on the transforming linear DNA fragment.

Competing Interest Statement The authors have no conflicts of interest to declare that are relevant to the content of this chapter.

References

1. Kurtzman CP (2005) Description of *Komagataella phaffii* sp. nov. and the transfer of *Pichia pseudopastoris* to the methylotrophic yeast genus *Komagataella*. Int J Syst Evol Microbiol 55(2):973–976
2. Kurtzman CP (2009) Biotechnological strains of *Komagataella* (*Pichia*) *pastoris* are *Komagataella phaffii* as determined from multigene sequence analysis. J Ind Microbiol Biotechnol 36(11):1435–1438
3. Cregg J, Shen S, Johnson M, Waterham HR (1998) Classical genetic manipulation. In: Higgins DR, Cregg JM (eds) *Pichia* protocols, Methods Mol Biol, vol 103. Humana Press, Totowa, NJ, pp 17–26
4. Tolstorukov II, Cregg JM (2007) Classical genetics. In: Cregg JM (ed) *Pichia* Protocols, Second edn, Methods Mol Biol, vol 389. Humana Press, Totowa, NJ, pp 189–201
5. Hanson SJ, Byrne KP, Wolfe KH (2014) Mating-type switching by chromosomal inversion in methylotrophic yeasts suggests an origin for the three-locus *Saccharomyces cerevisiae* system. Proc Natl Acad Sci USA 111:E4851. https://doi.org/10.1073/pnas.1416014111
6. Tolstorukov II, Dutova TA, Benevolensky SV, Soom YO (1977) Hybridization and genetic analysis of methanol-utilizing yeast *Pichia pinus*. Genetika (Soviet Genetics) 13(2): 322–329
7. Tolstorukov I, Burkett T (1991) Comparative analysis of genomic structures of methylotrophic yeasts. Paper presented in the 15th international specialized symposium on yeasts: "regulation of metabolism and biotechnology", Riga, 4-8 July 1991
8. Gleeson MA, Sudbery PE (1988) Genetic analysis in the methylotrophic yeast *Hansenula polymorpha*. Yeast 4:293–303
9. Tolstorukov II, Benevolensky SV (1978) Study of the mechanisms of mating and self-diploidization in haploid yeast *Pichia pinus*. I Bipolarity of mating. Genetika (Soviet Genetics) 14(3):519–526
10. Tolstorukov II, Benevolensky SV (1980) Study of the mechanisms of mating and self-diploidization in haploid yeast *Pichia pinus*. II. Mutations on the mating type locus. Genetika (Soviet Genetics) 16(8):1335–1341
11. Tolstorukov II, Benevolensky SV (1980) Study of the mechanisms of mating and self-diploidization in haploid yeast *Pichia pinus*. III Study of heterothallic mutants. Genetika (Soviet Genetics) 16(8):1342–1349
12. Tolstorukov II, Benevolensky SV (1982) Genetic control of cell type and complex organization of the mating type locus in the yeast *Pichia pinus*. Curr Genet 5(1):137–142
13. Wolfe KH, Butler G (2022) Mating-type switching in budding yeasts, from flip/flop inversion to cassette mechanisms. Microbiol Mol Biol Rev 86(2):7–21
14. De Schutter K, Lin YC, Tiels P et al (2009) Genome sequence of the recombinant protein production host *Pichia pastoris*. Nat Biotechnol 27(6):561–566
15. Mattanovich D, Callewaert N, Rouzé P et al (2009) Open access to sequence: browsing the *Pichia pastoris* genome. Microb Cell Factories 8:53. https://doi.org/10.1186/1475-2859-8-53

16. Mattanovich D, Graf A, Stadlmann J et al (2009) Genome, secretome and glucose transport highlight unique features of the protein production host *Pichia pastoris*. Microb Cell Factories 8:29. https://doi.org/10.1186/1475-2859-8-29
17. Ravin NV, Eldarov MA, Kadnikov VV et al (2013) Genome sequence and analysis of methylotrophic yeast *Hansenula polymorpha* DL1. BMC Genomics 14:837
18. Naumov G, Tolstorukov I (1973) Comparative genetics of yeast. X. Reidentification of mutators of mating types in Saccharomyces. Genetika (Soviet Genetics) 9(1):82–91
19. Harashima S, Nogi Y, Oshima Y (1974) The genetic system controlling homothallism in Saccharomyces yeasts. Genetics 77(4):639–650
20. Hicks JB, Herskowitz I (1976) Interconversion of yeast mating types I. Direct observations of the action of the homothallism (HO) gene. Genetics 83(2):245–258
21. Hicks J, Strathern JN (1977) Interconversion of mating type in *S. cerevisiae* and the Cassette model for gene transfer. Brookhaven Symp Biol 29:233–242
22. Strathern JN, Blair LC, Herskowitz I (1979) Healing of *mat* mutations and control of mating type interconversion by the mating type locus in *Saccharomyces cerevisiae*. Proc Natl Acad Sci USA 76(7):3425–3429
23. Haber JE (2012) Mating-type genes and MAT switching in *Saccharomyces cerevisiae*. Genetics 191(1):33–64
24. Strathern JN, Klar AJ, Hicks JB et al (1982) Homothallic switching of yeast mating type cassettes is initiated by a double-stranded cut in the *MAT* locus. Cell 31(1):183–192
25. Latham J, Kovacevich B, McNeill P et al. (2007) The development and optimization of *Pichia pastoris* as a platform for the manufacture of full-length therapeutic antibodies. Paper #4215, SIM, 2007
26. Cregg J, Latham J, Litton M et al (2011) Methods of synthesizing heteromultimeric polypeptides in yeast using a haploid mating strategy. US Patent 7,927,863, 19 Apr 2011
27. Lundbeck Seattle Biopharmaceuticals - Wikipedia
28. Chen M-T, Lin S, Shandil I et al (2012) Generation of diploid *Pichia pastoris* strains by mating and their application for recombinant protein production. Microb Cell Factories 11:91. https://doi.org/10.1186/1475-2859-11-91
29. Shaheen HH, Prinz B, Chen MT et al (2013) A dual-mode surface display system for the maturation and production of monoclonal antibodies in glyco-engineered *Pichia pastoris*. PLoS One 8. https://doi.org/10.1371/journal.pone.0070190
30. Wegner EH (1983) Biochemical conversions by yeast fermentation at high cell densities. US patent 4,414,329, 8 Nov 1983
31. Sturmberger L, Chappell T, Geier M et al (2016) Refined *Pichia pastoris* reference genome sequence. J Biotechnol 235:121–131
32. Banerjee H, Verma M (2000) Search for a novel killer toxin in yeast *Pichia pastoris*. Plasmid 43(2):181–183
33. Küberl A, Schneider J, Thallinger GG et al (2011) High-quality genome sequence of *Pichia pastoris* CBS7435. J Biotechnol 154(4):312–320
34. Renuse S, Madugundu AK, Kumar P et al (2014) Proteomic analysis and genome annotation of *Pichia pastoris*, a recombinant protein expression host. Proteomics 14(23–24):2769–2779
35. Cregg JM, Madden KR (1989) Use of site-specific recombination to regenerate selectable markers. Mol Gen Genet 219(1–2):320–323
36. Boeke JD, LaCroute F, Fink GR (1984) A positive selection for mutants lacking orotidine-5′-phosphate decarboxylase activity in yeast: 5-fluoro-orotic acid resistance. Mol Gen Genet 197(2):345–346
37. Boeke JD, Trueheart J, Natsoulis G, Fink GR (1987) 5-Fluoroorotic acid as a selective agent in yeast molecular genetics. Methods Enzymol 154:164–175
38. Gehlsen KR, Chappell TG (2017) *Pichia pastoris* strains for producing predominantly homogeneous glycan structure. US patent 9,617,550 B2, 11 Apr 2017
39. Biogrammatics, Inc. www.biogrammatics.com
40. Cregg JM, Tolstorukov I, Kusari A et al (2009) Expression in the yeast *Pichia pastoris*. Methods Enzymol 463:169–189
41. Thermo Fisher (http://www.thermofisher.com)
42. San FJ, Sung P, Klein H (2008) Mechanism of eukaryotic homologous recombination. Annu Rev Biochem 77:229–257
43. Lieber MR (2010) The mechanism of double-strand DNA break repair by the nonhomologous DNA end-joining pathway. Annu Rev Biochem 79:181–211

44. Hua SB, Qiu M, Chan E et al (1997) Minimum length of sequence homology required for in vivo cloning by homologous recombination in yeast. Plasmid 38(2):91–96
45. Nett JH, Gerngross TU (2003) Cloning and disruption of the Pp*URA5* gene and construction of a set of integration vectors for the stable genetic modification of *Pichia pastoris*. Yeast 20(15):1279–1290
46. Nett JH, Hodel N, Rausch S, Wildt S (2005) Cloning and disruption of the Pichia pastoris ARG1, ARG2, ARG3, HIS1, HIS2, HIS5, HIS6 genes and their use as auxotrophic markers. Yeast 22(4):295–304
47. Näätsaari L, Mistlberger B, Ruth C et al (2012) Deletion of the *Pichia pastoris* KU70 homologue facilitates platform strain generation for gene expression and synthetic biology. PLoS One 7(6):e39720. https://doi.org/10.1371/journal.pone.0039720
48. Choi B-K, Bobrowicz P, Davidson RC et al (2003) Use of combinatorial genetic libraries to humanize N-linked glycosylation in the yeast *Pichia pastoris*. PNAS 100(9):5022–5027
49. Vervecken W, Kaigorodov V, Callewaert N et al (2004) *In vivo* synthesis of mammalian-like, hybrid-type N-glycans in *Pichia pastoris*. Appl Environ Microbiol 70(5):2639–2646
50. Krainer FW, Gmeiner C, Neutsch L et al (2013) Knockout of an endogenous mannosyltransferase increases the homogeneity of glycoproteins produced in *Pichia pastoris*. Sci Rep 3:3279. https://doi.org/10.1038/srep03279
51. Chen Z, Sun H, Li P et al (2013) Enhancement of the gene targeting efficiency of non-conventional yeasts by increasing genetic redundancy. PLoS One 8:e57952. https://doi.org/10.1371/journal.pone.0057952
52. Pan R, Zhang J, Shen WL et al (2011) Sequential deletion of *Pichia pastoris* genes by a self-excisable cassette. FEMS Yeast Res 11(3):292–298
53. Spohner SC, Müller H, Quitmann H, Czermak P (2015) Expression of enzymes for the usage in food and feed industry with *Pichia pastoris*. J Biotechnol 202:118. https://doi.org/10.1016/j.jbiotec.2015.01.027
54. Cregg JM, Barringer KJ, Hessler AY, Madden KR (1985) *Pichia pastoris* as a host system for transformations. Mol Cell Biol 5(12):3376–3385
55. Gleeson MA, White CE, Meininger DP, Komives EA (1998) Generation of protease-deficient strains and their use in heterologous protein expression. Methods Mol Biol 103:81–94
56. Lin Cereghino GP, Lin CJ, Sunga AJ, at al. (2001) New selectable marker/auxotrophic host strain combinations for molecular genetic manipulation of *Pichia pastoris*. Gene 263(1–2):159–169

Chapter 28

Pichia pastoris Random Mutagenesis and Screening

Christine Winkler, Joan Lin-Cereghino, and Geoff P. Lin-Cereghino

Abstract

Random mutagenesis provides a valuable tool to elucidate gene function and to draw connections between gene products and molecular pathways. This chapter provides two protocols for the random mutagenesis of *Komagataella phaffii*, along with methods to screen for the resulting mutant phenotypes. Genes are first disrupted by the insertion of a non-homologous DNA cassette. In the next step, primers designed to anneal to the cassette enable the identification of the mutagenized locus. We applied our system in *K. phaffii* to enhance recombinant protein secretion of horseradish peroxidase (HRP) and β-galactosidase, respectively. The subsequent screening identified unexpected genes that were later confirmed to benefit recombinant protein secretion, validating the substantive potential of the methodology.

Key words *Pichia pastoris*, *Komagataella phaffii*, Insertional mutagenesis, Genome walking, Restriction enzyme-mediated integration (REMI), Protein secretion

1 Introduction

Generating mutants is a strategy often used to identify gene products involved in cellular mechanisms. The process can be divided into three steps: (a) generating a mutation in the genomic DNA; (b) using a screen or selection to recognize those cells with a phenotypic change in the desired molecular mechanism; and (c) determining the locus with the mutation. Although chemical agents have been used successfully as mutagens in *Komagataella phaffii* (previously *Pichia pastoris)* [1], these are problematic for several reasons. These agents are usually strong carcinogens, posing a threat to the user. In addition, even if the desired phenotype is acquired, the identification of the mutated locus is difficult and time-consuming, especially if cloning by complementation is not straightforward. Therefore, new methods of mutagenesis are needed for *K. phaffii*, especially to make the recovery of the mutant gene easier. This chapter provides two alternative approaches for the generation of mutant strains, along with ways to rapidly screen

Xavier Garcia-Ortega et al. (eds.), *Pichia pastoris: Methods and Protocols*, Methods in Molecular Biology, vol. 2697,
https://doi.org/10.1007/978-1-0716-4779-0_28,

or select these cells for the desired phenotype and identify the responsible gene.

Both approaches involve insertional mutagenesis, in which the mutation of interest is physically marked by the integration of a non-homologous sequence of DNA. The advantages of insertional mutagenesis include that these integrations are usually (a) single copy so that only one locus in the cell is mutated; (b) stable, meaning that there is little rearrangement of the DNA in and around the insertion site after many generations; and (c) conservative in that none of the genomic DNA is deleted or duplicated around the insertion site [2]. Both experimental approaches described in this chapter begin with random integration of a selectable marker gene into the *K. phaffii* genome to generate strains with enhanced secretion efficiency. Typically, random integration can be performed with any DNA that encodes for a selectable marker gene and can be supported by the concept of restriction enzyme-mediated integration (REMI). However, the first strategy involves cultivation in a multi-well format and a colorimetric assay to identify colonies with the desired mutant phenotype, while the second one singles out beneficial mutations through growth on selective media. The gene associated with the anticipated phenotype is then identified by PCR-based genome walking in the first method and by plasmid rescue in the second method. Because the *K. phaffii* genome has been sequenced and is readily available through Bioinformatics Online Genome Annotation System (BOGAS) and GenBank (http://bioinformatics.psb.ugent.be/orcae/ and http://www.ncbi.nlm.nih.gov/genbank/), these methodologies have allowed for the facile identification of genes involved in the regulation of protein secretion. The users should select the protocol that suits their needs the best. The knowledge obtained from these techniques can be used to understand the basic biology of *K. phaffii* and make this yeast an even better system for recombinant protein production.

2 Materials

2.1 Preparation of Competent Cells for Electroporation

1. *K. phaffii* strain (choice depends on the marker gene located in the transformed DNA).
2. YPD: 1% yeast extract, 2% peptone, 1% dextrose; all media are sterilized before use.
3. YPD containing 0.02 M HEPES.
4. 1 M dithiothreitol (DTT)
5. 1 M sorbitol (sterile-filtered)
6. Sterile water (ice cold).

2.2 Mutagenesis of HRP-Secreting Strain by Random Integration of Zeocin Resistance Cassette

2.2.1 Construction of the Mutagenesis Starting Strain

1. *K. phaffii* strain GS115 *his4* (Thermo Fisher Scientific, Carlsbad, CA).
2. Vector pPICtoGAPαHRP (*see* **Note 1**).
3. *Sal*I restriction enzyme.
4. Gel and PCR Clean-Up Kit.
5. Minimal dextrose (MD) agar: 1.34% YNB (w/o amino acids), 4×10^{-5}% biotin, 2% glucose, and 2% agar.

2.2.2 Amplification and Transformation of the Mutagenesis Cassette

1. Vector pGAPZA (Life Technologies, Carlsbad, CA).
2. Primer TEFfw, 5′-CCCACACACCATAGCTTCAAAATG-3′.
3. Primer CYC1rev, 5′-AGCTTGCAAATTAAAGCCTTCGAG-3′.
4. High-Fidelity polymerase.
5. Gel and PCR Clean-Up Kit.
6. BioRad MicroPulser electroporation apparatus.
7. 2 mm gap electroporation cuvettes.
8. 1 M sorbitol.
9. YPD.
10. YPD + Zeocin agar: 1% yeast extract, 2% peptone, 2% dextrose, 2% agar, and 100 μg/mL Zeocin (InvivoGen, San Diego, CA).

2.2.3 Glycerol Stocks of Mutant Library

1. 96-well deep-well plates (DWP) (Bel-Art Scienceware, Pequannock, NJ).
2. YPD + Zeocin: 1% yeast extract, 2% peptone, 2% dextrose, and 100 μg/mL Zeocin (InvivoGen, San Diego, CA).
3. 50% (w/v) glycerol.

2.2.4 Pre- and Main-Culture

1. DWP (Bel-Art Scienceware, Pequannock, NJ).
2. Multi-channel pipette.
3. DWP shakers (Infors HT Multitron, Bottmingen-Basel, Switzerland).
4. BYPD: 1% yeast extract, 2% peptone, 2% dextrose, 200 mM potassium phosphate buffer, pH 6.0.

2.2.5 HRP Activity Screen in Culture Supernatant

1. 50 mM sodium acetate buffer, pH 4.5.
2. 20× ABTS stock: 550 mg ABTS dissolved in 50 mL of 50 mM sodium acetate buffer, pH 4.5 (to be stored at 4 °C).
3. 30% (w/w) hydrogen peroxide (H_2O_2).
4. 96-well microtiter plates, polystyrene, flat-bottom.
5. Spectrophotometer for 96-well plates.
6. Multi-channel pipette.

2.2.6 Isolation of Genomic DNA

1. DWP (Bel-Art Scienceware, Pequannock, NJ).
2. DWP shaker (Infors HT Multitron, Bottmingen-Basel, Switzerland).
3. YPD.
4. Glass beads, 0.25–0.5 mm.
5. Lysis buffer: 2% Triton X-100, 1% SDS, 100 mM NaCl, 10 mM Tris, pH 8.0, 1 mM EDTA.
6. Phenol:chloroform: isoamylalcohol (25:24:1).
7. Disruptor Genie SI-D238 (Scientific Industries, Bohemia, NY).
8. TE buffer: 10 mM Tris, 1 mM EDTA, pH 8.0.
9. 95% EtOH.
10. 70% EtOH.
11. RNAse A, 10 mg/mL.
12. 4 M ammonium acetate.
13. Nuclease-free water.

2.2.7 Template-Blocking PCR

1. Primer CSF27, 5′- GACGCGTAATACGACTCACTA TAGGGA-3′.
2. Primer CSR30, 5′- ATCTCCCTATAGTGAGTCGTAT TACGCGTC-3′.
3. Primer CP, 5′- ACGCGTAATACGACTCACTATAGGGA GATC-3′.
4. Primer GSPTEFa, 5′-TTCCAAACCTTTAGTACGGGTAAT TAACGACAC-3′.
5. Primer GSPTEFb, 5′-GCTGTGCTTGGGTGTTTTGAAGT GGT-3′.
6. Primer GSPCYC1a, 5′- GAGTTAGACAACCTGAAGTC TAGGTCCCTA-3′.
7. Primer GSPCYC1b, 5′- GTACAGACGCGTGTACGCATG TAACATTATAC-3′.
8. Wizard SV Gel and PCR Clean-Up System (Promega, Madison, WI).
9. 50 mM $MgCl_2$.
10. 1 M Tris-HCl, pH 8.
11. *Bgl*II, *Bam*HI, or *Sau*3AI restriction enzymes.
12. ddGTP, 10 mM.
13. Klenow fragment, 10 U/μL.
14. T4 DNA ligase (5 U/μL).
15. Maxima Hot Start Green PCR Master Mix (Thermo Scientific, Waltham, MA).

2.3 Mutagenesis of β-Galactosidase-Secreting Strain by Restriction Enzyme-Mediated Integration (REMI)

2.3.1 Random Mutagenesis by REMI

1. pGAPZα B (Life Technologies, Carlsbad, CA).
2. pREMI-Z (NCBI accession number AF282723).
3. *Bam*HI restriction enzyme.
4. Zymo Research Clean and Concentrator Kit (Irvine, CA).
5. Electroporator (model ECM 630, BTX Harvard Apparatus, Holliston, MA).
6. 2 mm gap electroporation cuvettes.
7. 1 M sorbitol.
8. YPD.
9. YPD + Zeocin agar: 1% yeast extract, 2% peptone, 1% dextrose, 2% agar, and 100 μg/mL Zeocin (Life Technologies, Carlsbad, CA).

2.3.2 Screening/Selection of Mutants

1. YN + lactose + histidine plates: 0.17% yeast nitrogen base, 0.5% ammonium sulfate, 2% agar supplemented with 1.0% lactose and 50 μg/mL histidine, adjusted to pH 6.7 with 1 M sodium phosphate.
2. YN + methanol + histidine plates: 0.17% yeast nitrogen base, 0.5% ammonium sulfate, 2% agar supplemented with 0.5% methanol and 50 μg/mL histidine.
3. X-Gal.

2.3.3 Isolation of Genomic DNA

1. YPD.
2. Yeast Geno-DNA-Template DNA Extraction Kit (G-Biosciences, St. Louis, MO).
3. Chloroform.
4. Ethanol, 95%.
5. Ethanol, 70%.
6. TE buffer: 10 mM Tris, 1 mM EDTA, pH 8.0.

2.3.4 Plasmid Rescue from Mutant Strains

1. *Eco*RI or *Hin*dIII restriction enzymes.
2. QIAquick PCR Clean-up Kit (Qiagen, Valencia, CA).
3. T4 DNA ligase.
4. One Shot MAX Efficiency DH10B competent *E. coli* cells (Life Technologies, Carlsbad, CA).
5. Lennox Broth (LB), 0.5% yeast extract, 1% glucose, 0.5% NaCl.
6. Lennox Broth (LB) + Zeocin plates 0.5% yeast extract, 1% glucose, 0.5% NaCl, 2% agar supplemented with 25 μg/mL Zeocin.
7. QIAprep Spin Miniprep Kit (Qiagen, Valencia, CA).
8. *Nco*I and *Eco*RV restriction enzymes.

2.3.5 Sequencing and Bioinformatic Analysis

1. M13 Forward (−20) primer: 5′- GTAAAACGACGGCCAG-3′.
2. M13 Reverse primer: 5′- GGAAACAGCTATGAC-3′.

3 Methods

3.1 Preparation of Competent Cells for Electroporation

Both methods for mutagenesis require *K. phaffii* cells with a high transformation efficiency. Although a fairly rapid "condensed" procedure for this purpose has been developed [3], a more lengthy, conventional method, modified from a protocol described by Cregg [4], is used in order to produce cells that will give rise to a greater number of transformants. The following protocol should generate cells with a transformation efficiency of 10^4–10^5/μg DNA (*see* **Note 2**).

1. Inoculate 10 mL of YPD with a fresh *K. phaffii* colony from a plate and grow overnight with shaking at 30 °C (*see* **Note 3**).
2. The next morning, use the overnight culture to inoculate 500 mL YPD in a 2.8 L baffled flask to an OD_{600} of approximately 0.1.
3. Grow the 500 mL of cells to an OD_{600} of approximately 1.0 in a 30 °C shaker (*see* **Note 4**).
4. Pour the culture into sterile bottles and pellet by centrifugation for 10 min (perform this and all further centrifugation steps at 4000× *g* and 4 °C).
5. Resuspend the cell pellet gently in 100 mL YPD containing 0.02 M HEPES.
6. Add 2.5 mL of 1 M DTT dropwise while swirling the cells.
7. Incubate for 15 min with shaking at 30 °C.
8. Add ice-cold water to 500 mL.
9. Pellet by centrifugation for 10 min.
10. Resuspend the cells in 500 mL of ice-cold water.
11. Pellet by centrifugation for 10 min.
12. Resuspend cells in 250 mL of ice-cold water.
13. Pellet cells by centrifugation for 10 min.
14. Resuspend cells in 20 mL of ice-cold 1 M sorbitol and transfer them to a 50 mL conical tube.
15. Pellet by centrifugation for 10 min.
16. Resuspend the cells gently in 1.5–2.0 mL of ice-cold 1 M sorbitol by flicking and inverting the tube. Do not vortex.
17. Aliquot the cells in convenient volumes (50 or 100 μL) into chilled 1.5 mL centrifuge tubes.

18. Wrap tubes of cells in several layers of paper towels and place them inside a small Styrofoam box. Place the box in −80 °C freezer to allow the cells to freeze slowly.
19. When needed, thaw tubes of cells on ice.

3.2 Mutagenesis of HRP-Secreting Strain by Random Integration of Zeocin Resistance Cassette

This protocol describes an efficient insertion mutagenesis method for random targeting of genes in *K. phaffii*. Mutant colonies are generated by integration of a linear Zeocin resistance expression cassette, previously amplified from plasmid pGAPZA (Life Technologies) by PCR (*see* **Note 5**). We demonstrate the feasibility of this method by mutagenizing strains secreting recombinant horseradish peroxidase (HRP), with subsequent screening for improved secretion to the culture supernatant. To simplify deep-well plate cultivation in the screening step, we chose to express HRP from the constitutive *GAP* promoter. However, in our experience, the results obtained from screening with P_{GAP} could be successfully transferred to P_{AOX1}. The cassette insertion locus in mutants with enhanced secretory capacity is later identified by PCR-based genome walking.

3.2.1 Construction of Parental Strain for Mutagenesis

1. Linearize plasmid pPICtoGAPαHRP (Fig. 1) by restriction with *Sal*I, which cuts at a unique restriction site in the *HIS4* gene (*see* **Note 1**).
2. Confirm complete digestion on agarose gel before purifying the linear plasmid with the Promega Wizard SV Gel and PCR Clean-Up System.
3. Transform GS115 cells with 600 ng of purified DNA. Use a BioRad MicroPulser electroporation apparatus with "Pic" settings (2 mm gap cuvette, 2.0 kV pulse) for transformation.
4. Immediately after transformation, add 500 μL of 1 M sorbitol and 500 μL of YPD and allow cells to recover at 28 °C for 2 h.
5. Select transformants by plating the transformation mixture on MD plates for selection of histidine prototrophy. Streak to obtain single colonies.
6. Check several transformants for secretion of HRP employing the ABTS assay described in Subheading 3.3.3.
7. Pick one HRP-secreting strain, which displays an HRP secretion level that is representative of a majority of strains, as the starting strain for mutagenesis.

3.2.2 Amplification and Transformation with Mutagenesis Cassette

1. Amplify mutagenesis cassette (Sh ble gene with respective promoter and transcription terminator) from template pGAPZA DNA or any other similar vector containing such selection marker expression cassette with primers TEFfw and CYC1rev (Fig. 2). For Phusion High-Fidelity DNA polymerase, use the following cycling parameters: 98 °C/30 s – (98 °C/5 s – 70 °C/20 s – 72 °C/20 s) × 35–72 °C/7 min – 4 °C/∞.

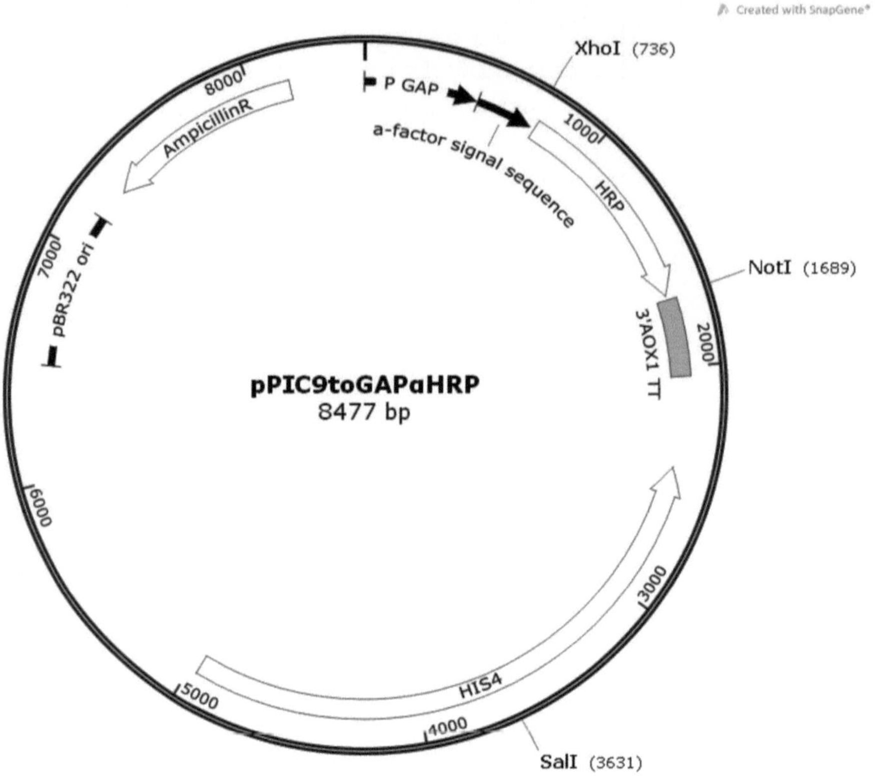

Fig. 1 Expression vector pPICtoGAPαHRP

Fig. 2 Schematic drawing of Zeocin resistance mutagenesis cassette. The 1172 bp fragment composed of eukaryotic promoter P_{TEF1}, prokaryotic promoter P_{EM7}, *She ble* open reading frame, and *CYC1* transcription terminator was amplified by PCR with the indicated primers

2. Load the PCR product on an agarose gel. Excise and purify the 1172 bp band using the Promega Wizard SV Gel and PCR Clean-Up System.
3. Use 1–2 μg of purified PCR product to transform 80 μL of electrocompetent cells of the mutagenesis starting strain. Prepare the electrocompetent cells as described in Subheading 3.1. Use a BioRad MicroPulser electroporation apparatus with "Pic" settings (2 mm gap cuvette, 2.0 kV pulse) for transformation.
4. Immediately after electroporation, add 500 μL of 1 M sorbitol and 500 μL of YPD and allow cells to recover at 28 °C for 2 h.

5. Plate different aliquots of transformation mixture on YPD + Zeocin agar plates, starting with 50 μL portions (*see* **Note 6**). It is preferred to have a low density of transformants on the plate to enable picking of single colonies (*see* **Note** 7). Incubate cells for 3 days at 28 °C.
6. Repeat **steps 1–5** until a sufficient number of transformants is obtained. We screened around 3000 colonies; this number can be adjusted depending on the anticipated secretion enhancement and/or screening effort and throughput.

3.2.3 Glycerol Stocks of Mutant Library

1. Fill an adequate number of 96-well DWPs with 250 μL YPD + Zeocin per well, except for column 4. This column is reserved for reference strains and should be filled with 250 μL YPD w/o Zeocin per well (*see* **Note 8**). The Zeocin is added to prevent contamination and to confirm the presence of the marker cassette in the genome. Using multi-channel pipettes is highly recommended for all pipetting steps involving 96-well plates.
2. Inoculate wells using sterile toothpicks. Column 4 is reserved for sterile controls, negative control GS115 (*his4* Zeo^-), and mutagenesis starting strain GS115 + pPICtoGAPαHRP (*HIS4* Zeo^-). All other wells are inoculated with mutagenesis cassette transformants (*HIS4* Zeo^+).
3. Incubate the DWPs for 48 h at 28 °C, 320 rpm, 80% humidity on a shaker (*see* **Note 9**).
4. Fill two 96-well microtiter plates per DWP with 100 μL of 50% glycerol per well. One will be your working plate, while the other will serve as the backup to retrieve identified hits. Transfer 100 μL of cell culture per well from the DWP to each microtiter plate and mix carefully by pipetting up and down. Freeze at −80 °C.

3.3 Deep-Well Plate Cultivation and HRP Activity Screening

Mutants are cultivated on buffered YPD (BYPD), since the reporter protein, HRP, is expressed from P_{GAP}. Alternatively, an induction protocol for expression from P_{AOX1} is described in the "Notes" section. The HRP activity screen is based on the protocol described in Morawski et al. [5].

3.3.1 Pre-culture

1. Thaw "working plates" of the glycerol stocks prepared in Subheading 3.2.3.
2. Fill 96-well DWPs with 250 μL BYPD per well.
3. Inoculate BYPD with 10 μL of glycerol stock.
4. Grow cells for 48 h (*see* **Note 9**) at 28 °C, 320 rpm, and 80% humidity.

3.3.2 Main-Culture

1. Fill 96-well DWPs with 500 μL BYPD per well (*see* **Note 10**).
2. Inoculate wells with 10 μL of pre-culture.
3. Grow cells for 48 h (*see* **Note 11**) at 28 °C, 320 rpm, and 80% humidity.
4. Prior to the activity screen, measure OD_{600} after diluting 1:20 (*see* **Note 12**).

3.3.3 HRP Activity Screen in Culture Supernatant

1. Prepare 1× ABTS screening solution: per screened DWP, mix 19 mL sodium acetate buffer, 1 mL of 20× ABTS stock, and 1.75 μL of 30% (w/w) H_2O_2 (*see* **Note 13**).
2. Centrifuge DWPs for 10 min at 1500× *g*, 22 °C.
3. Transfer 15 μL of supernatant into 96-well microtiter plates; be careful not to touch the cell pellet.
4. Immediately before analysis, add 140 μL of 1× ABTS screening solution per well to start the reaction. The uncolored solution will instantly start to turn turquoise.
5. Measure absorption kinetics at 405 nm in a spectrophotometer for 2–5 min.
6. Calculate activity/OD_{600} and retrieve the best hits from the "backup plate" of the glycerol stocks prepared in Subheading 3.2.3. We selected all the strains that had at least double the activity of the wild-type control.
7. Repeat cultivation and screening steps for best hits in subsequent rounds of rescreening.

3.4 PCR-Based Genome Walking

The described protocol for the isolation of genomic DNA (gDNA) is based on the method devised by Hoffman and Winston [6]. To avoid nonspecific amplification in the genome walking step, we adapted the template-blocking PCR protocol of Bae and Sohn [7]. In this protocol, the 3′ ends of the restricted genomic DNA fragments are blocked with dideoxynucleoside triphosphate (ddNTPs) prior to the ligation of appropriate adaptor cassettes (Fig. 3). This step prevents the generation of additional binding sites for the adaptor primer, which could be formed by filling in the recessed ends of the adaptor-ligated restriction fragments with DNA polymerase, ultimately resulting in nonspecific amplification. The unknown gDNA sequence between the adaptor and Zeocin^R-mutagenesis cassette is PCR-amplified and sequenced using primers that bind in these two elements. Subsequently, the resulting sequences, representing the mutagenesis cassette insertion loci, are mapped using NCBI BLAST.

3.4.1 Isolation of Genomic DNA

1. Fill the DWP with 600 μL YPD per well. Inoculate wells with the best hit *K. phaffii* mutant strains.
2. Grow cells 24–36 h at 28 °C with 320 rpm, and 80% humidity.

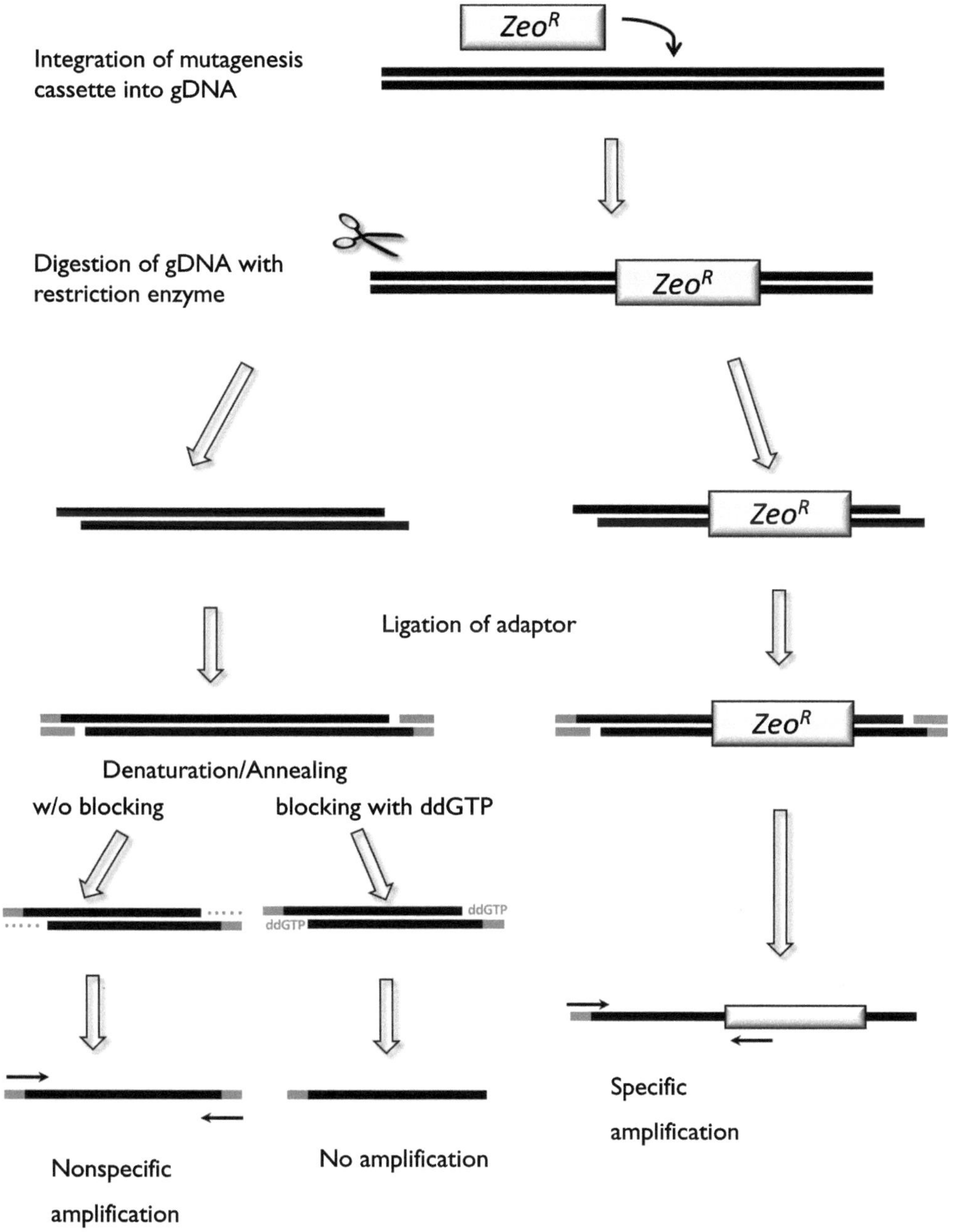

Fig. 3 Template-Blocking PCR scheme: Upon transformation, the mutagenesis cassette integrates randomly into the genome. Isolated gDNA is then cut with restriction enzymes, yielding fragments that contain the mutagenesis cassette and fragments that do not. In the next step, the adaptor is ligated to the complementary sticky ends of the restricted gDNA. If the fragment contains the mutagenesis cassette, the genomic sequence between the adaptor and the mutagenesis cassette can be amplified using primers that bind to these two elements. Through rounds of denaturation and annealing, DNA polymerase can potentially generate additional binding sites for the adaptor primer by filling in recessive ends, ultimately resulting in nonspecific amplification products of fragments that do not contain the mutagenesis cassette. This process is prevented when the 3′ ends of the restricted gDNA are blocked with dideoxynucleoside triphosphate (ddNTPs) prior to the ligation of appropriate adaptor cassettes

3. Transfer the culture volumes to 1.5 mL microcentrifuge tubes (*see* **Note 14**).
4. Spin for 1 min at max. speed (>13,000× *g*) in a tabletop centrifuge at 4 °C.
5. Decant supernatant and add approximately 0.3 g of glass beads, 150 μL of lysis buffer, and 150 μL of phenol/chloroform/isoamyl alcohol to the cell pellet.
6. Vortex for at least 8 min using the Disrupter Genie (*see* **Note 15**).
7. Add 150 μL of TE buffer and spin tubes for 5 min at max. speed.
8. Transfer the upper aqueous phase to a fresh 1.5 mL microcentrifuge tube. Precipitate DNA by adding 1 mL of ice-cold EtOH and incubating at −20 °C for 30 min.
9. Spin the tubes for 5 min at max. speed and decant the supernatants.
10. Dry the pellets at 60 °C until the EtOH has completely evaporated.
11. Resuspend the pellets in 400 μL of TE buffer and add 5 μL of RNAse A to degrade the RNA. Incubate for 2 h at 37 °C.
12. Precipitate the gDNA by adding 10 μL of 4 M ammonium acetate and 1 mL of ice-cold EtOH.
13. Spin 5 min at max. speed to pellet the gDNA. Remove the supernatants and wash the pellets with 1 mL of 70% EtOH. Spin again.
14. Dry pellets to remove any residual EtOH.
15. Dissolve each pellet in 50 μL of nuclease free, sterile water.

3.4.2 Template-Blocking PCR

1. A double-stranded adaptor cassette is generated by annealing the two primers CSF27 and CSR30. This cassette possesses a three-base (GAT) overhang at its 5′ end that is complementary to one-base-filled *Bam*HI-, *Bgl*II-, and *Sau*3AI-digested fragments, respectively. To assemble the adaptor, mix 20 μL of 100 μM primer CSF27, 20 μL of 100 μM primer CSR30, 10 μL of 50 mM $MgCl_2$, and 1.25 μL of 1 M Tris-HCl, pH 8. Heat to 100 °C for 5 min and allow to slowly cool down.
2. Digest 2 μg of gDNA with either *Bgl*II, *Bam*HI, or *Sau*3AI at 37 °C overnight in a final volume of 34 μL.
3. Add 0.8 μL ddGTP and 1 μL of Klenow fragment to the reaction mixture. Incubate for 3 h at 37 °C.
4. Purify the Klenow-treated DNA using the Promega Wizard SV Gel and PCR Clean-Up System, following the manufacturer's instructions. Elute in 55 μL of nuclease-free water.

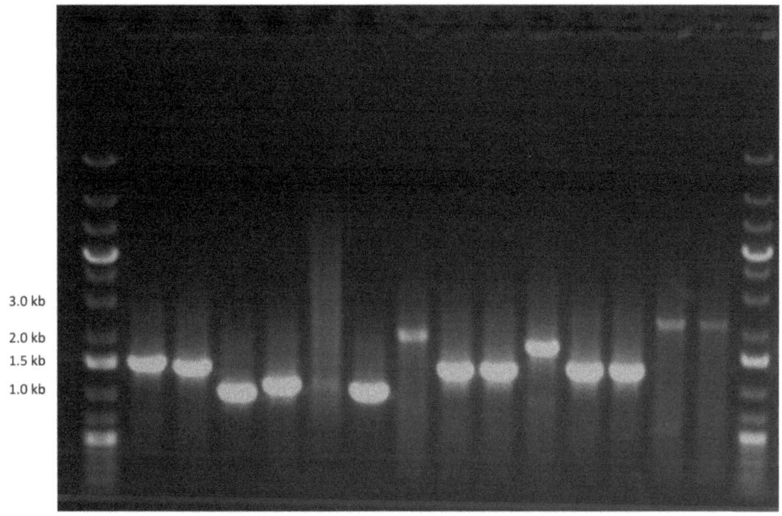

Fig. 4 A typical result of Template-Blocking PCR

5. Mix the eluted DNA with 1 μL of the adaptor (prepared in **step 1**), 6 μL of T4 ligase buffer, and 1 μL of T4 ligase. Incubate overnight at 16 °C for efficient ligation.
6. Purify the DNA again using the Promega Wizard SV Gel and PCR Clean-Up System. Elute in 40 μL TE buffer.
7. Perform template-blocking PCR in forward and reverse directions with primer pairs CP + GSPTEFa and CP + GSPCYC1a, respectively. As a template, use the restricted gDNA with ligated adaptor produced in **steps 5** and **6**. We use Thermo Scientific's Maxima Hot Start Green PCR Master Mix and the following cycling parameters for the PCR according to the manufacturer's instructions: 95 °C/4 min – (95 °C/30 s – 60 °C/50 s – 72 °C/4 min) × 35 cycles – 72 °C/10 min – 4 °C/∞. An example result is shown in Fig. 4.
8. Load a fraction of the PCR on an agarose gel. If the PCR resulted in a band, the product of the template-blocking PCR can be used as a template for a second, "nested" PCR reaction with the nested primer pairs CP + GSPTEFb and CP + GSPCYC1b, respectively (cycling parameters: 95 °C/4 min – (95 °C/30 s – 60 °C/50 s – 72 °C/1 min per kb, depending on the largest fragment to be amplified) × 40 cycles – 72 °C/10 min – 4 °C/∞).
9. Load the full volume of the nested PCR on an agarose gel. Extract the product band and purify DNA using the Promega Wizard SV Gel and PCR Clean-Up System.
10. Sequence the band employing the same primers that were used for amplification.

3.4.3 Bioinformatic Analysis

1. Identify the integration locus of the mutagenesis cassette by using NCBI blastn. Compare the obtained sequencing results to the genome sequence of the respective *K. phaffii* reference strain (e.g., *K. phaffii* GS115) in case GS115, X33, KM71H, or related strains are used as parental strains for mutagenesis, or the CBS7435 genome sequence in case the parental strain was made by transforming an NRLL Y-11430 or Y-4290 platform-derived strain with the reporter gene expression cassette (*see* **Note 16**).
2. Use Blastp to compare the corresponding amino acid sequence of the disrupted ORF (or ORFs in the vicinity of the integration site, if the cassette had integrated into an intergenic region) to proteins of other related yeasts. This step is recommended since the annotation of proteins in *K. phaffii* is sometimes incomplete or incorrect.

We confirmed the beneficial effect of gene disruptions identified in the screening by generating targeted knockouts of selected genes. The enhanced secretory performance of these knockout strains relative to the wild-type control in repeated rounds of screening is depicted in Fig. 5.

3.5 Mutagenesis of β-Galactosidase-Secreting Strain by Restriction Enzyme-Mediated Integration (REMI)

A second example and alternative strategy for generating mutations in the electrocompetent *K. phaffii* cells is based on restriction enzyme-mediated integration (REMI), a technique which has been used in many organisms [8, 9]. This method involves a) the random integration of a linearized plasmid containing a selectable marker into the *K. phaffii* genome; b) the identification of mutant colonies with a desired phenotype; and c) the retrieval of the circularized plasmid from the genomic DNA of the mutant strain and analysis of its DNA sequence to determine the disrupted gene (Fig. 6).

Although the details may differ depending on the microbe used, the integration event can be generally viewed in three steps: (a) during transformation, the restriction enzyme enters the nucleus along with a linearized plasmid whose ends have been generated by the same enzyme; (b) the restriction enzyme digests the host chromosome at its specific recognition site; and (c) the complementary ends of the chromosome and the linearized plasmid anneal and are consequently ligated by cellular enzymes to create a non-homologous integration [2] (*see* **Note 17**). The REMI protocol has been described previously [10]. The plasmid used in this procedure, pREMI-Z plasmid (a generous gift of Ben Glick, University of Chicago), is a 1985 bp vector containing a Zeocin resistance gene under the control of the *Saccharomyces cerevisiae TEF* and bacterial *EM7* promoters, Col E1 origin of replication, and a unique *Bam*HI site, allowing for transformation of *E. coli* and *K. phaffii*.

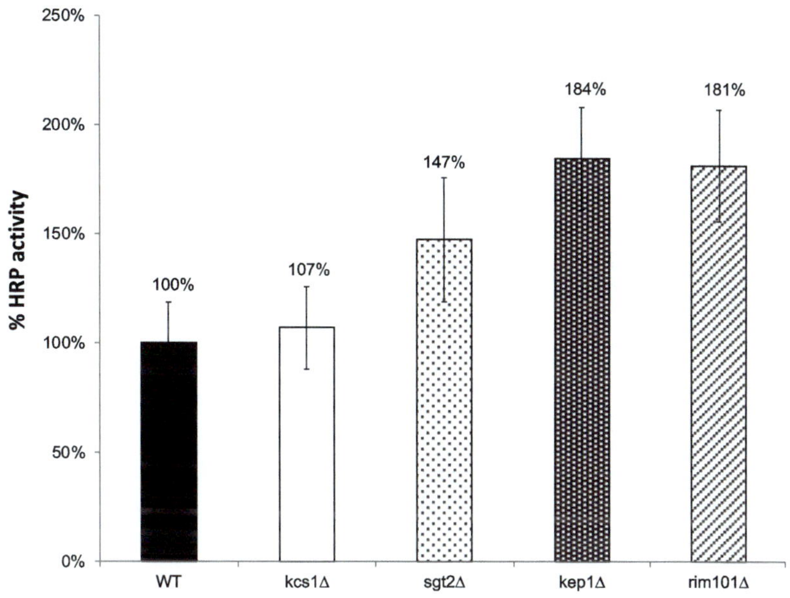

Fig. 5 Quantification of secreted HRP in culture supernatants of selected knockout strains. The genes *KCS1*, *SGT2*, *KEP1*, and *RIM101* had previously been identified in a screening for enhanced secretion of HRP, following random mutagenesis of the expression strain. Relative HRP activity upon expression from the *GAP* promoter. WT control and mutant strains expressing HRP from P_{GAP} were grown on BYPD (2% glucose) in DWP for 35 h before analysis. Peroxidase activity in supernatants was measured with the ABTS assay. Results represent the mean of four biologically independent experiments with 12 technical replicates per experiment

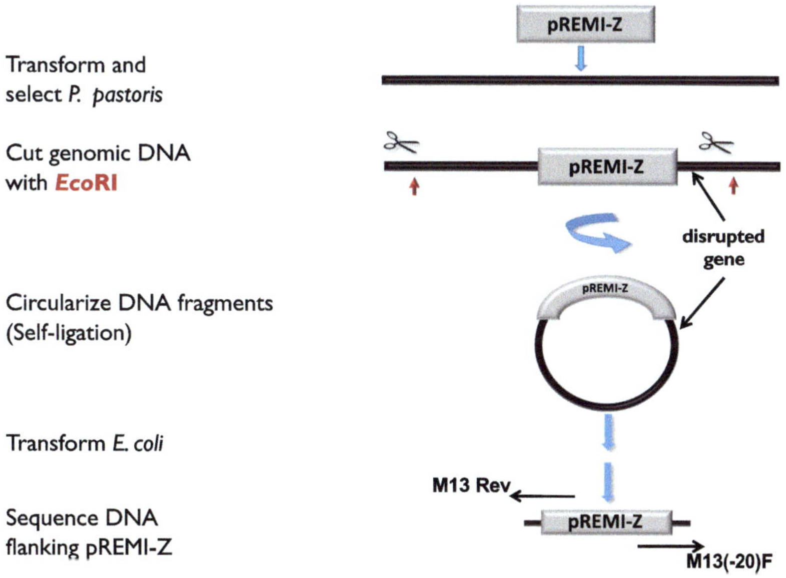

Fig. 6 Key steps in the REMI method

3.5.1 Random Mutagenesis by REMI

1. Digest 5 μg of pREMI-Z with *Bam*HI. A small portion of the reaction mixture should be run on a 1% agarose gel to confirm linearization.
2. Purify the digested plasmid to remove salts and enzymes using a Zymo Research Clean and Concentrator Kit.
3. Quantify the purified plasmid by absorbance at 260 nm.
4. Mix approximately 1 μg of the *Bam*HI-linearized plasmid with 50 μL of freshly thawed, electrocompetent cells and 1 unit of *Bam*HI in a 2 mm gap cuvette (*see* **Note 18**).
5. After incubating the cuvette on ice for 5 min, electroporate using standard conditions (1.5 kV, 50 uF, 200 Ω) with a BTX electroporation generator.
6. Immediately after electroporation, add a mixture of 500 μL of 1 M sorbitol and 500 μL of YPD to the cuvette. Transfer the cell suspension to a 1.5 mL microfuge tube, allowing it to recover for 1 h in a shaking incubator at 30 °C.
7. Plate the transformation mixture in 100 μL aliquots on YPD + Zeocin plates. Plates should be incubated for 2–3 days until colonies appear (expect approximately 500–1000 colonies/plate) (*see* **Note 19**).

3.5.2 Screening/Selection of Mutants

With the goal of isolating strains with enhanced secretion, a plasmid encoding an α-mating factor (α-MF)-β-galactosidase fusion was expressed from the *GAP* promoter and integrated into *K. phaffii* strain yDT39 *his4 met2* [11]. The plasmid contained *MET2* as a selectable marker. While the α-MF acts as a secretion signal, β-galactosidase serves to convert lactose, which cannot be utilized by *K. phaffii*, into galactose and glucose, which can be used metabolically. In wild-type cells, the protein is retained within the cell. Thus, when these cells are plated on medium containing lactose as a sole carbon source, they fail to grow because the enzyme is not efficiently secreted. These cells were made electrocompetent and mutagenized by the REMI method, as described above. If the strains harboring pREMI-Z insertion mutations allowed export of the β-galactosidase protein, they would be expected to grow on lactose (Fig. 7). This selection was performed as follows:

1. Replica-plate colonies from YPD+Zeocin plates (**step** 7 of Subheading 3.5.1.) to YN + lactose+histidine plates. A non-mutagenized, wild-type strain containing an integrated pGAPZα B plasmid (which carries the Zeocin resistance gene) should be used as a negative control (*see* **Note 20**).
2. Incubate the YN + lactose+histidine plates at 30 °C for 3–4 days. Supersecreting colonies should grow significantly better than the negative control strain (*see* **Note 21**).

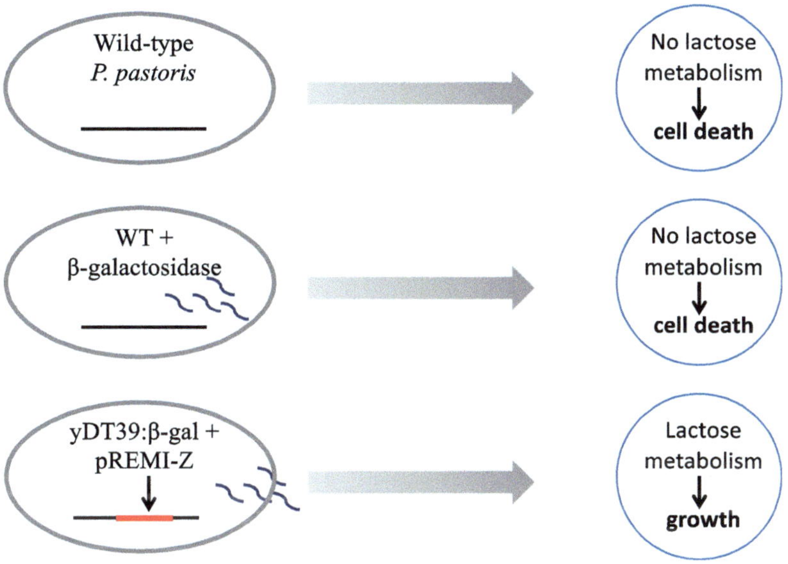

Fig. 7 Strategy to isolate strains that supersecrete β-galactosidase

3. Restreak the putative lactose-utilizing colonies on YN + lactose +histidine plates to confirm robust growth; however, mutant strains will grow at different rates.
4. Transfer colonies from lactose plates to YN + methanol+histidine plates supplemented with X-Gal (80 μg/mL) to confirm β-galactosidase secretion, as evidenced by the formation of a blue halo around the colony [12] (*see* **Note 22**).
5. Other reporter constructs (containing the G418 resistance gene or a prototrophic selectable marker gene if the strain carries auxotrophies) can be transformed into each mutant strain and assayed to see if the strain gives the expected phenotype. For instance, an α-MF-horseradish peroxidase (HRP) fusion can be expressed to see if its secretion is enhanced in a mutant strain compared to a wild-type strain [11].

3.6 Identification of Mutagenized Locus by Plasmid Rescue

As a first step, genomic DNA is isolated from each mutant using components of the Yeast Geno-DNA-Template DNA Extraction Kit (G-Biosciences, St. Louis, MO) with some modifications of the manufacturer's protocol. The genomic DNA of each strain should be analyzed by Southern blotting, using a probe generated from the pREMI-Z plasmid to confirm that the chromosomal DNA of each strain contains only one pREMI-Z insertion. A detailed description of the Southern blot method would go beyond the scope of this article, but standard protocols can be found in several sources [14]. Strains displaying a single band on the Southern are then subjected to plasmid rescue. The genomic DNA of these mutant strains is digested with a single restriction enzyme to generate fragments with self-annealing ends. The fragments will be allowed to anneal and be ligated. Only those circularized fragments containing the pREMI-Z sequence and flanking genomic DNA should be able to transform *E. coli* and confer Zeocin resistance (Fig. 6).

3.6.1 Isolation of Genomic DNA

1. Grow *K. phaffii* colonies in 3 mL of YPD in 50 mL conical tubes in a 30 °C shaker overnight.
2. The next day, pellet 1.5 mL of the liquid cultures in a microfuge tube.
3. After the supernatant is removed, add 150 μL of Yeast Suspension Buffer, 5 μL of LongLife Zymolyase, and 1 μL of β-mercaptoethanol to each cell pellet. The reagents, as well as those in the later steps, are provided by the Yeast Geno-DNA-Template DNA Extraction Kit.
4. Resuspend each pellet and incubate in a 37 °C water bath for 1 h, occasionally flicking the tubes to mix the samples.
5. Centrifuge the samples at 16,000× *g* for 5 min and discard the supernatant. Resuspend the pellets in 150 μL of sterile water.
6. Add 400 μL of lysis buffer and invert several times before the addition of 5 μL of LongLife Proteinase K solution.
7. Incubate at 60 °C for 3 h with mixing by inversion every 30 min (*see* **Note 23**).
8. After the samples cool to room temperature, add 200 μL of chloroform and vortex well.
9. Centrifuge for 5 min at 16,000× *g*.
10. Remove the upper aqueous phase and transfer it to a new microfuge tube.
11. Add 60 μL of DNA Stripping Solution. Invert several times and incubate at 60 °C for 10 min.
12. Cool samples to room temperature and add 120 μL of precipitation solution. Mix by inversion.
13. Centrifuge samples for 6 min at 16,000× *g*.
14. Transfer the supernatant to a new tube and add 700 μL of 95% ethanol.
15. Invert each tube 20 times to precipitate the genomic DNA (avoid vortexing, which will shear the DNA), and then centrifuge for 1 min at 16,000× *g* to pellet the DNA.
16. Remove the ethanol and add 1 mL of 70% ethanol to wash the pellet.
17. Invert and flick the tube to dislodge the pellet from the bottom.

18. After centrifugation at 16,000× *g* for 1 min, discard the 70% ethanol and leave each tube open and inverted for 5 min in a 37 °C oven to remove any residual ethanol (*see* **Note 24**).
19. Add 200 μL of TE buffer and 1 μL of LongLife RNAse to each sample. The DNA should be allowed to rehydrate for 30 min at 37 °C with intermittent inversion (no vortexing).
20. Determine the DNA concentration by measuring absorption at 260 nm. Typically, the samples contain 1–2 μg/μL of DNA.
21. Store at −80 °C.

3.6.2 Plasmid Rescue from Mutant Strains

1. Digest approximately 3 μg of genomic DNA with either *Eco*RI or *Hin*dIII for 3 h at 37 °C.
2. Purify the fragments with the QIAGEN QIAquick PCR Clean-up Kit and elute in 30 μL (*see* **Note 25**).
3. Incubate 1 μg of each purified, digested DNA with T4 DNA ligase and appropriate reaction buffer in a total volume of 20 μL overnight at room temperature.
4. Transform 3 μL of the ligation mix into One Shot MAX Efficiency DH10B competent *E. coli* cells according to the manufacturer's instructions. These cells are especially suitable for transformation by large plasmids.
5. Plate the transformation reaction on LB (low salt) + 25 μg/mL Zeocin solidified medium and incubate overnight at 37 °C.
6. Grow up at least six of the resulting colonies in liquid LB + 25 μg/mL Zeocin medium for 18 h at 37 °C (*see* **Note 26**).
7. Isolate the plasmids using QIAprep Spin Miniprep Kit and follow the special directions for elution of large plasmids.
8. Digest the plasmids with a restriction enzyme(s) that recognizes a site in pREMI-Z (i.e., *Nco*I and *Eco*RV) and compare the resulting restriction patterns on an agarose gel (Fig. 8 and *see* **Note 27**).

3.6.3 Sequencing and Bioinformatic Analysis

1. Use the M13 Forward (−20) primer and M13 Reverse primer to sequence the flanking genomic DNA in the rescued plasmids. These primer sites flank both sides of the original *Bam*HI site that is located in pREMI-Z (Fig. 6). At least 4 plasmids originating from the plasmid rescue of each mutant strain should be sequenced.
2. The National Center for Biotechnology Information (NCBI) (http://www.ncbi.nlm. nih.gov/) and the BOGAS (http://bioinformaticspsb.ugent.be/webtools/bogas) should be used to determine the gene disrupted in the mutant strains by the pREMI-Z sequence based on the identity between the

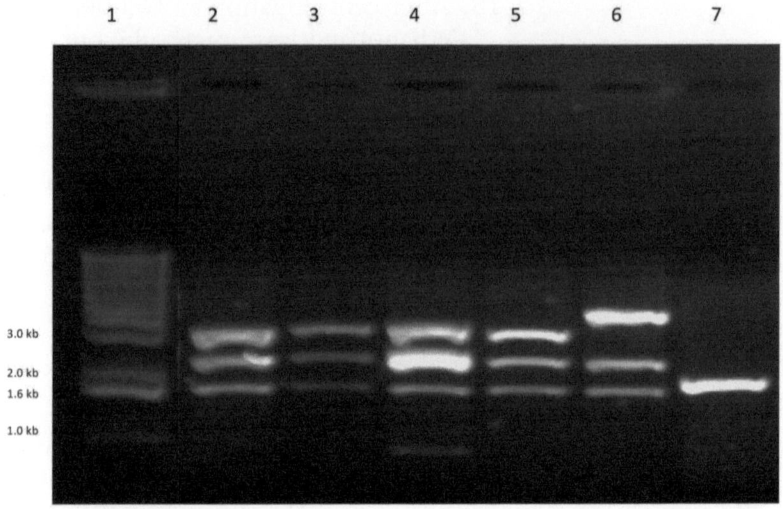

Fig. 8 Restriction analysis of plasmids isolated from colonies resulting from a plasmid rescue. Plasmids were digested with *Nco*I and *Eco*RV and then resolved on a 0.8% agarose gel. Lane 1: 1 kb molecular weight marker; lanes 2–6: plasmids isolated from colonies; lane 7: pREMI-Z. Lanes 2, 3, and 5 contain the pREMI-Z plasmid with the same flanking genomic DNA, while the plasmids in lanes 4 and 6 contain an additional insert along with this flanking genomic DNA

sequenced DNA and the published *K. phaffii* genome. The genes identified by this selection strategy have been previously described [11].

3. Amino acid sequence homology to other cloned genes in the database should be analyzed with the BLAST program. Expert Protein Analysis System (ExPASy) at http://www.expasy.org/ can be used to analyze the predicted amino acid sequences of the identified genes for motifs, such as kinase, DNA binding domains, and other structural analyses, which will provide clues to the function of the *K. phaffii* gene product.

4 Notes

1. We constructed the plasmid pPICtoGAPαHRP (Fig. 1), a chimera of the two vectors pPIC9 and pGAPZαA, which combines the two features *GAP* promoter and *HIS4* selection marker. The vector has the HRP gene inserted between *Xho*I and *Not*I restriction sites.
2. Electrocompetent cells can also be prepared by the condensed protocol, but the lower transformation efficiency of these cells will require many more electroporation reactions to achieve the same number of transformants.

3. Make sure that the strain is sensitive to Zeocin and has auxotrophic markers that will permit transformation with plasmids later.
4. Cells can be harvested at any OD_{600} between 0.7 and 1.3; however, an OD_{600} outside this range will lead to lower transformation efficiencies.
5. We also tested the applicability of a *KanMX6* marker cassette for mutagenesis, conferring resistance to Geneticin. We observed random integration into genomic DNA, similar to the Zeocin resistance cassette. However, elevated levels of transformation background on YPD plates containing 300 μg/mL Geneticin made us proceed with the Zeocin resistance marker instead.
6. We advise spreading the transformation mixture on plates by using glass beads, 2.85–3.3 mm (e.g., Roth).
7. If available, a colony picking robot can facilitate the picking procedure. Spread 500 μL of transformation mixture on agar prepared in appropriate bioassay trays. Use the picking head recommended by the supplier for picking the yeast colonies for the DWP.
8. Several thousand colonies have to be picked and screened to theoretically cover all nonessential genes of *K. phaffii*. Therefore, we recommend dividing the picking procedure into several working days, depending on the available capacities, e.g., automated pickers, DWP shakers, etc. The transformation plates should be stored at 4 °C during this time.
9. It is crucial for the outcome of the later screening that cells grow to a similar cell density. Grow them to saturation and possibly control OD_{600} after diluting 20-fold. Eventually, extend the incubation time until all wells show the same density.
10. In case of expression from P_{AOX1}, pre-grow cells for 32 h in 200 μL of $BMG_{lucose}Y$ (1% yeast extract, 2% peptone, 2% glucose, 0.2 M potassium phosphate buffer, pH 6.0, 1.34% YNB, 4×10^{-5}% biotin) per well. Start the induction by adding 200 μL of BMMY2 (1% yeast extract, 2% peptone, 2% methanol, 0.2 M potassium phosphate buffer, pH 6.0, 1.34% YNB, 4×10^{-5}% biotin). After another 12, 24, and 36 h, add 50 μL of BMMY10 (as BMMY2, but 10% methanol) to keep the methanol concentration at ~1% final concentration. Harvest cells 48 h after the start of induction [13].
11. Cultivation time can be adjusted depending on the expression level of the secreted protein.

12. Gene disruptions can cause reduced growth rates. Measuring OD_{600} is important, since altered HRP activity in the culture supernatant could stem from altered growth behavior.
13. 1× ABTS screening solution should always be prepared fresh. Keep on ice and protect from light until use in the assay.
14. Volumes of >200 μL are easily transferred from the DWPs by using a 200 μL-tip on top of a 1000 μL-tip.
15. Vortexing without the Disrupter Genie results in reduced cell lysis and, subsequently, reduced yields.
16. The genome sequence and protein annotation available in the NCBI database differ significantly between the sequences of GS115-derived strains, which were made by random mutagenesis, and the CBS7435/NRLL Y-11430-related strains. Due to existing differences and some remaining uncertainties, it can therefore be beneficial to compare results with both strains.
17. While the activity of the restriction enzyme has been shown to be required for this process in *Saccharomyces cerevisiae*, the restriction enzyme is not needed for REMI of some fungal species, such as *Ustilago maydis*. It is not clear if the addition of the restriction enzyme enhances the transformation efficiency with the linearized plasmid in *K. phaffii*.
18. Insufficient enzyme leads to lower transformation efficiency, while excess enzyme has been associated with genetic damage and multiple insertion events [2].
19. To analyze sufficient colonies, 5 electroporation reactions were initially performed, yielding approximately 50 plates in total.
20. If another *K. phaffii* strain is used, which has other nutritional auxotrophies, make sure to include the necessary amino acids in the growth medium.
21. To save time and effort, electroporated cells (**step # 6** above) can, instead, be allowed to recover in 1 mL of 1 M sorbitol for 1 h and then be plated directly onto YN + lactose+histidine plates.
22. β-galactosidase, which is secreted by cells, metabolizes the X-Gal in the medium to form a blue product on the periphery of the colony (blue halo). Colonies that do not show the supersecretion phenotype appear blue but show no blue halo.
23. This is the most crucial step to attain high yields of DNA. The samples should become clear over time as the cells lyse. If not, vortex them mildly to accelerate the process. By the end of 3 h, all solids should be gone, and the tube contents should be translucent and viscous. If not, continue incubation.

24. Excess ethanol will inhibit the digestion of the genomic DNA in the later steps, but the pellet should not be completely dried, or it will be tough to dissolve.
25. The fragments will be circularized and ligated, and then transformed into competent *E. coli*. If the resulting circularized pREMI-Z and flanking DNA are too large, the ligation product will be unable to transform the *E. coli*. Therefore, digesting the genomic DNA with each enzyme separately increases the chances that at least one recircularized plasmid will be small enough for transformation. Any restriction enzyme that is not methylation sensitive and does not cut within pREMI-Z can potentially be used for this purpose.
26. In some cases, presumably because the plasmid is so large, less than six transformant colonies will be produced.
27. One would expect that all plasmids originating from the same transformation reaction would be the same and therefore produce identical restriction patterns. However, extensive restriction analysis has indicated that a minority (<25%) of the plasmids may contain a random genomic fragment that was inserted prior to circularization or show rearrangement of the flanking genomic and plasmid DNA (Fig. 8). These plasmids should be ignored. In rare cases where the plasmid restriction patterns fall into two distinct groups, then representatives of both groups should be pursued for sequencing.

Competing Interest Statement The authors have no conflicts of interest to declare that are relevant to the content of this chapter.

References

1. Tolstorukov I, Cregg JM (2007) Classical genetics. In: Cregg JM (ed) Pichia protocols, 2nd edn. Humana Press, New Jersey, pp 189–201
2. Riggle PJ, Kumamoto CA (1998) Genetic analysis in fungi using restriction-enzyme-mediated integration. Curr Opin Microbiol 1(4):395–399. Epub 1999/03/06
3. Lin-Cereghino J, Wong WW, Xiong S et al (2005) Condensed protocol for competent cell preparation and transformation of the methylotrophic yeast *Pichia pastoris*. BioTechniques 38(1):44, 6, 8. Epub 2005/02/01. PMID: 15679083; PMCID: PMC2504082
4. Cregg JM, Russell KA (1998) Transformation. In: Higgins DR, Cregg JM (eds) Pichia protocols. Humana Press, New Jersey, pp 27–39
5. Morawski B, Lin Z, Cirino P et al (2000) Functional expression of horseradish peroxidase in *Saccharomyces cerevisiae* and *Pichia pastoris*. Protein Eng 13:377–384. https://doi.org/10.1093/protein/13.5.377
6. Hoffman CS, Winston F (1987) A ten-minute DNA preparation from yeast efficiently releases autonomous plasmids for transformaion of *Escherichia coli*. Gene 57:267–272. https://doi.org/10.1016/0378-1119(87)90131-4

7. Bae J-H, Sohn J-H (2010) Template-blocking-PCR: an advanced PCR technique for genome walking. Anal Biochem 398:112–116. https://doi.org/10.1016/j.ab.2009.11.003
8. Manivasakam P, Schiestl RH (1998) Nonhomologous end joining during restriction enzyme-mediated DNA integration in *Saccharomyces cerevisiae*. Mol Cell Biol 18(3):1736–1745. Epub 1998/03/06. PMID: 9488490; PMCID: PMC108888
9. Kuspa A (2006) Restriction enzyme-mediated integration (REMI) mutagenesis. Methods Mol Biol 346:201–209. Epub 2006/09/08
10. Schroder LA, Glick BS, Dunn WA (2007) Identification of pexophagy genes by restriction enzyme-mediated integration. Methods Mol Biol 389:203–218. Epub 2007/10/24
11. Larsen S, Weaver J, de Sa Campos K et al (2013) Mutant strains of *Pichia pastoris* with enhanced secretion of recombinant proteins. Biotechnol Lett 35:1925. https://doi.org/10.1007/s10529-013-1290-7
12. Staley CA, Huang A, Nattestad M et al (2012) Analysis of the 5′ untranslated region (5'UTR) of the alcohol oxidase 1 (*AOX1*) gene in recombinant protein expression in *Pichia pastoris*. Gene 496(2):118–127. Epub 2012/01/31. PMID: 22285974; PMCID: PMC3464313
13. Weis R, Luiten R, Skranc W et al (2004) Reliable high-throughput screening with *Pichia pastoris* by limiting yeast cell death phenomena. FEMS Yeast Res 5:179–189. https://doi.org/10.1016/j.femsyr.2004.06.016
14. Sambrook J, Fritsch EF, Maniatis T (1989) Molecular cloning: a laboratory handbook, 2nd edn. Cold Spring Harbor Laboratory Press

Chapter 29

Cryopreservation of *Pichia pastoris* Strains for Long-Term Storage

Pierre D. M. Reinprecht and Katharina Ebner

Abstract

This chapter describes easy and cheap methods for the long-term storage of wild-type and engineered *Komagataella phaffii* strains. When performed correctly, the outlined cryopreservation protocols allow almost indefinite storage of cells without losing their genetic integrity or viability.

Key words *Pichia pastoris*, *Komagataella phaffii*, Cryopreservation, Cryoconservation, Long-term storage, Glycerol stock, Freeze-drying

1 Introduction

Cryoconservation (i.e., the freezing and storage of samples at low temperatures) has been and is still the method of choice for long-term preservation of viable *Komagataella phaffii* (reclassified from *Pichia pastoris)* strains. Several techniques for cryoconservation, differing in the type of cryoprotective agent, freezing program, and final storage temperature, were developed [1]. Two methods—namely freeze-drying of the cells under vacuum conditions (lyophilization) and freezing at low temperatures using glycerol as a cryoprotectant—stood out and became standard methods of choice for the preservation of *K. phaffii* strains. In general, cryopreservation is also more efficient than previously typically used cultures on agar slants, which were stored for 1–5 years in cold rooms but needed laborious, regularly repeated cultivations. Thus, this chapter focuses on the preparation of glycerol stocks and the lyophilization of *K. phaffii* strains and provides detailed protocols for both methods.

Although freeze-drying is more labor-intensive and requires advanced equipment, compared to the rather simple preparation of glycerol stocks, lyophilized samples are better suited for international shipping as they are less temperature sensitive [2]. The major

Xavier Garcia-Ortega et al. (eds.), *Pichia pastoris: Methods and Protocols*, Methods in Molecular Biology, vol. 2697, https://doi.org/10.1007/978-1-0716-4779-0_29,

downside of both methods is the loss of cell viability. Approximately 90% of the cells do not survive the freezing process, and the viability further decreases at the same rate with every freeze–thaw cycle due to the formation of ice crystals and the cellular freeze–thaw stress response [3]. As a countermeasure, glycerol is added to the samples as a cryoprotectant. The glycerol interferes with the formation of hydrogen bonds between H_2O molecules, thus preventing the formation of ice crystals and increasing the number of viable cells [4]. A final glycerol concentration of approximately 30% (w/v) was identified as optimal for these purposes. Another decisive factor for the quality of the cryosamples is the freezing program. Although vitrification (i.e., rapid cooling under the formation of amorphous ice) as applied in specimen preparation for analytical purposes is one option to avoid the formation of ice crystals, the technique is not suitable as an everyday application in the lab. A more feasible strategy to prevent the excessive formation of ice crystals is to apply lower cooling rates. In theory, the optimum cooling rate is in the range from 5 to 180 °C min^{-1}, but controlled conditions like that are not possible without special equipment [3]. Therefore, a compromise between applicability and the number of viable cells must be found. Regarding the final storage temperature, the use of liquid nitrogen is recognized as the best practice for long-term storage, since temperatures fall below the glass transition of water at approximately −135 °C. A more convenient way for long-term preservation that is less maintenance-intensive and allows quicker access to stored samples but still ensures almost indefinite storage, is found by keeping the samples at −80 °C in an electrically cooled low temperature freezer [5].

2 Materials

For cryopreservation, the following materials, which are routinely available in every laboratory equipped for molecular biology, are needed.

2.1 Equipment

1. Autoclave.
2. Temperature-controlled shaking incubator (28–30 °C).
3. Freezer, −80 °C.
4. Tabletop centrifuge with rotor for reaction tubes.
5. Sterile, 15 mL screwcap plastic tubes or similar cultivation vials.
6. Sterile plastic Petri dishes (94 × 16 mm).
7. Screwcap cryovials (2 mL, temperature-resistant to −80 °C).
8. 96-deep-well plate and 96-well microtiter plate
9. Sterile, self-sticking aluminum foil.
10. Freeze dryer.

2.2 Media and Solutions

1. YPD broth: 1% (w/v) yeast extract, 2% (w/v) peptone, 2% (w/v) glucose.
2. YPD agar: 1% (w/v) yeast extract, 2% (w/v) peptone, 2% (w/v) glucose, 1.5%(w/v) agar-agar.
3. 30% (w/v) sterile glycerol.
4. 50% (w/v) sterile glycerol.
5. MTG solution: 10% skim milk powder, 10% trehalose, and 5% sodium glutamate. Sterilize for 5 min at 121 °C. Vortex thoroughly after autoclaving to homogenize the solution and store at 4 °C (*see* **Note 1**).

3 Methods

3.1 Single-Tube Glycerol Stocks

The following 5-step protocol describes how to prepare glycerol stocks for the preservation of *K. phaffii* from liquid culture (*see* Subheading 3.1.1) or cells grown on agar plates (*see* Subheading 3.1.2).

3.1.1 Liquid Culture

1. Prepare 2 mL of sterile YPD media in a sterile 15 mL tube or comparable container.
2. Use a sterile inoculation loop or a toothpick to pick a fresh single colony of the desired strain from an agar plate containing selective media and inoculate the media.
3. Grow the liquid culture overnight at 28–30 °C with shaking (*see* **Note 2**).
4. Transfer 600 μL of the liquid culture into a sterile 2 mL cryovial, add 900 μL of 50% (w/v) glycerol and mix thoroughly (final glycerol concentration should be approx. 30%) (*see* **Note 3**).
5. Freeze the tubes at −80 °C for storage (*see* **Note 4**).

3.1.2 Cells Grown on Agar Plates

1. Prepare fresh YPD-agar plates.
2. Streak out the cells on the agar plates.
3. Incubate the agar plates for 48 h at 28–30 °C.
4. Pour approximately 2 mL of 30% (w/v) glycerol on the agar plate, scrape off the cell material with a sterile spatula, and transfer 1.8 mL of the cell suspension into a sterile 2 mL cryovial.
5. Freeze the tubes at −80 °C for storage (*see* **Note 4**).

3.2 Glycerol Stocks in 96-well Format

To increase the sample throughput, the previously explained method for the preparation of single-tube glycerol stocks was adapted, enabling the simultaneous preparation of up to 96 glycerol stocks in microtiter plate format.

1. Prepare a sterile 96-deep-well plate with 250 μL of sterile YPD media in each well.
2. Pick fresh single colonies of the desired strains and use them to inoculate the media. One colony per well.
3. Grow the cultures overnight at 28 °C under constant shaking.
4. Transfer 80 μL of the culture from each well into a fresh, sterile 96-well microtiter plate, and add 120 μL of 50% (w/v) glycerol.
5. Seal the plate with self-sticking, sterile aluminum foil and freeze at −80 °C for storage.

3.3 Freeze-Drying (Lyophilization)

Freeze-drying is one of the preferred methods for the preparation of viable cell material, which should be shipped to other labs and also for reliable backup cultures to avoid total loss of strain collections in case of technical problems with the supply of electricity (e. g, in case of blackouts or similar situations).

1. Inoculate 2 mL of sterile YPD media in a sterile 15 mL tube or comparable container with a fresh single colony of the desired strain from an agar plate containing selective media.
2. Grow the culture overnight at 28 °C under constant shaking (*see* **Note 2**).
3. Transfer 1.8 mL of the cell suspension into a sterile 2 mL screwcap cryovial and centrifuge at 900 xg and room temperature for 5 min.
4. Discard the supernatant.
5. Add 50 μL of the MTG solution to the cell pellet.
6. Resuspend the pellet by gently mixing by pipetting up and down.
7. Freeze the cells at −80 °C for 24 h (*see* **Note 4**).
8. Unscrew the caps of the cryovial until they are hardly attached.
9. Place the frozen tubes in a freeze dryer and lyophilize at 0.38 mbar until a cell powder is formed.
10. Store the lyophilized cells at 4 °C in a dark environment to prevent UV-induced damage.

3.4 Resurrection of Stored Cells: Depending on the Method Chosen for the Preservation of the Cells, Different Strategies are Applied for Resurrection

3.4.1 Resurrection from Glycerol -Stocks

1. Take out the tube from the −80 °C freezer and place it into a cryorack. Make sure the cells stay frozen.
2. Scrap off some of the frozen cells with a sterile inoculation loop.
3. Streak out the biomass on an agar plate containing the appropriate media and selection conditions.
4. Immediately return the glycerol stock to the −80 °C freezer (*see* **Note 5**)

3.4.2 Resurrection of Lyophilized Cells

1. Add 1 mL of sterile YPD media to the cryovial.
2. Resuspend the cells thoroughly (*see* **Note 6**).
3. Plate 100 μL of a 1:1000 dilution on an appropriate agar plate with selective conditions and incubate at 28 °C until colonies are observed (*see* **Note 7**).

4 Notes

1. The MTG solution increases the viability of cells stored via freeze-drying by two orders of magnitude.
2. To ensure sufficient oxygen transfer, it is recommended to not completely tighten the screw cap of the tubes used for cultivation. Instead, just close the tube loosely and tape the cap for fixation. Furthermore, it is recommended to place the tube in the shaking incubator at an angle of approximately 45° to expand the surface area for oxygen transfer..
3. 100% glycerol is extremely viscous and difficult to handle; hence, the use of a 50% (w/v) solution is recommended.
4. It is recommended to place the samples in a styrofoam box prior to freezing to enable lower cooling rates to increase the viability of the cells.
5. It is crucial to ensure that the cells do not completely thaw during the resurrection process, since with every freeze–thaw cycle, the viability of the cells decreases massively. Therefore, the use of a cooling rack and quick handling is indispensable.
6. If cells are stored via freeze-drying, it is recommended to resurrect the whole glycerol stock and not to remove a small portion of the lyophilized cells and re-freeze to maintain the main stock. In contrast to the glycerol stocks, this method is not intended for long-term storage but for cheap and easy postal distribution since the cells can withstand room temperature conditions for at least 60 days.

7. Due to changing storage conditions during shipping, it might be necessary to plate a lower dilution (1:100 or 1:10) of the resurrected cells.

Competing Interest Statement

The authors have no conflicts of interest to declare that are relevant to the content of this chapter.

References

1. Bond C (2007) Cryopreservation of yeast cultures. In: Day JG, Stacey GN (eds) Cryopreservation and freeze-drying protocols. Humana Press, Totowa, pp 109–117. https://doi.org/10.1007/978-1-59745-362-2_7
2. Bond C (2007) Freeze-drying of yeast cultures. Methods Mol Biol (Clifton, N.J.) 368:99–107. https://doi.org/10.1007/978-1-59745-362-2_6
3. Cabrera E, Welch LC, Robinson MR, Sturgeon CM, Crow MM, Segarra VA (2020) Cryopreservation and the freeze–thaw stress response in yeast. Genes 11(8). https://doi.org/10.3390/genes11080835
4. Koh CM (2013) Storage of bacteria and yeast. Methods Enzymol 533:15–21. https://doi.org/10.1016/B978-0-12-420067-8.00002-7
5. Gujjari P, Muldrow T, Zhou JJ (2010) Effect of cryopreservation protocols on the phenotypic stability of yeast. Cryo Letters 31(3):261–267

Part VI

Upstream Bioprocessing

Chapter 30

Adapting *Komagataella phaffii* Cultivation Media to REACH Regulations

Julian Kopp, Alexander Pekarsky, and Oliver Spadiut

Abstract

REACH stands for "Registration, Evaluation, Authorization, and Restriction of Chemicals" and classifies certain substances as harmful to human health and the environment. Boric acid and cobalt dichloride, commonly used as boron (B) and cobalt (Co) sources in cultivation media for *Komagataella phaffii*, represent such harmful substances. However, B and Co support cellular growth, and yeast harbors a large number of co-containing enzymes. Even though cell physiology (up to 90 g/L) was not affected by the removal of boric acid and cobalt dichloride from the media, recombinant protein production varied when excluding these components from the cultivation media to comply with REACH regulations. Here, we describe how to analyze and potentially adapt a defined cultivation medium for *K. phaffii* bioreactor cultivations concerning current REACH regulations.

Key words *Pichia pastoris*, *Komagataella phaffii*, REACH, boron, Cobalt, Fermentation, Boron leakage, Glass vessel, Stainless steel

1 Introduction

Recombinant protein production with the methylotrophic yeast *Komagataella phaffii*, commonly known as *Pichia pastoris*, is widely performed both in academia and industry. Since 2006, the Registration, Evaluation, Authorization, and Restriction of Chemicals (REACH) regulation has targeted the reduction of hazardous chemicals in manufacturing processes to protect human health and the environment [1]. An important and widespread chemically defined medium for *K. phaffii* is Basal Salt Medium (BSM) (e.g., [2–9]), which is supplemented with a *Pichia* Trace Metal (PTM) solution that contains various metal salts, boric acid, and biotin [10, 11]. In fact, all published chemically defined media for *K. phaffii* comprise a PTM solution, which usually contains boric acid as a boron (B) source and cobalt dichloride as a cobalt (Co) source (e.g., [2–9, 12–15]). However, both substances are classified as "substances

Xavier Garcia-Ortega et al. (eds.), *Pichia pastoris: Methods and Protocols*, Methods in Molecular Biology, vol. 2697, https://doi.org/10.1007/978-1-0716-4779-0_30,

of very high concern" by REACH, which directs "progressively replacing these substances with less dangerous substances" [16]. This urges depletion or at least reduction of B and Co from *K. phaffii* cultivation media. However, the substitution of boric acid and cobalt dichloride is tricky because organic B compounds raise environmental questions [17], and the substitution of Co salts through vitamin B12 is hindered by the inability of *K. phaffii* to metabolize this vitamin [18]. One potential approach to reduce or even substitute the addition of boric acid could be the use of borosilicate glassware that reportedly leaks B over time [19, 20].

This chapter shows how to analyze and adapt cultivation media for certain trace elements. We demonstrate that both trace elements boron (B) and cobalt (Co), which are classified as "substances of very high concern" by REACH, can be excluded from the PTM1 solution in defined cultivation media for *K. phaffii* once pre-cultures are performed in glass vessels. The borosilicate glassware leaks sufficient B into the pre-culture to support cellular growth even in subsequent induced high cell density fermentations up to 90 g DCW L^{-1}. Even though the physiology and growth of *K. phaffii* are not affected by excluding B and Co from the defined cultivation medium, we, however, recommend analyzing potential effects on recombinant protein production as well as on the quality of the recombinant product.

2 Materials

Prepare all media and solutions with analytical-grade reagents and deionized water (=MQ).

2.1 Equipment

1. Thermostated shaker.
2. Standard 1 L baffled borosilicate glass Erlenmeyer flask for pre-cultures.
3. Plastic or standard borosilicate glass Erlenmeyer shake flasks for pre-cultures.
4. Lab-scale stainless-steel bioreactor for batch and fed-batch fermentations.
5. Spectrophotometer, centrifuge, and dry oven for sample preparation.
6. iCAP 6500 series spectrometer (Thermo Scientific, USA) coupled to an ASX-520 autosampler (Cetac, USA) for inductively coupled plasma optical emission spectrometry (ICP-OES) measurements. Instrumental parameters are described in Table 1.

Samples are automatically measured in triplicate by the device. The instrument is equipped with a standard sample introduction set

Table 1
ICP-OES instrumental parameters and analytical wavelengths in nm

Parameter	Value	Parameter	Value
RF power	1200 W	Nebulizer flow	0.80 L min^{-1}
Sample flow	0.80 mL min^{-1}	Auxiliary flow	0.8 L min^{-1}
Observation height	12 mm	Coolant flow	15 L min^{-1}
Integration time	10 s for 1:500 samples, 25 s for 1:10 samples		
Number of replicates	3		
Background correction	constant shift from the analytical line		
Element	Wavelength I (nm)		Wavelength II (nm)
B	246.773		209.893
Ca	422.673		393.366
Co	238.892		228.616
Cu	324.754		324.754
Fe	259.940		239.562
K	766.490		769.896
Mg	285.213		202.582
Mn	257.610		260.569
Mo	202.030		204.598
Zn	213.856		202.548
Eu	381.967		411.970

consisting of a concentric nebulizer and a cyclonic spray chamber. Two sensitive and non-interfering emission lines per element are used: one line for quantification and the second one for quality control.

2.2 Pre-culture (see Note 1)

1. Weigh YNB w/o amino acids and ammonium sulfate, $(NH_4)_2SO_4$, biotin, and C-source in a beaker.
2. Add 200 mL of 1.0 M potassium phosphate buffer (pH 6.0).
3. Dissolve by stirring and set the volume with distilled water to 1 L.
4. Filter-sterilize through a 0.2 μm cut-off filter into a sterile shake flask—either made from glass or from plastic.
5. If the strain of interest carries a resistance gene, add the respective antibiotics according to the specific resistance markers present in the strain (e.g., Zeocine®, Kanamycin) to an appropriate concentration (e.g., 100 μg mL^{-1} medium).
6. Store at 4 °C until further use.

2.3 Batch Media

1. BSM per L: 51.6 g glycerol, 10.79 ml H_3PO_4 85%, 0.18 g $CaSO_4 \times 2H_2O$, 13.62 g K_2SO_4, 4.48 g $MgSO_4 \times 7H_2O$, 4.13 g KOH and 0.3 mL of antifoam (e.g., Struktol J-673) in MQ.
2. Trace metal solution (PTM1) per L: 3.84 g $CuSO_4$, 0.08 g NaI, 3 g $MnSO_4 \times H_2O$, **0.02 g boric acid, 0.91 g $CoCl_2 \times 6H_2O$,** 0.2 g $Na_2MoO_4 \times 2H_2O$, 20 g $ZnCl_2$, 65 g $FeSO_4 \times 7H_2O$, 5 ml H_2SO_4, 0.2 g D-Biotin in MQ.
3. For experiments without B and Co, a PTM1 solution is prepared without the respective substances (*see* **Note 2**).
4. The base solution is to set the pH: 12.5% (v/v) NH_4OH (*see* **Note 3**).

2.4 Medium for Fed-Batch Culture

Different carbon sources can be used according to the application. The glucose and the glycerol feed can be sterilized by autoclaving prior to PTM1 addition; the methanol feed is sterile filtered through a 0.2 μm cut-off filter into a sterile flask in order to avoid methanol evaporation.

1. 500 g L^{-1} glucose monohydrate, 12 mL PTM1, 0.3 mL antifoam (Struktol J673).
2. Glycerol feed per L: 750 g glycerol, 12 mL PTM1, and 0.3 mL antifoam (Struktol J673).
3. Methanol feed per L: pure methanol, 4 mL PTM1, and 0.3 mL antifoam (Struktol J673) (*see* **Note 4**).

2.5 Materials Required for ICP-OES

1. 1%-HNO_3 (v/v) for matrix adjustment: prepare an adequate amount of MQ in a beaker, and add nitric acid of p.a. grade purity (Merck, Darmstadt, Germany) at 10 mL L^{-1} to reach the target concentration.
2. Dissolve europium (Specpure ICP standard, Alfa Aesar, Kandel, Germany) at a concentration of 1 mg L^{-1} in 1% HNO_3 as an internal standard.
3. 0.5 M potassium phosphate buffer at pH 5.0: weigh in 13.07 g K_2HPO_4 and 57.84 g·KH_2PO_4 and dissolve in 0.9 L MQ, adjust pH with phosphoric acid to 5, and fill up to 1 L in a beaker (*see* **Note 5**).

3 Methods

3.1 Fermenters Inoculation and Run

Specific materials to run continuous and fed-batch cultures are detailed in Chapters 31, 32, and 33 of this volume.

However, since the presence of trace components included in the pre-culture cultivation broth, key details are provided below:

1. Start a pre-culture in 100 mL of YNB medium in 1 L baffled glass Erlenmeyer flasks at 200–230 rpm and 30 °C for maximum of 24 h and to an OD_{600} ~ 3.0 (to guarantee good aeration, only 1/10 of the total volume of the flask is filled with medium).
2. Aseptically transfer the pre-culture into a sterile inoculation flask (e.g., a vessel providing a connection to the bioreactor). The inoculum is 10% of the final volume in the bioreactor to ensure reproducible batch phases.
3. Run the cultures according to the desired strategy.

3.2 ICP-OES Measurement

1. The cultivation sample solutions are diluted 1:10 with 1% HNO_3 (v/v) for matrix adjustment; thereby, europium with a concentration level of 1.0 mg L^{-1} is added as an internal standard to all samples.
2. ICP-OES for simultaneous, multi-element analysis is carried out on an iCAP 6500 series spectrometer coupled to an ASX-520 autosampler (*see* Table 1).
3. Ca, K, or Mg are analyzed using a sample dilution of 1:500 (m/m) since these elements are known to be present in higher concentrations in cell-free cultivation samples.
4. Quantification is done via external calibration with aqueous standard solutions using europium as an internal standard. Use high-purity water by purifying de-ionized water.
5. A certified stock solution of europium is used for method development and quantification of sample signals. Prepare working solutions by dilution of stock solutions with 1.0% nitric acid (v/v) prior to use.

3.3 Leakage of Trace Elements from Different Cultivation Vessels

1. Analyze the element leakage for the different materials, which are commonly used as vessels for *K. phaffii* pre-cultures:
   (a) standard borosilicate glass Erlenmeyer shake flasks,
   (b) standard borosilicate plastic Erlenmeyer shake flasks,
2. Element leakage is also analyzed in a stainless-steel bioreactor, which is typically used for controlled bioreactor cultivations, especially in an industrial environment.
3. The respective vessels must be vigorously cleaned with high-purity water (MQ) obtained from a Milli-Q system and then autoclaved with MQ.
4. After the autoclaved MQ is discarded, the vessels are again cleaned with fresh MQ and then filled with 0.5 M potassium phosphate buffer at pH 5.0 (*see* **Note 6**).
5. A defined volume of 0.5 M potassium phosphate buffer (0.05 L in glass Erlenmeyer flasks; 0.16 L in plastic; 7.20 L in steel

Table 2
ICP-OES results from elemental leakage experiments in mg L^{-1}

	B	Fe	Mn	Co	Cu	Zn	Mo	Mg	Ca
values are given at t_0/t_{142} in mg L^{-1}									
Glass 1	**0.30/1.36**	0.05/0.18	0.00/0.00	**< d/< d**	< d/< d	< d/< d	< d/< d	0.01/0.02	< d/0.16
Glass 2	**0.29/1.12**	0.07/0.08	0.00/0.00	**< d/< d**	< d/< d	< d/< d	< d/< d	0.04/0.16	< d/< d
Plastic 1	**0.30/0.30**	0.11/0.08	0.00/0.00	**< d/< d**	0.03/0.06	0.04/0.08	< d/< d	0.01/0.02	0.16/0.30
Plastic 2	**0.29/0.27**	0.06/0.10	0.00/0.00	**< d/< d**	0.03/0.07	0.02/< d	< d/< d	0.04/0.03	0.20/0.36
Steel 1	**0.24/0.21**	0.06/0.14	0.00/0.07	**< d/< d**	< d/< d	0.03/< d	< d/< d	0.34/2.89	0.62/0.32
Steel 2	**0.23/0.20**	0.05/0.12	0.00/0.07	**< d/< d**	< d/< d	0.03/< d	< d/< d	0.33/2.84	0.71/0.34

Experiments were performed in duplicates in borosilicate Erlenmeyer glass (Glass), or Utra Yield™ plastic (Plastic) shake flasks, which are commonly used for pre-cultures of *K. phaffii*, and in a stainless-steel bioreactor (Steel) used for controlled batch and fed-batch cultivations

< *d* below the limit of detection

bioreactor) is added to the respective vessels, and an initial sample (t_0) is taken.

6. Autoclave the filled vessels for 20 min at 121 °C and subsequently incubate at 30 °C under shaking (230 rpm) or stirring (300 rpm) for 142 h under sterile conditions. Additionally, the bioreactor is aerated with 0.3 vvm dried air.
7. After a significant time (in the described example 142 h), sample the respective vessels (t_{142}) and measure the final volume (e.g., 0.043 ± 0.002 L in glass; 0.135 ± 0.005 L in plastic; 6.58 ± 0.10 L in steel). This is necessary to account for water loss during the process.
8. Samples are stored in Eppendorf tubes until ICP-OES analysis. Experiments are performed in duplicates.
9. It was found that the concentrations of most measured trace elements did not change during such a 142 h lasting experiment (*see* Table 2).
   (a) Notably, the B concentration increased on average by 0.945 mg L^{-1} coming from borosilicate glass Erlenmeyer shake flasks.
   (b) Further, the magnesium concentration increased on average by 2.53 mg L^{-1} from stainless steel.
   (c) Taking the respective volumes before and after the experiments into account, the calculated leakage rates were 7.33 $\times 10^{-3}$ mg L^{-1} h^{-1} for B in glassware and 1.76 $\times 10^{-2}$ mg L^{-1} h^{-1} for magnesium in the stainless-steel vessel.

3.4 Elemental Media Composition

1. The respective concentrations for the single PTM1 trace elements that are added to BSM are shown in Table 3. Usually addition of PTM1 to BSM causes precipitate formation at a pH between 5 and 6 (1).
2. Thus, ICP-OES analyses can be performed to investigate the real initial concentration of each trace element in BSM at the start of cultivation and to demonstrate which elements are most affected by precipitation (Table 3). Samples were taken from BSM in the stainless-steel bioreactor directly after inoculation with the pre-culture.
3. The results clearly show that the elements Mn, Co, Cu, and especially Fe and Zn precipitate when PTM1 is added to BSM. Elements K and Ca were higher than those added via PTM1, which results from the addition of the pre-culture to the BSM cultivation medium.
4. It is especially interesting to see that B concentration increased extensively to 4.90 mg L^{-1} compared to the added 0.02 mg L^{-1}. This additional B results from pre-cultures that were

Table 3
Comparison of added and measured elemental concentrations in cultivation medium BSM and PTM1 containing B and Co

	B	Fe	Mn	Co	Cu	Zn	Mo	Mg	Ca	K
[mg L^{-1}]										
Added	**0.02**	56.6	4.2	**1.0**	6.6	41.6	0.3	442	41.9	5933
Measured	**4.90 ± 0.57**	< d	2.3 ± 0.2	**0.6 ± 0.07**	1.8 ± 0.3	5.5 ± 0.8	0.3 ± 0.1	468 ± 25.8	48.2 ± 6.8	8333 ± 535

Measured values were generated through ICP-OES directly after inoculation with the pre-culture
< *d* below the limit of detection

performed in glass Erlenmeyer shake flasks and from the base addition from glass bottles.

5. This hypothesis was confirmed when we inoculated with pre-cultures originating from plastic shake flasks and repeated the ICP-OES analyses. When the initial sample was measured after inoculation, we found only 0.06 mg L^{-1} B in the medium (instead of 4.90 mg L^{-1}).
6. Still, the three-fold higher B content (0.02 mg L^{-1} vs. 0.06 mg L^{-1}) should be attributed to the base addition from glass bottles.
7. Experiments indicated that the removal of B to PTM1 does not impact cellular growth.
8. Additionally, when using glass flasks or glass bioreactors, enough B is present in the media to avoid variations in recombinant protein production [21].
9. Experiments indicated that the removal of Co to PTM1 is not necessary to achieve cell concentrations up to 90 g/L; still, Co addition might be necessary when recombinant proteins are being produced, as Co limitation might affect recombinant protein production [21].

4 Notes

1. For pre-culture, buffered minimal dextrose/glycerol (BMD/BMG) medium (1.34% Yeast Nitrogen Base YNB, 4×10^{-5}% biotin, 200 mM potassium phosphate buffer pH 6.0, and 2% glucose) can be alternatively used as a medium for pre-cultures.
2. Filter-sterilize PTM solution through a 0.2 μm cut-off filter into a sterile bottle wrapped in aluminum foil and stored at 4 °C.
3. For the base titration, the following materials are required: base (12.5 NH_4OH), 0.25 M KHP, bromothymol blue (indicator), burette, and beaker with a magnetic stirrer. Add 2 mL of the base to the beaker (dilute base 1:10), and use the burette to add 0.25 M KHP. At the point of equivalence, the color of the indicator will turn from blue to gray and then to green. Calculate the molarity of the base according to Eq. 1:

$$\text{molarity\_base} = f \frac{\text{molarity\_KHP} \times \text{base\_consumption (KHP)}}{\text{volume\_base(diluted/before titration)}} \quad (1)$$

molarity_base [mol L^{-1}]

molarity_KHP [mol L^{-1}]

base_consumption (KHP) [mL]

volume_base [mL]

f = dilution factor of the base before titration

4. Different carbon sources can be used according to the application. The glucose and the glycerol feed can be sterilized by autoclaving prior to PTM1 addition; the methanol feed is sterile filtered through a 0.2 μm cut-off filter into a sterile flask in order to avoid methanol evaporation.
5. 0.5 M potassium phosphate buffer at pH 5.0 was used for element leakage studies because *K. phaffii* cultivations are performed at a pH value of 5 in a phosphate-buffered environment.
6. This buffer is chosen instead of a growth medium to improve the detectability of leaking elements in subsequent analytics and because *K. phaffii* cultivations are usually performed between pH 5 and 6 under high ionic concentrations [11].

Competing Interest Statement

The authors have no conflicts of interest to declare that are relevant to the content of this chapter.

References

1. European Chemicals Agency. Accessed 29th april 2024. https://echa.europa.eu/regulations/reach/understanding-reach
2. Min C-K, Lee J-W, Chung K-H, Park H-W (2010) Control of specific growth rate to enhance the production of a novel disintegrin, saxatilin, in recombinant Pichia pastoris. J Biosci Bioeng 110(3):314–319. https://doi.org/10.1016/j.jbiosc.2010.03.013
3. Bhatacharya P, Pandey G, Mukherjee KJ (2007) Production and purification of recombinant human granulocyte–macrophage colony stimulating factor (GM-CSF) from high cell density cultures of Pichia pastoris. Bioprocess Biosyst Eng 30(5):305–312. https://doi.org/10.1007/s00449-007-0124-1
4. Hang H-f et al (2008) A simple unstructured model-based control for efficient expression of recombinant porcine insulin precursor by Pichia pastoris. Korean J Chem Eng 25(5):1065–1069. https://doi.org/10.1007/s11814-008-0174-3
5. Yamawaki S et al (2007) Production of Single-Chain Variable Fragment Antibody (scFv) in fed-batch and continuous culture of pichia pastoris by two different methanol feeding methods. J Biosci Bioeng 104(5):403–407. https://doi.org/10.1263/jbb.104.403
6. Paulova L et al (2012) Use of a mixture of glucose and methanol as substrates for the production of recombinant trypsinogen in continuous cultures with Pichia pastoris Mut+. J Biotechnol 157(1):180–188. https://doi.org/10.1016/j.jbiotec.2011.10.010
7. Dietzsch C, Spadiut O, Herwig C (2011) A fast approach to determine a fed batch feeding profile for recombinant Pichia pastoris strains. Microb Cell Factories 10(1):85–85. https://doi.org/10.1186/1475-2859-10-85
8. Krainer FW et al (2015) Optimizing cofactor availability for the production of recombinant heme peroxidase in Pichia pastoris. Microb Cell Factories 14(1):4–4. https://doi.org/10.1186/s12934-014-0187-z
9. Krainer FW et al (2013) Knockout of an endogenous mannosyltransferase increases the homogeneity of glycoproteins produced in Pichia pastoris. Sci Rep 3:3279–3279. https://doi.org/10.1038/srep03279

10. Hitzman DO, Wegner EH (1976) Methanol foam fermentation to single cell protein by microorganisms. Accessed 29th April 2024, in: US3982998A Google Patents
11. Stratton J, Chiruvolu V, Meagher MM (1998) High cell-density fermentation. In: Higgins DR, Cregg JM (eds) Pichia protocols. Humana Press, Totowa, pp 107–120. https://doi.org/10.1385/0-89603-421-6:107
12. Maurer M, Kühleitner M, Gasser B, Mattanovich D (2006) Versatile modeling and optimization of fed batch processes for the production of secreted heterologous proteins with Pichia pastoris. Microb Cell Factories 5(1):37. https://doi.org/10.1186/1475-2859-5-37
13. Matthews CB, Kuo A, Love KR, Love JC (2018) Development of a general defined medium for Pichia pastoris. Biotechnol Bioeng 115:103–113. https://doi.org/10.1002/bit.26440
14. Prielhofer R et al (2013) Induction without methanol: novel regulated promoters enable high-level expression in Pichia pastoris. Microb Cell Factories 12(1):5. https://doi.org/10.1186/1475-2859-12-5
15. Laroche Y et al (1994) High-level secretion and very efficient isotopic labeling of tick anticoagulant peptide (TAP) expressed in the methylotrophic yeast, Pichia pastoris. Bio/Technology 12(11):1119–1124. https://doi.org/10.1038/nbt1194-1119
16. European Chemicals Agency, Accessed 29th April 2024. https://echa.europa.eu/substances-of-very-high-concern-identification-explained
17. Schoderboeck L et al (2011) Effects assessment: boron compounds in the aquatic environment. Chemosphere 82(3):483–487. https://doi.org/10.1016/j.chemosphere.2010.10.031
18. Zhang Y, Rodionov DA, Gelfand MS, Gladyshev VN (2009) Comparative genomic analyses of nickel, cobalt and vitamin B12 utilization. BMC Genomics 10(1):78. https://doi.org/10.1186/1471-2164-10-78
19. Jensen WB (2006) The origin of pyrex. J Chem Educ 83(5):692. https://doi.org/10.1021/ed083p692
20. Green GH, Blincoe C, Weeth HJ (1976) Boron contamination from borosilicate glass. J Agric Food Chem 24(6):1245–1246. https://doi.org/10.1021/jf60208a016
21. Pekarsky A, Mihalyi S, Weiss M, Limbeck A, Spadiut O (2020) Depletion of boric acid and cobalt from cultivation media: impact on recombinant protein production with Komagataella phaffii. Bioengineering 7(4):161. https://doi.org/10.3390/bioengineering7040161

Chapter 31

Fed-Batch Operational Strategies for Efficient and Robust *Pichia Pastoris* Cultivations Achieving High Cell Densities

Albert Sales-Vallverdú, Miguel Angel Nieto-Taype, Arnau Gasset, José Luis Montesinos-Seguí, Francisco Valero, and Xavier Garcia-Ortega

Abstract

Once a *Pichia pastoris (Komagataella phaffii)* producer clone is constructed for recombinant protein production (RPP), a proper design of a fermentation strategy becomes essential to maximize yields and productivity. Within the development of bioprocess technology, several strategies have been proposed based on a broad range of different operational conditions. Such is the case of the fed-batch operation mode, which has become the most extended in the industry. Concretely, μ-stat fed-batch has been widely used since it allows the achievement of pseudo-steady-state conditions for specific rates during the fed-batch phase. This approach allows for reaching and maintaining optimal conditions, obtaining relevant improvements in yields and productivity parameters toward robust and reproducible bioprocesses.

Additionally, in this chapter is described the implementation of advanced cultivation strategies based on the application of stress-based approaches, such as carbon-starving and oxygen-limiting conditions, which have shown to be promising methodologies with outstanding results.

Key words *Pichia pastoris*, *Komagataella phaffii*, Heterologous expression, Fed-batch, Carbon-starving, Hypoxia, Feeding profile

1 Introduction

During the development of a bioprocess that will end up becoming a biomanufacturing process, not only the strain engineering but also the bioprocess engineering has been displayed to exert a crucial role in working at the optimal conditions of the expression systems as well as during the scaling-up process from lab bench to the fermenter at industrial level. It usually harnesses the intrinsic traits of an engineered clone through a comprehensive characterization and selection of the best operational conditions with the aim of maximizing yields and both specific and volumetric productivities [1, 2].

Xavier Garcia-Ortega et al. (eds.), *Pichia pastoris: Methods and Protocols*, Methods in Molecular Biology, vol. 2697, https://doi.org/10.1007/978-1-0716-4779-0_31,

In this context, fed-batch cultures are considered the most common strategy extended at the industrial scale due to their advantages over chemostat and batch cultures. Some of them have the capability to reach high cell densities (which allows for accumulating very high product concentrations) without substrate inhibition and catabolite repression [3, 4]. Concerning the operational view, these cultures are divided into two well-defined phases: batch and fed-batch, respectively. During the first one, cells are grown under non-nutrient-limiting conditions with the aim of generating enough biomass for the productive stage. Later, when the initial substrate is depleted, the fed-batch phase is initiated by the addition of feeding solution into the bioreactor in order to allow cellular growth and/or protein expression, which are often coupled [5]. Consequently, the selection of an optimal feeding strategy becomes crucial to enhance recombinant expression, which is strongly associated with the corresponding promoter regulation.

From the vast number of feeding strategies described in the literature (Fig. 1), the use of pulses (lower process control), pre-defined profiles (constant, linear, exponential), or closed-loop control profiles has been extensively reported on *Komagataella phaffii* (reclassified from *Pichia pastoris*) fed-batch cultures [6–10]. Although substrate addition by pulses is considered the

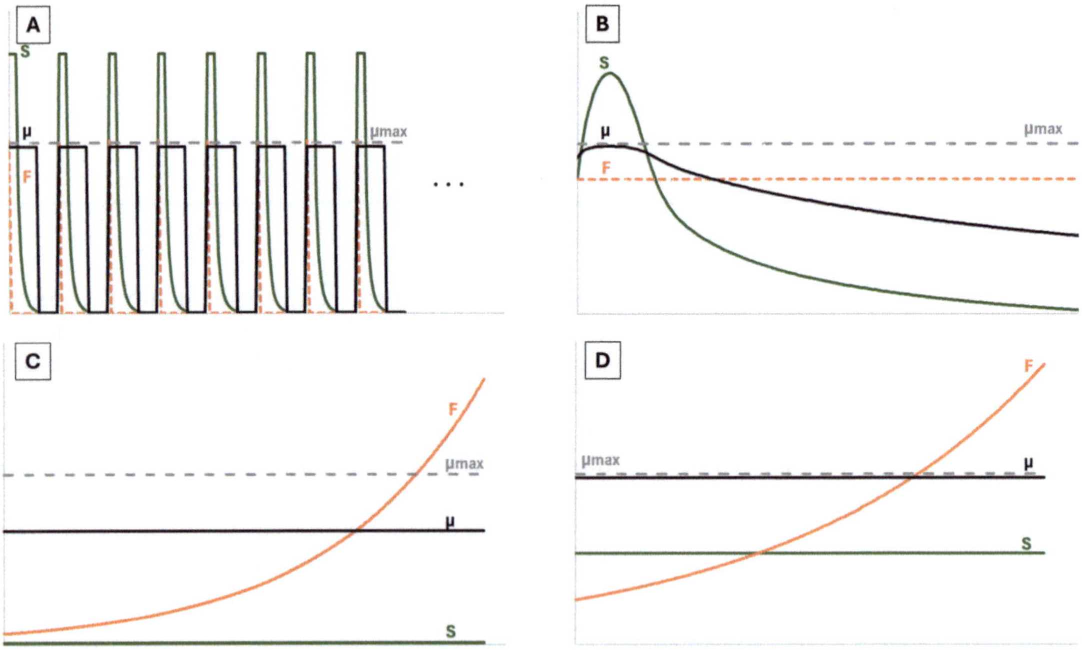

Fig. 1 Representation of different feeding strategies implemented for *Komagataella phaffii* fed-batch fermentations. Pulses (**a**), constant (**b**), exponential preprogrammed profile (μ-stat) (**c**), and closed-loop control (**d**) for constant substrate concentration. Feeding profile (F, orange line), substrate concentration (S, green line), and specific growth rate (μ, black line) time evolutions are presented for each strategy

simplest and easiest approach to implement, the eventual variability among equivalent fermentation runs might negatively affect the performance of the reproducibility and robustness of a bioprocess. On the other hand, closed-loop addition has been studied with relevant production improvements, but the high requirements in terms of software and reliable specialized substrate sensors make its implementation at the industrial scale more challenging [11].

In this sense, the implementation of preprogrammed exponential feeding profiles has been widely extended since, when working at carbon-limiting conditions or equivalent, it allows maintaining the pseudo-stationary state conditions at a constant specific growth rate (μ-stat) under carbon-limited conditions. Interestingly, this strategy has been reported to be successful for both episomal and integrative plasmid-based expression systems [12]. The specific growth rate (μ) reflects, to a certain extent, the metabolic state of the cell, including the equilibrium between the various steps until the product is secreted, as a balance of the different processes involved during the protein synthesis, folding, and secretion. Accordingly, optimized strategies based on the production kinetics profile (the relationship between specific production rate and specific growth rate) have been presented in the literature as a robust tool for bioprocess optimization with outstanding results in several studies [13–16]. The characterization of the production kinetics could be carried out both in C-limited fed-batch cultures and/or chemostats, as described in Chapter 33 of this book.

In the last years, fermentation strategies have been continually evolving with more complex and sophisticated approaches. In this sense, innovative strategies such as the introduction of carbon-starving (C-starving) and oxygen limitation (hypoxia) conditions have been proposed.

C-starving-based strategies attract interest in bioprocesses based on microbial fermentation at both lab and industrial scales due to the simplicity and boosting effect on recombinant protein production (RPP) for some *K. phaffii* expression systems. Specifically, Garcia-Ortega et al. in 2016 described the effect of carbon-starvation periods on μ-stat fed-batch cultures of Fab fragment recombinant P_{GAP}-driven expression (glycolytic constitutive growth-coupled promoter) [17]. In this study, superior yields of around 44–46% in comparison with standard conditions were observed.

Furthermore, oxygen limitation conditions, or hypoxia, also have proven to offer notable beneficial effects for P_{GAP}-driven RPP [18, 19]. Thus, physiological events implied by this fermentation strategy have been exhaustively investigated [18, 20–23]. Later studies have adapted the implementation of hypoxic conditions throughout the respiratory quotient (RQ) control with successful results [24–26]. This strategy relies on the analysis of the exhaust gas rather than measurements in the fermentation broth itself, thus avoiding the potential issues of submerged sensor-dependent measurements in large-scale bioreactors.

Interestingly, at industrial-scale bioprocesses with very high cell density, oxygen transfer is usually one of the most crucial limiting factors. In this sense, the application of both advanced strategies can also prove beneficial. Both strategies are easy to implement on large-scale bioprocesses since they depend on an externally applied limitation. Even though C-starving and hypoxia conditions have been studied with the *GAP* promoter, it does not mean they are restricted only to this glycolytic and constitutive promoter. In the last years, an increasing number of alternative promoters (native, engineered, and synthetic) have been presented in the literature with a great variety of regulation patterns where C-starving applications could be extended. One example is the case of de-repressible promoters, which are induced under carbon-limited conditions [27–30]. Hypoxia implementation with new development promoters could significantly reduce oxygen consumption, which is usually the most limiting bottleneck in *Pichia* large-scale bioprocesses.

This chapter describes in detail the steps to establish μ-stat fed-batch cultivations and the implementation of C-starving and hypoxia conditions on these cultures. Thereby, it will serve as a starting point to promote the research on these promising strategies as part of bioprocess engineering.

2 Materials

2.1 Culture Media

2.1.1 Inoculum Medium

1. Yeast extract peptone glycerol medium (YPG): 1% yeast extract, 2% peptone, and 2% glycerol. Medium is sterilized by autoclaving (*see* **Note 1**).
2. Buffered minimal glycerol medium (BMG): 100 mM phosphate buffer pH 6, 1.34% yeast nitrogen base without amino acids (YNB), 0.4 mg L^{-1} biotin, and 2% glycerol (*see* **Note 2**).

2.1.2 Batch Medium

1. *Pichia Batch Medium (PBM)*: it is composed of 40.0 g L^{-1} glycerol (*see* **Note 3**), 1.80 g L^{-1} citric acid, and 12.60 g L^{-1} $(NH_4)_2HPO_4$, 0.50 g L^{-1} $MgSO_4{\cdot}7H_2O$, 0.90 g L^{-1} KCl, 0.02 g L^{-1} $CaCl_2{\cdot}2H_2O$, 0.05 mL L^{-1} antifoam (Glanapon 2000kz as an example), 5 mL L^{-1} of PTM1 solution (described in Subheading 2.1.3), and 2 mL of biotin solution (described in Subheading 2.1.4) (*see* **Note 4**).
2. Dissolve all reagents except PTM1 trace salts and biotin solutions in the appropriate volume of distilled water and adjust to the desired pH (4–7) with the appropriate amount of HCl or NaOH (*see* **Note 5**).
3. Autoclave the medium during bioreactor sterilization. Later, add PTM1 and biotin solutions under sterile conditions (*see*

Note 6) when the medium is cooled down (room temperature).

Alternatively, *Basal salts medium (BSM)* was described by Invitrogen for *K. phaffii* fermentation in the respective *Pichia* Fermentation Process Guidelines [31]. This medium contains higher salts concentration in comparison with the described PBM. Its utilization will be a decision of the operator according to cultivation requirements. Its composition is detailed below.

1. *Basal salts medium (BSM)*: 26.7 mL L^{-1} 85% H_3PO_4, 0.93 g L^{-1} $CaSO_4$, 18.2 g L^{-1} K_2SO_4, 14.9 g L^{-1} $MgSO_4{\cdot}7H_2O$, 4.13 g L^{-1} KOH, 40.0 g L^{-1} glycerol (*see* **Note 4**), 4.35 mL L^{-1} of PTM1 solution, and 4.35 mL L^{-1} of biotin solution.
2. Dissolve all reagents and sterilize the solution by autoclaving inside the bioreactor.
3. Once the temperature is around room temperature, adjust to the desired pH with 28% NH_4OH and add 4.35 mL of PTM1 solution and 4.35 mL of 0.02% biotin solution under sterile conditions.

Another alternative is the one described by Matthews et al. (2018), named *Rich defined medium (RDM)* [32]. This medium is further supplemented with amino acids, vitamins, and lipids for the optimal growth of the yeast.

1. *Rich defined medium (RDM)*: composed of 40.0 mL L^{-1} glycerol, 12.0 g L^{-1} KH_2PO_4, 4.7 g L^{-1} $MgSO_4{\cdot}7\text{-}H_2O$, 0.36 g L^{-1} $CaCl_2{\cdot}2H_2O$, 1.65 g L^{-1} $(NH_4)_2SO_4$, 3.37 g L^{-1} KOH (*see* **Note 4**), 1.74 g L^{-1} glutamine, 1.46 g L^{-1} arginine, 1 mL L^{-1} vitamins solution mix (1000×) (described in Subheading 2.1.5), 10 mL L^{-1} lipids mix (described in Subheading 2.1.5), and 4.35 mL L^{-1} PTM1 solution.
2. Dissolve all reagents and adjust to the desired pH.
3. Sterilize the solution by autoclaving, and once the temperature is around room temperature, add the amino acids, the vitamins, the lipids, and the PTM1 solution.

2.1.3 PTM1 Trace Salts Solution

1. *PTM1 trace salts is* composed per liter of 5.0 mL L^{-1} H_2SO_4, 65.0 g L^{-1} $FeSO_4{\cdot}7\ H_2O$, 20.0 g L^{-1} $ZnCL_2$, 6.0 g L^{-1} $CuSO_4{\cdot}5H2O$, 3.36 g L^{-1} $MnSO_4{\cdot}H_2O$, 0.82 g L^{-1} $CoCl_2{\cdot}6H_2O$, 0.2 g L^{-1} $Na_2MoO_4{\cdot}2H_2O$, 0.08 g L^{-1} NaI, and 0.02 g L^{-1} H_3BO_3.
2. PTM1 trace salts solution is prepared by dissolving each reagent in the same order as is presented in the previous paragraph (*see* **Note 4**).

3. Sterilize the solution by 0.2 μm filtration.
4. Store the solution covered with aluminum foil at 4 °C.

2.1.4 Biotin Solution 0.02% (0.2 g L^{-1})

1. For 1 L, dissolve 200 mg of biotin. It will be necessary to have strong agitation without heating for a prolonged period for complete solubilization.
2. Sterilize by 0.2 μm filtration and store at 4 °C.

2.1.5 Additional Solutions for Rich Defined Medium

1. *Vitamin solution mix 1000*× is composed of 50 mg L^{-1} biotin, 200 mg L^{-1} p-aminobenzoic acid, 1 g L^{-1} nicotinic acid, 1 g L^{-1} Ca-pantothenate, 1 g L^{-1} pyridoxine-HCl, 1 g L^{-1} thiamine-HCl, and 25 g L^{-1} myo-inositol.
2. *Lipids solution mix* is supplied by Sigma–Aldrich with the following reference: Sigma–Aldrich (L0288–100ML). However, other alternative solutions by other suppliers might also be used.

2.1.6 Fed-batch Medium

According to the cultivation strategy, all the micronutrients required by the culture to reach high cell concentrations in the range of 100 g L^{-1} dry cell weight (DCW) might be included initially in the batch medium, such as when using *basal salts medium (BSM)*. In this case, the feeding medium usually only includes the carbon source (glucose, glycerol, or methanol) together with the PTM1 and biotin solution.

Alternatively, in order to reduce the initial quantity of salts in the batch medium, another alternative cultivation strategy is based on the addition of the key salts together with the carbon source in a more complex fed-batch feeding medium.

In this chapter, a *Pichia Fed-Batch Feeding Medium (PFFM)* composition complementary to the *Pichia* Batch Medium *(PBM)* is presented below:

1. *Pichia Fed-batch Feeding Medium (PFFM)*: contains 400 g L^{-1} of glycerol/glucose (*see* **Notes** 7 and **8**), 6.45 g L^{-1} $MgSO_4 \cdot 7H_2O$, 10 g L^{-1} KCl, 0.35 g L^{-1} $CaCl_2 \cdot 2\ H_2O$, 15 mL L^{-1} PTM1, 6 mL L^{-1} of biotin (0.2 g L^{-1}), and 0.2 mL L^{-1} antifoam (Glanapon 2000 kz as example).
2. Medium is prepared and sterilized by autoclaving in separate bottles as follows:
3. A bottle: glycerol/glucose and antifoam.
4. B bottle: $MgSO_4 \cdot 7H_2O$ and KCl.
5. C bottle: $CaCl_2 \cdot 2H_2O$.
6. After sterilization, cool down the solutions to room temperature.

7. Once bottles are cooled down, combine the contents of the three bottles into bottle "A" under sterile conditions. Subsequently, introduce the PTM1 solution and the biotin solution.

2.2 Consumable and Disposable Material

2.2.1 Inoculum

1. Sterile baffled shake flasks covered with aluminum foil or screw caps with venting filters can be used.
2. Laminar flow cabinet.
3. Orbital shaker incubator with temperature and humidity (optional) control.
4. A spectrophotometer able to read at a 600 nm wavelength.
5. High-speed, fixed-angle centrifuge with temperature control.
6. Fixed-angle rotor for bottles.
7. Sterile polycarbonate centrifugation bottles (or flasks) suitable for centrifugation.
8. Sterilized distilled water.

2.2.2 Fermentation

1. Stirred tank (glass or stainless-steel vessel) bioreactor (volume defined by operator) coupled to a DCU unit controller for stirring, air/oxygen inlet, DO, pH, temperature, and feeding addition (*see* **Note 9**).
2. Bioreactor requirements include pH, DO, and temperature inline sensors, an aeration system (ring sparger is recommended) with an exhaust gas condenser, two peristaltic pumps, and a cooling/heating system for jacket temperature control (alternatively, a cold finger can be used).
3. Three inlet ports (inoculum, pH control solution, and feeding) and an outlet port for sampling. Two ports compatible with probes are also required (*see* **Note 10**).
4. Chiller with a recirculation system. This equipment is connected with recirculating cooled-down water to the condenser and the temperature cooling system. Highly recommended.
5. Online off-gas analyzer (*see* **Note 11**).
6. Air, O_2, and N_2 supply.
7. Sterilized glass borosilicate bottles with two-port caps.
8. Sampling system (allowing sampling without the risk of culture broth contamination.)
9. Silicone cured/Norprene tubes, connectors, and clamps suitable for autoclaving.
10. 0.2 μm hydrophilic filters for sterilization of thermolabile solutions (PMT1 trace salts and biotin) (*see* **Note 12**).
11. 0.2 μm hydrophobic filters for gas venting and base/inoculum bottle assembly.

2.2.3 Feeding System

1. Glass bottle with two-port cap.
2. External/internal high volumetric precision pump for feeding addition. An additional controller (via DCU or external software) should be able to run customized additional profiles.
3. Sterilized glass borosilicate bottles with two-port stainless-steel caps.
4. 0.2 μm hydrophilic filters for sterilization of thermolabile solutions (PMT1 trace salts and biotin).
5. 0.2 μm hydrophobic filters for the feeding bottle assembly.
6. A magnetic stirrer to make sure that the fed-batch feeding medium is homogenized (antifoam might present solubility issues), and a weighing scale ensures that the addition profile of feeding is being accurately implemented.
7. Tubing (silicone and Norprene tubes), connectors, and clamps compatible with sterilization by autoclaving (*see* **Note 13**).

3 Methods

3.1 Fermenter Setup

A general scheme of fed-batch setup is represented in Fig. 2.

1. Assemble the vessel and the lid of the bioreactor. Tighten the screws firmly to avoid air leakage and potential contamination.

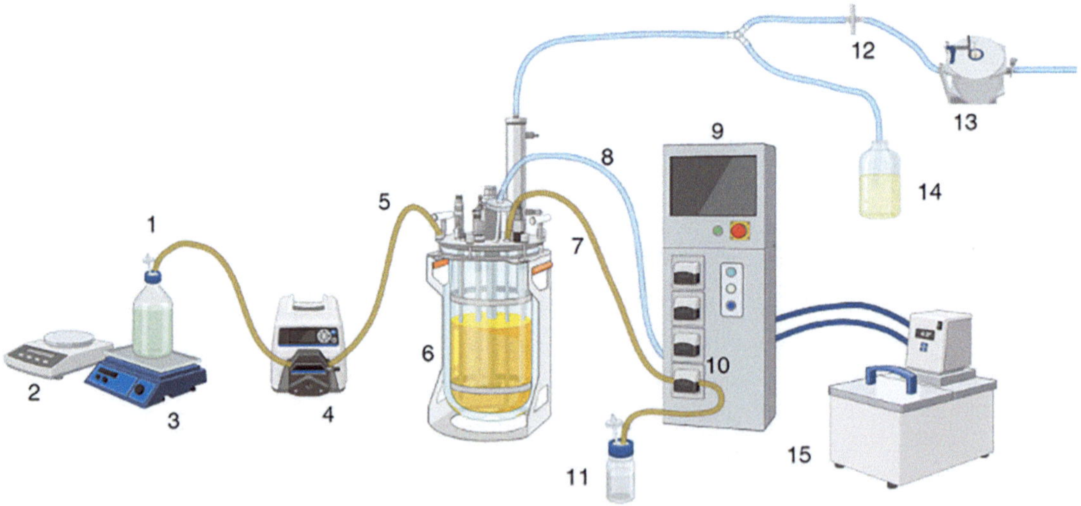

Fig. 2 Fermentation setup for fed-batch cultures. Each component is numbered as follows: (1) Feeding bottle; (2) Weighing scale; (3) Magnetic stirrer; (4) Feeding pump; (5) Feeding addition port; (6) Fermenter vessel; (7) Base addition port; (8) Venting port with filter; (9) Fermenter control unit; (10) Base pump; (11) Base bottle; (12) Off-gas filter; (13) Gas analyzer; (14) Kill tank; (15) Chiller. (Created with Biorender.com)

2. Prepare all the tubing systems for:
   (a) Sparger port: includes a 0.2 μm pore air filter for air venting into the bioreactor.
   (b) Sampling port: includes a tube connecting the sampling dip tube and the sampling recipient. The system used must avoid the risk of culture contamination.
   (c) Base control port: includes a base-tolerant tube connecting the inlet port and an empty glass bottle with a two-port cap (*see* Fig. 3). One port is destined for pressure equilibration, incorporating a 0.2 μm pore air filter, and the other is for base transfer to the bioreactor. The tube must be suitable for the use of the DCU's base peristaltic pump, incorporating a Norprene tube where the tubing system comes into contact with the pump, for better mechanical resistance and higher precision.
   (d) Inoculum port: includes a tube from the port to a glass bottle with a two-port cap. One of them is for the air filter (pressure compensation), and the other is for the inoculum transfer (*see* Fig. 3).
   (e) Feeding port: includes a tube connected to a dead end, closed with two clamps to maintain sterility of the tube. Following the sterilization of the fermenter and the

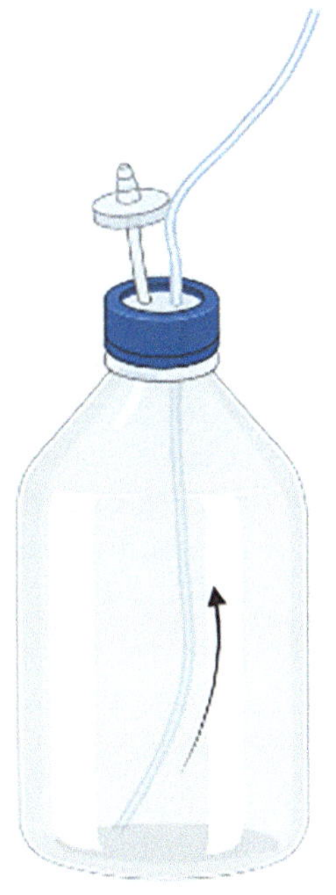

Fig. 3 Connection structure for bottles containing a two-port cap. On an empty glass bottle is assembled a screw cap with a head metal piece connected with a sterile air filter and the tubing for each system. (Created with Biorender.com)

feeding, this setup will facilitate the connection to the feeding system.

(f) Outlet gas: the outgas condenser outlet must be connected to a tubing system that allows the bifurcation of two streams of the outlet air. One will be redirected to an air filter (or a biological waste tank) (*see* **Note 14**), and the other stream will pass through the gas analyzer protected by another air filter. For the tubing preparation before the sterilization, incorporate an air filter at the outlet tube of the condenser. The tubing system with the bifurcation, which will carry the air outlet flow, does not need to be sterilized.

3. Clamp all tubing systems and cover the air filters with aluminum foil.
4. Add the batch medium into the bioreactor. Recommended using a suitable funnel.
5. Perform pH probe calibration by using a two-point method (recommended). Then incorporate pH and DO probes into their respective ports covered with aluminum foil on the top. Also, cover the top of the motor coupling adaptor.
6. Autoclave the bioreactor containing the batch medium with at least one port opened to avoid overpressure events that can damage the vessel.
7. After the sterilization process, close the open port as fast as possible. At this moment, perform a visual inspection to ensure that no tubing system has been disconnected, which could result in an air (or/and medium) leak. Also, make sure that the screws on the fermenter lid are tightly fastened after autoclaving. An airtightness test could be performed at this point by removing the bioreactor from the autoclave and letting the air get into the fermenter, penetrating through the air filter of the sparger (remove the clamp and the aluminum from this system). If the bioreactor is completely airtight, the differential pressure caused by a differential of temperatures will be equilibrated, causing bubbles in the media from the sparger. In case the media does not bubble, identify the air leak, make sure it is well closed, and repeat the sterilization process.
8. Remove the clamps from the required tubing systems. Also, remove the aluminum foil and connect the probes and stirrer to the DCU.
9. Activate the chiller and establish connections between the water-recirculating systems, the condenser, and the bioreactor jacket. Subsequently, insert the temperature probe and initiate the temperature controller in the DCU, configuring the working temperature as the set point (*see* **Note 15**).

10. Turn on the agitation controller at the desired set point.
11. Turn on the airflow. Then, detach the air filter from the condenser outlet, and rapidly connect the available tube to the bifurcated tubing system of the outgas. This bifurcated system now links the outgas coming from the condenser to the sterile filter (or biological waste tank) and the outgas analyzer (*See* **Note 16**).
12. Once the fermenter has cooled down, introduce the PTM1 and biotin solutions through the septum.
13. Prepare a glass bottle with NH_4OH 15–30% and mimicking sterile conditions (*see* **Note 17**), change caps with the sterile empty bottle of the pH control tubing system. Then, purge the tube and activate the pH controller in the DCU, configuring the working pH value as the set point. Make sure that the related pump is in control mode.
14. Calibrate de DO probe with a two-point (0 and 100%) method, especially if the process requires work in the low DO control range (oxygen limitation as an example).
15. If necessary, sterilize and calibrate the feeding addition system (*See* **Note 18**).
16. Simulating sterile conditions, connect the tubing system from the feeding port of the fermenter with the feeding solution. Remove the clamps and purge the tubing system with the feeding solution.
17. Measure the initial weight (*see* **Note 19**) of the feeding solution and place it on a magnetic stirrer for agitation.

3.2 Inoculum Preparation

The purpose of the inoculum culture is to generate enough biomass for bioreactor inoculation. Thus, it becomes essential to have a prior characterization of growth kinetics from the working strain under flask conditions, determining its μ_{max} and the OD (optical density) range corresponding to the exponential growth phase of the microorganism. Within this range is when the inoculum culture is intended to be used, thus avoiding long latency (lag) phases after the bioreactor's inoculation.

The following exponential growth formula, derived from the mass balance of a batch process, will assist in calculating the timings and working volumes for the preparation of the inoculum (*see* **Note 20**)

$$X = X_0 e^{\mu t} \tag{1}$$

1. Previously to the inoculum preparation, measure the μ_{max} of the strain on the inoculum medium and at the desired conditions. In the same experiment, determine the OD range where the cultivation grows exponentially.

2. Under sterile conditions (in a laminar flow cabin), introduce the inoculum medium into the shake flasks. The working volume of the media should be 10–15% of the total volume of the shake flask. Ensure to include the selection factor if necessary.
3. Following Eq. 1 and knowing X (the targeted OD), μ, and the desired cultivation time, X_0 (initial OD) can be determined. Calculate the volume from the cryovial cell bank (V_0) necessary using the formula $V_0C_0 = VC$, considering that V is the inoculum medium volume in the shake flask, C is the initial OD (X_0) of the cultivation, and C_0 is the OD from the cell bank cryovial.
4. Take one vial of the cell bank cryovial from the −80° freezer and thaw it on ice (*see* **Note 21**).
5. In sterile conditions, add the calculated volume from the cryovial to the shake flask. Homogenize by gently shaking.
6. Incubate the shake flask in an orbital incubator for the expected time and at the conditions (temperature and agitation rate) at which the strain was characterized.

3.3 Batch Phase

In accordance with the anticipated timeline established during the prior inoculum preparation phase, it is expected to find the strain culture in the middle of exponential growth, having accumulated substantial biomass suitable for fermenter inoculation. To smoothly move on to the next batch phase of the bioprocess, it's crucial to carefully follow these steps:

1. In sterile conditions, take a sample of the inoculum culture and measure the OD_{600} with the spectrophotometer.
2. Using the aforementioned equation ($V_0C_0 = VC$), determine the required inoculum volume to attain the desired cellular concentration in the bioreactor.
3. Consider V_0 as the volume of the inoculum being utilized, where C_0 denotes the cell concentration determined in **step 1**. V represents the operational volume within the fermenter, and C is the predetermined target cell concentration.
4. Quantify the determined inoculum volume using a sterile test tube and transfer it into the sterile centrifuge bottle. Centrifuge for 5 min at 6000 xg.
5. Under sterile conditions, discard the supernatant and resuspend the pellet with sterile distilled water. Transfer the resuspended inoculum to a sterile glass bottle.
6. Identify the empty bottle attached to the inoculum tubing system and, simulating sterile conditions, exchange the caps with the bottle containing the inoculum.
7. Inoculate the fermenter. Inoculation of the fermenter involves the transfer of the inoculum from the bottle to the bioreactor.

This can be accomplished by utilizing a peristaltic pump or by applying overpressure inside the bottle by applying overpressure through the vent filter on the inoculum bottle.

8. Make sure all controllers are turned on and the correct set-points have been successfully introduced.
9. To regulate DO, the cascade control of dissolved oxygen is commonly employed. In *K. phaffii* fermentations, a set point of 30% is the value considered ideal to avoid oxygen limitations in the culture.
10. Once the culture is adapted to bioreactor conditions, it is expected that biomass grows at a constant rate (μ_{max}) following exponential growth. The consumption rate of substrate and oxygen also follows the same trend. If the oxygen transfer conditions of the system do not change, the DO value will also decrease until it reaches the set-point value. At this point, the controller will activate, first increasing the agitation and then (if necessary) supplementing the air inlet with pure oxygen.
11. CO_2 production and O_2 consumption rates will have an exponential evolution; therefore, the detection of their concentrations in the fermenter's gas outlet should also follow this trend.
12. The completion of the batch phase occurs when the main carbon source, in this case glycerol, is depleted. At this point, a typical spike in DO is observed, attributed to the metabolic arrest. If the cascade control of the DO remains activated, this event is accompanied by a decrease in agitation rate (*see* **Note 22**).

3.4 Fed-batch Phase

This section is focused on the procedure to establish C-limiting fed-batch cultivations based on a preprogrammed exponential feeding profile to work at a constant specific growth rate.

Following the anticipated timeline established during the prior fermenter setup and the batch phase, at this point, the addition system is calibrated, and the feeding is connected to the inlet port. After the main carbon source of the batch phase has been depleted, and once we have passed the transition phase (if necessary), the system is ready to commence the fed-batch phase.

1. Calculate the addition volumes of the preprogrammed exponential feeding phase by using the following formulas:

$$F_t = F_0 \cdot e^{\mu \cdot t} \tag{2}$$

$$F_0 = \frac{X_0 \cdot V_0 \cdot \mu}{\left(S_0 \cdot Y_{X/S}\right)} \tag{3}$$

where F_t and F_0 are the feeding flow rates at time $= t$ (h) and time $= 0$ (h), respectively (L h^{-1}), X_0 is the biomass concentration (g L^{-1}), and V_0 is the culture broth volume (L) at the beginning of the fed-batch phase, μ is the selected nominal specific growth rate (h^{-1}) (*see* **Note 23**), t is the feeding phase time (h), S_0 is the substrate concentration in the feeding tank (g L^{-1}), and $Y_{X/S}$ is the overall biomass-to-substrate yield ($g_X\ g_S^{-1}$) (*see* **Note 24**).

2. Configure the corresponding software controlling the addition system and start the preprogrammed addition of the feeding media. To identify and avoid deviations from the designed feeding profile, it is helpful to check the feeding solution bottle weight during the fed-batch phase. If possible, using a closed-loop gravimetric control of the feeding addition system using a scale contributes to decreasing eventual deviations.
3. During the fed-batch phase, it is important to verify that carbon-limiting conditions are maintained throughout the fed-batch phase and, therefore, ensure that the main carbon source (in this case, glucose) does not accumulate in the fermentation culture.
4. Take samples along the cultivation to follow the bioprocess evolution.
5. At the end of the culture, carefully harvest the broth and determine the final cultivation volume, since it will importantly impact the determination of the key performance indicators (KPI) described in Subheading 3.6 (*see* **Note 25**).
6. Determine the carbon source concentration of the feeding solution, since it will also be key for the KPI calculations.

3.5 Advanced Strategies for Implementing Stress Conditions

3.5.1 Starving Conditions

The industrial biomanufacturing processes in large-scale fermentation facilities require the design and implementation of simple cultivation strategies that are robust and scalable, which is key for the tech transfer between different fermentation facilities as well as during the bioprocess scale-up. As it was indicated in the introduction, the implementation of carbon restriction stress has shown beneficial effects for recombinant expression under P_{GAP} regulation [17]. Equivalent results have been reported for other expression systems, such as those based on the promoter P_{DH} [33]. This strategy might also boost the titers and yields for other expression systems.

Thus, designing carbon-starving strategies in order to generate repeated cycles of carbon starving and preprogrammed exponential feeding periods might boost the production titers (Fig. 4). Since μ-stat fed-batch cultures work under carbon-limiting conditions, carbon source accumulation is not expected in the culture broth. Therefore, when the feeding addition pump is stopped, the carbon source is immediately depleted, and a carbon-starving event is triggered in the bioreactor. In this sense, initial conditions are recovered when feeding is reconnected.

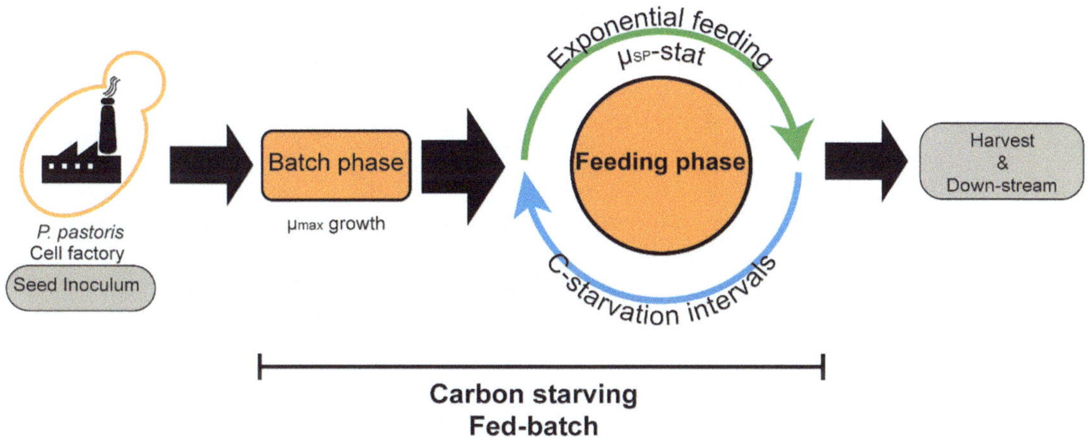

Fig. 4 General workflow of C-starving implementation. The feeding phase introduces C-starving intervals within a preprogrammed μ_{SP}-stat feeding profile to enhance RPP

After the evaluation of the results, the condition that led to the best yield and productivity was the implementation of 1.5 h of feeding followed by 0.5 h of starvation, which is detailed below (*see* **Note 26**):

1. Start preprogrammed exponential addition at a constant $\mu = 0.15\ h^{-1}$ and leave running for 1.5 h.
2. Stop feeding the pump for a time lapse of 0.5 h, triggering carbon-starvation conditions.
3. Later, restart feeding addition from the last volume dispensed before carbon-starving induction and continue with the exponential addition.
4. Repeat cycles from **Steps 4** to **6** during all fermentation until a biomass concentration of around 100 g L^{-1}. Feeding profile adaptation for carbon-starving can be observed in Fig. 5 (*see* **Note 27**).
5. Withdrawn several samples between each carbon-starving cycle in order to characterize product synthesis and biomass concentration.

3.5.2 Oxygen-limited Conditions

As described in the introduction, oxygen supply is often a limiting factor in large-scale cultivations. However, it has been proved that the implementation of controlled hypoxic conditions in the adequate range can boost RPP in *K. phaffii* when using the expression system based on the *GAP* promoter.

Applying hypoxic conditions in a dynamic culture growing at an exponential growth rate implies performing repetitive control actions to maintain the same level of oxygen limitation over time [25, 26]. Since oxygen consumption is coupled with substrate consumption, these control actions may have the same non-linear

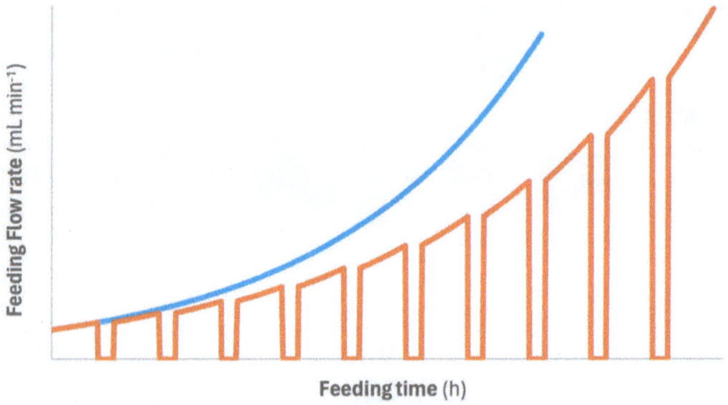

Fig. 5 Feeding time profile for fed-batch cultivation under standard (blue line) and carbon-starving (red line) conditions using a set point $\mu = 0.15\ h^{-1}$. For carbon-starving implementation, feeding was stopped every 1.5 h per 0.5 h along the preprogrammed feeding profile

behavior as substrate addition. Two operational parameters with a high impact on the oxygen transfer rate (*OTR*) can be manipulated to control the respiratory quotient (*RQ*), which is a clear physiological indicator of the oxygen limitation level: agitation rate and gas composition, this last by mixing pure oxygen or pure nitrogen with the inlet air. However, from an industrial point of view, the use of pure gases is not attractive and should be relegated to the point when hypoxia can no longer be controlled with agitation rate (i.e., when the agitation rate minimum or maximum limits have been reached).

To establish consistent oxygen-limiting conditions throughout a dynamic culture, such as in fed-batch operation mode, the following actions can be implemented:

1. Implement an online measurement of *RQ* with high accuracy and precision.
2. When the fed-batch phase has started, gradually reduce the agitation until reaching the desired *RQ*. The optimal *RQ* for each bioprocess should be determined, establishing a relationship with the bioprocess productivity (*see* **Note 28**).
3. If the agitation should be reduced below the minimum level to reach the desired *RQ*, pure nitrogen can be mixed with the inlet air to reduce the oxygen transfer to the culture.
4. As the feeding phase progresses, an increase in the agitation rate will be needed, in accordance with the substrate addition increase. In this sense, an automated control strategy is very useful (*see* **Note 29**).
5. If the maximum agitation rate is reached, pure oxygen can be mixed with the inlet air to increase the oxygen transfer. However, a reduction in the agitation rate may be necessary to keep the *RQ* set point.

3.6 Key Performance Indicators Determination (KPI)

Currently, fed-batch is the most useful operational strategy for the biomanufacturing of RPP and (bio)chemicals. To assess the efficacy of the process, as KPI, it is desirable to determine production rates and yields that are transferable between systems of different volumes or scales, and even with different biomass concentrations. Furthermore, the fact that the process is dynamic and the volume is not constant over time increases complexity in the determination of such KPI in comparison with continuous cultivations in which the volume is constant (*see* **Note 30**). Determination is detailed in Table 1.

4 Notes

1. The autoclave sterilization process requires exposure for thirty minutes at a temperature of 121 °C.
2. The use of products from animal sources can be avoided during inoculum preparation by using a defined buffered minimal glycerol medium (BMG).
3. The utilization of glucose as the main carbon source remains an option; however, the use of glycerol is recommended as it reduces by-product generation and exhibits a superior biomass-to-substrate yield.
4. The preparation sequence should remain consistent with the previously described protocol. This approach will enhance the dissolution of all components of the medium owing to the acidic pH.
5. *Pichia pastoris* exhibits optimal growth within a pH range of 4 to 7. The working pH will be determined based on product stability considerations.
6. Sterile conditions are defined as procedures in which sterility can be achieved. Such as procedures conducted in a laminar flow cabin or the fusion of tubing systems using specialized equipment, for example.
7. In this protocol, glucose is the primary carbon source described due to its potential for fermentation under oxygen-limiting conditions. Other conventional carbon sources, such as glycerol or methanol, can be utilized in the feeding media.
8. For the preparation of the fed-batch medium, the procedure remains consistent for glucose or glycerol. However, when methanol serves as the primary carbon source, it must be separately prepared and introduced into the fermenter. This precaution is taken to prevent the precipitation of salts within the medium.

Table 1
KPI determinations for fed-batch cultures

Variable		Units	Experimental parameter or variable	Equation
Specific growth rate	μ	h^{-1}	Biomass concentration (X, $g_X\ L^{-1}$) Bioreactor working volume (V, L) Fed-batch time (Δt, h)	$\mu_{mean} = \frac{\Delta(\ln(XV))}{\Delta t}$ (4)
Specific substrate uptake rate	q_S	$g_S\ g_x^{-1}\ h^{-1}$	Volume of feeding added (V_{feed}, L) Feeding substrate concentration (S_{feed}, g L^{-1}) Substrate concentration in culture broth (S, g L^{-1}) Biomass concentration (X, $g_X\ L^{-1}$) Bioreactor working volume (V, L) Fed-batch time (Δt, h)	$q_{s\,mean} = \frac{V_{feed} S_{feed} - \Delta(SV)}{(XV)_{mean} \cdot \Delta t}$ (5)
Overall biomass-to-substrate yield	$Y_{X/S}$	$g_X\ g_S^{-1}$	Volume of feeding added (V_{feed}, L) Feeding substrate concentration (S_{feed}, g L^{-1}) Substrate concentration in culture broth (S, g L^{-1}) Biomass concentration (X, $g_X\ L^{-1}$) Bioreactor working volume (V, L)	$Y_{X/S\,mean} = \frac{\Delta(XV)}{V_{feed} S_{feed} - \Delta(SV)}$ (6)
Specific product formation rate	q_P	PU $g_x^{-1}\ h^{-1}$	Product concentration (P, PU) Bioreactor working volume (V, L) Biomass concentration (X, $g_X\ L^{-1}$) Fed-batch time (Δt, h)	$q_{P\,mean} = \frac{\Delta(PV)}{(XV)_{mean} \cdot \Delta t}$ (7)
Specific productivity	Q_e	PU $g_x^{-1}\ h^{-1}$	Product concentration (P, PU) Bioreactor working volume (V, L) Final biomass concentration (X_f, $g_X\ L^{-1}$) Final bioreactor working volume (V_f, L) Fed-batch time (Δt, h)	$Q_e = \frac{\Delta(PV)}{(XV)_f \cdot \Delta t}$ (8)
Overall product to biomass yield	$Y_{P/X}$	PU g_X^{-1}	Product concentration (P, PU) Bioreactor working volume (V, L) Biomass concentration (X, $g_X\ L^{-1}$)	$Y_{P/X_{mean}} = \frac{\Delta(PV)}{\Delta(XV)}$ (9)
Volumetric productivity	Q_V	PU $L^{-1}\ h^{-1}$	Product concentration (P, PU) Bioreactor working volume (V, L) Final bioreactor working volume (V_f, L) Fed-batch time (Δt, h)	$Q_v = \frac{\Delta(PV)}{V_f \cdot \Delta t}$ (10)

Specific oxygen uptake rate	q_{O2}	mol O_2 g_x^{-1} h^{-1}	Oxygen uptake rate (OUR) (mol O_2 L^{-1} h^{-1}) Bioreactor working volume (V, L) Sampling time (dt, h) Fed-batch time (Δt, h) Biomass concentration (X, g_X L^{-1}) Inlet gas flow rate ($F_{g,in}$, mol h^{-1}) Inert factor (f_{inert}), needed if the off-gas flow rate is not measured. Calculated through inert gas balance Inlet O_2 molar fraction ($X_{O2,in}$) Outlet O_2 molar fraction ($X_{O2,out}$) Inlet CO_2 molar fraction ($X_{CO2,in}$) Outlet O_2 molar fraction ($X_{CO2,out}$) Inlet H_2O molar fraction ($X_{H2O,in}$) Outlet H_2O molar fraction ($X_{H2O,out}$)	$q_{O_{2mean}} = \frac{\sum OUR \cdot V \cdot dt}{(XV)_{mean} \cdot \Delta t}$ (11) $OUR = -\frac{F_{g,in}\left(f_{inert} \cdot x_{O_2,out} - x_{O_2,in}\right)}{V}$ (12) $f_{inert} = \frac{1 - x_{O_2,in} - x_{CO_2,in} - x_{H_2O,in}}{1 - x_{O_2,out} - x_{CO_2,out} - x_{H_2O,out}}$ (13)
Specific carbon dioxide production rate	q_{CO2}	mol CO_2 g_x^{-1} h^{-1}	Carbon dioxide production rate (CPR) (mol CO_2 L^{-1} h^{-1}) Bioreactor working volume (V, L) Sampling time (dt, h) Fed-batch time (Δt, h) Biomass concentration (X, g_X L^{-1}) Inlet gas flow rate ($F_{g,in}$, mol h^{-1}) Inert factor (f_{inert}), needed if the off-gas flow rate is not measured. Calculated through inert gas balance Inlet O_2 molar fraction ($X_{O2,in}$) Outlet O_2 molar fraction ($X_{O2,out}$) Inlet CO_2 molar fraction ($X_{CO2,in}$) Outlet O_2 molar fraction ($X_{CO2,out}$) Inlet H_2O molar fraction ($X_{H2O,in}$) Outlet H_2O molar fraction ($X_{H2O,out}$)	$q_{CO_{2mean}} = \frac{\sum CPR \cdot V \cdot dt}{(XV)_{mean} \cdot \Delta t}$ (14) $CPR = \frac{F_{g,in}\left(f_{inert} \cdot x_{CO_2,out} - x_{CO_2,in}\right)}{V}$ (15) $f_{inert} = \frac{1 - x_{O_2,in} - x_{CO_2,in} - x_{H_2O,in}}{1 - x_{O_2,out} - x_{CO_2,out} - x_{H_2O,out}}$ (16)
Respiratory quotient	RQ	mol CO_2 mol O_2^{-1}	Carbon dioxide production rate (CPR) (mol CO_2 L^{-1} h^{-1}) Oxygen uptake rate (OUR) (mol O_2 L^{-1} h^{-1})	$RQ = \frac{CPR}{OUR}$ (17)

9. Although glass fermenters are commonly used at the laboratory scale, stainless-steel fermenters also exist. With this type of fermenter, certain procedures described in this protocol (which is focused on glass fermenters) will differ. The main differences are as follows:
   (a) Sterilization of the equipment is achieved through steam recirculation, often with a Sterilization in Place (SIP) station.
   (b) Most tubing systems are fixed pipelines through which steam also circulates for sterilization. While fewer tubing systems are required for the addition and extraction of various fermentation components (inoculum, base, sampling, air inlet, etc.), a much deeper understanding of the equipment at the level of valves and pipelines will be necessary.
10. The use of the volatiles is recommended when using methanol as the main carbon source or when applying hypoxic conditions to a glucose-fed batch, since it will allow for monitoring of the methanol accumulation or the ethanol production, respectively.
11. The use of an off-gas analyzer is highly recommended when applying standard fermentative conditions, and it is essential when applying hypoxic conditions.
12. For the sterilization of media components via filtration, hydrophilic polyether sulfone (PES) membrane filters, featuring a pore size of 0.2 μm, are utilized.
13. If the addition system involves a peristaltic pump, the tubing used in this system must be suitable for such a pump to ensure optimal precision.
14. In certain instances, contingent upon the degree of biological hazard inherent in the bioprocess, the utilization of filters at the fermenter's gas outlet is avoided to prevent potential overpressure in case of filter saturation. In these cases, the air outgas is redirected to a biological waste tank containing 1% bleach (or 1 M NaOH or 70% ethanol).
15. In certain equipment, the temperature probe is also autoclaved; therefore, it must be installed before the sterilization process.
16. At this stage of the protocol, the bioreactor can be maintained in a standby mode pending inoculation, spanning several hours to days, if required. This period could be useful as a sterility assay, allowing for the detection of any biological contamination through monitoring changes in the pH, DO levels, and/or turbidity within the culture medium.

17. Semi-sterile conditions are defined as circumstances in which complete sterility cannot be fully guaranteed, but procedures are implemented to minimize the risks of contamination. To achieve this, actions are taken promptly, and the working environment, as well as the utilized materials, are sprayed with an alcoholic solution.
18. There is specific equipment wherein sterilization can only be achieved through chemical methods. This process entails the recirculation, by the equipment, of solutions comprising a strong acid, a potent base, and ethanol (sequentially), interspersed with sterile water between each solution.
19. Monitoring the weight evolution of the feeding solution throughout the fed-batch phase enables verification that the correct quantity is being added in accordance with the preprogrammed exponential addition. Measurement of solution density is imperative for this assessment.
20. Detailed information about the origin of this formula is given in Chapter 33 in this volume.
21. A gradual thawing process is advised to optimize the cellular viability of the WCB, thereby minimizing the lag phase during inoculum cultivation. It is recommended to immerse the WCB in ice for ten minutes and subsequently initiate the thawing process at ambient temperature.
22. Upon depletion of the main carbon source of the batch phase, it is recommended to initiate a transition phase of up to two hours. During this timeframe, the by-products resulting from yeast growth in the batch phase, predominantly organic acids, are metabolized. Subsequently, a slight elevation in the culture's pH is anticipated.

    If methanol is utilized as the primary carbon source in the fed-batch process, this transition phase assumes a distinct character. After observing the aforementioned DO spike, add sterilely the required amount of methanol to achieve 1 g L^{-1} in the culture broth (first methanol pulse). Wait until observing a second DO spike (it may take from 1 to 4 h) and repeat the process, but increase the concentration to 2 g L^{-1} of methanol in the fermenter. After adding the second methanol pulse, a rapid decrease in DO should be observed, indicating that biomass is already adapted to methanol assimilation. When observing that the second methanol pulse is consumed, indicated by a third DO spike, the fed-batch phase can start.
23. The targeted nominal specific growth rate during the fed-batch phase must be a value below de μ_{max} of the strain with that specific substrate, characterized during a batch phase.
24. Substrate-to-biomass yield values must be empirically determined or bibliographically obtained.

25. The final cultivation volume can be determined by weighing it into a tared container and considering culture broth density. As an alternative, the cultivation volume can be measured in a measurement cylinder.
26. Carbon-starving cycles are designed by varying the preprogrammed feeding periods with carbon-starvation intervals (pumping stop). So, the pumping stop should be enough to increase protein synthesis without triggering cellular damage and proteolysis processes. Internal experiments have displayed that after 90 min of carbon source depletion (glucose), no product losses have been observed during the starving periods. Results are further reported in the literature [17].
27. It is recommended to extend the feeding phase as much as possible to harness high biomass concentrations in combination with the C-starving boosting effect. In this sense, when more cells are affected by starvation, it is expected that higher product increments than those observed in the initial stages will occur.
28. Different *RQ* conditions should be evaluated in order to determine the optimal ones to enhance RPP. Thus, the usage of continuous operation mode is strongly recommended for bioprocess characterization before working under fed-batch conditions [25].
29. Online calculations based on first-principle models (mass balances) linked with Supervisory Control And Data Acquisition (SCADA) allow to maintain the target *RQ* control based on *CPR* and *OUR* online measurements [24].
30. Exponential addition should be considered for total volume estimation. Since it follows a preprogrammed exponential profile, volumes dispensed per minute should be added progressively in order to estimate the total volume in the bioreactor and contrast with the weight online measurement.

Acknowledgments

The project PID2022-136936OB-I00 by MCIN/AEI/10.13039/501100011033/FEDER, UE of the Spanish Ministry of Science and Innovation, funded this work. The group is a member of the 2021 SGR 00143 Department of Research and Universities of the Generalitat de Catalunya. We kindly thank the Christian Doppler Gesellschaft and the industrial partners Novo Nordisk, BioGrammatics, SeSam Biotech, ZETA, and SolarBiotech for funding the Christian Doppler Laboratory for Innovative *P. pastoris* Host and Vector Systems, in which MAN, AG, and XGO are contributing as members of the UAB team.

Competing Interest Statement The authors have no conflicts of interest to declare that are relevant to the content of this chapter.

References

1. Looser V, Bruhlmann B, Bumbak F et al (2015) Cultivation strategies to enhance productivity of *Pichia pastoris*: a review. Biotechnol Adv 33:1177–1193. https://doi.org/10.1016/j.biotechadv.2015.05.008
2. García-Ortega X, Cámara E, Ferrer P et al (2019) Rational development of bioprocess engineering strategies for recombinant protein production in *Pichia pastoris* (*Komagataella phaffii*) using the methanol-free *GAP* promoter. Where do we stand? New Biotechnol 53:24–34. https://doi.org/10.1016/j.nbt.2019.06.002
3. Çalık P, Hoxha B, Çalık G, Özdamar TH (2018) Hybrid fed-batch bioreactor operation design: control of substrate uptake enhances recombinant protein production in high-cell-density fermentations. J Chem Technol Biotechnol 93:3326–3335. https://doi.org/10.1002/jctb.5696
4. Potvin G, Ahmad A, Zhang Z (2012) Bioprocess engineering aspects of heterologous protein production in *Pichia pastoris*: a review. Biochem Eng J 64:91–105. https://doi.org/10.1016/j.bej.2010.07.017
5. Buchetics M, Dragosits M, Maurer M et al (2011) Reverse engineering of protein secretion by uncoupling of cell cycle phases from growth. Biotechnol Bioeng 108:2403–2412. https://doi.org/10.1002/bit.23198
6. Mears L, Stocks SM, Sin G, Gernaey KV (2017) A review of control strategies for manipulating the feed rate in fed-batch fermentation processes. J Biotechnol 245:34–46. https://doi.org/10.1016/j.jbiotec.2017.01.008
7. Ponte X, Barrigón JM, Maurer M et al (2018) Towards optimal substrate feeding for heterologous protein production in *Pichia pastoris* (*Komagataella spp*) fed-batch processes under P_{AOX1} control: a modeling aided approach. J Chem Technol Biotechnol 93:3208–3218. https://doi.org/10.1002/jctb.5677
8. Maurer M, Kühleitner M, Gasser B, Mattanovich D (2006) Versatile modeling and optimization of fed batch processes for the production of secreted heterologous proteins with *Pichia pastoris*. Microb Cell Factories 5:37. https://doi.org/10.1186/1475-2859-5-37
9. Garcia-Ortega X, Ferrer P, Montesinos JL, Valero F (2013) Fed-batch operational strategies for recombinant Fab production with *Pichia pastoris* using the constitutive *GAP* promoter. Biochem Eng J 79:172–181. https://doi.org/10.1016/j.bej.2013.07.013
10. Garrigós-Martínez J, Weninger A, Montesinos-Seguí JL (2021) Scalable production and application of *Pichia pastoris* whole cell catalysts expressing human cytochrome P450 2C9. Microb Cell Fact 20(1):90. https://doi.org/10.1186/s12934-021-01577-4
11. De Brabander P, Uitterhaegen E, Delmulle T et al (2023) Challenges and progress towards industrial recombinant protein production in yeasts: a review. Biotechnol Adv 64:108121. https://doi.org/10.1016/j.biotechadv.2023.108121
12. Weiss F, Requena-Moreno G, Pichler C et al (2024) Scalable protein production by *Komagataella phaffii* enabled by ARS plasmids and carbon source-based selection. Microb Cell Factories 23:1–16. https://doi.org/10.1186/s12934-024-02368-3
13. Nieto-Taype MA, Garrigós-Martínez J, Sánchez-Farrando M et al (2020) Rationale-based selection of optimal operating strategies and gene dosage impact on recombinant protein production in *Komagataella phaffii* (*Pichia pastoris*). Microb Biotechnol 13:315–327. https://doi.org/10.1111/1751-7915.13498
14. Garrigós-Martínez J, Nieto-Taype MA, Gasset-Franch A et al (2019) Specific growth rate governs *AOX1* gene expression, affecting the production kinetics of *Pichia pastoris* (*Komagataella phaffii*) P_{AOX1}-driven recombinant producer strains with different target gene dosage. Microb Cell Factories 18:187. https://doi.org/10.1186/s12934-019-1240-8
15. Garrigós-Martínez J, Vuoristo K, Nieto-Taype MA et al (2021) Bioprocess performance analysis of novel methanol-independent promoters for recombinant protein production with *Pichia pastoris*. Microb Cell Factories 20:74. https://doi.org/10.1186/s12934-021-01564-9
16. Rebnegger C, Graf AB, Valli M et al (2014) In *Pichia pastoris*, growth rate regulates protein synthesis and secretion, mating and stress

response. Biotechnol J 9:511–525. https://doi.org/10.1002/biot.201300334

17. Garcia-Ortega X, Adelantado N, Ferrer P et al (2016) A step forward to improve recombinant protein production in *Pichia pastoris*: from specific growth rate effect on protein secretion to carbon-starving conditions as advanced strategy. Process Biochem 51:681–691. https://doi.org/10.1016/j.procbio.2016.02.018
18. Baumann K, Maurer M, Dragosits M et al (2008) Hypoxic fed-batch cultivation of *Pichia pastoris* increases specific and volumetric productivity of recombinant proteins. Biotechnol Bioeng 100:177–183. https://doi.org/10.1002/bit.21763
19. Garcia-Ortega X, Valero F, Montesinos-Seguí JL (2017) Physiological state as transferable operating criterion to improve recombinant protein production in *Pichia pastoris* through oxygen limitation. J Chem Technol Biotechnol 92:2573–2582. https://doi.org/10.1002/jctb.5272
20. Carnicer M, Baumann K, Toplitz I et al (2009) Macromolecular and elemental composition analysis and extracellular metabolite balances of *Pichia pastoris* growing at different oxygen levels. Microb Cell Factories 8:65. https://doi.org/10.1186/1475-2859-8-65
21. Baumann K, Dato L, Graf AB et al (2011) The impact of oxygen on the transcriptome of recombinant *S. cerevisiae* and *P. pastoris* - a comparative analysis. BMC Genomics 12:218. https://doi.org/10.1186/1471-2164-12-218
22. Baumann K, Carnicer M, Dragosits M et al (2010) A multi-level study of recombinant *Pichia pastoris* in different oxygen conditions. BMC Syst Biol 4:141. https://doi.org/10.1186/1752-0509-4-141
23. Adelantado N, Tarazona P, Grillitsch K et al (2017) The effect of hypoxia on the lipidome of recombinant *Pichia pastoris*. Microb Cell Factories 16:86. https://doi.org/10.1186/s12934-017-0699-4
24. Sales-Vallverdú A, Gasset A, Requena-Moreno G et al (2024) Synergic kinetic and physiological control to improve the efficiency of *Komagataella phaffii* recom tools.lifetechnologies.com/content/sfs/manuals binant protein production bioprocesses. Microb Biotechnol 17:e14411. https://doi.org/10.1111/1751-7915.14411
25. Gasset A, Garcia-Ortega X, Garrigós-Martínez J et al (2022) Innovative bioprocess strategies combining physiological control and strain engineering of *Pichia pastoris* to improve recombinant protein production. Front Bioeng Biotechnol 10:818434. https://doi.org/10.3389/FBIOE.2022.818434
26. Gasset A, Van Wijngaarden J, Mirabent F et al (2024) Continuous Process Verification 4.0 application in upstream: adaptiveness implementation managed by AI in the hypoxic bioprocess of the *Pichia pastoris* cell factory. Front Bioeng Biotechnol 12:1439638. https://doi.org/10.3389/fbioe.2024.1439638
27. Prielhofer R, Reichinger M, Wagner N et al (2018) Superior protein titers in half the fermentation time: promoter and process engineering for the glucose-regulated *GTH1* promoter of *Pichia pastoris*. Biotechnol Bioeng 115:2479–2488. https://doi.org/10.1002/bit.26800
28. Fischer JE, Hatzl A-M, Weninger A et al (2019) Methanol independent expression by *Pichia pastoris* employing de-repression technologies. J Vis Exp e58589. https://doi.org/10.3791/58589
29. Vogl T, Sturmberger L, Kickenweiz T et al (2016) A toolbox of diverse promoters related to methanol utilization: functionally verified parts for heterologous pathway expression in *Pichia pastoris*. ACS Synth Biol 5:172–186. https://doi.org/10.1021/acssynbio.5b00199
30. Vogl T, Kickenweiz T, Pitzer J et al (2018) Engineered bidirectional promoters enable rapid multi-gene co-expression optimization. Nat Commun 9:3589. https://doi.org/10.1038/s41467-018-05915-w
31. Invitrogen Corporation (2002) *Pichia* fermentation process guidelines. https://tools.thermofisher.com/content/sfs/manuals/pichiaferm_prot.pdf
32. Matthews CB, Kuo A, Love KR, Love JC (2018) Development of a general defined medium for *Pichia pastoris*. Biotechnol Bioeng 115:103–113. https://doi.org/10.1002/bit.26440
33. Bernat-Camps N, Ebner K, Schusterbauer V et al (2023) Enabling growth-decoupled *Komagataella phaffii* recombinant protein production based on the methanol-free P_{DH} promoter. Front Bioeng Biotechnol 11:1130583. https://doi.org/10.3389/fbioe.2023.1130583

Chapter 32

A Guide to Fed-Batch Cultivation of *Pichia pastoris* on Laboratory Scale

Marcel Straumann, Hana Raschmanová, Valentina Haag, Iwo Zamora, and Karin Kovar

Abstract

Fed-batch is the most commonly used cultivation mode for industrial production of recombinant proteins with *Pichia pastoris*. On a laboratory scale, fed-batch culture provides a way to control the specific biomass growth rates at any pre-set value, allowing the conditions of biomass growth and recombinant product formation to be systematically studied.

In this chapter, we present an accessible and versatile approach for designing, performing, and evaluating a fed-batch cultivation in laboratory-scale stirred tank bioreactors.

Key words *Pichia pastoris*, *Komagataella phaffii*, Fed-batch culture, Bioreactor, Fermentation, Bioprocess design and development, Feed rate, Specific growth rate, Specific productivity, Data evaluation

1 Introduction

The widely recognized goal of industrial or laboratory bioreactor cultivation is to deliver as much product as possible in as short a time and as small a volume as possible [1], implying both the highest possible product titers (c_p) and volumetric productivities (r_p). This is achieved under appropriate microbial culture conditions (i.e., concerning the optimum supply of nutrients as well as controlling dissolved oxygen concentrations, pH, and temperature), which are controlled (or pre-set) and monitored in bioreactors. On a laboratory scale, fed-batch culture also provides a way to control biomass growth and thus allow a study of product formation kinetics, the relationship between specific productivity (q_p) and specific biomass growth rate (μ).

As the volumetric productivity of heterologous protein production in *Pichia pastoris* is dependent on biomass produced, control of the specific biomass growth rate is crucial [2, 3]. A typical

Xavier Garcia-Ortega et al. (eds.), *Pichia pastoris: Methods and Protocols*, Methods in Molecular Biology, vol. 2697, https://doi.org/10.1007/978-1-0716-4779-0_32,

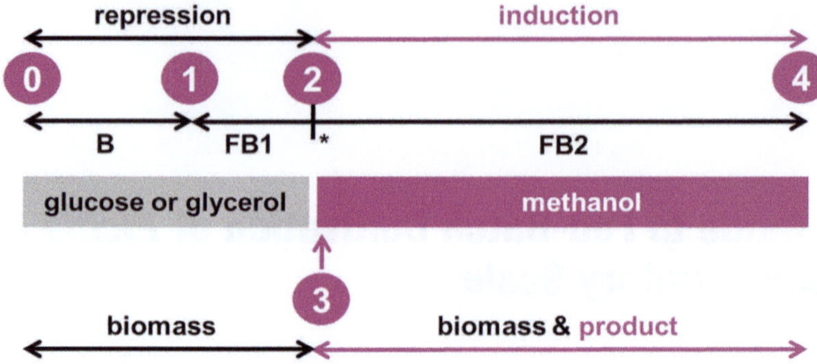

Fig. 1 Phases of a bioprocess with *Pichia pastoris* on a bioreactor scale. Glucose or glycerol is used to grow the biomass; they also repress the alcohol oxidase promoter 1 (P_{AOX1}) so that no recombinant product is produced. Then, when the glucose or glycerol is completely consumed (i.e., depleted from the culture broth) and methanol is added, the P_{AOX1} promoter is induced and the heterologous protein is produced. At the same time, methanol serves as a carbon source for biomass growth. Typically, fed-batch bioreactor cultivation with *P. pastoris* comprises three main phases, which start or end (numbers 1, 2, 3, and 4) and have distinct phases: *Phase B:* initial batch phase with glucose or glycerol as a carbon source, which starts at time point (**0**), characterized by initial concentrations of substrate and biomass, s_0 and x_0, respectively. *Phase FB1:* biomass growth in a fed-batch mode with glucose or glycerol, which starts at time point (**1**), characterized by concentrations of biomass at the end of the batch phase (x_{end}), zero substrate concentration in the supernatant, and an initial feed (F_0), which increases exponentially with time to maintain a pre-set specific biomass growth rate (μ). *Phase FB2:* a production fed-batch phase with methanol starts at time point (**2**), at which the current biomass concentration and productivity optimum determine the design of the methanol feed.Optionally a short intermediate phase * (asterisk) can be implemented during which neither the growth substrate (glucose or glycerol) nor methanol is fed continuously. At this time point (3), for instance, an initial pulse of methanol ($<$ 1 g/L final concentration in bioreactor) allows the culture to completely deplete the growth substrate and its metabolites while adapting to methanol
The harvest point (4) is typically reached either at the limit of the working volume V_{end}, when oxygen delivery is unable to maintain $pO_2 > 20\%$, or at the limit of cooling capacity, or due to loss of productivity. Alternatively, the bioprocess may be arbitrarily terminated for any operational reason

bioreactor cultivation with *Pichia pastoris* comprises three phases (Figs. 1, 2, and 3):

- Phase B: an initial phase of biomass growth in batch mode with glucose or glycerol as a carbon source,
- Phase FB1: a biomass growth phase in fed-batch mode with glucose or glycerol,
- Phase FB2: a production phase in fed-batch mode with methanol.

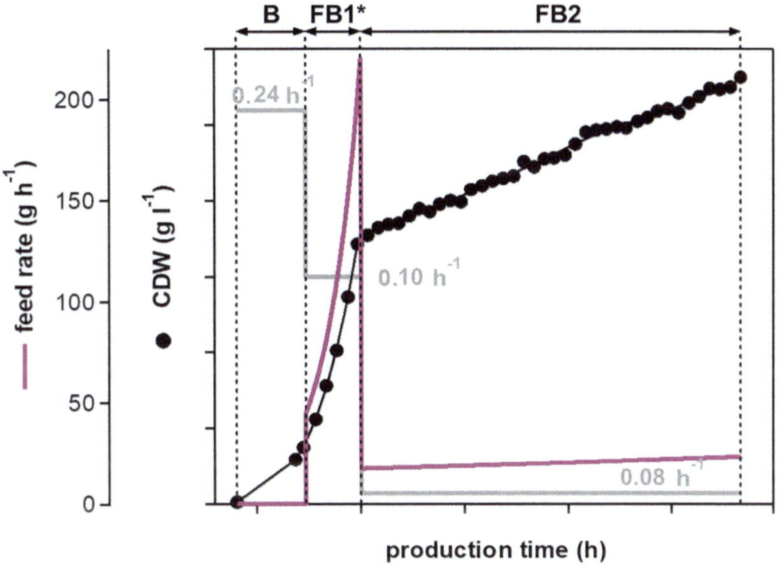

Fig. 2 Biomass growth in fed-batch cultures with *Pichia*. As long as substrate is supplied, biomass is produced over the entire process. Black circles represent cell dry weight (CDW). With these values, the maximum specific growth rate in batch culture can be determined, as well as pre-set specific growth rates in fed-batch phases can be verified. The specific growth rate with methanol during the production phase is typically maintained at a low value (i.e., not higher than 80% of the maximum specific growth rate on methanol), in order to avoid methanol accumulation and misfolding of proteins

Note that bioreactor cultivation (i.e., upstream bioprocessing) is a specialized field with its own jargon (*see* **Note 1**) and nomenclature in Fig. 4.

During initial batch cultivation (phase B), the *Pichia* strain under investigation is basically characterized with respect to its growth behavior using the specified substrate. This includes determination of the maximum specific biomass growth rate (μ_{max}), the maximum biomass to substrate yield ($Y_{x/s,max}$), and the maximum specific substrate utilization rate ($q_{s,max}$).

Fed-batch mode (i.e., phases FB1 and FB2) follows when the carbon source and eventually its metabolites are completely depleted in the batch or previous fed-batch. Feeding can be carried out in different modes (Fig. 5). The carbon source (i.e., glycerol, glucose, or methanol, or mixtures thereof [2]) can be added to the bioreactor in pulses or continuously, preferably at an exponential rate or alternatively at a constant, linearly increasing, or decreasing rate.

Exponential substrate feeding allows one to keep the specific biomass growth rate constant. We recommend the decoupling of the fed-batch phase for biomass growth (phase FB1) from the fed-batch phase for recombinant protein production (phase FB2).

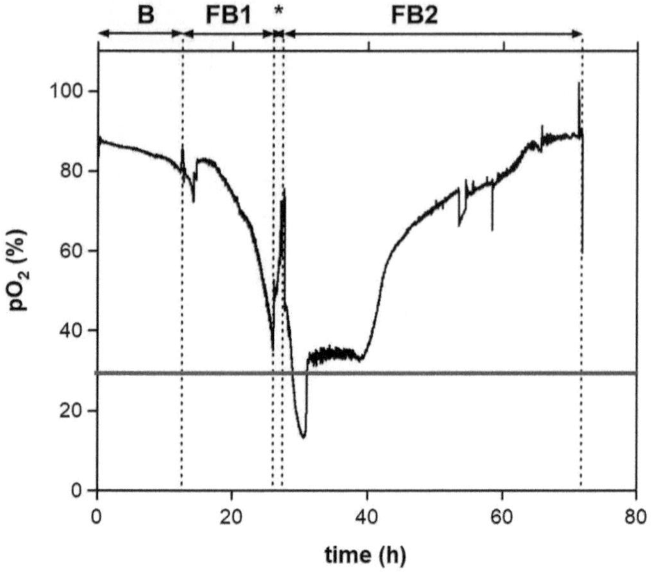

Fig. 3 Time course of dissolved oxygen concentration. As cells grow and oxidatively catabolize the supplied substrate, oxygen is consumed, and the culture broth becomes acidic, which is again compensated by adding ammonia. If a dissolved oxygen concentration (pO_2) sensor is available and pO_2 is not controlled (i.e., all settings are held constant over the process/phase duration), biomass growth is indicated by changes in the pO_2 value over time. For example, if the substrate is depleted at the end of batch cultivation, an abrupt increase in pO_2 will be observed. Such typical time histories can be used advantageously for process monitoring and control, as well as comparison between processes

Although this strategy does not necessarily lead to the maximum possible amount of product, it often yields high amounts of product in a comparably short period of time (*see* **Note 2**).

Moreover, from several fed-batch cultivations performed at different exponential feeding rates, production kinetics can be determined; these represent the relationship between specific biomass growth rate and specific productivity (Fig. 6).

Submerged cultivation of *Pichia pastoris* in sterile containers (bioreactors or fermenters) has become both a valuable research tool and a mature manufacturing technology. In this chapter, the concepts of bioprocess design with *Pichia*—as founded through extensive practical experience—are presented in such a way that bioreactor cultivations can be straightforwardly reproduced in any laboratory appropriately equipped.

c_p	mg L^{-1}	product titre, product concentration (in supernatant)
$F(t)$	g h^{-1}	feed rate changing with time
F_0	g h^{-1}	initial feed rate
M	kg	mass
m_s	g g^{-1} h^{-1}	specific maintenance rate
μ	h^{-1}	specific growth rate
μ_{max}	h^{-1}	maximum specific growth rate
q_p	mg g^{-1} h^{-1}	specific productivity (specific rate of product formation)
$q_{s,max}$	g g^{-1} h^{-1}	maximum specific substrate utilisation rate
r_p	mg L^{-1} h^{-1}	volumetric productivity
ρ	kg L^{-1}	density
s	g L^{-1}	(residual) concentration of substrate in the reactor
s_0	g L^{-1}	initial substrate concentration
t	h	process time
V	L	working volume
V_0	L	initial working volume
V_{end}	L	final culture volume (of entire bioreactor)
V_s	L	volume of supernatant
w_{in}	g g^{-1}	mass fraction of substrate in the feed solution
x	g L^{-1}	concentration of biomass
x_0	g L^{-1}	initial biomass concentration
x_{end}	g L^{-1}	biomass concentration at end of a process phase
$Y_{p/x}$	mg g^{-1}	observed yield (product/biomass)
$Y_{x/s}$	g g^{-1}	observed yield (biomass/substrate)
$Y_{x/s,max}$	g g^{-1}	maximum yield (biomass/substrate)

Fig. 4 Nomenclature. All abbreviations used in this chapter are listed in alphabetical order

2 Materials

Due to the potentially harmful materials and conditions used during bioreactor cultivations (e.g., concentrated acid, elevated pressure in a glass vessel), good laboratory practice and safety guidelines, as well as user manuals and other instructions provided by suppliers, should be followed; *see* Subheading 3.1.

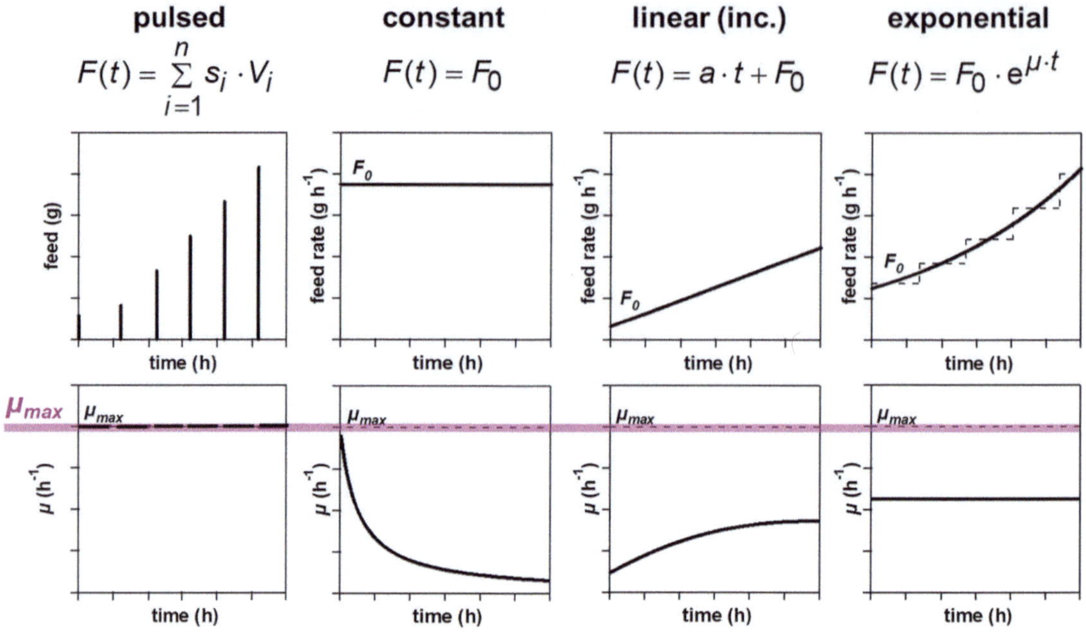

Fig. 5 Feeding regimes (profiles). A carbon source can be fed into a bioreactor in several defined ways: (**a**) as pulses, (**b**) at a constant rate, (**c**) at a linearly increasing rate, or (**d**) at an exponentially increasing rate. This chapter focuses on the design of exponential feed rate cultures, which allow the specific growth rate of the biomass to be maintained constant over time

2.1 Inoculum, Media, and Stock Solutions

1. *P. pastoris* cryopreserved stock cultures stored at −80 °C.
2. Cultivation medium BMGY (Buffered Complex Medium with Glycerol) containing per liter:

   10 g glycerol, 10 g yeast extract, 20 g peptone, 13.4 g yeast nitrogen base (YNB), 0.4 mg biotin, and 100 mM potassium phosphate buffer (pH 6.0). Dissolve 10 g of glycerol, 10 g of yeast extract, and 20 g of peptone in 798 mL of deionized water and sterilize it by autoclaving. When cooled to ambient temperature, add 100 mL of sterile 10x concentrated YNB (10x YNB), 100 mL of sterile 1 M potassium phosphate buffer (pH 6.0), and 2 mL of sterile 500x concentrated biotin (500x Biotin).
3. Optionally, add a solution of antibiotics to maintain selective pressure on the recombinant strain.
4. Stock solution 10x YNB (13.4% yeast nitrogen base with ammonium sulfate without amino acids): 134 g yeast nitrogen base (YNB) with ammonium sulfate and without amino acids per liter. Dissolve 134 g of YNB in 1000 mL of deionized water and filter sterilize.
5. Stock solution of 1 M potassium phosphate buffer (pH 6.0), 1 liter:

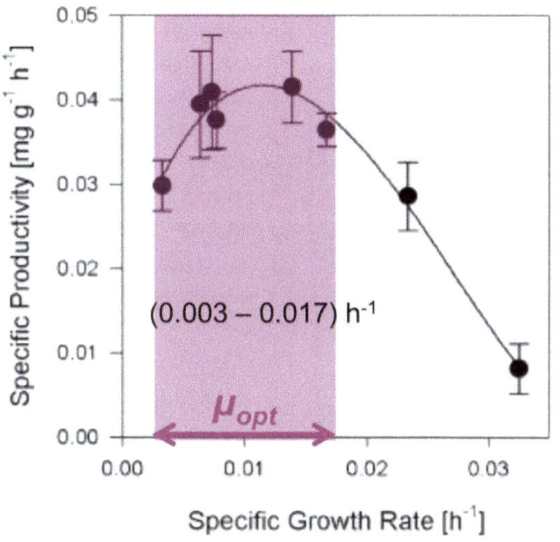

Fig. 6 Production kinetics. Production kinetics describes the relationship between the specific growth rate of biomass (μ) and specific productivity (q_p). A bell-shaped production kinetics curve is typical for proteins whose production is induced by methanol under the alcohol oxidase promoter (P_{AOX1}). Each point on this graph is obtained from a single fed-batch cultivation; the specific growth rate is maintained at a constant value over the whole production phase, and the respective specific productivity is calculated and plotted against the specific growth rate. The underlying data were adapted from Potgieter et al. (2010) [3]

Mix 132 mL of 1 M K_2HPO_4 and 868 mL of 1 M KH_2PO_4, adjust pH to 6.0 using phosphoric acid (H_3PO_4) or potassium hydroxide (KOH), and sterilize this solution by autoclaving.

6. Stock solution 500x Biotin (0.02% Biotin):

   20 mg of biotin per 100 mL of deionized water, i.e., dissolve 20 mg of biotin in 100 mL of deionized water and filter sterilize.

2.2 Bioreactor, Accessories, and Media

1. The laboratory bioreactor unit is equipped with a control unit and an appropriate bioprocess software platform to collect and evaluate data, as well as advanced control (Fig. 7). Make sure to use a bioreactor with a microbial configuration including a Rushton impeller, baffles, aeration via sparger, independent air and oxygen lines, pH control via acid and base pump, temperature control by active heating and cooling, foam detection, and antifoam control.
2. pO_2 cascade control is optional via stirrer speed, aeration rate, and oxygen content of aerated gas as suited for high-cell-density *Pichia* cultures.

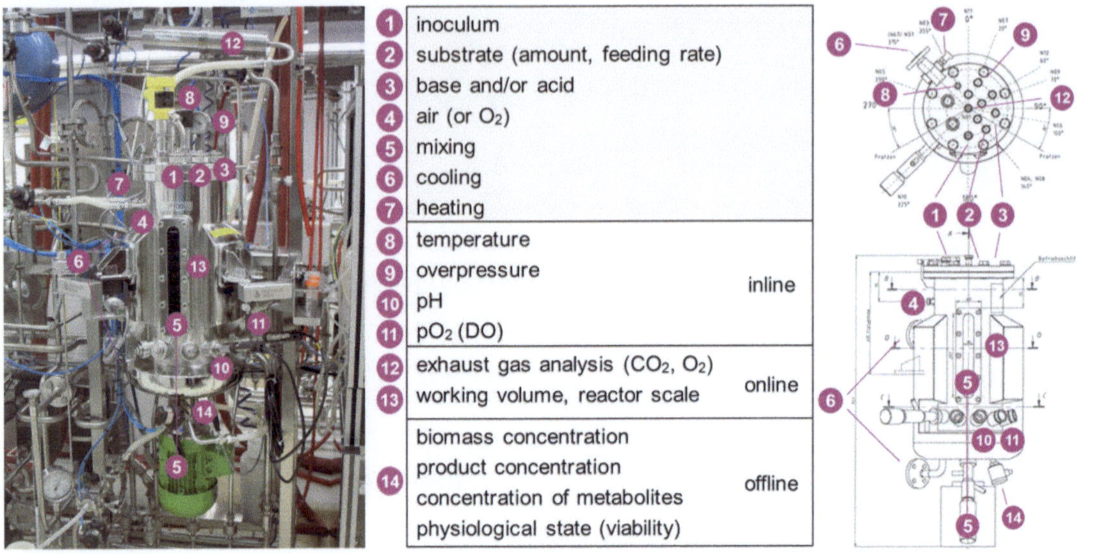

Fig. 7 Laboratory-scale bioreactor with external accessories and probes. The bioreactor lid has ports for attaching various external vessels, e.g., bottles with inoculum, substrates, or solutions for pH adjustment (1–3). A gas inlet (4) is typically part of the lid, and gas is introduced into the liquid at the bottom of the bioreactor vessel, being released from under an impeller (5). The heating/cooling jacket and probe for monitoring temperature (6 and 7) are crucial to control heat transfer. Various other probes/sensors can be attached to the bioreactor: (8–11)—pressure, pH and dissolved oxygen (pO_2) are monitored inline (12–13)—exhaust gases (CO_2, O_2), working volume, or weight of the liquid in the vessel is optionally monitored online (12–13). Samples are taken via sampling port (14) to determine concentrations of biomass, substrates, product, or metabolites, or for any additional offline analyses

3. External bioreactor accessories: Sampler port, scales (optional), pumps, flasks or containers for feeding solutions, pH correction, antifoam agent, 0.22 μm filters, silicone hoses, hose clamps, needles, and septa (*see* **Notes 3** and **4**).
4. Sensors for temperature, pH, pO_2, and foam (*see* **Note 5**).
5. Defined mineral medium used for batch [4] culture in bioreactor (phase B) containing per liter:

   30 g of glycerol (or an appropriate amount of another carbon source), 2.86 g K_2SO_4, 0.64 g KOH, 2.32 g $MgSO_4{\cdot}7\,H_2O$, 0.17 g $CaSO_4{\cdot}2\,H_2O$, 0.60 g EDTA disodium dehydrate, 7.23 g 85% H_3PO_4, 0.22 g NaCl and 0.1 mL polypropylene glycol (PPG) (*see* **Note 6**). Dissolve the components in deionized water (total volume 992 mL) and sterilize by autoclaving. Add 4.35 mL of filter-sterilized PTM1 stock solution and 4.35 mL of filter-sterilized 500x Biotin.
6. PTM1 stock solution: 5.0 mL 96% H_2SO_4, 3.84 g $CuSO_4$, 0.08 g NaI, 3.0 g $MnSO_4{\cdot}H_2O$, 0.2 g $Na_2MoO_4{\cdot}2\,H_2O$, 0.02 g H_3BO_3, 0.92 g $CoCl_2{\cdot}6\,H_2O$, 20.0 g $ZnCl_2$, and

65.0 g $FeSO_4 \cdot 7\ H_2O$ per liter. Dissolve the components in deionized water and filter sterilize the solution.

7. Solutions for adjusting pH in the bioreactor: 8.5% H_3PO_3 (acid) and 25% NH_4OH (base) (see **Note 7**).
8. Feeding solution for growth fed-batch (phase FB1) containing per kilogram:

   588 g glycerol, 12 mL of filter-sterilized PTM1 solution, and 12 mL of filter-sterilized 500x Biotin per 1000 g of feed solution. Dissolve 588 g of glycerol in 388 g of deionized water and sterilize by autoclaving. Add 12 mL of filter-sterilized PTM1 solution and 12 mL of filter-sterilized 500x Biotin.
9. Feeding solution for production fed-batch (phase FB2) containing per kilogram:

   976 g methanol, 12 mL of filter-sterilized PTM1 solution, and 12 mL of filter-sterilized biotin solution per 1000 g of feed solution. Do not sterilize methanol.
10. Spectrophotometer (600 nm).
11. Oven at 105 °C and tared 1.5 ml microtubes (dried at 105 °C for >48 h) for biomass sampling.
12. Autoclave for 20 min at 121 °C to sterilize volumes of several liters.

3 Methods

3.1 General Instructions

1. Prepare all solutions using demineralized water at room temperature.
2. Store all sterilized cultivation media and stock solutions at 4 °C.
3. Use a laminar (air) flow cabin for sterile and aseptic work, e.g., mixing cultivation media, handling the *P. pastoris* culture, and the like.
4. Strictly follow the manufacturer's instructions when calibrating and maintaining probes (sensors), scales, and pumps.
5. Diligently follow all waste disposal regulations when disposing of waste materials.
6. Use caution and safety glasses when working with and/or around pressurized bioreactors.

Methods 3.2 and 3.3 can be applied for any bioreactor cultivation with *P. pastoris*, i.e., batch, fed-batch, or continuous cultures.

3.2 Preparation of Inoculum

1. Thaw the stock culture of your *P. pastoris* strain at room temperature.
2. Vortex the thawed stock culture briefly and transfer it to 100 mL of fresh sterile BMGY in a 500 mL Erlenmeyer flask (see **Note 8**).
3. Incubate the first pre-culture at 30 °C for 12–24 h.
4. Check the grown first pre-culture under a microscope for potential contamination.
5. If the first pre-culture is reasonable (i.e., not contaminated and growing well), use it to inoculate fresh sterile BMGY at a 10% inoculation ratio. For example, mix 50 mL of the first pre-culture with 450 mL of fresh, sterile BMGY to obtain 500 mL of a second pre-culture (in a 2 L Erlenmeyer flask).
6. Incubate the second pre-culture at 30 °C for 12 h. After 12 h, this culture should be in the exponential growth phase with an optical density of >1 (*see* **Notes 9** and **10**).
7. Examine the second pre-culture under a microscope for possible contamination prior to further use.

3.3 Bioreactor Assembly and Preparation

1. Prepare all transfer vessels prior to bioreactor sterilization (*see* **Note 7**). Depending on the size of the reactor, such (empty) vessels are needed for:
   (a) Defined mineral medium used in the batch culture.
   (b) PTM1 solution.
   (c) Biotin solution.
   (d) Feed solution for fed-batch with biomass growth.
   (e) Methanol feed solution.
   (f) Base.
   (g) Acid.
   (h) Inoculum.
   (i) Antifoam solution.
2. If it is possible to sterilize several liters (in a closed glass or plastic container) in an autoclave, sterilize the batch medium outside the bioreactor. This will avoid volume and, thus, concentration changes due to evaporation or steam condensation.
3. Calibrate the pH probe with reference buffers at pH 4 and 7. Place the pH probe in the appropriate port of the bioreactor.
4. Assemble the pO_2 probe and all other probes and sensors in the bioreactor.
5. Fill the bioreactor to one-third of its volume with water or with medium. All probe heads must be covered with liquid.

6. Install the septa needed. If there are not enough ports, combine the two liquid-transfer assemblies with a Y-connector.
7. Install the pressure sensor, close the bioreactor completely, and perform a pressure test, i.e., at 0.4 bar overpressure, the pressure should not change more than 5% over 10 min after all vents are closed.
8. Start sterilization at 121 °C for at least 20 min. If not automated, keep the air inlet valve open at temperatures below 100 °C.
9. Refill and mix sterile-filtered PTM1 and biotin solutions in a sterile manner in a transfer vessel. Alternatively, if the batch medium is sterilized in an autoclave, transfer the sterile PTM1 and biotin solutions directly into the sterile medium.
10. When the temperature of the water in the bioreactor vessel decreases below 30 °C, replace the water with autoclaved medium and complement it with PTM1 and biotin, if not done before. If the bioreactor is on a scale, tare this scale when the bioreactor vessel is empty.
11. Fill a sterilized transfer vessel with 25% ammonia solution, i.e., prepared with sterilized water, but not sterilized. Weigh this vessel and attach it to the bioreactor via a septum.
12. Weigh the vessel with 8.5% phosphoric acid, previously autoclaved. Attach this vessel to the bioreactor via a septum.
13. Apply a stirrer speed of 800–1200 rpm and keep it constant.
14. Set temperature control to 30 °C; test temperature probe and heating/cooling system operation.
15. Start the pH control at 5.5 and check that the pH is adjusted with the base solution.
16. Start aeration with 2–3 liters of air per liter of batch medium and keep this setting constant, i.e., we recommend not applying a pO_2 cascade control.
17. Calibrate the 0% point of the pO_2 probe by sparging N_2 in the bioreactor. If no N_2 is available, calibrate this point during sterilization. Calibrate the 100% point at maximum stirrer speed, aeration rate, and oxygen content of aerated gas.
18. Apply 0.5 bar overpressure to maintain pO_2 above 20%. If this is not possible, supplement the air with pure oxygen.
19. When all cultivation conditions are set, the bioreactor is ready for inoculation.

3.4 Batch Cultivation and Strain Characterization

1. Attach the inoculum vessel to the bioreactor and transfer its entire content aseptically into the bioreactor, using a peristaltic pump or overpressure. Document the time of the inoculation.

Table 1
Kinetic and stoichiometric characteristics of *Pichia pastoris* Mut+ and MutS strains

	Methanol			Glycerol		
	μ_{max}	$Y_{x/s}$	m_s	μ_{max}	$Y_{x/s}$	m_s
	(h^{-1})	$(g\,g^{-1})$	$(g\,g^{-1}\,h^{-1})$	(h^{-1})	$(g\,g^{-1})$	$(g\,g^{-1}\,h^{-1})$
Mut^+	0.028–0.154	0.15–0.38	0.011–0.030	0.18–0.24	0.43–0.61	0.009–0.032
Mut^S	0.011–0.035	0.30	0.026	0.093–0.180	0.500–0.619	n.a.
approx.	0.014	0.42	0.014	0.24	0.61	0.009

Values were determined from cultivation of recombinant strains at 30 °C; $Y_{x/s}$ (biomass to substrate yield) and m_s (specific maintenance rate) are based on cell dry weight in grams; data adapted from [2]. approx.—these values should be used in calculation examples in Subheading 3.5

2. Allow the cultivation to run. Biomass growth is indicated by a decrease in dissolved oxygen (pO_2) concentration over time (Fig. 3).
3. Take samples regularly, for strain characterization every 2–3 h (*see* **Note 11**). For a basic strain characterization, analyze the biomass concentration both as optical density and gravimetrically (*see* **Note 12**) and the substrate concentration by HPLC. Check the culture regularly under the microscope for eventual contamination.
4. Document the time of the end of the batch cultivation.

   The end of the batch, i.e., complete consumption of the carbon source (glycerol, glucose, or methanol), is indicated by a sudden increase in the pO_2 signal. When using exponentially growing inoculum, the batch usually takes 12–16 h.
5. Collect a final sample at the end of cultivation. Measure the weight of the culture broth to account for volume loss due to evaporation and sampling, if necessary (*see* **Note 13**). Measure the amount of acid and base consumed. Terminate the control of the cultivation process in the control unit.
6. Pump out and inactivate the culture if not intended for further use or analysis. Fill the bioreactor vessel with water and sterilize the bioreactor.
7. A basic strain characterization includes determination of maximum specific biomass growth rate (μ_{max}) and maximum yield of biomass to substrate ($Y_{x/s,max}$) with all carbon sources, i.e., glucose, glycerol, and methanol. Experimental determination of the specific maintenance rate (m_s) is very demanding, requiring several fed-batch cultures. We recommend using m_s values from the scientific literature (Table 1).

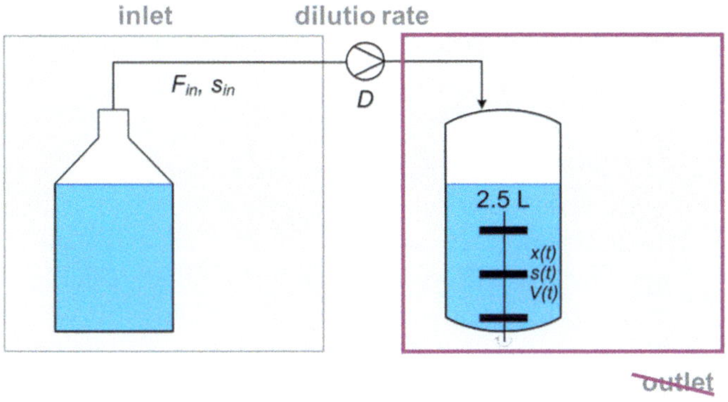

mass change = inlet – outlet + reaction

$$\frac{d}{dt}(c_i \cdot V) = F_{in}(t) \cdot c_{i,in}(t) - F_{out}(t) \cdot c_{i,out}(t) + r_i(t) \cdot V(t)$$

Fig. 8 Mass balance: The concept of fed-batch cultivation. During fed-batch cultivation, the substrate (inlet) is pumped into the bioreactor vessel, and no mass (outlet) is pumped out. This implies an increasing culture volume (V) over time; losses by sampling and evaporation are not considered. To take volume changes into account, masses of, for example, biomass (in g) or product (typically in mg) must be used in all calculations and data evaluations, not concentration values (in g/L or mg/L)

8. If you have not characterized your *Pichia* strain (*see* **Note 14**), use the values from Table 1 as a feasible starting point approximation.

3.5 Example of a Calculation for the Design of a Fed-Batch Culture

Most importantly, as the culture volume is changing over time, use the mass ($x \cdot V$ in g) and not the concentrations (in g/L) in all calculations and data evaluations below (Fig. 8).

The conditions of this exemplary fed-batch culture are as follows:

Batch cultures with glycerol and methanol are conducted to determine μ_{max} and $Y_{x/s,max}$ for each of these substrates. The values for m_s are taken from the literature (Table 1). For glycerol and methanol, these are μ_{max} of 0.24^{-1}/h, and $0.14-^{-1}$/h $Y_{x/s,max}$ of 0.61 g/g and 0.42 g/g, and m_s of 0.009 g/g/h and 0.014 g/g/h, respectively.

Batch (phase B) starts at an initial volume of 6 L, which corresponds to 1/3 of the maximum working volume of 18 L. This initial volume includes 0.6 L of inoculum, and 30 g/L of glycerol is used as the substrate. The batch phase ends when glycerol is depleted. At a starting volume of 6 L, glycerol feed is added (phase FB1) exponentially over 8 hours in order to maintain a constant μ of 0.1^{-1}/h. Subsequently, methanol feed (phase FB2) is added exponentially until the maximum working volume has been reached, enabling a constant μ of 0.08^{-1}/h and the expression of a recombinant protein.

1. Determine (*see* **Note 15**) the maximum specific growth rate of biomass (μ_{max}, $^{-1}$/h) and the maximum biomass/substrate yield ($Y_{x/s,max}$, g/g) for the *Pichia* strain under investigation, i.e., for this calculation, we use values given above.

If you determine the biomass concentrations (x) experimentally (e.g., gravimetrically or as optical density) at defined time points over the time course of a cultivation, calculate (*see* **Note 15**) the specific growth rates as follows: First, calculate mass from the biomass concentrations by multiplying by working volume (V). Subsequently, calculate logarithmic values of the biomass, $\ln(x \cdot V)$, and plot them on the y-axis against time points (t in hours) on the x-axis. Linear regression is applied, and the slope of the linear part corresponds to the specific growth rate (μ in $^{-1}$/h). For batch cultivation (with an excess of substrate during cultivation), the maximum specific growth rate (μ_{max} in $^{-1}$/h) is determined.

2. Calculate the biomass concentration x_{end} achieved when all substrate has been consumed.

Provided the batch culture started with 30 g/L of glycerol (s_0) and 1.7 g/L of biomass (x_0), according to Eqs. (32.1a) and (32.1b), batch growth ends at 20 g/L and, in addition to the 1.7 g/L inoculum, 18.3 g/L of biomass is produced from 30 g/L of glycerol.

$$x_{end} = s_0 \cdot Y_{x/s,\,max} + x_0 \ \text{(in g/L)} \tag{32.1a}$$

$$x_{end} = 30\ \text{g/L} \cdot 0.61\ \text{g/g} + 1.7\ \text{g/L} = 20\ \text{g/L} \tag{32.1b}$$

3. Calculate the initial feeding rate F_0 for the fed-batch with glycerol as a substrate (phase FB1).

Biomass concentration of 20 g/L (x_{end}, Eq. 32.1b) at the end of the batch (phase B) is then used to calculate the initial feeding rate (Eqs. 32.2a and 32.2b) using glycerol as a substrate (phase FB1). In Eqs. (32.2a) and (32.2b), μ is the pre-set value of specific growth rate (0.1^{-1}/h), V_0 is the initial volume (6 L), i.e., volume at the end of the batch, and w_{in} (g/g) is the mass fraction of substrate in feed solution, i.e., in this example 0.588 g/g.

$$F_0 = \left(\frac{\mu}{Y_{x/s,\,max}} + m_s\right) \cdot \frac{V_0 \cdot x_0}{w_{in}} \ \text{(in g/h)} \tag{32.2a}$$

$$F_0 = \left(\frac{0.1\ ^{-1}/\text{h}}{0.61\ \text{g/g}} + 0.009\ \text{g/g/h}\right) \cdot \frac{6\ L \cdot 20\ \text{g/L}}{0.588\ \text{g/g}} = 35.3\ \text{g/h} \tag{32.2b}$$

4. Plot $F(t)$ in Excel, i.e., using Eqs. (32.3a) and (32.3b) and the exponential function (i.e., =EXP()), calculate $F(t)$ values (between feed start and probable end time) in a column of

this spreadsheet with, for example, a time interval of 0.05 h. The feeding rate starts at the initial feed rate F_0 and changes with time $F(t)$, whereas biomass grows with a constant μ value of 0.1^{-1}/h.

$$F(t) = F_0 \cdot e^{\mu \cdot t} \text{ (in g/h)} \tag{32.3a}$$

$$F(t) = 35.3 \text{ g/h} \cdot e^{0.1 \text{ 1/h} \cdot t} \text{ (in g/h)} \tag{32.3b}$$

5. Calculate the feed integral (i.e., glycerol added to the bioreactor), the biomass volume, and the reactor volume as a function of time (Eq. 32.4a). Use another row in the Excel sheet. The density of the feed solution (ρ) with 0.588 g/g glycerol is 1.26 g/mL.

For example, after 8 h of feed addition (phase FB2), 432.6 g of glycerol was added (Eq. 32.4b), which corresponds to 735.7 g of glycerol feed (0.588 g/g) added. This corresponds to a reactor volume of 6.584 L (Eqs. 32.5a and 32.5b), resulting in a biomass concentration of 58.3 g L^{-1} (Eqs. 32.6a and 32.6b).

$$\int_{t0}^{t} F(t)dt = F_0 \int_{t0}^{t} e^{\mu \cdot t} dt = F_0 \cdot \frac{1}{\mu} \cdot \left(e^{t} - e^{t0}\right) \text{ (in g)} \tag{32.4a}$$

$$\int_{0}^{8} F(t)dt = \int_{0}^{8} 35.3 \cdot e^{0.1 \cdot t} dt = 35.3 \cdot \frac{1}{0.1} \cdot \left(e^{0.8} - e^{0}\right)$$
$$= 432.6 \text{ g} \tag{32.4b}$$

$$V(t) = V_0 + \frac{\int_{0}^{t} F(t)dt}{w_{in} \cdot \rho} \text{ (in L)} \tag{32.5a}$$

$$V(8) = 6 \text{ L} + \frac{432.6 \text{ g}}{0.588 \text{ g/g} \cdot 1.26 \text{ g/mL}} = 6 \text{ L} + 0.584 \text{ L}$$
$$= 6.584 \text{ L} \tag{32.5b}$$

$$x_{end} = \frac{\int_{0}^{8} F(t)dt \cdot Y_{x/s} + x_0 \cdot V_0}{V_{end}} \text{ (in g/L)} \tag{32.6a}$$

$$x_{end} = \frac{432.6 \text{ g} \cdot 0.61 \text{ g/g} + 20 \text{ g/L} \cdot 6 \text{ L}}{6.584 \text{ L}}$$
$$= \frac{263.9 \text{ g} + 120 \text{ g}}{6.584 \text{ L}} = 58.3 \text{ g/L} \tag{32.6b}$$

6. Repeat all calculations for the methanol feed, using $w_{in} = 0.976$ g/g (Eqs. 32.7a and 32.7b).

$$F_0 = \left(\frac{\mu}{Y_{x/s,\,max}} + m_s\right) \cdot \frac{V_0 \cdot x_0}{w_{in}} \text{ (in g/h)} \tag{32.7a}$$

$$F_0 = \left(\frac{0.08^{-1}/\text{h}}{0.42\ \text{g/g}} + 0.014\ \text{g/g/h}\right) \cdot \frac{6.584\ L \cdot 58.3\ \text{g/L}}{0.976\ \text{g/g}}$$

$$= 80.4\ \text{g/h} \qquad (32.7\text{b})$$

The density of the feed solution with 0.976 g/g methanol is 0.792 g/mL. Using the above equations (Eqs. 32.4a, 32.4b, 32.5a, 32.5b, 32.6a and 32.6b): After 15 h in total, 2332 g of methanol corresponding to 2389 g of methanol feed is added, the end volume of the cultivation after phase FB2 is 9.6 L, and the theoretical achievable biomass concentration after 15 h is 142 g/L. The biomass concentration might be lower depending on the proportion of substrate utilized for the product formation.

3.6 Performing Fed-Batch Cultivation: A Summary

1. Plan your start with one-third of the total bioreactor volume for batch medium, one-third for all feed media together, and one-third to compensate for aeration, stirring, and foam formation.

Typically, the duration of the cultivation process is limited by the size (volume) of the bioreactor.

2. Set the specific growth rate of the growth fed-batch (e.g., Fig. 2), which should not exceed 80% of the maximum specific growth rate of biomass (μ_{max}), in order not to accumulate substrate.
3. Plan your fed-batch cultivation in an Excel spreadsheet before you start, i.e., perform all calculations according to the example given in Subheading 3.5 of this protocol.
4. Assemble your media and bioreactor according to the instructions in Subheadings 3.1, 3.2, 3.3, and 3.4.
5. If sensors are available, use the pO_2 or pH signal to start the growth fed-batch (phase FB1). For instance, every growth fed-batch begins with the end of the initial batch cultivation (Fig. 1, *see* **Note 16**).
6. Verify the pre-set specific growth rates for fed-batch using (Eq. 32.8, see Fig. 2):

$$\mu(t) = \frac{\ln(x \cdot V) - \ln(x_0 \cdot V_0)}{t - t_0} \quad (\text{in } 1/\text{h}) \qquad (32.8)$$

7. Start the methanol feed directly after the growth fed-batch has ended and measure the methanol concentration. The culture may need some time to adapt its metabolism to methanol. If accumulation of methanol is observed, consider a 2–3 h adaptation period with a lower feed rate.
8. Take regular samples (for biomass, substrate, and product concentrations) during the whole calculation, i.e., a minimum of

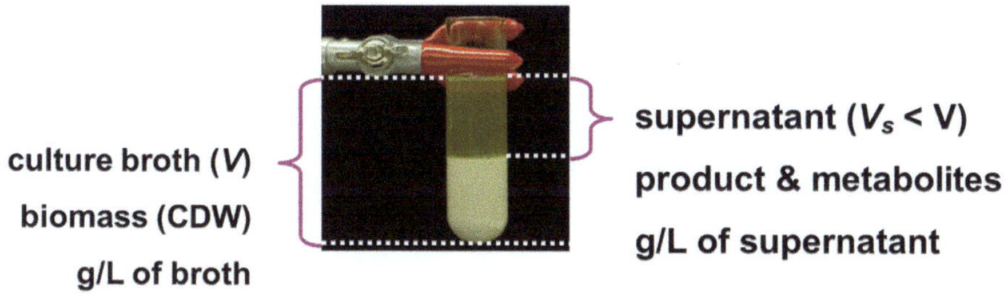

Fig. 9 Volume of supernatant in high-cell-density *P. pastoris* culture. The difference between the total volume of culture broth (*V*) and supernatant volume (Vs) is significant in high-cell-density *P. pastoris* cultures. The biomass concentration (e.g., cell dry weight, CDW) is experimentally determined in the whole culture broth volume, while products and metabolites are typically assessed in the centrifuged supernatant. Given the potentially massive difference between the total broth volume and the supernatant volume, the estimated concentration of product(s) and metabolites must be recalculated to the total broth volume

5–7 samples per phase, and, at the end, a last sample or harvest the whole culture volume for further use.

9. At the end of the cultivation, determine the weight of the culture broth and the amount of acid and base used. Use these values and your calculations to account for volume losses due to evaporation and sampling.
10. Sterilize the bioreactor filled with water again and disassemble it for thorough cleaning without using detergents.
11. Determine the amount of product produced in the entire bioreactor ($c_{p,end} \cdot V_s$ in g) from the concentration in the last sample ($c_{p,end}$) and the volume of the supernatant (V_s), i.e., the liquid in which the secreted product is dissolved (Fig. 9).

For this, calculate the approximate volume of supernatant (Fig. 7) according to equation (32.9) and the density of culture broth according to equation (32.10), as well as the wet biomass (g/L) c according to equation (32.11).

ρ_{broth} is the density of culture broth (kg/L), $M_{x,\text{ WCW}}$ is the mass of wet cells (g/L), and $\rho_{\text{supernatant}}$ is the density of supernatant that is approximately 1.03 kg/L [3, 4].

$$V_s(t) = \frac{\left(V(\mathrm{t})\cdot\rho_{\text{broth}}\right) - \frac{M_{x,\text{WCW}}}{1000}}{\rho_{\text{supernatant}}} \text{ (in L)} \tag{32.9}$$

x_{WCW} is the concentration of wet biomass cell weight (g/L).

$$\rho_{\text{broth}} \approx 0.000132\cdot x_{\text{WCW}} + 1.03 \text{ (in kg/L)} \tag{32.10}$$

$$\text{WCW} \approx 4.2\cdot\text{CDW (in g/L)} \tag{32.11}$$

12. Determine the product to biomass yield for fed-batch according to equation (Eq. 32.12), where V_s (L) is the volume of supernatant, i.e., the volume of cell-free liquid after centrifugation:

$$\Upsilon_{p/x} = \frac{\Delta(c_p \cdot V_S)}{\Delta(x \cdot V)} \text{ (in mg/g)} \tag{32.12}$$

13. Calculate the specific productivity in fed-batch according to equation (Eq. 32.13):

$$q_p(\mu) = \mu \cdot \Upsilon_{p/x} \text{(in mg/g/h)} \tag{32.13}$$

4 Notes

1. The field of bioprocessing, i.e., cultivation of cells in a bioreactor, has its own technical language, which can contribute to the understanding of characteristics in use. For example, the terms "fermenter" or "fermentation" are commonly used as synonyms for "bioreactor" or "cultivation," respectively, even though the bioprocess is not fermentative by definition.

   The concept of "volumetric" and "specific" rates is commonly applied, where "specific" stands for a value of the concentration change per unit time, which is divided by the respective biomass concentration, i.e., concentrations at the time of determination.

   Final product concentration c_p (at the bioreactor harvest) is named titer.

   Yield ($\Upsilon_{p/s}$, $\Upsilon_{x/s}$) refers to the ratio of substance produced (i.e., product or biomass) to substrate consumed. Yield is also commonly used as the ratio of product produced to biomass produced ($\Upsilon_{p/x}$).

2. Volumetric productivity r_x (in g/g/h) and yield $\Upsilon_{p/s}$ (in mg product to g substrate) are useful measures of bioprocess economy. The highest possible values are targeted.

   In addition, the yield $\Upsilon_{x/s}$ expresses how much biomass is built from one gram of substrate.

   $\Upsilon_{x/s}$ for glycerol (0.5–0.6 g/g) and glucose (0.7 g/g) are higher than $\Upsilon_{x/s}$ for methanol (0.2–0.4 g/g).

3. Use silicon hoses where possible, except when inserting any hose into peristaltic pumps; then use the more robust Norprene (e.g., Merck) or PharMed® BPT hoses. For assembly of transfer vessels, always cut a triangle out of the hose in the vessel so it cannot generate a vacuum at the vessel wall and prevent liquid flow. Prior to sterilization, always test the assembly of all transfer and feed vessels in order to prevent too short or defective hoses.

4. The suitability of external accessories for bioreactors, which are required for fed-batch cultivation, is best discussed with your bioreactor supplier, as well as having some such parts as a backup.

5. Additional sensors can be included. These are not crucial for the method described in this chapter but can provide additional information for process monitoring and advanced process control, for example, a turbidimetric probe (i.e., referring to cell growth), analyzers of exhaust gases (e.g., oxygen, carbon dioxide), automated analyzers of substrates (i.e., methanol), metabolites and the product, or foam and conductivity probes.
6. To avoid a malfunction due to foaming, add 0.1% PPG 2000 to the medium, especially if foaming is a recognized problem in a shake flask pre-culture. In addition, a foam sensor is recommended to automatically turn on/off the antifoam dosing during cultivation. As PPG is a surfactant, it may interfere with recombinant protein secretion and/or oxygen transfer. Therefore, monitor the antifoam dosage and try to keep it to the minimum necessary.
7. Do not autoclave ammonium solutions which are used for pH adjustment. This solution is sterile by definition and would evaporate at elevated temperatures. Do not autoclave methanol solutions due to explosion risk. If in doubt, a methanol solution can be sterilized by microfiltration (0.22 μm).
8. Since good aeration is important for the cultivation of *P. pastoris*, use baffled Erlenmeyer flasks and an air-permeable cover.
9. Measure the absorbance at 600 nm. To accurately measure an absorbance higher than 1.0, dilute culture samples with PBS or 0.9% NaCl before measurement.
10. Optionally, add a solution of antibiotics to maintain selective pressure on the recombinant strain. Note that with sterile *Pichia* cultures and robust recombinant constructs, these antibiotics are not needed, and their addition would not be economically feasible.
11. Take about 10 mL sample volume in total: 1 mL of sample for OD measurement and microscopic check of septicity of the culture, twice 1 to 2 mL for gravimetric determination of biomass, and twice 1 mL for HPLC analysis.
12. For accurate gravimetric determination of biomass concentration, especially when biomass concentration is low, label, dry (105 °C, 24 h), and weigh sample tubes before use.
13. The medium evaporates during cultivation, especially when the evaporation rate of methanol is high.
14. Growth characteristics for two carbon substrates can be determined within one cultivation process. For example, when the batch with glycerol is finished (pO_2 has risen rapidly), instead of ending the process and starting a completely new batch cultivation, you can start a subsequent second batch directly.

For example, you can pulse methanol by adding 5 mL (of methanol) per liter to the previous batch culture.

15. For more on the general basics of determining the specific growth rate from data points, see Potgieter et al. (2010) [3] and Looser et al. (2015) [2].

16. Prefill all hoses before starting any feeds, especially if the feeding is gravimetrically controlled. The same applies to other hoses, e.g., acid, base, and antifoam.

Acknowledgments

This chapter compiles most experimental experiences that our team has gathered and is organized in a conceptual and didactic manner over about 20 years. For accomplishing this, we would like to express our deepest gratitude to invaluable support by staff and students at the Zurich University of Applied Sciences (ZHAW, Wädenswil, Switzerland), and the Institute of Chemical Technology (VSCHT, Prague, Czech Republic) and during the "Hands-on Pichia" courses with TU Graz/ACIB (Austria) and partners, as well as to our numerous industrial collaborators.
Special mention is for the long-standing productive partnerships with Lonza AG and Infors HT. Their commitment made the development of our know-how on *Pichia* possible and was also supported by Innosuisse funds (formerly KTI) for projects 7403.3;3 LSPP-LS and 18163.1 PFLS-LS.

References

1. Meyer H, Schmidhalter DR (2014) Industrial scale suspension culture of living cells. Wiley-VCH Verlag GmbH & Co KGaA. https://doi.org/10.1002/9783527683321
2. Looser V, Brühlmann B, Bumbak F et al (2015) Cultivation strategies to enhance productivity of *Pichia pastoris*: a review. Biotechnol Adv 33: 1177–1193. https://doi.org/10.1016/j.biotechadv.2015.05.008
3. Potgieter TI, Kersey SD, Mallem MR et al (2010) Antibody expression kinetics in glycoengineered *Pichia pastoris*. Biotechnol Bioeng 106:918–927. https://doi.org/10.1002/bit.22756
4. Hyka P, Züllig T, Ruth C et al (2010) Combined use of fluorescent dyes and flow cytometry to quantify the physiological state of *Pichia pastoris* during the production of heterologous proteins in high-cell-density fed-batch cultures. Appl Environ Microbiol 76:4486–4496. https://doi.org/10.1128/aem.02475-09

Chapter 33

Pichia pastoris Continuous Cultivations for Strain Characterization and Continuous Biomanufacturing

Núria Bernat-Camps, Arnau Gasset, José Luis Montesinos-Seguí, Francisco Valero, and Xavier Garcia-Ortega

Abstract

Continuous cultivation provides physiological and product-related data of the recombinant-producing strain, which is crucial for bioprocess design and optimization. In particular, continuous cultivation enables the study of one condition at a time, being also a suitable mode for systems biology. Besides, regulatory and technological bases are being set at the industrial scale to achieve continuous biomanufacturing in the near future. Therefore, this chapter will describe the process of conducting carbon-limiting continuous cultivations, also known as chemostat cultivations, the most used continuous strategy for *Pichia pastoris* (*Komagataella phaffii*). In this protocol, design equations and setup descriptions can be found in order to carry out chemostat cultivations in a benchtop bioreactor. Although the process is described for *K. phaffii* using glucose, glycerol, or methanol as the sole carbon source, the protocol can be adapted to other substrates and microorganisms.

Key words *Pichia pastoris*, *Komagataella phaffii*, Chemostat cultivation, Recombinant protein production, Physiological characterization

1 Introduction

Recombinant proteins are used worldwide in several different applications, whose market is expected to increase to USD 13.04 billion by 2030 [1]. The use of yeasts as a cell factory is gaining great interest due to their advantages over higher eukaryotes and bacteria [2, 3]. Among the different yeast hosts, *Komagataella phaffii* (reclassified from *Pichia pastoris*) has been widely selected for recombinant protein production (RPP) since the 2000s, surpassing conventional and nonconventional yeasts [4].

Integrated continuous biomanufacturing is rising as a key approach for RPP, especially in the biopharma market. Continuous fermentation means constantly adding fresh medium to the bioreactor while the broth is removed, achieving steady-state conditions.

Xavier Garcia-Ortega et al. (eds.), *Pichia pastoris: Methods and Protocols*, Methods in Molecular Biology, vol. 2697, https://doi.org/10.1007/978-1-0716-4779-0_33,

This approach ensures a robust and reproducible process, besides allowing a constant product recovery. Furthermore, metabolites and product accumulation are avoided, a potential disadvantage that is present in batch or fed-batch operations when toxic or inhibitory levels are reached. However, although some concerns are being addressed to successfully implement continuous fermentations at an industrial scale, the fed-batch mode is still widely preferred [5–9]. Nevertheless, continuous cultivations can be a crucial source of knowledge for rational optimization of fed-batch processes [2].

Kinetic and physiological characterization of the producing strain is critical to rationally design and optimize bioprocesses. Accordingly, continuous cultivation is the most suitable operating mode to study these parameters since steady-state conditions can be reached and maintained over time. Chemostat cultivation is the most used continuous operation in *K. phaffii*, which is based on continuously adding fresh medium in growth-limiting conditions, allowing for setting a desired specific growth rate over time [2]. In this way, specific growth rate (μ), media composition, pH, and hypoxic conditions, among others, can be easily studied to enhance the RPP [2, 10]. In particular, the relationship between the specific production rate (q_p) and μ, named production kinetics, has been reported to be a key and protein-specific parameter for bioprocess optimization [2]. This relationship reflects several physiological interactions and steps, from gene expression to protein folding and secretion, making some bottlenecks potentially undetectable [11]. However, chemostat cultivations are also a powerful tool for systems biology research. The study of transcriptomic, proteomic, metabolomic, and fluxomic in continuous cultures allows an understanding of the cell behavior of the biological system. Consequently, the knowledge achieved at the molecular level can be used to enhance the RPP [2].

As alternatives to the chemostat cultures, other continuous cultivation strategies have been described for *K. phaffii* and other microbial cell factories in the literature, for example: retentostats [12, 13], perfusion [14, 15], or turbidostats [16]. However, this chapter will be focused only on the performance of chemostat cultivations.

In *K. phaffii*, several promoters with different regulations are being used for RPP. From the promoter of the *alcohol oxidase 1* gene (P_{AOX1}), which is strongly induced by methanol, to the *glyceraldehyde 3-phosphate dehydrogenase* gene promoter (P_{GAP}), considered constitutive and strong on glucose or glycerol [17]. Therefore, depending on the promoter used, the carbon source might differ.

Due to the importance of continuous cultivations in current bioprocess optimization and future biomanufacturing, a complete protocol to carry out chemostat cultivations is described in this chapter. Briefly, a chemostat process starts by preparing the inoculum and setting up the bioreactor. When the inoculum cultivation has grown and is still at the exponential phase, the bioreactor is inoculated, and thus the batch phase starts. After approximately 24 hours, a sharp spike in the dissolved oxygen (DO) is noticed, meaning that the carbon source has been depleted. Then, the chemostat phase can start, by feeding fresh medium and taking broth out in order to keep the bioreactor volume constant, obtaining the following mass balance for the biomass:

No biomass addition to the system (sterile feed) $\rightarrow$ input = 0.

Steady-state conditions $\rightarrow$ accumulation = 0

$$\text{Input} + \text{generation} = \text{accumulation} + \text{output},$$

$$r_X V = X F_{\text{out}}; r_X = \mu X,$$

$$\mu X V = X F_{\text{out}},$$

$$\mu = \frac{F_{\text{out}}}{V}; D = \frac{F_{\text{out}}}{V},$$

$$\mu = D. \tag{1}$$

where r_X is the volumetric growth rate (g_X L $^{-1}$ h^{-1}), V the bioreactor working volume (L), X the biomass concentration (g_X L^{-1}), F_{out} the outlet volumetric flow rate (L h^{-1}), μ the specific growth rate (h^{-1}) and D the dilution rate (h^{-1}). To achieve chemostat continuous operation considering similar densities of the solutions involved, the total inlet and outlet flow rates should be the same, so the V is maintained constant over time. Therefore, a desired μ is set by adjusting the D, which can only be varied by the volumetric flow rate. Along the chemostat cultivation, small outlet volumetric flow rate variations can take place due to base and antifoam addition, gas mixing, sampling, etc. Base addition for pH control is the main contributor to the variation of total outlet flow with respect to the inlet flow, since *K. phaffii* continuously produces acids while it grows. Therefore, the total volumetric flow rate can be estimated as:

$$F_{\text{feed}} + F_{\text{base}} = F_{\text{out}}, \tag{2}$$

where F_{feed} is the feeding flow rate, F_{base} is the mean base flow rate, and F_{out} is the total outlet flow rate, all of them are volumetric in L h^{-1}. Equation (2) should be adapted when solution densities are considerably different between them, giving the following one when considering the volume constant:

$$F_{\text{feed}} \frac{\rho_{\text{feed}}}{\rho_{\text{out}}} + F_{\text{base}} \frac{\rho_{\text{base}}}{\rho_{\text{out}}} = F_{\text{out}}, \tag{3}$$

where ρ_{feed}, ρ_{base}, and ρ_{out} are the densities of the feed, base, and outlet solutions, respectively, in Kg L^{-1}. To robustly set the set-point D, the overall flow rate of the system is primarily governed by the feeding rate, which is added by a high-precision peristaltic pump. Regarding the outlet flow rate, the outlet pipe must be set at the height of the working volume during the chemostat cultivation, with the pump at a higher nominal flow rate. Hence, the liquid added by the feed and base pumps will come out as soon as the broth level reaches the outlet pipe.

When designing an experiment, bear in mind that the D must be lower than the μ_{max} of the producing strain, and rigorously lower than the critical dilution rate (Dc, where $Dc = \mu_{max}$ Se/(Ks + Se) for a *Monod* growth kinetics), otherwise the bioreactor will wash out. Accordingly, chemostat cultivations with glucose or glycerol usually are performed in the μ range of 0.01 to 0.15 h^{-1}, while those with methanol go from 0.01 to 0.10 h^{-1} [18–20]. However, the μ_{max} of the producing strain should be always determined because it might be protein-dependent, causing even a significant decrease compared to the WT μ_{max} [21].

Environmental stressful conditions have been reported to be indirectly beneficial for enhancing RPP, due to their impact on global metabolism regulation [22]. As shown in some works, applying a certain degree of oxygen limitation can boost RPP, especially when working with glucose as a carbon source [23–25]. To apply oxygen-limiting conditions in chemostat cultivations, the oxygen transfer from the gas flow to the culture broth should be reduced, whose speed is defined by the oxygen transfer rate (*OTR*):

$$\mathrm{OTR} = k_L a\left(\mathrm{O_2}^{\mathrm{Sat}} - \mathrm{O_2}\right), \tag{4}$$

where *OTR* is the oxygen transfer rate ($\mathrm{mols_{O2}\,L^{-1}\,h^{-1}}$), $k_L a$ is the volumetric mass transfer coefficient (h^{-1} or s^{-1}), O_2^{Sat} is the saturation oxygen concentration in the culture broth ($\mathrm{mols_{O2}\,L^{-1}}$) and O_2 the real oxygen concentration ($\mathrm{mols_{O2}\,L^{-1}}$). Mainly, two parameters can be manipulated to modify the *OTR*: agitation rate (affecting $k_L a$) and oxygen fraction in the inlet gas (affecting O_2^{Sat}). As mentioned previously, the agitation rate is intended to be constant along the chemostat cultivation since it affects the broth volume. Then, oxygen limitation can be achieved by mixing oxygen or nitrogen with air in the inlet gas.

This protocol aims to describe the procedures to carry out chemostat cultivations with *K. phaffii* on glucose, glycerol, or methanol as sole carbon and energy sources, using benchtop bioreactors. In these cultivations, different μ and basic operating conditions can be studied for systems biology or strain characterization. Besides, specific instructions for hypoxic conditions are given.

2 Materials

2.1 Inoculum

1. 1 L baffled shake flasks covered with aluminum foil, fabric, or filter cap and autoclaved.
2. Laminar flow cabin.
3. Complex yeast extract peptone glycerol (YPG) media: 1% yeast extract, 2% peptone, and 2% glycerol. To prepare 1 L, mix 10 g of yeast extract, 20 g of peptone, and 20 g of glycerol in 900 mL of distilled water. Once dissolved, fill up to 1 L with distilled water and autoclave at 121 °C for 30 min (*see* **Note 1** for alternative media).
4. Orbital shaker incubator (preferably with a working range from 25 to 30 °C).
5. Spectrophotometer suitable for 600 nm wavelength measurements.
6. 250–500 mL centrifuge bottle (conical) filled with water, to prevent plastic deformation, and sterilized by autoclaving.
7. Centrifuge equipped with a rotor adapter for 250–500 mL conical bottles.

2.2 Main Fermentation

1. Glass or stainless-steel bioreactor coupled to the control unit with a software control system (*see* **Note 2**). The bioreactor must have at least three inlet ports (inoculum, pH control solution, and feeding) and two outlet ports with long pipes (sampling and chemostat outlet). The system must have at least two peristaltic pumps (pH base-based control and chemostat outlet) and one variable and high-precision peristaltic pump (feeding solution) (*see* **Note 3**).
2. Off-gas condenser and a temperature vessel control system.
3. Chiller is highly recommended to cool down the water or refrigerant stream to both the condenser and the vessel, especially at high cell density cultures.
4. Temperature, pH, and dissolved oxygen probes (*see* **Note 4**).
5. Off-gas analyzer (recommended) (*see* **Note 5**).
6. Pressure sensor for bioreactor headspace pressure control (*see* **Note 6**).
7. A continuous supply of air (mixtures with pure oxygen or nitrogen may also be necessary) (*see* **Note 7**).
8. Small bottles with 2-port caps.
9. Sampling system.
10. Silicone, peristaltic pump and base-tolerant tubing, 0.2 μm hydrophobic air filters, connectors, and clamps compatible with autoclave sterilization.

11. *Pichia Batch Medium (PBM):* For 1 L preparation, dissolve in 900 mL distilled water 1.80 g citric acid, 40.0 g glycerol, 12.60 g $(NH_4)_2HPO_4$, 0.50 g $MgSO_4{\cdot}7\ H_2O$, 0.90 g KCl, 0.02 g $CaCl_2{\cdot}2\ H_2O$, and 0.05 mL antifoam Glanapon 2000kz (*see* **Note 8**) [26]. Adjust the pH to the set-point using HCl 25% (v/v) solution (*see* **Note 9**). Fill up to 1 L with distilled water, homogenize, and measure the volume that will be used for inoculating, which will depend on the inoculation system (*see* **Note 10**). Since this protocol describes inoculation through a bottle, measure 93.4 mL of medium using a test tube (*see* **Note 11**). Transfer the 93.4 mL to a small bottle and introduce the rest of the medium into the bioreactor. Sterilize both the bioreactor and the small bottle by autoclaving. Once the temperature has decreased to the set-point cultivation temperature, add 4.6 mL PTM1 and 2 mL biotin 0.2 g L^{-1} under sterile conditions through the bioreactor septum (*see* **Note 12** for alternative media).
12. *PTM1 trace salts solution*: For 1 L, mix 5.0 mL H_2SO_4, 65.0 g $FeSO_4{\cdot}7\ H_2O$, 20.0 g $ZnCL_2$, 6.0 g $CuSO_4{\cdot}5\ H_2O$, 3.36 g $MnSO_4{\cdot}H_2O$, 0.82 g $CoCL_2{\cdot}6\ H_2O$, 0.2 g $Na_2MoO_4{\cdot}2\ H_2O$, 0.08 g NaI, 0.02 g H_3BO_3 with distilled water [26]. The solution must be filter-sterilized (0.22 μm nominal pore size) and stored at 4 °C covered with aluminum foil.
13. Biotin 0.2 g L^{-1}: Completely dissolve 200 mg biotin in 1 L of distilled water. Sterilize by microfiltration (0.22 μm nominal pore size) and store at 4 °C.
14. Base control: NH_4OH solution at 15% (v/v). No sterilization is needed. Prepare it in a gas extraction hood.

2.3 Feeding System

1. Glass or plastic bottle (5 to 20 L) with 3-port cap. The bottle must have precise volume marks.
2. High-performance precision tubing and silicone tubing, 0.2 μm hydrophobic air filters, connectors, and clamps compatible with autoclave sterilization.
3. Hydrophilic polyethersulfone membrane filter capsule with 0.22 μm nominal pore size (Sartorius Sartopore 25445307H8--00—A or equivalent).
4. Collecting autoclavable bottles for the chemostat outlet.
5. *Pichia Continuous Feeding Medium (PCFM)*: Carbon source can vary depending on the scope of the cultivation between glycerol, glucose, or methanol (or even mixtures of them), which will give different biomass concentrations (*see* **Note 13**). The recipe for 1 L is: 0.9 g citric acid, 50.0 g of carbon source, 4.35 g $(NH_4)_2HPO_4$, 0.65 g $MgSO_4{\cdot}7\ H_2O$, 1.70 g KCl, 0.01 g $CaCl_2{\cdot}2\ H_2O$, 1.6 mL PTM1, 1 mL biotin 0.2 g L^{-1}, and 0.2 mL antifoam Glanapon 2000kz in distilled

water [26]. The pH should be lowered to around 5 to facilitate salt dissolution. Sterilization by microfiltration as described in methods (*see* **Note 14**).

3 Methods

3.1 Flowchart for the Chemostat cultivation and Bioreactor Setup

Figure 1a represents the flow chart for carrying out chemostat cultivations, from inoculum preparation and bioreactor setup to chemostat initiation. However, some actions must be performed before starting the process, like setting up the bioreactor in the correct configuration to allow for an accurate performance. In particular, the outlet pipe must be calibrated at the correct height before starting the bioreactor preparation for autoclaving.

1. Introduce into the bioreactor all the probes that will be used during the fermentation, as well as pipes or other pieces that affect the working volume.
2. Accurately measure 1 L of water, or the selected working volume during chemostat cultivation, and introduce it into the bioreactor (*see* **Note 2**).
3. Set all working conditions for agitation speed, temperature, gas flow, and overpressure (*see* **Note 6**) that will be maintained constant during the chemostat phase.

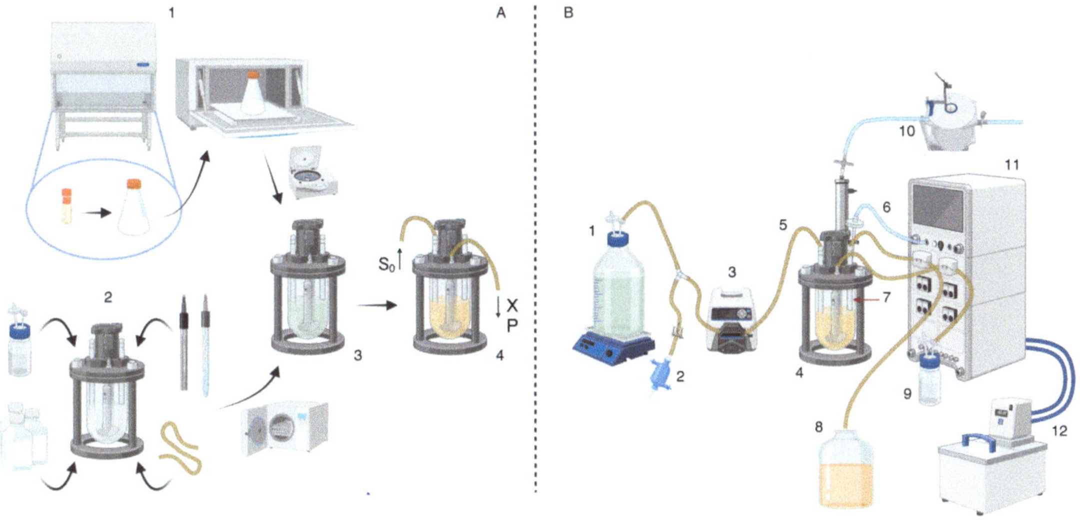

Fig. 1 (**a**) Flow-chart of a chemostat cultivation: (1) inoculum preparation, (2) bioreactor set-up, (3) bioreactor inoculation and batch initiation, and (4) chemostat initiation. (**b**) Scheme of a chemostat cultivation setup; (1) feeding bottle, (2) membrane filter capsule for medium sterilization, (3) high-precision peristaltic pump, (4) bioreactor, (5) feeding tube, (6) gas-flow tube, (7) outlet pipe calibrated to the cultivation heigh, (8) collecting bottle of the outlet broth, (9) base-containing bottle for pH control, (10) gas analyzer, (11) control unit, and (12) chiller

4. Connect a silicone tube to the outlet pipe and place it in the peristaltic outlet pump (change to the correct tubing with connectors if it is necessary), securing all connections with cable ties.
5. Set the outlet pipe at a high position, but close enough to the water level, and turn on the outlet pump. Slowly decrease the outlet pipe until the first drop of water goes through the silicone tube.
6. Tightly screw the outlet pipe in that position, concluding its calibration.
7. Then, the bioreactor can be setup as normally for a standard fermentation.
8. Calibrate the control's unit peristaltic pumps as required for the software system.
9. Prepare all connections and tubing for:
   (a) Sparger port: connect a silicone tube from the port to a 0.2 μm pore hydrophobic air filter.
   (b) Sampling port (outlet): connect the silicone tubing from the sampling pipe, which must reach the bottom of the bioreactor, to the sampling recipient. The system used must allow taking samples without risk of contamination.
   (c) Base control port (inlet): Connect a base-tolerant tube from the bioreactor port to an empty bottle closed with a 2-port cap. One cap port is used to transfer the liquid, while the other allows the pressure equilibration through a 0.2 μm pore hydrophobic air filter. The tube must be long enough and suitable for use in the control's unit base peristaltic pump.
   (d) Feeding port (inlet): connect a silicone tube to the bioreactor port, add a connector to the other end, and close it with two clamps, creating a dead end. After sterilization, it will be used to connect the feed system.
   (e) Inoculating port (inlet): connect the bioreactor port to an empty bottle closed with a 2-port cap with tubing (silicone and peristaltic pump), same setup as in **step 9(c)** (*see* **Note 15**).
   (f) Chemostat outlet port (outlet): connect a long silicone tube from the bioreactor port (already calibrated to the correct height) to a peristaltic pump tubing with a dead-end setup.
10. Calibrate the pH probe by performing a 2-point calibration and introduce it into the bioreactor, together with the DO probe.

11. Clamp all tubes and cover the air filters with aluminum foil to prevent them from getting wet.
12. Prepare the batch medium and preserve 93.4 mL in a small bottle (it will be used later for inoculating the bioreactor), transferring the rest volume into the bioreactor.
13. Autoclave the bioreactor, the small bottle with medium, and the 250–500 mL centrifuge bottle filled with water for 30 min at 121 °C, leaving at least one port open in the bioreactor.
14. Once the sterilization process has finished, open the autoclave and close the bioreactor open port as fast as possible. As a sterility control check, take out the bioreactor from the autoclave and remove the clamp and the aluminum foil from the sparger air filter. Immediately afterward, air should come in to equilibrate pressures, indicated by bubbles in the medium. If no bubbling occurs, then the bioreactor might have another opening where air is getting in. In this case, find the bridge, solve it, and autoclave it again with the same procedure.
15. Turn on the chiller and connect the water or refrigerant stream to both the condenser and bioreactor temperature control system.
16. Introduce the temperature sensor and connect all probes.
17. Start controlling the temperature at the set-point and turn on the agitation speed and the gas flow. Connect the off-gas to the pressure sensor and set the selected overpressure (*see* **Note 6**).
18. Add biotin and PTM1 trace solutions through the septum when the bioreactor temperature has decreased to the temperature set-point (*see* **Note 16**).
19. Prepare a glass bottle with NH_4OH 15% (v/v) in the extraction hood. Go to the bioreactor and quickly change caps with the sterile empty bottle connected to the bioreactor, which was prepared for pH control.
20. Activate pH control for the selected set-point, making sure all related devices (pumps) are in the automatic position (if applicable).
21. Once all set-point values have been reached, calibrate the DO probe (*see* **Note 17**).
22. Connect the off-gas, after the pressure sensor, to the gas analyzer (if applicable). Always put a 0.2 μm pore hydrophobic air filter before the gas analyzer.
23. Wait until inoculation must be performed.

3.2 Inoculum Preparation

The inoculum is a small-scale cultivation performed in shake flasks, whose aim is to obtain enough biomass in the exponential phase to inoculate the bioreactor. It is important to characterize the growth kinetics of the target strain, mainly to know the μ_{max} and the OD_{600} range where the cells are still growing exponentially. Having this information, it is possible to calculate when and at what OD_{600} the inoculum should start, using the biomass mass balance of a batch (*see* **Note 18**):

$$\cancel{\text{Input}} + \text{generation} = \cancel{\text{accumulation}} + \text{output},$$

$$r_X V = \frac{\mathrm{d}(XV)}{\mathrm{d}t},$$

$$r_X V = V\frac{\mathrm{d}X}{\mathrm{d}t} + X\frac{\mathrm{d}V}{\mathrm{d}t}; \text{where } V \sim \text{const.and } r_X = \mu X,$$

$$\mu X = \frac{\mathrm{d}X}{\mathrm{d}t}; \mu t = \ln\frac{X}{X_0},$$

$$X = X_0 e^{\mu t}. \tag{5}$$

1. Before starting the inoculum preparation, the μ_{max} of the strain on YPG at the desired operating conditions should be estimated. Otherwise, a bibliographical value can be used (*see* **Note 19**). In the same experiment, determine the OD_{600} range where the cultivation grows exponentially.
2. In the laminar flow cabin, add 100 mL of sterile YPG to a 1 L baffled shake-flask, under sterile conditions. Add antibiotics or supplements if needed due to resistance markers or auxotrophies (*see* **Note 20**).
3. Take the working cell bank (WCB) from the −80 °C ultrafreezer and wait until it is thawed. If agar plates are used instead of WCB as a starting point, make an additional seeding cultivation at a small scale (5–50 mL).
4. In the laminar flow cabin, inoculate the 1 L shake-flask with 0.5–2% (v/v) of WCB and homogenize by gently shaking. The volume to inoculate will depend on the operator's preferences, which can set the cultivation time and calculate the initial OD_{600} by taking Eq. (5).
5. Incubate the shake-flask in an orbital shaker incubator for the selected time and at the desired temperature, ideally the same one at the μ_{max} was determined (*see* **Note 19**).

3.3 Batch Phase

Figure 1a represents the flow chart for carrying out chemostat cultivations, from inoculum preparation and bioreactor setup to chemostat initiation. However, some actions must be performed before starting the process, like setting up the bioreactor in the correct configuration to allow for an accurate performance. In particular, the outlet pipe must be calibrated at the correct height before starting the bioreactor preparation for autoclaving.

1. Introduce into the bioreactor all the probes that will be used during the fermentation, as well as pipes or other pieces that affect the working volume.
2. Accurately measure 1 L of water, or the selected working volume during chemostat cultivation, and introduce it into the bioreactor (*see* **Note 2**).
3. Set all working conditions for agitation speed, temperature, gas flow, and overpressure (*see* **Note 6**) that will be maintained constant during the chemostat phase.
4. Connect a silicone tube to the outlet pipe and place it in the peristaltic outlet pump (change to the correct tubing with connectors if it is necessary), securing all connections with cable ties.
5. Set the outlet pipe at a high position, but close enough to the water level, and turn on the outlet pump. Slowly decrease the outlet pipe until the first drop of water goes through the silicone tube.
6. Tightly screw the outlet pipe in that position, concluding its calibration.
7. Then, the bioreactor can be setup as normally for a standard fermentation.
8. Calibrate the control's unit peristaltic pumps as required for the software system.
9. Prepare all connections and tubing for:
   (a) Sparger port: connect a silicone tube from the port to a 0.2 μm pore hydrophobic air filter.
   (b) Sampling port (outlet): connect the silicone tubing from the sampling pipe, which must reach the bottom of the bioreactor, to the sampling recipient. The system used must allow taking samples without risk of contamination.
   (c) Base control port (inlet): Connect a base-tolerant tube from the bioreactor port to an empty bottle closed with a 2-port cap. One cap port is used to transfer the liquid while the other allows the pressure equilibration through a 0.2 μm pore hydrophobic air filter. The tube must be long enough and suitable for use in the control's unit base peristaltic pump.
   (d) Feeding port (inlet): connect a silicone tube to the bioreactor port, add a connector to the other end, and close it with two clamps, creating a dead end. After sterilization, it will be used to connect the feed system.

(e) Inoculating port (inlet): connect the bioreactor port to an empty bottle closed with a 2-port cap with tubing (silicone and peristaltic pump), same setup as in **step 9(c)** (*see* **Note 15**).

(f) Chemostat outlet port (outlet): connect a long silicone tube from the bioreactor port (already calibrated to the correct height) to a peristaltic pump tubing with a dead-end setup.

10. Calibrate the pH probe by performing a 2-point calibration and introduce it into the bioreactor, together with the DO probe.
11. Clamp all tubes and cover the air filters with aluminum foil to prevent them from getting wet.
12. Prepare the batch medium and preserve 93.4 mL in a small bottle (it will be used later for inoculating the bioreactor), transferring the rest volume into the bioreactor.
13. Autoclave the bioreactor, the small bottle with medium, and the 250–500 mL centrifuge bottle filled with water for 30 min at 121 °C, leaving at least one port open in the bioreactor.
14. Once the sterilization process has finished, open the autoclave and close the bioreactor open port as fast as possible. As a sterility control check, take out the bioreactor from the autoclave and remove the clamp and the aluminum foil from the sparger air filter. Immediately afterwards, air should come in to equilibrate pressures, indicated by bubbles in the medium. If no bubbling occurs, then the bioreactor might have another opening where air is getting in. In this case, find the bridge, solve it, and autoclave it again with the same procedure.
15. Turn on the chiller and connect the water or refrigerant stream to both the condenser and bioreactor temperature control system.
16. Introduce the temperature sensor and connect all probes.
17. Start controlling the temperature at the set-point and turn on the agitation speed and the gas flow. Connect the off-gas to the pressure sensor and set the selected overpressure (*see* **Note 6**).
18. Add biotin and PTM1 trace solutions through the septum when the bioreactor temperature has decreased to the temperature set-point (*see* **Note 16**).
19. Prepare a glass bottle with NH_4OH 15% (v/v) in the extraction hood. Go to the bioreactor and quickly change caps with the sterile empty bottle connected to the bioreactor, which was prepared for pH control.

20. Activate pH control for the selected set-point, making sure all related devices (pumps) are in the automatic position (if applicable).
21. Once all set-point values have been reached, calibrate the DO probe (*see* **Note 17**).
22. Connect the off-gas, after the pressure sensor, to the gas analyzer (if applicable). Always put a 0.2 μm pore hydrophobic air filter before the gas analyzer.
23. Wait until inoculation must be performed.

3.4 Chemostat Phase: Study of μ and Key Working Conditions

In Fig. 1b, a scheme of the chemostat setup is represented, including the feeding system and the outlet configuration. Since the appealing aspect of chemostat cultivations is the capability to study different conditions in just one setup, the feed medium is usually prepared in large volumes. Therefore, sterilization by autoclaving is not feasible; in contrast, the microfiltration process is described in the following section. To save time, all these processes can be conducted while the batch phase is running.

1. Calibrate the high-precision peristaltic pump with the selected tubing, according to the software used.
2. Prepare the feeding system as shown in Fig. 1b. Take the 5–20 L bottle with a 3-port cap with the following configuration:
   (a) Port 1—sterile filter: Use a short tube to connect the port with a 0.2 μm pore hydrophobic air filter to equilibrate pressures.
   (b) Port 2—feed filtration: Connect the port with a membrane filter capsule of 0.2 μm pore size via the correct tubing size. Make sure the filter is placed in the right orientation. Add a small tube to the other side of the membrane filter capsule.
   (c) Port 3—feed addition: Connect tubes to both sides of the cap, one reaching the bottom of the bottle and the other one long enough to reach the high-precision peristaltic pump. Depending on the flow rate set-point, a specific tubing might be used (*see* **Note 26**). At the end of the tubing after the peristaltic pump, prepare a dead-end with a connector to, eventually connect the system to the bioreactor after the sterilization.
3. Add around 2 L of distilled water to the 5–20 L bottle and dilute the antifoam (volume specified in Subheading 2.3). Add a magnet stirrer to the bottle (*see* **Note 27**).
4. Close all the tubes with clamps and cover the filters with aluminum foil.

5. Autoclave the whole system for 30 min at 121 °C.
6. In the meantime, dissolve all the feed medium components described in Subheading 2.3 in 2–5 L of distilled water, including the carbon source and the trace salts, and biotin solutions (*see* **Note 28**).
7. When the feeding system has been sterilized and cooled down, start filtering the medium solution through the microfilter capsule:
   (a) First, extend the nonsterile side of the filter with more tubing, to be able to use a peristaltic pump.
   (b) Make sure all clamps from the 3-port-cap have been removed except for the feed addition tube.
   (c) Place the nonsterile tube from the filter capsule inside the medium solution and start pumping the solution. Immediately afterward, open the air exhaust valves of the filter capsule to fill it, displacing the air inside.
   (d) Once the capsule is filled and all air removed, close the air exhaust valves and increase the pumping rate.
   (e) When the medium has been totally filtered, add distilled water to the container as many times as needed to reach the feed volume set-point. Use the volume marks on the bottle (*see* **Note 29**).
   (f) Afterward, close the tube with clamps and remove the filter capsule. Homogenize the feed solution by gently shaking it and leaving it stirring with the magnet inside.
   (g) To dry the filter capsule, first open both valves and drain all water remaining in the capsule chamber. Then, remove the water inside the filter chamber by flowing air through the inlet tube with the valves opened. Close each valve alternatively to increase the pressure inside the capsule chamber and push the water inside through the outlet tube (*see* **Note 30**).
   (h) Repeat the process until no water is going out of the outlet tube of the filter. Then, leave the filter capsule valves open with an air stream for approximately half an hour, until the filter is completely dried (*see* **Note 31**).
8. Connect the feeding system to the bioreactor designated tube with all clamps still closed, removing the dead end and using the sterile connectors.
9. Insert the feeding tubing in the high-precision peristaltic pump and remove the clamps of the circuit to allow an eventual feeding flow to the bioreactor (*see* **Note 32**).
10. Insert the outlet tubing into the peristaltic pump and connect the end to the collecting bottle. This bottle does not have to be sterile.

11. Wait until the batch phase finishes.

    When the batch phase is finished, as indicated by the DO peak, the chemostat cultivation can start. Waiting about 30–90 min before starting the chemostat mode is recommended to make sure all carbon sources and potential by-products are totally consumed. No further step is required for glucose or glycerol feed (*see* **Note 33** for methanol feed).

12. Change the operating parameters (if applicable) to chemostat set points. DO control must be turned off, and agitation speed and gas flow kept constant (at the outlet pipe calibrating conditions).

13. Turn on the outlet peristaltic pump.

14. Set the feeding flow rate according to the calibration of the high-precision peristaltic pump and start the addition (*see* **Note 34**). The set-point should be about 2% lower than the goal *D* (*see* **Note 35**).

15. Leave the system to equilibrate for some minutes and make sure DO is not lower than 20% (air saturation). Make air mixtures with pure O_2 in case DO cannot be maintained, but always maintain the same total gas flow (*see* **Note 36**).

16. Measure the outlet flow rate with a tared bottle, test tube, or flow meter. In case the outlet flow rate is higher or lower than the set-point *D*, adjust the feeding flow rate and repeat the process. The chemostat cultivation starts when the goal outlet flow rate is achieved.

17. Wait until the third residence time to consider that the cultivation has probably reached a steady state (*see* **Note 25**). The residence time is defined as the time needed to change the whole volume of the bioreactor:

$$\tau = \frac{V}{F_{out}} = \frac{1}{D}, \tag{6}$$

    where τ is the residence time (h), F_{out} is the total outlet volumetric flow rate (L h^{-1}), and *D is* the dilution rate (h^{-1}).

18. Verify that steady-state conditions have been reached by taking samples at the 3rd, 4th' and 5th residence time (*see* **Note 37**). To do so, measure and compare key parameters (substrate and product concentration, biomass concentration, DO, and off-gas CO_2 and O_2 concentration) to evaluate cultivation stability. Store samples at −20 °C for further analyses, if applicable.

19. If the cultivation is stable, the study under these conditions can be concluded.

20. Start the next cultivation condition, varying the feeding flow rate again or some other parameter. Repeat steps from the 14th to the 18th until cultivation stabilization (*see* **Note 38**).

21. To stop the cultivation, turn off all the pumps (feeding, base, and outlet), remove the overpressure, and disconnect the controllers.
22. Carefully harvest the broth and measure the volume to know the actual working volume (*see* **Note 39**).
23. Remove all probes and prepare the bioreactor for killing off. Fill the vessel with water and autoclave at 121 °C for 30 min. Once finished, wait until the bioreactor cools down to clean it.
24. Take a sample of the feed bottle to measure the carbon source concentration (if applicable) and clean the system.

3.5 Chemostat Phase: Study of Hypoxic Conditions

To study oxygen limitation (hypoxia) in chemostat cultivations, all working conditions are fixed despite the *OTR*, which is reduced by mixing pure nitrogen in the inlet gas flow. Be aware that oxygen-limiting conditions can reduce the D_C; therefore, select an appropriate D to avoid washing out the cultivation.

1. Make sure the bioreactor can perform gas mixing of air and nitrogen. Otherwise, replace the biorector's oxygen supply tube with a nitrogen supply tube.
2. Follow the protocol until Subheading 3.4, **step 16**.
3. Once the outlet flow rate is set to goal D, reduce the *OTR* by gradually increasing the nitrogen flow rate and decreasing the air flow rate, accordingly, maintaining the total gas flow rate constant.
4. Measure substrate concentration in the outlet broth to make sure the bioreactor is not washing out (substrate concentration should be close to zero).
5. Measure the O_2 and CO_2 concentrations in the inlet gas since minimal deviations of the nominal gas flow rates can have a great impact on oxygen fraction.
6. When the steady state is reached (*see* Subheading 3.4), start a new condition modifying the gas flow rate to apply a different degree of oxygen limitation.
7. To compare the different oxygen-limiting conditions between them, properly measure the respiratory quotient (*RQ*, *see* Subheading 3.6), since DO will be zero or close to zero in all hypoxic conditions. For a correct *RQ* calculation, humidity in both the inlet and the outlet gas should be measured [24, 25].

3.6 Variable Calculations

The main bioprocess variables are summarized in the following table, including their equations and the experimental parameter or calculated variable needed. More calculations and demonstrations can be found elsewhere [23].

Variable		Units	Experimental parameter or variable	Equation
Dilution rate	D	h^{-1}	Bioreactor working volume (V, L) Outlet volumetric flow rate (F_{out}, L h^{-1})	$D = \frac{F_{out}}{V}$
Specific growth rate	μ	h^{-1}	Bioreactor working volume (V, L) Outlet flow rate (F_{out}, L h^{-1})	$\mu = D = \frac{F_{out}}{V}$
Specific substrate uptake rate	q_S	$g_S\ g_x^{-1}\ h^{-1}$	Inlet substrate concentration (S_{in}, g_S L^{-1}) Inlet feeding volumetric flow rate (F_{in}, L h^{-1}) Outlet substrate concentration (S_{out}, g_S L^{-1}) Outlet volumetric flow rate (F_{out}, L h^{-1}) Biomass concentration (X, g_X L^{-1}) Bioreactor working volume (V, L)	$q_s = \frac{\frac{S_{in}}{X}F_{in} - \frac{S_{out}}{X}F_{out}}{V}$
Overall biomass-to-substrate yield	$Y_{X/S}$	$g_X\ g_S^{-1}$	Outlet volumetric flow rate (F_{out}, L h^{-1}) Biomass concentration (X, g_X L^{-1}) Inlet substrate concentration (S_{in}, g_S L^{-1}) Inlet feeding volumetric flow rate (F_{in}, L h^{-1}) Outlet substrate concentration (S_{out}, g_S L^{-1})	$Y_{X/S} = \frac{F_{out}\ X}{F_{in}S_{in} - F_{out}S_{out}}$
Specific product formation rate	q_P	PU $g_x^{-1}\ h^{-1}$	Dilution rate (D, h^{-1}) Protein concentration (P, PU L^{-1}) Biomass concentration (X, g_X L^{-1})	$q_P = \frac{D\ P}{X}$
Overall product to biomass yield	$Y_{P/X}$	PU g_X^{-1}	Protein concentration (P, PU L^{-1}) Biomass concentration (X, g_X L^{-1})	$Y_{P/X} = \frac{P}{X}$
Volumetric productivity	Q_V	PU $L^{-1}\ h^{-1}$	Dilution rate (D, h^{1}) Protein concentration (P, PU L^{-1})	$Q_V = D\ P$
Specific oxygen uptake rate	q_{O2}	mol $O_2\ g_x^{-1}\ h^{-1}$	Oxygen uptake rate (OUR, mol O_2 L^{-1} h^{-1}) Biomass concentration (X, g_X L^{-1}) Bioreactor working volume (V, L) Inlet gas molar flow rate ($F_{g,in}$, mol h^{-1}) Inert factor (f_{inert}), needed when the off-gas flow rate is not measured. Calculated through inert gas balance. Inlet O_2 molar fraction ($X_{O2,in}$)	$q_{O2} = \frac{OUR}{X}$ $OUR = -\frac{F_{g,in}(f_{inert} x_{O_2,out} - x_{O_2,in})}{V}$ $f_{inert} = \frac{1 - x_{O_2,in} - x_{CO_2,in} - x_{H_2O,in}}{1 - x_{O_2,out} - x_{CO_2,out} - x_{H_2O,out}}$

(continued)

Variable		Units	Experimental parameter or variable	Equation
			Outlet O_2 molar fraction ($X_{O2,out}$) Inlet CO_2 molar fraction ($X_{CO2,in}$) Outlet CO_2 molar fraction ($X_{CO2,out}$)	
Specific carbon dioxide production rate	q_{CO2}	mol CO_2 g_x^{-1} h^{-1}	Carbon dioxide production rate (CPR, mol CO_2 L^{-1} h^{-1}) Biomass concentration (X, g_X L^{-1}) Bioreactor working volume (V, L) Inlet gas molar flow rate ($F_{g,in}$, mol h^{-1}) Inert factor (f_{inert}), needed when the off-gas flow rate is not measured. Calculated through inert gas balance Inlet CO_2 molar fraction ($X_{CO2,in}$) Outlet CO_2 molar fraction ($X_{CO2,out}$) Inlet O_2 molar fraction ($X_{O2,in}$) Outlet O_2 molar fraction ($X_{O2,out}$)	$q_{CO2} = \frac{CPR}{X}$ $CPR = \frac{F_{g,in}\left(f_{inert} \cdot x_{CO_2,out} - x_{CO_2,in}\right)}{V}$ $f_{inert} = \frac{1 - x_{O_2,in} - x_{CO_2,in} - x_{H_2O,in}}{1 - x_{O_2,out} - x_{CO_2,out} - x_{H_2O,out}}$
Respiratory quotient	RQ	mol CO_2 mol O_2^{-1}	Carbon dioxide production rate (CPR, mol CO_2 L^{-1} h^{-1}) Oxygen uptake rate (OUR, mol O_2 L^{-1} h^{-1})	$RQ = \frac{CPR}{OUR}$

4 Notes

1. The use of products from animal sources can be avoided during inoculum preparation by using defined buffered minimal glycerol medium (BMG): 100 mM phosphate buffer pH 6, 1.34% yeast nitrogen base without amino acids (YNB), 0.4 mg L^{-1} biotin, and 1% glycerol. To prepare 1 L, autoclave 700 mL of distilled water and add 100 mL of phosphate buffer 1 M pH 6 (autoclaved), 100 mL YNB 10× (sterilized by filtration), 2 mL biotin 500× (0.2 g L^{-1}, sterilized by filtration), and 100 mL glycerol 10× (autoclaved) with distilled water.
2. To make the chemostat more user-friendly, the working volume (V) normally used is 1 L. Therefore, $\mu = D = F_{\text{out}}/V$ is equal to F_{out}, making the setup easier and faster. Considering this, 2 L glass bioreactors are recommended.
3. The current bioreactor's control unit usually has three constant-speed peristaltic pumps (base, acid, and antifoam) and one speed-variable peristaltic pump (feed). For the chemostat outlet, the acid or antifoam pump can be used if the bioreactor software allows it. In case a peristaltic pump is missing, an external peristaltic pump can be used.
4. The use of a volatile probe is recommended when using methanol as a carbon source or when applying hypoxic conditions to a glucose chemostat cultivation, since it will allow for monitoring the methanol accumulation or the ethanol production, respectively.
5. An off-gas analyzer is highly recommended in chemostat cultivations since it provides online data of the physiological state of the cultivation.
6. Since the risk of contamination in long-term fermentations is high, working with overpressure is recommended. Usually, chemostat cultivations are operated with 0.2 bar of overpressure (therefore 1.2 bar inside the bioreactor). When working under overpressure, remember to calibrate the height of the outlet pipe with this overpressure, since the volume occupied by the water might be slightly different.
7. Enriching the inlet gas flow with pure oxygen might be necessary at high D, especially when working with methanol or glycerol, due to their higher degree of reduction. Alternatively, nitrogen supplementation might be necessary for applying hypoxic conditions to the chemostat cultivation. Ensure that an installation of both gases is available when designing the experiments.
8. During batch culture medium preparation, it is recommended to first weigh the glycerol and add the water and the salts. Citric

acid should be added before salts, so the pH is decreased, and salts can be better dissolved.

9. The pH and sometimes temperature must be set depending on the protein or molecule produced. It is important to check the pH where the protein is active and stable, always avoiding working near its isoelectric point since it can cause protein aggregation.
10. The medium is filled up to 993.4 mL because, after sterilization, PTM1 trace salts and biotin solutions will be added, which will make the final volume 1000 mL. Consider this when preparing batch medium volumes different than 1000 mL.
11. If inoculation is performed through the septum using a syringe and a needle, the volume used to resuspend the inoculum can be lower. Therefore, recalculate the volume to be stored, considering that the volume used is normally between 20 and 50 mL.
12. *Basal salt medium (BSM)* from Invitrogen [27] or a defined medium described by Jungo et al. (2007) [28] can be used or adapted for the batch phase.
13. At the steady state, the biomass concentration and the substrate consumption can be estimated as:

$$r_X V = X F_{\text{out}}; r_S V = S_0 F_{\text{in}} - S F_{\text{out}},$$

$$r_x = r_S \Upsilon_{X/S}; X F_{\text{out}} = (S_0 F_{\text{in}} - S F_{\text{out}}) \Upsilon_{X/S}.$$

Then, assuming that $S = 0$ (accumulation $= 0$) and $F_{\text{in}} \approx F_{\text{out}}$:

$$X F_{\text{out}} = S_0 F_{\text{out}} \Upsilon_{X/S}; X = S_0 \Upsilon_{X/S},$$

where r_S is the volumetric substrate consumption rate ($g_S\ L^{-1}\ h^{-1}$), S_0 and S are the substrate concentrations in the feeding tank and in the culture broth, respectively ($g_S\ L^{-1}$), F_{in} is the inlet volumetric flow rate ($L\ h^{-1}$), and $\Upsilon_{X/S}$ is the biomass-to-substrate yield ($g_X\ g_S{}^{-1}$). Considering that $\Upsilon_{X/S}$ is substrate- and strain-dependent, substrate concentration can be adapted to achieve a goal biomass concentration.

14. Alternatively, chemostat media described by Jungo; et al. (2007) [28] can be used or adapted, especially when using methanol as a carbon source.
15. When inoculating through the septum, no inoculating port or further preparation is needed.
16. Measure the solutions in the laminar flow cabin with a sterile syringe and couple a needle. When injecting the solution through the septum, make sure it goes directly to the medium

and not to the walls. If there are some remaining PTM1 trace salts solution on the bioreactor walls, it might be oxidized at some point. However, it has no observed effect on cell growth.

17. When adding the trace salts to the bioreactor, the DO drops due to the oxidation of the metals. Make sure it has been stabilized before calibrating the DO probe.
18. Inoculating the bioreactor when the cultivation is still in the exponential phase avoids a long lag phase at the beginning of the fermentation.
19. Growth rate depends on cultivating temperature; the higher T, the faster μ. Usually, *K. phaffii* is cultivated in a range of 25–30 °C. Bear in mind that the μ_{max} will describe the growth kinetics only under the conditions it was determined.
20. The inoculum shake flasks must have baffles, and the medium volume should not be above 10% of the total flask volume, to ensure good aeration.
21. The optimal range of a spectrophotometer for OD_{600} determinations can be considered between 0.2 and 0.8. Dilute the sample, if necessary. Check for linearity.
22. Make sure the conical bottle for inoculum centrifugation can handle the centrifugation speed.
23. In case the bioreactor is inoculated with a syringe through the septum, resuspend the pellet with the dedicated medium volume collected during medium preparation, which should be between 20 and 50 mL.
24. When inoculating via a syringe, transfer the resuspended pellet to a syringe coupled with a needle. Go to the bioreactor and, after cleaning all surfaces and gloves with ethanol, punch the septum of the bioreactor with the syringe and inoculate the entire volume.
25. No foam formation should be detected during the cultivation because batch and chemostat media already contain antifoam. If high foaming is observed, add a few drops of 1:10 diluted and sterile (by autoclaving) antifoam (Glanapon 2000kz) through the septum. Repeat the process in case foaming persists, always avoiding adding more antifoam than required.
26. Since the same chemostat cultivation setup might be used to study different μ, it is recommended to split the feeding tube in two with a "Y" connector, so two different diameters of tubes suitable for the high-precision peristaltic pump can be used. In this way, all μ ranges will be covered with the same setup.

27. Antifoam is autoclaved instead of filtered because it would block the filter capsule. Alternatively, glycerol could also be dissolved together with antifoam in the feeding bottle to avoid viscosity problems in the filter capsule.
28. When preparing a glycerol feed media, first weigh the amount of glycerol and then dissolve it in water. For glucose feed, dissolve stepwise the amount of glucose in stirred warm water. Alternatively, methanol or mixtures can also be used for the chemostat feed. In this case, simply dissolve in water the volume of pure methanol needed.
29. If precipitation inside the filter capsule is noticed, pump HCl solution (2 M is recommended) to redissolve the salts. This problem should be avoided by decreasing the feeding medium pH to 5.
30. The filter capsule should be already clean because the last step of medium sterilization is filtering just water. In case some dirtiness is observed, rinse the filter with water and proceed to dry it.
31. Complete dryness of the filter capsule is crucial to avoid mold growing in the filter.
32. If working with overpressure and in case the feeding pipe reaches the fermentation broth, make special attention to always have the tubes clamped to prevent the broth from being pushed into the feeding system. Therefore, before removing the clamps, the high-precision peristaltic pump must be already setup. If a mistake occurs, and the broth comes into contact with the feeding system (tubing or even reaches the feeding bottle), the fermentation process must be restarted because the carbon source concentration entering the bioreactor will be variable and indeterminable.
33. Methanol feeding requires a transition step to adapt the culture to methanol metabolism, so the cells can consume it as soon as the feeding starts, preventing accumulations at the beginning that can be fatal. A methanol probe can be helpful in this phase. If it is not available, the transition phase can be followed by changes in DO, off-gas composition, pH, and temperature. In both cases, setting the agitation speed too high and a constant rpm is highly recommended. To start the transition phase, add a 1 g L^{-1} methanol pulse and wait until total depletion. The consumption of methanol can be followed by the methanol probe; otherwise, a change in pH, temperature, and DO trend will be noticed when methanol is consumed. In particular, pH will slightly increase, the temperature might slightly decrease due to the controller's inertia, and DO will rise faster. Then, add a second methanol pulse of about 2 g L^{-1}. A rapid decrease in DO should be observed, meaning that the cultivation

already has the metabolic machinery to properly consume methanol. In this case, the chemostat phase can be started; otherwise, add a new methanol pulse of 2 g L^{-1}.

34. The usual D range for glucose and glycerol is 0.01–0.15 h^{-1}. Lower D can be performed, although the total cultivation time will be very long, while higher D is not recommended to prevent bioreactor washout. With methanol, the maximum D might be 0.10 h^{-1} or lower. Therefore, for highly engineered strains or under restrictive conditions (and growing on methanol), the μ_{max} must be determined since it can greatly decrease.
35. From experience, the base flow rate accounts for approximately 2% of the outlet flow rate. Therefore, the feeding flow rate should be set about 2% lower than the D set-point, so the addition of the base flow rate will give the goal outlet flow rate.
36. Depending on the D goal, air mixtures of pure oxygen and air might be needed at high μ to avoid oxygen-limiting conditions. To make the process easier and cheaper, the carbon source concentration can be reduced, and so will be the biomass. In this way, air mixtures will be avoided. However, some accuracy in the parameters recorded might be lost, mainly in the off-gas analysis when working at low μ due to the lower consumption and production of O_2 and CO_2, respectively.
37. When working with overpressure, be especially careful taking samples, independently of the sampling system used. Must not pull back the plunger of the syringe, the pressure inside the bioreactor will push the plunger as soon as the clamp is removed. Just make slight resistance with the thumb against the plunger, until withdrawing the goal sample volume.
38. This process can be repeated as many times depending on the volume of feed prepared. For instance, if the working volume is 1 L, 5 L of feed will be needed to complete one condition. To handle more easily the process and avoid potential problems due to the long-term cultivation, it is recommended to perform a maximum of 4 conditions in the same setup.
39. The final cultivation volume can be determined by weighing it into a tared beaker and then measuring its density, using a densitometer or calculating it with a volumetric flask. As an alternative, the cultivation volume can be measured in a test tube.

Acknowledgments

This work was funded by Project PID2022-136936OB-I00 by MCIN /AEI /10.13039/501100011033 /FEDER, UE of the Spanish Ministry of Science and Innovation. The group is a member of the 2021 SGR 00143 Department of Research and Universities of Generalitat de Catalunya. Núria Bernat-Camps acknowledges the Austrian Centre of Industrial Biotechnology (acib) for funding her scholarship through the COMET program. Additionally, we kindly thank the Christian Doppler Gesellschaft and the industrial partners Novo Nordisk, BioGrammatics, SeSam Biotech, ZETA, and SolarBiotech for funding the Christian Doppler Laboratory for Innovative *P. pastoris* host and vector systems, in which NBC, AG, and XGO are contributing as members of the UAB team.

Competing Interest Statement The authors have no conflicts of interest to declare that are relevant to the content of this chapter.

References

1. (2023) Global recombinant proteins market - Forecast 2023–2030. In: Research and Markets. https://www.researchandmarkets.com/report/protein-drugs
2. Nieto-Taype MA, Garcia-Ortega X, Albiol J et al (2020) Continuous cultivation as a tool toward the rational bioprocess development with *Pichia Pastoris* cell factory. Front Bioeng Biotechnol 8. https://doi.org/10.3389/FBIOE.2020.00632
3. Bernauer L, Radkohl A, Lehmayer LGK, Emmerstorfer-Augustin A (2021) *Komagataella phaffii* as emerging model organism in fundamental research. Front Microbiol 11. https://doi.org/10.3389/FMICB.2020.607028
4. De Brabander P, Uitterhaegen E, Delmulle T et al (2023) Challenges and progress towards industrial recombinant protein production in yeasts: a review. Biotechnol Adv 64:108121. https://doi.org/10.1016/j.biotechadv.2023.108121
5. Nasr MM, Krumme M, Matsuda Y et al (2017) Regulatory perspectives on continuous pharmaceutical manufacturing: moving from theory to practice: September 26–27, 2016, international symposium on the continuous manufacturing of pharmaceuticals. J Pharma Sci. Elsevier B.V. 106:3199–3206
6. Kumar A, Udugama IA, Gargalo CL, Gernaey KV (2020) Why is batch processing still dominating the biologics landscape? Towards an integrated continuous bioprocessing alternative. PRO 8:1–19. https://doi.org/10.3390/pr8121641
7. Badr S, Okamura K, Takahashi N et al (2021) Integrated design of biopharmaceutical manufacturing processes: operation modes and process configurations for monoclonal antibody production. Comput Chem Eng 153:107422. https://doi.org/10.1016/j.compchemeng.2021.107422
8. Narayanan H, Sponchioni M, Morbidelli M (2022) Integration and digitalization in the manufacturing of therapeutic proteins. Chem Eng Sci 248:117159. https://doi.org/10.1016/j.ces.2021.117159
9. Garrigós-Martínez J, Weninger A, Montesinos-Seguí JL (2021) Scalable production and application of *Pichia pastoris* whole cell catalysts expressing human cytochrome P450 2C9. Microb Cell Fact 20(1):90. https://doi.org/10.1186/s12934-021-01577-4
10. García-Ortega X, Cámara E, Ferrer P et al (2019) Rational development of bioprocess engineering strategies for recombinant protein production in *Pichia pastoris (Komagataella phaffii)* using the methanol-free GAP promoter. Where do we stand? New Biotechnol 53:24–34. https://doi.org/10.1016/J.NBT.2019.06.002

11. Looser V, Bruhlmann B, Bumbak F et al (2014) Cultivation strategies to enhance productivity of *Pichia pastoris*: a review. Biotechnol Adv 33:1177–1193. https://doi.org/10.1016/j.biotechadv.2015.05.008
12. Coltman BL, Rebnegger C, Gasser B, Zanghellini J (2024) Characterising the metabolic rewiring of extremely slow growing *Komagataella phaffii*. Microb Biotechnol 17:e14386. https://doi.org/10.1111/1751-7915.14386
13. Rebnegger C, Vos T, Graf AB et al (2016) *Pichia pastoris* exhibits high viability and a low maintenance energy requirement at near-zero specific growth rates. Appl Environ Microbiol 82:4570–4583. https://doi.org/10.1128/AEM.00638-16
14. Crowell LE, Lu AE, Love KR et al (2018) On-demand manufacturing of clinical-quality biopharmaceuticals. Nat Biotechnol 36:988. https://doi.org/10.1038/nbt.4262
15. Matthews CB, Kuo A, Love KR, Love JC (2018) Development of a general defined medium for *Pichia pastoris*. Biotechnol Bioeng 115:103–113. https://doi.org/10.1002/bit.26440
16. Mcgeachy AM, Meacham ZA, Ingolia NT (2019) An accessible continuous-culture turbidostat for pooled analysis of complex libraries. https://doi.org/10.1021/acssynbio.8b00529
17. Vogl T, Glieder A (2013) Regulation of *Pichia pastoris* promoters and its consequences for protein production. New Biotechnol 30:385–404. https://doi.org/10.1016/J.NBT.2012.11.010
18. Nieto-Taype MA, Garrigós-Martínez J, Sánchez-Farrando M et al (2020) Rationale-based selection of optimal operating strategies and gene dosage impact on recombinant protein production in *Komagataella phaffii (Pichia pastoris)*. Microb Biotechnol 13:315–327. https://doi.org/10.1111/1751-7915.13498
19. Garrigós-Martínez J, Nieto-Taype MA, Gasset-Franch A et al (2019) Specific growth rate governs AOX1 gene expression, affecting the production kinetics of *Pichia pastoris (Komagataella phaffii)* P_{AOX1} -driven recombinant producer strains with different target gene dosage. Microb Cell Factories 18:187. https://doi.org/10.1186/s12934-019-1240-8
20. Garrigós-Martínez J, Vuoristo K, Nieto-Taype MA et al (2021) Bioprocess performance analysis of novel methanol-independent promoters for recombinant protein production with *Pichia pastoris*. Microb Cell Factories 20:74. https://doi.org/10.1186/s12934-021-01564-9
21. López-Fernández J, Benaiges MD, Valero F (2021) Constitutive expression in *Komagataella phaffii* of mature *Rhizopus oryzae* lipase jointly with its truncated prosequence improves production and the biocatalyst operational stability. Catalysts 11:1192. https://doi.org/10.3390/CATAL11101192/S1
22. Chen X, Li C, Liu H (2021) Enhanced recombinant protein production under special environmental stress. Front Microbiol 12. https://doi.org/10.3389/fmicb.2021.630814
23. Garcia-Ortega X, Valero F, Montesinos-Seguí JL (2017) Physiological state as transferable operating criterion to improve recombinant protein production in *Pichia pastoris* through oxygen limitation. J Chem Technol Biotechnol 92:2573–2582. https://doi.org/10.1002/jctb.5272
24. Gasset A, Garcia-Ortega X, Garrigós-Martínez J et al (2022) Innovative bioprocess strategies combining physiological control and strain engineering of *Pichia pastoris* to improve recombinant protein production. Front Bioeng Biotechnol 10. https://doi.org/10.3389/fbioe.2022.818434
25. Gasset A, Van Wijngaarden J, Mirabent F et al (2024) Continuous Process Verification 4.0 application in upstream: adaptiveness implementation managed by AI in the hypoxic bioprocess of the *Pichia pastoris* cell factory. Front Bioeng Biotechnol 12:1439638. https://doi.org/10.3389/fbioe.2024.1439638
26. Baumann K, Maurer M, Dragosits M et al (2008) Hypoxic fed-batch cultivation of *Pichia pastoris* increases specific and volumetric productivity of recombinant proteins. Biotechnol Bioeng 100:177–183. https://doi.org/10.1002/bit.21763
27. Barrero JJ, Pagazartaundua A, Glick BS et al (2021) Bioreactor-scale cell performance and protein production can be substantially increased by using a secretion signal that drives co-translational translocation in *Pichia pastoris*. New Biotechnol 60:95. https://doi.org/10.1016/J.NBT.2020.09.001
28. Jungo C, Marison I, von Stockar U (2007) Mixed feeds of glycerol and methanol can improve the performance of *Pichia pastoris* cultures: a quantitative study based on concentration gradients in transient continuous cultures. J Biotechnol 128:824–837. https://doi.org/10.1016/j.jbiotec.2006.12.024

Chapter 34

Best Practices for Protein Expression in *Pichia pastoris* Combining Design of Experiments with an Automated Platform for Parallelized Cultivation in Small-Scale Stirred Tank Bioreactors

Holger Morschett, Joan Cortada-Garcia, Hauke Holm, Núria Adelantado, and Karlheinz Flicker

Abstract

This chapter provides practical guidance for studying the expression of proteins in the yeast *Komagataella phaffii* by using a commercial automated platform for parallelized cultivation in single-use stirred tank bioreactors. In addition to discussing the advantages and disadvantages of such a platform regarding scale, automatability, and single-use vessels, the main focus is on the presentation of a flexibly adaptable, operator-centric workflow for the use of the cultivation system. In particular, it includes highlighting typical pitfalls and sources of error in handling and culminates in the presentation of best practices to support the proper execution of such experiments. To complement, a generic guide to its use for strain characterization and process development using statistical design of experiments is suggested based on the parallelization capabilities of the system.

Key words *Pichia pastoris*, *Komagataella phaffii*, Fed-batch, Single-use, Parallelized cultivation, Process development, Automation, Optimization, Design of experiments

1 Introduction

Komagataella phaffii (reclassified from *Pichia pastoris*) represents a major workhorse for the microbial production of pharmaceutically relevant biologics. However, biological and process diversity requires efficient strategies for their experimental assessment. While microscale systems like BioLector® XT (Beckman Coulter) or bioREACTOR 8/48 (2mag) have emerged as powerful tools for early-stage research [1–8], their scale-related constraints (e.g., mass transfer and analytical accessibility) limit prediction power toward bench scale and beyond. Consequently, the classic stirred tank bioreactor (STR) remains the gold standard for scalable bioprocess

Xavier Garcia-Ortega et al. (eds.), *Pichia pastoris: Methods and Protocols*, Methods in Molecular Biology, vol. 2697,
https://doi.org/10.1007/978-1-0716-4779-0_34,

development to this day. Nonetheless, a general trend toward reduced STR working volumes and parallelized operation is observed. While a multitude of such systems became available, for example, DASGIP®/DASBOX® Parallel Bioreactor System (Eppendorf), Multifors® (Infors HT), and BioXplorer (H.E.L.), this chapter focuses on the Ambr® 250 High Throughput (Sartorius). Although many general aspects can be abstracted to any (parallelized) bioreactor system, details on instrument use are tailored to the Ambr® 250 High Throughput, which combines the following key design features:

1. Geometric similarity to large-scale STRs for simplified scale-up (e.g., as per power input, stirrer tip speed, mass transfer coefficient, or maximum gas transfer rates) [9, 10].
2. Sub-liter nominal volume (ca. 250 mL) as a sweet spot between material/media savings and sampling capabilities while occupying a reasonable in-lab footprint.
3. Full instrumentation for process monitoring and control comparable to larger systems (temperature, pH, dissolved oxygen (DO), off-gas analysis, and feeding).
4. An adequate degree of parallelization (12 or 24 STRs) for strain/process studies, with or without using statistical experimental design in biological replicates.
5. Robotic integration for walk-away automated sampling/dosing regardless of typical working hours for reduced manual handling errors and improved data quality and traceability.
6. Ready-to-use single-use (SU) consumables for minimal changeover time and contamination risk.

With cultivation strategies designed to fit the physiology of the expression system, special emphasis needs to be put on the applied promoter system:

Offering tight control and strong induction capabilities, the expression system based on the *AOX1* promoter (P_{AOX1}) represents probably the widespread system described for *K. phaffii* recombinant protein production. For a comprehensive description of this promoter and its engineered derivatives, the reader is referred to literature [11–14]. As a common feature, they are strictly repressed by glucose/glycerol and strongly induced in the presence of methanol. However, the technical realization of such processes is potentially challenging, as analytical tools for strict control of methanol levels [15–20] as well as effective explosion hazard measures are required [21].

As an alternative, the methanol-free P_{GAP}/P_{GTH}-derived expression systems have gained increased popularity over the past years as they avoid flammable reagents, typically reduce process times, and provide easier handling due to simplified cultivation regimes. Although induction strength varies with the carbon source used, the P_{GAP}-based system features strictly growth-coupled expression [22]. The P_{GTH}-based system is tightly regulated, and expression takes place under glucose-limiting conditions either by restricting glucose feeding or by substituting glucose for glycerol. This induces expression of the protein of interest while maintaining growth capabilities. However, expression is directly coupled with biomass formation, where oxygen supply and heat dissipation become the main (technical) limitations [23–27]. Therefore, selecting equipment with high mass and heat transfer capabilities is essential.

Both briefly described promoter systems represent industrially relevant examples, being complemented by engineered variations and alternatives as reviewed by Weinhandl et al. [28]. In summary, the promoter system defines the general process regime and equipment.

Producer strain characterization and process parameter optimization can be highly streamlined using small-scale parallelized systems, particularly if combined with a design of experiments (DoE) approach instead of studying the different factors in isolation (one factor at a time), which can lead to "getting stuck" in local optima. When different strains are investigated, different genetic constructs are compared, such as different promoters, transcription terminators, codon optimization sequences, gene copy numbers, or integration loci, to name a few [29]. Typical process parameters investigated are the feeding strategy, cultivation medium composition, the type of inducer and induction strength, and the pH and temperature during protein production.

Subheading 3.1 will provide general guidance on using a DoE approach for process optimization using a parallelized cultivation system, while Subheading 3.2 and 3.3 will delve into how to conduct experiments with such systems, particularly focusing on the aforementioned bioreactor platform. For further details about fed-batch and continuous cultures, please see Chapters 31, 32, and 33 in this volume.

2 Materials

Prepare all media and stock solutions from deionized water at room temperature. It is recommended to provide about half of the required amount of water, dissolve the components with constant stirring, and only then fill up to the final volume unless stated otherwise. In the case of multicomponent solutions, dissolve the individual components in order of appearance. Each component is only added after the previous one has completely dissolved. Where indicated, sterilize solutions by autoclaving (21 min at 121 °C in

1 bar steam-saturated atmosphere). Unless otherwise stated, store at room temperature.

2.1 Preculture Medium

The use of "yeast peptone base" medium based on [30] is recommended.

1. For a 1 L preparation, dissolve 10 g yeast extract, 20 g peptone, and 20 g glycerol.
2. Sterilize by autoclaving.

2.2 Batch Medium

1. The usage of a freshly prepared defined batch medium used by [31] is recommended (*see* **Notes 1** and **2**).
2. For a 1 L preparation dissolve 40 g glycerol anhydrous, 12.4 g $NH_4H_2PO_4$, 0.5 g $MgSO_4 \cdot 7\ H_2O$, 0.9 g KCl, 0.022 g $CaCl_2 \cdot 2\ H_2O$, and 2 g $C_6H_8O_7 \cdot H_2O$ (citric acid monohydrate).
3. Sterilize by autoclaving.
4. Aseptic addition of 4.6 mL/L trace elements solution (*see* Subheading 2.4) and pH adjustment is done later directly in the bioreactors.

2.3 Feed Solution

1. A feed solution based on [31] is recommended (*see* **Notes 3–5**).
2. The feed solution is prepared fresh prior to use in a 500 mL Duran® bottle with feeding lines and pump connectors.
3. For a 1 L preparation dissolve 6.45 g KCl, 0.35 g $CaCl_2 \cdot 2\ H_2O$, 10 g $MgSO_4 \cdot 7\ H_2O$, and 495 g glucose monohydrate in approximately 275 mL water.
4. Fill to 0.988 L.
5. Sterilize by autoclaving.
6. After cooling to room temperature aseptically add 12 mL trace elements stock solution (*see* Subheading 2.4).

2.4 Trace Elements Solution

1. A trace elements solution based on [31] is recommended.
2. For a 1 L preparation, dissolve 0.2 g biotin, 9.15 g H_2SO_4 (95–98%), 6 g $CuSO_4 \cdot 5\ H_2O$, 0.08 g NaI, 3.36 g $MnSO_4 \cdot H_2O$, 0.2 g $Na_2MoO_4 \cdot 2\ H_2O$, 0.02 g H_3BO_3, 2.05 mL of a 400 g/L $CoCl_2 \cdot 7\ H_2O$ stock solution, 20 g $ZnCl_2$, and 65 g $FeSO_4 \cdot 7\ H_2O$ (*see* **Note 6**).
3. Sterilize by autoclaving.
4. The trace elements solution can be kept for up to 6 weeks at 4 °C.

2.5 Titration Agents

1. Titration agents are prepared in 500 mL Duran® bottles with feeding lines and pump connectors (*see* **Note 5**).
2. Prepare a 10% (w/w) H_3PO_4 solution by dilution of concentrated (typically 85% (w/w)) H_3PO_4 in water. Always add water first, then the acid.
3. For the NH_3 solution, prepare a bottle with half the target volume of water.
4. Autoclave both bottles.
5. Complete the NH_3 solution by aseptic addition of 25% (w/v) NH_3 to obtain a final concentration of 12.5% (w/v) (*see* **Notes 7** and **8**).

2.6 Cleaning-in-Place Agents & Solutions

Cleaning-in-place (CIP) agents and solutions are all to be prepared in Duran® bottles and autoclaved.

1. One empty 1 L bottle with a feeding line and a pump connector.
2. Two 1 L bottles filled with 900 mL water, a feeding line, and a pump connector.
3. Two 500 mL bottles containing 400 mL water, a feeding line, and a pump connector. After autoclaving—and right before CIP—aseptically add 100 mL 30% (w/v) H_2O_2 for a final concentration of 6% (w/v).
4. One 1 L bottle with 700 mL water, a feeding line, and a pump connector. After autoclaving, aseptically add 300 mL absolute ethanol for a final concentration of 70% (v/v).
5. Two 2 L bottles without feeding line or pump connector, each containing 1.8 L water.
6. Two 1 L bottles without feeding line or pump connector, containing 1 L H_2O_2 6% (w/v) prepared as **step 3** above.
7. Wipes soaked with either 70% (v/v) ethanol or 99.9% (v/v) isopropanol for cleaning surfaces.

2.7 Antifoam Solution

1. Fill one empty 175 mL Nalgene® bottle with approximately 100 mL pure AF204 (Sigma-Aldrich) antifoam agent (*see* **Note 9**).
2. Sterilize by autoclaving.

2.8 Bioreactor Instrumentation

The parallelized cultivation methods described in this chapter are detailed to be run in an Ambr® 250 High Throughput automated cultivation platform (Sartorius), but equivalent methods can be implemented in the alternative systems mentioned in Subheading 1.

The bespoke platform carries either 12 or 24 bays fitting 250 mL nominal volume SU STRs each featuring an agitator shaft with two Rushton impellers, options for headspace and/or submersed aeration and four syringe pump-driven addition lines. Inline pH readout is obtained via an electrochemical probe while DO is measured by an optical sensor spot [32] (*see* **Note 10**). Oxygen and carbon dioxide concentration in the off-gas of the individual vessels is recorded. An additional option for inline biomass estimation using light scattering is available but was not used in this context (*see* **Note 11**). The instrument is embedded into a class two biosafety cabinet and also including a liquid handler capable of dosing and sampling actions using disposable tips (e.g., inoculation, induction, and sampling). Samples are deposited on chilled deck nests (*see* **Note 12**). The instrument comes with a set of dedicated software to plan, schedule, and execute experiments and visualize the results.

3 Methods

3.1 Parallelization and Experimental Design

As previously introduced (*see* Subheading 1), DoE is a methodology that can improve (bio)process understanding while making better use of resources, particularly in complex multifactorial systems with interactions between parameters. However, it requires care when setting up the experimental design, as a smart design is considered more resource-efficient, while a suboptimal one can come at a higher cost for appropriate design space coverage. The reader is directed to specialized literature [33–36] to learn more about the topic, but general guidance on this matter is given below:

1. Start defining the experimental purpose and the response variables to be investigated (*see* **Note 13**).
2. Identify factors and levels to be studied as per the following guidance:
   (a) Fishbone diagrams are often useful to identify relevant factors for the response variables.
   (b) Levels are often defined from prior knowledge of the system (biological and technical), the experimental purpose, and familiarization runs (*see* point Subheading 3.1 below).
   (c) The reader is encouraged to use these literature examples to familiarize themselves with practical DoE methodology and how multiple factors and their interactions can be studied.
   (d) Examples of factors included in *K. phaffii* DoE optimization studies comprise:
      - pH, temperature, and inducer concentration [37–39];
      - pH, temperature, DO, and induction time [40];

- pH, temperature, salt supplementation at induction, and batch glycerol concentration [41];
- pH, temperature, aeration, and agitation [42].

3. Determine the number of experiments that can be executed based on the available time and economic resources (*see* **Note 14**). Depending on the scope of the study, one or multiple of the following stages might be followed:
   (a) Familiarization: particularly important when little information is known about the system (e.g., to define the factor levels). Typically, a small amount of resources (e.g., 10%) is invested in this stage.
   (b) Screening/sensitivity analysis: to define which factors are most important for the response variables. This is commonly assessed by model fitting.
   (c) Optimization: based on the screening results, a new experimental space is defined to locate and characterize the optimal region for the response variables.
   (d) Robustness testing: particularly important before a process is transferred into manufacturing or a product is released. This step allows to identify factors that lead to higher process variability and quantifies their influence on critical quality attributes.

   Spread all available resources in the aforementioned stages required for the purpose of the study.
4. Once the number of experiments has been defined and allocated to the individual stages, design the exact experiments to be conducted.
   (a) For screening experiments, factorial or fractional factorial designs are most commonly used. However, due to constraints in resources or the design space, alternative designs might be preferred—e.g., Plackett-Burman, Rechtschaffner, or L-designs.
   (b) For optimization experiments, response surface methodology (RSM) designs such as central composite designs (CCDs) and Box–Behnken designs are most typical.
   (c) Non-standard designs—such as D-optimal designs—might be preferred for both screening and optimization (*see* **Notes 15** and **16**).
5. The degree of parallelization (12 or 24 parallel STR cultivations in this case) comes with direct implications on experimental design considerations:
   (a) Choosing the appropriate number of replicates is essential. Regarding strain comparisons, it is strictly recommended to conduct experiments at least in biological

duplicates for a basic impression on reproducibility while more replicates are required for proper statistical analysis. In that regard, evaluating statistical power is worth considering [43, 44].

(b) For DoE approaches, it is typically relevant to have center or reference points in replicate. However, the allocation of experimental resources should be evaluated carefully when it comes to replicating conditions versus integrating additional ones [33].

(c) Depending on the complexity of the design, splitting the total number of experiments into different runs (blocks) might be necessary. In such cases, it is important to assess the comparability of the individual blocks, e.g., by randomization, homogeneity testing, and blocking of variables [33, 45].

(d) Changing continuous parameters at runtime allows for testing multiple conditions (e.g., feed rates) during a single cultivation enabling a so-called intensified DoE (iDoE) approach [46, 47]. However, such approach requires careful evaluation to make sure the individual phases do not influence each other.

To account for these considerations in an efficient manner, the reader is recommended to use a DoE software, such as the dedicated Design of Experiments software of the instrument. Alternative dedicated DoE software options are also available, such as JMP® (JMP Statistical Discovery) and Design-Expert® (StatEase).

3.2 Preculture

1. Fill 100 mL preculture medium (*see* Subheading 2.1) into a threefold baffled 500 mL shake flask.
2. Inoculate to an initial optical density (OD_{600}) of 0.15.
3. Incubate the flask for 20 ± 2 h in a shaking incubator at 30 °C, and 170 rpm orbital shaking at 25 mm shaking diameter.
4. Measure the OD_{600} 10–20 min before the scheduled inoculation time of the main culture to predict when the target OD_{600} will be reached (typically ca. 40–55). Return the flasks to the incubator until the desired inoculation time (*see* **Note 17**).

3.3 Preparation of the Culture

All main cultivations are executed using the bioreactor system described in Subheading 2.8. Key details are also indicated in the notes section (*see* **Note 18**)

1. Define the experimental control script in the "Definition" software as per the experiment's requirements. Table 1 provides an exemplary configuration for reference (*see* **Note 19**).

Table 1
Typical process parameters for *K. phaffii* cultivation. Individual settings may need to be adjusted as per strain's needs

Parameter	Typical setpoint	Typical range	Comments/details
Aeration with pressurized air in v/v/min	1	1	Corrected for volume changes
Additional aeration with pressurized oxygen in mL/min	0	0–150	As per cascaded DO control loop, only used when stirrer at maximum stirring speed
DO in %	30	≥30	Regulated as per cascaded DO control loop – Level 1 stirrer speed: Range 1000–4500/min, P term 25, I term 300 s, D term 0 s – Level 2 (additional oxygen flow): Range 0–150 mL/min, P term 0.5, I term 70 s, D term 0 s
pH	6.0	4.0–6.8	Regulated as per pH control loop – Base: 0.05 dead band, 0–4 mL/min flow rate, P term 100, I term 0 s, D term 0 s – Acid: 0.05 dead band, 0–4 mL/min flow rate, P term 70, I term 0 s, D term 0 s
Stirring speed in 1/min	≥1000	1000–4500	As per cascaded DO control loop
Temperature in °C	30	25–30	Might be changed for specific process phases
Working volume in mL	90	90–220	Corrected for volume changes. Given range reflects starting to maximal working volume (*see* **Note 27**)
Calibration settings 100% DO and ambient off-gas	n/a	n/a	– Settings: 4500/min stirring, 1 v/v/min airflow, 0 mL/min additional oxygen airflow, 30 °C – Medium complete including antifoam – DO sensor spot reading is on – Calibrate DO spot and off-gas sensor – Delay time: 1 h (time during which readings must be stable to allow calibration) – Acceptance ranges for stable reading: ±0.2 mV oxygen sensor raw reading, ±0.2° DO sensor spot phase, ±0.2 °C off-gas temperature, ±0.2% expected air saturation, ±0.2 °C temperature – Max wait time can be defined, e.g., 10 h
Calibration settings pH	n/a	n/a	Conditions: – Temperature 30 °C – pH of the medium measured offline and introduced into the pH calibration step – Timings: 5 min minimum wait, 60 min maximum wait, 2 min delay time Max allowed changes in readings: ±0.03 raw pH reading, ±0.05 °C culture temperature

(continued)

Table 1
(continued)

Parameter	Typical setpoint	Typical range	Comments/details
Experiment phases	n/a	n/a	Triggered in order of appearance 1. Start: Startup of all control loops and definition of setpoints (except feed rate). DO 100%, off-gas and pH calibration & offset correction 2. Inoculation: Manual or automatic inoculation of the bioreactors and timestamp for cultivation start 3. Batch: Feed start trigger evaluation, potentially sampling 4. Feed: Feeding, time-triggered sampling 5. End: After last sampling event. Stops all pumps, minimum stirring, cooldown to 12 °C
Feed start	n/a	n/a	1st condition: CER >20 mmol/L/h 2nd condition: CER drop by more than 25% compared to peak value 2nd condition only checked after 1st evaluated true
Feed regime	n/a	n/a	– Step finishes immediately, under "option" choose the general feed regime (e.g., linear). If different feed regimes are to be combined (e.g., exponential feed followed by constant feed) choose option "profile" and define the parameters and timings under "edit profiles" (*see* **Notes 28** and **29**) – Adjust for sampling: Flow rate and volume (sampling since inoculation) (*see* **Note 30**) – Priming option: First move liquid to end
Sampling	n/a	n/a	– Typical sampling volume: 4.5 mL, adjust according to analytical needs (*see* **Note 31**) – Sample into 15 mL conical tubes (24× 15 mL racks in cooled positions). Define the sample destinations for each sample individually (*see* **Note 12**) – Schedule samples along phases of interest (*see* **Note 32**)
Foam control	on	n/a	– Settings: Bolus pipette mode (*see* **Note 33**), 0.1 mL bolus volume, 5 mL maximum volume added, 5 min dead time between boli, condition "foam sensor" reading >100, 2 min delay time

2. Have the following prepared:
   (a) Batch medium (*see* Subheading 2.2).
   (b) Feed solution (*see* Subheading 2.3).
   (c) Titration agents (*see* Subheading 2.5).

   (d) CIP solutions (*see* Subheading 2.6).
   (e) Antifoam solution (*see* Subheading 2.7).
   (f) Two autoclaved measuring cylinders covered with aluminum foil (to fill the bioreactors).
   (g) An autoclaved lid for a 175 mL bottle wrapped in aluminum foil (for antifoam).
3. Start the "Runtime" software that controls the instrument.
4. Sanitize the wash pump circuit by replacing the wash pump fluid with 6% (w/v) H_2O_2 and running it for 30–60 min. Change back to sterile water afterward. The wash pump needs to be stopped when swapping solutions.
5. Check that all tubing is correctly installed.
6. Start the CIP protocol "regular sanitization" for all tubing needed during the run.
7. Prepare the liquid handler:
   (a) Check that the tip waste bag is empty, replace if not.
   (b) Clean the liquid handler tools using an ethanol or isopropanol wipe.
   (c) Replace the drip catcher pad.
8. Fill the bioreactors with 95 mL batch medium each using the sterile measuring cylinders from **Preparation step 2.6**. Take an offline pH measurement from leftover batch medium.
9. Install the bioreactors in their respective bays and connect pH probes and stirrer headplates.
10. Enter the actual starting volumes, in this case 95 mL, in the software.
11. In the experiment control script, step "manual sample": Enter the volume that needs to be sampled to reach the actual batch starting volume of 90 mL, in this case 5 mL. This sample will later be used for pH offset correction (*see* **Preparation step 22**).
12. Verify that all bioreactors are working properly by checking the online pH and DO. In case of defects, replace affected bioreactors.
13. Aseptically add 90 μL (0.1% (v/v) antifoam agent to each bioreactor (*see* **Note 9**).
14. Provide the pH of the batch medium measured in **Preparation step 8** in the "calibrate pH" step of the experiment control script.
15. Switch on the chiller.
16. Verify that the parameters for DO calibration are correct (as per Table 1).

17. Once the CIP procedure is finished, connect the titration agent and feed reservoir bottles to their respective addition lines and prime them using the "fill tube lines" step applying the following settings: 7.5 mL per line, 100 mL/h flow rate (*see* **Note 20**).
18. Connect the liquid manifolds to the individual bioreactors (step "connect bioreactors").
19. Prime the addition ports using the "prime reactor inlets" step applying the following settings: 140 μL per port, 8 mL/h flow rate. Confirm successful priming by visual inspection.
20. Acknowledge "Enable automatic pumping" option to switch them from manual to automatic mode.
21. Next, the instrument will automatically execute a 1-point DO sensor spot calibration at 100% air saturation as per experiment definition. With the exemplary configuration from Table 1, this will take approximately 1 h and the off-gas module will automatically be calibrated alongside (*see* **Note 21**).
22. The automated 1-point pH calibration as per experiment definition will take place right afterward. With the exemplary configuration from Table 1, the instrument allows for a 14 h equilibration time for the pH probes and subsequently calibrates to the offline pH value from **Preparation step 8** (*see* **Note 21**).
23. Once the pH calibration has completed, the instrument will switch on the pH control loop for 2 h to titrate the cultivation medium to the target value. It will switch off again after this period to avoid oscillation effects (*see* **Note 21**).
24. After calibration and pH titration, manually sample the individual bioreactors to bring them to their target initial batch volume (90 mL in this case). Take offline pH measurements of all samples. Enter the measured values in the "Runtime" software in the "pH offsets" tab for a 1-point adjustment of the individual pH probes to the current values.
25. Load the labware onto the worktable of the instrument:
    (a) Three racks of 10 mL tips (24 tips each) for up to 6 samples from 12 bays.
    (b) One rack of 1 mL tips (96 tips each) for antifoam dosage.
    (c) Up to 4 racks of sample containers. The authors had a good experience with 15 mL conical tubes for sampling (24 per rack), but other tubes or deep-well plates also work.
    (d) One 175 mL Nalgene® bottle with sterile antifoam (for automated dosage) with corresponding bottle lid (can be removed by the liquid handler to aspirate from the bottle).

26. Have a final check of the instrument:
    (a) Are all parameters at setpoint?
    (b) Are the titration agents and feed connected correctly?
    (c) Is the communication between the instrument and the computer up and running?
    (d) Are all materials required on the worktable?

    Preparation is now completed, and bioreactors are ready for inoculation (*see* **Note 22**).

3.4 Cultivation Run

1. Continue from here, once both the instrument is ready (Subheading 3.3, **Preparation step 26** completed) and the precultures are ready (Subheading 3.2, **step 4** completed).
2. It is recommended to inoculate the main cultures to a defined starting biomass concentration, typically by OD_{600} for reasons of practicability. Determine the OD_{600} of the individual precultures and calculate the respective volumes required to inoculate the main cultures to the desired starting biomass concentration.
3. Inoculate the individual main cultures with the volumes calculated before, e.g., by using serological pipettes (*see* **Notes 23** and **24**). Having done so, acknowledging the "manual inoculation" checkbox will transition the control script into the next phase and start the inoculation timer.
4. From this point on, the main cultures are run fully automated according to the pre-defined process. In addition to the control of process parameters, feeding, and any antifoam dosing, this also includes sampling in accordance with the defined sampling regime. The generated samples are provided on the worktable in the sample containers (*see* **Note 12**).

3.5 Post-use Cleaning

1. Once the main cultivation and potential harvest are finished, uninstall the bioreactors using the "unload bioreactors" step (*see* **Note 25**):
    (a) Unscrew stirrer headplates.
    (b) Disconnect pH cables and put rubber caps back on the sockets.
    (c) Disconnect liquid manifolds and put the wash caps back on.
    (d) Discard the bioreactors as per effective biological waste disposal procedure.
2. Sanitize the wash pump circuit by replacing the wash pump fluid with 6% (w/v) H_2O_2 and running it for 30–60 min. Change back to sterile water afterward.
3. Run the post-use CIP procedure for all tubing used during the run using the default "regular sterilization with Ethanol" step.

4. Clean and tidy up the instrument:
   (a) Clean all surfaces inside the laminar flow hood using a wipe soaked with 70% (v/v) ethanol or 99.9% (v/v) isopropanol (*see* **Note 26**).
   (b) Unload potentially remaining consumables, empty racks, waste, drip capture pad, etc. and dispose as per effective (biological) waste disposal procedure.
   (c) Check the off-gas filter and replace it if it is dirty or contaminated.
   (d) Check the pump lines for crystals or precipitate and exchange as required.

3.6 Exemplary Process Data

Exemplary process data for glucose-regulated heterologous expression of a nanobody in *K. phaffii* applying both a linear and an exponential feeding regime are shown in Fig. 1. All parameters are normalized to respective maxima with the process time relative to feed start (negative values correspond to the batch phase while positive values correspond to the feed/expression phases). The respective feed rates are shown for easy visual discrimination of the individual process phases.

The following observations can be made from Fig. 1:

1. During the batch phase, the biomass increases under conditions of excess substrate supply (Fig. 1a, b), leading to an exponential increase in CTR and OTR (Fig. 1c, d), as well as titration base consumption (Fig. 1a, b).
2. After a sudden decrease in DO due to the inoculation of the bioreactors with the seed culture, the DO in the culture liquid decreases exponentially until the setpoint is reached. Beyond this point, the stirring speed and supplementation with pure oxygen are adjusted by the cascaded DO control loop to maintain a DO above the setpoint (Fig. 1c, d).
3. Product formation is not observed until the beginning of the feeding phase (Fig. 1a, b), as expression remains repressed along batch substrate excess phases.
4. The end of the batch phase is characterized by a sharp increase in the DO signal together with a corresponding drop in the CTR and OTR (Fig. 1c, d). These readings serve to trigger the start of the feed phase, leading to a further increase in the biomass concentration and the initiation of target protein expression.
5. The CTR, OTR, and titration base consumption also increase during the feeding phase, with a profile that is dictated by the feeding regime in an approximate linear and exponential fashion, respectively, corresponding to the feeding regime.

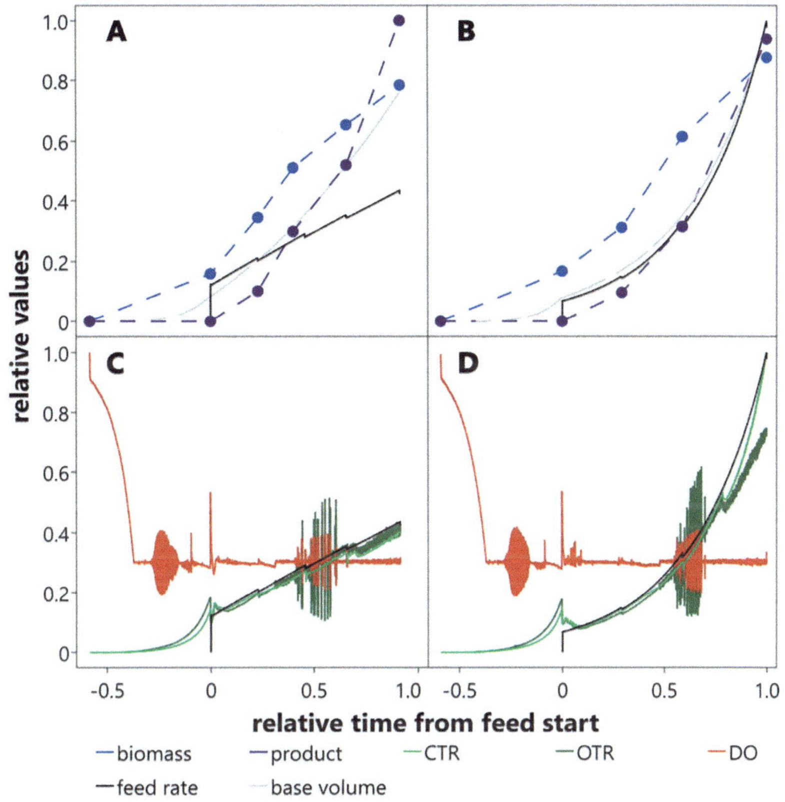

Fig. 1 Heterologous expression of a nanobody in *K. phaffii* using two different feeding regimes (left and right column). Time series of biomass, extracellular product and cumulative base addition are shown as examples of (**a**) linear and (**b**) exponential feeding regimes. Resulting courses of DO, carbon dioxide and oxygen transfer rates are illustrated for (**c**) linear and (**d**) exponential feedings. All subplots display time series of feed rates for illustration purposes. All timescales are normalized to time of feed start

6. From distinct drops in the CTR, OTR, and feed flow rate signals (e.g., relative time 0.23, 0.46, 0.66, and 0.91 for the linear feed process), it can be observed how the process control software adjusts the feed rate based on volume corrections upon sample withdrawal from the individual bioreactors (*see* **Note 30**).
7. The merits of evaluating different feed regimes are clearly illustrated. Namely, in this example, a linear feeding regime translates to comparable target product titers with less production of biomass (which needs to be separated, inactivated, and discarded) and a lower consumption of feed (carbon source), titration base, and oxygen. This translates into higher product yields and a lower risk of incurring mass and/or heat transfer limitations during scale-up.

4 Notes

1. To improve batch-to-batch reproducibility and minimize variations due to raw materials used, defined media are generally favored [48]. Depending on existing strain auxotrophies, minimal media might require specific supplementation. Depending on the targeted biomass, concentration limitations in biotin and/or trace elements might occur. If these are observed, additional supplementation is recommended.
2. A more recent version of the batch medium features an increase in $NH_4H_2PO_4$ to 20.8 g/L and KCl to 1.6 g/L as well as an additional post-sterilization addition of 8 mL/L of a 0.1 g/L biotin stock solution. Biotin is still added to the trace elements solution [49].
3. A more recent version of the feed solution features an increase in KCl to 8 g/L, $MgSO_4 \cdot 7\ H_2O$ to 22 g/L, and a reduction of $CaCl_2 \cdot 2\ H_2O$ to 0.058 g/L with no additional supplementation of trace elements solution [49].
4. To ensure reproducible and precise feeding performance, the viscosity of feed fluids needs consideration. The maximum supported concentrations for glycerol and glucose feed solutions are 75% (w/v) and 68% (w/v), respectively. The authors had good experiences with ≤70% (w/v) glycerol and ≤ 50% (w/v) glucose.
5. Priming of addition lines requires 7.5 mL per line. Therefore, the authors recommend preparing at least 200 mL more of the respective solution than is required for the actual cultivation process.
6. $CoCl_2 \cdot 7\ H_2O$ (CAS no. 7791-13-1) is defined as a substance of very high concern according to the European Chemicals Agency and thus requires handling with appropriate care [50].
7. For safety reasons, the NH_3 is not to be autoclaved but prepared by aseptic addition.
8. Maximum concentration of NH_3 for titration purposes should not exceed 12.5% (w/v) to avoid damage to the tubing material and dosing inaccuracy due to gassing-out.
9. Antifoam can modulate membrane permeability and thus impact target protein secretion [51]. Therefore, antifoam use (type and amount) should be harmonized across processes as far as possible. The authors had good experiences with AF204 (Sigma-Aldrich) and PPG2000.
10. DO is measured by means of an oxygen-sensitive fluorescent dye, which is inherently subject to optical interference [52]. In suspected cases, it is recommended to check any biogenic

fluorescence (especially where fluorescent reporter proteins are used) against the wavelengths relevant for the DO measurement.

11. While inline biomass estimation via light scattering can reduce hands-on efforts, implementation of appropriate calibration models (potentially dependent on physiological state), dynamic range, and potential interference by dispersed bubbles need consideration.
12. The liquid handler integrated into the bioreactor platform is only capable of acquiring and storing (typically at 4 °C) samples, but not further processing them. They might no longer be representative if stored for too long. If the implementation of quenching protocols [53–55] is not an option, sample processing should be scheduled as timely as possible.
13. While this step appears straightforward at first look, it is strongly recommended to properly define the research question before moving forward, as its revision during the study can easily turn any data acquired as of that point futile. Key aspects are controllable inputs as well as reliably quantifiable response variables relating to the research question.
14. It is recommended to start defining how many experiments can be afforded before defining a detailed experimental plan.
15. The most common DoE designs for screening purposes are two-level full factorial and fractional factorial designs, requiring 2^k and 2^{k-p} number of experiments, respectively (where k is the number of factors being studied and p is the fraction studied). In Plackett–Burman designs, the number of experiments is a multiple of 4, providing more flexibility at the cost of a higher level of confounding of main effects with two-factor interactions. Therefore, Plackett–Burman designs might be more suitable for screening experiments of large main effects when resources are limited (e.g., when only 12 or 24 experiments can be run). Less common alternatives are Rechtschaffner and L-designs. Rechtschaffner designs are saturated two-level fractional factorials that are efficient in situations where two-factor interaction estimates are important. L-designs represent three- or mixed-level fractional factorial designs that can be useful if—for example—three different strains are to be tested, or if quadratic behavior is anticipated. CCDs are the most common RSM designs and can be considered as augmented (fractional) factorial designs. Box–Behnken RSM designs are recommended when the high and low levels of the k studied factors are considered "too extreme" (e.g., if high pH combined with high temperature seems unreasonable) and the focus is placed in the level midpoints instead. With DoE software readily available, computer-generated designs such as optimal designs

have emerged as interesting alternatives. They aim at the best possible subset of experiments selected from a candidate set of designs for selected factor settings, thereby allowing to span the largest possible design space with least experimental effort. D-optimal designs can be useful due to their flexibility and ability to support constrained experimental regions and complex problems, a typical challenge to bioprocesses. For more details, the reader is referred to specialized literature [34–36].

16. While empirical approaches have routinely been used to fit DoE-derived data with purely statistical models (e.g., linear, quadratic, and polynomic), a paradigm shift toward so-called model-based DoEs (mDoE) has been observed for a few years now. Here, data fitting has initially been done with mechanistic models that allow for extraction of descriptive features. As a more recent development, hybrid models, which combine data-driven and mechanistic components, are becoming increasingly popular as they combine the specific advantages of both approaches [46, 56, 57].
17. To minimize lag phase duration in main cultivation, precultures should always be harvested during exponential growth phase.
18. The sequence of individual preparation steps can be adjusted to adapt to the general working environment. In any case, it is essential to complete preparation before harvest of the precultures.
19. The authors recommend that experiment templates are created to minimize risk of misconfiguration during setup.
20. Before starting the priming procedure, it should be double-checked that all connections are tight and do not leak. If bubbles are observed in the tubing after priming is completed, priming should be repeated.
21. The authors recommend scheduling the instrument preparation in such a way that these steps are automatically executed by the instrument overnight.
22. It is recommended to deactivate the "wash pump flow," "wash pump running," and "airflow" (laminar flow hood) alarms before starting the main cultivation. In the event of an alarm, these will stop pumps and the liquid handler which may result in loss of the cultivation run.
23. Using its liquid handler, the bioreactor platform also supports automatic inoculation from up to 3 stirred 175 mL Nalgene® bottles that can be placed on the worktable. However, for more than three different precultures, manual exchange of the bottles is required. Also, the automated inoculation procedure is

comparably slow. Therefore, the authors typically prefer manual inoculation.

24. In the most simplistic approach, untreated precultures are used for inoculation. In case, one might want to minimize carryover of preculture medium into main culture, it is recommended to wash the inoculum by (multiple) cycles of centrifugation and resuspension in buffer (e.g., phosphate buffered saline). In such case, it is highly recommended to measure OD after the final resuspension step and use this value to calculate the volume for inoculation.
25. It is essential to first stop any ongoing operation of the system by using the "unload" function before starting the disassembly. Trying to disassemble the stirrer headplate while stirrers are still running can lead to damage of the instrument.
26. Do not wipe the inside of the bays to avoid scratching the measurement optics. Only do so if unavoidable due to contamination or spill.
27. While the minimum initial volume specified is 115 mL, the authors had good experiences with 90 mL starting volume.
28. Compared to most other microbial processes, *K. phaffii* cultivation processes may feature extraordinary oxygen transfer and/or heat dissipation demands or even feed pump rates beyond the capabilities of the instrumentation while particularly exponential regimes can be challenging. This may (temporarily) lead to oxygen limitation, observable either by DO levels around 0% or plateau phases in OTR over time [58], and temperature rise beyond the desired setpoint. These phenomena typically have a detrimental effect on process performance and should be avoided. In the first instance, respective control loops should be checked. If all controlled parameters are already at maximum, the instrument's capacities have been reached and restricting carbon supply to the bioreactors might need to be considered. If this has not (yet) happened, the control loop settings might not be aggressive enough to follow the process kinetics, and their optimization should be considered.
29. Maximum specific growth rates of *K. phaffii* are in the range of 0.2/h [31, 59]. To minimize the risk of overfeeding no profiles that translate into μ >0.18/h should be employed.
30. In contrast to larger bioreactors, the relative fraction of volume removed by sampling is comparably high (e.g., a 5 mL sample from 100 mL working volume already translates into a 5% change). Thus, feed rates should be adjusted accordingly to

avoid overfeeding. This can be configured in the instrument control software using eq. (1):

$$f(t)_{\mathrm{corr}} = f(t) \cdot X(t)_{\mathrm{cell}} \tag{1}$$

where $f(t)_{corr}$ in mL/h is the feed rate after correction, $f(t)$ in mL/h is the feed rate prior to correction, and $X(t)_{cell}$ is the relative fraction of biomass remaining at time t. $X(t)_{cell}$ is initially instantiated to 1 and is reduced upon sampling throughout the process. For example, if 10 mL is sampled from an initial volume of 100 mL, it is assumed that $X(t)_{cell}$ is thereby reduced from 1 to 0.9 which translates into a 10% reduction of the actual feed rate at this point in time. Since this estimation is based on the amount of biomass removed, it is restricted to sampling event corrections and does not take into account volume loss due to evaporation or additions such as titration or antifoam agents, and inducers.

31. During sampling, dispersed gas bubbles are aspirated together with culture broth, decreasing the obtained against the desired volume. This can be compensated by correcting with the actual gas hold-up. Gas hold-up can be estimated by correlation models that need to be implemented by the operator. While parametrization is mostly process specific, literature provides different models that can be utilized as a baseline [60, 61]. Such corrections need careful parametrization, as wrong estimates will cause a difference between actual and assumed culture broth volume, translating to an incorrect feed rate adjustment upon sampling events if enabled. In particular, for late samples with typically high biomass concentrations and elevated viscosity, a tendency of sampling too little liquid volume has frequently been observed.
32. Sampling schedules are defined to meet the experiment's purpose (e.g., estimation of specific rates or other key performance indicators during certain process phases), but should be kept at the lowest frequency feasible to minimize interference with the ongoing process.
33. Alternatively, if high antifoam needs are expected, one of the instrument's addition lines can be used to feed diluted antifoam.

Acknowledgments

The authors wish to thank Samantha Capewell (Lonza AG) and Mechael Kanovsky for language editing.

Competing Interest Statement The authors have no conflicts of interest to declare that are relevant to the content of this chapter.

References

1. Mattanovich D, Borth N (2006) Applications of cell sorting in biotechnology. Microb Cell Factories 5:12. https://doi.org/10.1186/1475-2859-5-12
2. Bahnemann J, Grünberger A (2022) Microfluidics in biotechnology: overview and status quo. In: Bahnemann J, Grünberger A (eds) Microfluidics in biotechnology, 1st edn. Springer Nature, Cham. https://doi.org/10.1007/10_2022_206
3. Duetz WA (2007) Microtiter plates as mini-bioreactors: miniaturization of fermentation methods. Trends Microbiol 15:469–475. https://doi.org/10.1016/j.tim.2007.09.004
4. Hemmerich J, Noack S, Wiechert W, Oldiges M (2018) Microbioreactor systems for accelerated bioprocess development. Biotechnol J 13: 1700141. https://doi.org/10.1002/biot.201700141
5. Hemmerich J, Adelantado N, Barrigón JM, Ponte X, Hörmann A, Ferrer P, Kensy F, Valero F (2014) Comprehensive clone screening and evaluation of fed-batch strategies in a micro-bioreactor and lab scale stirred tank bioreactor system: application on *Pichia pastoris* producing *Rhizopus oryzae* lipase. Microb Cell Factories 13:36. https://doi.org/10.1186/1475-2859-13-36
6. Back A, Rossignol T, Krier F, Nicaud J-M, Dhulster P (2016) High-throughput fermentation screening for the yeast *Yarrowia lipolytica* with real-time monitoring of biomass and lipid production. Microb Cell Factories 15: 147. https://doi.org/10.1186/s12934-016-0546-z
7. Müller C, Bakkes PJ, Lenz P, Waffenschmidt V, Helleckes LM, Jaeger K-E, Wiechert W, Knapp A, Freudl R, Oldiges M (2022) Accelerated strain construction and characterization of *C. glutamicum* protein secretion by laboratory automation. Appl Microbiol Biotechnol 106:4481–4497. https://doi.org/10.1007/s00253-022-12017-7
8. Helleckes LM, Müller C, Griesbach T, Waffenschmidt V, Moch M, Osthege M, Wiechert W, Oldiges M (2023) Explore or exploit? A model-based screening strategy for PETase secretion by *Corynebacterium glutamicum*. Biotechnol Bioeng 120:139–153. https://doi.org/10.1002/bit.28261
9. de Souza Vandenberghe LP, Herrmann LW, de Oliveira Penha R, de Mello AFM, Martínez-Burgos WJ, Junior AIM, de Souza Kirnev PC, de Carvalho JC, Soccol CR (2022) Engineering aspects for scale-up of bioreactors. In: Sirohi R, Pandey A, Taherzadeh MJ, Larroche C (eds) Current developments in biotechnology and bioengineering, 1st edn. Elsevier, Amsterdam. https://doi.org/10.1016/B978-0-323-91167-2.00002-2
10. Gill NK, Appleton M, Baganz F, Lye GJ (2008) Quantification of power consumption and oxygen transfer characteristics of a stirred miniature bioreactor for predictive fermentation scale-up. Biotechnol Bioeng 100:1144–1155. https://doi.org/10.1002/bit.21852
11. Yurimoto H, Sakai Y (2019) Methylotrophic yeasts: Current understanding of their C1-metabolism and its regulation by sensing methanol for survival on plant leaves. Curr Issues Mol Biol 33:197–210. https://doi.org/10.21775/cimb.033.197
12. Hartner FS, Ruth C, Langenegger D, Johnson SN, Hyka P, Lin-Cereghino GP, Lin-Cereghino J, Kovar K, Cregg JM, Glieder A (2008) Promoter library designed for fine-tuned gene expression in *Pichia pastoris*. Nucleic Acids Res 36:e76. https://doi.org/10.1093/nar/gkn369
13. Vogl T (2022) Engineering of promoters for gene expression in *Pichia pastoris*. Methods Mol Biol 2513:153–177. https://doi.org/10.1007/978-1-0716-2399-2_10
14. Vogl T, Glieder A (2013) Regulation of *Pichia pastoris* promoters and its consequences for protein production. New Biotechnol 30:385–404. https://doi.org/10.1016/j.nbt.2012.11.010
15. Stratton J, Chiruvolu V, Meagher M (1998) High cell-density fermentation. Methods Mol Biol 103:107–120. https://doi.org/10.1385/0-89603-421-6:107
16. Chongchittapiban P, Borg J, Waiprib Y, Pimsamarn J, Tongta A (2016) On-line methanol sensor system development for recombinant human serum albumin production by *Pichia pastoris*. Afr J Biotechnol 15:2374–2383. https://doi.org/10.5897/AJB2015.15122
17. Zhou X-S, Fan W-M, Zhang Y-X (2002) Effects of different methanol feeding strategy on hirudin production in high-density fermentation by recombinant *Pichia pastoris*. Chin J Biotechnol 18:348–351
18. Li P, Anumanthan A, Gao X-G, Ilangovan K, Suzara VV, Düzgüneş N, Renugopalakrishnan V (2007) Expression of recombinant proteins in *Pichia pastoris*. Appl Biochem Biotechnol 142:105–124. https://doi.org/10.1007/s12010-007-0003-x

19. Allampalli P, Rathinavelu S, Mohan N, Sivaprakasam S (2022) Deployment of metabolic heat rate based soft sensor for estimation and control of specific growth rate in glycoengineered *Pichia pastoris* for human interferon alpha 2b production. J Biotechnol 359:194–206. https://doi.org/10.1016/j.jbiotec.2022.10.006
20. Pla IA, Damasceno LM, Vannelli T, Ritter G, Batt CA, Shuler ML (2006) Evaluation of Mut^+ and Mut^S *Pichia pastoris* phenotypes for high level extracellular scFv expression under feedback control of the methanol concentration. Biotechnol Prog 22:881–888. https://doi.org/10.1021/bp060012+
21. Mattanovich D, Jungo C, Wenger J, Dabros M, Maurer M (2014) Yeast suspension culture. In: Meyer HP, Schmidhalter D (eds) Industrial scale suspension culture of living cells, 1st edn. Wiley, Weinheim. https://doi.org/10.1002/9783527683321.ch02
22. Waterham HR, Digan ME, Koutz PJ, Lair SV, Cregg JM (1997) Isolation of the *Pichia pastoris* glyceraldehyde-3-phosphate dehydrogenase gene and regulation and use of its promoter. Gene 186:37–44. https://doi.org/10.1016/S0378-1119(96)00675-0
23. Chisti Y, Moo-Young M (1996) Bioprocess intensification through bioreactor engineering. Chem Eng Res Des 74:575–583
24. Takors R (2012) Scale-up of microbial processes: Impacts, tools and open questions. J Biotechnol 160:3–9. https://doi.org/10.1016/j.jbiotec.2011.12.010
25. Chmiel H (2011) Bioprozesstechnik. Spektrum Akademischer Verlag, Heidelberg. https://doi.org/10.1007/978-3-8274-2477-8
26. Noorman H (2011) An industrial perspective on bioreactor scale-down: What we can learn from combined large-scale bioprocess and model fluid studies. Biotechnol J 6:934–943. https://doi.org/10.1002/biot.201000406
27. Doran PM (2013) Reactor engineering. In: Doran PM (ed) Bioprocess engineering principles, 2nd edn. Academic, London. https://doi.org/10.1016/B978-0-12-220851-5.00014-9
28. Weinhandl K, Winkler M, Glieder A, Camattari A (2014) Carbon source dependent promoters in yeasts. Microb Cell Factories 13:1–17. https://doi.org/10.1186/1475-2859-13-5
29. Gao J, Jiang L, Lian J (2021) Development of synthetic biology tools to engineer *Pichia pastoris* as a chassis for the production of natural products. Synth Syst Biotechnol 6:110–119. https://doi.org/10.1016/j.synbio.2021.04.005
30. Gou XH, Liu YY, Chen QL, Tang JJ, Liu DY, Zou L, Wu XY, Wang W (2012) High level expression of bikunin in *Pichia pastoris* by fusion of human serum albumin. AMB Express 2:14. https://doi.org/10.1186/2191-0855-2-14
31. Maurer M, Kühleitner M, Gasser B, Mattanovich D (2006) Versatile modeling and optimization of fed batch processes for the production of secreted heterologous proteins with *Pichia pastoris*. Microb Cell Factories 5:37. https://doi.org/10.1186/1475-2859-5-37
32. John GT, Klimant I, Wittmann C, Heinzle E (2003) Integrated optical sensing of dissolved oxygen in microtiter plates: A novel tool for microbial cultivation. Biotechnol Bioeng 81:829–836. https://doi.org/10.1002/bit.10534
33. Montgomery DC (2017) Design and analysis of experiments. Wiley, Arizona State University
34. Eriksson L, Johansson E, Kettaneh-Wold N, Wikström C, Wold S (2008) Design of experiments: Principles and applications. Umetrics Academy, Umeå
35. NIST/SEMATECH (2012) e-Handbook of statistical methods. http://www.itl.nist.gov/div898/handbook/. Accessed 12 Apr 2024
36. Coffey T (2018) Design of experiments (DoE) for process development. In: Statistics for biotechnology process development, 1st edn. Chapman and Hall/CRC, New York
37. Jafari R, Sundström BE, Holm P (2011) Optimization of production of the anti-keratin 8 single-chain Fv TS1-218 in *Pichia pastoris* using Design of Experiments. Microb Cell Factories 10:34. https://doi.org/10.1186/1475-2859-10-34
38. Fricke J, Pohlmann K, Tatge F, Lang R, Faber B, Luttmann R (2011) A multi-bioreactor system for optimal production of malaria vaccines with *Pichia pastoris*. Biotechnol J 6:437–451. https://doi.org/10.1002/biot.201000385
39. Fricke J, Pohlmann K, Jonescheit NA, Ellert A, Joksch B, Luttmann R (2013) Designing a fully automated multi-bioreactor plant for fast DoE optimization of pharmaceutical protein production. Biotechnol J 8:738–747. https://doi.org/10.1002/biot.201200190
40. Harms J, Wang X, Kim T, Yang X, Rathore AS (2008) Defining process design space for biotech products: Case study of *Pichia pastoris* fermentation. Biotechnol Prog 24:655–662. https://doi.org/10.1021/bp070338y
41. Lin H, Kim T, Xiong F, Yang X (2007) Enhancing the production of Fc fusion protein in fed-batch fermentation of *Pichia pastoris* by

Design of Experiments. Biotechnol Prog 23: 621–625. https://doi.org/10.1021/bp0603199

42. Palmerín-Carreño D, Martínez-Alarcón D, Dena-Beltrán JL, Vega-Rojas LJ, Blanco-Labra A, Escobedo-Reyes A, García-Gasca T (2021) Optimization of a recombinant lectin production in *Pichia pastoris* using crude glycerol in a fed-batch system. Processes 9:876. https://doi.org/10.3390/pr9050876
43. Blainey P, Krzywinski M, Altman N (2014) Replication. Nat Methods 11:879–880. https://doi.org/10.1038/nmeth.3091
44. Goulet M-A, Cousineau D (2019) The power of replicated measures to increase statistical power. Adv Methods Pract Psychol Sci 2:199–213. https://doi.org/10.1177/2515245919849434
45. Box GEP, Hunter JS, Hunter WG (2005) Statistics for experimenters: Design, innovation, and discovery. Wiley, Hoboken
46. Bayer B, Striedner G, Duerkop M (2020) Hybrid modeling and intensified DoE: an approach to accelerate upstream process characterization. Biotechnol J 15:2000121. https://doi.org/10.1002/biot.202000121
47. von Stosch M, Willis MJ (2017) Intensified design of experiments for upstream bioreactors. Eng Life Sci 17:1173–1184. https://doi.org/10.1002/elsc.201600037
48. Matthews CB, Kuo A, Love KR, Love JC (2018) Development of a general defined medium for *Pichia pastoris*. Biotechnol Bioeng 115:103–113. https://doi.org/10.1002/bit.26440
49. Prielhofer R, Maurer M, Klein J, Wenger J, Kiziak C, Gasser B, Mattanovich D (2013) Induction without methanol: novel regulated promoters enable high-level expression in *Pichia pastoris*. Microb Cell Factories 12:5. https://doi.org/10.1186/1475-2859-12-5
50. European Chemicals Agency (2022) Substance infocard cobalt (II) chloride hexahydrate. https://echa.europa.eu/substance-information/-/substanceinfo/100.119.108. Accessed 30 Jan 2023
51. Routledge SJ, Hewitt CJ, Bora N, Bill RM (2011) Antifoam addition to shake flask cultures of recombinant *Pichia pastoris* increases yield. Microb Cell Factories 10:17. https://doi.org/10.1186/1475-2859-10-17
52. Kunze M, Roth S, Gartz E, Büchs J (2014) Pitfalls in optical on-line monitoring for high-throughput screening of microbial systems. Microb Cell Factories 13:53. https://doi.org/10.1186/1475-2859-13-53
53. Cruz Bournazou MN, Barz T, Nickel DB, Lopez Cárdenas DC, Glauche F, Knepper A, Neubauer P (2017) Online optimal experimental re-design in robotic parallel fed-batch cultivation facilities. Biotechnol Bioeng 114: 610–619. https://doi.org/10.1002/bit.26192
54. Baumschlager A, Aoki SK, Khammash M (2017) Dynamic blue light-inducible T7 RNA polymerases (Opto-T7RNAPs) for precise spatiotemporal gene expression control. ACS Synth Biol 6:2157–2167. https://doi.org/10.1021/acssynbio.7b00169
55. Lanier LL, Warner NL (1981) Paraformaldehyde fixation of hematopoietic cells for quantitative flow cytometry (FACS) analysis. J Immunol Methods 47:25–30. https://doi.org/10.1016/0022-1759(81)90253-2
56. Tsopanoglou A, Jiménez del Val I (2021) Moving towards an era of hybrid modelling: advantages and challenges of coupling mechanistic and data-driven models for upstream pharmaceutical bioprocesses. Curr Opin Chem Eng 32:100691. https://doi.org/10.1016/j.coche.2021.100691
57. Narayanan H, Sokolov M, Morbidelli M, Butté A (2019) A new generation of predictive models: The added value of hybrid models for manufacturing processes of therapeutic proteins. Biotechnol Bioeng 116:2540–2549. https://doi.org/10.1002/bit.27097
58. Anderlei T, Büchs J (2001) Device for sterile online measurement of the oxygen transfer rate in shaking flasks. Biochem Eng J 7:157–162. https://doi.org/10.1016/S1369-703X(00)00116-9
59. Garcia-Ortega X, Ferrer P, Montesinos JL, Valero F (2013) Fed-batch operational strategies for recombinant Fab production with *Pichia pastoris* using the constitutive GAP promoter. Biochem Eng J 79:172–181. https://doi.org/10.1016/j.bej.2013.07.013
60. Kudrewizki F (1982) Zur Berechnung des Gasgehaltes in begasten Rührkesseln. Chem Tech 5:247–249
61. Ganguly SK, Thakare A, Patil DP, Tibrewal A, Dethe A, Majumder CB, Ray A (2020) Estimation of gas holdup using the gassed to ungassed power ratio of an oxygen–water system in a stirred and sparged tank contactor. ACS Omega 5:28929–28941. https://doi.org/10.1021/acsomega.0c02292

Chapter 35

Kinetic Modeling of *Pichia pastoris* Bioreactor Systems Under Different Operating Conditions: Batch, Fed-Batch, and Chemostat

Julio Berrios, Carlos Martínez, and Patrick Fickers

Abstract

Mathematical models describing the behavior of *Komagataella phaffii* in bioreactors are a powerful tool for process optimization and control. In this chapter, models accounting for different operation modes, namely batch, fed-batch, and continuous culture (i.e., chemostat) under different substrate regimes, have been considered through illustrative examples. Consideration has been given to the dynamics of both intracellular and extracellular components. Simulations highlighting specific variable behavior for each operation mode, such as biomass in fed-batch cultures and steady-state productivity in chemostats, have been included. In particular, the use of methanol with supplementary substrates, such as glycerol or sorbitol, and the integration of intracellular enzyme alcohol oxidase synthesis have been explored. The chapter provides a systematic protocol guiding readers to construct such kinetic models.

Key words *Pichia pastoris*, *Komagataella phaffii*, Modeling, Co-feeding, Methanol, Mass balance, Bioreactor dynamics

Nomenclature

μ	Specific growth rate (h^{-1})
D	Dilution rate (h^{-1})
F	Flow ($L\ h^{-1}$)
k_{Aox}	Aox formation kinetic constant ($mg_{Aox}\ g_X^{-1}\ h^{-1}$)
k_{deg}	Recombinant protein degradation rate (h^{-1})
k_{ev}	Methanol evaporation constant (h^{-1})
K_I	Cell growth Inhibition constant of methanol ($g_M\ L^{-1}$)
K_s	Half saturation constant ($g_M\ L^{-1}$)
P	Recombinant protein (product) concentration ($g\ L^{-1}$)
q	Specific consumption/production rate ($g_S\ g_X^{-1}\ h^{-1}$)
S	Substrate concentration ($g_S\ L^{-1}$ or $mg_S\ L^{-1}$)
E	Intracellular enzyme concentration ($mg_{Aox}\ L^{-1}$)
V	Volume (L)
X	Biomass concentration ($g_X\ L^{-1}$)

Xavier Garcia-Ortega et al. (eds.), *Pichia pastoris: Methods and Protocols*, Methods in Molecular Biology, vol. 2697, https://doi.org/10.1007/978-1-0716-4779-0_35,

$Y_{x/s}$ Yield of substrate S into biomass X ($g_x\, g_S^{-1}$)
vvm volume of air/gas per volume of culture medium per minute ($V_{gas}\, Vl_{iq}\, min^{-1}$)

Subscripts

0 Feeding or Inlet to the bioreactor
G Glycerol
M Methanol
SOR Sorbitol
Aox Alcohol oxidase
max Maximum

1 Introduction

In recent decades, *Komagataella phaffii* (previously named *Pichia pastoris*) has gained significant attention in both fundamental and applied research [1–3]. The increasing number of industrial applications based on this yeast, particularly in the field of recombinant protein production, have underscored the need for quantitative tools for bioprocess characterization, control, and optimization. Mathematical models have emerged as an effective solution to meet this objective. In this context, a mathematical model is a formal representation of yeast cells growing in a bioreactor system under specific operating conditions. From a biotechnological point of view, the main objective of constructing such mathematical models is to establish a predictive framework for understanding the behavior of cells in culture under specific culture conditions for better control and optimization of a process system. Various mathematical models, including intracellular and extracellular levels (i.e., intracellular and extracellular components), have been developed for *K. phaffii* [4, 5]. In this regard, this chapter provides the fundamentals of modeling *K. phaffii* growth in bioreactor systems, focusing on models that are based on mass balances and kinetic parameters. It is important to note that the scope herein is limited to the formulation of mathematical equations and model simulation. Parameter estimation and validation with data collection were not considered. *See* Chapters 31, 32 and 33 in this volume for further technical details about how to run continuous and fed-batch cultivations.

Mass and energy balances are the basis for modeling the cell behavior in a bioreactor. For *K. phaffii* production system, most models have been established using mass balances of biomass and limiting nutrients [4]. These models represent the dynamics of relevant variables for cell growth and recombinant protein

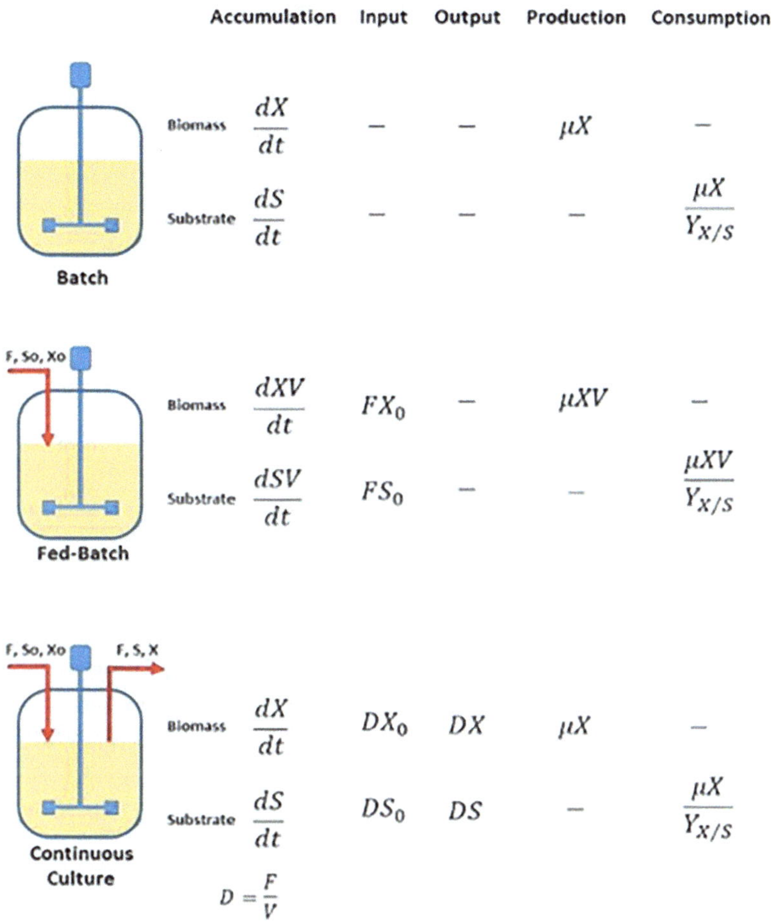

Fig. 1 Mass balance equations for biomass (*X*) and substrate (*S*) in a bioreactor system. Three operating modes are shown: batch, fed-batch, and continuous culture. Note that the mass balances for the fed-batch reactor are based in terms of biomass (g) and not biomass concentration (g L^{-1}) (*see* **Note 1**). As a low cell death rate is assumed in those processes, the cell death rate was neglected in the mass balance calculation

production. *Accumulation* rate is a key concept for setting up these balances and, as a general approach, it considers four terms as follows:

$$\text{Accumulation} = \text{input} - \text{output} + \text{production} - \text{consumption} \quad (1)$$

The terms *input* and *output* in a bioreactor system are the feeding of fresh medium or the removal of spent medium, depending on the operation mode considered (Fig. 1). The *production* is normally related to biomass formation or product synthesis, including recombinant proteins; whereas the *consumption* describes nutrient uptake or degradation (catabolism), cell death, or any process that leads to the transformation of a medium component into a different one.

To illustrate these concepts, three case studies of *P. pastoris* grown under batch, fed-batch, and continuous (i.e., chemostat) culture modes using methanol as the main carbon source are developed below. As a co-substrate, glycerol or sorbitol was considered. Each case relies on mass balances with different considerations tailored to specific operational features.

1.1 Batch Culture of K. phaffii Growing on Methanol as the Sole Carbon Source

The mass balance of a bioreactor operating in batch mode is presented. X denotes the biomass concentration and S_M, the methanol concentration. Then

$$\frac{dX}{dt} = \mu_M X \tag{2}$$

$$\frac{dS_M}{dt} = -q_M X \tag{3}$$

where μ_M represents the specific growth rate and q_M the uptake rate of methanol. Following Eq. (1), input and output are negligible in the batch reactor as it is a closed system. The cell growth is described by μX, and methanol consumption by $-q_M X$.

The description of this simple model is completed by specifying the forms of μ and q_M. If it is assumed that methanol is below its inhibitory concentration, the specific growth rate follows a Monod function, that is,

$$\mu_M(S_M) = \mu_{\max,M} \frac{S_M}{K_{S,M} + S_M} \tag{4}$$

where $\mu_{\max,\,M}$ is the maximum specific growth rate and $K_{S,\,M}$ is the half-saturation constant. The uptake rate of methanol is described by

$$q_M = \frac{\mu}{Y_{X/M}} \tag{5}$$

A simulation of batch culture based on these equations is presented in Subheading 3.1.

1.2 Fed-Batch Culture of K. phaffii Growing in on Methanol and Glycerol

Bioprocess operated in fed-batch mode for recombinant protein production may include the use of a co-substrate such as glycerol together with methanol [6]. Therefore, the consumption of those two carbon sources must be considered in the mass balance equations. Methanol acts as both an inducer and a source of carbon and energy, while glycerol, which also provides carbon and energy, reduces methanol consumption. Such a co-feeding process improves the volumetric productivity of the recombinant protein. For a fed-batch process with two carbon sources, the mass balance is given by the following four-dimensional system:

$$\frac{dXV}{dt} = \mu XV \tag{6}$$

$$\frac{dS_G V}{dt} = FS_{G0} - q_G XV \tag{7}$$

$$\frac{dS_M V}{dt} = FS_{M0} - q_M XV - k_{ev} S_M V \tag{8}$$

$$\frac{dV}{dt} = F \tag{9}$$

In contrast to the batch process of Subheading 1.1, this process accounts for the input flow rate F of carbon source and the feeding solutions, in this case for glycerol (S_{G0}) and methanol (S_{M0}). The terms q_G and q_M account for the specific uptake rates of glycerol and methanol, respectively, and $k_{ev}S_MV$ corresponds to a volatilization term (*see* **Note 2**). Equation (9) describes the increase of volume (V) due to the substrate solution feeding during the culture.

Dual carbon source supply in *K. phaffii* cultures has been modeled in different ways (e.g., [7, 8]). In this case, methanol and glycerol are assimilated simultaneously in fed-batch cultures, and the growth of *K. phaffii* should be described considering the contribution of both carbon sources to cell growth. Thus, a simple model was developed, with the specific growth rate described by the sum of two distinct contributions of the carbon sources, methanol and glycerol:

$$\mu = \mu_G + \mu_M \tag{10}$$

$$\mu_G = \mu_{\max,G} \frac{S_G}{K_{S,G} + S_G} \tag{11}$$

$$\mu_M = \mu_{\max,M} \frac{S_M}{K_{S,M} + S_M + \frac{S_M^2}{K_I}} \tag{12}$$

According to this model, μ comprises both the carbon contribution from glycerol (μ_G) and methanol (μ_M), as described in Eq. (10). In addition, the inhibitory effect of methanol is incorporated into μ_M, introducing an inhibition constant K_I.

The uptake rate for methanol is given by Eq. (5). Similarly, the uptake rate for glycerol is given by

$$q_G = \frac{\mu_G}{Y_{X/G}} \tag{13}$$

A simulation of a two-substrate fed-batch modeling is shown in Subheading 3.2.

1.3 Continuous Culture of *K. phaffii* Growing on Methanol and Sorbitol Considering Intracellular Dynamics

A further step in developing growth models is the integration of intracellular metabolic processes, giving more accuracy and complexity to the model. This approach requires identifying and selecting the cellular processes that are relevant to the variables that are being studied (Fig. 2). To exemplify this modeling approach, a variation of the model developed by Çelik et al. [9] has been

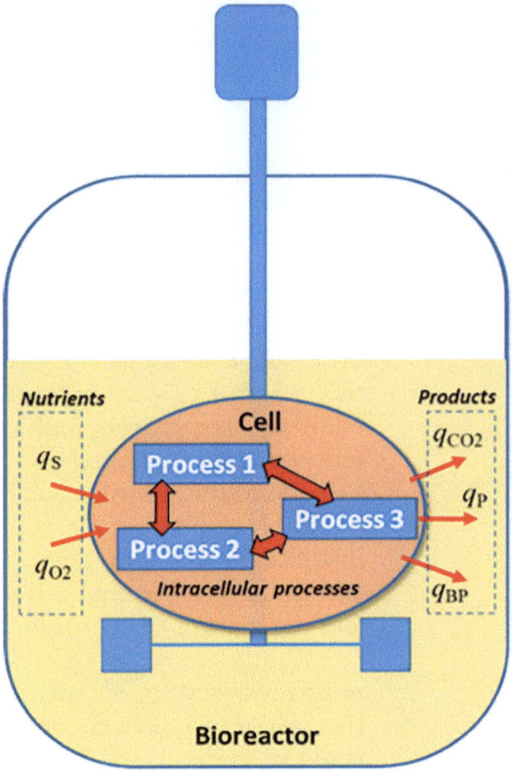

Fig. 2 General representation of an integrated model considering the bioreactor and metabolic processes. Intracellular processes consider the relevant conversions (i.e. by metabolic pathways) that will be considered in the model (Process 1, 2, 3, etc). These processes interact with the bioreactor-level variables through the exchange of nutrients and products by rates of nutrient consumption, production, cell growth, allowing the integration at both levels. BP: by-products

considered, in which the presence of alcohol oxidase enzyme (Aox) and cell maintenance are explicitly taken into account.

For this, a continuous bioreactor in which *K. phaffii* grows on two substrates (methanol and sorbitol) has been considered. The variables X, S_M, and S_{SOR} denote the concentrations of yeast, methanol, and sorbitol, respectively. It is assumed that cells produce a recombinant protein, whose concentration is denoted by P. According to this model, the production of P relies on the activity of the enzyme alcohol oxidase, Aox (concentration within the cell is represented by EAox), whose synthesis is induced by methanol. Thus, the production of Aox is a required step in the conversion of methanol into metabolites and biomass components such as proteins by *K. phaffii*. The balance equations are the following:

$$\frac{dX}{dt} = -DX + \mu X \tag{14}$$

$$\frac{dS_M}{dt} = DS_{M,0} - q_M X - DS_M \tag{15}$$

$$\frac{dS_{\mathrm{Sor}}}{dt} = DS_{\mathrm{Sor},0} - q_{\mathrm{Sor}} X - DS_{\mathrm{Sor}} \tag{16}$$

$$\frac{dE_{\mathrm{Aox}}}{dt} = q_{\mathrm{Aox}} X - \mu E_{\mathrm{Aox}} \tag{17}$$

$$\frac{dP}{dt} = q_P X - DP - k_{\mathrm{deg}} P \tag{18}$$

The mass balance for Aox was made inside the cell, therefore Eq. (17) does not include a term for the reactor dilution, but a term for "cell dilution" in the form μE_{AOX} . This term represents the effect of metabolite or enzyme dilution in the cell compartment since the total volume of the cytosol increases as the cell population grows (*see* **Note 3**). In specific experimental conditions, such as low growth rates, this dilution effect can be considered neglectable [10]. The product equation (Eq. (18)) is described by three terms: the production rate of recombinant protein ($q_p X$), the dilution due to medium feeding rate (DP), and the product degradation ($k_{\mathrm{deg}} P$) which considers a kinetic constant (k_{deg}).

The specific growth rate (μ) is defined as the sum of the specific growth rate on methanol (μ_M) and sorbitol (μ_{SOR}), separately. That is,

$$\mu = \mu_M + \mu_{\mathrm{SOR}}, \tag{19}$$

where μ_M is defined by Eq. (4) and μ_{SOR} is given by

$$\mu_{\mathrm{SOR}} = \frac{\mu_{\max,\mathrm{SOR}} S_{\mathrm{SOR}}}{K_{S,\mathrm{SOR}} + S_{\mathrm{SOR}}}. \tag{20}$$

The specific uptake rate q_M is given by (5) and q_{SOR} is given by

$$q_{\mathrm{SOR}} = \frac{\mu_{\mathrm{SOR}}}{Y_{X/\mathrm{SOR}}} + m_{\mathrm{SOR}}, \tag{21}$$

where m_S is a term accounting for the maintenance on sorbitol. The specific protein production rate is given by

$$q_P = k_p \frac{E_{\mathrm{Aox}}}{X}, \tag{22}$$

where k_p is a kinetic positive constant. Finally, to facilitate the description of the processes, it is assumed that the specific product degradation rate (k_{deg}) is constant and that the enzyme formation rate is proportional to the uptake rate of methanol, that is,

$$q_{\mathrm{Aox}} = k_{\mathrm{Aox}} q_M. \tag{23}$$

More accurate descriptions of these terms can be found in Joseph et al. [11]. A simulation of a continuous culture based on these equations is shown in Subheading 3.3.

2 Materials

2.1 Biological Knowledge of the Growth Kinetics of K. phaffii

Mathematical models for bioreactors primarily rely on mass balance equations that incorporate terms linked to the growth of specific microorganisms. It is crucial to understand which variables significantly influence *K. phaffii* growth and how they impact the overall growth dynamics. This comprehension is imperative for constructing a well-founded model. Thus, a thorough reference to the existing literature on *K. phaffii* becomes essential, providing knowledge to justify the model development.

2.2 Computational Tools for Solving Ordinary Differential Equations

Ordinary differential equations (ODEs) are typically addressed numerically rather than through explicit solutions. Various computational tools have been developed for this purpose, offering efficient numerical solutions. Notable examples include the deSolve package in R [12], the ode solver in Matlab [13], and the Scipy. integrate package available in Python [14].

2.3 Experimental Data from Bioreactor Cultures

These data typically include biomass, methanol, and co-substrate concentrations at different time points of the culture.

3 Methods

A general protocol for modeling and simulating *K. phaffii* growth in bioreactors:

1. *Define the system boundaries.* For example, in the cases exposed in Subheadings 1.1 and 1.2, the boundary around the bioreactor system was considered, whereas in Subheading 1.3 cell boundaries were also considered for process modeling.
2. *Identify key variables.* Enumerate the essential variables representing both bioreactor and metabolic processes. These may include the concentration of yeast cells, substrate(s) concentration, product(s) concentration, or intracellular metabolite(s).
3. *Establish interactions between sub-systems.* If employing a multilevel modeling approach (i.e., incorporating intracellular and extracellular substances in **step 2**), establish interactions between the extracellular and intracellular sub-systems. Identify substance transfers or conversions between these sub-systems.
4. *Formulate mass balance equations.* Develop mass balance equations for each identified variable, considering inflows, outflows, generation, and consumption within the system boundaries.

5. *Incorporate reaction kinetics.* Describe specific growth rates, uptake rates, production rates, and metabolic process rates.
6. *Provide parameter values.* Assign values to kinetic parameters, drawing from literature or experimental data.
7. *Perform the dynamic simulation.* Perform simulations of the formulated model using computational tools.

Following this 7-step protocol, three different examples are presented in Subheadings 3.1, 3.2, and 3.3, considering the systems and conditions described above in Subheadings 1.1, 1.2, and 1.3, respectively.

3.1 Simulation of a Batch Culture with Methanol as Carbon Source

The abovementioned protocol applied a batch culture presented in Subheading 1.1 to simulate the profile as different methanol initial concentrations. Batch culture is the simplest mode to operate a culture in a reactor. Thus, the protocol considerations in this case for each step are as follows:

1. The system boundaries are the batch reactor.
2. The key variables are yeast biomass (X) and methanol (S_M) concentration.
3. There are no intracellular variables.
4. Mass balances are provided by Eqs. (2) and (3). In batch reactors, there are no inflows or outflows, therefore mass balances only consider the consumption of methanol by yeast (q_M).
5. The production and consumption rates are defined in Eqs. (4) and (5).
6. Parameters of this example are provided in the caption of Fig. 3.
7. Simulations are performed with the library deSolve of R (*see* **Note 4**). They are graphically shown in Fig. 3.

3.2 Simulation of a Fed-Batch Culture with Two Carbon Sources

The protocol was now used to construct a model of a fed-batch culture based on Subheading 1.2 using methanol and glycerol as the main substrate and co-substrate, respectively. During this type of culture, external components (e.g., nutrients, inducer, etc.) are added at a specific time point, for a specific period, and at a specified feeding rate. Protocol considerations for each step are:

1. The system boundaries are the fed-batch reactor.
2. The key variables are yeast (X), methanol (S_M), and glycerol (S_G) concentration. Since volume (V) is not constant, it is also added as a variable.
3. There are no intracellular variables.

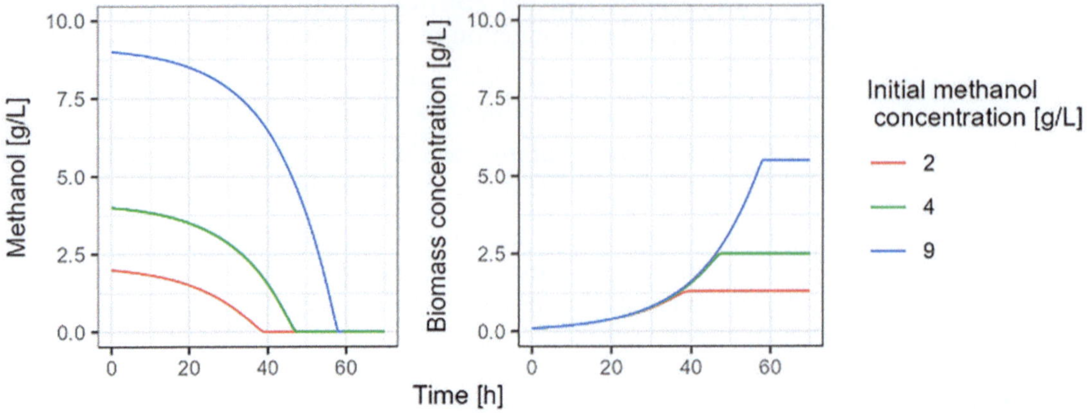

Fig. 3 Simulation of the batch model described in Subheading 1.1 at different initial concentrations of methanol (2, 4, and 9 g L^{-1}). The yield coefficient $Y_{X/M} = 0.6$ gX/g is taken from Canales et al. [7], and the kinetic parameters $\mu_{max} = 0.07$ h^{-1}, and $K_M = 0.05$ g L^{-1} are from Joseph et al. [11]. In all cases, all methanol is completely consumed by the cells, causing the biomass growth to stop. Depending on the methanol concentration, the culture time ranges between about 40 to 60 h

4. Mass balances are provided by Eqs. (6), (7), (8), and (9). They are based on the following processes:
   (a) Inflows of methanol and glycerol.
   (b) Volatilization of methanol.
   (c) Production of yeast biomass through the consumption of methanol and glycerol.
5. The specific growth rate is defined by Eqs. (10), (11), and (12). The consumption rates are defined by Eqs. (5) and (13).
6. Parameters are provided in the caption of Fig. 4.
7. Simulations are performed with the library deSolve of R (*see* **Note 4**). They are graphically shown in Fig. 4.

3.3 Simulation of a Continuous Culture Integrating Bioreactor and Intracellular Dynamics

The protocol has been applied to construct a model as described in Subheading 1.3 using methanol and sorbitol as the main substrate and co-substrate, respectively. In a continuous culture, the nutrient medium flows in at the same rate it exits, such that the volume of the reactor is constant. The protocol is:

1. The system boundaries are the continuous reactor.
2. The key variables are the yeast cell concentration (X), the extracellular concentrations of methanol (S_M), sorbitol (S_{SOR}), and recombinant protein (P), and the intracellular concentration of alcohol oxidase (E_{Aox}).
3. Alcohol oxidase (Aox) is intracellular.

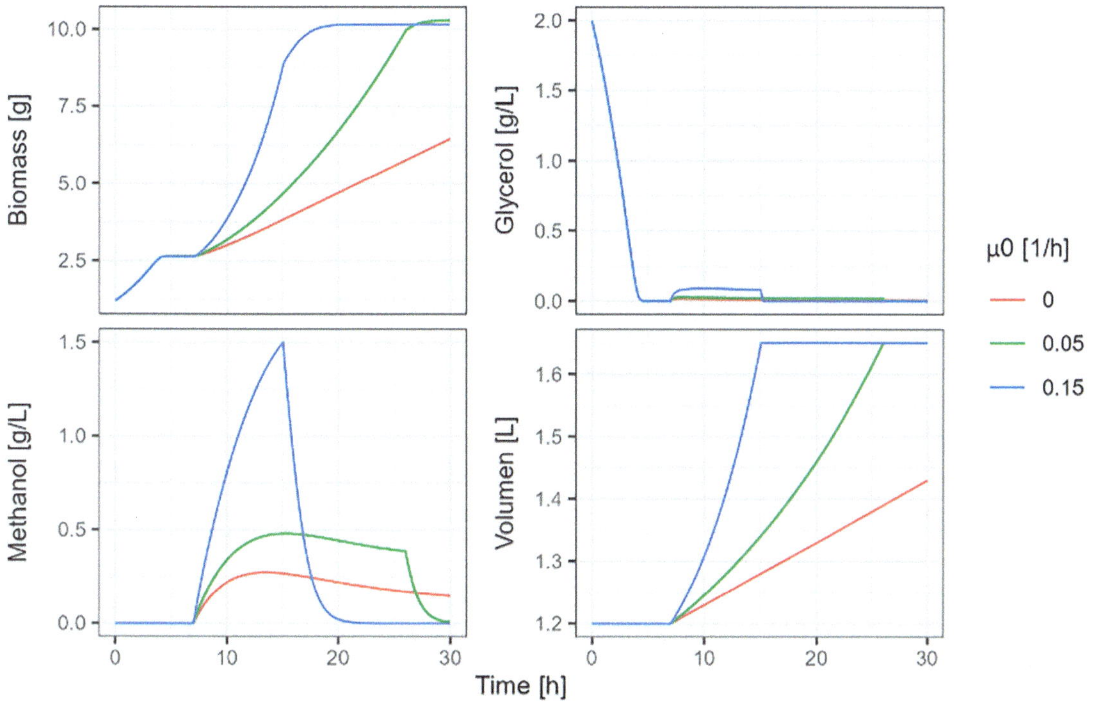

Fig. 4 Simulation of a fed-batch process with linear ($\mu_0 = 0$; red) and exponential (green and blue) feeding based on Subheading 1.2. After 8 h of batch phase, a mixture of methanol and glycerol was fed during 22 h with a feed rate defined as $F = F_0 \exp(\mu_0 t)$, with $F_0 = 0.01\ h^{-1}$. Kinetic parameters were taken from Canales et al. [7]

4. Mass balances are provided by Eqs. (14), (15), (16), (17), and (18). They are based on the following processes:
   (a) Inflow and outflow of methanol and sorbitol.
   (b) Production of yeast cells through the consumption of methanol and sorbitol.
   (c) Production and degradation of the recombinant protein, and its dilution due to inflow of substrates.
   (d) Production of Aox inside the cell and intracellular dilution due to population growth.
5. The specific growth rate is defined through Eqs. (4), (19), and (20). The uptake rates are defined through Eqs. (5) and (21). The production rate of recombinant protein is defined by Eq. (22), and its specific degradation rate is assumed constant. The production rate of Aox is defined by Eq. (23).
6. The caption of Fig. 5 provides references for the parameters of this example.
7. Simulations are performed with the library deSolve of R (*see* **Note 4**) and shown in Figs. 5 and 6.

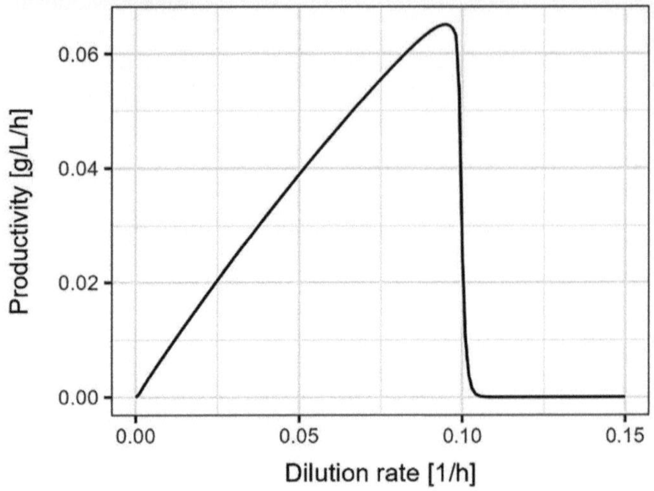

Fig. 5 Productivity of the recombinant protein at steady-state with different dilution rates in a continuous culture of *K. phaffii*, according to Subheading 1.3. The productivity is then determined by multiplying the steady state value of the product by the dilution rate (i.e., $D \cdot P$). Parameters associated with sorbitol and Aox are taken from Çelik et al. [9]. Parameters associated with methanol are described in the caption of Fig. 4. The input concentrations are $S_{M,\,0} = 10$ g L^{-1} and $S_{SOR,\,0} = 10$ g L^{-1}. The volume is 1 L and the dilution rate is 0.03 h^{-1}. The degradation rate is 0.5 h^{-1} and $k_{AOX} = 0.05$ mg g^{-1} h^{-1}

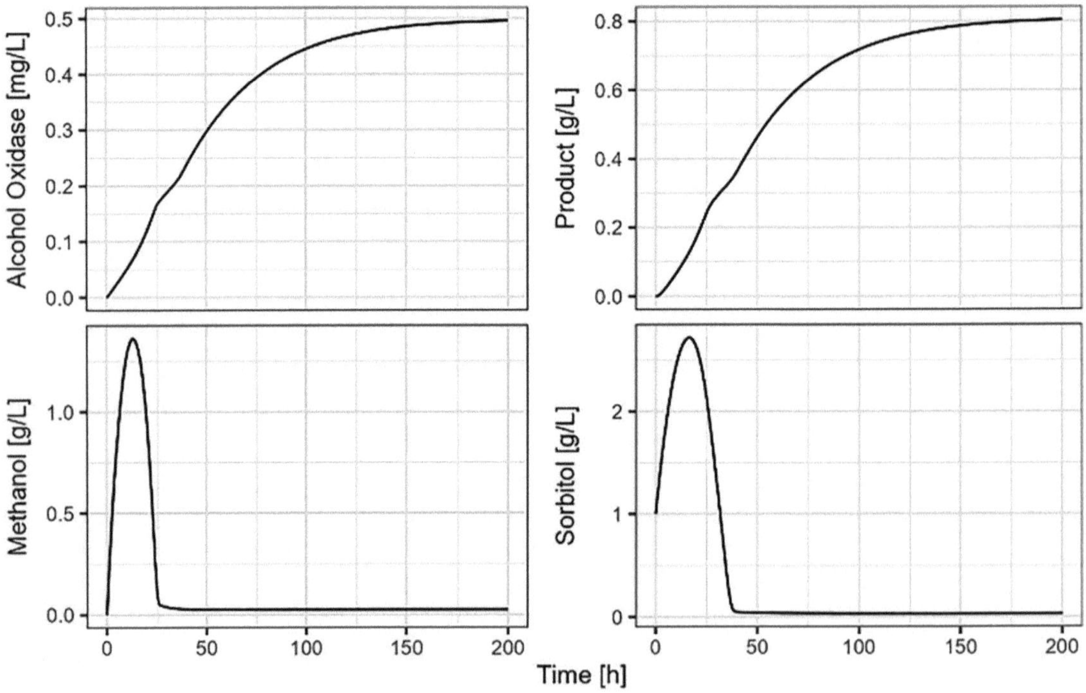

Fig. 6 Simulation of the dynamics of a continuous culture of *P. pastoris* based on equations from Subheading 1.3. A mix of sorbitol and methanol is fed with $D = 0.03$ h^{-1}, as detailed in the caption of Fig. 5. Profile for Aox is intracellular whereas product (recombinant protein), methanol, and sorbitol are at the bioreactor level (culture medium)

4 Notes

1. Mass balance equations are generally stated in terms of mass as done for the fed-batch reactor in Fig. 1, and an additional equation describes volume variation. However, when a reactor is operated in batch or continuous mode, and the volume remains constant, equations can be simplified.
2. This model also accounts for methanol evaporation. Although methanol evaporation can occur at the usual bioreactor operation temperature and aeration rate, it is normally considered negligible in models (e.g., [15]). The first-order constant for methanol volatilization can range between 0.02 and 0.04/h (i.e., the same magnitude order as μ) for aeration rates between 1 and 2 vvm [7]. If the volatilization is considered, then the term $k_{ev}S_MV$ should be added (as shown in Eq. (7)) to complete the mass balance.
3. In contrast to extracellular concentrations, which are quantified in relation to the reactor volume, the concentration of alcohol oxidase represents the mass of the enzymes per unit volume of the cellular cytosol.
4. deSolve is an R package designed for solving initial value problems of a system of first-order differential equations [11]. All models were implemented in R version 4.3.2 using the method ode45 with time steps set at 0.01 h.

Acknowledgments

Authors wish to acknowledge FONDECYT Regular project 1191196 and ANILLO Regular de Ciencia y Tecnología project ACT210068, Agencia Nacional de Investigación y Desarrollo (ANID), Chile; and Wallonie-Bruxelles International through the Cooperation bilateral Belgique-Chili project SUB/2019/435787.

Competing Interest Statement The authors have no conflicts of interest to declare that are relevant to the content of this chapter.

References

1. Bernauer L, Radkohl A, Lehmayer LGK, Emmerstorfer-Augustin A (2021) Komagataella phaffii as emerging model organism in fundamental research. Front Microbiol 11. https://doi.org/10.3389/fmicb.2020.607028
2. Patra P, Das M, Kundu P, Ghosh A (2021) Recent advances in systems and synthetic biology approaches for developing novel cell-factories in non-conventional yeasts. Biotechnol Adv 47:107695. https://doi.org/10.1016/j.biotechadv.2021.107695

3. Peña DA, Gasser B, Zanghellini J et al (2018) Metabolic engineering of Pichia pastoris. Metab Eng 50:2–15. https://doi.org/10.1016/j.ymben.2018.04.017
4. Theron CW, Berrios J, Delvigne F, Fickers P (2018) Integrating metabolic modeling and population heterogeneity analysis into optimizing recombinant protein production by Komagataella (Pichia) pastoris. Appl Microbiol Biotechnol 102:63–80. https://doi.org/10.1007/s00253-017-8612-y
5. Ergün BG, Berrios J, Binay B, Fickers P (2021) Recombinant protein production in Pichia pastoris: from transcriptionally redesigned strains to bioprocess optimization and metabolic modelling. FEMS Yeast Res 21:foab057. https://doi.org/10.1093/femsyr/foab057
6. Berrios J, Flores M-O, Díaz-Barrera A et al (2017) A comparative study of glycerol and sorbitol as co-substrates in methanol-induced cultures of Pichia pastoris: temperature effect and scale-up simulation. J Ind Microbiol Biotechnol 44:407–411. https://doi.org/10.1007/s10295-016-1895-7
7. Canales C, Altamirano C, Berrios J (2018) The growth of Pichia pastoris Mut+ on methanol–glycerol mixtures fits to interactive dual-limited kinetics: model development and application to optimised fed-batch operation for heterologous protein production. Bioprocess Biosyst Eng 41:1827–1838. https://doi.org/10.1007/s00449-018-2005-1
8. Isidro IA, Portela RM, Clemente JJ et al (2016) Hybrid metabolic flux analysis and recombinant protein prediction in Pichia pastoris X-33 cultures expressing a single-chain antibody fragment. Bioprocess Biosyst Eng 39:1351–1363. https://doi.org/10.1007/s00449-016-1611-z
9. Çelik E, Çalık P, Oliver SG (2009) A structured kinetic model for recombinant protein production by Mut+ strain of *Pichia pastoris*. Chem Eng Sci 64:5028–5035. https://doi.org/10.1016/j.ces.2009.08.009
10. Niu H, Jost L, Pirlot N et al (2013) A quantitative study of methanol/sorbitol co-feeding process of a Pichia pastoris Mut+/pAOX1-lacZ strain. Microb Cell Factories 12:33. https://doi.org/10.1186/1475-2859-12-33
11. Joseph JA, Akkermans S, Van Impe JF (2023) Macroscopic modeling of the growth and substrate consumption of wild type and genetically modified Pichia pastoris. Biotechnol J 18: e2300164. https://doi.org/10.1002/biot.202300164
12. Soetaert K, Petzoldt T, Setzer RW (2010) Solving differential equations in R: package deSolve. J Stat Softw 33:1–25. https://doi.org/10.18637/jss.v033.i09
13. Shampine LF, Gladwell I, Thompson S (2003) Solving ODEs with MATLAB. Cambridge University Press, Cambridge
14. Sundnes J (2024) Solving ordinary differential equations in python. Springer Nature, Cham
15. Mendoza Muñoz DF, Algecira Enciso NA, Córdoba Ruiz H, Barrera Avellaneda LA (2008) A simple structured model for recombinant IDShr protein production in Pichia pastoris. Biotechnol Lett 30:1727–1734. https://doi.org/10.1007/s10529-008-9750-1

Chapter 36

Online Monitoring and Control of Methanol Concentration in *Pichia Pastoris* Fermentations

Francisco Valero, José Luis Montesinos-Seguí, and Arnau Gasset

Abstract

Culture broth methanol concentration is the key parameter in the production of recombinant proteins by using *Pichia pastoris (Komagataella phaffii)* as a cell factory under the regulation of the methanol-inducible Alcohol Oxidase 1 (*AOX1*) promoter (P_{AOX1}) or other alternative promoters induced directly or indirectly by methanol. Accordingly, the implementation of an efficient methanol monitoring and control system is critical to guarantee the proper methanol concentration toward achieving an efficient and robust production while assuring reproducible bioprocessing conditions.

Key words *Pichia pastoris*, *Komagataella phaffii*, Methanol monitoring and control, Calibration curve, Closed-loop control, Bioprocess control

1 Introduction

The alcohol oxidase 1 (*AOX1*) promoter (P_{AOX1}) and related or alternative methanol-inducible promoters are still the most widely used and efficient promoters for recombinant protein production (RPP) in the methylotrophic yeast *Komagataella phaffii* (formerly *Pichia pastoris*). Thus, the monitoring and control of methanol concentration in the culture broth can significantly contribute to the design and implementation of efficient bioprocesses, enhancing their production rates and yields as well as ensuring quality, robustness, and reproducibility [1]. However, some operational strategies can also be implemented without online methanol determination with satisfactory results. As an example, among others, a methanol-limited fed-batch operation can be carried out based on a pre-programmed exponential methanol addition to maintain a constant specific growth rate [2]. Further details about this fed-batch cultivation strategy are described Chapter 32. Alternatively, another cultivation strategy based on the so-called dissolved oxygen (DO) spike method, which relies on the methanol feeding to be

Xavier Garcia-Ortega et al. (eds.), *Pichia pastoris: Methods and Protocols*, Methods in Molecular Biology, vol. 2697,
https://doi.org/10.1007/978-1-0716-4779-0_36,

dependent on dissolved oxygen in the culture broth, has been described in the literature [3]. This second alternative strategy is based on the fact that when the methanol is depleted while being the sole carbon source in the culture broth, the oxygen uptake rate (OUR) of the culture will decrease immediately, and accordingly, a sudden peak will be observed in the DO profile. Taking into consideration that methanol might become inhibitory for cell growth at a concentration above 10 g L^{-1}, it is important to be aware that neither of the two strategies guarantees the non-accumulation of cytotoxic and inhibitory methanol concentrations. This fact is relatively risky, especially when mixed feeding strategies are implemented, in which the methanol is fed jointly with other carbon sources to increase protein production rate and productivity [4].

Different analytical approaches have been used to monitor methanol concentration. These include the use of Fourier Transform Mid-InfraRed Spectroscopy (FT-MIRS) [5] as well as the use of automated Sequential Injection Analysis (SIA) [6]. However, these approaches are difficult to set up and require expensive instrumentation, and accordingly, at present, none of them is often used as a technology benchmark by the *Pichia* community. On the other hand, currently, the most used methanol monitoring methods are based on the detection of the methanol vapor concentration either in the exhaust gas from the bioreactor [7], or alternatively, in the vapor phase diffusing across a membrane from culture broth [8].

The first of them has important drawbacks mainly related to the need for precise pressure control on the head of the fermenter, as well as to the addition of a gas flow rate control system. Additionally, preventive actions in the exhaust gas line to avoid foam, liquids, or condensates that could damage mass flow measuring and control equipment and the methanol sensor are necessary. Moreover, different extents of interferences may be produced, depending on the sensing element and gas composition, with the presence of other volatile compounds, especially when ammonium solution is added from the head of the fermenters and not directly into the culture broth to control the pH of the bioprocess [9].

Consequently, the most applied approach is the use of an in-line probe based on a silicone rubber membrane. It is inserted in the bioreactor, ensuring that the membrane is fully submerged in the culture broth. The silicone membrane is continuously perfused with a clean, dry air supply (ideally synthetic gas, not coming from a standard compressor, since humidity can affect the signal of the sensor), diffusing across the gas-permeable membrane the vapor phase (methanol or other volatiles) at a partial pressure governed by the concentration in the bulk liquid phase. The vapor phase is carried to the sensing element, for instance, a tin dioxide (SnO_2) semiconductor, whose conductivity depends on the methanol concentration in the gas phase.

Some suppliers offer methanol analyzers for *Pichia*'s bioprocesses based on the mentioned method:

- http://www.biotechnologie-kempe.de
- https://frings.com

Raven Biotech used to be a reference for online methanol sensors in the *Pichia community,* but it stopped its activity in 2024.

In this chapter, the methodology and maintenance tasks to use this kind of sensor for online monitoring are described in detail, especially the first one (from Raven Biotech) as a reference, because the others on the list work with the same principle and similar setup. Additionally, some types of methanol closed-loop control systems are also presented with a generic setup for a methanol analysis system.

2 Materials

The described in-line methanol concentration monitoring system can be implemented in glass or stainless-steel fermentation vessels of different sizes and suppliers.

Detailed protocols to perform *K. phaffii* cultivations for both fed-batch and continuous operational modes are widely described in Chapters 31 and 33 in this volume.

The following equipment is required for an efficient methanol concentration determination:

1. In-line methanol sensor.
2. The methanol analyzer suppliers usually offer adapters to the standard fermenter ports for glass and stainless-steel vessels.
   (a) For glass vessels, the probes are usually inserted in the headplate using an adapter for 12 mm diameter probes.
   (b) For stainless-steel vessels, the probes are usually inserted in a lower standard port using an adapter for 25 mm diameter probes.
3. Spare parts of the methanol probe and sensor for maintenance tasks (*see* **Note 1**).
4. Non-permeable (Norprene® as an example) tubing for the connection between the methanol probe and the sensing element.
5. Dry air supply line (ideally synthetic gas).
6. For real-time monitoring and data storage, a PC with adequate connectivity, including software that allows storage of data in real time, is required.

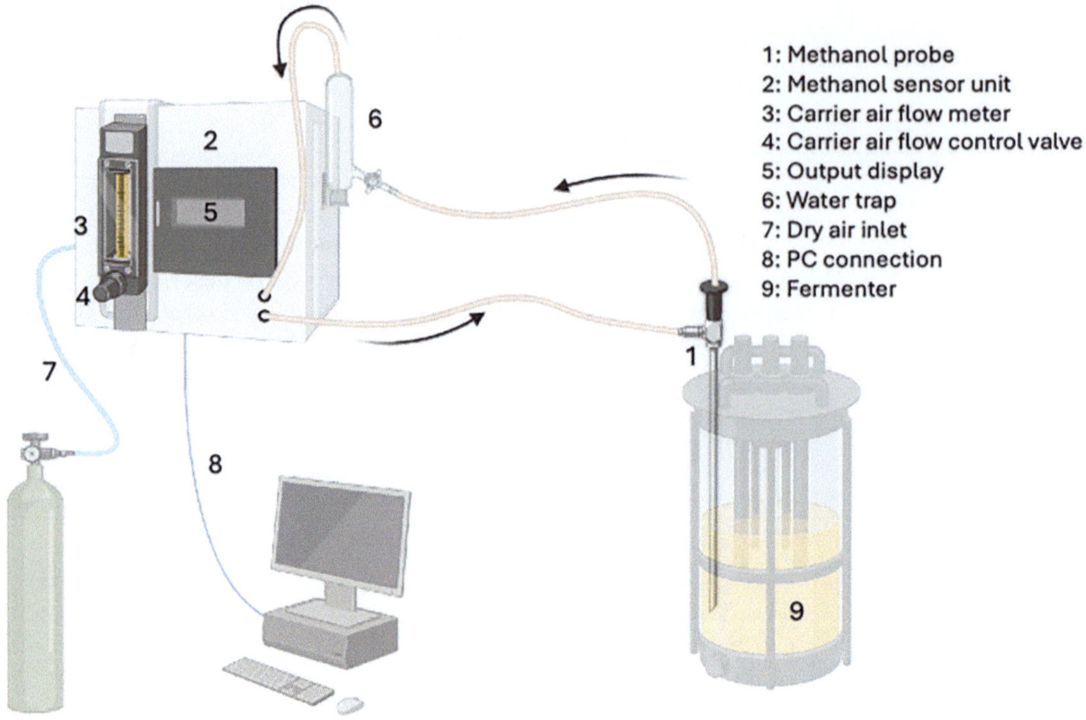

Fig. 1 Setup of methanol analysis equipment. (Created with Biorender.com)

7. For a closed-loop control implementation, in addition to the PC and software mentioned in the previous point, a pumping system controlled by the PC, able to add the methanol amount determined by the control law, will be required.
8. HPLC or other reference analysis equipment capable of measuring methanol concentration for a correct system calibration.

In Fig. 1, a generic setup for a methanol monitoring system is presented.

3 Methods

3.1 Methanol Probe Setup

1. Before the bioprocess starts, the probe connection, tubing, and silicone membrane must be inspected for any signs of moisture or deterioration.
2. The replacement of the silicone membrane can be done on your own, although it takes a little practice (*see* **Note 1**).
3. The probe, utilizing a silicone rubber membrane, must be inserted into the bioreactor, ensuring that it is fully submerged in the culture broth.

4. The inlet and outlet ports of the probe must be connected to the sensor through a non-permeable tube.
5. The air pressure for the regulated air supply should not exceed 15 psi. It is recommended that the use of synthetic air is better than air from a compressor to ensure dry, clean, and oil-free air. The inlet airflow has to be fixed close to 12 $cm^3\ min^{-1}$.
6. Check the operation manual to perform the preliminary test before using it in the monitoring and optional control of the bioprocess.
7. The probe can be sterilized following a standard sterilization bioreactor protocol.
8. For sterilization, the probe tubing must be disconnected from the sensor. To avoid overpressure generation inside the probe chamber, the tubing must not be clamped. However, to avoid humidity entering the probe during the sterilization process, the ends of the tubes should be covered with aluminum foil.

3.2 Sensing Element and Fermenter Setup

1. The sensor output is a voltage signal corresponding to methanol concentration. The sensor is normally calibrated to monitor methanol concentration in the range of 0–1% v/v. The system is not designed typically to measure volatile organic vapors exceeding 2% v/v (*see* **Note 2**). The tin dioxide (SnO_2) semiconductor could be easily found in a microelectronics store (*see* **Note 3**).
2. The condenser of the bioreactor must be well-dimensioned and work with cooling water or other refrigerant below 10 °C to minimize the possible stripping of methanol.
3. With the same goal of minimizing the stripping of methanol, the substrate solution containing methanol must be added directly to the fermentation broth, not in the headspace of the fermenter, since a significant fraction of methanol is then evaporated through the exhaust gas together with the airflow coming from the sparger.
4. It is also worth noting that, if the silicone membrane suffers any damage during the fermentation, culture broth can enter the sensor pipes and reach the tin dioxide sensor, which should then be replaced since the components from the culture medium can spoil the sensor.

   To avoid it, a water/condensate trap could be installed between the probe outlet and the sensor inlet.
5. In some cases, a closed-loop control system to maintain a constant methanol concentration in the culture broth can be applied to improve bioprocess yields and productivity.

3.3 Calibration of the Equipment

It is recommended to use at least the same fermentation medium for the calibration of the methanol analyzer since it can impact the measurements.

1. Once the temperature is constant, a pulse of methanol at the lowest concentration (0.1%) is added.
2. Once the signal of the sensor has been stabilized (5–10 min), a new pulse of methanol has to be added.
3. The increased methanol concentration will depend on the range of methanol concentration expected to be measured.
4. Normally, five-point calibration curves are enough to satisfactorily correlate sensor output (millivolts) with methanol concentration (g L^{-1}).
5. The calibration response curve follows an exponential response. Thus, at high methanol concentrations, a low change in voltage signal can be produced by a high change in methanol concentration, which leads to a very low sensor sensibility for higher methanol concentrations (Fig. 2).
6. When methanol is not added at the beginning of the culture, online calibration is also possible, for instance, at the beginning of the induction phase, taking advantage of the fact that *Pichia*'s metabolism has not yet adapted to the consumption of methanol and thus following the same protocol previously explained. However, after *Pichia*'s metabolism adaptation to methanol consumption, it must be considered that the signal reached after the stabilization of the sensor can be underestimated due to the consumption of methanol and the response time of the overall system.

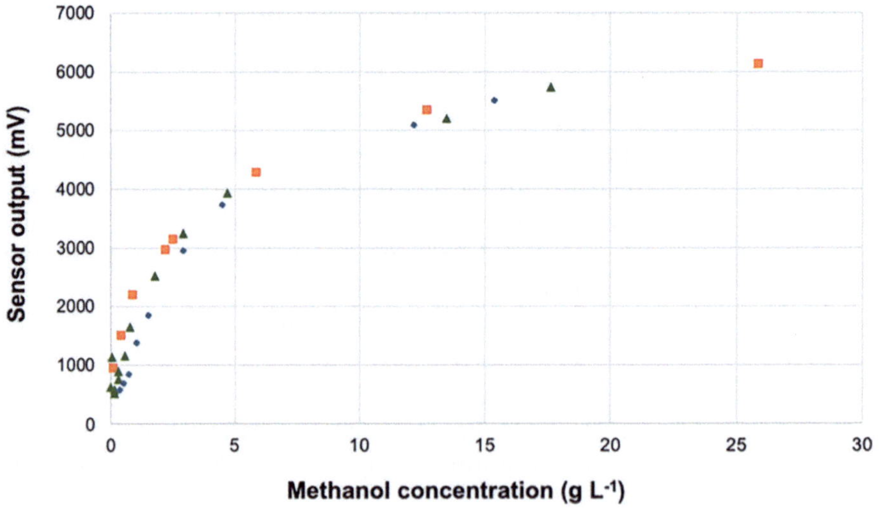

Fig. 2 Methanol calibration curves. Sensor output (mV) vs. methanol concentration (g L^{-1})

7. If methanol control is to be implemented, considering only one-point calibration might be enough when a single methanol set-point is applied.
8. To correctly determine the methanol concentration reached in each pulse calibration, a comparison with other reference analytical methods, for instance, HPLC or GC, is recommended.
9. Usually, during the course of a standard culture (5 days), a recalibration should not be necessary; however, a periodic check of the carrier flow rate and the baseline voltage is recommended.

3.4 Closed-Loop Methanol Control

Most commercial methanol measurement devices incorporate an automatic on-off control system, which governs methanol addition to maintain the methanol concentration at the desired set-point. This scheme is rather simple and can be enough for the Mut^s phenotype, where the methanol consumption rate (r_S) is moderate to low. However, for the Mut^+ phenotype, where r_S *is* much higher, the methanol concentration may present important fluctuations around its set-point. To avoid this problem, more complex control algorithms have been implemented considering feedback control strategies [10, 11]:

1. Proportional-integral (PI), proportional-integral-derivative (PID), and heuristics control [12].
2. Model-based predictive control algorithms can even improve the control strategy [2, 13]. In Fig. 3, the results obtained for a methanol non-limited fed-batch culture controlled at a methanol set-point of 10 g L^{-1}, with a model-based predictive control combined with a PI feedback control, for the production of *Rhizopus oryzae* lipase (Rol) with a Mut^+ phenotype, are presented.

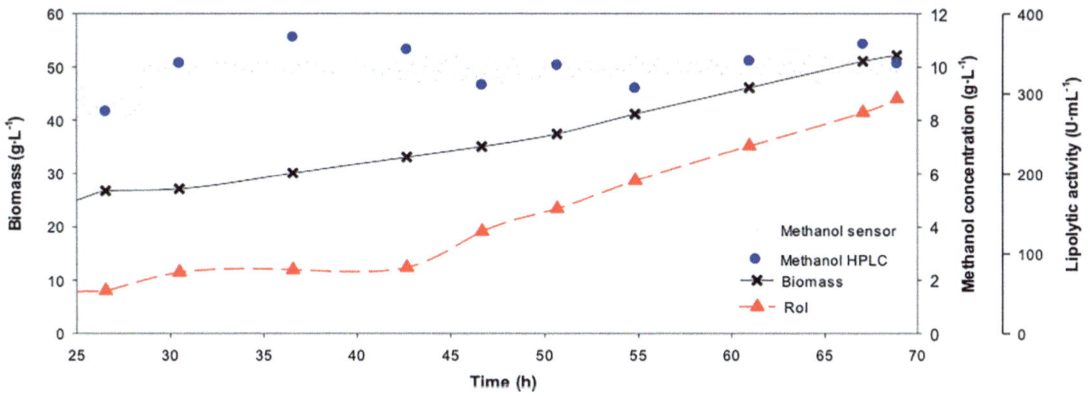

Fig. 3 Example of the performance of an online methanol monitoring and control system for a methanol set-point of 10 g L^{-1} in a fed-batch fermentation producing Rol

3. In addition, apart from controlling the methanol concentration at a constant value, these model-based controllers can be used to implement optimal substrate trajectories [14].

4 Notes

1. The weakest element of the methanol monitoring system is the semi-permeable silicone membrane through which the volatile compounds can diffuse. If any damage to the silicone membrane is detected, it should be replaced with a new one. The methodology depends on the model of the sensor employed, but in general, it requires disassembling the probe cap, changing the membrane, ensuring that it is well secured, and tightly assembling the probe cap again. It is convenient to have spare membranes.
2. The measurement range for other models can reach 6% v/v (https://frings.com/wp-content/uploads/2020/01/Flyer_ALCOSENS.pdf).
3. The sensor employed for methanol detection is usually the Figaro TGS822 sensor (https://www.figarosensor.com/product/sensor/).

Acknowledgments

The project PID2022-136936OB-I00 funded by MCIN /AEI / 10.13039/501100011033 /FEDER, UE of the Spanish Ministry of Science and Innovation, funded this work. The group is a member of the 2021 SGR 00143 Department of Research and Universities of the Generalitat de Catalunya.

References

1. Cos O, Ramon R, Montesinos JL, Valero F (2006) A simple model-based control for *Pichia pastoris* allows a more efficient heterologous protein production bioprocess. Biotechnol Bioeng 95:145–154. https://doi.org/10.1002/bit.21005
2. Barrigón JM, Montesinos JL, Valero F (2013) Searching the best operational strategies for *Rhizopus oryzae* lipase production in *Pichia pastoris* Mut+ phenotype: methanol limited or methanol non-limited fed-batch cultures? Biochem Eng J 75:47–54. https://doi.org/10.1016/J.BEJ.2013.03.018
3. Zhang W, Bevins MA, Plantz BA et al (2000) Modeling *Pichia pastoris* growth on methanol and optimizing the production of a recombinant protein, the heavy-chain fragment C of botulinum neurotoxin, serotype A. Biotechnol Bioeng 70:1–8. https://doi.org/10.1002/1097-0290(20001005)70:1<1::AID-BIT1>3.0.CO;2-Y
4. Arnau C, Casas C, Valero F (2011) The effect of glycerol mixed substrate on the heterologous production of a *Rhizopus oryzae* lipase in *Pichia pastoris* system. Biochem Eng J 57:30–37. https://doi.org/10.1016/j.bej.2011.08.004
5. Crowley J, Mccarthy B, Nunn NS et al (2000) Monitoring a recombinant *Pichia pastoris* fed batch process using Fourier transform mid-infrared spectroscopy (FT-MIRS).

Biotechnol Lett 22:1907–1912. https://doi.org/10.1023/A:1026784101162/METRICS

6. Surribas A, Cos O, Montesinos JL, Valero F (2003) On-line monitoring of the methanol concentration in *Pichia pastoris* cultures producing an heterologous lipase by sequential injection analysis. Biotechnol Lett 25:1795–1800. https://doi.org/10.1023/A:1026228518418
7. Guarna MM, Lesnicki GJ, Tam BM et al (1997) On-line monitoring and control of methanol concentration in shake-flask cultures of *Pichia pastoris*. Biotechnol Bioeng 56:279–286. https://doi.org/10.1002/(SICI)1097-0290(19971105)56:3
8. Hellwig S, Emde F, Raven NPG et al (2001) Analysis of single-chain antibody production in *Pichia pastoris* using on-line methanol control in fed-batch and mixed-feed fermentations. Biotechnol Bioeng 74:344–352. https://doi.org/10.1002/bit.1125
9. Ramon R, Feliu JX, Cos O et al (2004) Improving the monitoring of methanol concentration during high cell density fermentation of *Pichia pastoris*. Biotechnol Lett 26:1447–1452
10. Pla IA, Damasceno LM, Vannelli T et al (2006) Evaluation of Mut + and Mut S *Pichia pastoris* phenotypes for high level extracellular scFv expression under feedback control of the methanol concentration. https://doi.org/10.1021/BP060012
11. Surribas A, Stahn R, Montesinos JL et al (2007) Production of a *Rhizopus oryzae* lipase from *Pichia pastoris* using alternative operational strategies. J Biotechnol 130:291–299. https://doi.org/10.1016/j.jbiotec.2007.04.009
12. Potvin G, Ahmad A, Zhang Z (2012) Bioprocess engineering aspects of heterologous protein production in *Pichia pastoris*: a review. Biochem Eng J 64:91–105. https://doi.org/10.1016/j.bej.2010.07.017
13. Allampalli SSP, Sekhar S, Sivaprakasam S (2024) Enhanced production of human interferon α2b in glycoengineered *Pichia pastoris* by robust control of methanol feeding and implications of various control strategies. Biochem Eng J 201:109152. https://doi.org/10.1016/J.BEJ.2023.109152
14. Ponte X, Barrigón JM, Maurer M et al (2018) Towards optimal substrate feeding for heterologous protein production in *Pichia pastoris (Komagataella spp)* fed-batch processes under P_{AOX1} control: a modeling aided approach. J Chem Technol Biotechnol 93:3208. https://doi.org/10.1002/jctb.5677

Chapter 37

Off-Gas Balancing for Bioprocess Monitoring *Pichia pastoris* Bioprocesses

Lisa Michel, Gesine Cornelissen, and Christian Kaiser

Abstract

The observation of aerobic cultivations via their oxygen demand is an important part of any management of a biotechnological process. In addition to process monitoring, exhaust gas composition is used for process control or process and bioreactor characterization. Oxygen and carbon dioxide (O_2, CO_2) evolution in the exhaust gas flow are usually the key components that allow obtaining important real-time insights about the cultivations, which are very valuable for the online monitoring and control of bioprocesses. They are commonly described by their mole fractions, x_{O_2} and x_{CO_2}, respectively, which can be easily measured by implementing commercially available off-gas analyzers in the exhaust gas.

Key words *Pichia pastoris*, *Komagataella phaffii*, Bioprocess monitoring and control, Fed-batch cultivation, Off-gas balancing, Oxygen transfer coefficient, Oxygen transfer rate, Process analytical technology

Nomenclature

CER	:CO_2 evolution rate [g/L/h]
c_{OL}	:mass concentration of O_2 in V_L [g/L]
c_{OL100}	:saturation concentration of O_2 in V_L during DO-calibration [g/l]
F_{nG}	:aeration rate under normalized conditions [L/h]
H_{nO_2}	:Henry coefficient of O_2 under normalized conditions [Nm/kg]
H_{O_2}	:Henry coefficient of O_2 [Nm/kg]
K_{iO_2}	:Henry coefficient i (i = 1–4) [°C^{-i}]
k_La	:volumetric O_2 transfer coefficient [/h]
M_{O_2}	:molecular mass of O_2 [32 g/mol]
M_{CO_2}	:molecular mass of CO_2 [44 g/mol]
N_{St}	:stirrer speed [/min]
OTR	:volumetric O_2 transfer rate [g/L/h]
OUR	:volumetric O_2 uptake rate [g/L/h]
p_G	:pressure in the gas phase [N/m]
p_{Gcal}	:pressure in the gas phase during *DO*-calibration [N/m]
pH	:pH value [−]

Xavier Garcia-Ortega et al. (eds.), *Pichia pastoris: Methods and Protocols*, Methods in Molecular Biology, vol. 2697, https://doi.org/10.1007/978-1-0716-4779-0_37,

DO	:dissolved O_2 tension [%]
q_{CO_2}	:cell-specific CO_2 evolution rate [g/g/L/h]
Q_{CO_2}	:volumetric CO_2 discharge rate [g/L/h]
q_{O_2}	:cell specific O_2 uptake rate [g/g/L/h]
Q_{O_2}	:volumetric O_2 supply rate [g/L/h]
RQ	:respiratory quotient [−]
ϑ_L	:temperature of V_L [°C]
V_L	:liquid volume [L]
V_{nM}	:mole volume under normalized conditions [22.41 L/mol]
X	:bio dry weight in V_L [g_x /L]
x_{CO_2}	:CO_2-mole fraction in the gas phase [%]
x_{CGcal}	$x_{CAIR} = 0.0003 \Rightarrow 100$ % CO_2-saturation of AIR
x_{CGin}	:CO_2-mole fraction at gas inlet [−]
x_{CGout}	:CO_2-mole fraction at gas outlet [−]
x_{O_2}	:O_2-mole fraction in the gas phase [%]
x_{OG}	:O_2-mole fraction in the gas phase [−]
x_{OGcal}	:O_2-mole fraction in the gas phase during DO-100 %-calibration [−]
x_{OGcal}	:$x_{OAIR} = 0.2094 \Rightarrow 100$ % O_2-saturation of AIR
x_{OGcal}	:1 (pure O_2)$\Rightarrow$ 100 % O_2-saturation
x_{OGin}	:O_2-mole fraction at gas inlet [−]
x_{OGout}	:O_2-mole fraction at gas outlet [−]
$y_{X/Owl}$	:O_2 yield coefficient at constant μ [−]
μ	:cell specific growth rate [/h]

1 Introduction

Off-gas analysis and balancing of microbial fermentations allow for obtaining important real-time insights about the cultivations, which are very valuable for the online monitoring and control of bioprocesses. Suitable exhaust gas analysis systems are available on the market at low cost. Alternative off-gas analyzers are offered by different vendors. Maihak off-gas analyzer consisting of an Oxygor 6 N analyzer and a Defor analyzer for oxygen and carbon dioxide (O_2, CO_2). It is important to mention that this system requires an additional dehumidifier. The cost-effective BlueSens systems Blue-Vary and BlueInOne (BlueSens gas sensor GmbH) can be used without an additional dehumidifier since this analyzer also measures relative humidity and gas pressure besides the mole fractions of O_2 and CO_2 in the outlet gas.

Additionally, mass spectrometry has been frequently applied to monitor the O_2 and CO_2 content in the off-gas of fermentations, but its cost and skills required for handling it are significantly higher.

Off-gas balancing enables a real-time observation of the respiratory activity of cells. It allows to generate conclusions about substrate consumption, metabolic shifts, as well as usually indicates deviations from the planned course of the process, e.g., due to nutrient supply or mass transfer limitation [1–3]. Consequently, it is an advantageous tool during process development and manufacturing regarding process reproducibility, process understanding, and early anomaly detection. Thus, in the context of process analytical technology (PAT) and quality by design off-gas analysis is a powerful mechanism to give control to industrial manufacturing processes as they can be considered critical process parameters [4–6]. Some parameters, such as oxygen transfer rate (OTR), and maximum mass transfer coefficient (k_La) require a deep characterization for successful process scale-up to production scale [7].

The performance of aerobic cultivations with the need for a continuous supply of O_2 is influenced significantly by the mass-transfer phenomena between the gas and liquid phases. In large production-scale reactors, usually, the O_2 transfer to the liquid phase is the limiting factor to transfer efficiently the bioprocess developed and smaller scales.

Many operating variables can affect the O_2 transfer in bioreactors, and the interactions between these variables are complex: bubble size distribution, residence time, medium composition, O_2 solubility, cell concentration, temperature, viscosity, and power input [8, 9]. Due to the large number of influencing factors that cannot be determined with any degree of accuracy during the process, the mass balancing for off-gas analysis is essentially reduced to one simplified approach.

The mass transfer rate between gas and liquid phase is proportional to the concentration gradient in the interfacing area and to the volumetric mass transfer coefficient. For O_2 and CO_2, the major mass transfer resistance is located at the liquid side and the mass transfer, either by adsorption or desorption, can be calculated by the two-film theory [10].

In the following chapter, the mathematical model for off-gas balancing and its application to *Komagataella phaffii* (previously *Pichia pastoris*) fed-batch cultivation processes is outlined. This includes the calculation of key cultivation parameters such as the oxygen uptake rate (OUR), carbon dioxide evolution rate (CER), respiration quotient (RQ), and mass transfer coefficient (k_La).

2 Materials

Off-gas analysis is commonly used for cultivations performed in bioreactors, however, some devices able to be implemented in shake flask cultures can also be found in the market (*see* **Note 1**).

2.1 Bioreactor Setup

Considering that the main contribution of this chapter is focused on gas analysis, the materials and methods for performing cultivations in bioreactors will be skipped. In this sense, other chapters of this Volume are specifically focused on continuous cultivation (Chapter 33) and fed-batch cultivations (Chapters 31 and 32).

Additionally, for adequate off-gas balancing accurate volume determination of the liquid reaction volume (V_L) throughout the whole fermentation process is a pre-requisite (*see* **Note 2**).

2.2 Off-Gas Analyzer Setup

When external analyzers are employed, it is mandatory to create the right conditions for measurement. The sensors should be protected by correctly integrating them into the system (Fig. 1) (*see* **Note 3**).

1. The use of thermal mass flow controllers (TMFC) is recommended to ensure that your system provides consistent air and O_2 flow rates (*see* **Note 4**).
2. The use of a sterile barrier of the gas outlet is recommended to protect the off-gas analyzer. It can be achieved by installing the supplied adapter in the exhaust pipe and attaching the sensor head to the adapter, in the case of the BCP Series from

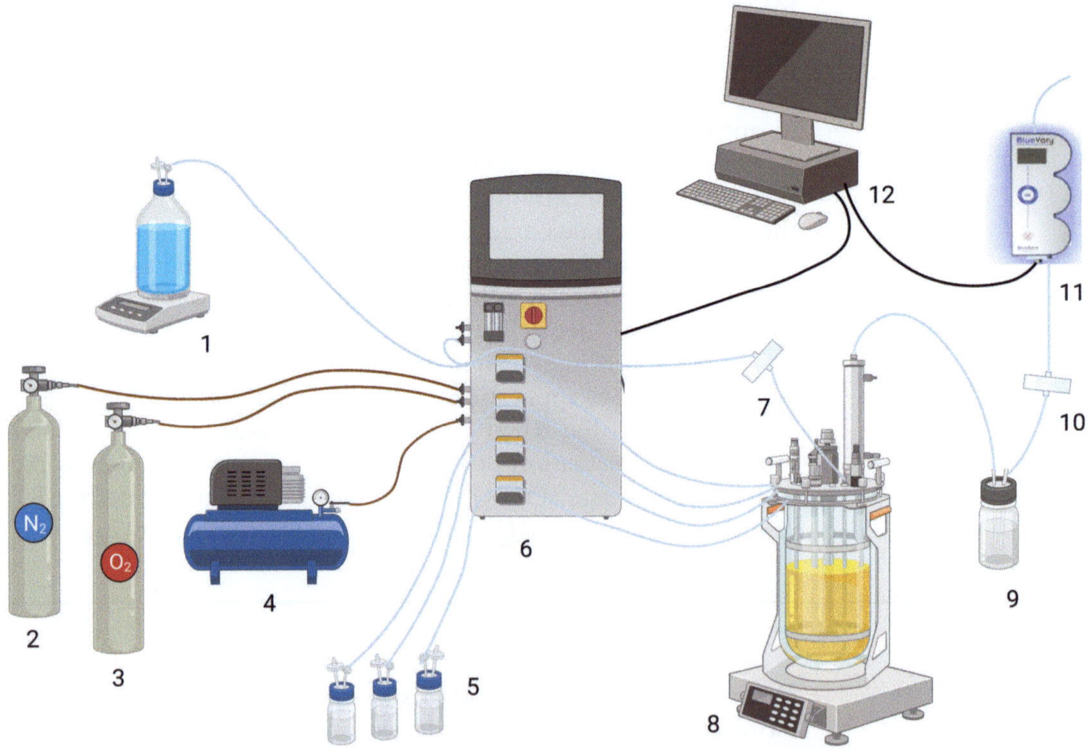

Fig. 1 Recommended bioreactor setup: 1. Feedreservoir, 2. nitrogen supply, 3. oxygen supply, 4. air compressor, 5. acid, base, and antifoam, 6. supply tower with digital control unit (DCU), 7. air filter gas inlet, 8. bioreactor, 9. foam trap, 10. air filter gas outlet, 11. gas analyzer. Image created with Biorender

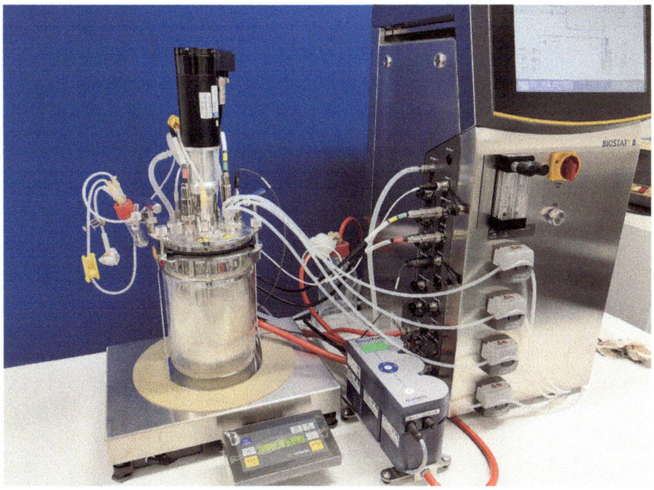

Fig. 2 Autoclavable small scale bioreactor with 2 L working volume equipped with off-gas analyser BlueVary (BlueSens gas sensor GmbH)

BlueSens. Regarding for BlueInOne from BlueSens sensors, both are already pre-assembled in the flow adapter (*see* **Note 5**).

3. Required tubing to keep the distance between the bioreactor headspace and the exhaust gas analyzer as short as possible (*see* **Note 6**).
4. Examples of the integration of those off-gas analyzers are shown in Figs. 2 and 3.

2.3 Software for Proper Monitoring and On-Line Data Calculation

A real-time connectivity system between the sensor and the process control system allows online bioprocess monitoring and control (*see* **Note 7**).

1. BlueSens gas sensor GmbH provides the software BlueVis, which collects all the data generated by the off-gas analyzers and can be used as a process control system itself.
2. Equivalent software to read the data from other gas analyzer systems such as Maihak of MS can be used.
3. The data are transferred in real-time also to other process control systems (SCADA), e.g., MFCS/win, Dasware®, Eve® or alternative software, e.g., MatLab is offered by an OPC Server/Client.
4. Online calculation methods for gas balancing. The most relevant are described in detail in Subheading 3.

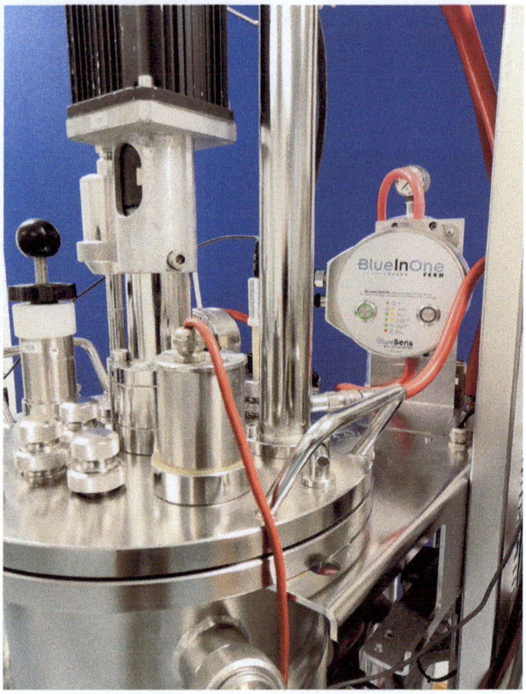

Fig. 3 In-situ sterilizable pilot scale bioreactor with 20 L working volume equipped with BlueInOne off-gas analyser (BlueSens gas sensor GmbH)

3 Methods

3.1 Calibration Before Its Use

One-point calibration is recommended at least once a month or after each disconnection of the sensor to maintain a reliable measurement.

1. Expose the sensor for at least 30 min to an appropriate and specified gas flow. Note that any changes in the gas mixture will cause errors in the measurements (*see* **Note 8**).
2. If your sensor was exposed to high CO_2 concentrations in previous cultivations flush your sensor with air for several hours (*see* **Note 9**).
3. Additional gas mixtures can be further used to improve the calibration accuracy since the 1-point calibration might be limited (especially in terms of CO_2) (*see* **Note 10**).

3.2 k_La Determination Based on Gas Analysis

The oxygen transfer capacity, namely the O_2 transfer coefficient (k_La), is a key parameter for successful cultivation development, process transfers, and for further scale-up. This parameter is a useful method for bioreactor characterization, and its determination can be carried out by applying off-gas analyses during your running processes (*see* **Note 11**).

Alternative methods have been described in the literature. Some are based on physico-chemical methods, thus not requiring biological consumption of O_2 in a growing culture. Others are considered biological since they are based on cellular respiration in a cultivation system. Among those, the k_La can be calculated based on dissolved oxygen (DO) values, a method which is usually called static gassing out method [7, 11, 12]. Alternatively, it can be determined from experimentally gained off-gas data following the next steps (*see* **Note 12**).

1. Calculate the O_2 supply rate (Q_{O_2}) based on in the aeration rate (F_{nG}), the actual liquid volume (V_L) in the bioreactor, the molar weight of O_2 (M_{O_2}) as well as the molar standard volume (V_{nM}) the mole fractions of O_2 and CO_2 in the inlet- and outlet gas flow (x_{OGin}, x_{CGin}, x_{OGout}, x_{CGout}) using Eq. (1) (*see* **Note 13**).

$$Q_{O_2}(t) = \frac{F_{nG}(t) \cdot M_{O_2}}{V_{nM} \cdot V_L(t)} \cdot \frac{x_{OGin}(t) \cdot [1 - x_{CGout}(t)] - x_{OGout}(t) \cdot [1 - x_{CGin}(t)]}{1 - x_{OGout}(t) - x_{CGout}(t)} \tag{1}$$

2. Identify the temperature-dependent capability of water to take up O_2 using the Henry coefficient (H_{O_2}) by setting in the temperature of the liquid phase (ϑ_L) as well as the theoretical Henry coefficient for O_2 at 0 °C under normalized conditions (H_{nO_2}) into Eq. (2) (*see* **Note 14**).

$$\mathrm{H_{O_2}(t)} = \frac{\mathrm{H_{nO_2}}}{1 + \sum_{i=1}^{4} \mathrm{K_{iO_2}} \cdot \theta_L(t)} \tag{2}$$

3. Use Eq. (3) to determine the O_2 saturation concentration (c_{OL100}) at the time point of calibration using air.

$$c_{OL100}(t) = \frac{p_{Gcal} \cdot x_{OGcal}}{H_{O_2}(t)} \tag{3}$$

4. Calculate the actual O_2 concentration in the liquid phase (c_{OL}), which is determined from the actual relative dissolved oxygen tension measurement in the liquid phase (DO) and depends on the underlying calibration conditions (c_{OL100}) by filling in the obtained values in Eq. (4) (*see* **Note 15**).

$$c_{OL}(t) = \frac{DO(t)}{100\%} \cdot c_{OL100}(t) \tag{4}$$

5. Calculate the k_La-values of your process by using Eq. (5).

$$k_La(t) = \frac{Q_{O2}(t)}{\frac{p_G(t) \cdot x_{OG}(t)}{H_{O2}(t)} - c_{OL}(t)} \tag{5}$$

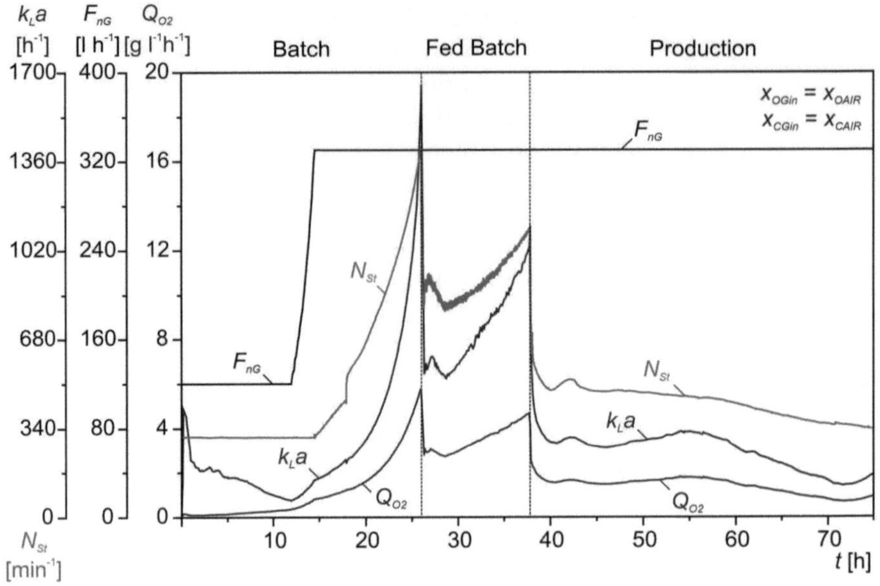

Fig. 4 Analysis of O_2 transfer control with oxygen supply rate (Q_{O_2}), total aeration rate (F_{nG}), stirrer speed (N_{St}), and oxygen transfer coefficient (k_La)

Figure 4 an application example and the course of k_La. Even F_{nG} and N_{St} were constant at the beginning of the process, the k_La first decreases. This is an often observed phenomenon. With increasing F_{nG} and N_{St} from $t = 12$ h on an exponentially increasing k_La can be monitored. At the end of the batch phase, the stirrer speed almost reached its maximum of 1500/min. The volumetric O_2 need at this time point was 5.8 g/L/h in this 5 L bioreactor, which leads to a total amount of almost 30 g/h O_2 need. The k_La values 1650/h at the end of the batch phase. The maximum O_2 transfer capacity was valued at 8.7 g/L/h at the end of the batch phase and could have been increased to 41.5 g/L/h by the usage of pure O_2 [13].

3.3 On-Line Determination of O_2-Related Respiratory Parameters (OUR, qO_2)

1. Key respiratory parameters and their use as soft sensors in real-time are often used to improve the monitoring and control of bioprocesses. These parameters are also very helpful to characterize strain and bioprocesses, as well as to implement Process Analytic Technologies (PAT) (*see* **Note 16**).
2. *OUR determination*: The oxygen uptake rate depends on the (micro)organism, the physiological state, and the fermentation conditions. Therefore, online monitoring of this process variable can provide valuable insights into the process [14].
3. The O_2 accumulation in the bioreactor as well as the convective O_2 input and output are negligible in relation to the O_2 transfer from the gas to the liquid phase. Consequently, the O_2 mass flow into the bioreactor, the mass transfer flow from the gas

into the liquid phase, and the mass transfer flow into the cell can be equated. Then one has $Q_{O_2} = \text{OTR} = \text{OUR}$.

4. This means that in a quasi-steady state O_2 transfer equilibrium, the O_2 supply rate (Q_{O_2}) corresponds to the OUR.
5. Use the gas balancing method for the online determination of the OUR. With Eq. 1 we provide a formula that can be directly implemented into your control system to calculate the OUR in real-time (*see* **Note 17**).
6. q_{O_2} *determination*: Complimentarily, the OUR can be calculated in its specific form, (q_{O_2}), which describes the O_2 consumption per unit of time and cell or biomass.
7. Derive the q_{O_2} from the former calculated OUR by dividing the value by the corresponding biomass concentration, usually used in terms of dry cell weight (DCW) (X, g_X/L) (*see* **Note 18**).
8. q_{O_2} is characteristic of every microorganism and provides important information about cell metabolism and cell physiology, on the basis of which, for example, feed strategies or process control can be developed [15].

3.4 On-Line Determination of CO_2-Related Respiratory Parameters (CER, qCO_2, RQ)

1. *CER determination*: The carbon dioxide evolution rate (*CER*) can be used to identify different growth phases based on the correlation of CO_2 production and microbial growth kinetics.
2. Use the Eq. (7) to implement on-line monitoring of the *CER* during your bioprocess (*see* **Note 19**).

$$\text{CER}(t) = \frac{F_{nG}(t)\cdot M_{CO2}}{V_{nM}\cdot V_L(t)} \cdot \frac{x_{\text{CGout}}(t)\cdot[1 - x_{\text{OGin}}(t)] - x_{\text{CGin}}(t)\cdot[1 - x_{\text{OGout}}(t)]}{1 - x_{\text{OGout}}(t) - x_{\text{CGout}}(t)} \tag{7}$$

3. q_{CO_2} determination: Calculate the cell-specific carbon evolution rate (q_{CO_2}) by dividing the *CER* by the current biomass concentration, usually used in terms of dry cell weight (DCW) (X, g_X/L).
4. *RQ determination*: The respiratory quotient or respiratory coefficient (*RQ*) is a measure of the ratio between O_2 consumed, and CO_2, produced by the cells of the system.
5. Complete your online monitoring setup with Eq. (8) to enable an online calculation of the *RQ*.

$$\text{RQ}(t) = \frac{\text{CER}(t)}{\text{OUR}(t)} \tag{8}$$

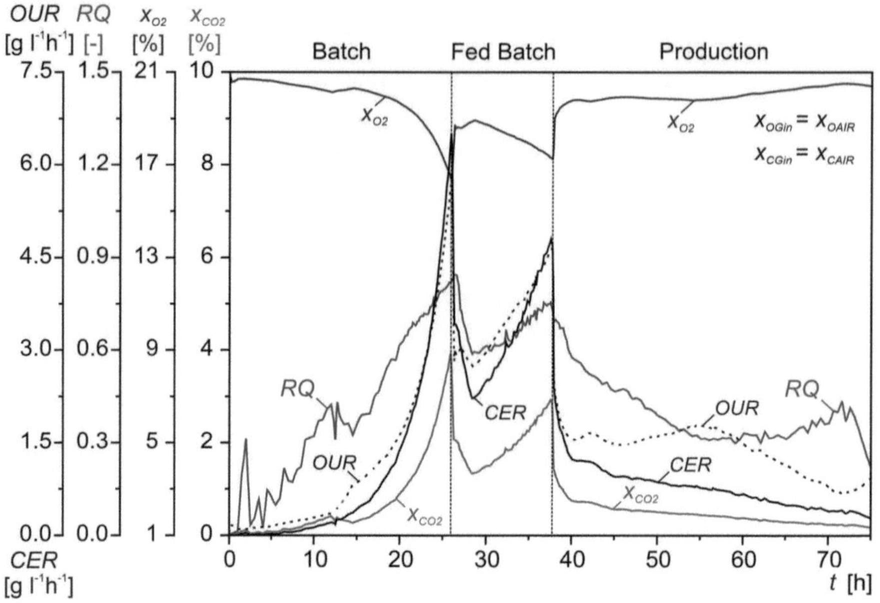

Fig. 5 Course of O_2 mole fraction in the off gas (x_{O_2}), CO_2 mole fraction in the off gas (x_{CO_2}), *OUR*, *CER*, and respiratory quotient (*RQ*)

Figure 5 shows the courses of off-gas analysis (x_{O_2} and x_{CO_2}) and the results of the above-described off-gas balancing (OUR, CER, and RQ). The cell activity showed higher levels during growth on glycerol than during growth on methanol. The volumetric *OUR* and *CER* increased exponentially and reached almost 6 and 7 g/L/h at the end of the batch phase and lower values at the end of the glycerol-based fed-batch phase, respectively. The decreasing activity of the cells during protein production can also be seen from decreasing rates. The course of the *RQ* shows a typical behavior. Due to CO_2 buffering and maybe a small offset in O_2 off-gas analysis, the *RQ* starts at zero and converges toward 0.6 in this cultivation. At the beginning of the cultivation, the calculation is rather difficult due to very small changes in x_{O_2} and x_{CO_2} because of a low cell density. During the glycerol phases, approximately 0.9 g CO_2 were produced per gram of used-up O_2 compared to only 0.4 g CO_2 produced per gram of used-up O_2 during growth on methanol during protein production, where the O_2 need is proportionately higher.

3.5 Interpretation of Metabolic Changes Obtained Based on Gas Analysis

By continuously monitoring the composition of the exhaust gas flow, valuable information about cellular metabolic changes is always available.

3.5.1 Carbon Source Depletion (End of Batch)

1. The values of the exhaust gas analysis can be used for the detection of the batch end.

2. The carbon source depletion at the end of the batch phase leads to a strong decrease in O_2 demand.
3. This is noticeable by a sudden increase in the mole fraction of O_2 (x_{O_2}) and a simultaneous steep drop in the mole fraction of CO_2 (x_{CO_2}) (Fig. 3).
4. By the start of the fed batch phase, again the exponentially increasing and decreasing courses of the mole fractions can be observed.
5. This information can be used to automate the bioprocess flow. For example, by transferring the exhaust gas values to a toolbox, the end of the batch can be automatically recognized via previously-stored conditions and the feed pump can be started. This means that the fed-batch is only started when the substrate provided has been completely used up.

3.5.2 Identification of the Carbon Source Consumed Based on RQ

1. The RQ is defined as the ratio of the number of CO_2 molecules produced to the number of O_2 molecules consumed by an organism.
2. Under normoxic conditions, the amount of CO_2 produced and O_2 consumed, and consequently also the *RQ*, depends on which C source is metabolized by the organisms. This is very useful because we can identify the substrate consumed by the organism by *RQ* monitoring.
3. In *Pichia* cultivations on glucose, *RQ* values of around 1, whereas on glycerol or methanol, the *RQ* is around 0.6 (*see* **Note 20**).
4. Differences in *RQ* values along a culture might indicate which main substrate is being consumed or whether undesirable other metabolic processes are taking place, such as the breakdown of by-products and the switch from respiratory to fermentative processes.
5. An example could be the metabolic shift from respiratory to respiro-fermentative pathways in O_2-limiting conditions, thus producing ethanol as a by-product.
   - When growing on glucose, the *RQ* would shift from values around 1, to values significantly higher than 1 (values up to 1.8 have been reported) [16].
   - Consequently, by-product formation of Ethanol during aerobic cultivations can be detected by an increased *RQ*-value during batch and fed-batch phases.

3.5.3 Online-Specific Growth Rate Determination Based on Gas Analysis

1. Results from off-gas balancing can be used to estimate the cell-specific growth rate (μ) online. At this point, we describe how a successful online determination of μ can be implemented based on the measured x_{CO_2} (*see* **Note 21**).

2. To get the μ based on the CO_2 production create a semi-logarithmic plot with $\ln(x_{CO_2})$ (y-axis) over the time course of the cultivation (t) (x-axis).
3. Draw a straight line through the plotted $\ln(x_{CO_2})$-values.
4. Determine the slope of the straight line. μ corresponds to the calculated slope.

4 Notes

1. Commercially available systems are, for example, the Kuhner Tom system from Kuhner Shaker, but a good alternative is also the BCP series from BlueSens, which can be used both on a shake flask scale and later the bioreactor.
2. The most accurate method to determine the liquid reaction volume V_L throughout the whole fermentation process is to convert the liquid reaction mass (m_L) with the corresponding density. The density of water can be used as a first approximation. To determine m_L, the reactor is placed on a scale and the scale signal is transferred to the process control system as a further process variable. Other options are the use of quantitative-level probes or mathematical balancing.
3. The cost-effective BlueSens systems BlueVary and BlueInOne (BlueSens gas sensor GmbH) can be used without an additional dehumidifier and measure relative humidity and gas pressure besides the mole fractions of O_2, and CO_2 in the outlet gas.
4. Usually, the gas composition at the bioreactor inlet is not measured but assumed constant if using air, or calculated when using air mixtures. Accordingly, it is recommended to use thermal mass flow controllers (TMFC) since they are less prone to errors compared to manually operated rotameters. Most bioreactor systems today are equipped with TMFCs or can be retrofitted at a reasonable cost.
5. Install the analyzer behind the exhaust cooler as well as the sterile filter of the exhaust line. If your exhaust cooler does not provide a sterile position your sensor needs to be protected from hot steam during sterilization. For example, the SGP-1 sample gas pump from BlueSens can be used to cool down the gas, collect the water, and thus protect the sensor. Additionally, a foam trap can be connected between the exhaust gas filter and the gas analyzers to protect the latter from liquids.
6. The residence time distribution is significantly influenced by the distance, flow velocities, and installations and therefore has a significant influence on the precision of the measurement results.

7. When using the BCP-Series from BlueSens easily up to 12 sensor heads can be connected via the *BaCCOm12* connection box to the corresponding process system.
8. You can minimize the effect of dilution by accumulating water molecules in the dry process gas, by flushing the O_2 sensors using humid process gas (20.97 Vol. % O_2) at process temperature instead of ambient air during the calibration process.
9. You can save specified gas when normal fresh air is used during the first flushing time.
10. Due to the relatively low proportion of CO_2 in the air, it can be advantageous to use special calibration gases to increase the accuracy of the measurement.
11. Non-optimized *DO*-controller parameters, resulting in strong fluctuations of the manipulated value (e.g., stirrer speed, head pressure, aeration rate, or molar fraction of O_2 in the inlet gas). Using common PI-controllers more stable regulation can be achieved in most cases by reducing the controller gain and/or the integral part.
12. Use calculation software that can handle large amounts of data well, such as MatLab from MathWorks or Microsoft Excel. A good freely available software alternative is Scilab.
13. Whilst the aeration rate (F_{nG}) and the actual liquid volume (V_L) are process-specific variables of your process setup, the molar mass of O_2 (M_{O_2} = 32.0 g/mol) and the molar standard volume (V_{nM} = 22.41 L) are fixed natural constants.
14. The theoretical Henry coefficient for O_2 at 0 °C under normalized conditions, H_{nO_2} = 1.448E06 Nm/kg, as well as the four constants, $K_1O_2 = -2.723\text{E-}02\ °C^{-1}$, $K_2O_2 = 5.627\text{E-}04/°C$, $K_3O_2 = -6.597\text{E-}06/°C$, $K_4O_2 = 3.283\text{E-}08/°C$, of the modified Bunsen's law [17, 18].
15. This method is not suitable for very small bioreactors, as the difference between the O_2 input and output rate is very small due to the short contact time.
16. At low O_2 consumption and thus low CO_2 evolution rates, due to low cell concentrations (X), e.g., close to inoculation and/or low respiratory cell activity, relative errors of calculated O_2 balance values are high. In those process phases, results should be critically scrutinized. For accurate cell mass estimations in such process phases, e.g., turbidity or capacity probes should be used instead of off-gas balances [19–21].
17. Strong fluctuations in the *DO* tension might impact the accuracy of off-gas balances significantly. The reasons can be many. The main pitfalls are: (a) Use of lid probes in combination with low stirrer speed resulting in temporary adhesion of gas bubbles at the tip of the probe. Potential solutions are the usage of

side ports for *DO*-probes or higher minimum stirrer speed values in case the former is not feasible e.g. in benchtop bioreactor systems. (b) Discontinuous substrate feed rates in fed-batch or continuous processes might result in alternating cell activity. This can be caused by too low remaining feed rates in small-scale fermentations and/or working at the lower end of the working range of the feeding pump used. Solutions could be the use of lower substrate concentration in the feed solution (with the negative effect of a higher resulting dilution rate of the cultivation broth) or adjustment of the working range of the feeding system (e.g., different tubing, pump, etc.).

18. You can get the dry cell weight (X) online by using a turbidity probe. Offline the X can be determined by measuring the optical density at 600 nm (OD_{600}) or by determining the dry cell weight (DCW) in g/L.
19. Whilst the aeration rate (F_{nG}) and the actual liquid volume (V_L) are process-specific variables of your process setup, the molar mass of CO_2 ($M_{CO_2} = 44.0$ g/mol) and the molar standard volume ($V_{nM} = 22.41$ L) are fixed natural constants.
20. At the beginning of the batch phase, a strong deviation between the expected *RQ* and the one calculated online can often be observed. Only after a certain time does the *RQ* reach the expected value. This is because the measurement of the O_2 content in the exhaust gas flow (x_{OGout}) is very inaccurate at the beginning, due to the slight drop in value during this period. It is therefore recommended to monitor the much more accurate CO_2 content in the exhaust gas (x_{CGout}) during this period.
21. A complimentary method for μ-calculation based on O_2 uses Eq. 9.

$$\hat{\mu}(t) = \frac{\mathrm{OTR(t) \cdot V_L(t)}}{\frac{V_{L1} \cdot c_{XL1}}{y_{X/Ow1}} + \int_{t1}^{t} (OTR(\tau) \cdot V_L(\tau)) d\tau} \tag{9}$$

Acknowledgments

We hereby acknowledge the Department of Biotechnology, Faculty of Life Sciences, Hamburg University of Applied Sciences for providing us with all the facilities and financial support required to perform the practical work. Lisa Michel was kindly funded by Pro Exzellenzia 4.0 and the graduation program of the Hamburg University of Applied Sciences. The Hamburg University of Applied Sciences funds the PharmCycle project in which this subproject is integrated.

The authors Gesine Cornelissen and Christian Kaiser would like to thank their mentor Reiner Luttmann for the introduction to the mathematical description and analysis of biotechnological processes.

Competing Interest Statement The authors have no conflicts of interest to declare that are relevant to the content of this chapter.

References

1. Heyman B, Tulke H, Putri SP, Fukusaki E, Büchs J (2020) Online monitoring of the respiratory quotient reveals metabolic phases during microaerobic 2,3-butanediol production with *bacillus Licheniformis*. Eng Life Sci 20(3–4): 133–144. https://doi.org/10.1002/elsc.201900121
2. Dobrescu AC, Veras HCT, Varrone C, Knudsen JD (2021) Novel propagation strategy of saccharomyces cerevisiae for enhanced xylose metabolism during fermentation on softwood hydrolysate. Fermentation 7(4):288. https://doi.org/10.3390/fermentation7040288
3. Zavec D, Gasser B, Mattanovich D (2020) Characterization of methanol utilization negative *Pichia Pastoris* for secreted protein production: new cultivation strategies for current and future applications. Biotechnol Bioeng 117(5): 1394–1405. https://doi.org/10.1002/bit.27303
4. Streefland M, Martens DE, Coen Beuvery E, Wijffels RH (2013) Process analytical technology (PAT) tools for the cultivation step in biopharmaceutical production. Eng Life Sci 13(3): 212–223. https://doi.org/10.1002/elsc.201200025
5. Helgers H, Schmidt A, Lohmann LJ, Vetter FL, Juckers A, Jensch C, Mouellef M, Zobel-Roos S, Strube J (2021) Towards autonomous operation by advanced process control—process analytical Technology for Continuous Biologics Antibody Manufacturing. Processes 9(1):172. https://doi.org/10.3390/pr9010172
6. Gillespie C, Wasalathanthri DP, Ritz DB, Zhou G, Davis KA, Wucherpfennig T, Hazelwood N (2022) Systematic assessment of process analytical Technologies for Biologics. Biotechnol Bioeng 119(2):423–434. https://doi.org/10.1002/bit.27990
7. Garcia-Ochoa F, Gomez E (2009) Bioreactor scale-up and oxygen transfer rate in microbial processes: an overview. Biotechnol Adv 27(2): 153–176. https://doi.org/10.1016/j.biotechadv.2008.10.006
8. Mersmann A, Kind M, Stichlmair J (2005) Thermische Verfahrenstechnik: Grundlagen und Methoden. 2., Wesentlich erw. und aktualis. Aufl. Verfahrenstechnik. Springer, Berlin/Heidelberg/New York
9. Schügerl K (1993) Comparison of different bioreactor performances. Bioprocess Eng 9(5):215–223. https://doi.org/10.1007/BF00369405
10. Yagi H, Yoshida F (1977) Desorption of carbon dioxide from fermentation broth. Biotechnol Bioeng 19(6):801–819. https://doi.org/10.1002/bit.260190603
11. Cornelissen G (2004) Integrierte Bioprozessentwicklung zur Herstellung pharmakologischer wirksamer Proteine mit Pichia pastoris. Fortschritt-Berichte VDI Reihe 17, Biotechnik, Medizintechnik 249. VDI-Verl, Düsseldorf
12. Damiani AL, Kim MH, Wang J (2014) An improved dynamic method to measure k L a in bioreactors. Biotechnol Bioeng 111(10): 2120–2125. https://doi.org/10.1002/bit.25258
13. Petříček R, Labík L, Moucha T, Rejl FJ, Valenz L, Haidl J (2017) Volumetric mass transfer coefficient in fermenters: scale-up study in viscous liquids. Chem Eng Technol 40(5):878–888. https://doi.org/10.1002/ceat.201600585
14. Garcia-Ochoa F, Gomez E, Santos VE, Merchuk JC (2010) Oxygen uptake rate in microbial processes: an overview. Biochem Eng J 49(3):289–307. https://doi.org/10.1016/j.bej.2010.01.011
15. Goudar CT, Piret JM, Konstantinov KB (2011) Estimating cell specific oxygen uptake and carbon dioxide production rates for mammalian cells in perfusion culture. Biotechnol Prog 27(5):1347–1357. https://doi.org/10.1002/btpr.646

16. Garcia-Ortega X, Valero F, Montesinos-Seguí JL (2017) Physiological state as transferable operating criterion to improve recombinant protein production in Pichia pastoris through oxygen limitation. J Chem Technol Biotechnol 92:2573–2582. https://doi.org/10.1002/jctb.5272
17. Schumpe A, Quicker G, Deckwer W-D (1982) Gas solubilities in microbial culture media. In: Reaction engineering, Advances in biochemical engineering/biotechnology, vol 24. Springer Berlin Heidelberg, Berlin/Heidelberg, pp 1–38. https://doi.org/10.1007/3-540-11699-0_9
18. Wicke E (1969) Landolt-Börnstein: Zahlenwerte Und Funktionen Aus Physik, Chemie, Astronomie, Geophysik Und Technik, 6. Auflage. Band II: Eigenschaften Der Materie in Ihren Aggregatzuständen; 5. Teil/Bandteil b: Transportphänomene II, Kinetik, Homogene Gasgleichgewichte. Herausgegeben von Kl. Schäfer. Springer-Verlag, Berlin-Heidelberg-New York 1968. XII, 397 Seiten. Preis: DM230,—. Ber Bunsenges Phys Chem 73(3):325–326. https://doi.org/10.1002/bbpc.19690730322
19. Grigs O, Bolmanis E, Galvanauskas V (2021) Application of in-situ and soft-sensors for estimation of recombinant P. Pastoris GS115 biomass concentration: a case analysis of HBcAg (Mut+) and HBsAg (MutS) production processes under varying conditions. Sensors 21(4):1268. https://doi.org/10.3390/s21041268
20. Horta ACL, Da Silva AJ, Sargo CR, Cavalcanti-Montaño ID, Galeano-Suarez ID, Velez AM, Santos MP, Gonçalves VM, Giordano RC, Zangirolami TC (2015) On-line monitoring of biomass concentration based on a capacitance sensor: assessing the methodology for different bacteria and yeast high cell density fed-batch cultures. Braz J Chem Eng 32(4): 821–829. https://doi.org/10.1590/0104-6632.20150324s00003534
21. Kaiser C, Pototzki T, Ellert A, Luttmann R (2008) Applications of PAT-process analytical Technology in Recombinant Protein Processes with *Escherichia Coli*. Eng Life Sci 8(2): 132–138. https://doi.org/10.1002/elsc.200720232

Chapter 38

A Single-Use Technology Platform: How to Set Up Bioprocesses with *Pichia pastoris*

Melanie Ottinger and Jason Brown

Abstract

Among the many advantages of single-use technologies (SUTs), a low contamination rate - typically around 1% - is one of the most desirable. The unique, highly demanding performance requirements of *Pichia pastoris* bioprocesses (i.e., high-performance mixing and both high oxygen demand and heat transfers) have so far limited the transfer of such bioprocesses from traditional (glass or stainless steel) stirred-tank bioreactors to plastic-made single-use systems.

In this chapter, we present a comprehensive guide to cultivating *Pichia pastoris* using the HyPerforma™ Single-Use Fermentor (SUF), a truly scalable single-use stirred tank bioreactor shown to perform comparably to conventional systems. While detailed user and software manuals are available from the manufacturer, our protocol provides additional methods and optimizations tailored to the specific needs of *Pichia* bioprocesses - improving the overall efficiency and effectiveness of lab-scale SUF operations.

Key words *Pichia pastoris*, *Komagataella phaffii*, Single-use technologies, Single-use fermenter

1 Introduction

Single-use technology (SUT) has revolutionized the biotechnology industry, particularly in mammalian and insect cell applications. Single-use bioreactors (SUB), as well as vessels for storage, sampling, freezing, or mixing (i.e., flexible bags made from plastic material instead of glass or stainless steel), are well established in biopharmaceutical production. Single-use bioreactors (SUB) (*see* **Note 1**) are now established as the standard for current good manufacturing practices (cGMP) in the production of mammalian cell cultures [1, 2].

1.1 Why to Use (or Not) SUT for Bioprocesses with Pichia Pastoris?

As *Pichia pastoris* is a promising system for the production of recombinant proteins, capable of achieving complex, humanized glycosylation patterns together with high cell densities and robust and efficient protein production with inexpensive mineral medium,

Xavier Garcia-Ortega et al. (eds.), *Pichia pastoris: Methods and Protocols*, Methods in Molecular Biology, vol. 2697,
https://doi.org/10.1007/978-1-0716-4779-0_38,

it is increasingly replacing mammalian cell systems [3]. Given these advantages, there is great interest in culturing *Pichia* under such demanding safety conditions as SUT allows (*see* **Notes 2** and **3**). Despite benefits such as a reduced risk of cross-contamination, minimal maintenance, faster turnaround between runs, and increased flexibility with greater equipment utilization, SUT for microbial cells (including *Pichia* recombinant methylotrophic yeast) faces significant technological challenges, e.g., achieving adequate mixing and control of both temperature and dissolved oxygen levels.

The HyPerforma™ S.U.F. represents a truly scalable single-use stirred tank system (Figs. 1, 2, and 3), designed to be an alternative to conventional stirred tank aerobic fermenters (made from glass or stainless steel) currently used in microbial applications. For example, Ablynx, B.V., acquired by Sanofi, has reported using *P. pastoris* for large-scale manufacturing of several Nanobody® products [4] and has expanded its capacity with multiple 300 L SUFs.

Fig. 1 HyPerforma Enhanced Single-Use Fermentor (e.S.U.F.) hardware. Engineering improvements of S.U.F. were implemented on hardware as well as the BioProcess Container (BPC) (30 L and 300 L), to match the oxygen delivery and cooling capacity of the stainless-steel fermenters. The cooling jacket surface area was increased by 25% on the sides and bottom, with an additional 10% in gas hold-up volume area for the 300 L S.U.F. to support adequate cooling of the production process. More technical details are discussed directly in the supplier's manual: Thermo Fisher Scientific (2021). HyPerforma Single-Use Fermentor (S.U.F.) User's Guide, DOC0031, Revision H, October 2021

Fig. 2 Enhanced parabolic impeller used with e.S.U.F. A plastic (polyethylene) impeller with a bearing/seal assembly linking to an external mixer drive is the key element in the single-use design. The BPCs for 30 L and 300 L were redesigned to have enhanced impellers that increase, by at least four (4) times, oxygen delivery from spargers. As a result, experimental comparisons showed that oxygen supplementation was reduced by 70–93% vs. the original S.U.F. and is comparable with the performance of stainless-steel fermenters for the same strain and procedure

1.2 How to Design a Single-Use Facility?

While biopharmaceutical manufacturers are ultimately responsible for ensuring the suitability of single-use equipment, suppliers must also ensure regulatory compliance of disposable materials. Some developers and manufacturers of biopharmaceuticals are critical of SUT due to potential interactions with cells or proteins. Extractables and leachables (*see* **Note 4**), which derive from additives and are formed during bag manufacturing or gamma sterilization, pose a concern [5, 6].

The biopharmaceutical manufacturing approach should therefore include:

(a) Risk mitigation through chemical compatibility, clearance steps, and pre-flush.

(b) Chemical safety assessment involving extractables profiling, a controlled extractables study, sum parameters, unknown compounds, a simulated use extractables study, a leachables study, an exposure scenario, and a toxicological risk assessment.

Defining the type of production (multi-product or dedicated product) is crucial for designing a single-use facility. These systems

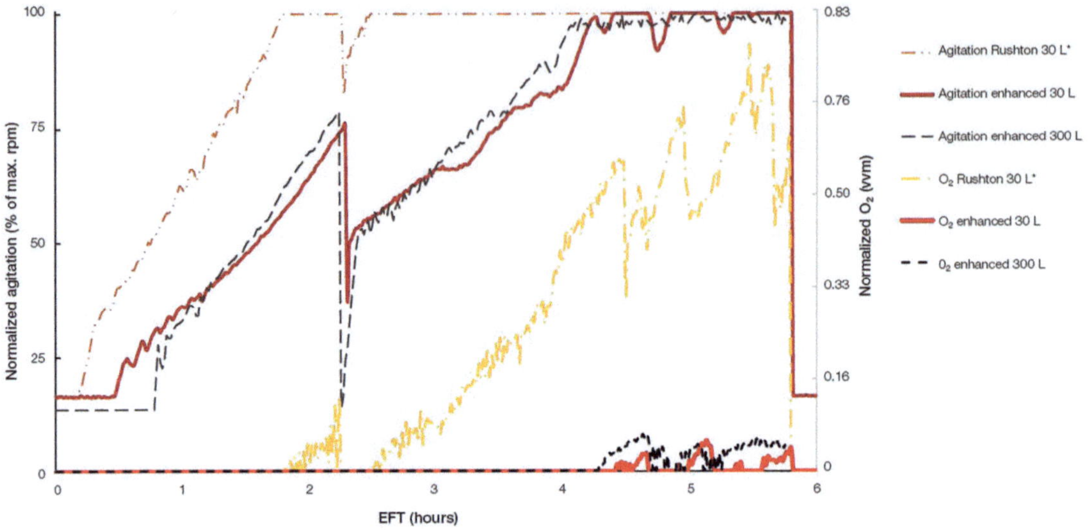

Fig. 3 Comparison of HyPerforma S.U.F. with a Rushton and enhanced impeller. Data showing oxygen (O_2) supplementation for the enhanced impeller, at 30 L and 300 L scales, compared to the original Rushton impeller in a 30 L HyPerforma S.U.F. The HyPerforma S.U.F. with Rushton impeller was set to 10% dissolved oxygen (DO), but with the HyPerforma e.S.U.F., the DO control was more precise and could be set to 5% DO. The enhanced impeller in the HyPerforma e.S.U.F. reduced the oxygen supplementation by 93%. Many cultures need systems capable of high oxygen transfer (kLa $> 600\ h^{-1}$), requiring gas flows of two (2) vessel volumes per minute (vvm). Single-use systems have constrained gas flow rates due to pressure safety limits of the single-use container, i.e., internal pressure <0.04 bar. This required a low operating backpressure to decrease the gas flow capacity of the exhaust filters [2, 8]

require different considerations than traditional stainless-steel systems, particularly in transferring process solutions like medium and buffer. Containers for single-use solution storage should be placed nearby, often outside core production suites, with transfer lines being fed through walls to minimize personnel traffic.

Single-use technology heavily relies on the procurement of raw materials and design, focusing on single-use consumables. The bag and tubing assembly must balance functionality and cost-efficiency. Testing and validating single-use hardware follows the same approach as stainless-steel systems, but qualification of consumables and vendor quality is critical. The quality of single-use bags and assemblies is determined during manufacturing and testing at the vendor site. Collaboration between bag manufacturers and film or resin suppliers is essential. A quality-by-design (QbD) approach involves selecting appropriate resins, defining film extrusion design space, and validating the welding process. Users must ensure that suppliers maintain high-quality controls. Leading suppliers are part of the Bio-Process Systems Alliance (BPSA), which promotes single-use manufacturing standards and guidance.

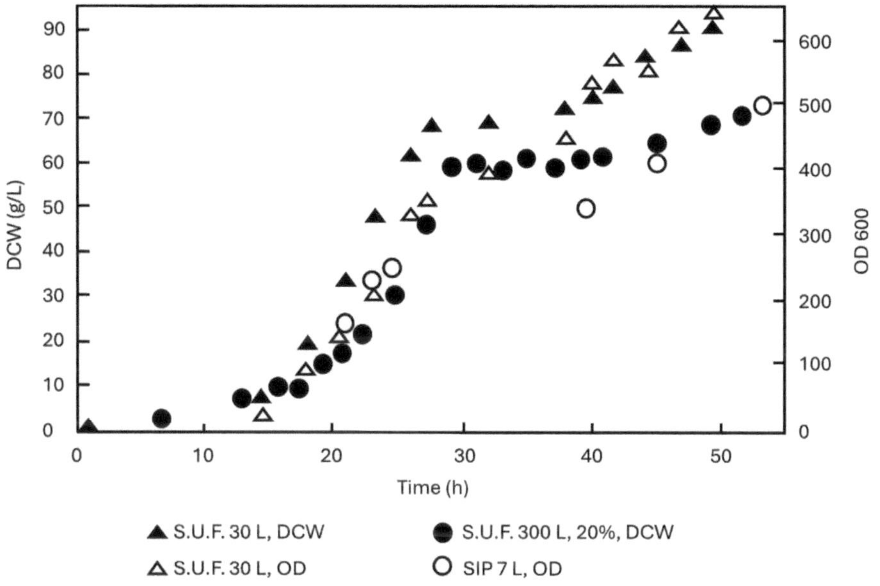

Fig. 4 Time course of biomass concentration in a fed-batch with *Pichia pastoris*. Dry cell weight (DCW) and optical density (OD600) for *Pichia pastoris* cultures from the 7 L SIP and S.U.F. fermenters are compared. The bioprocess started in BSM medium containing 2% glycerol, then glycerol was fed exponentially for 4 h, followed by 24 h of methanol feed. Production of recombinant protein (cutinase) was apparent at 6 h after induction with methanol. The maximum gas flow required was five times the liquid volume during expression culture, whereas with e.S.U.F., supplementation with oxygen was reduced by >70%. Finally, 2.1 g/L cutinase was produced, which was achieved with a similar maximum OD600 of 659, corresponding to a wet and dry cell weight (WCW and DCW) of 303 g/L and 95.6 g/L, respectively

1.3 Why This Protocol?

This protocol was established from experience using several fed-batch cultures of *P. pastoris* strain GS115 expressing cutinase, a 22 kDa protein with glycerol and methanol. They were performed in HyPerforma 30 L and 300 L S.U.F., and the final concentrations of biomass and product were similar to those typically achieved with *P. pastoris* in a Sartorius™ 10 L (7 L working volume) stainless-steel Sterilize-in-Place (SIP) bioreactor at our pilot-scale protein production laboratory (Fig. 4).

This protocol is intended for laboratory practitioners who wish to transfer their established bioprocesses with *Pichia* to an SUF system because they are working with biopharmaceutical applications with special safety requirements. The focus is therefore on the handling and practical operation of this new type of fermenter , and principles described here are also applicable to other experimental designs with microorganisms.

2 Materials

The following materials were routinely used for cutinase production in SUF, exemplified in this protocol.

2.1 Strain and Media

1. *P. pastoris* strain GS115 expressing cutinase.
2. All culture media commonly used with *Pichia pastoris*, i.e., YPD medium (yeast extract–peptone–dextrose, containing Zeocin™ antibiotic), BMGY (buffered glycerol complex medium), and BSM medium (basal salt medium).

2.2 Fermentor and Laboratory Equipment

1. HyPerforma™ S.U.F. of 30 L volume that consists of the following primary components:
   (a) Outer support container with standard water jacket heating system (Fig. 1).
   (b) S.U.F. or e.S.U.F. BioProcess Container (BPC) (Fig. 5).
   (c) Rushton impellers with standard S.U.F. or an enhanced parabolic impeller with e.S.U.F. (Figs. 2 and 3).
   (d) Impellers coupled directly to a shaft through a bearing to the drive motor.

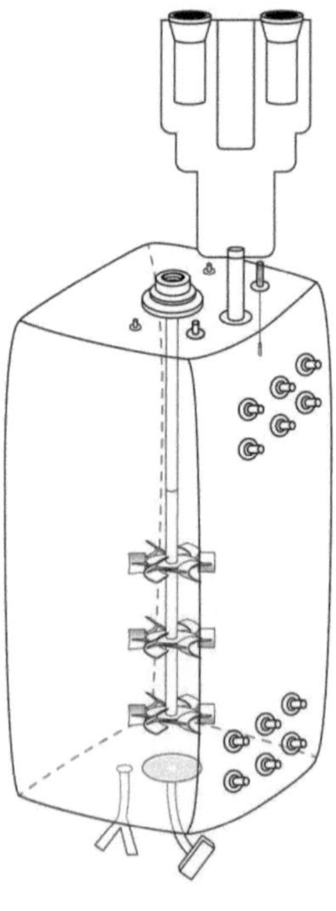

Fig. 5 Illustration of the e.S.U.F. BioProcess Container (BPC, 30 L). BioProcess Containers (BPCs) are supplied pre-sterilized by irradiation. A key element in the single-use design is the plastic (polyethylene) impeller with a bearing/seal assembly linking to an external mixer drive

(e) Direct drive agitation mixing assembly with an AC motor.
(f) 4 baffles.
(g) Drilled-hole sparger (DHS) with ring-like design.
(h) Exhaust management and condenser systems designed to provide reliable off-gassing and to help minimize liquid loss during the fermentation process while extending filter life.
(i) Exhaust gas vent filter heater and backup filter automated pinch clamp (to extend filter life).
(j) Turnkey system with HyPerforma G3 process control system or electrical control panel (E-Box) for agitation with open-architectural control.

2. Microscope.
3. Safety cabinet class II.
4. Centrifuge (The Thermo Scientific™ Sorvall™ BIOS 16 Centrifuge with Thermo Scientific™ CentriPAK™ BPC system).
5. Water bath.
6. Magnetic stirrer.
7. Incubators.
8. pH meter.
9. Pipettes.
10. 0.2 μm filters.
11. Sample vials (recommend vacuum sample tubes with blood sample device so no culture is dripped on the floor as it is with typical syringe samples, or several 50 mL Labtainer Pro BioProcess Containers built into the sample line set).
12. Shake flasks.
13. Medium storage bags.
14. Peristaltic pumps.
15. Balances.
16. Tubes and clamps.
17. Autoclave.
18. Welder.
19. Single-Use Mixer DS 300 (Thermo Scientific™ HyPerforma™) for medium preparation.

3 Methods

In this protocol, we describe best practices for working with *P. pastoris* in the HyPerforma 30 L. Users of this protocol will preferably be trained on single-use equipment and will be able to develop their own comprehensive operating protocols.

Cutinase (22 kDa protein) production in the 30 L HyPerforma S.U.F. includes inoculum production as well as preliminary production of media and multiple steps to prepare the S.U.F. for fed-batch cultures grown subsequently with glycerol and methanol; this includes preparation of the bioprocess containers (BPCs), including assemblies, their filling, and post-use disposal.

All culture medium recipes correspond to the compositions given in "*Pichia Fermentation Process Guidelines*" (or purchased from a supplier, *see* **Note 2**) and include YPD, BMGY, and BSM.

3.1 Inoculum Production in Baffled Flasks

1. Prepare the YPD medium (containing Zeocin™ antibiotic) for the initial seed culture. The liquid volume should be a maximum of 10–30% of the total flask volume.
2. Start the initial seed culture 40 h prior to S.U.F. inoculation from a single colony transferred from a YPD agar plate to YPD liquid medium.
3. Prepare the BMGY medium for 10% of the initial S.U.F. volume.
4. 20 h prior to S.U.F. inoculation, transfer 8 mL of the initial seed culture to each liter of BMGY medium.
5. Operate the culture at 30 °C and 250 rpm for 16 h up to an optical density of 2. 6. That is, OD at 600 nm. To accurately determine OD_{600} >1.0, dilute the sample ten-fold before reading.

3.2 Medium Preparation

1. Prepare all BSM components.
2. Hydrate the medium components in 25 L batches using the Thermo Scientific™ HyPerforma™ Single-Use Mixer DS 300.
3. Filter the hydrated medium through a 0.45 μm filter, followed by a gamma-sterilized 2″ 0.2 μm PVDF filter.

3.3 Unpacking: Preparation of BPC

1. Store the BPC in its original packaging, under ambient conditions (2–30 °C).
2. One day before inoculation, prepare the S.U.F. Before unpacking the BPC, ensure that a safety cutter or other similar tool is available.
3. For unpacking, position the box with the top side facing up. Acceptable carriers and surfaces have smooth edges and rounded corners.
4. Add the BPC restraint system to the S.U.F. with BPC assemblies to keep the BPC flat and, thus, prevent its damage. Remove the BPC from the restraint system inside the box.

5. Inspect the outer polybag for any damage. If there is no damage, use a finger to punch a hole through the end of the polybag and manually tear it open, using both hands.
6. Place the BPC on a proper workspace surface. Unfold the chamber either all the way open or in segments to inspect for damage or abnormalities. Normal folding marks, light surface wrinkles, and light stress marks are acceptable.
7. Check for particulate risk according to the screening criteria provided by the supplier (*see* **Note 5**).
8. Inspect the tubing for signs of damage or abnormalities (*see* **Notes 6** and **2**).
9. Inspect the connection points for the presence of a retainer type (such as cable ties or BarbLock™ retainers) and proper fit.
10. Inspect connectors for missing caps or plugs on the ends.
11. Ensure that aseptic, steam-through, and steam-to connectors are in the correct position and have not been activated.
12. Inspect filters and verify the flow direction (*see* **Note 2**).

3.4 Preparation and Sterilization of the Probe Assemblies for the BPC

Options for single-use probes are available for DO, pH, pressure, foam sense, optical density, and capacitance. In case you prefer conventional probes, the following procedure describes the use of the probe assembly.

1. Select the appropriate probe. Verify the presence of a Teflon™ support ring and O-ring on the probe and visually inspect them for damage.
2. Perform any required probe maintenance and calibrate the pH probe.
3. Insert the probe into the probe assembly through the threaded adaptor.
4. Verify that the probe tip is not touching the membrane of the aseptic connector (more than 6.35 mm, i.e., 1/4 in. gap) before threading it into the probe adaptor.
5. Hand-tighten the adapter and verify that the probe tip is not touching the membrane.
6. Place the probe assembly with the probe into the autoclave tray for probe kits.
7. Autoclave the probe assembly using a validated sterilization cycle, approximately 30 min at 122 °C.
8. Allow enough time for probe assemblies to cool completely before connecting to the BPC for probe insertion.
9. When stored properly, the autoclaved probe assemblies can be stored dry for short periods of time (less than 24 h) without loss of sensor longevity, performance, or sterility.

3.5 BPC Installation for 30 L Units

1. Confirm that the correct volume BPC is being used, i.e., each outer support container is designed for a specific size S.U.F. BPC (can be customized).
2. Before starting to load the BPC, verify that the S.U.F. hardware is stationary, with the casters locked into place.
3. Open the S.U.F. door and load the BPC into the stainless-steel outer support container. Orient the BPC with the bearing port up (before removing the cap) and toward the motor drive, with the connector probe ports facing the bottom access cutout.
4. Place the bearing port into the bearing port receiver. Close the door and the clamp.
5. Route the bottom ports through the opening in the bottom of the outer support container, and route the sparge line and bottom drain line through the bottom access opening.
6. Open the door of the condenser unit and feed the condenser portion of the bag through the condenser. Insert the vent filter (s) into the filter bracket. Straighten and smooth the condenser bag in the condenser unit, and close and latch the door of the condenser. Ensure that the condenser BPC is not too tight or too loose in the condenser hardware unit.
7. Secure filter heaters on the filters. The vent filter heater controller is preset to 55 °C.
8. Connect the incoming gas line to the sparger filter and inflate the BPC with air through the sparger. Always inflate the BPC with air prior to adding medium (*see* **Note 2**).
9. To reach the desired pressure, obstruct the exhaust flow path by gently pressing on the chamber above the 1 in. port/tube or cover the exhaust exit port with your hand until pressure reaches between 14 and 34 mbar (0.2–0.5 psi).
10. Check valves during typical use, i.e., for the sparge line, it is not uncommon for some fluid to bypass check valves. Elevate the filter above the liquid level in the S.U.F. and secure it to prevent exposing the filter to liquid. Inflation time is approximately one to 5 min; do not exceed 25 slpm or 0.034 bar (0.5 psi) internal pressure for 30 L BPCs.
11. Use the bottom cutouts located at the base of the support container as a reference to align the hanging tabs on the BPC. Attach the two hanging tabs on the bottom of the BPC to the tab holders (pins) on the SUF.
12. Align the row of connector ports and probes, and then close the S.U.F. door. Verify that all port clamps are closed and located as close as possible to the body of the BPC.
13. Allow for drive shaft insertion and fill the BPC with air for approximately 10–20 min (times will vary based on flow rate,

inlet pressure, and container volume). To prevent damaging the BPC, the BPC must be partially inflated before you begin to insert the drive shaft. Close the access door of the S.U.F. before fully inflating the BPC.

14. Connect the medium ground clip to the stainless-steel insert in the sample line of the BPC. This grounds the medium inside the BPC and helps to eliminate electrostatic charge.
15. Install the exhaust filter pinch clamp on the vent filter post. Check to ensure that the blue light on the release button is illuminated, which indicates that power to the unit is enabled.
16. Before inserting the drive shaft, remove the latch pin from the motor assembly and unscrew the threaded cap covering the pass-through of the motor. Use two hands to load the drive shaft through the top of the motor assembly; a slight back-and-forth twisting motion will aid in insertion and prevent stretching of the impeller tubing. In order to directly couple the drive shaft to the motor drive, place the threaded cap on the hollow pass-through and hand-tighten in a clockwise direction. Then tighten the cap by placing the spanner wrench on the hollow pass-through and tighten the cap using the supplied torque wrench. Close the safety access cover and insert the latch pin.
17. Turn off the air supply once the drive shaft has been inserted.
18. Position and close a bar clamp on the bottom drain line as close as possible to the BPC port.
19. Before inserting the RTD or a selected temperature sensor into the thermowell, place a small amount (0.5 mL) of conductive glycerol in the thermowell to aid in heat transfer. Secure by twisting the Luer lock collar. The thermowell will stretch slightly when the RTD is seated.
20. Connect the pressure sensor to the monitoring device.
21. Before beginning probe insertion, become familiar with the CPC AseptiQuik or Kleenpak connector. Attach probe clips onto the outer support container above the probe assembly. Plastic probe clips slide on with firm pressure. Install the pre-sterilized sensor and probe kit using aseptic connection methods. The aseptic connection is completed prior to the bellows being collapsed. Insert the probe by collapsing the bellows. Position the probe clip in the desired horizontal location. Lift the probe and set it into the probe clip. Verify that the probe remains at the proper insertion depth and angle as the bellows expand to rest freely in the probe clip.
22. Plug the foam sensor wire into the port at the top of the BPC and attach the ground clip to the stainless-steel insert in the RTD/sampling port.

3.6 Medium Fill to the BPC

1. Always inflate the BPC with air prior to adding medium. Ensure a good fit of the BPC to the outer support container/jacket. This will reduce heat-up and cool-down times.
2. Select the desired line set from the BPC for fluid introduction and make an aseptic connection (either tubing welder, quick connect, or tri-clamp), and begin the liquid fill.
3. After approximately 10% of the projected volume has been added, check the position of the BPC in the outer support container, particularly the sparger and the drain line. Adjust the position if necessary for a proper fit.
4. Fill to the desired liquid volume, i.e., 20–100% of the rated volume.
5. The oxygen transfer rate is best with all 3 impellers submerged, so starting at 22 L for batch culture and, after fed-batch culture, finishing at 30 L of liquid working volume is recommended.

3.7 Process Strategy: Design and Implementation

The process strategy comprises a series of steps ensuring the performance of specific actions in a specific order at specific times.

1. Write a process strategy, i.e., this often involves translating an existing process from shake flasks, glass, or stainless-steel bioreactors into the S.U.F., whereas some compromises may be necessary to account for the flexible single-use technology.
2. Implement this strategy into the Thermo Scientific TruBio Bioprocess Control Software (*see* **Note 7**).
3. Control temperature at 30 °C in batch and fed-batch with glycerol, i.e., before methanol induction. For the fed-batch with methanol, set and control the temperature to 28 °C.
4. Set the pH controller to 5.0 ± 0.1 and maintain this pH during S.U.F. cultivation by automatically adding 28% NH_4OH and 43% H_2SO_4.
5. Set the DO controller to 30% and maintain this level with a cascade controller (*see* **Note 8**).
6. Check that the S.U.F., which is previously filled with medium, is operating at the targeted steady equilibrated state and has achieved the desired temperature and pH. This S.U.F. is then ready for inoculation with *P. pastoris* strain GS115.
7. Connect the inoculum addition line, i.e., at the end of the seed culture vessel, and equip it with the proper connectors/tubing.
8. Transfer the inoculum to the S.U.F. Typically, this is done with the tubing connection process (aseptic Luer lock connection or tube welding) and a peristaltic pump. Pump the desired volume of seed cells into the S.U.F.

9. Once glycerol in the batch culture is consumed, as indicated by a peak in the percentage of dissolved oxygen (DO), add a 50% glycerol feeding solution and incubate for 4 h (*see* Fig. 3).
10. Upon completion of the glycerol feed, a second peak in DO is noted. Then, begin adding a 99% methanol induction feed.
11. The methanol feed is increased (or decreased) as needed for optimal protein expression for another 24 h (*see* Figs. 3 and 4).
12. Take samples regularly for dry cell weight (DCW) and protein analysis. Protein concentrations are estimated using a densitometric method in which samples are compared against BSA standards and gel images are analyzed using TotalLab™ software.
13. After the working volume is filled or the planned process duration is accomplished, dispense and harvest the culture as follows:
    (a) Connect the bottom drain tubing set to the intended transfer line.
    (b) Open the clamp positioned at the bottom drain port.
    (c) Begin to drain by using the peristaltic pump.
    (d) When approximately three to five liters remain in the BPC, lift the BPC at the top hanging tabs located opposite the bottom drain (this will pool medium toward the drain).
    (e) Hold the bottom drain line near the floor while lifting the exhaust filter side of the BPC. This will facilitate draining the final liter of harvest medium.
14. Shutdown the bioprocess and dispose of the BPC post-use (*see* **Note 9**).

4 Notes

1. All terms "fermenter," "fermentor," or "bioreactor" can refer to the vessel used in cultivation. "Fermentor" is more specifically used to describe the equipment (apparatus, vessel) used in the microbial fermentation process. Single-use fermentors (SUF) or single-use bioreactors (SUB) for cell culture processes are today the norm. Their acceptance in microbial fermentations is still rare, becoming more common at a smaller scale of up to 300 L.

    Based on the long-accepted traditional stirred tank reactor (STR) made of glass or stainless steel with Clean-in-Place (CIP) and Sterilization In Place (SIP) features, the SUF mimics their scalability (height-to-diameter ratio, mixer design, and location) and operating parameters, but with the unique advantage of being a single-use device.

2. All design features and hardware specifications can be found on the supplier's website when using the search function with the product name "HyPerforma Enhanced Single-Use Fermentor":

   https://www.thermofisher.com

   The following Application Notes (Thermo Fisher Scientific and former Invitrogen) and Guidelines are relevant in the context of this chapter:

   (a) Cultivation of aggressive microbial cultures of Pichia and Bacillus using the HyPerforma 30 L and 300 L Single-Use Fermentors. (2016)

   https://assets.thermofisher.com/TFS-Assets/BPD/Application-Notes/suf-cultivation-aggressive-microbial-cultures-app-note.pdf

   (b) Bioprocessing solutions

   https://assets.thermofisher.com/TFS-Assets/BPD/brochures/BPD-Product-Capabilities-Catalog.pdf

   (c) Enhanced single-use fermentor for improved cooling and oxygen mass transfer. (2021)

   https://assets.thermofisher.com/TFS-Assets/BPD/Application-Notes/application-note-enhanced-single-use-fermentor-improved-cooling-oxygen-mass-transfer.pdf

   (d) HyPerforma 5:1 Single-Use Bioreactor (SUB) User's Guide DOC0022, Revision H, February 2021

   https://assets.thermofisher.com/TFS-Assets/BPD/manuals/5to1-single-use-bioreactor-users-guide.pdf

   (e) Standard BioProcess Container Unpacking and Inspection Guide, Revision G, June 2022

   https://assets.thermofisher.com/TFS-Assets/BPD/Product-Guides/bioprocess-container-unpacking-inspection-guide.pdf

   (f) *Pichia* Fermentation Process Guidelines (by Invitrogen Life Technologies, 2002)

   https://tools.thermofisher.com/content/sfs/manuals/pichiaferm_prot.pdf

3. Plastic containers (BPCs) of SUB or SUFs are provided pre-sterilized by the supplier using gamma irradiation and are replaced after each batch, thereby reducing cleaning and cleaning validation requirements. The eliminated need for CIP/SIP saves on the one hand time, in the form of faster product changeover procedures, and on the other hand labor, plus savings from lower energy and water consumption. The lack

of complicated utilities also lowers the capital investment as well as the cost and effort for qualification and maintenance. The lower risk of cross-contamination, since the product flow path is discarded and replaced after each batch, is contributing to the operational speed. Therefore, single-use facilities have significantly shorter downtimes, thus increasing efficiency and batch output per year. Furthermore, the simple reconfiguration and assembly of mobile and disposable equipment allow optimal customization of processes and increase the flexibility of the production site. Due to its advantages, single-use equipment is, at present, widely used in mammalian bioprocesses outside of cell culture, including harvest, filtration, and buffer/medium preparation. Disposable filtration and clarification are well established, and single-use membrane adsorbers are beginning to replace polishing columns [2].

4. Extractables are compounds that originate from components in a solution under extreme conditions, whereas leachables originate under typical operational conditions. Suppliers should provide generic extractables data.

   For further studies, the DECHEMA publication "Recommendation for Leachables Studies – Standardized cell culture test for the early identification of critical films" and the ASTM Standard Guide for Cell Culture Growth Assessment of Single-Use Material (E3231 – 19) are recommended.

5. The unintentional presence of particulates (loose or mobile matter) can potentially affect patients, cell growth, and/or drug products. Major classifications for SUT particulate risk are:
   (a) Location of particle, i.e., product contact surface, exterior of the SUT, or embedded in the SUT material,
   (b) Particle characteristics, i.e., intrinsic or extrinsic,
   (c) Size of the particle, appearance of the particle, e.g., shape, texture, or color,
   (d) Count of particles, i.e., extent of particle contamination and number of particles per square unit, i.e., TAPPI size estimation chart T 564 is used by industrial manufacturers to measure the size of spots, defects, or inclusions.

6. After irradiation, a tube may be set in a slightly bent or kinked position as a result of the packaging operation. Most tubing kinks are cosmetic and may be opened manually by pinching the wall of the tubing to release the kink without impacting product integrity.

7. Thermo Scientific TruBio Bioprocess Control Software, powered by DeltaV™ distributed control system from Emerson™ is an open, fully configurable software solution designed to

provide a flexible system for the control of bioprocesses. The software has been developed according to GAMP5 methods and conforms to regulatory requirements for use in current good manufacturing practice (cGMP)-compliant processes.

8. The cascade DO controller of the 20% DO setpoint works as follows: First, the stirrer is ramped up, then air is added, followed by pure oxygen. The cascade minimum set points are increased manually to avoid dropping the DO below 20–10%.
9. Appropriate post-process treatment depends on the characteristics of the product generated during the manufacturing process, as well as geography and waste management preferences [7]. Those biopharmaceutical manufacturers located near waste-to-energy facilities or recyclers that handle infectious waste should carefully consider these options over landfilling or incineration without energy recovery.

Acknowledgments

We thank our colleagues Dmitrij Bugajev and Juozas Šiurkus from Thermo Fisher Scientific Baltics for providing experimental data. We acknowledge Prof. Dr. Karin Kovar (daspool, Wädenswil, Switzerland) for her technical and conceptual assistance, many helpful discussions, and support.

Competing Interest Statement The authors have no conflicts of interest to declare that are relevant to the content of this chapter.

References

1. Schmidhalter DR, Elzner S, Schmid R (2019) Progress in the development of single-use solutions in antibody–drug conjugate (ADC) manufacturing. In: Single-use technology in biopharmaceutical manufacture, pp 303–310. https://onlinelibrary.wiley.com/doi/book/10.1002/9781119477891
2. Jones N (2015) Single-use processing for microbial fermentations. Bioprocess Int 13(4):56–62. https://www.bioprocessintl.com/microbial-cell-culture/single-use-processing-for-microbial-fermentations
3. Love KR, Dalvie NC, Love JC (2018) The yeast stands alone: the future of protein biologic production. Curr Opin Biotechnol 53:50–58. https://doi.org/10.1016/j.copbio.2017.12.010
4. Webpage: https://pichia.com/science-center/commercialized-products/
5. Eibl R, Steiger N, Fritz C, Eisenkrätzer D, Bär J, Müller D, Eibl D (2014) Standardized cell culture test for the early identification of critical films for CHO cell lines and chemically defined culture media. DECHEMA, Frankfurt aM. ISBN, 978-3-89746-149-9. https://dechema.de/dechema_media/Downloads/Positionspapiere/SingleUse_Leachables_2014_en.pdf
6. Sobańtka A, Weiner C (2019) Extractables/leachables from single-use equipment: considerations from a (bio) pharmaceutical manufacturer. In: Single-use technology in biopharmaceutical manufacture, pp 143–158. https://onlinelibrary.wiley.com/doi/book/10.1002/9781119477891

7. Whitford WG, Petrich MA, Flanagan WP (2019) Environmental impacts of single-use systems. In: Single-use technology in biopharmaceutical manufacture, pp 169–179. https://onlinelibrary.wiley.com/doi/book/10.1002/9781119477891
8. Schirmer C, Dreher T, Leupold M, Glaser R, Castan A, Brown J et al (2019) Recommendation for biological evaluation of bioreactor performance for microbial processes. DECHEMA, Gesellschaft für Chemische Technik und Biotechnologie eV. ISBN 978-3-89746-222-9. https://dechema.de/dechema_media/Downloads/Positionspapiere/SingleUse_Microbial_ezl+2Ed+Sept2019.pdf

Chapter 39

A *Pichia pastoris* Cell-Free Protein Synthesis Platform

Rochelle Aw, Farzana Alam, Alex J. Spice, and Karen M. Polizzi

Abstract

Cell-free protein synthesis is an in vitro transcription and translation process that takes advantage of the fact that cell growth is dissociated from protein production. After cell cultures have grown to the optimal cell density, the cells are lysed while preserving the cellular machinery so that they can be used in an in vitro protein synthesis reaction. The system uses three main components: cell lysate, a reaction mix, and plasmid DNA. Protein production can be performed in hours on a lab bench, expanding the possibility to evaluate multiple plasmids at once without the need to transform and select positively expressing clones.

The use of *Komagataella phaffii* to generate the lysate for cell-free protein synthesis has been shown with a number of different strains, including a ribosome-overexpressing strain, *FHL1*. The high cell densities result in a highly active lysate that has been shown to be capable of producing a variety of products, including human serum albumin and virus-like particles (VLPs). Furthermore, as a eukaryotic platform, the system is capable of performing post-transcriptional modifications such as creating disulfide bonds without the need for additional folding chaperones or supplements. Here, we describe the methods required for generating cell lysate, creating a reaction mix that has been optimized through design of experiments, optimizing the concentration of DNA template, and using a reporter protein to monitor expression yields. Harnessing the potential of CFPS using *K. phaffii* can lead to faster prototyping of vectors before strain construction, facilitating the rapid identification of optimal protein variants. It can also be used to produce proteins that are toxic to living cells, providing an alternative source of proteins for experimentation.

Key words *Pichia pastoris*, *Komagataella phaffii*, Cell-free protein synthesis, In vitro transcription-translation, Rapid prototyping, Synthetic biology, Eukaryotic cell-free system

1 Introduction

Komagataella phaffii (reclassified from *Pichia pastoris*) has proven itself as a successful recombinant expression platform with the approval of multiple products by the Food and Drug Administration (FDA), including a recently approved monoclonal antibody [1, 2]. Despite this, there are products that can be difficult to produce in *K. phaffii*, and prototyping of strains can be laborious

Authors Rochelle Aw and Farzana Alam have equally contributed to this chapter.

Xavier Garcia-Ortega et al. (eds.), *Pichia pastoris: Methods and Protocols*, Methods in Molecular Biology, vol. 2697,
https://doi.org/10.1007/978-1-0716-4779-0_39,

due to the length of time needed for transformation and expression. The popularity of cell-free protein synthesis (CFPS), whereby active cell lysates are produced containing the cellular machinery required for protein synthesis, opens up novel avenues for the production of proteins at a small scale [3, 4]. CFPS was first used in the 1960s by Nirenberg and Matthaei to decipher the genetic code through the use of *Escherichia coli* extracts [5] and has in recent years expanded to include a product that has been given Fast Track designation by the FDA [6]. CFPS is advantageous over in vivo protein production as products can be made in hours as opposed to days, and it is an open reaction that allows fine-tuning of processes, which can be particularly important for downstream processing. Furthermore, the production scale is small, and therefore it is easy to compare multiple conditions or targets at one time, and it overcomes issues such as secretion saturation or the limitations of producing toxic products.

In theory, any cell type can be used for CFPS, with many already reported including *E. coli*, wheat germ [7], rabbit reticulocyte [8], Chinese hamster ovary cells [9, 10], *Streptomyces venezuelae* [11], *Bacillus megaterium* [12], tobacco plants [13], as well as several yeast species including *Saccharomyces cerevisiae* [14], *Aspergillus niger* [15], *Neurospora crassa* [16], and *P. pastoris* [17, 18]. Multiple strains of *K. phaffii* have been tested, including the wild-type strain X33, two strains optimized for ribosome overexpression, *FHL1* and *HMO1,* and a protease-deficient strain, SMD1163 [17, 18]. A variety of products have also been made to date using the platform including reporter proteins such as luciferase, GFP, human serum albumin (up to 116 μg mL^{-1}) [19], and vaccine products such as Hepatitis B core antigen virus-like particles with higher reported yields (6.5 μg mL^{-1}) than other CFPS platforms [20].

CFPS is composed of three necessary parts for functionality: the cell lysate, the reaction mix, and the DNA template. We have performed extensive optimization of the harvest point in order to finally balance ribosome content and cell density [17]. Additionally, using a definitive screening design (DSD), we have optimized the reaction mix, resulting in a 3.5-fold improvement in yield [19].

Here we describe methods for the generation of each of these components, including increasing the versatility of this platform by creating a system where sonication can be used for cell disruption, making it more accessible to research labs. This protocol utilizes an improved reaction mix that was informed by a Design of Experiment (DoE) approach [19].

2 Materials

2.1 Preparation of Pichia pastoris lysate

1. Single colonies from a fresh YPD agar plate at 30 °C.
2. YPD liquid broth: 1% yeast extract, 2% peptone, 2% dextrose.
3. YPD agar: 1% yeast extract, 2% peptone, 2% dextrose, 2% agar; add antibiotic as appropriate for strain selection.
4. Buffer A: 20 mM HEPES-KOH (pH 7.4), 100 mM potassium acetate, 2 mM magnesium acetate. Prepare it fresh and keep ice cold.
5. Dry ice (optional)
6. Methanol (optional)
7. Lysis buffer A: 20 mM HEPES-KOH, 100 mM potassium acetate, 2 mM magnesium acetate, 2 mM DTT, 0.5 mM PMSF. Prepare fresh and keep it ice cold.
8. Bradford assay kit (or another suitable protein quantification kit)
9. 1 liter baffled glass flask
10. Orbital shaking incubator, 30 °C
11. UV/visible spectrophotometer
12. 1.5 mL cuvettes
13. 15 mL centrifuge tubes
14. 50 mL centrifuge tubes
15. High-speed refrigerated centrifuge
16. Sonicator (e.g., Fisherbrand™ Model 120 Sonic Dismembrator) with a 1/8-inch probe.
17. 10 mL serological pipettes
18. A syringe with an 18-gauge needle
19. 3.5 K MWCO Slide-A-Lyzer™ G2 dialysis cassette (Thermo-Fisher Scientific, Paisley, UK)
20. 1.5 mL microcentrifuge tubes

2.2 Reaction Mix

1. DNA template (plasmid) prepared by extraction with Qiagen Plasmid Maxi, Mega, or Giga Kit (Qiagen, Crawley, UK). DNA template resuspended in TE buffer: 100 mM Tris-Cl (pH 8.0), 10 mM EDTA (pH 8.0).
2. 1 M HEPES-KOH (pH 7.4), filter sterilized
3. 1 M magnesium glutamate, filter sterilized
4. 1 M dithiothreitol (DTT)
5. RTS Amino Acid Mix (biotechrabbit GmbH, Hennisdorf, Germany)

6. NTP solution set, 100 mM each (ThermoFisher Scientific)
7. 500 mM creatine phosphate, filter sterilized
8. 4 mg/mL creatine phosphokinase (Sigma-Aldrich, St. Louis, MI) in 25 mM glycylglycine (pH 7.5), filter sterilized
9. 40 U/μL murine RNase inhibitor (New England Biolabs, Ipswich, MA)
10. T7 polymerase, HC (200 U/μl, ThermoFisher Scientific)
11. Nuclease-free H_2O (ThermoFisher Scientific)
12. Cell lysate
13. 1.5 mL microcentrifuge tubes
14. Microcentrifuge
15. Water bath or heat block, 60 °C
16. Incubator or water bath, 21 °C

2.3 Determining Luciferase Production from Cell-Free Reactions

1. CFPS reaction product
2. 200 mM tricine, pH 7.6
3. 25 mM ATP, made fresh and stored on ice
4. 50 mM $MgSO_4$
5. 10 mM $MgCO_3$
6. 10 mM EDTA, pH 7.6
7. 500 mM DTT, made fresh and stored on ice
8. 25 mM D-luciferin or luciferin salt must be stored at −20 °C and protected from light
9. 1.5 mL microcentrifuge tubes
10. 96-well, half-area, flat-bottom white microtiter plates (Grenier Bio, Gloucestershire, UK; cat no. 675075)
11. Plate reader for luminescence

3 Methods

3.1 Preparation of K. phaffii Lysate Using a SONICATOR

The protocol described here is for the production of cell lysate using a sonicator to disrupt the cell wall, but under conditions that have been optimized to maintain the integrity of the cellular machinery (Fig. 1). This is an amended protocol that aims to improve the accessibility of this protocol to laboratories where high-pressure cell-disruptors are not available. The yield for CFPS may be lower when using lysate produced with a sonicator than when produced using a high-pressure cell homogenizer due to the reduced total protein concentration achieved after lysis. *See* **Note 1** for details on the use of a high-pressure cell homogenizer for the production of cell lysate. Glass beads have also been used for cell

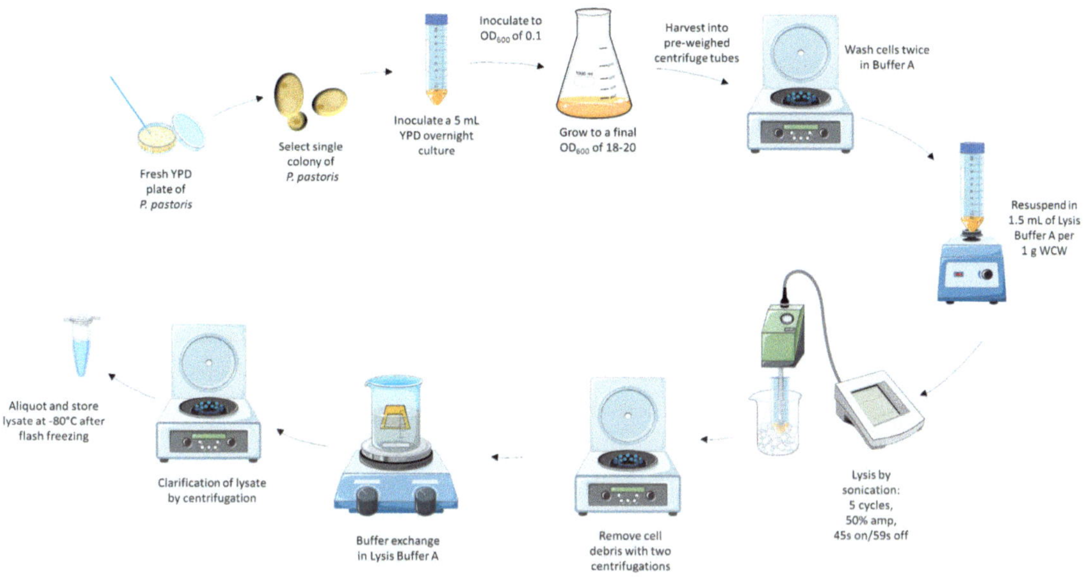

Fig. 1 Schematic representation of the cell-free extract preparation process. Icons adapted from BioIcons (https:/bioicons.com/) and Servier (https:/smart.servier.com/), modified and licensed under CC-BY 3.0 Unported https:/creativecommons.org/licenses/by/3.0/

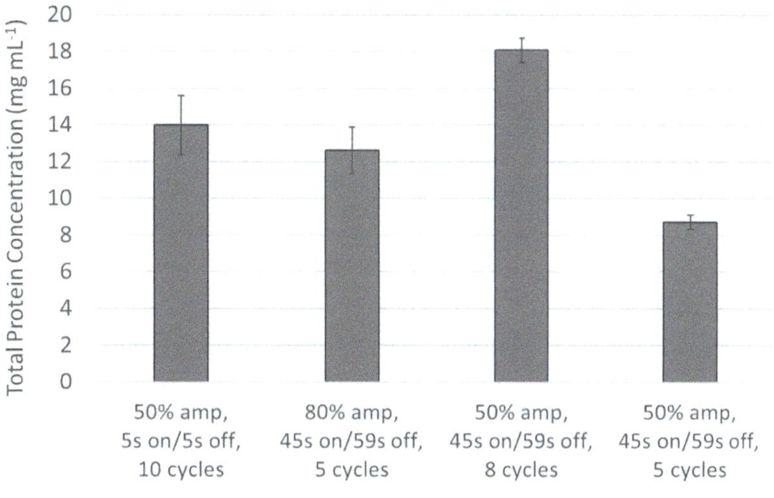

Fig. 2 Protein concentrations achieved from sonication under four different cycle conditions. Concentrations were measured in triplicate using Pierce™ Coomassie Plus (Bradford) Assay Kit

lysis, and the ratio of beads to cells and cycles has been optimized [18].

1. A fresh YPD agar plate should be prepared from glycerol stocks prior to the preparation of the cell lysate (*see* **Note 2**) and left to grow for 3–5 days at 30 °C. If a genetically engineered strain is being used, maintain antibiotic selection where applicable.

2. Pick a single colony and inoculate into 5 mL of YPD medium (maintaining antibiotic selection if applicable) in a 50 mL centrifuge tube. Incubate overnight at 30 °C with shaking at 250 rpm.
3. Measure the OD_{600} of the overnight culture using a spectrophotometer.
4. Subculture the overnight culture into a sterilized 1 L glass baffled flask containing 200 mL of YPD medium to an OD_{600} of 0.1 (*see* **Note 3**).
5. Incubate the culture in a shaking incubator at 250 rpm at 30 °C until an OD_{600} of 18–20 has been reached (*see* **Note 4**).
6. Pre-cool four 50 mL centrifuge tubes on ice and record the weight of one of the tubes (*see* **Note 5**).
7. Once the desired OD_{600} has been reached, the culture should be split equally between the four pre-cooled tubes. All following steps should be performed on ice and using a refrigerated centrifuge (*see* **Note 6**).
8. Centrifuge for 10 min at 3000× *g* at 4 °C.
9. Remove the supernatant and resuspend the pellet from each tube in 12.5 mL of ice-cold buffer A. Vortex or pipette to resuspend the cells, ensuring that the cells are regularly incubated on ice to reduce heat generation. Once fully resuspended, combine the four tubes into the pre-weighed 50 mL centrifuge tube.
10. Centrifuge for 10 min at 3000× *g* at 4 °C.
11. Remove the supernatant and wash the cells in 20 mL of ice-cold buffer A.
12. Centrifuge for 10 min at 3000× *g* at 4 °C.
13. Repeat **steps 11** and **12**.
14. After the final centrifugation, blot the 50 mL centrifuge tube on a towel to remove excess buffer.
15. Weigh the 50 mL centrifuge tube containing the combined cell pellet in order to calculate the wet cell weight.
16. Either continue with cell lysate preparation or flash freeze the pellet in a dry ice and methanol bath prior to storage at −80 °C until ready to continue (*see* **Note 7**).
17. Once ready to proceed with cell lysis, if the pellet has been frozen, ensure that it is fully defrosted before starting. The pellet should be resuspended in 1.5 mL ice-cold lysis buffer A per 1 g of wet cell weight, and if necessary, transfer it to a 15 mL centrifuge tube (*see* **Notes 8** and **9**).
18. Lyse the cells using a sonicator with a 1/8-inch probe (*see* **Note 10**). Use 50% amplitude with 45 s on and 59 s off for

5 cycles (total lysis time 3 min 45 s) (*see* **Notes 11** and **12**). It is essential that the tube is kept on ice at all times during sonication.

19. A sample should be collected post-lysis for total protein analysis at the end of the procedure.
20. Centrifuge the tube for 30 min at 18,000× *g*, 4 °C.
21. Transfer the supernatant to a fresh 15 or 50 mL centrifuge tube.
22. Repeat **step 20**.
23. Pipette the supernatant into a 3.5 K MWCO Slide-A-Lyzer™ G2 dialysis cassette following the manufacturer's guidelines for correct rehydration prior to use (*see* **Note 13**).
24. Incubate the dialysis cassette in at least 50 volumes of lysis buffer A for 30 mins at 4 °C.
25. Repeat **step 24** a further three times.
26. Remove the sample from the dialysis cassette using either an 18-gauge needle or a serological pipette as described in the manufacturer's guidelines and transfer to a 15 or 50 mL centrifuge tube.
27. Centrifuge the tube for 1 h at 18,000× *g*, 4 °C.
28. Pre-cool 1.5 mL microcentrifuge tubes at −80 °C for a minimum of 30 min.
29. Once the centrifugation is complete, immediately aliquot the supernatant into the pre-cooled 1.5 mL microcentrifuge tube. Aliquots should be single-use (*see* **Note 14**).
30. Aliquots should be flash frozen immediately using a dry ice and methanol bath before being stored in a −80 °C freezer (*see* **Note 15**).
31. Alongside the aliquot from **step 19**, a second aliquot should be retained in order to test the total protein concentration using a Bradford assay (or other protein determination assay). Protein concentrations of greater than 8 mg mL^{-1} should be achieved (*see* **Note 16** and Fig. 2).

3.2 Preparation of the Reaction Mix and Running the CFPS Reaction

The protocol described here is for the second component of the CFPS reaction, which is the reaction mix. The reaction mix here is used for coupled transcription and translation, although it is possible to perform a linked reaction where in vitro transcription is performed first and the cell extract is only used for in vitro translation, which can be particularly useful for troubleshooting if the system is not functional. As the reaction mix involves small volumes, we recommend making a fresh master mix on the day and aliquoting it out when ready to use. During the preparation of cell lysate, all steps should be carried out on ice, and frozen

Table 1
Composition of the optimized reaction mix for coupled transcription and translation

Component	Volume (μL)
1 M HEPES-KOH, pH 7.4	3
1 M magnesium glutamate	0.3
100 mM DTT	1
RTS amino acid mix (0.9 mM each of 19 amino acids, 0.75 mM leucine; *see* **Note 19**)	7.5
100 mM ATT	0.5
100 mM GTP	0.5
100 mM UTP	0.5
100 mM CTP	9.5
500 mM creatine phosphate (*see* **Note 20**)	4.5
4 mg mL^{-1} creatine phosphokinase	3.38
40 U μL^{-1} RNase inhibitor	0.25
200 U μL^{-1} T7 polymerase	0.5
0.8 nM to 40 nM Plasmid DNA or 1 μL water	Variable
Water	to 25 μL

components should be fully thawed prior to use. To prepare the DNA required for the reaction, we recommend using the Qiagen Plasmid Maxi Prep kit and eluting in TE buffer (*see* **Note 17**).

1. Combine the components in the exact order detailed in Table 1 (*see* **Note 18**). The volumes listed are for 1 reaction; to create a master mix, multiply this by $n + 2$ (where n is the total number of desired reactions). Once mixed, the tubes are flicked and briefly centrifuged; vortexing is not recommended.
2. Pipette 25 μL of the reaction mix into individual tubes for each reaction.
3. Add 25 μL of *K. phaffii* lysate.
4. Flick the components, do not vortex, and briefly spin in a microcentrifuge.
5. Incubate the mix at 21 °C in either a water bath or in a static incubator (*see* **Notes 22** and **23**).
6. Reaction time will vary, although on average it will take 2–8 h. More complex proteins, for instance, those that require post-translational modifications such as disulfide bonds or particularly large proteins, may require a longer reaction time.
7. Once protein synthesis has finished, the reaction products should be analyzed immediately (*see* **Note 24**). Certain

products can be frozen for later analysis, but this will be protein-specific and will require optimization.

3.3 DNA Concentration Titration to Improve Yield

The reaction mix has been modified using design of experiments to improve the yield of the reaction [19]; however, it is also possible to adjust the DNA concentration in order to increase yield further. While previous research in *K. phaffii* has shown that there is a saturation of the positive effects of increasing DNA concentration [18], our research has shown that with our system, increasing DNA concentration to at least 40 nM improves production, *see* Fig. 3. The optimal DNA concentration may be product-specific, and a titration should be carried out to establish the optimum for the application.

1. Prepare DNA with either Qiagen Maxi Prep, Qiagen Mega Prep, or Qiagen Giga Prep kit according to the manufacturer's notes and resuspend in TE buffer.
2. Dilute DNA according to the range of concentrations required (*see* **Note 25**).
3. Perform coupled in vitro transcription and translation according to Subheading 3.2.
4. Analyze the protein using your method of choice.

Our recommendation for first-time users is to produce a reporter protein such as luciferase and monitor production using a

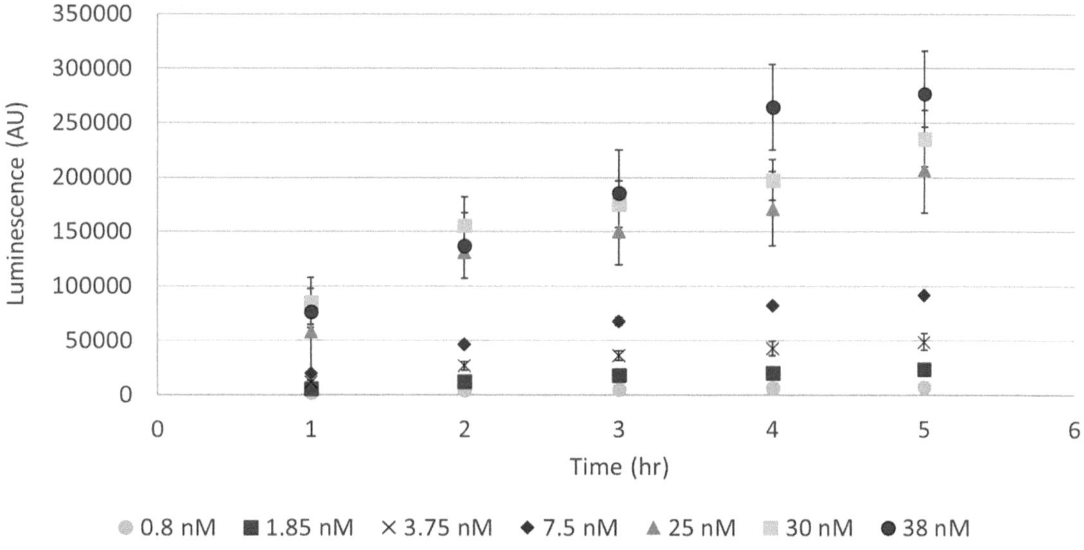

Fig. 3 DNA concentration has an impact on protein production. CFPS reactions of DNA titrated from 0.8 μM to 38 μM concentration in a standard CFPS reaction. Error bars represent the standard deviation of the mean of three biological replicates for the raw data, and two for the negative control, and are calculated using Gaussian error propagation

Table 2
Components needed to make the luciferase assay buffer for 15 reactions

Component	Volume (μL)
200 mM tricine	50
25 mM ATP	5
50 mM $MgSO_4$	26.7
10 mM $MgCO_3$	53.4
10 mM EDTA	5
500 mM DTT	17
25 mM D-luciferin	5
Water	337.9

3.4 Determining Luciferase Production from Cell-Free Reactions

luminescence assay (*see* **Note 26**). For all troubleshooting, an easy-to-visualize reporter protein is recommended so that issues with quantification of the product are not the limiting factor. We use a homemade luciferase assay buffer to ensure that the system is affordable [21]. The components, with the exception of D-luciferin, DTT, and ATP, may be made in advance and stored at room temperature. D-luciferin should be stored at −20 °C and protected from light, and for each reaction, ATP and DTT should be prepared fresh. We recommend taking samples every hour to determine the CFPS production profile (*see* **Note 27**).

1. Make a master mix for the luciferase assay buffer for 15 reactions (Table 2). Combine the components on ice protected in a foil-covered 1.5 mL microcentrifuge tube (*see* **Note 28**):

1. Use a 96-well half-area flat-bottom white microtiter plate to measure the luminescence. Every hour, mix 30 μL of luciferase assay buffer and 5 μL of the CFPS reaction.
2. Measure the luminescence on a plate reader that is capable of detecting luminescence. The reading should be taken immediately after combining the luciferase assay buffer with the CFPS reaction product (*see* **Note 29**).
3. To calculate the production of luciferase, the negative control must be subtracted, and Gaussian error propagation used to calculate the error bars from the triplicate reactions, *see* Fig. 4. A commercial standard of luciferase can be used to create a standard curve to enable the quantification of protein yield. Alternatively, the luminescence units produced are sufficient for comparing different conditions for their effect on yield.

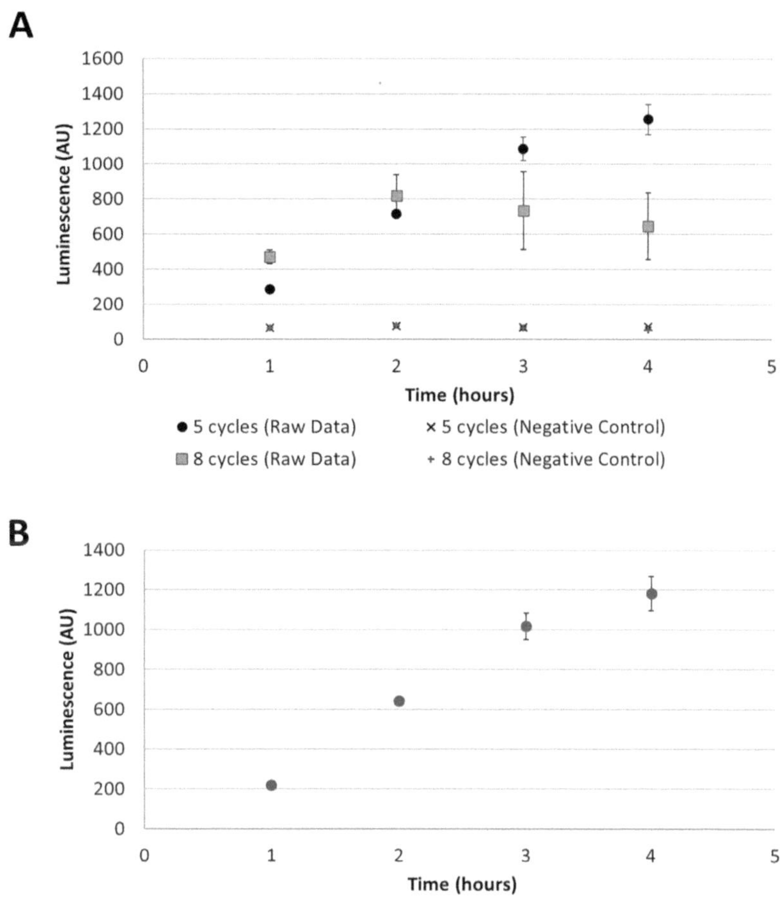

Fig. 4 An example time course of luciferase production from a coupled in vitro transcription/translation protocol. (**a**) Example data using two different sonication lysis protocols from the extract derived from the *FHL1* strain. Both sonication methods use the cycle conditions of 50% amplitude, 45 s on, 59 s off, and differ only in the number of cycles and therefore total lysis time, for either 5 cycles or 8 cycles. The reaction uses DNA containing the Cricket Paralysis Virus (CrPV) IRES to allow for cap-independent mRNA transcription. Luciferase production was measured over a 4-h period, and a negative control was included that replaced the DNA with water. (**b**) Background corrected data from the final selected sonication protocol of 50% amplitude, 45 s on, 59 s off for 5 cycles. Error bars represent the standard deviation of the mean of three experimental replicates for the raw data, and two for the negative control, and are calculated using Gaussian error propagation

4 Notes

1. A high-pressure cell homogenizer has been widely used for the production of cell lysate from *K. phaffii*, and detailed method papers describe this process in full [22, 23]. A higher protein concentration is obtained using high-pressure disruption

compared to sonication, but this equipment is not always available in all laboratories.

2. Fresh YPD agar plates, no older than 2 weeks (stored at 4 °C), should be used for the preparation of cell lysate. Streaking out strains from glycerol stocks ensures that there are limited passages to reduce the likelihood of mutations.
3. The method described here has specifically been designed to prepare a lysate from a 200 mL culture. It is possible to adjust the scale (both scaling up and down), but the entire method for lysate processing must then be adjusted accordingly. Due to the importance of keeping the lysate as cold as possible, it is suggested to split larger cultures into 200 mL batches.
4. The harvest point is critical for active cell lysates, as there is a tradeoff between active ribosomes and the culture phase [17]. The length of time that the culture is grown for will be dependent on the strain that is being utilized. Slower-growing strains may need to be grown for longer, although this may have an impact on protein synthesis. It is important to harvest the strains during the mid-exponential phase, which may vary for each strain. For instance, in previous work using SMD1163 cultures were harvested after 18 h, but only an OD_{600} of 6 was achieved, which will have an impact on the volume of lysate that can be made at one time [18]. Different strains will require optimization to determine the optimal harvest point.
5. Pre-weighing an empty centrifuge tube is pivotal in order to calculate the mass of cells. This step will ensure that the correct volume of lysis buffer is used and will ensure the lysate is neither too concentrated nor too dilute, which is a critical attribute in CFPS.
6. All steps should be performed on ice to ensure that the cell lysate remains as cold as possible.
7. This is a natural stopping point within the protocol that allows the users to freeze the prepared pellet or continue with the lysis protocol. Due to the length of time it takes to grow and wash the cells, we would recommend splitting the protocol over 2 days, especially for first-time users. It is critical that the pellet is flash frozen; we recommend using a dry ice and methanol bath; however, it is possible to use an alternative such as liquid nitrogen. To ensure that the pellet has fully frozen, hold the tubes in the dry ice and methanol bath until a color change can be observed throughout the entire pellet. Additionally, it may be possible to see a raised point at the center of the pellet. The pellet must be stored at −80 °C; however, prior to use, it should be thoroughly defrosted on ice, which may take at least 30 min.

8. If the user is interested in using a high-pressure cell disruptor, only 1 mL of lysis buffer A should be used per 1 g of wet cell weight.
9. The size of the pellet and the sonicator equipment will determine which size of centrifuge tube to use. Normally, if the cell pellet is resuspended in 5 mL of lysis buffer A, a 15 mL centrifuge tube will be used. If greater than 10 mL of lysis buffer A is used, then a 50 mL centrifuge tube will be used. This will also be determined by the manufacturer's requirements for the probe size being used. If necessary, split the lysate into batches and process separately.
10. The sonication protocol has been optimized for this particular probe size. If a probe with an alternative diameter is used, the user must optimize the process to suit their equipment.
11. We have tested multiple sonication cycle conditions and initially determined the protein concentration as a preliminary indicator of successful lysis. This is a critical quality attribute, and it has been suggested that a concentration of 15–25 mg L^{-1} is required for successful CFPS [24]. A high protein lysate concentration does not guarantee a high-yield CFPS reaction, as the tradeoff between efficient lysis and maintaining a cold temperature is essential. *See* Figs. 2 and 4a, where a higher protein concentration did not equate to higher CFPS production. See **Note 16** for more information on protein concentration.
12. If the user is using a high-pressure cell disruptor, two passes at 30,000 psi are used. Please refer to Aw et al. [22] and Spice et al. [23] for further details.
13. An 18-gauge needle is required to ensure that the air has been fully removed from the cassette prior to submersion in lysis buffer A.
14. Single-use aliquots should be used, as freeze-thaw cycles greatly diminish the activity of the cell lysate and are not recommended. Based on our workflows, a set of six runs (three samples and three controls) uses 175 μL of lysate. Therefore, this is our chosen aliquot volume.
15. The stability of the lysate has been observed for up to 1 year at −80 °C.
16. While for other systems it has been suggested that concentrations of 15–25 mg mL^{-1} of protein are required for successful CFPS reactions, we have found that lower concentrations still result in active lysate when prepared using our method. If low protein yields are reported, it is important to determine whether this occurs after lysis (the sample from **step 19**) or after dialysis and clarification (the sample from **step 31**). In our

experience, low total protein concentration is due to poor lysis. This can be optimized by altering the sonication cycle or altering the volume of lysis buffer A that is used for resuspending the cells. It is important to note that increasing lysis may lead to an increase in the sample temperature that will negatively affect lysate activity, and therefore, this is not a definitive determinant of a successful lysate preparation.

17. Maxi preps result in higher quality DNA than using kits such as the Qiagen Mini Prep kit and affect the overall yield of the CFPS reaction. The plasmid that we use contains a T7 promoter, internal ribosome entry site (IRES), and Kozak sequence upstream of the gene of interest, a polyA tail to improve RNA stability, and a T7 transcription terminator. Our system utilizes the Cricket Paralysis Virus (CrPV) IRES; however, alternative IRESs have been used in other eukaryotic systems, such as the EMCV IRES [25], although any alternative IRES will need to be characterized in this system.
18. The order in which the components are added to a master mix can have an effect on the functionality of the system. All components except for the lysate are combined into a 1.5 mL microcentrifuge tube and are mixed by flicking 10 times. This is followed by a brief centrifugation. Individual replicates are then mixed into separate tubes.
19. We have found that using a commercial amino acid mix (e.g., RTS Amino Acid Sampler from biotechrabbit, GmbH) results in higher yields than using a homemade amino acid mix. To ensure the amino acid mix is fully dissolved, once thawed, it is heated to 60 °C for 5 min prior to use. It is then cooled to room temperature, and it is important not to return the dissolved mixture to ice prior to use to prevent precipitation.
20. Creatine phosphate and creatine phosphokinase are the components that are most affected by the freeze-thaw cycle. This is observed by a rapid drop-off in the efficiency of the CFPS reaction. Therefore, we recommend aliquoting both of these components into single-use aliquots.
21. In addition to running triplicates of the reaction, we suggest running triplicates of a negative control. While our preferred method is to replace the plasmid with water, it is possible to create a negative control that lacks the T7 RNA polymerase and maintains the plasmid instead, although this will require setting up a second master mix for the reaction mix (minus the T7 RNA polymerase).
22. Our system has been optimized to work at 21 °C, although it has been shown for alternative *K. phaffii* CFPS platforms that temperatures of 30 °C improve yields [18]. However, we have found this to be detrimental to the yield of our CFPS reactions

and predict that the difference may be strain-dependent and could be an aspect that requires optimization based on the strain chosen.

23. Unlike other CFPS systems, shaking the reaction prevents protein synthesis. This has implications for the production of products that require oxygen to fold, such as GFP, and therefore should be taken into consideration. Alternatively, samples from the reaction can be incubated for 2 h in water to allow the GFP to fold prior to measuring fluorescence [18].

24. Methods to investigate the success of a reaction will depend on the product being produced. To establish a CFPS platform, we recommend working with luciferase as a reporter (*see* Subheading 3.3. Determining luciferase production from cell-free reactions). For well-expressed proteins, it is possible to visualize the protein using Western blots, although generally yields are not high enough for visualization by SDS-PAGE gels alone (due to the high concentration of host cell proteins in the lysate). It is also possible to purify the protein directly from the CFPS reaction for downstream processing and further analysis.

25. We use a range of DNA concentrations from 0.8 nM to 40 nM; however, depending on the difficulty of extracting DNA and the importance of cost-effectiveness, the amount of DNA can be decreased and still result in protein synthesis.

26. Amino acid sequence for the firefly luciferase reporter protein

    >tr|Q27758|Q27758_PHOPY Firefly luciferase OS=Photinus pyralis OX=7054 GN=luc PE=4 V=1

    MEDAKNIKKGPAPFYPLEDGTAGEQLHKAMK RYALVPGTIAFTDAHIEVNITYAEYFEMSVRLAEAMK RYGLNTNHRIVVCSENSLQFFMPVLGALFIGVAVAPAN DIYNERELLNSMNISQPTVVFVSKKGLQKILNVQKKL PIIQKIIIMDSKTDYQGFQSMYTFVTSHLPPGFNEYDFV PESFDRDKTIALIMNSSGSTGSPKGVALPHRTACVRF SHARDPIFGNQIIPDTAILSVVPFHHGFGMFTTLGY LICGFRVVLMYRFEEELFLRSLQDYKIQSALLVPTLFSF FAKSTLIDKYDLSNLHEIASGGAPLSKEVGEAVAKRFHLP GIRQGYGLTETTSAILITPEGDDKPGAVGKVVPF FEAKVVDLDTGKTLGVNQRGELCVRGPMIMSGYVND PEATNALIDKDGWLHSGDIAYWDEDEHFFIVDRLKSLI KYKGCQVAPAELESILLQHPNIFDAGVAGLPGDDAGEL PAAVVVLEHGKTMTEKEIVDYVASQVT TAKKLRGGVVFVDEVPKGLTGKLDARKIREILI KAKKGGKSKL

27. The profile of protein production may vary depending on the user producing the cell lysate. We have observed differences in lag time, as well as the length of time of production time, although total yields are often similar [17, 19]. Therefore,

measuring every hour in triplicate helps to establish differences. It is also a good indication of the reliability of the reaction, as inconsistent results across the triplicates may be indicative of poor mixing of the reaction mix and cell lysate or potentially evaporation within the CFPS reaction tube. If a noticeable decrease in activity between experiments is observed, this suggests that fresh reagents should be prepared, in particular creatine phosphate and creatine phosphokinase as previously detailed in **Note 20**.

28. D-luciferin is light sensitive and therefore should be protected from light sources. The reaction volume is 500 μL, sufficient to measure 15 reactions at one time, and can be scaled up or down accordingly. It is preferred to make fresh luciferase assay buffer before each time point.
29. It is common for the initial reading of the luciferase assay to be significantly higher than other readings within the 20-min window. This is due to the nature of the assay and the instant photon generation and exponential decay. To prevent variability between users, particularly if a large number of samples are assayed, it is possible that differing speeds in pipetting can lead to variation. Therefore, we recommend averaging the luminescence values from the second time point until the end of the run, which should be after the initial decay of the luminescence signal [26].

Acknowledgments

This research is funded by the Department of Health and Social Care using UK Aid funding and is managed by the Engineering and Physical Sciences Research Council (EPSRC, grant number: EP/R013764/1). The views expressed in this publication are those of the author(s) and not necessarily those of the Department of Health and Social Care. A.S. would also like to thank the EPSRC Centre for Innovative Manufacturing in Emergent Macromolecular Therapies (EP/I033270/1) for Ph.D. studentship funding. F.A. would like to thank the Prime Minister's Fellowship, Bangladesh, for Ph.D. studentship funding and Jagannath University, Bangladesh, for supporting her study leave.

Competing Interest Statement The authors have no conflicts of interest to declare that are relevant to the content of this chapter.

References

1. Yuan H, Silberstein SD (2021) Eptinezumab. In: Maassen van den Brink A, Martelletti P (eds) Monoclonal antibodies in headache: from bench to patient. Springer International Publishing, Cham, pp 109–119
2. Ahmad M et al (2014) Protein expression in *Pichia pastoris*: recent achievements and perspectives for heterologous protein production. Appl Microbiol Biotechnol 98(12): 5301–5317. https://doi.org/10.1007/s00253-014-5732-5
3. Ogonah OW, Polizzi KM, Bracewell DG (2017) Cell free protein synthesis: a viable option for stratified medicines manufacturing? Curr Opin Chem Eng 18:77–83. https://doi.org/10.1016/j.coche.2017.10.003
4. Carlson ED et al (2012) Cell-free protein synthesis: applications come of age. Biotechnol Adv 30(5):1185–1194. https://doi.org/10.1016/j.biotechadv.2011.09.016
5. Nirenberg MW, Matthaei JH (1961) The dependence of cell-free protein synthesis in E. coli upon naturally occurring or synthetic polyribonucleotides. Proc Natl Acad Sci USA 47:1588–1602
6. Wilding KM et al (2019) Endotoxin-free *E. coli*-based cell-free protein synthesis: pre-expression endotoxin removal approaches for on-demand cancer therapeutic production. Biotechnol J 14(3):1800271. https://doi.org/10.1002/biot.201800271
7. Anderson CW, Straus JW, Dudock BS (1983) Preparation of a cell-free protein-synthesizing system from wheat germ. Methods Enzymol 101:635–644. https://doi.org/10.1016/B978-0-12-765560-4.50044-7
8. Jackson RJ, Hunt T (1983) Preparation and use of nuclease-treated rabbit reticulocyte lysates for the translation of eukaryotic messenger RNA. Methods Enzymol 96:50–74
9. Brodel AK, Sonnabend A, Kubick S (2014) Cell-free protein expression based on extracts from CHO cells. Biotechnol Bioeng 111(1): 25–36. https://doi.org/10.1002/bit.25013
10. Heide C et al (2021) Design, development and optimization of a functional mammalian cell-free protein synthesis platform. Front Bioeng Biotechnol 8. https://doi.org/10.3389/fbioe.2020.604091
11. Moore SJ et al (2017) *Streptomyces venezuelae* TX-TL – a next generation cell-free synthetic biology tool. Biotechnol J 12(4):1600678. https://doi.org/10.1002/biot.201600678
12. Moore SJ et al (2018) Rapid acquisition and model-based analysis of cell-free transcription–translation reactions from nonmodel bacteria. Proc Natl Acad Sci USA 115(19):E4340. https://doi.org/10.1073/pnas.1715806115
13. Buntru M et al (2014) Tobacco BY-2 cell-free lysate: an alternative and highly-productive plant-based in vitro translation system. BMC Biotechnol 14(1):37. https://doi.org/10.1186/1472-6750-14-37
14. Hodgman CE, Jewett MC (2013) Optimized extract preparation methods and reaction conditions for improved yeast cell-free protein synthesis. Biotechnol Bioeng 110(10): 2643–2654. https://doi.org/10.1002/bit.24942
15. Schramm M et al (2022) Cell-Free Protein Synthesis with Fungal Lysates for the Rapid Production of Unspecific Peroxygenases. Antioxidants 11(2):284. https://doi.org/10.3390/antiox11020284
16. Wu C et al (2017) The cell free protein synthesis system from the model filamentous fungus Neurospora crassa. Methods 137:11. https://doi.org/10.1016/j.ymeth.2017.12.003
17. Aw R, Polizzi KM (2019) Biosensor-assisted engineering of a high-yield Pichia pastoris cell-free protein synthesis platform. Biotechnol Bioeng 116(3):656–666. https://doi.org/10.1002/bit.26901
18. Zhang L, Liu W-Q, Li J (2020) Establishing a eukaryotic pichia pastoris cell-free protein synthesis system. Front Bioeng Biotechnol 8(536). https://doi.org/10.3389/fbioe.2020.00536
19. Spice AJ et al (2020) Improving the reaction mix of a Pichia pastoris cell-free system using a design of experiments approach to minimise experimental effort. Synth Syst Biotechnol 5(3):137–144. https://doi.org/10.1016/j.synbio.2020.06.003
20. Spice AJ et al (2020) Synthesis and assembly of Hepatitis B virus-like particles in a *Pichia pastoris* cell-free system. Front Bioeng Biotechnol 8:72–72. https://doi.org/10.3389/fbioe.2020.00072
21. Siebring-van Olst E et al (2013) Affordable luciferase reporter assay for cell-based high-throughput screening. J Biomol Screen 18(4): 453–461. https://doi.org/10.1177/1087057112465184
22. Aw R, Spice AJ, Polizzi KM (2020) Methods for expression of recombinant proteins using a *Pichia pastoris* cell-free system. Curr Protoc Protein Sci 102(1):e115. https://doi.org/10.1002/cpps.115

23. Spice AJ, Aw R, Polizzi KM (2022) Cell-free protein synthesis using pichia pastoris. In: Karim AS, Jewett MC (eds) Cell-free gene expression: methods and protocols. Springer US, New York, pp 75–88

24. Fujiwara K, Doi N (2016) Biochemical preparation of cell extract for cell-free protein synthesis without physical disruption. PLoS One 11(4):e0154614. https://doi.org/10.1371/journal.pone.0154614

25. Makrydaki E et al (2021) Chapter 18: Cell-free protein synthesis using Chinese hamster ovary cells. In: Kelman Z, O'Dell WB (eds) Meth enzymol, vol 659. Academic, pp 411–435

26. Gan R, Jewett MC (2014) A combined cell-free transcription-translation system from *Saccharomyces cerevisiae* for rapid and robust protein synthesis. Biotechnol J 9(5):641–651. https://doi.org/10.1002/biot.201300545

Part VII

Downstream Processing and Product Analysis

Chapter 40

Clarification of *Pichia pastoris* Broth by TFF with Ceramic Membranes

Fuensanta Verdú-Navarro, Juan A. Moreno-Cid, Ana G. Del Hierro, Xavier Garcia-Ortega, and Ruth Ordóñez

Abstract

In industrial processes using *Komagataella phaffii*, the most commonly used technology in the downstream processes to clarify the culture broths is continuous centrifugation. However, currently, tangential flow filtration (TFF) is considered an interesting scalable strategy to clarify broths from microbial fermentation processes at high cell concentrations. TFF presents numerous advantages over centrifugation depending on the bioprocess workflow, scale, and nature of the final products. In this chapter, we describe the methodology to achieve biomass clarification by microfiltration of *K. phaffii* fermentation broth using ceramic membranes.

Key words Komagataella phaffii, Tangential Flow Filtration, Microfiltration, Ceramic, Clarification

1 Introduction

Downstream processes for clarifying broth typically involve centrifugation, depth filtration, or tangential flow filtration (TFF) [1]. In the case of *Komagataella phaffii* (previously *Pichia pastoris*), continuous centrifugation is currently the most commonly employed strategy at an industrial scale [2]. However, in addition to the high costs of the industrial centrifugation equipment, this method often needs an additional step of depth filtration to guarantee a clear and transparent broth, free of cell debris, before proceeding with subsequent downstream processes [3]. Consequently, new approaches using continuous microfiltration technologies are being studied and compared with those traditionally used. Whenever TFF has been chosen for clarifying this type of broth, hollow fiber membranes, and cassettes, to a lesser extent, have been the preferred option [4].

Both membrane systems are characterized by their compacity, allowing high filtration areas per volume. Hollow fiber membranes

Xavier Garcia-Ortega et al. (eds.), *Pichia pastoris: Methods and Protocols*, Methods in Molecular Biology, vol. 2697,
https://doi.org/10.1007/978-1-0716-4779-0_40,

are offered in different polymeric materials (polyethersulfone or PES, polysulfone or PS, polypropylene or PP, polyvinylidene fluoride or PVDF, mixed cellulose ester or ME, and regenerated cellulose or RC) and are the best option when facing shear-sensitive cells, however, they are very limited to high viscous media. Cassettes are also manufactured in polymeric materials, but mainly in PES and RC [5]. Although cassettes are more compact, they are not as gentle with the cells as hollow fibers are and are fouled more easily, that is the reason for limiting their use in cell clarification and for being focused on biomolecules concentration. The key when selecting membrane material is how the substances in the media and the target product interact with it (fouling, extractables, etc). PES and RC are both hydrophilic with low protein binding characteristics, but generally, PES membranes provide higher permeate fluxes. RC, however, has broad solvent compatibility and guarantees a low extractable level, which makes it the preferred material when sample analyses rely on HPLC techniques.

Despite all the advantages of polymeric materials, one important disadvantage is that they are considered consumables, which could turn a process economically unfeasible, especially with low-added-value final products. Alternatively, ceramic tubular membranes are ideal to face the mentioned drawback since they present low protein binding properties and the ability to tolerate extreme pH. Additionally, their lifetime of more than 10 years [4], if they are cleaned properly after each filtration batch, also represents a very positive feature. An additional key benefit of ceramic tubular membranes, versus already mentioned hollow fiber modules, is their good capacity to manage viscous media (up to 300–400 cP). This is an important advantage to clarify the very high cell concentration of *K. phaffii* fed-batch cultivation broth, which can reach up to 120 g L^{-1} of dry cell weight (DCW). Accordingly, these very highly dense and viscous cell suspensions (that can even reach up to 200–300 cP after 2X concentration) usually might clog the filtration membranes quickly. Their main disadvantage is that they require working at high cross-flow velocities to get representative permeate flux, which is not recommended for shear-sensitive downstream processes, but this is not the case for most of *K. phaffii* bioprocesses.

After this first clarification, in most cases where the target protein is secreted, the next step of the DSP will be an ultrafiltration/diafiltration step followed by chromatography purification, if needed a high purity. Details about these methods can be found in Chapters 42 and 43. Otherwise, if the target protein is expressed intracellularly, the biomass will have to be disrupted. An efficient method for high-pressure homogenization is detailed in Chapter 41, as well as in the literature [6].

In this chapter, a protocol suggested by the authors for harvesting high cell density *K. phaffii* broths utilizing a 0.4 μm pore-size tubular ceramic membrane in a tangential flow filtration system is detailed. This material and filtration cut-off offer very good performance for highly viscous *K. phaffii* fermentation broths, which is often the case in fed-batch cultures. Alternatively, a hollow fiber or cassette system using a similar filtration cut-off could also be used.

Additionally, a protocol for cleaning and recovering the membrane for reuse is also detailed.

The authors have developed the method using the system M1 from BIONET®, which can control the retentate flow automatically by changing the speed of the pump accordingly. Equivalent automated TFF systems are also offered by alternative vendors such as Sartorius or Cytiva.

2 Materials

2.1 Cell Suspensions

1. As a working material, a *Komagataella phaffii* fermentation broth was used. The key physicochemical properties were: cell concentration 110 g L^{-1} of dry cell weight (DCW), at pH 6.3, and viscosity 10.5 cP.
2. Details about how to obtain high cell density culture in fed-batch cultures performing alternative culture strategies can be found in Chapters 31 and 32, as well as in the literature [7, 8].

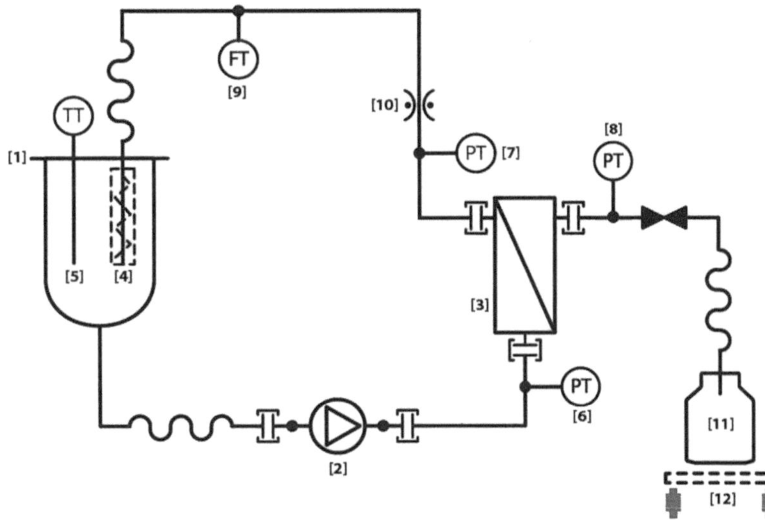

Fig. 1 Tangential Flow Filtration system. Main components: (1) feed vessel; (2) recirculation pump; (3) membrane module; (4) heat exchanger; (5) temperature transmitter; (6) feed pressure transmitter; (7) retentate pressure transmitter; (8) permeate pressure transmitter; (9) flow transmitter; (10) backpressure pinch valve; (11) permeate reservoir; and (12) scale

2.2 Filtration System

1. Automated TFF system (M1 from BIONET®). In Fig. 1, a scheme of a typical TFF system used for *K. phaffii* fermentation broth clarification is shown.
2. A heat exchanger is connected to an external chilling/heating system in charge of supplying heating and cooling capacity for DSP and/or CIP operations.
3. Ceramic (α-Al_2O_3) membranes with a pore diameter of 0.4 μm and a channel diameter of 2.7 mm. From Atech Innovations GmbH.
4. Raw chemicals for pH adjustment and membrane cleaning: NaOH 50%wt, NaClO 15%wt (as free chlorine), and HNO_3 60%wt.
5. Demineralized water at room temperature (RT) is used for membrane conditioning and buffer exchange in the diafiltration stage.

2.3 Sampling and Analysis

Take samples along the clarification/diafiltration process to determine the yield of protein recovery (*see* **Note 1**). In the case of feed and retentate samples, a centrifugation step is needed to separate the biomass from the supernatant with the target protein.

1. Centrifugation of feed and retentate samples should be performed at 9000 rpm (around 9000 g) for 20 min at RT. The rotor should be suitable for 50 mL tubes and high-speed.
2. Total protein determination kit requires the following materials:
   - Pierce Coomassie (Bradford) or BCA-based Protein Assay Kit. Different alternatives are available from different vendors.
   - A spectrophotometer device for cuvettes or 96 well-plates able to quantify colorimetric reactions at a wavelength of 595 nm.
3. For protein electrophoresis, SDS-PAGE (sodium dodecyl sulfate-polyacrylamide gel electrophoresis) with Coomassie staining at the desired polyacrylamide concentration.

3 Methods

Details about the abbreviations are detailed in the Chapter List of Abbreviations.

3.1 Conditioning

1. Before clarification, broth pH is adjusted to 4.7 with HNO_3. This pH reduction is beneficial for increasing protein recovery yield (*see* **Note 2**).

2. Fill the system with demineralized water and open the permeate and backpressure valves completely.
3. Adjust water pH to that of fermentation broth (*see* **Note 3**). The volume of water used for conditioning has to be at least 30% of the hold-up volume of the system to guarantee flow stability.
4. Start pumping in recirculation mode (permeate sent back to the feed vessel) at a cross-flow velocity (CFV) of 5 m s^{-1} and keep the temperature under control at 25 °C (*see* **Note 4**).
5. Once the temperature and CFV are achieved, adjust the backpressure valve if necessary to ensure permeate generation.
6. Keep the system in recirculation mode for 5 min.

3.2 Cell Clarification

1. After membrane conditioning, add the fermentation broth and keep the system in recirculation mode until setting a transmembrane pressure (TMP) of 0.4–0.8 barg (*see* Eq. (1)).

$$\mathrm{TMP} = \frac{P_f + P_r}{2} - P_p \tag{1}$$

2. Keep the system in recirculation mode for 5 min.
3. After this time, put the system in concentration mode, sending the permeate to an external vessel.
4. Start registering the permeate weight with time to determine permeate flow, which can be transformed into permeate flux with Eq. (2).

$$J = \frac{Q_p}{A} \tag{2}$$

5. Clarification is done by keeping the previous variables (CFV, temperature, and TMP) constant until achieving approximately 2X of the concentration factor (*see* Eq. (3)).

$$CF = \frac{V_f}{V_r} \tag{3}$$

3.3 Diafiltration

1. Demineralized water (adjusted at the pH of the broth) is used for diafiltration (*see* **Note 5**), which starts once the cells are concentrated 2X in the clarification stage. The amount of water is calculated following Eq. (4).

$$V_{DF} = DV \cdot V_o \tag{4}$$

2. It is recommended to work between 3 and 5 DV in batch mode, combined with a step of membrane surface cleaning.

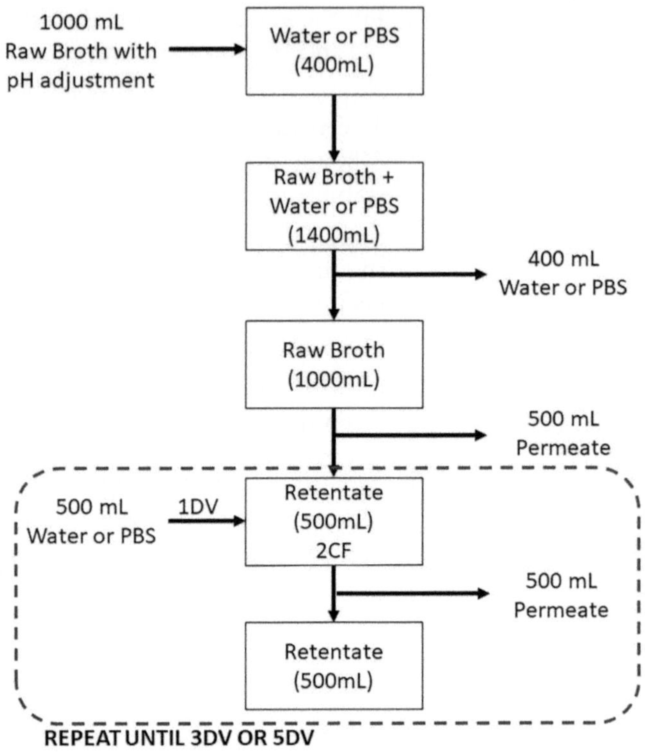

Fig. 2 Scheme of conditioning, clarification, and diafiltration stages

The procedure suggested by the authors includes the following steps:

(a) Open the backpressure valve completely.

(b) Close the permeate valve.

(c) Add the first volume (1DV) of demineralized water.

(d) Recirculate for 5 min, keeping the same CFV and temperature.

(e) Open the permeate, adjust TMP to the same value used in the clarification stage and start collecting permeate.

3. Figure 2 represents a scheme with all the steps involved in the clarification process.

3.4 Harvest

Once the diafiltration stage is finished and before harvesting.

1. Open the backpressure valve completely.
2. Close the permeate valve.
3. Keep the system in recirculation for 5 min at the same CFV and temperature used in the clarification stage (*see* **Note 6**).
4. After this time, open the drainage valve and collect all the retentate (exhaust yeast).

3.5 Cleaning in Place (CIP)

Proceed with the following cleaning protocol to recover the reference membrane water flux:

1. Rinse the feed and retentate (permeate close) with demineralized water for 5 min at RT and at the same CFV used for clarification/diafiltration stages.
2. Recirculate through the feed and retentate (permeate close) NaOH (2.5%wt) solution at 85 °C (*see* **Note 7**) for 30 min at the same CFV used for clarification/diafiltration stages.
3. Rinse the feed and retentate (permeate close) with demineralized water for 5 min at the same CFV used for clarification/diafiltration stages.
4. Recirculate through the feed and retentate (permeate close) NaOH (2.5%wt) + NaClO (100 mg/L free chlorine) solution at 85 °C for 10 min at the same CFV used for clarification/diafiltration stages.
5. Rinse the feed, retentate, and permeate with demineralized water until neutrality at the same CFV used for clarification/diafiltration stages and at the lowest TMP that allows permeate flow.

4 Notes

1. To determine the yield of protein recovery, it is possible to use different analytical methods: enzymatic activity, protein quantification, protein electrophoresis, etc. Choose the best option for your protein.
2. This pH is the specific isoelectric point for this protein. pH adjustment may vary depending on the protein of interest.
3. Demineralized water can be exchanged for a specific buffer (i.e. PBS) if required. The buffer has to be similar in pH and ionic strength to that of the sample.
4. Some of the filtration conditions may vary depending on the protein of interest, i.e. temperature. Knowing the optimal conditions of the target protein prior to the filtration to avoid denaturation is a must.
5. Diafiltration or washing stage is necessary to optimize the recovery of the target protein.
6. This procedure is recommended to detach as much yeast as possible from the membrane to facilitate subsequent CIP and membrane recovery.
7. When setting the rinsing temperature, avoid temperature gradients higher than 30 °C to avoid membrane shocks.

Acknowledgments

Part of this work was possible thanks to professors Francisco Valero and José Luis Montesinos, of the Bioprocess Engineering and Applied Biocatalysis Group of the Autonomous University of Barcelona (UAB), for the collaboration initiated for the development of different works with the friendly and scientific transfer of the strain used.

Competing Interest Statement The authors FVN, JMC, ADH, and RO are employees of Bionet, a company engaged in the commercialization of bioprocessing equipment and offering related services.

The author XGO has no conflicts of interest to declare that are relevant to the content of this chapter.

Appendix

Here below are listed all the abbreviations mentioned in the chapter.

TMP	Transmembrane pressure, barg
P_f	Inlet or feed pressure, barg
P_r	Outlet or retentate pressure, barg
P_p	Permeate pressure, barg
J	Permeate flux, L h^{-1} m^{-2}
Q_p	Permeate flow, L h^{-1}
A	Membrane area, m^2
CF	Concentration factor
V_f	Initial volume, L
V_r	Final (or retentate) volume, L
V_{DF}	Buffer (or water) volume, L
DV	Diafiltration factor (also called Diavolume)
V_o	Diafiltration starting volume, L

References

1. Kalyanpur M (2002) Downstream processing in the biotechnology industry. Mol Biotechnol 22: 87–98. https://doi.org/10.1385/MB:22:1:087
2. Jahic M, Veide A, Charoenrat T, Teeri T, Enfors S-O (2006) Process technology for production and recovery of heterologous proteins with Pichia pastoris. Biotechnol Prog 22:1465–1473. https://doi.org/10.1021/bp060171t
3. Yavorsky DP, McGee S (2002) Selection and sizing of clarification depth filters. Genet Eng News 22:44–45
4. Wang A, Lewus R, Rathore AS (2006) Comparison of different options for harvest of a therapeutic protein product from high cell density yeast fermentation broth. Biotechnol Bioeng 94:91–104. https://doi.org/10.1002/bit.20816
5. van Reis R, Zydney A (2007) Bioprocess membrane technology. J Membr Sci 297:16–50. https://doi.org/10.1016/j.memsci.2007.02.045
6. García-Ortega X, Reyes C, Montesinos JL, Valero F (2015) Overall key performance indicator to optimizing operation of high-pressure homogenizers for a reliable quantification of intracellular components in Pichia pastoris. Front Bioeng Biotechnol 3:1–9. https://doi.org/10.3389/fbioe.2015.00107
7. Garrigós-Martínez J, Weninger A, Montesinos-Seguí JL et al (2021) Scalable production and application of Pichia pastoris whole cell catalysts expressing human cytochrome P450 2C9. Microb Cell Factories 20:90. https://doi.org/10.1186/s12934-021-01577-4
8. Gasset A, Garcia-Ortega X, Garrigós-Martínez J et al (2022) Innovative bioprocess strategies combining physiological control and strain engineering of Pichia pastoris to improve recombinant protein production. Front Bioeng Biotechnol 10:1–12. https://doi.org/10.3389/fbioe.2022.818434

Chapter 41

Pichia pastoris Cell Disruption by High-Pressure Homogenization (HPH)

Antoni Casablancas, Xavier Garcia-Ortega, and Francisco Valero

Abstract

Cell disruption is a key unit operation when an intracellular target product has to be recovered for further downstream processing. A wide variety of chemical, biological, physical, and mechanical methods are available to release biological products from the cell factories. Among them, high-pressure homogenization (HPH) is recognized as one of the most promising methods due to its efficiency, reproducibility, scalability, and relative easiness. In this chapter, a guide to applying HPH in the cell factory *Komagataella phaffii* is presented.

Key words *Komagataella phaffii*, *Pichia pastoris*, High-pressure homogenization, Cell disruption, Critical process parameter (CPP), Key performance indicators (KPI)

1 Introduction

Currently, biopharmaceuticals, organic acids, antibiotics, enzymes, amino acids, and other products are produced in cell factories using bioprocess technology [1]. Bioprocesses based on bacteria such as *Escherichia coli* usually have limitations as a protein production host due to their general inability to perform post-translational modifications and secrete recombinant proteins. These constraints restrict the number of biological products that can be produced in a soluble and active form. Accordingly, to access the target intracellular product, cells must be disrupted. [2]. On the other hand, *Komagataella phaffii* (reclassified from *Pichia pastoris*) can secrete the target product into the culture broth while secreting very low levels of native proteins, facilitating the downstream processes, which is considered one of the main advantages of this microbial cell factory [3]. Nevertheless, some bioproducts are still expressed intracellularly in this cell factory if the secretion is not efficient. Furthermore, high levels of target bioproduct expression can lead to saturation or overloading of the secretory pathways, provoking the intracellular

Xavier Garcia-Ortega et al. (eds.), *Pichia pastoris: Methods and Protocols*, Methods in Molecular Biology, vol. 2697, https://doi.org/10.1007/978-1-0716-4779-0_41,

accumulation of the bioproduct and, in the worst cases, its intracellular degradation [4]. Thus, for the recovery of the intracellular bioproduct or to study the effect of secretory pathway saturation on the bioprocess efficiency in *K. phaffii*, the reliable quantification and recovery of the intracellular product of interest accumulated in the cells is mandatory [5].

Existing principles for cell disruption are based on chemical, biological/enzymatical, physical, and mechanical methods. Eggenreich et al. summarized the most common cell disruption techniques, indicating their advantages and disadvantages, focusing on the cell factory *Escherichia coli* [6], and Gauterio et al. on yeasts [7]. Among them, high-pressure homogenization (HPH) methods are considered the most widely used due to their best performance [8].

When comparing different methods, it is key to define the critical process parameter (CPP) to determine the key performance indicators (KPI) and assess the required critical quality attribute (CQA). Classical KPIs are product purity, lysis efficiency, product titer, and product activity [9].

In *K. phaffii,* as well as other microbes, the selection of the carbon source used during cell growth has an important impact on the cell wall properties. When glucose and glycerol are used as substrates, the cell wall structure is similar to other yeasts and, accordingly, does not require extraordinarily demanding disruption methods. However, when *K. phaffii* grows on methanol, the thickness of the cell walls has been described to increase notably, and consequently, the cell disruption needs more severe methods and conditions [10].

To determine the performance of the selected cell disruption method, a decrease in optical density (OD) or a release of total protein after cell disruption are the most common methods used. Similarly, intracellular alkaline phosphatase activity is a measure of the total protein release but also an indicator of preservation of enzyme activity along the disruption. García-Ortega et al. compared several disruption methods based on different principles (mechanical, chemical, physical…), specifically the performance of cell disruption methods for *K. phaffii* based on high-pressure homogenization (HPH), bead milling, and cell permeabilization using N,N-dimethyltetradecylamine. In conclusion, HPH proved to be more effective in terms of OD decrease, total protein released, and alkaline phosphatase activity released [11]. The optimal conditions of HPH to release intracellular levels of Fab2F5 have also been reported [12].

According to the results obtained both for *E. coli* and *K. phaffii*, HPH turned out to be the best method from the point of view of efficiency, reproducibility, scalability, and method complexity. Hence, in this chapter, a brief guide to implementing an intracellular cell bioproduct recovery based on HPH at the lab and pilot plant scale is presented.

2 Materials

2.1 Microorganism

Suspensions of a *K. phaffii* X33 (Mut$^+$) strain under the regulation of the methanol-inducible *AOX1* promoter (P_{AOX1}) growing on mixed feeds of glucose and methanol were obtained from steady-state chemostat cultures described elsewhere [11]. Additionally, the strain *K. phaffii* X33 (Mut$^+$) expressing both light and heavy chain genes of the human Fab 2F5 under the control of the constitutive *GAP* promoter was included in the comparison since the effect of the carbon source is known to have an important effect on the cell wall thickness [12].

2.2 Sample Preparation Previous to Disruption

1. 20 mL Falcon tubes.
2. Refrigerated centrifuge for biomass separation. Centrifugation standard conditions are 4200 g, 4 °C for 10–15 min.
3. Phosphate buffered saline (PBS; pH = 7.4): 137 mM NaCl; 2.7 mM KCl; 8.1 mM $Na_2HPO_4{\cdot}2H_2O$; 1.76 KH_2PO_4.
4. UV-visible spectrophotometer.

2.3 Disruptor

1. Ice bath.
2. For lab-scale cell disruptions, benchtop HPH with working pressure in the range of 2–3 kbars.

   The One-Shot Cell Disrupter (Constant Systems Ltd., Warwickshire, UK) can be used at a lab scale, with 8 mL being a volume of the disruption samples.
3. For pilot plant scale cell disruptions, continuous HPH, connected to a chilling system, with working pressure in the range of 2–3 kbars.

   As an example, the homogenizer TS Series Cabinet Disruption System (Constant Systems Ltd., Warwickshire, UK) can be used, in which 250 mL of the volume can be disrupted.
4. For both equipment mentioned in points 3 and 4, alternative equipment offering similar performance can be found in the market.

2.4 Additional Material for Analysis

1. Total protein determination analysis kit, which could be based on Bradford, BCA, Lowry, etc.
2. Tris–HCl buffer for alkaline phosphatase activity assay (pH = 8.8) (Tris–HCl 1 M; MgCl 0.001 M).
3. Substrate for alkaline phosphatase activity assay: p-nitrophenyl phosphate (PNPB, 0.04 M).

4. UV-visible spectrophotometer.
5. Equipment to perform SDS-PAGE analysis that allows to assess the protein purity and concentration estimation.

3 Methods

To optimize the conditions of the HPH, it is recommended to use at least the following critical process parameter (CPP):

1. Initial optical density at 600 nm (OD_{600}).
2. Working pressure and number of cycles.

To evaluate HPH performance, it is recommended to use at least the following key performance indexes (KPI), whose analytical procedures are described in Subheading 3.4 of this chapter:

1. OD_{600} decrease.
2. Total protein released.
3. Product purity.
4. Product titter.
5. Alkaline phosphatase activity release (this parameter is related to the preservation of enzymatic activity by the effect of HPH).

The correlation between the CPP and the KPI achieved was reported by Garcia-Ortega et al. [11].

3.1 Sample Preparation

A typical sample volume for HPH lab scale is 8 mL, while a minimum of around 200 mL is usual for a continuous pilot plant scale HPH.

1. Before cell disruption, centrifuge the sample of cell suspensions from the bioprocess for 10 min at 4200 g, 4 °C (washing process) (*see* **Note 1**).
2. Eliminate supernatant and resuspend in PBS.
3. Repeat the washing process 3 times.
4. After washing, cells were resuspended in PBS to achieve the recommended initial OD of the HPH.
5. Keep the cells at 4 °C, and the suspension is disrupted.

3.2 Lab-Scale Cell Disruption (8 mL)

1. The disruptor's sample reservoir is kept in a freezer for 30 min before the disruption.
2. Repeat this action every 3 or 4 disruption cycles (*see* **Note 2**).
3. Fix the optimal work pressure in the range of 1–2.7 kbar (*see* **Note 3**).
4. Introduce the sample at the recommended initial OD.

5. Repeating the process for the number of optimal cycles (*see* Note 4).
6. Determine the final OD_{600} of the sample obtained.
7. Centrifuge the sample of cell suspension for 15 min at 17500 g, 4 °C.
8. Discard the cell debris.
9. Reservoir of the supernatant to analyze the KPIs.

3.3 Pilot-Scale Cell Disruption (up to 10 L).

1. A thermostatic reservoir is used to keep the sample cooled during the process, and another thermostatic reservoir is used to recover the processed sample.
2. The sample must be adjusted to the recommended initial OD.
3. Fix the optimal working pressure (*see* **Note 3**).
4. Start the cell disruption cycle. The equipment will keep pumping samples from the thermostatic reservoir to the disruption chamber until the thermostatic reservoir is empty.
5. Transfer the processed sample to the sample thermostatic reservoir and start the disruption process again. Repeat the process as many times as the number of optimal cycles.
6. Determine the final OD of the sample obtained.
7. Centrifuge the sample of cell suspension for 15 min at 17500 g, 4 °C.
8. Discard the cell debris.
9. Reserve the supernatant to analyze the KPIs.

3.4 Analytical Methods

3.4.1 Optical Density (OD)

OD at 600 nm can be used as a direct measure of the cell integrity of the samples; hence, a relative decrease in OD was related to the proportion of cells disrupted. Spectrophotometric analyses taken in triplicate are recommended. Typical standard deviation is below 5% (*see* **Note 5**).

3.4.2 Total Protein Released (TPR)

TPR was considered a suitable indirect performance indicator to be correlated with disruption efficiency. It was determined by Bradford assay at 595 nm using Coomassie PlusTM protein assay reagent (Pierce Rockford, IL, USA) using bovine albumin as a standard. Spectrophotometric analyses taken in triplicate are recommended. The typical standard deviation is below 5%.

3.4.3 Alkaline Phosphatase Activity Released (APAR)

As an intracellular enzyme, the APAR gives not only an indication of the protein released but also the preservation of enzymatic activity, which was considered a reliable indirect performance index. The protocol was adapted from the literature [13].

1. First, in the spectrophotometry cuvette, add 8 mL of buffer (Tris–HCl 1 M, $MgCl_2$ 0.001 M, pH = 8.8).

2. Second, add 0.1 mL of p-nitrophenyl phosphate 0.04 M.
3. Third, add 0.1 mL of the sample to analyze.
4. Mixing the cuvette and introducing it into the spectrophotometer at a controlled temperature of 30 °C.
5. Read the absorbance at 410 nm for 10 min and calculate the slope (Abs/min) (*see* **Note 6**).
6. Activity units (AU) of alkaline phosphatase (AP) are defined as the quantity of AP required to release 1 μmol of p-nitrophenol per minute under the specified assay conditions.
7. Spectrophotometric analyses taken in triplicate are recommended. Typical standard deviation is below 4% (*see* **Note 7**).

4 Notes

1. The clean suspensions are recommended to be vortexed vigorously for homogenizing the samples and dispersing any cellular aggregates.
2. The idea is to avoid an increase in temperature, which can denature the product of interest.
3. The best working pressure and maximum number of cycles are a compromise between the efficiency of the disruption and the stress resistance of the bioproduct. *In Komogataella's* disruption, it is very important to characterize the OD decrease or alkaline phosphatase activity released after each cycle. Although at certain CPP, the process could be effective for an *E. coli* or a *K. phaffii* growing on glucose, it may not be so for *K. phaffii* grown on methanol. It is due to an increase in the cell wall thickness, so the monitoring of one of these parameters is key to ensuring the proper functioning of the HPH for the specific expression system.
4. All the samples were kept on ice during disruption steps to avoid possible endogenous protease activity.
5. To quantify the efficiency of the cell disruption, the OD decrease (ODD) indicator is used. ODD is determined as a normalized quotient between the pre-homogenized (OD_0) and the post-homogenized (OD_H) values (Eq. 1)

$$ODD = 1 - \frac{OD_H}{OD_0} \tag{1}$$

6. Discharged in the first 2 min. If the slope is not linear, repeat the test, diluting the sample.
7. The effect of the initial OD_{600} can affect the values of TPR and APAR. Since this parameter is directly related to the total

amount of biomass that will be disrupted, to be able to compare disruption results between samples with different initial biomass content, these performance indicators were always normalized with the pre-homogenized OD_{600} of the samples, and hence, using the corresponding specific values; specific TPR (mg OD^{-1} L^{-1}); specific APAR (AU OD^{-1} L^{-1}).

Acknowledgments

The project PID2022-136936OB-I00 by MCIN /AEI / 10.13039/501100011033 /FEDER, UE of the Spanish Ministry of Science and Innovation, funded this work. The group is a member of the 2021 SGR 00143 Department of Research and Universities of the Generalitat de Catalunya.

Competing Interest Statement The authors have no conflicts of interest to declare that are relevant to the content of this chapter.

References

1. García-Ortega X, Ferrer P, Montesinos JL, Valero F (2013) Fed-batch operational strategies for recombinant Fab production with *Pichia pastoris* under glyceraldehyde-3-phospahte dehydrogenase promoter: from carbon source metabolism to bioreactor operational parameters. Biochem Eng J 79:172–118. https://doi.org/10.1016/j.bej.2013.07.013
2. Kotzsch A, Vernet E, Hammarström M, Berthelsen J, Weigelt J, Gräslund S, Sundström M (2011) A secretory system for bacterial production of high-profile protein targets. Prot Sci 20:597–609. https://doi.org/10.1002/pro.593
3. Demain AL, Vaishnav P (2009) Production of recombinant proteins by microbes and higher organisms. Biotechnol Adv 27:297–306. https://doi.org/10.1016/j.biotechadv.2009.01.008
4. Kuo TC, Shaw JF, Lee GC (2015) Improvement in the secretory expression of recombinant *Candida rugosa* lipase in *Pichia pastoris*. Process Biochem 50:2137–2143. https://doi.org/10.1016/j.procbio.2015.09.013
5. Balasundaram B, Harrison S, Bracewell DG (2009) Advances in product release strategies and impact on bioprocess design. Trends Biotecnol 27:447–485. https://doi.org/10.1016/j.tibtech.2009.04.004
6. Eggenreich B, Rajamanickam V, Wurm DJ, Fricke J, Herwig C, Spadiut O (2017) A combination of HPLC and automated data analysis for monitoring the efficiency of high-pressure homogenization. Microb Cell Fact 16:134. https://doi.org/10.1186/s12934-017-0749-y
7. Gauterio GV, Maciel da Silva R, Karraz FC, Coelho MAZ, Riberio BD, Lesmes AC (2023) Cell disruption and permeabilization methods for obtaining yeast bioproducts. Cleaner Chem Eng 6:100112. https://doi.org/10.1016/j.clce.2023.100112
8. Inguva P, Graselli S, Heng PWS (2024) High pressure homogeneization – an update on its usage and understanding. Chem End Res Design 202:284–302. https://doi.org/10.1016/j.cherd.2023.12.026
9. Pekarsky A, Spadiut O, Rajamanickam V, Wurm J (2018) A fast and simple approach to optimize the unit operation high pressure homogenization – a case study for a soluble therapeutic protein in *E. coli*. Prep Biochem Biotechnol 49:74–81. https://doi.org/10.1080/10826068.2018.1536988
10. Canales M, Buxadó JA, Heynngnezz L, Enríquez A (1998) Mechanical disruption of *Pichia pastoris* yeasts to recover the recombinant glycoprotein Bm86. Enzyme Microb Technol 23:58–63. https://doi.org/10.1016/S0141-0229(98)00012-X
11. García-Ortega X, Reyes C, Montesinos JL, Valero F (2015) Overall key performance

indicator to optimizing operation of high-pressure homogenizers for a reliable quantification of intracellular components in *Pichia pastoris*. Front Bioeng Biotechnol 3:1–9. https://doi.org/10.3389/fbioe.2015.00107

12. García-Ortega X, Adelantado N, Ferrer P, Montesinos JL, Valero F (2016) A step forward to improve recombinant protein production in *Pichia pastoris*: from specific growth rate effect on protein secretion to carbon-starving conditions as an advanced strategy. Process Biochem 51:681–691. https://doi.org/10.1016/j.procbio.2016.02.018

13. Berstine EG, Hooper ML, Grandchamp S, Ephrussi B (1973) Alkaline phosphatase activity in mouse teratoma. Proc Natl Acad Sci USA 70:3899–3903. https://doi.org/10.1073/pnas.70.12.3899

Chapter 42

Concentration and Purification of Proteins Produced by *Pichia pastoris* by TFF with Ceramic Membranes

Fuensanta Verdú-Navarro, Juan A. Moreno-Cid, Ana G. Del Hierro, Xavier Garcia-Ortega, and Ruth Ordóñez

Abstract

Tangential flow filtration is a common strategy to concentrate proteins. In the case of proteins produced by *Komagataella phaffii*, filtration is commonly based on cassette membranes. However, there is an alternative that is more affordable, durable, and resistant: ceramic membranes. This chapter describes the methodology to concentrate and purify *K. phaffii* secreted recombinant proteins using ceramic membranes as an alternative to cassettes in a tangential flow filtration system.

Key words *Komagataella phaffii*, *Pichia pastoris*, Tangential flow filtration, Ultrafiltration, Ceramic membrane, Protein, Concentration, Purification

1 Introduction

The microorganism *Komagataella phaffii* (previously called *Pichia pastoris)* holds significant importance as a yeast strain in industrial protein synthesis. One of these main features is the ability to carry out post-translational modifications and secrete the products of interest into the extracellular medium, keeping the concentration of endogenous proteins at low levels, which allows to simplify notably the downstream processing (DSP) [1, 2].

Tangential Flow Filtration (TFF) with ultrafiltration (UF) membranes is a common strategy employed for protein concentration [3, 4]. Eventually, it could also be combined with chromatography when high-purity degrees are required [5]. In these cases, when it is considered essential and cannot be skipped, its yield can be improved if the filtration process is combined with a coarse purification stage called diafiltration, in which the desired buffer can be exchanged by the product at the same time impurities are forced to flow with the permeate fraction [6].

Xavier Garcia-Ortega et al. (eds.), *Pichia pastoris: Methods and Protocols*, Methods in Molecular Biology, vol. 2697,
https://doi.org/10.1007/978-1-0716-4779-0_42,

Since the background proteins, of all proteins secreted to the culture broth by *K. phaffii,* are low and often found at low concentrations, the purity of the protein of interest concentrated by UF can be considered sufficient in numerous cases out of the biopharmaceutical applications, thus simplifying enormously the DSP operations. The *Pichia* secretome, which describes the native proteins secreted by wild-type *K. phaffii* strains, is published in the literature detailing the proteins usually found in the broths [7]. This might help to identify eventual contaminant proteins in the process and final product.

In the specific case of extracellular proteins released by *K. phaffii,* UF/DF methods based on cassette membranes [8, 9] are the most standard alternative used. Cassettes are manufactured in polymeric materials (Polyethersulfone or PES and Regenerated Cellulose or RC)[10]. PES and RC are both considered hydrophilic with low protein binding characteristics, but generally, PES membranes provide higher permeate fluxes. RC, however, has broad solvent compatibility and guarantees a low extractable level, which makes it the preferred material when sample analyses rely on HPLC techniques.

Despite all the advantages of polymeric materials, one important disadvantage is that cassettes lose integrity batch after batch and might be considered single-use consumables, which could turn a process economically unfeasible, especially with low-value-added final products. Ceramic membranes are a good alternative to consider. With a lifetime of more than 10 years (if they are cleaned properly after each filtration batch) and the ability to tolerate extreme pH, ceramic tubular membranes are ideal to face this drawback [11, 12].

In this chapter, a protocol to concentrate and purify recombinant proteins produced by *K. phaffii* using a 15 kDa pore-size ceramic membrane in a TFF system is detailed. This example is suitable for retaining and purifying proteins with a molecular weight (MW) higher than 45 kDa since they will be kept in the retentate fraction, while smaller molecules will go through the permeate, thus eliminating low-MW contaminants. Further recommendations about how to select the adequate membrane cut-off are included in the notes section (*See* **Note 1**).

In case the target protein is expressed intracellularly, an additional cell disruption step would be required. Details can be found in Chapter 41, as well as in the literature [13].

Additionally, a protocol for cleaning and recovering the membrane for reuse is also detailed.

The authors have developed the method using the system M1 from BIONET®, which can control the retentate flow automatically by changing the speed of the pump accordingly. Equivalent automated TFF systems are also offered by alternative vendors such as Sartorius or Cytiva.

2 Materials

2.1 Clarified Supernatants

1. As a working material, *Komagataella phaffii* clarified supernatants obtained from high-cell-density cultures were used. Details can be found in Chapters 31 and 32, as well as in the literature [14, 15].
2. This clarification is often obtained after a combination of centrifugation and microfiltration steps (0.4 μm pore size). Alternatively, as described in Chapter 40, a suitable clarification can be obtained through a single-step TFF using hollow fiber or ceramic tubular membranes.

2.2 Filtration System

1. Automated TFF system (M1 from BIONET®). In Fig. 1, a scheme of a typical TFF system used for protein concentration is shown.
2. A heat exchanger is connected to an external chilling/heating system in charge of supplying heating and cooling capacity for DSP and/or Cleaning in Place (CIP) operations.
3. Ceramic (α-Al_2O_3) membrane with 15 kDa pore diameter (*see* **Note 1**) and a channel diameter of 2.7 mm. From Atech Innovations GmbH.
4. Raw chemicals for pH adjustment and membrane cleaning: NaOH 50% wt., HNO_3 60% wt.

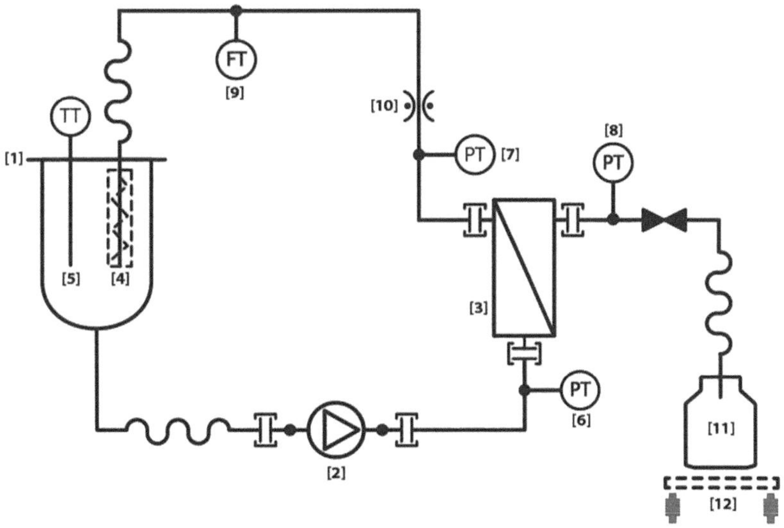

Fig. 1 Tangential Flow Filtration system. Main components: (1) Feed vessel; (2) recirculation pump; (3) membrane module; (4) heat exchanger; (5) temperature transmitter; (6) feed pressure transmitter; (7) retentate pressure transmitter; (8) permeate pressure transmitter; (9) flow transmitter; (10) backpressure pinch valve; (11) permeate reservoir; (12) scale

5. Demineralized water at room temperature (RT) is used for membrane conditioning and buffer exchange in the diafiltration stage.

2.3 Sampling and Analysis

It is important to take samples and analyze the protein content along the concentration-diafiltration process to determine the yield of protein recovery (*see* **Note 2**).

1. Total protein determination kit:
   - Pierce Coomassie (Bradford) or BCA-based Protein Assay Kit. Different alternatives are available from different vendors.
   - A spectrophotometer device for cuvettes or 96-well plates, able to quantify colorimetric reactions at a wavelength of 595 nm.
2. For protein electrophoresis, sodium dodecyl sulfate-polyacrylamide gel electrophoresis (SDS-PAGE) with Coomassie staining at the desired polyacrylamide concentration.

3 Methods

Details about the abbreviations are detailed in the chapter list of abbreviations.

3.1 Conditioning

1. Fill the system with demineralized water and open the permeate and backpressure valves completely.
2. Adjust water pH (*see* **Note 3**) to that of permeate (*see* **Note 4**). The volume of water used for conditioning has to be at least 30% of the holdup volume of the system to guarantee flow stability.
3. Start pumping in recirculation mode (permeate sent back to the feed vessel) at a cross-flow velocity (CFV) of 4 m s^{-1} (or a similar range according to membrane saturation tests previously performed) and keep the temperature under control at 25 °C (*see* **Note 5**).
4. Once the temperature and CFV are achieved, adjust the backpressure valve if necessary to ensure permeate generation.
5. Keep the system in recirculation mode for 5 min.

3.2 Protein Concentration

1. After membrane conditioning, add the permeate obtained in the biomass clarification stage (it contains the protein of interest) and keep the system in recirculation mode until setting a transmembrane pressure (TMP) of 3 barg (*see* Eq. (1)).

$$\text{TMP} = \frac{P_f + P_r}{2} - P_p \tag{1}$$

2. Keep the system in recirculation mode for 5 min.
3. After this time, put the system in concentration mode, sending the permeate to an external vessel.
4. Start registering the permeate weight with time to determine permeate flow, which can be transformed into permeate flux with Eq. (2).

$$J = \frac{Q_p}{A} \tag{2}$$

5. Concentration is done by keeping the previous variables (CFV, temperature, and TMP) constant until achieving approximately 10X of concentration factor (*see* **Note 6**) (*see* Eq. (3)).

$$\mathrm{CF} = \frac{V_f}{V_r} \tag{3}$$

3.3 Diafiltration

1. Diafiltration (*see* **Note 7**) starts once the protein achieves the desired concentration factor. The amount of buffer is calculated following Eq. (4).

$$V_{\mathrm{DF}} = \mathrm{DV} \cdot V_o \tag{4}$$

2. It is recommended to work between 2 and 4 DV in batch mode, combined with a step of membrane surface cleaning (*see* **Note 8**). This is the procedure:
   (a) Open the backpressure valve completely.
   (b) Close the permeate valve.
   (c) Add the first volume (1DV) of demineralized water.
   (d) Recirculate for 5 min, keeping the same CFV and temperature.
   (e) Open the permeate, adjust the TMP to the same value used in the concentration stage, and start collecting permeate.
3. Figure 2 represents a scheme with all the steps involved in the concentration and purification process.

3.4 Recovering Product

Once the diafiltration stage is finished, and before recovering the product:

1. Open the backpressure valve completely.
2. Close the permeate valve.
3. Keep the system in recirculation for 5 min at the same CFV and temperature used in the concentration and purification stage (*see* **Note 9**).

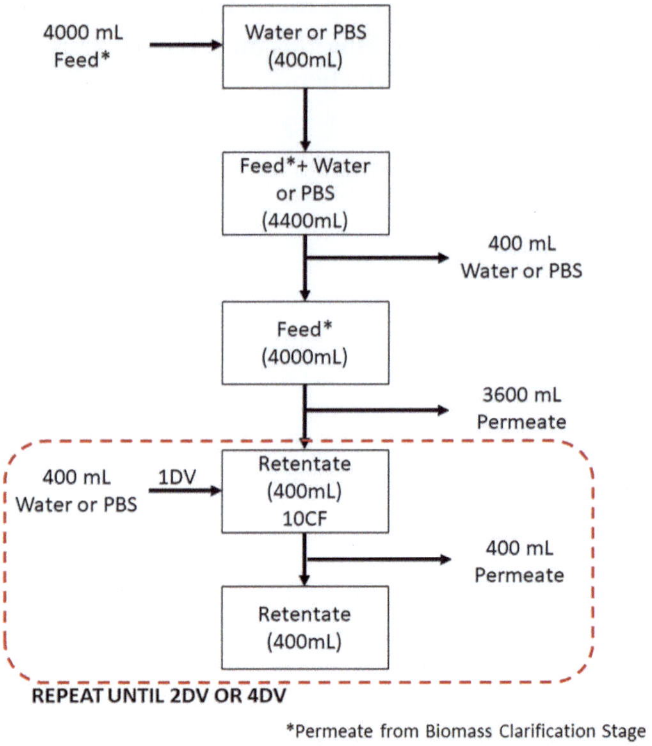

Fig. 2 Scheme of conditioning, concentration, and diafiltration stages

4. After this time, open the drainage valve and collect all the retentate (concentrated protein).

3.5 Cleaning in Place (CIP)

Proceed with the following cleaning protocol to recover the reference membrane water flux:

1. Rinse the feed and retentate (permeate close) with demineralized water for 5 min at RT and at the same CFV used for concentration/diafiltration stages.
2. Recirculate through the feed and retentate (permeate close) NaOH (2%wt) solution at 85 °C (*see* **Note 10**) for 30 min at the same CFV used for concentration/diafiltration stages.
3. Rinse the feed, retentate, and permeate with demineralized water until neutrality at the same CFV used for concentration/diafiltration stages and at the lowest TMP that allows permeate flow.

4 Notes

1. Selection of the optimal filtration cut-off is equally important, whether to concentrate and retain proteins completely or to let them pass into the permeate fraction while other unwanted

substances are retained. In the first option, as is the case with *K. phaffii's* secreted proteins, a general rule is to choose a membrane with a nominal molecular weight cut-off (NMWC) that is 3–5 times smaller than the molecular weight (MW) of the target molecule to retain. However, if protein passage is desired, then an order of magnitude difference between the membrane's NMWC and the protein size is preferable.

2. To determine the yield of protein recovery, it is possible to use different analytical methods: enzymatic activity, protein quantification, protein electrophoresis, etc. Choose the best option for your protein.
3. Demineralized water can be exchanged for a specific buffer (i.e., PBS) if required. The buffer has to be similar in pH and ionic strength to that of the sample.
4. This pH is the specific isoelectric point for this protein. pH adjustment may vary depending on the protein of interest.
5. Some of the filtration conditions may vary depending on the protein of interest, i.e., temperature. Knowing the optimal conditions of the target protein before the filtration to avoid denaturation is a must.
6. The maximum concentration factor that can be achieved depends not only on having a good and stable permeate flux ($\geq$40 L h^{-1} m^{-2}) but also on the holdup volume of the installation. Do not overconcentrate the protein to avoid protein aggregation.
7. In this case, diafiltration has a double purpose: (a) increase protein purity, favoring impurities permeation; and (b) buffer exchange to a more conservative one.
8. The number of DVs is calculated to ensure that all the medium has been exchanged and maximum protein purity is achieved.
9. This procedure is recommended to recover as much as possible of the protein that could be attached to the membrane surface. Longer flushing times are not recommended to avoid prolonging (excessively) the shear stress on the product.
10. When setting the rinsing temperature, avoid temperature gradients higher than 30 °C to avoid membrane shocks.

Acknowledgments

Part of this work was possible thanks to professors Francisco Valero and José Luis Montesinos, of the Bioprocess Engineering and Applied Biocatalysis Group of the Autonomous University of Barcelona (UAB), for the collaboration initiated for the development of different works with the friendly and scientific transfer of the strain used.

Competing Interest Statement The authors FVN, JMC, ADH, and RO are employees of Bionet, a company engaged in the commercialization of bioprocessing equipment and offering related services.

The author XGO has no conflicts of interest to declare that are relevant to the content of this chapter.

Appendix

Below are listed all the abbreviations mentioned in the Chapter.

TMP	Transmembrane pressure, barg
P_f	Inlet or feed pressure, barg
P_r	Outlet or retentate pressure, barg
P_p	Permeate pressure, barg
J	Permeate flux, L h^{-1} m^{-2}
Q_p	Permeate flow, L h^{-1}
A	Membrane area, m^2
CF	Concentration factor
V_f	Initial volume, L
V_r	Final (or retentate) volume, L
V_{DF}	Buffer (or water) volume, L
DV	Diafiltration factor (also called Diavolume)
V_o	Diafiltration starting volume, L

References

1. Qin X, Qian J, Yao G, Zhuang Y, Zhang S, Chu J (2011) GAP promoter library for fine-tuning of gene expression in Pichia pastoris. Appl Environ Microbiol 77:3600–3608. https://doi.org/10.1128/AEM.02843-10
2. Meehl MA, Stadheim TA (2014) Biopharmaceutical discovery and production in yeast. Curr Opin Biotechnol 30:120–127. https://doi.org/10.1016/j.copbio.2014.06.007
3. Powell MJ, Timperman AT (2005) Quantitative analysis of protein recovery from dilute, large volume samples by tangential flow ultrafiltration. J Membr Sci 252:227–236. https://doi.org/10.1016/j.memsci.2004.12.024
4. Ghosh R (2008) Ultrafiltration-based protein bioseparation. In: Kumar Pabby A, Rizvi S, Maria Sastre A (eds) Handbook of membrane separations. CRC Press, pp 497–511
5. Chen S, Tan Y, Zhu Y, Sun L, Lin J, Zhang H (2022) Application of ultrafiltration and ion exchange separation technology for lysozyme separation and extraction. Fermentation 8:

297. https://doi.org/10.3390/fermentation8070297

6. Prapulla SG, Karanth NG (2014) Fermentation (industrial) | recovery of metabolites. In: Batt CA, Tortorello ML (eds) Encyclopedia of food microbiology, 2nd edn. Academic, Oxford, pp 822–833
7. Burgard J, Grünwald-Gruber C, Altmann F, Zanghellini J, Valli M, Mattanovich D, Gasser B (2020) The secretome of *Pichia pastoris* in fed-batch cultivations is largely independent of the carbon source but changes quantitatively over cultivation time. Microb Biotechnol 13: 479–494. https://doi.org/10.1111/1751-7915.13499
8. Bavaro T, Ubiali D, Brocca S, Rocchietti S, Nieto I, Pregnolato M, Lotti M, Terreni M (2010) Recombinant lipase from Candida rugosa for regioselective hydrolysis of peracetylated nucleosides. A comparison with commercial non-recombinant lipases. Biocatal Biotransform 28:108–116. https://doi.org/10.3109/10242420903497362
9. Brocca S, Secundo F, Ossola M, Alberghina L, Carrea G, Lotti M (2003) Sequence of the lid affects activity and specificity of Candida rugosa lipase isoenzymes. Protein Sci 12: 2312–2319. https://doi.org/10.1110/ps.0304003
10. van Reis R, Zydney A (2007) Bioprocess membrane technology. J Membr Sci 297:16–50. https://doi.org/10.1016/j.memsci.2007.02.045
11. Wang A, Lewus R, Rathore AS (2006) Comparison of different options for harvest of a therapeutic protein product from high cell density yeast fermentation broth. Biotechnol Bioeng 94:91–104. https://doi.org/10.1002/bit.20816
12. Birrenbach O, Faust F, Ebrahimi M, Fan R, Czermak P (2021) Recovery and purification of protein aggregates from cell lysates using ceramic membranes: fouling analysis and modeling of ultrafiltration. Front Chem Eng 3
13. García-Ortega X, Reyes C, Montesinos JL, Valero F (2015) Overall key performance indicator to optimizing operation of high-pressure homogenizers for a reliable quantification of intracellular components in Pichia pastoris. Front Bioeng Biotechnol 3:1–9. https://doi.org/10.3389/fbioe.2015.00107
14. Gasset A, Garcia-Ortega X, Garrigós-Martínez J et al (2022) Innovative bioprocess strategies combining physiological control and strain engineering of Pichia pastoris to improve recombinant protein production. Front Bioeng Biotechnol 10:1–12. https://doi.org/10.3389/fbioe.2022.818434
15. Garrigós-Martínez J, Weninger A, Montesinos-Seguí JL et al (2021) Scalable production and application of Pichia pastoris whole cell catalysts expressing human cytochrome P450 2C9. Microb Cell Fact 20:90. https://doi.org/10.1186/s12934-021-01577-4

Chapter 43

Protein Purification Protocols for Recombinant Enzymes Produced in *Pichia pastoris*

Merve Keser, Mikel Dolz, Javier Viña-Gonzalez, David Gonzalez-Perez, and Miguel Alcalde

Abstract

Komagataella phaffii is usually a microbial host of choice for eukaryotic enzyme expression and overproduction in fed-batch bioreactors. The design of a fast and reliable protein purification protocol is crucial to perform a detailed biochemical characterization of the recombinant enzyme. Here, we provide standardized methods for ion exchange chromatography (IEC) and immobilized metal affinity chromatography (IMAC) in the ÄKTA system that are broadly used for the purification of heterologous enzymes produced in *K. phaffii*.

Key words *Pichia pastoris*, *Komagataella phaffii*, Enzyme heterologous expression, Protein purification, Ion exchange chromatography (IEC), Immobilized metal affinity chromatography (IMAC), *Pichia pastoris*, Bioreactor

1 Introduction

Komagataella phaffii, previously known as *Pichia pastoris*, is widely used by both academia and industry due to its ability to secrete to the extracellular medium functionally folded while expressing lower amounts of native proteins in the culture broth, which significantly facilitates the downstream operations (DSP) [1]. Compared to the prokaryotic expression system *Escherichia coli*, the secretory pathway of *K. phaffii* allows enzymes to be directly released into the culture supernatant, avoiding tedious lysis steps [2]. Furthermore, *K. phaffii* can perform complex post-translational modifications, including disulfide-bound formation, N- and O-glycosylation, and N- and C-terminal processing [3].

In standard *Pichia* bioprocesses, once the upstream operations are completed, the DSP sequence usually starts with the product

Authors Merve Keser and Mikel Dolz have equally contributed to this chapter.

Xavier Garcia-Ortega et al. (eds.), *Pichia pastoris: Methods and Protocols*, Methods in Molecular Biology, vol. 2697,
https://doi.org/10.1007/978-1-0716-4779-0_43,

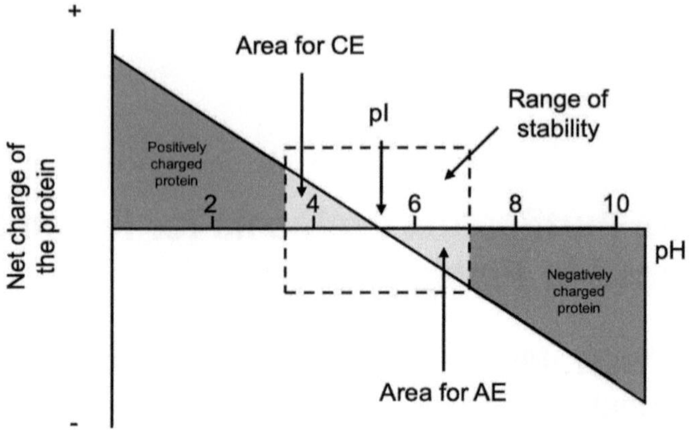

Fig. 1 Schematic representation of the relationship between the net charge of the protein and pH. The overlapping areas of stability and charge indicate the possible strategies to follow

recovery and concentration steps through clarification and ultra/diafiltration methods, whose details are described in Chapters 40 and 42. Subsequently, as a crucial part of the DSP, protein purification is considered the final step that allows the target protein to be characterized at the biochemical and structural level and/or achieve the required purity for its industrial or pharmaceutical applications. As *K. phaffii* secretes a fairly low number of endogenous proteins in comparison to the recombinant fraction [4], this facilitates the application of fast purification protocols.

Among the most common purification strategies for recombinant proteins produced by *K. phaffii are* spotlight ion exchange chromatography (IEC) and immobilized metal affinity chromatography (IMAC). IEC offers a high resolution with substantial loading capacity. An ion exchange column comprises a matrix of spherical particles (*e.g.*, agarose beads) that are functionalized with ionic groups, either negatively or positively charged, depending on the type of chromatography: anion exchange (AE) or cation exchange (CE). Anion exchanger matrices present positively charged functional groups (*e.g.*, quaternary ammonium, diethylaminoethyl, or diethylaminopropyl) which interact with and bind negatively charged proteins. In contrast, cation exchangers display negatively charged functional groups, such as sulfopropyl, methyl sulfonate or carboxymethyl, and therefore present affinity towards positively charged proteins. When performing an ionic exchange purification of a certain protein, it is critical to know its physicochemical characteristics, particularly its isoelectric point (pI) and its pH stability. Considering these parameters, an effective purification strategy can be designed (Fig. 1).

IMAC is a purification technique that exploits the strong interaction between histidine residues present in proteins and metal ions such as nickel, cobalt, zinc, or copper, which are immobilized onto a resin matrix. Typically, nickel-nitrilotriacetic acid (Ni-NTA) and cobalt-nitrilotriacetic acid (Co-NTA) serve as common ligands attached to the agarose resin in IMAC. Ni-NTA exhibits high binding affinity, while Co-NTA offers greater specificity and reduced background binding. Alternatively, iminodiacetic acid (IDA) serves as another ligand option, known for its versatility and cost-effectiveness, albeit with slightly lower metal ion binding affinity compared to NTA ligands. IMAC proves highly effective in purifying His-tagged proteins featuring 4-10 histidine residues either at the N- or C-termini of recombinant proteins. The increased number of histidine tags enhances binding to the metal ions, facilitating selective elution and yielding cleaner samples through stringent wash steps. Incorporating cleavage sites before the His-tags presents an alternative method to eliminate any potential interference from the tags in the final product functionality. The advantages of IMAC include its high specificity, user-friendly nature, and its capability to achieve both high purity and yield. Compared to non-his tagged proteins, IMAC offers an efficient purification solution, making it a popular choice in both research and industrial biotechnology settings.

This chapter provides general enzyme purification guidance to include IEC and IMAC-based steps, through which a high purity degree can be obtained in *K. phaffii*-based bioprocesses.

2 Materials

2.1 General Chemicals

All chemicals used were of reagent grade purity and sourced from Merck (New Jersey, USA) unless specified otherwise.

1. 70% Ethanol.
2. NaCl.
3. Imidazole.
4. di-Potassium hydrogen phosphate, anhydrous (K_2HPO_4).
5. Potassium phosphate monobasic (KH_2PO_4).
6. Ortho-phosphoric acid (H_3PO_4) (99%).
7. NaOH pellets.
8. ABTS (2,2′-azino-bis(3-ethylbenzothiazoline-6-sulfonic acid) (Panreac Applichem ITW Reagents, Darmstadt, Germany).
9. Protein assay dye reagent concentrate (Bio-Rad, California, USA).
10. Hydrogen peroxide (Bio-Rad, California, USA).
11. SeeBand protein staining solution (Gene Bio-Application LTD., Yavne, Israel).

2.2 Buffer Solutions

2.2.1 Buffer Solutions Ion Exchange Chromatography (IEC)

1. Binding buffer for anion exchange chromatography: Tris–HCl 20 mM, pH 7.8.
2. Elution buffer for anion exchange chromatography: Tris–HCl 20 mM, pH 7.8, NaCl 1 M.
3. Binding buffer for cation exchange chromatography: Sodium phosphate/citrate 20 mM, pH 3.3.
4. Elution buffer for cation exchange chromatography: Sodium phosphate/citrate 20 mM, pH 3.3, NaCl 1 M.

2.2.2 Buffer Solutions Immobilized Metal Affinity Chromatography (IMAC)

1. Stock solution: 1 M Potassium phosphate buffer (KPi) at pH 8.0 in ddH_2O.
2. Stock solution: 5 M NaCl in dH_2O.
3. Stock solution: 1 M Imidazole dH_2O.
4. Stock solution: 10 M NaOH, NaOH pellets dissolved in dH_2O.
5. Stock solution: 5 M H_3PO_4.
6. Binding buffer (Buffer A): 100 mM KPi pH 8.4 containing 300 mM NaCl (*see* **Note 1**).
7. Wash buffer: 100 mM KPi pH 8.4 containing 150 mM NaCl and 5 mM Imidazole (*see* **Note 2**).
8. Elution buffer (Buffer B): 100 mM KPi pH 8.4 containing 150 mM NaCl and 300 mM imidazole (*see* **Note 3**).

2.3 General Equipment

The authors of this chapter use the equipment described in this section. However, other equivalent alternatives may be offered by other vendors.

1. Äkta Pure (Cytiva, Massachusetts, USA).
2. Vertical polyacrylamide electrophoresis gel system (Bio-Rad, Mini-PROTEAN® Tetra System).
3. Tangential flow filtration system (Minimate™ EVO System, Cytiva, Massachusetts, USA) and/or hollow fiber cartridge tangential flow filtration (TFF) system, pore size 0.45 μm (Sartorius, Gottingen, Germany).
4. Steri-top Millipore Express PLUS 0.22 μm PES, 500 mL for vacuum pump (Millipore, Massachusetts, USA).
5. 0.22 μm Rotilabo-syringe filters, PVDF (Carl Roth GmbH, Karlsruhe, Germany).
6. Ultrafiltration system (Amicon® Stirred Cell 50 or 400 mL, Millipore, Massachusetts, USA).
7. 10 kDa pore-sized membrane (Millipore, Massachusetts, USA).

2.3.1 Columns for IEC

1. Cation-exchange Column (HiTrap SP FF Cytiva, Massachusetts, USA).
2. High-resolution 10 μM anion-exchange column (Biosuite Q column, Waters, Massachusetts, USA).

2.3.2 Columns for IMAC

1. His-trap fast flow (FF), 5 mL (Cytiva, Massachusetts, USA).

2.4 Working Material Before Purification Steps

The general workflow before starting the purification steps is described below.

1. *K. phaffii* expressed enzyme fermentation broth (*see* **Note 4**). Specific methods to perform fed-batch and continuous cultures are described in Chapters 31, 32, and 33.
2. Centrifugation for clarification steps. Alternatively, a method based on TFF is also described in Chapter 40.
3. TFF system for ultrafiltration and/or filtration systems to concentrate the target protein and exchange is buffer.

3 Methods

3.1 Downstream Steps

1. Harvest the fermentation broth and centrifuge at 4000 rpm for 20 min and discard the pellet (*see* **Note 5**).
2. Set up the tangential flow filtration and/or ultrafiltration systems according to the manufacturer's specifications (*see* **Note 6**).

3.2 IEC Purification Protocol

1. Dialyze the enzyme supernatant at least 20 times in the binding buffer using a TFF or ultrafiltration system. Depending on the pH of the binding buffer and the isoelectric point (pI) of the protein, the enzyme will be either in an anionic or cationic state, necessary for the protein to get attached to the column due to electrostatic interactions (*see* **Note 7**).
2. Centrifuge and filter the resulting rebuffered enzyme using a 0.22 μM Rotilabo syringe filter, or if the rebuffered concentrated enzyme volume is more than 50 mL, Steri-top Millipore Express 0.22 μM PES using a vacuum pump is recommended to remove impurities that may damage the column.
3. Equilibrate the purification instrument and column in the selected binding buffer (*see* **Note 8**).
4. Load the protein into the column.
5. Apply a gradient with increasing NaCl concentration to elute your protein of interest (*see* **Note 9**).

3.3 IMAC Purification Protocol

1. Rebuffer the supernatant using the TFF hollow fiber cartridge system using dH_2O according to the manufacturer's specifications (*see* **Note 6**).
2. Add 10% of 1 M KPi buffer pH 8.4 to the retentate sample in the previous step, along with 150 mM NaCl from the stock NaCl solution to bring the sample to the equilibration conditions (equilibration buffer, *see* buffer and stock solutions section). The sample must be in a final concentration of 100 mM KPi, 150 mM NaCl, at pH 8.4. Use 10 M NaOH or 5 M H_3PO_4 to adjust pH if necessary (*see* **Notes** 7 and **10**).
3. Attach the His-Trap 5 mL FF column to the injection module on the Äkta system according to the Äkta system user manual.
4. Equilibration: Equilibrate the column with 5 column volumes (CV) using buffer A (*see* Subheading 2.2.2).
5. Inject the rebuffered supernatant: Load the equilibrated enzyme broth onto the HisTrap column attached to Äkta using a superloop, or alternatively by using an inlet line directed towards the column. If using an inlet tube, place the rebuffered enzyme sample on ice. If using a superloop, fill the superloop with the rebuffered enzyme sample (*see* **Note 11**).
6. Wash the column: Using 3 CVs of wash buffer, wash the column to remove any unspecifically bound components from the column (*see* **Note 11**).
7. Elution: Elute the protein using Buffer B (*see* Subheading 2.2.2) in a linear or step gradient, which should be adjusted in each case.
8. Re-equilibration or cleaning in place (CIP): If running a new purification with the same conditions, repeat from **step 4** on going. If the purification run is finished, run a CIP method according to the manufacturer's instructions that are specified in the column manual.

3.4 Identification of Pure Enzyme and Polishing

1. Collect the elution profile in 0.5 mL fractions. Measure plate activity to identify the active fractions and select those fractions with the highest activity (Fig. 2) (*see* **Note 12**).
2. Removal of His-tag and dialysis for storage conditions: Continuing with downstream processing of the enzyme, if a cleavage tag was added before the His-tag, remove the His-tag by means of using specific proteases for the cleavage site designed according to the protease manufacturer's specifications. Whether using IEC or IMAC, dialyze the purified protein against an appropriate buffer in which the protein will be stable and store it in aliquots at 4 °C for short-term storage, or flash freezing in liquid nitrogen and storage at −80 °C for long-term (*see* **Note 13**).

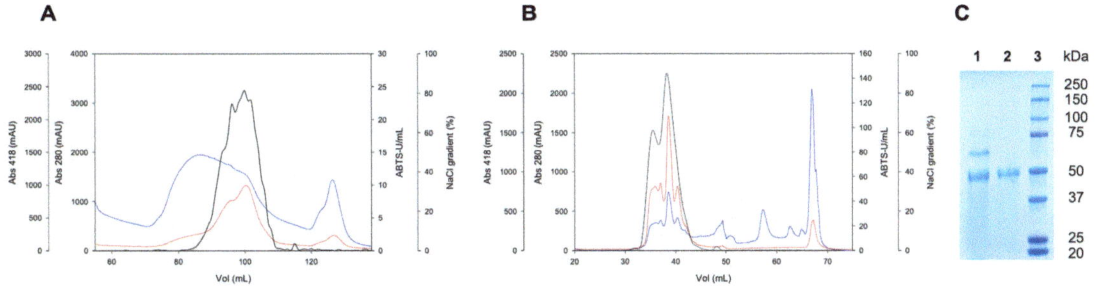

Fig. 2 Illustrative examples of purification chromatograms and SDS-Page protein gel. Representation of a cation exchange chromatography (**a**). Representation of an anion exchange chromatography, using a MonoQ column (**b**). MonoQ purification results in a more refined and polished purification—evidenced by the differentiated peaks of glycoforms. In red is represented the absorbance at 418 nm due to the presence of the heme group in the protein. In blue is represented the absorbance at 280 nm. In black is represented the activity of the different fractions (measured in ABTS units). The dotted line represents the NaCl gradient applied for the protein elution. SDS-Page gel with different elution fractions from the ÄKTA instrument (**c**). Lane 1 represents a fraction after the first step of CE, presenting some impurities. Lane 2 represents a more polished sample after an AE MonoQ chromatography step. Lane 3 represents the gel ladder with its corresponding molecular weight markers

3. Measure protein concentration at 595 nm using Bradford reagent [5] (*see* **Note 14**).
4. Load 6 μg of your protein of interest in an SDS-Page gel. Dye and reveal the SDS-Page gel (Fig. 2c) with SeeBand staining solution (or similar) to confirm the purity grade of your sample (*see* **Note 15**).

4 Notes

1. It is recommended to use soy peptone in the composition of the fermentation media to ensure higher yields, as peptone from casein causes inhibition in expression. Anti-foam is crucial for the fermentation to prevent the secreted enzyme from denaturing due to foaming and agitation. Antifoam from Struktol J673A is highly recommended as this has been shown to boost expression in fermentation studies [6].
2. Equilibration buffer is composed accordingly using the stock solutions made. This buffer does not contain imidazole to ensure the binding of the target protein within the supernatant to the column. If the target protein is unstable at pH 8.4, this value can be decreased to 7.5, which is 1 unit above the pKa of His to ensure the correct protonation stage.
3. Wash buffer contains a low concentration of imidazole to remove the unspecifically bound components from the media

as well as ensuring the cleanliness of the purified sample. Typically, from 5 to 25 mM imidazole can be used.

4. Elution using imidazole is generally in the range of 150–350 mM, however, depending on the amount of his tags used as well as the strength of the binding, this concentration can be increased. If a higher concentration of imidazole is used, a prolonged dialysis procedure must be carried out after purification to fully remove imidazole as not to oxidize the purified protein when placed in storage.
5. *K. phaffii* secretes the protein of interest into the extracellular medium. Therefore, after centrifugation, the recombinant enzyme is in the supernatant, and the pellet can be discarded. To prevent damage to ulterior devices, filtration (*e.g.*, glass fiber filters) of the collected supernatant is highly recommended.
6. Quality control of the rebuffered supernatant must include checking the volumetric activity, volume, and protein concentration of the flow-through and retentate, ensuring total loss of units does not exceed 10%. If losses are higher, repeat the step and assess the TFF system's functionality. TFF is recommended for large-volume supernatants and culture broths from bioreactors. Sequential steps of tangential flow filtration and ultrafiltration are advised: use TFF for larger volumes due to higher speeds, and ultrafiltration for dialysis to achieve lower volumes for easier loading into the purification instrument, using less binding buffer. Measure enzymatic activity to track the enzyme's location, and if flow-through activity exceeds 10%, replace the membrane or use one with a smaller pore cut-off for ultrafiltration. Creating a purification table with the yield of each step can help optimize the process (Table 1).
7. Before designing a purification strategy, it is crucial to conduct pH stability studies to determine the range of pHs in which the protocol can be implemented. Additionally, knowing the protein's isoelectric point (pI) will help establish the appropriate

Table 1
An illustrative example of a purification table.

Sample	Volume (mL)	U/mL	Total Units	Protein concentration (mg/mL)	Total protein (mg)	Specific activity (U/mg)	Purification factor	Purification yield (%)
Crude extract	920	1.80	1656	0.17	156.40	10.59	1.00	100
CE	30	22.75	682.5	0.40	9.60	56.87	5.37	41.21
AE	0.2	3265	653	10.075	1.00	312.72	29.53	39

purification procedure (Fig. 1). Whether using anion exchange (AE), cation exchange (CE) chromatography, or IMAC, the selected pH should be stable for the protein while altering its net charge. Prior determination of the target protein's pH stability is highly recommended.

8. To ensure all the protein binds to the column, flow through at least 10 column volumes. Monitor the conductivity of the mobile phase before and after protein loading to assess dialysis quality. Differences in conductivity indicate inadequate dialysis, which can lead to failed binding to the column.
9. Bound proteins are eluted by increasing the ionic strength of the buffer, typically using a gradient of NaCl. This method, especially useful for unknown samples, reduces the interaction between the matrix and bound proteins, causing them to elute. The gradient length and final salt concentration should be optimized for each enzyme based on its specific interaction with the column. This approach ensures efficient and controlled protein elution.
10. After adding the 10% stock solution of 1M KPi buffer at pH 8.4, the enzyme can be stored for later use.
11. Collect the flow through in case any mistakes or non-binding occurs, preventing any loss of enzyme, allowing for further troubleshooting if necessary.
12. To identify the fraction containing the enzyme of interest, observe several parameters. First, the enzyme should correspond to a peak in absorbance at 280 nm [7, 8]. If the enzyme bears a cofactor, measure the wavelength at which the cofactor absorbs (e.g., 418 nm for hemoproteins) to locate the enzyme throughout the fractions. The most recommended method is using colorimetric assays to detect the most active fractions [9]. Additionally, for quality control, run an activity assay throughout the purification process to prevent enzyme loss and ensure proper purification.
13. Different cleavage sites can be introduced in the design of the target protein construction in its N-terminal or C-terminal to ensure that the purification tags do not interfere with the activity or the stability of the enzyme. These cleavage sites to remove tags are targeted by proteases such as Factor Xa, TEV protease, HRV-3C protease, or thrombin, to name a few. It is recommended to use a linker peptide between the cleavage site and the gene of interest to prevent the digestion of the native enzyme [10]. Salt presence and high imidazole concentration might inactivate the enzyme; therefore, storage conditions of the purified protein should be tested immediately upon purification to determine the best protein stability.

14. For the sake of accuracy, composing an *ad hoc* calibration curve with triplicates of BSA standards is recommended.
15. In case the purity of the enzyme of interest is not enough, sequential steps modifying the pH—which may result in a different type of chromatography—can eliminate impurities. Additionally, the use of high-resolution columns (*see* Subheading 2.3.1) can lead to a more polished—and therefore purer sample (Fig. 2b, c). It is also possible that not only a unique peak in the purification chromatogram or a band in the SDS-Page gel appears showing the same activity (Fig. 2c). This might be derived from different isoforms with distinct glycosylation patterns. Employing deglycosidases as PNGase F [11–13] can be applied to the sample to remove these moieties.

Acknowledgments

This work was supported by the I+D+I PID2022-142074OB-I00-RAINBOW funded by the Ministerio de Ciencia e Innovación/Agencia Estatal de Investigación/AEI/doi: 10.13039/501100011033/. D.G.P. thanks to Comunidad de Madrid Atraccion de Talento Mod. 1 Project 2022-T1/BIO-238521/ECOCHEM.

Competing Interest Statement The authors have no conflicts of interest to declare that are relevant to the content of this chapter.

References

1. Cregg JM, Tolstorukov I, Kusari A et al (2009) Chapter 13 Expression in the yeast *Pichia pastoris*. In: Burgess R (ed) Methods in enzymology, vol 263. Elsevier, Berkeley, pp 169–189
2. Daly R, Hearn MTW (2005) Expression of heterologous proteins in *Pichia pastoris*: a useful experimental tool in protein engineering and production. J Mol Recognit 18:119–138. https://doi.org/10.1002/jmr.687
3. Bretthauer RK, Castellino FJ (1999) Glycosylation of *Pichia pastoris* -derived proteins. Biotech App Biochem 30:193–200. https://doi.org/10.1111/j.1470-8744.1999.tb00770
4. Burgard J, Grünwald-Gruber C, Altmann F et al (2020) The secretome of *Pichia pastoris* in fed-batch cultivations is largely independent of the carbon source but changes quantitatively over cultivation time. Microb Biotech 13:479–494. https://doi.org/10.1111/1751-7915.13499
5. Rosenberg IM (2005) Protein analysis and purification: benchtop techniques, 2nd edn. Birkhauser, Boston, pp 130–134
6. Routledge SJ, Hewitt CJ, Bora N, Bill RM (2011) Antifoam addition to shake flask cultures of recombinant *Pichia pastoris* increases yield. Microb Cell Fact 10:17. https://doi.org/10.1186/1475-2859-10-17
7. Warburg O, Christian W (1941) Isolierung und Kristallisation des Gärungsferments Enolase. Naturwis 29:589–590. https://doi.org/10.1007/BF01482279
8. Layne E (1957) Spectrophotometric and turbidimetric methods for measuring proteins. Methods Enzymol 3:447–454. https://doi.org/10.1016/S0076-6879(57)03413-8

9. Dolz M, Monterrey DT, Beltran-Nogal A et al (2023) The colors of peroxygenase activity: colorimetric high-throughput screening assays for directed evolution. In: Renata H (ed) Methods in enzymology: P450 enzymes, vol 693. Academic, Cambridge, MA, pp 73–109
10. Waugh DS (2011) An overview of enzymatic reagents for the removal of affinity tags. Prot Exp Purif 80:283–293. https://doi.org/10.1016/j.pep.2011.08.005
11. Maley F, Trimble RB, Tarentino AL, Plummer THJ (1989) Characterization of glycoproteins and their associated oligosaccharides through the use of endoglycosidases. Anal Biochem 180:195–204. https://doi.org/10.1016/0003-2697(89)90115-2
12. Tretter V, Altmann F, März L (1991) Peptide-N4-(N-acetyl-beta-glucosaminyl)asparagine amidase F cannot release glycans with fucose attached alpha 1—3 to the asparagine-linked N-acetylglucosamine residue. Eur J Biochem 199:647–652. https://doi.org/10.1111/j.1432-1033.1991.tb16166.x
13. Plummer THJ, Tarentino AL (1991) Purification of the oligosaccharide-cleaving enzymes of Flavobacterium meningosepticum. Glycobiology 1:257–263. https://doi.org/10.1093/glycob/1.3.257

Chapter 44

Characterization of Low Molecular Weight Variants (Fragmentation) in *Pichia pastoris* by CE-SDS

Rocío Cuevas Casado and Albert Font Inglés

Abstract

Komagataella phaffii is an efficient system for secreted recombinant protein expression. However, this host also expresses native proteases whose activity can degrade the target proteins, thus impacting native yield and quality. This proteolysis creates impurities considered critical quality attributes (CQAs) by regulatory agencies. Capillary electrophoresis sodium dodecyl sulfate (CE-SDS) is an ideal analytical method for characterizing these impurities due to its high resolution, sensitivity, and ability to separate protein fragments, making it valuable for the control strategy and characterization of a biopharmaceutical product.

Key words *Pichia pastoris*, *Komagataella phaffii*, Capillary electrophoresis, Sodium dodecyl sulfate, Critical quality attributes, High throughput, Biologics

1 Introduction

One advantage of using *Komagataella phaffii* (previously known as *Pichia pastoris*) as a recombinant protein expression system is its high productivity, the ability to reach high cell density, and the use of strong promoters that drive high levels of protein expression. These features make *Pichia pastoris* a popular choice for producing biologics efficiently and cost-effectively.

However, an important drawback when expressing biopharmaceutical products is their tendency to degrade proteins, totally or partially, due to their native protease activity. This proteolytic degradation has been observed to occur by different kinds of proteases, either internally or extracellularly. Therefore, during the manufacturing process, specifically upstream (USP), where the protein is overexpressed under controlled conditions (bioreactor), this event might happen, especially when growing cells in a defined medium as well as when using methanol for the *AOX1* promoter induction [1, 2].

Xavier Garcia-Ortega et al. (eds.), *Pichia pastoris: Methods and Protocols*, Methods in Molecular Biology, vol. 2697, https://doi.org/10.1007/978-1-0716-4779-0_44,

But this protein fragmentation might also occur, usually to a lesser extent [1, 2], through the downstream (DSP) operations, where the protein is purified by diverse filtration methods and/or single or multiple chromatographic steps. Usually at DSP, this degraded variant can be targeted and removed using specific developed chromatography steps during capture, intermediate, and/or polishing steps to obtain the desired impurity levels. Every purification step presents a purpose, and for that, the capture chromatography step aims to bind the specific product by the affinity of its physicochemical properties such as pI. In contrast, using a completely different chromatography approach, the intermediate and polishing steps remove impurities that present similar physicochemical behavior compared to specific products.

The degradation produced intrinsically by *K. phaffii* can significantly impact not only the yield but also the quality and stability of the biopharmaceutical products resulting in incomplete or non-functional products that could be unsuitable for therapeutic use [3, 4]. These degraded products can become impurities, which are considered critical quality attributes (CQAs) by regulatory agencies. For this reason, both the FDA and EMA require the full characterization of this biological product, which includes the determination of physico-chemical properties, biological activity, immuno-chemical properties, purity, and impurities, by appropriate techniques to allow a suitable specification to be established [5]. Complimentary, it is also required to monitor, along the whole manufacturing process, impurities species related to the product called product-related variants.

The effect of fragmentation on the function of the evaluated product depends on whether cleavage sites are observed in the active site of the protein, which might affect the mechanism of action of the molecule, and therefore reducing the activity of the product, or at other regions which do not compromise the product stability and/or efficacy.

As fragmentation events might be considered CQAs, they must be thoroughly characterized using orthogonal methods to ensure the safety and efficacy of the final biopharmaceutical product [6, 7]. Briefly, an orthogonal method is an additional method that provides very different selectivity to the primary method and gives information about the same quality attribute [8, 9].

In case of fragmentation, sodium dodecyl sulfate capillary electrophoresis (CE-SDS) is an ideal technique for characterizing these products, since it can be used in both process control and as part of the control strategy and release testing [10].

This technique, which is a highly sensitive method for good fragment separation, might be orthogonally evaluated alongside SEC-HPLC [11, 12], while intact mass (LC-MS) might provide a different insight by identifying the specific cleavage site [13, 14]. Complementarily, the presence of specific amino acids remaining from the secretion signal at the N-terminal protein region, commonly observed in *Pichia* products, such as EA or EAEA, can also be determined by LC-MS [4].

Prior to evaluating any results, all these methods require a previous qualification for performance assessment following ICH guidelines [15]. For method identification, only specificity should be assessed, while for method quantification (CE-SDS), precision, linearity, accuracy, and limit of quantification should be properly determined.

The CE-SDS might be run under reducing and nonreducing conditions. In case the protein does not present disulfide bonds, only nonreducing conditions might be evaluated. This chapter specifies the preparations for both conditions; however, only CE-SDS nonreducing results are shown.

2 Materials

2.1 Sample Preparation

1. Purified final product (filtered by 0.22 μm). Many chapters, for example, Chapters 40, 41, 42, 43, and this chapter, describe, in detail, the most used DSP operation for *Pichia* bioprocesses.
2. MilliQ water.
3. 1.5 mL Eppendorf tubes.
4. Benchtop vortex, centrifuge, and water bath (or similar) up to 70 °C for Eppendorf tubes.
5. Pipettes: 1000 μL and 200 μL.
6. Master mix including standard for different conditions (described in Subheading 3.2).
7. 250 mM Iodoacetamide (nonreducing conditions).
8. 100 mM DTT (reducing conditions).

2.2 Capillary Electrophoresis Equipment

Different vendors offer alternative equipment. The most frequently used are capillary electrophoresis equipment from SCIEX and Maurice Protein Simple. The authors used the equipment supplied by SCIEX to develop the method described in this chapter. The mentioned SCIEX instrument can be used for different applications, among them, the described CE-SDSD, which requires a specific kit.

1. Capillary electrophoresis instrument.
2. Kits from the manufacturer of this equipment are also available. In this case, SDS-MW kits can be purchased.
3. The detector required for the analysis should be a PDA (photodiode array) to register a wide wavelength range. For this analysis, at least registration at 210–220 nm is applied.

2.3 SDS-MW Kits

For this analysis, the author uses a specific kit containing:

1. Sample buffer.
2. 10 kDa Internal standard.
3. Molecular weight ladder.
4. Gel SDS buffer.
5. Acid solution (0.1 M HCl).
6. Basic solution (0.1 M NaOH).
7. Universal plastic vials.
8. Universal rubber vial caps-blue.
9. 200 μL micro vials.
10. Pre-installed cartridge containing bare fused silica capillary.

3 Methods

3.1 Equipment Setup

1. Place universal plastic vials as it is demonstrated in Fig. 1.
2. Add the required volume:
   - 1.5 mL of MilliQ water for "Water vials.
   - 1 mL MilliQ water for Waste vials.
   - 1.2 mL Gel buffer for Gel-R and Gel-S vials.
   - 1.4 mL NaOH and HCl for NaOH and HCl vials, respectively, and put the universal rubber vial caps on top (*see* **Note 1**).

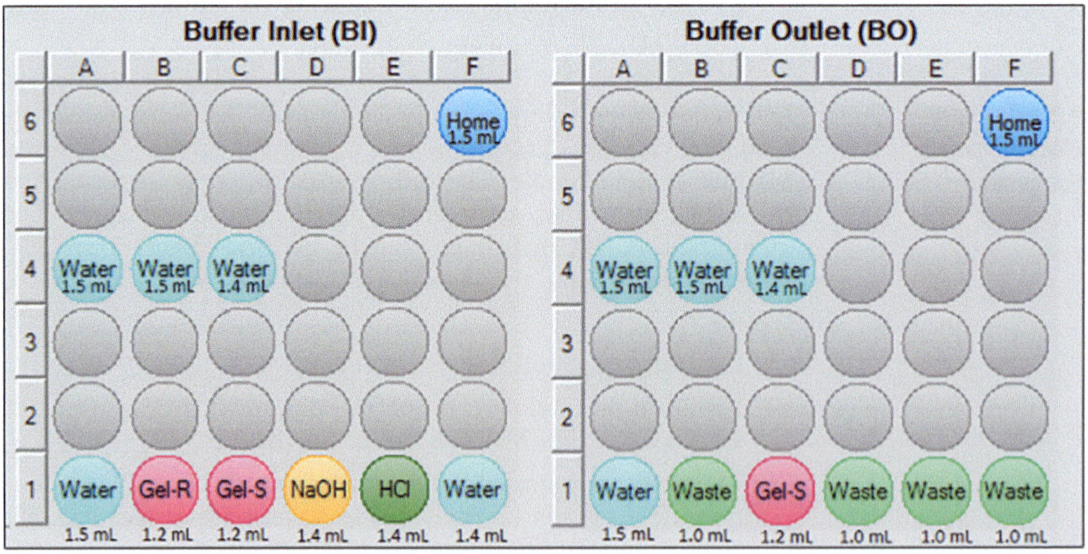

Fig. 1 Buffer tray layout for solution vials

3. Place the pre-installed capillary cartridge into the cartridge compartment and close the front panel.
4. Install the PDA detector (*see* **Note 2**).

3.2 Molecular Weight Ladder Preparation

1. In 1.5 mL Eppendorf tubes, add the following components:

Components	Volume (μL) $n = 1$
Sample buffer	85
DTT 100 mM	5
Molecular weight ladder	10

2. Mix thoroughly by vortexing and subsequently, spin it for 30 s at 3.000 rpm to remove air bubbles.
3. Heat the mixture in a water bath at 70 °C for 10 min.
4. Place the vial at room temperature to cool down for 2–3 min.
5. Transfer 98 μL into 200 μL micro vials and place them inside universal plastic vials.
6. Place these vials on an inlet sample tray and cover them with universal rubber vial caps-blue.

3.3 Sample Preparation

7. It is highly recommended to prepare samples along with the buffer presented in this sample as a negative control.
8. Dilute the sample down to 1.0 mg/mL in MilliQ water.
9. Prepare a master mix in 1.5 mL Eppendorf tubes following the tables:
   - Master mix for non-reducing conditions (*see* **Note 3**):

Components	Volume (μL) $n = 1$
Sample buffer	50
10 KDa Internal standard	2
IAM 250 mM	5

   - Master mix for reducing conditions:

Components	Volume (μL) $n = 1$
Sample buffer	50
10 KDa Internal standard	2
DTT 100 mM	5

10. Mix thoroughly. Spin for 30 s at 3.000 rpm to remove air bubbles.
11. Divide the master mix by adding 57 μL of master mix to a new 1.5 mL Eppendorf tubes by reverse pipetting.
12. Add 45 μL of sample.
13. Mix thoroughly by vortexing and subsequently, spin it for 30 s at 3.000 rpm to remove air bubbles.
14. Heat this mixture in a water bath at 70 °C for 10 min.
15. Transfer 98 μL into 200 μL micro vials and place them inside universal plastic vials.
16. Place these vials on an inlet sample tray and cover them with universal rubber vial caps blue.

3.4 Separation Method

3.4.1 Temperature and Registration Settings

1. Recommended Current: 300 μA.
2. Recommended Cartridge and sample storage temperature: 25 °C.
3. Recommended wavelength: 220 nm.

3.4.2 Before Injecting the Sample

1. Rinse with 0.1 M NaOH to clean the capillary surface.
2. Rinse with 0.1 M HCl to neutralize the silanol groups of the capillary.
3. Rinse with water to remove any acidic residue.
4. Fill the capillary with Gel SDS buffer.

3.4.3 Sample Injection

1. Inject by voltage: 5.0 kV for 20.0 s.
2. Separate by voltage: 15.0 kV for 30 min.
3. Press the Run button.

3.5 Data Analysis

3.5.1 Fragmentation Quantification

1. 10KDa internal standard, which is part of the master mix (*see* Subheading 3.2), appoints the starting point of integration (*see* **Note 4**)
2. To integrate the samples (Fig. 2 for full view profile and Fig. 3 as the zoom of the fragments region):
   (a) Load both the sample electropherogram profile and its corresponding buffer (negative control).
   (b) Align both profiles.
   (c) Integrate the peaks that are only present in the sample profile. In this case, the software presents specific icons for baseline integration (see Fig. 4 for an example of integration).
   (d) Report the %corrected area for every integrated peak.

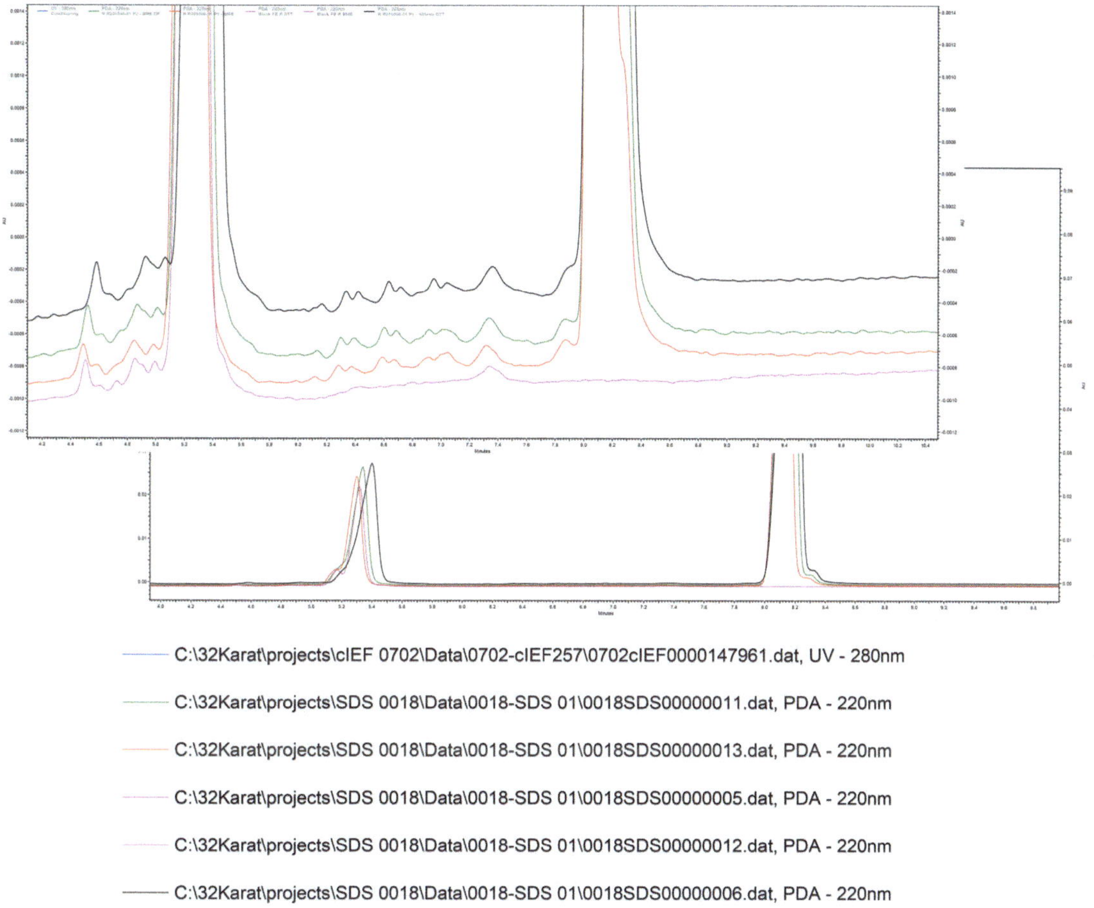

Fig. 2 Full view of the electrophoretic profile of a purified protein expressed in Pichia Pastoris by CE-SDS non-reducing

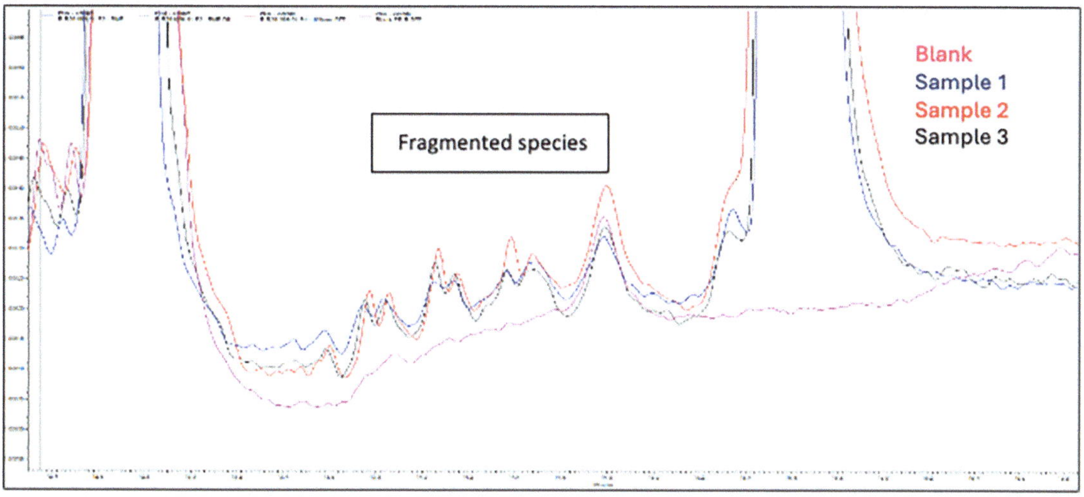

Fig. 3 Zoom view of the electrophoretic profile of a purified protein expressed in Pichia Pastoris by CE-SDS non-reducing

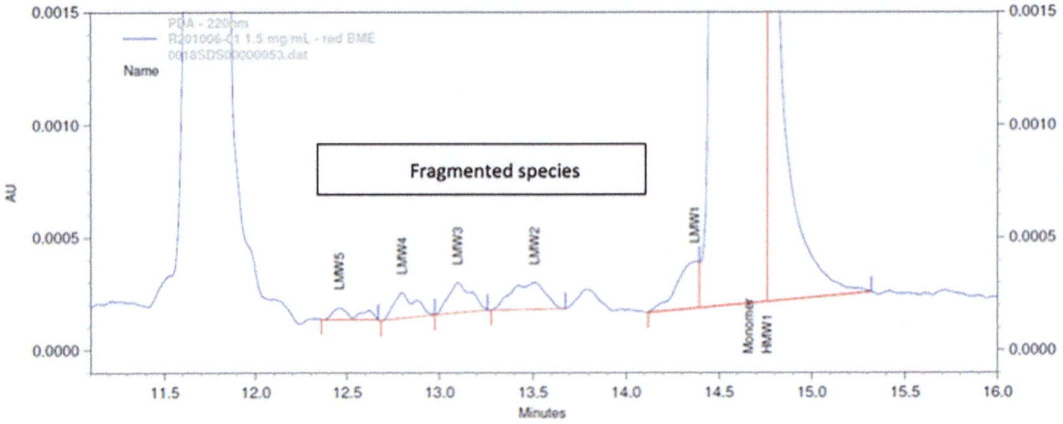

Fig. 4 Integration of the zoom view of the electrophoretic profile of a purified protein expressed in Pichia Pastoris by CE-SDS non-reducing

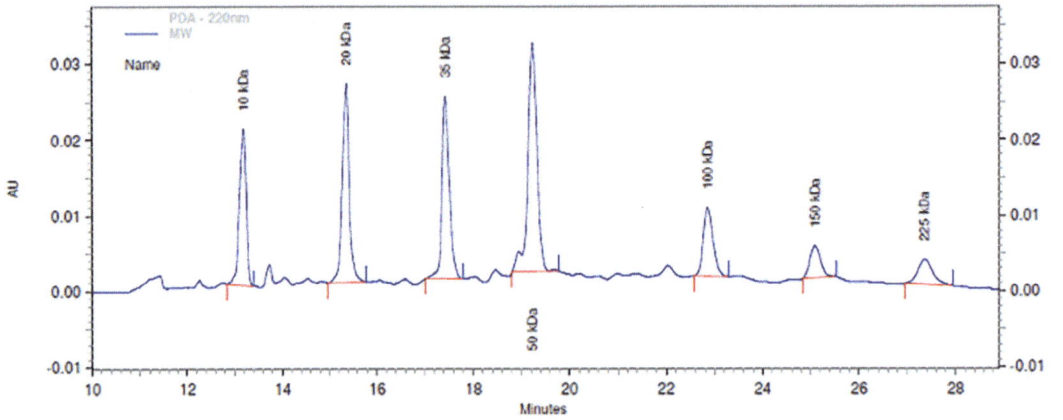

Fig. 5 Molecular weight ladder electrophoretic profile

3.5.2 Molecular Weight Determination for Fragmentation Species

1. Load the molecular weight ladder electropherogram profile and integrate 7 peaks as Fig. 5. The MW for every peak is: 10 kDa, 20 kDa, 35 kDa, 50 kDa, 100 kDa, 150 kDa, and 225 kDa.
2. In the analysis section of the software, write down the MW of every peak and its corresponding migration time to generate a quadratic curve (*see* Fig. 6).
3. Introduce the migration time of the 10KDa internal standard of every sample in the Reference peak (marked in red in Fig. 6).
4. Automatically, every MW of every integrated peak (Subheading 3.4.3) is reported (*see* **Note 5**).

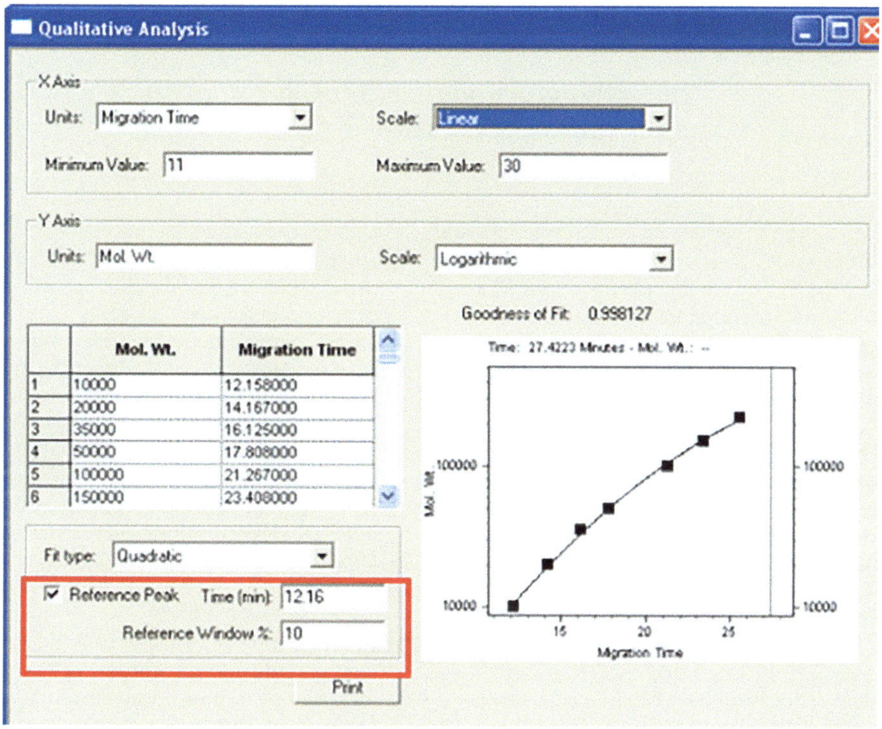

Fig. 6 Screenshot from qualitative analysis to determine the molecular weight for unknown fragmented peaks of samples

4 Notes

1. Before putting the rubber vial caps on top of the vials, pinch them with your fingers to reduce their rigidity and avoid broken capillaries.
2. It is advisable to calibrate the PDA detector before starting the analysis. The calibration procedure recommended by each vendor might present differences.
3. IAM solution should be prepared freshly before use, and it should be protected from light because Iodoacetamide is unstable and light sensitive.
4. SDS-MW gel is formulated to provide an effective sieving range of approximately 10–225 kDa.
5. Molecular weight determination of every fragment is obtained, and therefore, the degradation pattern can be elucidated by subtracting the molecular weight of the full-length protein from the molecular weight of every fragment. This technique can accurately assess differences between molecular weights higher than 1 kDa.

Competing Interest Statement

The authors have no conflicts of interest to declare that are relevant to the content of this chapter.

References

1. Garcia-Ortega X, Reyes C, Montesinos JL, Valero F (2015) Overall key performance indicator to optimizing operation of high-pressure homogenizers for a reliable quantification of intracellular components in Pichia pastoris. Front Bioeng Biotechnol 3:1–9. https://doi.org/10.3389/fbioe.2015.00107
2. Garrigós-Martínez J, Weninger A, Montesinos-Seguí JL et al (2021) Scalable production and application of Pichia pastoris whole cell catalysts expressing human cytochrome P450 2C9. Microb Cell Fact 20:90. https://doi.org/10.1186/s12934-021-01577-4
3. Cereghino JL, Cregg JM (2000) Heterologous protein expression in the methylotrophic yeast Pichia pastoris. FEMS Microbiol Rev 24:45–66. https://doi.org/10.1111/j.1574-6976.2000.tb00532.x
4. Ahmad M, Hirz M, Pichler H, Schwab H (2014) Protein expression in Pichia pastoris: recent achievements and perspectives for heterologous protein production. Appl Microbiol Biotechnol 98:5301–5317. https://doi.org/10.1007/s00253-014-5732-5
5. EMA (2018) Guideline-Requirements-Quality-Documentation-Concerning-Biological-Investigational-Medicinal_En-0. 44:1–21. Retrieved from EMA. https://www.ema.europa.eu/en/requirements-quality-documentation-concerning-biological-investigational-medicinal-products-clinical-trials-scientific-guideline
6. Food and Drug Administration (2018) Quality attribute considerations for chewable tablets. Final 271. Retrieved from FDA. https://www.fda.gov/files/drugs/published/Quality-Attribute-Considerations-for-Chewable-Tablets-Guidance-for-Industry.pdf
7. EMA (2014) Guideline on process validation for the manufacture of biotechnology-derived active substances and data to be provided in the regulatory submission. Committee for Medicinal Products for Human Use (CHMP) Guideline 44:Stability studies on rFVIIIFc drug substance. 1–14. Retrieved from EMA. https://www.ema.europa.eu/en/process-validation-manufacture-biotechnology-derived-active-substances-data-be-provided-regulatory-submission-scientific-guideline
8. Zheng F, Rasmussen HT (2009) Use of orthogonal methods during pharmaceutical development: case studies. Retrieved from LCGC Supplements Special Issues 27:16–21. https://www.chromatographyonline.com/view/use-orthogonal-methods-during-pharmaceutical-development-case-studies
9. World Health Organization (2013) Guidelines on the quality, safety and efficacy of biotherapeutic protein products prepared by recombinant DNA technology. Retrieved from World Health Organization. https://www.who.int/publications/m/item/recombinant-dna-annex-4-trs-no-987
10. EMA (2009) ICH topic Q 6 B specifications: test procedures and acceptance criteria for biotechnological/biological products. Retrieved from ICH. https://www.ema.europa.eu/en/ich-q6b-specifications-test-procedures-acceptance-criteria-biotechnological-biological-products-scientific-guideline
11. Fekete S, Beck A, Veuthey JL, Guillarme D (2014) Theory and practice of size exclusion chromatography for the analysis of protein aggregates. J Pharm Biomed Anal 101:161–173. https://doi.org/10.1016/j.jpba.2014.04.011
12. Striegel AM (2022) Size-exclusion chromatography: a twenty-first century perspective. Chromatographia 85:307–313. https://doi.org/10.1007/s10337-022-04143-1
13. Coradin M, Karch KR, Garcia BA (2017) Monitoring proteolytic processing events by quantitative mass spectrometry. Expert Rev Proteomics 14:409–418. https://doi.org/10.1080/14789450.2017.1316977
14. Grotefend S, Kaminski L, Wroblewitz S et al (2012) Protein quantitation using various modes of high-performance liquid chromatography. J Pharm Biomed Anal 71:127–138. https://doi.org/10.1016/j.jpba.2012.08.024
15. ICH Steering Committee (2023) ICH Q2 (R2) validation of analytical procedures: method. Retrieved from ICH. https://www.ema.europa.eu/en/ich-q2r2-validation-analytical-procedures-scientific-guideline

INDEX

Xavier Garcia-Ortega et al. (eds.), *Pichia pastoris: Methods and Protocols*, Methods in Molecular Biology, vol. 2697, https://doi.org/10.1007/978-1-0716-4779-0,

M

O

P

Q

R

S

T

U

Y

Z

MIX
Papier aus verantwortungsvollen Quellen
Paper from responsible sources
FSC® C105338

If you have any concerns about our products,
you can contact us on
ProductSafety@springernature.com

In case Publisher is established outside the EU,
the EU authorized representative is:
Springer Nature Customer Service Center GmbH
Europaplatz 3, 69115 Heidelberg, Germany

Printed by Libri Plureos GmbH
in Hamburg, Germany